Free Student Aid.

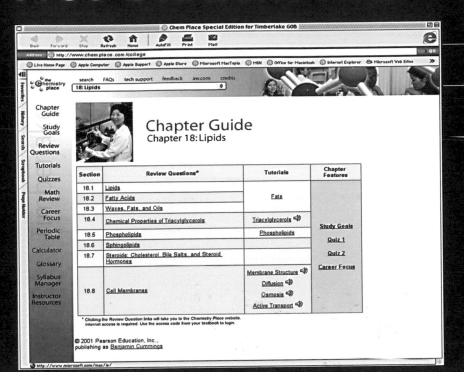

Log on.

Tune in.

Succeed.

To help you succeed in general, organic, and biological chemistry or organic and biological chemistry, your professor has arranged for you to enjoy access to great media resources on an interactive CD-ROM and using **The Chemistry Place™** for *General, Organic, and Biological Chemistry*. You'll find that these resources will enhance your understanding of chemistry.

Here's your personal ticket to success:

How to log on to www.chemplace.com/college:

1. Go to www.chemplace.com/college.
2. Click the *General, Organic, and Biological Chemistry* book cover.
3. Click "Register Here."
4. Enter your pre-assigned Access Code exactly as it appears below and click "Submit."
5. Complete the online registration form to create your own personal User ID and Password.
6. Once your personal ID and Password are confirmed by email, go back to www.chemplace.com/college, click the appropriate book cover, type in your new User ID and Password, and click "Enter."

Your Access Code is:

USABG-BUCKO-LIPPY-DIVED-FACTO-PSHAW

Detach this card and keep it handy. It's your ticket to valuable information.

Important: Please read the License Agreement located on the site before using **The Chemistry Place™** for *General, Organic, and Biological Chemistry* web site or CD-ROM. By using the web site or CD-ROM, you indicate that you have read, understood, and accepted the terms of this agreement.

Got technical questions?
For technical support, please visit www.aw.com/techsupport, send an email to online.support@pearsoned.com (for web site questions), or send an email to media.support@pearsoned.com (for CD-ROM questions) with a detailed description of your computer system and the technical problem. You can also call our tech support hotline at 1-800-677-6337 Monday-Friday, 8 a.m. to 5 p.m. CST.

What your system needs to use these media resources:

WINDOWS
250 MHz Windows-95/98/NT/2000
32 MB RAM installed, 64 preferred
800 x 600 screen resolution
Thousands of colors
4x CD-ROM drive
Browsers: Internet Explorer 5.0; Netscape 4.7
Plug-Ins: Shockwave Player, Flash Player, Chime
56K Modem

MACINTOSH
233 MHz G3
OS 8.1 or higher
32 MB RAM minimum
800 x 600 screen resolution
Thousands of colors
4x CD-ROM drive
Browsers: Internet Explorer 5.0; Netscape 4.7 (for CD-ROM, Netscape only)
Plug-Ins: Shockwave Player, Flash Player, Chime
56K Modem

0-8053-2983-8

Your User ID

_____

Your Password

_____

Organic and Biological

Chemistry

STRUCTURES OF LIFE

Karen C. Timberlake

Benjamin
Cummings

An imprint of Addison Wesley

San Francisco • Boston • New York
Capetown • Hong Kong • London • Madrid • Mexico City
Montreal • Munich • Paris • Singapore • Sydney • Tokyo • Toronto

Executive Editor	Ben Roberts
Acquisitions Editor	Maureen Kennedy
Project Editor	Claudia Herman
Production Editor	Jean Lake
Director of Marketing	Stacy Treco
Marketing Manager	Christy Lawrence
Market Development Manager	Chalon Bridges
Art Coordinators	Betty Gee, Side By Side Studios
	Anthony J. Asaro
	Jean Lake
Composition	Pre-Press
Manufacturing Supervisor	Vivian McDougal
Copyeditor	Cathy Cobb
Proofreader	Martha Ghent
Art Direction, Text and Cover Designer, Page Layout	Mark Ong, Side By Side Studios
Illustrators	Laura Brown, Carin Caine, Wen Chao, J.B. Woolsey Associates, Pre-Press, Side By Side Studios
Photo Research	Anthony J. Asaro
Text Photography	John Bagley, Richard Tauber
Cover Photographs	John Bagley, Richard Tauber
Cover Illustration	Blakely Kim

Library of Congress Cataloging-in-Publication Data
Timberlake, Karen.
General, organic, and biological chemistry : structures of life / Karen C. Timberlake.
p. cm.
Includes index.
ISBN 0-321-04283-2   OB version: ISBN 0-8053-2983-8
1. Chemistry.   I. Title.
QD33.2 .T56 2001
540—dc21
2001047274

The Benjamin/Cummings Publishing Company, Inc.
1301 Sansome Street
San Francisco, CA 94111

Trademarks

Brief Contents

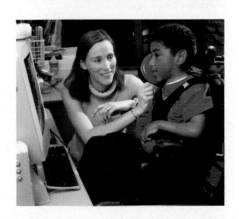

Contents

14 Alcohols, Phenols, Ethers, and Thiols 449

15 Aldehydes, Ketones, and Chiral Molecules 482

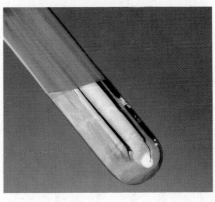

18 Lipids   575

19 Amines and Amides   614

20 Amino Acids and Proteins   641

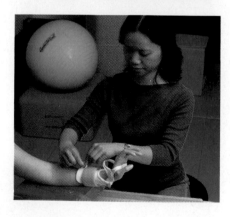

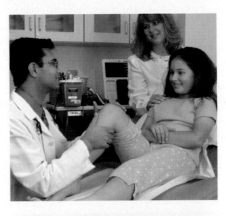

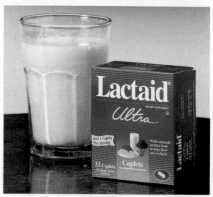

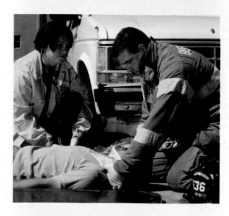

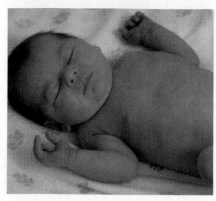

25 Metabolic Pathways for Lipids and Amino Acids 814

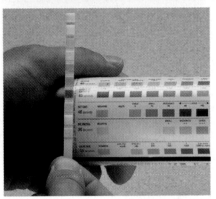

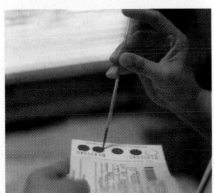

Applications and Activities

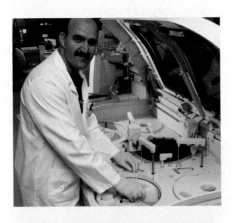

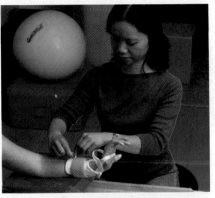

ENVIRONMENTAL NOTE

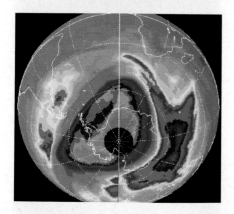

EXPLORE YOUR WORLD

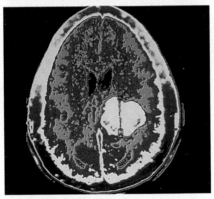

About the Author

Karen Timberlake is professor emeritus of chemistry at Los Angeles Valley College, where she taught chemistry for allied health and preparatory chemistry for 36 years. She received her bachelor's degree in chemistry from the University of Washington in 1962 and her master's degree in biochemistry from the University of California at Los Angeles in 1965. She has also taken graduate courses in science education during a sabbatical at the University of Northern Colorado.

Professor Timberlake has been writing chemistry textbooks for general, organic, and biological chemistry for 28 years. During that time, her name has become associated with the strategic use of pedagogical tools that promote student success in chemistry and the application of chemistry to real-life situations in health and medicine. More than one million students have learned chemistry using texts, laboratory manuals, and study guides written by Karen Timberlake. In addition to *Organic and Biological Chemistry: Structures of Life,* she is also the author of *General, Organic, and Biological Chemistry: Structures of Life, Chemistry: An Introduction to General, Organic, and Biological Chemistry,* and the accompanying *Study Guide with Solutions for Selected Problems, Laboratory Manual,* and *Essentials Laboratory Manual.*

Professor Timberlake belongs to numerous science and education organizations including the American Chemical Society (ACS) and the National Science Teachers Association (NSTA). In 1987, she was the Western Regional Winner of Excellence in College Chemistry Teaching Award given by the Chemical Manufacturers Association. She has participated in education grants for science teaching including the Los Angeles Collaborative for Teaching Excellence (LACTE) and the Title III grant at her college. She often speaks at conferences and education meetings on using student-centered teaching methods in chemistry to promote the learning success of students.

Her husband, Bill, is also a chemistry professor and contributed much to the writing of *Chemistry.* When the Professors Timberlake are not writing textbooks, they relax by hiking, traveling to Mexico and Europe, trying new restaurants, and playing lots of tennis. The Timberlakes' son, John, and daughter-in-law, Cindy, are marketing consultants for Internet businesses in Burlingame, California. The Timberlakes' grandson, Daniel, who was born in September, does not know what he wants to be yet.

To the Student

I hope that this textbook helps you discover exciting new ideas and gives you a rewarding experience as you develop an understanding and appreciation of the role of chemistry in your life. If you would like to share your experience with chemistry, or have questions and comments about this text, I would appreciate hearing from you.

Karen C. Timberlake
Email: khemist@aol.com

I dedicate this book to

- my husband for his patience, loving support, and preparation of late meals,

- to my son, John, daughter-in-law, Cindy, and grandson, Daniel, for the precious things in life, and

- to my wonderful students whose hard work and commitment puts purpose in this text.

The whole art of teaching is only the art of awakening the natural curiosity of young minds.
— ANATOLE FRANCE

One must learn by doing the thing; though you think you know it, you have no certainty until you try.
— SOPHOCLES

Discovery consists of seeing what everybody has seen and thinking what nobody has thought.
— ALBERT SZENT-GYORGI

Preface

Welcome to the first edition of *Organic and Biological Chemistry: Structures of Life*. My main objective in writing this text is to relate the structure and behavior of matter to its functions in health and life because all too often students find chemistry a series of facts to be memorized. This text is designed to be used in a one-year course that includes general, organic, and biological chemistry. It is written for students who are preparing for careers in health-related professions such as nursing, dietetics, respiratory therapy, or laboratory technology.

The Theme Emphasizes the Relationship Between Structures and Life

Throughout the text, the structures of inorganic, organic, and biochemical compounds and molecules are related to their function. The topic of stereochemistry and chiral organic molecules, which begins in Chapter 15 with Aldehydes, is revisited as an important concept in the understanding of the structures of carbohydrates, amino acids, and chiral drugs. The structures of molecules are related to their physical and chemical properties such as solubility in water, density, and boiling point. The structural levels of proteins are related to their function, while chemical processes that denature proteins emphasize the importance of structure to activity. Throughout the text, the atomic structure of matter is highlighted by macro-to-micro illustrations that relates the atomic level to the macroscopic structures of real-life materials. In this way the chemical concepts and structures of molecules are continuously related to the behavior and function of biomolecules in the body. My goal in this text is to provide a learning environment that makes the study of chemistry an engaging and positive experience. It is also my goal to help every student become a critical thinker by instilling in them the scientific concepts that form the basis for making important decisions about issues concerning health and the environment.

Pedagogical Features

Students are often challenged by the study of chemistry and have difficulty finding the relevance of chemistry to their career paths. Many features in this text illustrate how chemistry is connected to students' lives and their interest in allied health careers. Chapter openers begin with interviews with health care professionals who discuss their type of work. Health Notes and Environmental Notes in every chapter continually relate the chemistry to current topics in health, medicine, and research, which support the student's interest in a health career.

A comprehensive learning program in each chapter provides many learning tools. Learning Goals in every section preview the concepts the student is to learn. The Sample Problems in every section model successful problem-solving techniques for the student. In this text I have redesigned the typical linear process of reading a whole chapter and then doing problems into a more cyclic process that combines each topic with immediate applications using Explore Your World activities, Health and Environmental Notes, and immediate problem solving. Questions and Problems now follow each section to encourage the student to apply newly learned material immediately to the problem-solving experience. Students are encouraged to be active learners by working problems and thinking about real-life issues as they progress through each chapter. Additional Problems found at the end of the chapter integrate the topics in the entire chapter to promote further study and critical thinking. Answers for Study Checks and odd-numbered problems located at the end of the text give immediate feedback to problem solving. Chapter Reviews give a brief overview of the important concepts in each section of the chapter. Key Terms and Summary of Reactions remind students of the new vocabulary and new reactions presented in each chapter.

Throughout the text, a strong art program visually connects the real-life world of materials familiar to students with their atomic-level structure. Students realize that all of the things they see every day have an atomic level of organization and structure that relates to their behavior and function. Macro-to-micro illustrations from typical materials to their atomic level of organization continually relate structure to ordinary objects. Every figure throughout the text contains a question that requires the student to study that figure and relate the visual representation to the text. Molecular structures prevalent throughout the organic and biochemistry chapters aid in the visualization of the three-dimensional structures of organic and biochemical molecules. The plentiful use of three-dimensional structures takes the descriptions of molecules in the text to a visual level, which stimulates the imagination of the student. The use of color in diagrams, organic equations, and biochemical pathways clarifies and organizes the features of rather complex ideas.

Chapter Organization

In each textbook I write, I consider it essential to relate every chemical concept to real-life issues of health and environment. In this text I use the theme of Structures of Life to indicate the relationship between structures and function.

Although it is traditional to place the topics of organic and biochemistry in two separate sections, I have placed chapters on carbohydrates and lipids next to the organic chapters that discuss similar functional groups. Thus I begin the second part of the text with Chapter 11 with the shape of organic compounds, physical and chemical properties, and an overview of functional groups and constitutional isomers that form the structure of organic chemistry and form a basis for understanding the biomolecules of living systems. In Chapters 12 through 15, I describe the structures and nomenclature of saturated and unsaturated hydrocarbons, alcohols, thiols, ethers, aldehydes, and ketones. In each chapter, physical and chemical properties are related to the structure and functional groups in each family of organic compounds. In Chapter 15, I introduce the concept of chiral compounds and enantiomers, which I revisit in the chapters on carbohydrates and amino acids to show its importance in the nature of biomolecules and chiral drugs and their behavior in living systems. Chapter 16 integrates a biochemistry chapter, Carbohydrates, which contains the functional groups discussed in the previous chapters. To the student, the application of organic chemistry to biochemistry is often the most exciting part of their study of chemistry because of relationship to health and medicine. Chapter

17 discusses Carboxylic Acids and Esters, which is followed by Chapter 18 on Lipids, which contain the same functional groups in larger molecules such as triacylglycerols and glycerophospholipids. I discuss the role of these lipids and cholesterol in cell membranes as well as the lipids that function as bile salts and steroid hormones.

Chapters 19 and 20 continue with the physical and chemical properties of organic compounds, Amines and Amides, followed by related biomolecules, Amino Acids and Proteins. The importance of the structure of proteins from primary to quaternary is related to the shapes and activity of proteins. Chapter 21 relates the importance of the three-dimensional shape of proteins to their function as enzymes. The student learns that the shape of an enzyme is a factor in enzyme regulation and how end products might change the shape to increase or decrease the rate of an enzyme-catalyzed reaction. We also see that proteins change shape and lose function when subjected to pH changes and high temperatures. The important role of water-soluble vitamins as coenzymes is related to enzyme function. Chapter 22 describes the nucleic acids and their importance as biomolecules that store and direct information for cellular components, growth, and reproduction.

The final three chapters, Chapters 23, 24, and 25, discuss the metabolic pathways of biomolecules from the digestion of foodstuffs to the synthesis of ATP. Chapter 23 describes the stages of metabolism and the digestion of carbohydrates, our most important fuel. Then we see how the metabolic pathway of glycolysis degrades the glucose molecule to pyruvate. Under aerobic conditions, we learn that pyruvate is converted to acetyl CoA. We also look at how the storage molecule glycogen is replenished and how glucose is synthesized from noncarbohydrate sources. In Chapter 24, we look at the central role of the citric acid cycle in utilizing acetyl CoA to produce reduced coenzymes under aerobic conditions for the electron transport system and oxidative phosphorylation. Details on the structure and function of ATP synthase are included. Chapter 25 discusses the digestion of lipids and proteins and the metabolic pathways that convert fatty acids and amino acids to energy. We also learn how excess carbohydrates are converted to triacylglycerols in adipose tissue and how the intermediates of the citric acid cycle are converted to nonessential amino acids. Finally, we summarize the relationships between the catabolic and anabolic pathways in metabolism.

Because a chemistry course for allied health may be taught within different time frames, it may be difficult to cover all the chapters in this text. However, each chapter is now a complete package, which allows some chapters to be skipped or the order of presentation to be changed.

Feature Guide

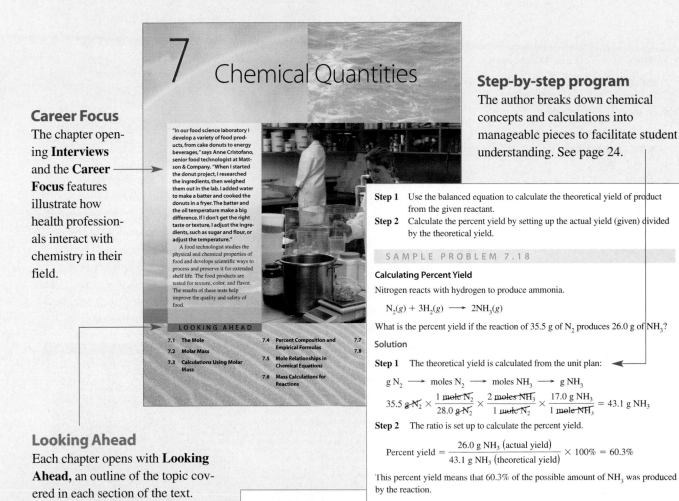

Career Focus

The chapter opening **Interviews** and the **Career Focus** features illustrate how health professionals interact with chemistry in their field.

Looking Ahead

Each chapter opens with **Looking Ahead,** an outline of the topic covered in each section of the text.

Learning Goals

Learning Goals at the beginning of each new section in a chapter give the outcome that can be expected after reading the material in the section and working the problems.

7 Chemical Quantities

"In our food science laboratory I develop a variety of food products, from cake donuts to energy beverages," says Anne Cristofano, senior food technologist at Mattson & Company. "When I started the donut project, I researched the ingredients, then weighed them out in the lab. I added water to make a batter and cooked the donuts in a fryer. The batter and the oil temperature make a big difference. If I don't get the right taste or texture, I adjust the ingredients, such as sugar and flour, or adjust the temperature."

A food technologist studies the physical and chemical properties of food and develops scientific ways to process and preserve it for extended shelf life. The food products are tested for texture, color, and flavor. The results of these tests help improve the quality and safety of food.

LOOKING AHEAD

7.1 The Mole
7.2 Molar Mass
7.3 Calculations Using Molar Mass
7.4 Percent Composition and Empirical Formulas
7.5 Mole Relationships in Chemical Equations
7.6 Mass Calculations for Reactions
7.7
7.8

Step-by-step program

The author breaks down chemical concepts and calculations into manageable pieces to facilitate student understanding. See page 24.

Step 1 Use the balanced equation to calculate the theoretical yield of product from the given reactant.

Step 2 Calculate the percent yield by setting up the actual yield (given) divided by the theoretical yield.

SAMPLE PROBLEM 7.18

Calculating Percent Yield

Nitrogen reacts with hydrogen to produce ammonia.

$$N_2(g) + 3H_2(g) \longrightarrow 2NH_3(g)$$

What is the percent yield if the reaction of 35.5 g of N_2 produces 26.0 g of NH_3?

Solution

Step 1 The theoretical yield is calculated from the unit plan:

$$g\,N_2 \longrightarrow moles\,N_2 \longrightarrow moles\,NH_3 \longrightarrow g\,NH_3$$

$$35.5\,\cancel{g\,N_2} \times \frac{1\,\cancel{mole\,N_2}}{28.0\,\cancel{g\,N_2}} \times \frac{2\,\cancel{moles\,NH_3}}{1\,\cancel{mole\,N_2}} \times \frac{17.0\,g\,NH_3}{1\,\cancel{mole\,NH_3}} = 43.1\,g\,NH_3$$

Step 2 The ratio is set up to calculate the percent yield.

$$\text{Percent yield} = \frac{26.0\,g\,NH_3\,(\text{actual yield})}{43.1\,g\,NH_3\,(\text{theoretical yield})} \times 100\% = 60.3\%$$

This percent yield means that 60.3% of the possible amount of NH_3 was produced by the reaction.

LEARNING GOAL

Draw the structure of the product from the reaction of a triacylglycerol with hydrogen, an acid or base, or an oxidizing agent.

18.4 Chemical Properties of Triacylglycerols

The chemical reactions of the triacylglycerols (fats and oils) are the same as we discussed for alkenes (Chapter 13), and carboxylic acids and esters (Chapter 17). We will look at the addition of hydrogen to the double bonds, the oxidation of double bonds, and the acid and base hydrolysis of the ester bonds of fats and oils.

Hydrogenation

The **hydrogenation** of unsaturated fats converts carbon–carbon double bonds to single bonds. The hydrogen gas is bubbled through the heated oil in the presence of a nickel catalyst.

$$-CH{=}CH- + \ H_2 \xrightarrow{Ni} \ -\overset{\overset{\displaystyle H}{|}}{\underset{\underset{\displaystyle H}{|}}{C}}-\overset{\overset{\displaystyle H}{|}}{\underset{\underset{\displaystyle H}{|}}{C}}-$$

For example, when hydrogen adds to all of the double bonds of triolein using a nickel catalyst, the product is the saturated fat tristearin.

Health Note

A rich array of **Health Notes** in each chapter apply chemical concepts to real-life issues and relevant topics of health and medicine. These topics include weight loss and weight gain, artificial fats, anabolic steroids, alcohol, genetic diseases, viruses, and cancer.

Environmental Note

Environmental Notes delve into issues such as global warming, radon, ozone depletion, acid rain, food irradiation, pheromones, and recycling of plastics.

HEALTH NOTE

Stored Fat and Obesity

The storage of fat is an important survival feature in the lives of many animals. In hibernating animals, large amounts of stored fat provide the energy for the entire hibernation period, which could be several months. In camels, such as dromedary camels, large amounts of food are stored in the camel's hump, which is actually a huge fat deposit. When food resources are low, the camel can survive months without food or water by utilizing the fat reserves in the hump. Migratory birds preparing to fly long distances also store large amounts of fat. Whales are kept warm by a layer of body fat called "blubber" under their skin, which can be as thick as 2 feet. Blubber also provides energy when whales must survive long periods of starvation. Penguins also have blubber, which protects them from the cold and provides energy when they are sitting on a nest of eggs.

Humans also have the capability to store large amounts of fat, although they do not hibernate or usually have to survive for long periods of time without food. When humans survived on sparse diets that were mostly vegetarian, the fat content was about 20%. Today, a typical diet includes more dairy products and foods with high fat levels, which increases the daily fat intake to as much as 60% of the diet. The U.S. Public Health Service now estimates that in the United States, more than one-third of adults are obese. Obesity is defined as a body weight that is more than 20 percent over an ideal weight. Obesity is a major factor in health problems such as diabetes, heart disease, high blood pressure, stroke, and gallstones as well as some cancers and forms of arthritis.

that certain pathways in lipid and carbohydrate metabolism may cause excessive weight gain in some people. In 1995, scientists discovered that a hormone called *leptin* is produced in fat cells. When fat cells are full, high levels of leptin signal the brain to limit the intake of food. When fat stores are low, leptin production decreases, which signals the brain to increase food intake. Some obese persons have high levels of leptin, which means that leptin did not cause them to decrease how much they ate.

Research on obesity has become a major research field. Scientists are studying differences in the rate of leptin production, degrees of resistance to leptin, and possible combinations of these factors. After a person has dieted and lost weight, the leptin level drops. This decrease in leptin may cause an increase in hunger, slow metabolism, and increased food intake, which starts the weight-gain cycle all over again. Currently, studies are being made to assess the safety of leptin therapy following a weight loss.

ENVIRONMENTAL NOTE

CFCs and Ozone Depletion

The compounds called chlorofluorocarbons (CFCs) were used as propellants for hairs sprays, paints, and as refrigerants in home and car air conditioners. Two widely used CFCs, Freon 11 (CCl_3F) and Freon 12 (CCl_2F_2), were developed in the 1920s as nontoxic refrigerants, which were safer than the sulfur dioxide and ammonia used at the time.

Freon 11   Freon 12

In the stratosphere, a layer of ozone (O_3) absorbs the ultraviolet (UV) radiation of the sun and acts as a protective shield for plants and animals on earth. Ozone is produced in the stratosphere when oxygen reacts with ultraviolet light and breaks into oxygen atoms that quickly combine with oxygen molecules to form ozone.

$$O_2 \xrightarrow{UV} O + O$$
$$O_2 + O \longrightarrow O_3$$

In the 1970s scientists became concerned that CFCs entering the atmosphere were accelerating the depletion of ozone and threatening the stability of the ozone layer. CFCs decompose in the upper atmosphere in the presence of UV light to produce highly reactive chlorine atoms.

$$CCl_3F \xrightarrow{UV\ light} CCl_2F + Cl$$

The reactive chlorine atoms catalyze the breakdown of ozone molecules.

$$Cl + O_3 \longrightarrow ClO + O_2$$
$$ClO + O_3 \longrightarrow Cl + 2O_2$$

It has been estimated that one chlorine atom can destroy as many as 100,000 ozone molecules. Normally, there is a balance between the ozone and oxygen in the atmosphere but the rapid destruction of ozone has upset that equilibrium.

Reports of polar ozone depletion over Antarctica in March 1985 prompted scientists to call for a freeze on the production of

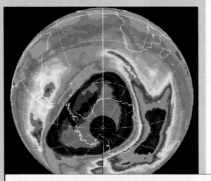

Explore Your World

Explore Your World includes hands-on activities that use everyday materials to encourage students to actively explore selected chemistry topics, either individually or in group-learning environments. Each activity is followed by questions to encourage critical thinking.

EXPLORE YOUR WORLD

Modeling Cis–Trans Isomers

Because cis–trans isomerism is not easy to imagine, here are some things you can do to understand the difference in rotation around a single bond compared to a double bond and how it affects groups that are attached to the carbon atoms in the double bond.

Using Your Hands and Fingers as Single and Double Bonds

Put the fingertips of your index fingers together. This is a model of a single bond. Consider the index fingers as a pair of carbon atoms and think of your thumbs and other fingers as other parts of a carbon chain. While your index fingers are touching, twist your hands and change the position of the thumbs relative to each other. Notice how the relationship of your other fingers changes.

Now place the tips of your index fingers and middle fingers together in a model of a double bond. As you did before, twist your hands to move the thumbs apart. What happens? Can you change the location of your thumbs relative to each other without breaking the double bond? The difficulty of moving your hands with two fingers touching represents the lack of rotation about a double bond. You have made a model of a cis isomer when both thumbs are on the same side. If you

Cis-hands (*cis*-thumbs/fingers)

Trans-hands (*trans*-thumbs/fingers)

turn one hand over so one thumb points down and one points up, you have made a model of a trans isomer.

Using Toothpicks and Gumdrops to Form Single and Double Bonds

Use toothpicks for bonds and gumdrops for atoms. Place a toothpick between two black gumdrops (carbon atoms). To each carbon atom attach three toothpicks with a yellow gumdrop on the other end. Rotate one of the gumdrop carbon atoms to show the movement of the attached atoms.

Trans-gumdrop isomer

Remove a toothpick and yellow gumdrop from each carbon atom. Place a second toothpick between the carbon atoms. Each carbon atom should now be attached to two noncarbon atoms. Make those two atoms different colors, such as yellow and red. First place the red gumdrops on the same side of the double bond, which makes a model of a cis isomer. Try to twist the double toothpicks. Can you do it? If you turn the whole molecule upside down, are the red gumdrops still on the same side?

Now switch one of the red gumdrops with the yellow gumdrop on the same carbon. The red gumdrops should be opposite or trans to each other. Twist the double bond again. Can you move the red gumdrops to the same side of the double bond? It should not happen without breaking the double bond. How does this illustrate two isomers of a double bond molecule when the atoms in the double bond are attached to different groups of atoms? If you attach two yellow or two red gumdrops to the same carbon atom, can you make cis–trans isomers? Why or why not?

Sample Problems appear throughout the text to immediately demonstrate the application of each new concept. The worked-out solution gives step-by-step explanations, provides a problem-solving model, and illustrates required calculations.

Study Checks at the end of each Sample Problem ask the student to work a similar problem.

SAMPLE PROBLEM 14.1

Classifying Alcohols

Classify each of the following alcohols as primary, secondary, or tertiary.

a. $CH_3—CH_2—CH_2—OH$

b. $CH_3—CH_2—\overset{\overset{\displaystyle OH}{|}}{\underset{\underset{\displaystyle CH_3}{|}}{C}}—CH_3$

c. OH

Solution

a. One alkyl group attached to the carbon atom bonded to the —OH makes this a primary alcohol.
b. Three alkyl groups attached to the carbon atom bonded to the —OH makes this a tertiary alcohol.
c. Two alkyl groups attached to the carbon atom bonded to the —OH makes this a secondary alcohol.

Study Check

Classify the following as primary, secondary, or tertiary:

H_3C OH

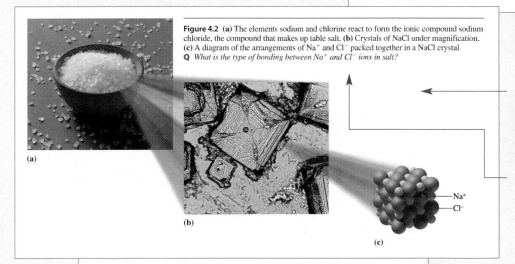

Figure 4.2 (a) The elements sodium and chlorine react to form the ionic compound sodium chloride, the compound that makes up table salt. (b) Crystals of NaCl under magnification. (c) A diagram of the arrangements of Na^+ and Cl^- packed together in a NaCl crystal.
Q *What is the type of bonding between Na^+ and Cl^- ions in salt?*

(a)

(b)

(c)

—Na^+
—Cl^-

Macro-to-micro Illustrations

Macro-to-micro illustrations of familiar objects and phenomena teach students to visualize the behavior of atoms and molecules.

Questions paired with figures challenge students to think critically about photos and illustrations.

QUESTIONS AND PROBLEMS

Amino Acids as Acids and Bases

20.13 Write the zwitterion of each of the following amino acids:
    a. glycine   **b.** cysteine   **c.** serine   **d.** alanine
20.14 Write the zwitterion of each of the following amino acids:
    a. phenylalanine   **b.** methionine   **c.** leucine   **d.** valine
20.15 Write the positive ion (acidic ion) of each of the amino acids in problem 20.13 at a pH below 1.0.
20.16 Write the negative ion (basic ion) of each of the amino acids in problem 20.13 at a pH above 12.0.
20.17 Would the following ions of valine exist at a pH above, below, or at pI?

a. $H_2N—\overset{\overset{\displaystyle CH}{|}}{\underset{\underset{\displaystyle CH_3 \;\; CH_3}{}}{}}—COO^-$ b. $\overset{+}{H_3N}—\overset{\overset{\displaystyle CH}{|}}{\underset{\underset{\displaystyle CH_3 \;\; CH_3}{}}{}}—COOH$ c. $\overset{+}{H_3N}—\overset{\overset{\displaystyle CH}{|}}{\underset{\underset{\displaystyle CH_3 \;\; CH_3}{}}{}}—COO^-$

20.18 Would the following ions of serine exist at a pH above, below, or at pI?

a. $\overset{+}{H_3N}—\overset{\overset{}{|}}{\underset{\underset{\displaystyle CH_2OH}{}}{CH}}—COO^-$ b. $\overset{+}{H_3N}—\overset{\overset{}{|}}{\underset{\underset{\displaystyle CH_2OH}{}}{CH}}—COOH$ c. $H_2N—\overset{\overset{}{|}}{\underset{\underset{\displaystyle CH_2OH}{}}{CH}}—COO^-$

Questions and Problems

Questions and Problems at the end of each section encourage students to immediately apply concepts and begin problem solving after learning a workable amount of information, instead of waiting until the end of the chapter to work problems.

Chapter Review

Chapter Reviews at the end of each chapter briefly review the main ideas in each section.

Chapter Review

8.1 Gases and Kinetic Theory

In a gas, particles are so far apart and moving so fast that their attractions are unimportant. A gas is described by the physical properties of pressure (P), volume (V), temperature (T), and amount in moles(n).

8.2 Gas Pressure

A gas exerts pressure, the force of the gas particles striking the surface of a container. Gas pressure is measured in units of torr, mm Hg, atm, and pascal.

8.3 Pressure and Volume (Boyle's Law)

The volume (V) of a gas changes inversely with the pressure (P) of the gas if there is no change in the amount and temperature: $P_1V_1 = P_2V_2$. This means that the pressure increases if volume decreases; pressure decreases if volume increases.

8.4 Temperature and Volume (Charles' Law)

The volume (V) of a gas is directly related to its Kelvin temperature (T) when there is no change in the amount and pressure of the gas:

Summary of Reactions

Organic and biological chemistry chapters contain additional summaries such as this **Summary of Reactions.**

Summary of Reactions

Hydrogenation

Alkene + H_2 $\xrightarrow{Pt}$ alkane

$CH_2=CH-CH_3 + H_2 \xrightarrow{Pt} CH_3-CH_2-CH_3$

Alkyne + $2H_2$ $\xrightarrow{Pt}$ Alkane

$CH_3-C\equiv CH + 2H_2 \xrightarrow{Pt} CH_3-CH_2-CH_3$

Halogenation

Alkene + Cl_2 (or Br_2) $\longrightarrow$ dihaloalkane

$CH_2=CH-CH_3 + Cl_2 \longrightarrow CH_2-CH-CH_3$ (with Cl, Cl substituents)

Hydrohalogenation

Alkene + HX $\longrightarrow$ haloalkane

$CH_2=CH-CH_3 + HCl \longrightarrow CH_3-CH-CH_3$ (with Cl substituent)
Markovnikov's rule

Hydration of Alkenes

Alkene + H—OH $\xrightarrow{H^+}$ alcohol

$CH_2=CH-CH_3 + H-OH \xrightarrow{H^+} CH_3-CH-CH_3$ (with OH substituent)
Markovnikov's rule

Key Terms

Key Terms appear in bold-faced type in a chapter and are listed as a glossary at the end of the chapter to encourage vocabulary review.

Key Terms

alcohols Organic compounds that contain the hydroxyl (—OH) functional group, attached to a carbon chain, R—OH.

alkoxy group A group that contains oxygen bonded to an alkyl group.

cyclic ethers Compounds that contain an oxygen atom in a carbon ring.

dehydration A reaction that removes water from an alcohol in the presence of an acid to form alkenes at high temperature, or ethers at lower temperatures.

oxidation The loss of two hydrogen atoms from a reactant to give a more oxidized compound, e.g., primary alcohols oxidize to aldehydes, secondary alcohols oxidize to ketones. An oxidation can also be the addition of an oxygen atom as in the oxidation of aldehydes to carboxylic acids.

phenol An organic compound that has an —OH group attached to a benzene ring.

primary (1°) alcohol An alcohol that has one alkyl group bonded to the alcohol carbon atom, R—CH$_2$OH.

Additional Problems

Additional Problems at the end of each chapter combine concepts from that chapter and previous chapters. These problems are more complex mathematically and require more critical thinking; several involve real-life relationships.

Answers to all **Study Checks** and odd-numbered problems are given at the end of the text.

Additional Problems

15.43 Describe the bonds between the carbon and oxygen in the carbonyl group.

15.44 Why does the C=O double bond have a dipole, while the C=C does not?

15.45 Write the constitutional isomers for the carbonyl compounds of C_4H_8O.

15.46 Write the constitutional isomers for the carbonyl compounds of $C_5H_{10}O$.

15.47 Give the IUPAC and common names (if any) for each of the following compounds:

a. [structure]   b. [structure]

15.49 Draw the condensed structural formulas of each of the following:
a. 3-methylcyclopentanone
b. 4-chlorobenzaldehyde
c. 3-chloropropionaldehyde
d. ethyl methyl ketone
e. 3-methylhexanal
f. 2-heptanone, an alarm pheromone of bees

15.50 Draw the condensed structural formulas of each of the following:
a. propionaldehyde
b. 2-chlorobutanal
c. 2-methylcyclohexanone
d. 3,5-dimethylhexanal
e. 3-bromocyclopentanone
f. *trans*-2-hexenal, an alarm pheromone of an ant

A strong media package supports learning and instructional needs at many levels.

The Chemistry Place, Special Edition for *General, Organic, and Biological Chemistry* contains many tools to help students prepare for exams. This web site includes review questions and quiz questions for each chapter, interactive tutorials, career resources, a glossary, an interactive periodic table, and a calculator.

You can also find the contents of the web site on the CD-ROM in your book.

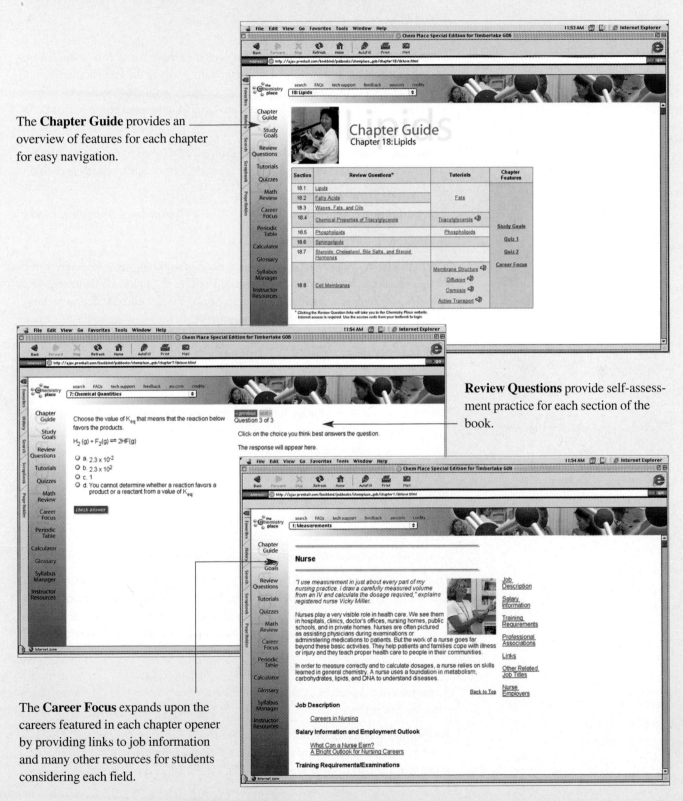

The **Chapter Guide** provides an overview of features for each chapter for easy navigation.

Review Questions provide self-assessment practice for each section of the book.

The **Career Focus** expands upon the careers featured in each chapter opener by providing links to job information and many other resources for students considering each field.

Instructional Package

Organic and Biological Chemistry: Structures of Life is the nucleus of an integrated teaching and learning package of support materials for both professors and students.

For Students

Student Study Guide and Selected Solutions for *Organic and Biological Chemistry* by Karen Timberlake. Keyed to the learning goals in the text and designed to promote active learning through a variety of exercises with answers as well as mastery exams. Also contains complete solutions to odd-numbered problems. (0-8053-2986-2)

Laboratory Manual for *Organic and Biological Chemistry* by Karen Timberlake. Contains 30 experiments for the standard course sequence of topics. (0-8053-2992-7)

Special Edition of The Chemistry Place™ for *Organic and Biological Chemistry*
www.chemplace.com/college
This web site, tailored specifically to *Organic and Biological Chemistry*, offers comprehensive and interactive teaching and learning resources for chemistry. Features include tutorials, simulations, practice quizzes, web links, career resources, and an interactive periodic table. The online syllabus manager allows instructors to post class notes and assignments on the Web. Packaged with every new copy of the book.

The Chemistry Tutor Center
www.aw.com/tutorcenter
Provides one-to-one tutoring by phone, fax, email, and/or the Internet during evening hours and weekends. Qualified college instructors answer questions and provide instruction regarding examples, exercises, and other content from the text.

For Instructors

Instructor's Manual and Complete Solutions by Karen Timberlake. Highlights chapter topics and includes suggestions for the laboratory. Also includes answers and solutions for all problems in the text. (0-8053-3562-5)

Online Instructor Resources by Karen Timberlake. The Chemistry Place, Special Edition for *Organic and Biological Chemistry,* includes many online resources for instructors. The Instructor's Manual highlights chapter topics and includes suggestions for the laboratory. Also includes answers and solutions for all problems in the text. The Instructor's Manual to the Laboratory Manual provides solutions to lab exercises. www.chemplace.com/college

Transparency Acetates. A set of 300 color transparencies and 50 transparency masters with problems from each chapter. (0-8053-2988-9)

Benjamin Cummings Digital Library CD-ROM *General, Organic, and Biological Chemistry: Structures of Life,* and *Organic and Biological Chemistry: Structures of Life.* This easy-to-search CD-ROM includes every figure and table from the book. The images can be exported into PowerPoint presentations, other presentation tools, and web sites. (0-8053-2985-4)

Test Bank by Lynn Carlson, University of Wisconsin, Parkside. Includes 50 multiple choice, 15 matching, and 10 short answer questions per chapter, approximately 2,100 questions total.

Acknowledgments

The preparation of this new text was a continuous effort of many people for a two-year period. As in my work on many editions of *Chemistry: An Introduction to General, Organic, and Biological Chemistry,* I am thankful for the support, encouragement, and dedication of many people who put in hours of tireless effort to produce a high quality book that provides an outstanding learning package. Once again the editorial team at Benjamin Cummings has done an excellent job. I appreciate the work of my Executive Editor, Ben Roberts, and Acquisitions Editor, Maureen Kennedy, who supported my vision of a new book with a new art program and learning package. My Project Editor, Claudia Herman, is like an angel who continually encouraged me on the development of this new text, while she skillfully coordinated reviews, art, website materials, interviews, and all the things it takes to make a book come together. Jean Lake, Production Editor, brilliantly coordinated all phases of the manuscript to the final pages of a beautiful book. Cathy Cobb, Copyeditor, precisely edited the manuscript to make sure the words were correct to help students learn chemistry. Betty Gee was absolutely invaluable in coordinating two photographers and four different artists as they created an art program of over seven hundred pieces.

I am especially proud of the new art program in this text, which lends beauty and understanding to chemistry. The photographs are the creative work of Mark Ong, Art Director, and photographers John Bagley and Richard Tauber, who took wonderfully vivid photographs that tell the story of the beauty of chemistry and its relationship to the structures of life. Photos of chemical compounds make students feel as though they are seeing a chemical in the laboratory. The macro-to-micro features that relate real-life photos to their atomic and molecular structure are a fantastic learning tool for students. There are new three-dimensional ball-and-stick models by J.B. Woolsey and Associates that provide students with the visual impressions and dimensions of organic and biochemical compounds. The entire art program is new and geared to the students' interest in allied health careers. Photos and interviews with health care professionals open each chapter and appear within chapters. I applaud Mark's creative eye as he art-directed and interviewed different professionals in the health care field. It was a big effort and is the keynote for every chapter. Thanks to the Production Coordinator, Tony Asaro, who gathered all the materials for many of the photos in the text, conducted many of the interviews, and agreed to be the "patient" in some of the photos.

It has been especially helpful to have accuracy reviewers for this new text. Without them, much would have gone unnoticed and uncorrected. Thanks to Professors Juliette Bryson and Yuri Zhorov, who read many galley and page proofs, for their valuable comments and corrections. Thanks to Martha Ghent for the hours of proofreading and to Lynn Carlson for her outstanding preparation of the Test Bank. Thanks to an enthusiastic and tireless Marketing Team, Christy Lawrence, Marketing Manager, Stacy Treco, Director of Marketing, and Chalon Bridges, Market Development Manager, whose interviews with professors provided valuable information for the direction of this book.

This text also reflects the contributions of many professors who took the time to review and edit the manuscript, provide outstanding comments, help, and suggestions. I am extremely grateful to an incredible group of peers for their careful assessment of all the new ideas for the text, suggested additions, corrections, changes, and deletions, and an incredible amount of feedback about the best direction for the text. In addition, I appreciate the time that health care professionals took to let us take photos and discuss their work with them. I admire and appreciate every one of you.

Reviewers

Peter Balanda, Ferris State University
Mark Benvenuto, University of Detroit Mercy
Tom Burkholder, Central Connecticut State University
G. Lynn Carlson, University of Wisconsin, Parkside
Ana Ciereszko, Miami-Dade Community College
Mark Chiu, Seton Hall University
Patricia Draves, University of Central Arkansas
Fabian Fang, California State University, Bakersfield
Don Glover, Bradley University
Sharon Kapica, County College of Morris
Robert Kolodny, Armstrong-Atlantic State
Patti Landers, University of Oklahoma
Richard Langley, Steven F. Austin State University
Tim Lubben, Northwestern College
Charlene McMahon, Carroll College
Melvin Merken, Worcester State College
Pamela Mork, Concordia College
Barbara Mowery, Thomas Nelson Community College
Shane Phillips, California State University, Stanislaus
Elizabeth Roberts-Kirchhoff, University of Detroit Mercy
Richard Sheardy, Seton Hall University
Kevin Siebenlist, Marquette University
Howard Silverstein, Georgia Perimeter College
Steve Socol, McHenry County College
John Sowa, Seton Hall University
Karen Wiechelman, University of Louisiana at Lafayette
Don Williams, Hope College
Suzanne Williams, Northern Michigan University

11 Introduction to Organic Chemistry

"The purpose of our research was to create a way to make Taxol," says Paul Wender, Francis W. Bergstrom Professor of organic chemistry and head of the Wender research group at Stanford University. "Taxol is a chemotherapy drug originally derived from the bark of the Pacific yew tree. However, removing the bark from yew trees destroys them, so we need a renewable resource. We worked out a synthesis that began with turpentine, which is both renewable and inexpensive. Initially, Taxol was used with patients who did not respond to chemotherapy. The first person to be treated was a woman who was diagnosed with terminal ovarian cancer and given three to six months to live. After a few treatments with Taxol, she was declared 98% disease–free. A drug like Taxol can save many lives, which is one reason that a study of organic chemistry is so important."

LOOKING AHEAD

www.chemplace.com/college
Visit the URL above or use the CD-ROM in the book for extra quizzing, interactive tutorials, and career resources.

Organic chemistry is the chemistry of carbon compounds. The element carbon has a special role in chemistry because it bonds with other carbon atoms to give a vast array of molecules. The variety of molecules is so great that we find organic compounds in many common products we use such as gasoline, medicine, shampoos, plastic bottles, and perfumes. The food we eat is composed of different organic compounds that supply us with fuel for energy and the carbon atoms needed to build and repair the cells of our bodies. Large organic molecules make up the proteins in hair and skin, the lipids in cell membranes and adipose tissue, and the DNA in cell nuclei.

Although many organic compounds occur in nature, chemists have synthesized even more. The cotton, wool, or silk in your clothes contain naturally occurring organic compounds, whereas materials such as polyester, nylon, or plastic have been synthesized through organic reactions. Sometimes it is convenient to synthesize a molecule in the lab even though that molecule is also found in nature. For example, vitamin C synthesized in a laboratory has the same structure as the vitamin C in oranges or lemons. In these next chapters, you will learn about the structures and reactions of organic molecules, which will provide a foundation for understanding the more complex molecules of biochemistry.

LEARNING GOAL

Identify the number of bonds for carbon and other atoms in organic molecules.

11.1 Organic Compounds

In Chapter 4, we learned that a carbon atom has four valence electrons. By sharing those valence electrons to achieve an octet, a carbon atom forms four covalent bonds. In all organic molecules, all carbon atoms always have four covalent bonds. This is important to remember when you write the structures of organic compounds.

$$\cdot\ddot{C}\cdot\ +\ 4H\cdot\ \longrightarrow\ H\!:\!\underset{\cdot\cdot}{\overset{\cdot\cdot}{C}}\!:\!H\ =\ H\!-\!\underset{\underset{H}{|}}{\overset{\overset{H}{|}}{C}}\!-\!H$$

Methane

In organic molecules with two or more carbon atoms, one or more of the valence electrons is used to form covalent bonds between carbon atoms.

$$H\!:\!\overset{H}{\underset{H}{\overset{\cdot\cdot}{C}}}\!:\!\overset{H}{\underset{H}{\overset{\cdot\cdot}{C}}}\!:\!H\ =\ H\!-\!\overset{\overset{H}{|}}{\underset{\underset{H}{|}}{C}}\!-\!\overset{\overset{H}{|}}{\underset{\underset{H}{|}}{C}}\!-\!H\qquad\text{carbon–carbon single bond}$$

Ethane

$$\overset{H}{\underset{H}{\overset{\cdot\cdot}{C}}}\!::\!\overset{H}{\underset{H}{\overset{\cdot\cdot}{C}}}\ =\ \overset{H}{\underset{H}{\diagup}}C\!=\!C\overset{H}{\underset{H}{\diagdown}}\qquad\text{carbon–carbon double bond}$$

Ethene (ethylene)

In organic compounds, carbon atoms are most likely to bond with hydrogen, oxygen, nitrogen, sulfur, and halogens such as chlorine. Hydrogen with one valence electron forms a single covalent bond. An octet is achieved by nitrogen forming three covalent bonds and oxygen and sulfur each forming two covalent bonds. The halogens with seven valence electrons form one covalent bond to complete octets. Table 11.1 lists the number of covalent bonds most often formed by elements found in organic compounds.

Table 11.1 Covalent Bonds for Elements in Organic Compounds

Element	Group	Covalent Bonds	Structure of Atoms
H	1A	1	H—
C	4A	4	—C—
N	5A	3	—N—
O, S	6A	2	—O— —S—
F, Cl, Br, I	7A	1	—X: (X = F, Cl, Br, I)

SAMPLE PROBLEM 11.1

Carbon Bonds

Complete the following structures by adding the correct number of hydrogen atoms:

a. C—C—C   **b.** C—O—C   **c.** 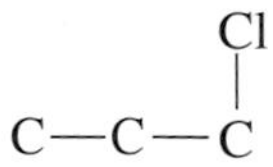 C—C—C with Cl

Solution

We need to add hydrogen atoms to give each carbon atom four covalent bonds, oxygen two bonds, and chlorine one bond.

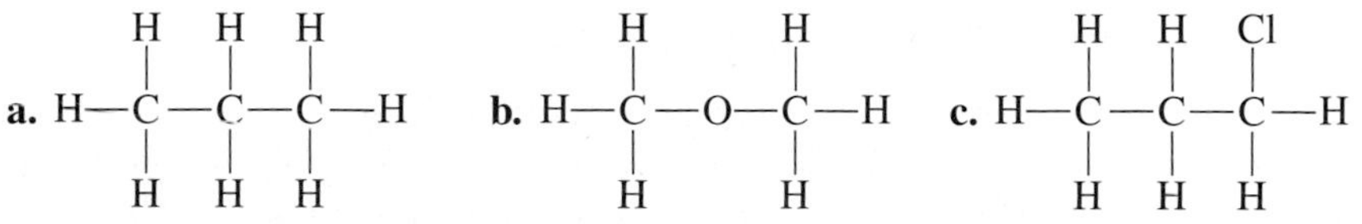

a.
```
    H  H  H
    |  |  |
H—C—C—C—H
    |  |  |
    H  H  H
```
b.
```
    H     H
    |     |
H—C—O—C—H
    |     |
    H     H
```
c.
```
    H  H  Cl
    |  |  |
H—C—C—C—H
    |  |  |
    H  H  H
```

Study Check

Complete the following structure by adding hydrogen atoms.
C—C—N

QUESTIONS AND PROBLEMS

Organic Compounds

11.1 Add hydrogen atoms to the following to give a correct structure:
  a. C—C—C   **b.** C—C—O
  c. C=C—C—N

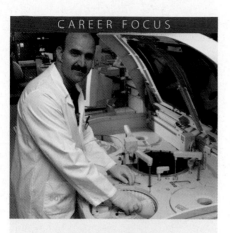

11.2 Add hydrogen atoms to the following to give a correct structure:

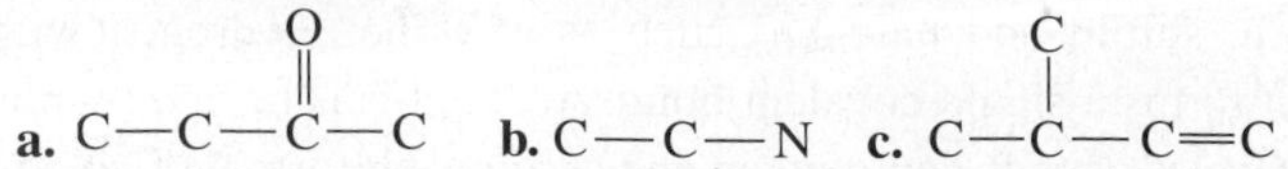

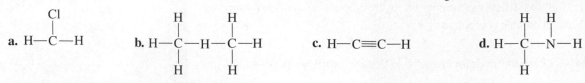

a. C—C—C—C   b. C—C—N   c. C—C—C≡C

11.3 Determine whether each of the following formulas is correct or incorrect:

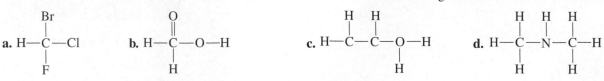

a. H—C—H   b. H—C—H—C—H   c. H—C≡C—H   d. H—C—N—H

11.4 Determine whether each of the following formulas is correct or incorrect:

a. H—C—Cl   b. H—C—O—H   c. H—C—C—O—H   d. H—C—N—C—H

11.2 The Tetrahedral Structure of Carbon

In most organic molecules, a carbon atom is bonded to four other atoms. The VSEPR theory (Chapter 4) predicts that when four bonds are arranged as far apart as possible, they have a tetrahedral shape. For example, in the simplest organic molecule methane, CH_4, the bonds to hydrogen are directed to the corners of a tetrahedron with bond angles of 109.5°. The three-dimensional structure of methane can be illustrated as a ball-and-stick model or a space-filling model.

Chemists often use a shorthand notation to represent the three-dimensional structure. Wedge-shaped bonds show bonds that come toward the reader, while dashed lines depict bonds that go back, away from the reader. (See Figure 11.1.) However, we also write the two-dimensional structure for convenience with the understanding that the bonds to the carbon atoms are in a tetrahedral shape.

Tetrahedral Shapes in Carbon Chains

When there are two or more carbon atoms in a molecule, each carbon retains the tetrahedral shape when it is bonded to four other atoms. For example, the ball-and-stick model of ethane C_2H_6 is based on two tetrahedra attached to each other. (See Figure 11.2.) In this structure, straight lines depict the atoms in the plane of the page, wedges indicate bonds coming forward, and dashed lines show bonds going back of the plane. All the bond angles are close to 109.5°.

SAMPLE PROBLEM 11.2

Structure of Carbon Compounds

In a propane molecule with 3 carbon atoms, what is the geometry around each carbon atom?

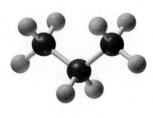

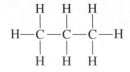

H—C—C—C—H

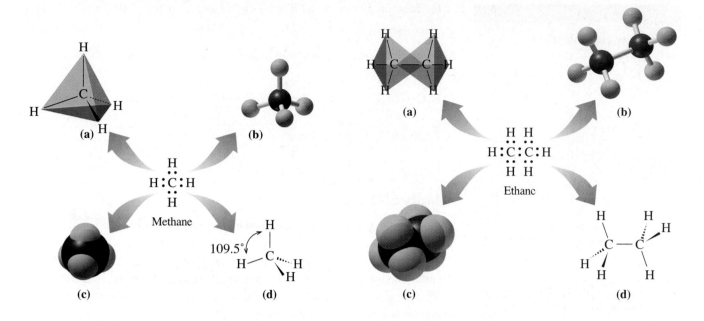

Figure 11.1 Three-dimensional representations of methane, CH₄:
(a) tetrahedron, **(b)** ball-and-stick model, **(c)** space-filling model,
(d) wedges and dashes.
Q *Why does methane have a tetrahedral shape and not a flat shape?*

Figure 11.2 Three-dimensional representations of ethane, C₂H₆:
(a) tetrahedral shape of each carbon, **(b)** ball-and-stick model, **(c)** space-filling model, **(d)** wedges and dashes.
Q *How is the tetrahedral shape maintained in a molecule with two carbon atoms?*

Solution

Each carbon atom is bonded to four other atoms, which gives a tetrahedral shape to each carbon in the molecule.

Study Check

Why does a carbon atom bonded to four bromine atoms have a tetrahedral shape?

QUESTIONS AND PROBLEMS

The Tetrahedral Structure of Carbon

11.5 Why is the structure of the CH_4 molecule three-dimensional rather than two-dimensional?

11.6 In the abbreviated three-dimensional structure of CH_3—CH_3, what do the wedges and dashes represent?

11.3 Polarity of Organic Molecules

LEARNING GOAL

Predict whether an organic molecule is polar or nonpolar.

A covalent bond between two carbon atoms is nonpolar because the electrons are shared equally. However, when a carbon atom bonds to a different atom, the bond is polar covalent. Most of the elements found in organic compounds are more electronegative than carbon, with the exception of hydrogen.

Solubility and Polarity of Hydrocarbons

Fill a small bowl about half full with water. Add a few drops of some mineral oil or a small amount of petroleum jelly and mix. Mineral oil, a laxative and lubricant, is a mixture of liquid hydrocarbons, and petroleum jelly or Vaseline is a colloidal mixture of solid and liquid hydrocarbons.

Questions

1. What do your observations tell you about the solubility of hydrocarbons in water?
2. If water is polar, would you classify mineral oil or petroleum jelly as polar or nonpolar substances?
3. From your observations, are these hydrocarbons more or less dense than water? Explain.

Element	F	O	N	Cl	Br	C	H
Electronegativity value	4.0	3.5	3.0	3.0	2.8	2.5	2.1

As the carbon bonds to atoms with higher electronegativity values, the polarity of the bonds increases.

$$H_3C—CH_3 \qquad H_3C—Br \qquad H_3C—NH_2 \qquad H_3C—OH$$

Nonpolar ⟶ Increasing polarity of covalent bonds

Polarity of Organic Molecules

The polarity of a molecule is determined by the polarity of its bonds and the structure of the molecule. A bond between identical atoms is nonpolar, which makes molecules such as H_2 and Cl_2 nonpolar. A polar bond forms between atoms of different electronegativity values, which makes a molecule such as HCl polar. However, when molecules consist of several polar bonds, the arrangement of the bonds determines whether it is a polar or nonpolar molecule. If a molecule contains a symmetrical arrangement of polar bonds so that the dipoles cancel out, the molecule is nonpolar. For example, CCl_4 is nonpolar because it has a tetrahedral structure that cancels out the C—Cl dipoles. However, if there are different atoms bonded to carbon, the dipoles are not likely to cancel each other, which makes the molecule polar. For example, the CH_3Cl molecule is polar because the C—Cl bond is much more polar than the C—H bonds. (See Figure 11.3.)

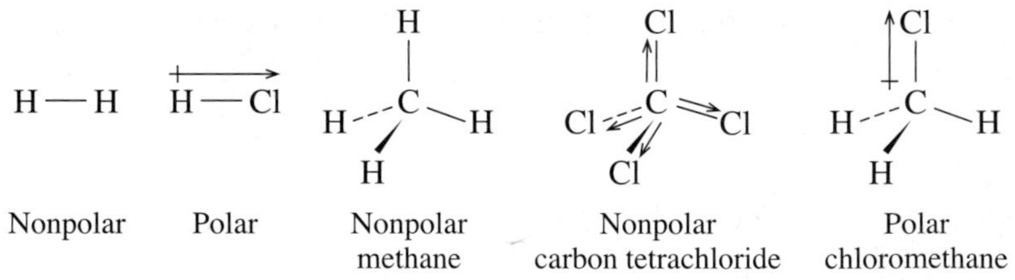

Figure 11.3 Effects of polarity of bonds on the polarity of molecules.
Q *Why are the CH_4 and CCl_4 molecules nonpolar?*

| Nonpolar | Polar | Nonpolar methane | Nonpolar carbon tetrachloride | Polar chloromethane |

SAMPLE PROBLEM 11.3

Polarity of Organic Molecules

Predict whether each of the following molecules will be polar or nonpolar.
a. F_2 **b.** CF_4 **c.** CCl_3F

Solution

a. A bond between two identical atoms such as F—F is nonpolar, which makes the F_2 molecule nonpolar.
b. The CF_4 molecule is nonpolar because the polarities of C—F bonds cancel out in the tetrahedral shape.
c. The CCl_3F molecule is polar because the C—F bond is more polar than the C—Cl bonds.

Study Check

Why is CH_3F a polar molecule?

QUESTIONS AND PROBLEMS

Polarity of Organic Molecules

11.7 Predict whether each of the following molecules will be polar or nonpolar:
  a. Br_2 **b.** CBr_4 **c.** $CHBr_3$ **d.** CH_3Br

11.8 Predict whether each of the following molecules will be polar or nonpolar:
  a. HI   **b.** CH_3OH **c.** CH_2BrF **d.** CCl_4

11.4 Properties of Organic Compounds

LEARNING GOAL

Identify properties characteristic of organic or inorganic compounds.

The polarities of molecules have a strong influence on the physical and chemical behavior of organic compounds. **Organic compounds** containing carbon and hydrogen have covalent bonds and form nonpolar molecules. As a result, the attractions between molecules are weak, which accounts for the low melting and boiling points of carbon compounds. Most organic compounds are not soluble in water. For example, vegetable oil, which is a mixture of organic compounds, does not dissolve in water, but floats on top. Small organic compounds with oxygen or nitrogen atoms are somewhat soluble because the electronegative atom forms hydrogen bonds with water.

Many of the organic compounds undergo combustion and burn vigorously in air. In contrast, many of the inorganic compounds, which we studied earlier, are ionic, which leads to high melting and boiling points. Inorganic compounds that are ionic or polar covalent are usually soluble in water and many produce ions and conduct electrical currents. Most inorganic substances do not burn in air. Table 11.2 contrasts some of the properties associated with organic and inorganic compounds such as propane, C_3H_8, and sodium chloride, NaCl. (See Figure 11.4.)

Table 11.2   Some Properties of Organic and Inorganic Compounds

Property	Organic	Example: C_3H_8	Inorganic	Example: NaCl
Bonding	Covalent	Covalent	Many are ionic, some covalent	Ionic
Polarity of bonds	Nonpolar, unless a more electronegative atom is present	Nonpolar	Most are ionic or polar covalent, a few are nonpolar covalent	Ionic
Melting point	Low	−188°C	High	801°C
Boiling point	Low	−42°C	High	1413°C
Flammability	High	Burns in air	Low	Does not burn
Solubility in water	Not soluble, unless a polar group is present	No	Most are soluble, unless nonpolar	Yes
Particles in solution	Molecules	Molecules	Many produce ions in water	Na^+Cl^-

Figure 11.4 Propane, C_3H_8, is an organic compound, whereas sodium chloride, NaCl, is an inorganic compound.
Q *Why is propane used as a fuel?*

SAMPLE PROBLEM 11.4

Properties of Organic Compounds

Indicate whether the following properties are characteristic of organic or inorganic compounds:

a. Not soluble in water   **b.** High melting point   **c.** Burns in air

Solution

a. Many organic compounds are not soluble in water
b. Inorganic compounds are most likely to have high melting points.
c. Organic compounds are most likely to be flammable.

Study Check

Octane floats on water. What type of compound is octane?

QUESTIONS AND PROBLEMS

Properties of Organic Molecules

11.9 Identify the following as formulas of organic or inorganic compounds:
   a. KCl   **b.** C_4H_{10} **c.** CH_3CH_2OH
   d. H_2SO_4 **e.** $CaCl_2$ **f.** CH_3CH_2Cl

11.10 Identify the following as formulas of organic or inorganic compounds:
   a. $C_6H_{12}O_6$ **b.** Na_2CO_3 **c.** I_2
   d. C_2H_5Cl **e.** $C_{10}H_{22}$ **f.** CH_4

11.11 Identify the following properties as most typical of organic or inorganic compounds:

a. soluble in water   **b.** low boiling point
c. burns in air   **d.** solid at room temperature

11.12 Identify the following properties as most typical of organic or inorganic compounds:

a. high melting point   **b.** gas at room temperature
c. covalent bonds   **d.** produces ions in water

11.5 Functional Groups

Classify organic molecules according to their functional groups.

Organic compounds number in the millions and more are synthesized every day. It might seem that the task of learning organic chemistry would be overwhelming. However, within this vast number of compounds, there are characteristic structural features called **functional groups,** which are a certain group of atoms that react in a predictable way. Compounds with the same functional group undergo similar chemical reactions. The identification of functional groups allows us to classify organic compounds according to their structure and to name compounds within each family. For now we will focus on recognizing the patterns of atoms that make up the functional groups.

Alkanes, Alkenes, and Alkynes

The **hydrocarbons,** which contain only carbon and hydrogen, are the simplest of the organic compounds. The **alkanes** contain only carbon–carbon single bonds. They are also called **saturated hydrocarbons** because they cannot add any more hydrogen atoms to the structure. The alkanes are not very reactive compared to compounds with functional groups, but they serve as a basic structure for the rest of the organic molecules.

Hydrocarbons with double or triple bonds are called **unsaturated hydrocarbons,** because they can add atoms of hydrogen, oxygen, or a halogen. The **alkenes** contain a functional group that is a double bond between two adjacent carbon atoms; **alkynes** contain a triple bond. The alkenes and the alkynes are much more reactive than the single-bonded alkanes.

As we proceed, we will write the structures in an abbreviated form called **condensed structural formulas.** In a condensed structural formula, the hydrogen atoms attached to each carbon are written adjacent to the symbol C for carbon. Thus, a CH_3— is the abbreviation for a carbon attached to three hydrogen atoms, whereas —CH_2— shows a carbon attached to two hydrogen atoms. The functional group is usually written separately so it can be easily recognized.

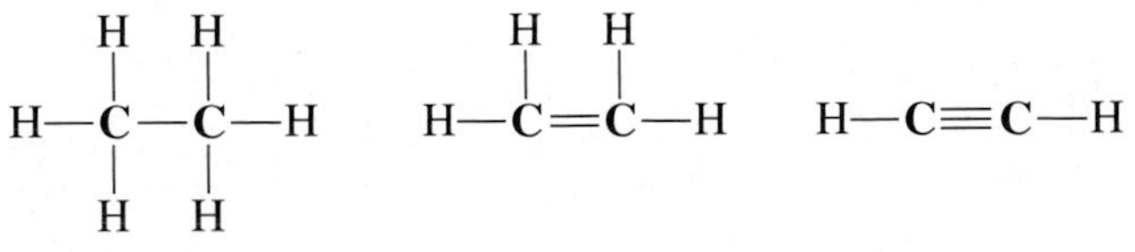

	An alkane	An alkene	An alkyne
Functional group	—C—C—	—C=C—	—C≡C—
Condensed structural formulas	CH_3—CH_3	CH_2=CH_2	HC≡CH

An alcohol

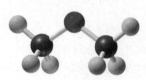

An ether

Alcohols and Ethers

The characteristic functional group in **alcohols** is the **hydroxyl (—OH) group** bonded to a carbon atom. In **ethers,** the characteristic structural feature is an oxygen atom bonded to two carbon atoms. The oxygen atom also has two unshared pairs of electrons, but they will not be shown in the structural formulas.

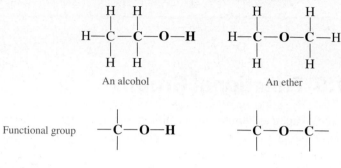

An alcohol   An ether

Functional group

Condensed structural formulas $CH_3—CH_2—OH$ $CH_3—O—CH_3$

Aldehydes and Ketones

The aldehydes and ketones contain a **carbonyl group** $(C{=}O)$, which is a carbon with a double bond to oxygen. In an **aldehyde,** the carbon atom of the carbonyl group is bonded to another carbon and one hydrogen atom. Only the simplest aldehyde, CH_2O, has a carbonyl group attached to two hydrogen atoms. In a **ketone,** the carbonyl group is bonded to two other carbon atoms.

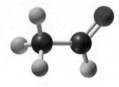

An aldehyde

A ketone

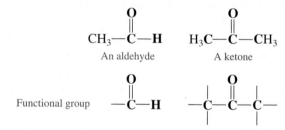

Functional group

SAMPLE PROBLEM 11.5

Classifying Organic Compounds

Classify the following organic compounds according to their functional groups:

a. $CH_3—CH_2—CH_2—OH$ **b.** $CH_3—CH{=}CH—CH_3$

c. 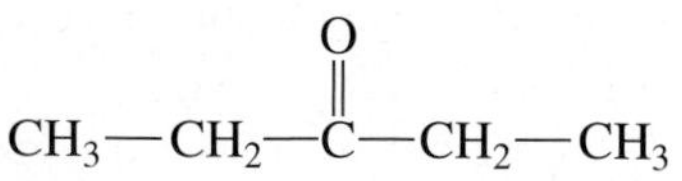 $CH_3—CH_2—\overset{\overset{\textstyle O}{\|}}{C}—CH_2—CH_3$

Solution

a. alcohol   **b.** alkene   **c.** ketone

Study Check

Why is $CH_3—CH_2—O—CH_3$ an ether, but $CH_3—\overset{\overset{\textstyle OH}{|}}{CH}—CH_3$ is an alcohol?

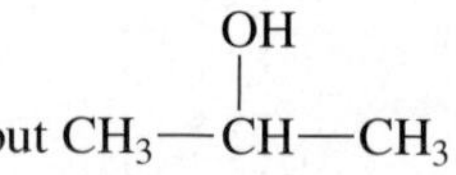

Carboxylic Acids and Esters

In **carboxylic acids,** the functional group is the *carboxyl group,* which is a combination of the *carbo*nyl and hydro*xyl* groups.

$$CH_3-\overset{\displaystyle O}{\overset{\|}{C}}-O-H \quad \text{or} \quad CH_3COOH \quad \text{or} \quad CH_3CO_2H$$

A carboxylic acid

Functional group $\quad -\overset{\displaystyle O}{\overset{\|}{C}}-O-H \quad \text{or} \quad -COOH \quad \text{or} \quad -CO_2H$

A carboxylic acid

An **ester** is similar to a carboxylic acid, except the oxygen of the carboxyl group is attached to a carbon and not to hydrogen.

$$CH_3-\overset{\displaystyle O}{\overset{\|}{C}}-O-CH_3 \quad \text{or} \quad CH_3COOCH_3 \quad \text{or} \quad CH_3CO_2CH_3$$

An ester

Functional group $\quad -\overset{\displaystyle O}{\overset{\|}{C}}-O-\overset{\displaystyle |}{\underset{\displaystyle |}{C}}- \quad \text{or} \quad -COOC- \quad \text{or} \quad -CO_2C-$

An ester

Amines

In **amines,** the central atom is a nitrogen atom. Amines are derivatives of ammonia, NH_3, in which carbon atoms replace one, two, or three of the hydrogen atoms. Amines are classified as primary, secondary, or tertiary amines according to the number of carbon groups bonded to the nitrogen atoms. A nitrogen atom has an unshared pair of electrons, but the lone pair will not be shown in the structural formulas.

A primary amine

NH_3	CH_3-NH_2	$CH_3-\underset{\displaystyle \underset{\displaystyle CH_3}{\|}}{NH}$	$CH_3-\underset{\displaystyle \underset{\displaystyle CH_3}{\|}}{N}-CH_3$
Ammonia	A primary amine (1°)	A secondary amine (2°)	A tertiary amine (3°)

A list of the common functional groups in organic compounds is shown in Table 11.3.

A secondary amine

SAMPLE PROBLEM 11.6

Identifying Functional Groups

Classify the following organic compounds according to their functional groups:

a. $CH_3-CH_2-NH-CH_3$ 　　　**b.** $CH_3CO_2CH_2CH_3$

c. $CH_3-CH_2-\overset{\displaystyle O}{\overset{\|}{C}}-OH$

Solution

a. amine 　　　**b.** ester 　　　**c.** carboxylic acid

Study Check

How does a primary amine differ from a secondary amine?

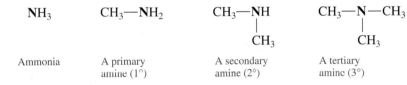

Table 11.3 Classification of Organic Compounds

Class	Example	Functional Group	Characteristic
Alkane	$H_3C{-}CH_3$		Carbon–carbon single bond
Alkene	$H_2C{=}CH_2$		Carbon–carbon double bond
Alkyne	$HC{\equiv}CH$	$-C{\equiv}C-$	Carbon–carbon triple bond
Aromatic	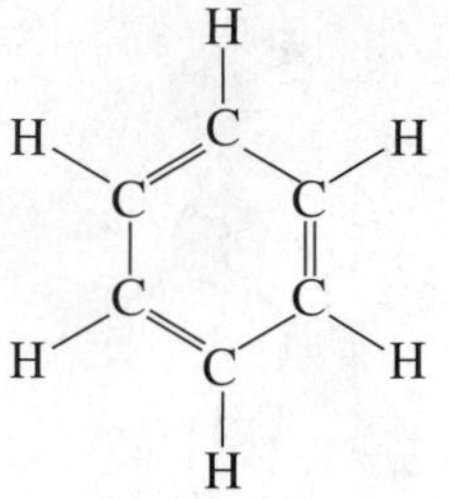	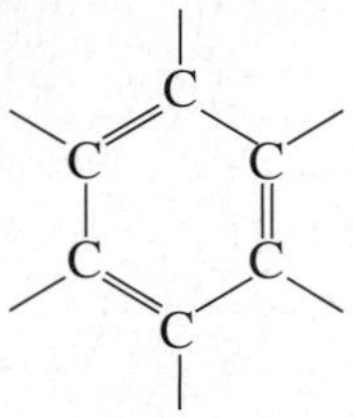	Six-atom ring
Haloalkane	$CH_3{-}Cl$	$-F, -Cl, -Br, -I$	One or more halogen atoms
Alcohol	$CH_3{-}CH_2{-}OH$	$-OH$	Hydroxyl group ($-OH$)
Ether	$H_3C{-}O{-}CH_3$	$-O-$	Oxygen atom bonded to two carbons
Thiol	$CH_3{-}SH$	$-SH$	A $-SH$ group bonded to carbon
Aldehyde			Carbonyl group (carbon-oxygen double bond) with $-H$
Ketone			Carbonyl group (carbon–oxygen double bond) between carbon atoms
Carboxylic acid			Carboxyl group (carbon oxygen double bond and $-OH$)
Ester			Carboxyl group with $-H$ replaced by a carbon
Amine	$CH_3{-}NH_2$		Nitrogen atom with one or more carbon groups
Amide			Carbonyl group bonded to nitrogen

Functional Groups in Familiar Compounds

The flavors and odors of foods and our uses for fuels and many household products can be attributed to the functional groups of organic compounds. As we discuss these familiar products, look for the functional groups we have described.

Hydrocarbons, particularly alkanes, are the compounds that make up natural gas, motor oils, gasoline, and other fuels. The coal and crude oil that provides these hydrocarbons was formed 200 million years ago from the pressures on buried organic matter deep in the earth. Because the hydrocarbons are flammable we use compounds such as methane, propane, and butane as energy sources

$$CH_4 \qquad CH_3-CH_2-CH_3$$
Methane   Propane

$$CH_3-CH_2-CH_2-CH_3$$
Butane

When high temperatures and pressures are used to join small alkenes, extremely long carbon chains called polymers are formed. You may recognize the names of some polymers such as Styrofoam, Teflon, polyethylene, and polystyrene, which are used to make plastic bottles, photographic film, plastic utensils, outdoor clothing, and plastic parts for the body. The polymer polyvinyl chloride (PVC) is used in plastic pipes, garden hoses, and shower curtains. Even in long polymers, every carbon atom that is bonded to four other atoms has a tetrahedral shape.

$$Many \quad H-C=C-Cl \longrightarrow$$

Polyvinyl chloride (PVC)

Perhaps you wrap food in a plastic wrap to keep it fresh. This polymer contains many chlorine atoms that help the plastic adhere to the containers.

Polyvinylidene chloride (Saran)

Ethyl alcohol is the alcohol found in alcoholic beverages. Isopropyl alcohol is another alcohol commonly used to disinfect skin before giving injections and to treat cuts.

$$CH_3-CH_2-OH \qquad CH_3-\overset{\overset{\textstyle OH}{|}}{C}H-CH_3$$
Ethyl alcohol   Isopropyl alcohol

Ketones and aldehydes are in many items we use or eat each day. Acetone or dimethyl ketone is produced in great amounts commercially. Acetone is used as an organic solvent because it dissolves a wide variety of organic substances. You may be familiar with acetone as fingernail polish remover. Ketones and alde-

hydes used in the food industry are found in flavorings such as vanilla, cinnamon, and spearmint. When we buy a small bottle of liquid flavorings, the aldehyde or ketone is dissolved in alcohol because the compounds are not very soluble in water. The aldehyde butyraldehyde adds a "buttery" taste to foods and margarine.

$$CH_3-CH_2-CH_2-\overset{\overset{\textstyle O}{||}}{C}-H$$
Butyraldehyde "butter" flavoring

The sour tastes of vinegar and fruit juices and the pain from ant stings are all due to carboxylic acids. Acetic acid is the carboxylic acid that makes up vinegar. Aspirin also contains a carboxylic acid group. Esters found in fruits produce the pleasant aromas and tastes of bananas, oranges, pears, and pineapples. Esters are also used as solvents in many household cleaners, polishes, and glues.

$$CH_3-\overset{\overset{\textstyle O}{||}}{C}-OH$$
(acetic acid in vinegar)

$$CH_3-\overset{\overset{\textstyle O}{||}}{C}-O-CH_2-CH_2-CH_3$$
Propyl acetate (pears)

$$CH_3-NH_2$$

$$CH_3-\overset{\overset{\textstyle O}{||}}{C}-O-CH_2-CH_2-CH_2-CH_2-CH_3$$
Pentyl acetate (bananas)

One of the characteristics of fish is their odor, which is due to amines. Amines produced when proteins decay have a particularly pungent and offensive odor.

$$H_2N-CH_2-CH_2-CH_2-CH_2-NH_2$$
Putrescine

$$H_2N-CH_2-CH_2-CH_2-CH_2-CH_2-NH_2$$
Cadaverine

Alkaloids are biologically active amines synthesized by plants to ward off insects and animals. Some typical alkaloids include caffeine, nicotine, histamine, and the decongestant epinephrine. Many are painkillers and hallucinogens such as morphine, LSD, marijuana, and cocaine. Certain parts of our neurons have receptor sites that respond to the various alkaloids. By modifying the structures of certain alkaloids to eliminate side effects, chemists have synthesized painkillers and drugs such as Novocain, codeine, and Valium.

QUESTIONS AND PROBLEMS

Functional Groups

11.13 Identify the class of compounds that contains each of the following functional groups.
  a. hydroxyl group attached to a carbon chain
  b. carbon–carbon double bond
  c. carbonyl group attached to two hydrogen atoms
  d. carboxyl group in which the hydrogen is replaced by a carbon atom

11.14 Identify the class of compounds that contains each of the following functional groups:
  a. a nitrogen atom attached to one or more carbon atoms
  b. carboxyl group
  c. oxygen atom bonded to two carbon atoms
  d. a carbonyl group between two carbon atoms

11.15 Classify the following molecules according to their functional groups. The possibilities are alcohol, ether, ketone, carboxylic acid, or amine:

a. $CH_3-CH_2-O-CH_2-CH_3$

b. $CH_3-\underset{\underset{OH}{|}}{CH}-CH_3$

c. $CH_3-\underset{\overset{O}{\|}}{C}-CH_2-CH_3$

d. $CH_3-CH_2-CH_2-COOH$

e. $CH_3-CH_2-NH_2$

11.16 Classify the following molecules according to their functional groups. The possibilities are alkene, aldehyde, carboxylic acid, ester, or amine:

a. $CH_3CH_2COOCH_2CH_3$

b. $CH_3-\underset{\underset{CH_3}{|}}{N}-CH_3$

c. $CH_3-CH_2-CH_2-\underset{\overset{O}{\|}}{C}-H$

d. $CH_3-CH_2-CH_2-COOH$

e. $CH_3-CH=CH-CH_3$

LEARNING GOAL

Write condensed structural formulas for constitutional isomers.

11.6 Constitutional Isomers

As we begin to look at the structure of organic compounds, we will find that several compounds can have the same molecular formula. This property called **isomerism** occurs when compounds have identical molecular formulas, but a different arrangement of atoms. There are several kinds of isomers in organic chemistry, but for now we will look at the **constitutional isomers** (or structural isomers) in which atoms of identical molecular formulas are connected in a different order.

With the exception of some small molecules such as CH_4 and C_2H_6, most organic molecules have isomers and that number increases as the number of carbon atoms increases. For example, there are 3 constitutional isomers with a molecular formula C_5H_{12}, 75 with a molecular formula of $C_{10}H_{22}$, and 4,327 with a molecular formula of $C_{15}H_{32}$. We are not going to write all of their structures, but it does demonstrate another reason for the large number of organic compounds.

We can illustrate this idea by looking at the constitutional isomers with molecular formula C_2H_6O. There are two different compounds with two different functional groups that have this formula. One isomer, ethyl alcohol, is a liquid at room temperature and boils at 78°C. The other isomer, dimethyl ether, is a gas at room temperature with a boiling point of $-24°C$. In the alcohol, the oxygen atom is connected to the end of the carbon atoms, but in the ether, it is between the carbon atoms. (See Figure 11.5.) This difference in the way the carbon and oxygen atoms connect strongly influences the physical and chemical behavior. Ethyl alcohol reacts vigorously with sodium metal producing hydrogen gas. Dimethyl ether does not react with sodium. We conclude that the difference in the structural formulas of these two constitutional isomers accounts for their markedly different physical and chemical properties.

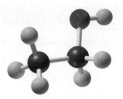

Ethyl alcohol

Dimethyl ether

	Ethyl alcohol	Dimethyl ether
Functional group	hydroxyl group (—OH)	oxygen atom (—O—)
Molecular formula	C_2H_6O	C_2H_6O
Condensed formula	CH_3—CH_2—OH	CH_3—O—CH_3
Molar mass	46	46
Room temperature	liquid	gas
Melting point	$-117°C$	$-138°C$
Boiling point	$78°C$	$-25°C$
Reaction with sodium metal	vigorous; produces H_2	none

Figure 11.5 The constitutional isomers of ethyl alcohol and dimethyl ether can be seen in their ball-and-stick models.
Q *What makes ethyl alcohol and dimethyl ether constitutional isomers?*

SAMPLE PROBLEM 11.7

Constitutional isomers

Indicate whether the following pair of molecules represents constitutional isomers or not.

Aldehyde (C_3H_6O)

Solution

These two compounds are constitutional isomers because they have identical molecular formulas C_3H_6O, but the oxygen atom forms a double bond to a different carbon atom in the chain.

CH_3—CH_2—$\overset{\overset{\textstyle O}{\|}}{C}$—H
Aldehyde (C_3H_6O)

CH_3—$\overset{\overset{\textstyle O}{\|}}{C}$—$CH_3$
Ketone (C_3H_6O)

Ketone (C_3H_6O)

Study Check

Would you expect the two compounds in Sample Problem 11.7 to have the same boiling point? Explain.

HEALTH NOTE

Organic Compounds in Living Organisms

Our investigation into the structure and variety of organic compounds will be the foundation for understanding of the structures and functions of large molecules of life. You may already be familiar with the major biomolecules of carbohydrates, lipids, proteins, and nucleic acids such as the DNA in the nuclei of our cells. Each of these compounds consists of many carbon atoms bonded to hydrogen, but also to oxygen and nitrogen. We might say that the four major elements of life are C, H, O, and N. The tetrahedral geometry for carbon that we described for methane is also true in structures of long chains of carbon atoms.

In plants, photosynthesis combines CO_2 and water to form monosaccharides such as glucose and polysaccharides such as starch and cellulose. Glucose has a six-carbon chain with an aldehyde group and several hydroxyl groups. A variation in the direction of a single ——OH group in the three-dimensional structures gives another monosaccharide galactose. Thus structure plays an important role in the understanding of carbohydrates. When glucose molecules join by ether links, long polymers of complex carbohydrates called polysaccharides are formed, which we know as starch and cellulose. A difference in the direction of the bond between the glucose molecules allows us to digest the starches in our cereals and bread, but not the cellulose in their packaging.

Lipids containing carbon, hydrogen, and oxygen make up a variety of components in our body but you are probably most familiar with body fat and steroids. Body fat is composed of fatty acids, which are carboxylic acids with long carbon chains, bonded to glycerol by ester bonds. Because most lipids are insoluble, they play

Nucleic acids, DNA in cells

Carbohydrates for energy

Lipids, body fats, cell walls, hormones

Protein in skin, hair, muscles, hemoglobin, antibodies, enzymes

a major role in the structure of cell walls, which separate cell contents from aqueous fluids outside the cells. Steroids, a group of lipids that contain four-carbon rings, include cholesterol, hormones, neurotransmitters, and vitamins. By modifying structures of steroids, chemists have developed drugs that mimic certain physiological behavior in the body. For example, changes in the structure of progesterone produced a compound that is now used in birth control pills.

Proteins are composed of carbon, hydrogen, oxygen, and nitrogen. These large polymers consist of small molecules called amino acids that have both amine and carboxylic acid functional groups. The order of amino acids in a protein determines the structure and function of that protein in the body. There are proteins such as hemoglobin that transport oxygen to our cells, contract our muscles, form antibodies, make up our hair, skin, and

enzymes that catalyze biological reactions. In an enzyme, a part of the protein structure binds and reacts only with compounds of a certain shape. Substances that have structures similar to that of the proper reactant can attach to the enzyme and inhibit enzyme action. In enzyme regulation, an end product binds to the enzyme in a way that alters protein structure and decreases enzyme activity.

Nucleic acids are long chains of nucleotides, which contain a carbohydrate called ribose, a nitrogen base, and a phosphate group. Our nucleic acids carry the messages for the production of all the proteins produced in cells. Each time a cell divides the DNA (deoxyribonucleic acid) is duplicated, which places the same instructions for protein synthesis in each new cell. In genetic diseases, mutations alter the nucleic acids causing a change in the proteins, which affect enzyme structure and function in the cell.

SAMPLE PROBLEM 11.8

Writing Constitutional Isomers

Write condensed structural formulas of two alcohols that are constitutional isomers of C_3H_8O.

Solution

Each formula contains three carbon atoms and the hydroxyl group (—OH). We can write two different alcohols by connecting the hydroxyl group to different carbon atoms.

$$\underset{\displaystyle C—C—C}{\overset{\displaystyle OH}{|}} \qquad C—C—C—OH$$

We complete the structural formula of the constitutional isomers by adding the correct number of hydrogen atoms.

$$\begin{array}{c} \text{OH} \\ | \\ \text{CH}_3\text{—CH—CH}_3 \end{array} \qquad \text{CH}_3\text{—CH}_2\text{—CH}_2\text{—OH}$$

Study Check

Draw a condensed structural formula of two amines that are constitutional isomers with the formula C_2H_7N.

QUESTIONS AND PROBLEMS

Constitutional Isomers

11.17 Indicate whether each of the following pairs represents constitutional isomers, identical compounds, or different compounds that are not constitutional isomers:

a. $\text{CH}_3\text{—CH}_2\text{—CH}_2\overset{\displaystyle \overset{O}{\|}}{\text{—C}}\text{—H}$ and $\text{CH}_3\text{—CH}_2\overset{\displaystyle \overset{O}{\|}}{\text{—C}}\text{—CH}_3$

b. $\text{CH}_3\text{—CH}_2\text{—OH}$ and $\text{HO—CH}_2\text{—CH}_3$

c. $\text{CH}_3\text{—CH}_2\text{—CH}_2\text{—CH}_3$ and $\begin{array}{c}\text{CH}_3\\|\\\text{CH}_3\text{—CH—CH}_3\end{array}$

d. $\text{CH}_3\text{—CH}_2\text{—CH}_2\text{—OH}$ and $\begin{array}{c}\text{OH}\\|\\\text{CH}_3\text{—CH—CH}_3\end{array}$

e. $\text{CH}_3\text{CH}_2\text{COOH}$ and $\text{CH}_3\text{COOCH}_3$

f. $\begin{array}{c}\text{OH}\\|\\\text{CH}_3\text{—CH—CH}_3\end{array}$ and $\text{CH}_3\overset{\displaystyle \overset{O}{\|}}{\text{—C}}\text{—CH}_3$

11.18 Indicate whether each of the following pairs of molecules represents constitutional isomers, identical compounds, or different compounds that are not constitutional isomers:

a. $\begin{array}{c}\text{CH}_3\\|\\\text{CH}_3\text{—CH}_2\text{—CH—CH}_2\text{—CH}_3\end{array}$ and $\begin{array}{c}\text{CH}_3\ \ \text{CH}_3\\|\ \ \ \ |\\\text{CH}_3\text{—CH—CH—CH}_3\end{array}$

b. $\text{CH}_3\text{—O—CH}_2\text{—CH}_3$ and $\text{CH}_3\text{—CH}_2\text{—O—CH}_3$

c. $\text{CH}_3\text{—CH}_2\overset{\displaystyle \overset{O}{\|}}{\text{—C}}\text{—H}$ and $\text{CH}_2\text{=CH—CH}_2\text{—OH}$

d. $\text{CH}_3\text{—CH}_2\text{—NH}_2$ and $\begin{array}{c}\text{H}\\|\\\text{CH}_3\text{—N—CH}_3\end{array}$

e. $\text{CH}_3\overset{\displaystyle \overset{O}{\|}}{\text{—C}}\text{—O—CH}_3$ and $\text{CH}_3\text{—CH}_2\overset{\displaystyle \overset{O}{\|}}{\text{—C}}\text{—O—H}$

f. $\text{CH}_3\text{—CH}_2\overset{\displaystyle \overset{O}{\|}}{\text{—C}}\text{—H}$ and $\text{CH}_3\text{—CH}_2\overset{\displaystyle \overset{O}{\|}}{\text{—C}}\text{—O—H}$

11.19 Draw condensed structural formulas of a carboxylic acid and two esters that have the molecular formula $C_3H_6O_2$.

11.20 Draw condensed structural formulas for two alcohols and two ethers with a molecular formula $C_4H_{10}O$.

Chapter Review

11.1 Organic Compounds

In organic compounds, carbon atoms share four valence electrons to form four covalent bonds. In organic molecules with two or more carbon atoms, additional valence electrons are used to form carbon–carbon covalent bonds. Although carbon is most often bonded to hydrogen, it can also bond to oxygen, nitrogen, sulfur, or a halogen such as chlorine or bromine.

11.2 The Tetrahedral Structure of Carbon

The VSEPR theory predicts that the bonds in a carbon atom bonded to four atoms are arranged as far apart as possible to give a tetrahedral shape. In the simplest organic molecule, methane, CH_4, the four bonds that bond hydrogen to the carbon atom are directed out to the corners of a tetrahedron with bond angles of $109.5°$.

11.3 Polarity of Organic Molecules

Polar bonds result when carbon is bonded to atoms with higher electronegativity values such as oxygen, nitrogen, or one of the halogens. If polar bonds are arranged symmetrically, as in a tetrahedron, their dipoles cancel, which makes the molecule nonpolar. A molecule of CCl_4 is nonpolar even though it consists of polar $C-Cl$ bonds. However, if one of the bonds is more polar ($C-Cl$) than the others as in CH_3Cl, the dipoles do not cancel, which makes the molecule polar.

11.4 Properties of Organic Compounds

Most organic compounds have covalent bonds and form nonpolar molecules. Often they have low melting points and low boiling points, are not very soluble in water, produce molecules in solutions, and burn vigorously in air. In contrast, many inorganic compounds are ionic or contain polar covalent bonds and form polar molecules. Many have high melting and boiling points, are usually soluble in water, produce ions in water, and do not burn in air.

11.5 Functional Groups

An organic molecule contains a characteristic group of atoms called a functional group that determines the molecule's family name and chemical reactivity. Functional groups are used to classify organic compounds, act as reactive sites in the molecule, and provide a system of naming for organic compounds. Some common functional groups include the hydroxyl group ($-OH$) in alcohols, the carbonyl group ($C=O$) in aldehydes and ketones, and a nitrogen atom ($-NH_2$) in amines.

11.6 Constitutional Isomers

Constitutional isomers (or structural isomers) are compounds that have the same molecular formula, but differ in the order that their atoms are connected. They have the same numbers of atoms, the same molar mass, but differ in their physical and chemical properties such as solubility, boiling points, and reactivity.

Key Terms

alcohols A class of organic compounds that contains the hydroxyl ($-OH$) group bonded to a carbon atom.

aldehydes A class of organic compounds that contains a carbonyl group ($C=O$) bonded to at least one hydrogen atom.

alkanes Hydrocarbons containing only single bonds between carbon atoms.

alkenes Hydrocarbons that contain carbon–carbon double bonds.

alkynes Hydrocarbons that contain carbon–carbon triple bonds.

amines A class of organic compounds that contains a nitrogen atom bonded to carbon atoms.

carbonyl group A functional group that contains a double bond between a carbon atom and an oxygen atom.

carboxylic acids A class of organic compounds that contains the functional group ($-COOH$).

condensed structural formulas An abbreviated structure in which the hydrogen atoms attached to each carbon atom are written adjacent to the symbol for carbon.

constitutional isomers Isomers having the same molecular formula but a different order of bonding in the structure.

esters A class of organic compounds that contains a $-COO-$ group with an oxygen atom bonded to carbon.

ethers A class of organic compounds that contains an oxygen atom bonded to two carbon atoms.

functional group A group of atoms that determine the physical and chemical properties and naming of a class of organic compounds.

hydrocarbons Organic compounds consisting of only carbon and hydrogen.

hydroxyl group The group of atoms ($-OH$) characteristic of alcohols.

isomerism The property of organic compounds in which identical molecular formulas have different arrangements of atoms.

ketones A class of organic compounds in which a carbonyl group is bonded to two carbon atoms.

organic compounds Compounds made of carbon that typically have covalent bonds, nonpolar molecules, low

melting and boiling points, are insoluble in water, and flammable.

saturated hydrocarbon A compound of carbon and hydrogen that contains only single carbon–carbon (C—C) bonds.

unsaturated hydrocarbon A compound of carbon and hydrogen that contains at least one double carbon–carbon (C=C) or triple (C≡C) bond.

Additional Problems

11.21 Compare organic and inorganic compounds in terms of:
   a. types of bonds   **b.** solubility in water
   c. melting points   **d.** flammability

11.22 Identify each of the following compounds as organic or inorganic:
   a. Na_2SO_4 **b.** CH_2=CH_2
   c. Cr_2O_3 **d.** $C_{12}H_{22}O_{11}$

11.23 Match the following physical and chemical properties with the compounds butane, C_4H_{10}, or potassium chloride, KCl.
   a. melts at $-138°C$ **b.** burns vigorously in air
   c. melts at $770°C$ **d.** produces ions in water
   e. is a gas at room temperature

11.24 Match the following physical and chemical properties with the compounds cyclohexane, C_6H_{12}, or calcium nitrate, $Ca(NO_3)_2$.
   a. contains only covalent bonds
   b. melts above $500°C$
   c. insoluble in water
   d. liquid at room temperature
   e. produces ions in water

11.25 Complete the following structures by adding the correct number of hydrogen atoms:
   a. C—C   **b.** C—C—O

   c. C—C—C (with O double bonded to third C)   **d.** C—O—C—C

11.26 Identify the errors in the following formulas:
   a. CH_4F **b.** CH_3—CH_2

   c. CH_3≡CH_3 **d.** CH_3CH=C—H (with O double bonded to C)

11.27 Distinguish between the following:
   a. a hydroxyl group and carbonyl group
   b. an alcohol and an ether
   c. a carboxylic acid and an ester

11.28 Classify the following as alkanes, alkenes, alkynes, alcohols, ethers, aldehydes, ketones, carboxylic acids, esters, or amines.
   a. CH_3NH_2

   b. CH_3—C—CH_3 (with O double bonded to C)

   c. CH_3—CH—CH_3 (with OH on middle C)
   d. CH_3—CH_2—CH_3
   e. CH_3—C≡CH
   f. CH_3—O—CH_2—CH_3

11.29 Identify the pairs of compounds that are constitutional isomers, identical compounds, or different compounds that are not constitutional isomers:

   a. CH_3—CH—Cl and Cl—CH_2—CH_2—Cl (with Cl on middle C of first)

   b. CH_3CH_2CH and CH_3CH_2COH (both with O double bonded to C)

   c. CH_3—O—$CH_2CH_2CH_3$ and CH_3CH_2—O—CH_2CH_3

   d. CH_3—O—CH_2CH_3 and CH_3—C—CH_3 (second with O double bonded to C)

   e. CH_3—C—O—CH_3 and CH_3—CH_2—C—OH (both with O double bonded to C)

   f. CH_3—C—CH_2—CH_3 and CH_3—CH_2—C—CH_3 (both with O double bonded to C)

   g. CH_3—CH_2—CH_2—CH_3 and CH_3—CH—CH_2—CH_3 (second with CH_3 on middle C)

11.30 Draw condensed structural formulas for the following constitutional isomers with a molecular formula C_3H_9N:

    a. two primary amines

    b. a secondary amine

    c. a tertiary amine

11.31 Predict whether each of the following molecules will be polar or nonpolar:

    a. CH_3F **b.** $CHCl_3$ **c.** CF_4

    d. CH_3-CH_2-Cl **e.** CH_3OH **f.** CH_2F_2

11.32 Write structural formulas of two esters and a carboxylic acid that are constitutional isomers with molecular formula $C_3H_6O_2$.

11.33 Match each of the following terms with the corresponding description:

hydrocarbon, alcohol, ether, aldehyde, ketone, carboxylic acid, ester, amine, saturated hydrocarbon, unsaturated hydrocarbon, functional group, tetrahedral, constitutional isomers

    a. An organic compound that contains a hydroxyl group bonded to a carbon.

    b. A hydrocarbon that contains one or more carbon–carbon double bonds.

    c. An organic compound in which the carbon of a carbonyl group is bonded to a hydrogen.

    d. A hydrocarbon that contains only carbon–carbon single bonds.

    e. An organic compound in which the carbon of a carbonyl group is bonded to a hydroxyl group.

    f. An organic compound that contains a nitrogen atom bonded to one or more carbon atoms.

    g. The three-dimensional shape of a carbon bonded to four hydrogen atoms.

    h. Organic compounds with identical molecular formulas that differ in the order the atoms are connected.

    i. An organic compound in which the hydrogen atom of a carboxyl group is replaced by a carbon atom.

    j. Organic compounds that contain only carbon and hydrogen atoms.

    k. An organic compound that contains an oxygen atom bonded to two carbon atoms.

    l. A hydrocarbon that contains a carbon–carbon triple bond.

    m. A charactistic group of atoms that make compounds behave and react in a particular way.

    n. An organic compound in which the carbonyl group is bonded to two carbon atoms.

12 Alkanes

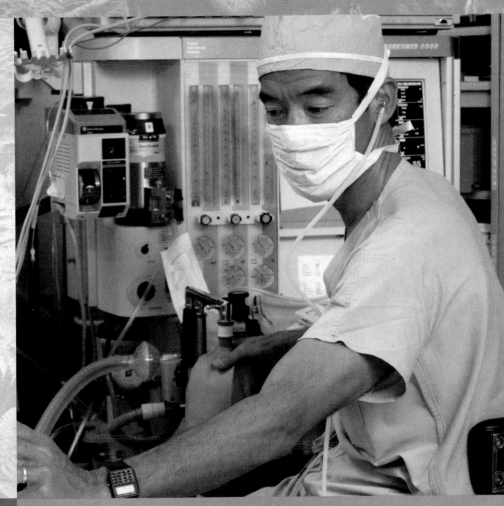

"During surgery, I work with the surgeon to provide a safe level of anesthetics that renders the patient free from pain," says Mark Noguchi, nurse anesthetist (CRNA), Kaiser Hospital. "We do spinal and epidural blocks as well as general anesthetics, which means the patient is totally asleep. We use a variety of pharmaceutical agents including halothane ($C_2HBrClF_3$), and bupivacain ($C_{18}H_{28}N_2O$), as well as muscle relaxants such as midazolam ($C_{18}H_{13}ClFN_3$) to achieve the results we want for the surgical situation. We also assess the patient's overall hemodynamic status. If blood is lost, we replace components such as plasma, platelets, and coagulation factors. We also monitor the heart rate and run EKGs to determine cardiac function."

LOOKING AHEAD

the **Chemistry** place

www.chemplace.com/college

Visit the URL above or use the CD-ROM in the book for extra quizzing, interactive tutorials, and career resources.

At the beginning of the nineteenth century, scientists classified chemi-cal compounds as inorganic and organic. An inorganic compound was a substance that was composed of minerals, and an organic com-pound was a substance that came from an organism, thus the use of the word "organic." It was thought that some type of "vital force," which could only be found in living cells, was required to synthesize an organic compound. This perception was shown to be incorrect in 1828, when the German chemist Friedrick Wöhler synthesized urea, a prod-uct of protein metabolism, by heating an inorganic compound ammo-nium cyanate.

$$NH_4CNO \xrightarrow{\text{Heat}} H_2N-\overset{\displaystyle \overset{O}{\|}}{C}-NH_2$$

Ammonium
cyanate
(inorganic)   Urea, organic

We now define organic chemistry as the study of hydrocarbons and their derivatives. As we go into the twenty-first century, we have iden-tified around ten million organic compounds. When carbon atoms bond, they form chains of carbon atoms that can be hundreds or thou-sands of atoms long. There are also many carbon compounds with identical molecular formulas, but different structures because they have different arrangements of atoms. In this chapter, we look at alka-nes and the ways their atoms are arranged in carbon chains, chains with carbon branches, and rings of carbon atoms.

12.1 Alkanes

In Chapter 11, we looked at several families of organic compounds. Among these are the hydrocarbons, compounds that contain only carbon and hydrogen. The **alkanes** are hydrocarbons that contain only carbon–carbon single bonds. The gen-eral formula for an alkane is C_nH_{2n+2}, where n is the number of carbon atoms. The number of hydrogen atoms is twice the number of carbon atoms plus two more. Some examples follow:

General Formula (C_nH_{2n+2}) *for Some Alkanes*	*C atoms*	*H atoms*
CH_4	1	$4 = (2 \times 1) + 2$
C_2H_6	2	$6 = (2 \times 2) + 2$
C_4H_{10}	4	$10 = (2 \times 4) + 2$
$C_{10}H_{22}$	10	$22 = (2 \times 10) + 2$

When relatively few organic compounds were known, names were assigned in a random fashion. Today the **IUPAC system** (International Union of Pure and Applied Chemistry) determines the protocol for naming organic compounds. The names for the first four continuous-chain saturated alkanes are methane, ethane, propane, and butane. Then the IUPAC system uses Greek prefixes such as *pent (5),* *hex (6),* and *hept (7)* to indicate the number of carbon atoms in the chain. The suf-fix *ane* indicates that the compounds are in the alkane family.

Table 12.1   IUPAC Names for the First Ten Continuous-Chain Alkanes

Number of Carbon Atoms	Prefix	Name	Molecular Formula	Condensed Structural Formula
1	Meth	Methane	CH_4	CH_4
2	Eth	Ethane	C_2H_6	CH_3-CH_3
3	Prop	Propane	C_3H_8	$CH_3-CH_2-CH_3$
4	But	Butane	C_4H_{10}	$CH_3-CH_2-CH_2-CH_3$
5	Pent	Pentane	C_5H_{12}	$CH_3-CH_2-CH_2-CH_2-CH_3$
6	Hex	Hexane	C_6H_{14}	$CH_3-CH_2-CH_2-CH_2-CH_2-CH_3$
7	Hept	Heptane	C_7H_{16}	$CH_3-CH_2-CH_2-CH_2-CH_2-CH_2-CH_3$
8	Oct	Octane	C_8H_{18}	$CH_3-CH_2-CH_2-CH_2-CH_2-CH_2-CH_2-CH_3$
9	Non	Nonane	C_9H_{20}	$CH_3-CH_2-CH_2-CH_2-CH_2-CH_2-CH_2-CH_2-CH_3$
10	Dec	Decane	$C_{10}H_{22}$	$CH_3-CH_2-CH_2-CH_2-CH_2-CH_2-CH_2-CH_2-CH_2-CH_3$

It is important to learn the prefixes in Table 12.1 because they indicate the same number of carbon atoms in the names of other kinds of organic compounds.

Structural Formulas

The **structural formulas** for an alkane show the order of bonding for the carbon atoms in the chain. In the **expanded structural formula**, all the individual atoms and their covalent bonds are drawn. In a **condensed structural formula**, each carbon atom is grouped with its bonded hydrogen atoms. A subscript indicates the number of hydrogen atoms bonded to each carbon atom. The C—H bonds are understood but not written individually.

$$
\begin{array}{c} H \\ | \\ H-C- \\ | \\ H \end{array} \text{ becomes } CH_3- \quad \text{and} \quad \begin{array}{c} H \\ | \\ -C- \\ | \\ H \end{array} \text{ becomes } -CH_2-
$$

In **continuous alkanes**, the carbon atoms are connected in a row. However, in chains of three carbon atoms or more, the carbon atoms do not actually lie in a straight line. The tetrahedral shape of carbon (Chapter 11) arranges the carbon bonds in a zigzag pattern, which is seen in the ball-and-stick model of hexane. (See Figure 12.1.) An abbreviated structure called the **line-bond formula** shows only the bonds from carbon to carbon. The ends of the lines and the corners where the lines meet are understood to be carbon atoms attached to the proper number of hydrogen atoms to give four bonds. In Table 12.2, we see how these structural formulas are written to represent the first three alkanes methane, ethane, and propane. Note that the molecular formula gives the total number of carbon and hydrogen atoms, but does not indicate the arrangement of atoms in the molecule.

Figure 12.1 A ball-and-stick model of hexane.

Q *Why do the carbon atoms in hexane appear to be arranged in a zigzag chain?*

Conformation of Alkanes

Another structural property of alkanes is the rotation of atoms around each carbon–carbon single bond. In an alkane, the groups attached to each carbon are

Table 12.2　Writing Structural Formulas for Some Alkanes

Alkane	Methane	Ethane	Propane
Molecular formula	CH_4	C_2H_6	C_3H_8
Structural formulas			
Expanded	$H-\underset{\underset{H}{\mid}}{\overset{\overset{H}{\mid}}{C}}-H$	$H-\underset{\underset{H}{\mid}}{\overset{\overset{H}{\mid}}{C}}-\underset{\underset{H}{\mid}}{\overset{\overset{H}{\mid}}{C}}-H$	$H-\underset{\underset{H}{\mid}}{\overset{\overset{H}{\mid}}{C}}-\underset{\underset{H}{\mid}}{\overset{\overset{H}{\mid}}{C}}-\underset{\underset{H}{\mid}}{\overset{\overset{H}{\mid}}{C}}-H$
Condensed	CH_4	CH_3-CH_3	$CH_3-CH_2-CH_3$ or $CH_3\diagdown^{CH_2}\diagup CH_3$
Line-bond		—	⋀

not in fixed positions, but rotate freely about the bond connecting the two carbon atoms. This motion is analogous to the independent rotation of the wheels of a toy car. The different arrangements that occur during the rotation about a single bond are called **conformations.**

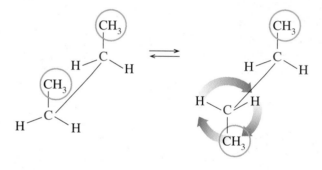

Suppose we could look down the center C—C bond in butane, C_4H_{10}, as it rotates. Sometimes the CH_3 groups line up in front of each other, and at other times they are opposite each other. (See Figure 12.2.) As the CH_3 groups turn around the single bond, many conformations are possible, some of which are shown in Figure 12.3. Although conformations constantly change during the rotation, the conformation in which the CH_3 groups are farthest apart is the most stable. The conformations of butane are depicted by a variety of two-dimensional structural formulas in Table 12.3. It is important to recognize that all of these structural formulas represent the same continuous chain of four carbon atoms.

Figure 12.2 As the CH_3 groups rotate freely around a carbon–carbon single bond in butane, they have different conformations.

Q *Why are the CH_3 groups in butane close to each other in one conformation and far apart from each other in another conformation?*

SAMPLE PROBLEM 12.1

Drawing Expanded, Condensed, and Line-Bond Structural Formulas

Draw expanded, condensed, and line-bond structural formulas for pentane.

Solution

The expanded structural formula for pentane, C_5H_{12}, is written with 5 carbon atoms bonded in a continuous chain connected to hydrogen atoms. In the condensed structural formula each carbon atom is grouped with its attached hydrogens. The line-bond formula shows only the bonds connecting the carbon atoms.

$H-\underset{\underset{H}{\mid}}{\overset{\overset{H}{\mid}}{C}}-\underset{\underset{H}{\mid}}{\overset{\overset{H}{\mid}}{C}}-\underset{\underset{H}{\mid}}{\overset{\overset{H}{\mid}}{C}}-\underset{\underset{H}{\mid}}{\overset{\overset{H}{\mid}}{C}}-\underset{\underset{H}{\mid}}{\overset{\overset{H}{\mid}}{C}}-H$　　　$CH_3-CH_2-CH_2-CH_2-CH_3$　　　⋁⋀⋁

Expanded formula　　　　　　　　　　Condensed　　　　　　　　　Line-bond

Table 12.3   Some Structural Formulas and Conformations for Butane C₄H₁₀

Expanded structural formula

$$H-\overset{\displaystyle H}{\underset{\displaystyle H}{C}}-\overset{\displaystyle H}{\underset{\displaystyle H}{C}}-\overset{\displaystyle H}{\underset{\displaystyle H}{C}}-\overset{\displaystyle H}{\underset{\displaystyle H}{C}}-H$$

Condensed structural formulas

CH₃—CH₂—CH₂—CH₃

CH₃—CH₂
      |
    CH₂—CH₃

CH₂—CH₂
 | |
CH₃ CH₃

CH₃
 |
CH₂—CH₂—CH₃

CH₃
 |
CH₂—CH₂
     |
    CH₃

CH₃,   CH₂,
   CH₂   CH₃

CH₃
 |
CH₂
 |
CH₂
 |
CH₃

Line-bond formula

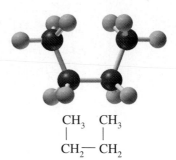

CH₃ CH₃
 | |
CH₂—CH₂

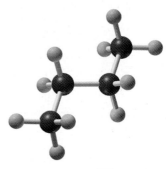

        CH₃
         |
CH₂—CH₂
 |
CH₃

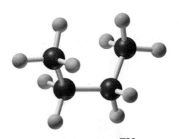

           CH₃
            |
CH₃—CH₂—CH₂

Figure 12.3 In the possible conformations of butane, the CH₃ groups may be close to each other or far apart, which is the most stable arrangement.
Q *Why is it possible for a butane molecule to have many conformations?*

Study Check

Draw the condensed structural formula for the following line-bond formula:

QUESTIONS AND PROBLEMS

Alkanes

12.1 Use the general formula for alkanes to determine the following:
   a. the molecular formula of an alkane with 7 carbon atoms
   b. the number of hydrogen atoms bonded to 5 carbon atoms
   c. the number of carbon atoms that are bonded to 10 hydrogen atoms

12.2 Use the general formula for alkanes to determine the following:
   a. the molecular formula of an alkane with 9 carbon atoms
   b. the number of hydrogen atoms bonded to 8 carbon atoms
   c. the number of carbon atoms that are bonded to 26 hydrogen atoms

12.3 Write the stated type of structural formula for a continuous chain of each of the following alkanes:
   a. an expanded structural formula for propane
   b. the condensed structural formula for hexane
   c. a line-bond formula for hexane

12.4 Write the stated type of structural formula for a continuous chain of each of the following alkanes:
   a. an expanded structural formula for butane
   b. the condensed structural formula for octane
   c. a line-bond formula for decane

12.5 Give the IUPAC name for each of the following alkanes.

a. CH_3
   CH_2—CH_2—CH_2
                CH_3

b. /\/\/\

c. CH_3—CH_2—CH_2
   CH_2
   CH_2—CH_3

12.6 Give the IUPAC name for each of the following alkanes.

a. CH_4

b. /\/\/\/\

c. CH_3
   CH_2
   CH_3
   CH_3

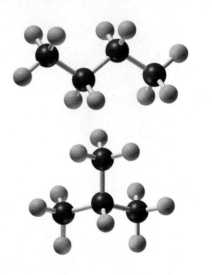

Figure 12.4 The constitutional isomers of C_4H_{10} have the same number and type of atoms, which are bonded in a different order.
Q *What makes these molecules constitutional isomers?*

LEARNING GOAL
Write the IUPAC names for alkanes.

12.2 IUPAC Naming System for Alkanes

When two molecules have the same molecular formula but different arrangements of atoms, they are called **constitutional isomers** (Chapter 11). For example, there are two different condensed structural formulas for alkanes with the molecular formula C_4H_{10}. (See Figure 12.4.) Although they contain the same type and number of atoms, the atoms are connected in a different order. In one molecule, the carbon atoms are in a continuous carbon chain. In the other, a carbon side group called a **branch** or **substituent** replaces a hydrogen atom on a carbon chain. An alkane with one or more branches is called a **branched-chain alkane.** Constitutional isomers have different properties such as boiling point, melting point, and density, as shown in Table 12.4.

SAMPLE PROBLEM 12.2

Constitutional Isomers

Identify each pair of structural formulas as constitutional isomers or conformations of the same molecule.

a. CH_3 CH_3
   CH_2—CH_2 and CH_2—CH_2—CH_3
                               CH_3

b. CH_3—CH—CH_2—CH_2—CH_3 and CH_3—CH—CH—CH_3
              CH_3 CH_3 CH_3

Solution

a. The structural formulas represent conformations of the same molecule because each has four atoms in a continuous chain.

b. These are constitutional isomers because the molecular formula C_6H_{14} is identi-

Table 12.4 Some Properties of Constitutional Isomers of C$_4$H$_{10}$

Condensed formula	$CH_3-CH_2-CH_2-CH_3$	$CH_3-\overset{\overset{\displaystyle CH_3}{\displaystyle |}}{CH}-CH_3$
Melting point	$-138°C$	$-159°C$
Boiling point	$-1°C$	$-12°C$
Density	0.58 g/mL	0.55 g/mL

cal but the atoms are bonded in a different order. One has one CH_3 side group attached to a five-carbon chain, and the other has two CH_3 side groups attached to a four-carbon chain.

Study Check

The following compound has the same molecular formula as the compounds in part b of the preceding Sample Problem. Is the following structural formula a constitutional isomer or identical to one of the molecules in part b?

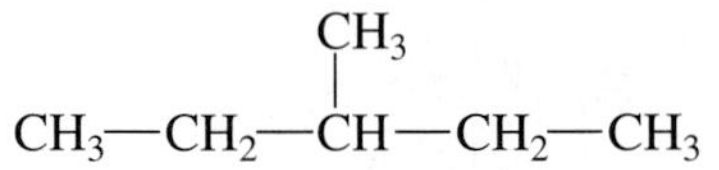

$$CH_3-CH_2-\overset{\overset{\displaystyle CH_3}{\displaystyle |}}{CH}-CH_2-CH_3$$

Classifying Carbon Atoms in Hydrocarbons

In a hydrocarbon, each carbon atom is classified according to the number of carbon atoms connected to it. A **primary** (1°) **carbon** is bonded to one other carbon atom. A **secondary** (2°) **carbon** has two carbon atoms attached to it. A **tertiary** (3°) **carbon** is bonded to three other carbon atoms.

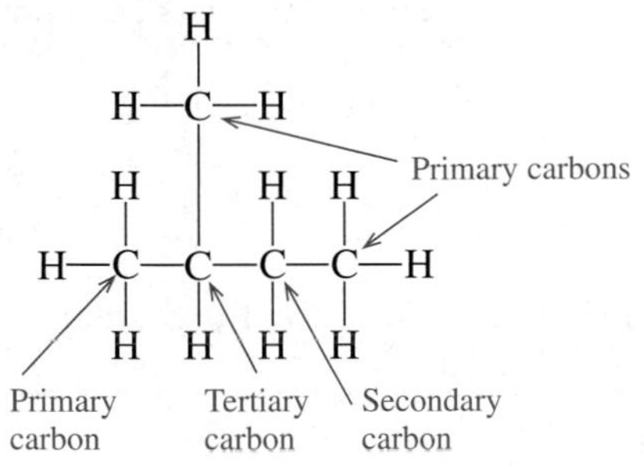

Alkyl Groups in Branched Alkanes

In the IUPAC names for branched-chain alkanes, the substituents are named as alkyl groups. An **alkyl group** is an alkane that is missing one hydrogen atom. The alkyl group is named by replacing the *ane* ending of the corresponding alkane name with *yl*. Alkyl groups do not exist on their own: they must be attached to something such as a carbon chain. Some of the common alkyl groups are illustrated in Table 12.5.

Some of the alkyl groups use common names such as isopropyl and isobutyl. The classification of the carbon in the alkyl group attached to a carbon chain is used in the names secondary-butyl (*sec*-butyl) and tertiary-butyl (*tert*-butyl).

Table 12.5   Names and Formulas of Some Common Alkyl Groups

Alkane	Name of Alkane	Corresponding Alkyl Group	Name of Alkyl Group
One carbon			
CH_4	Methane	CH_3—	Methyl
Two carbons			
CH_3—CH_3	Ethane	CH_3—CH_2—	Ethyl
Three carbons			
CH_3—CH_2—CH_3	Propane	CH_3—CH_2—CH_2—	Propyl
		CH_3—CH—CH_3	Isopropyl
Four carbons			
CH_3—CH_2—CH_2—CH_3	Butane	CH_3—CH_2—CH_2—CH_2—	Butyl
		CH_3—CH—CH_2—CH_3	sec-Butyl (secondary butyl)
CH_3—$\overset{\displaystyle CH_3}{CH}$—$CH_3$	Isobutane	CH_3—$\overset{\displaystyle CH_3}{CH}$—$CH_2$—	Isobutyl
		CH_3—$\overset{\displaystyle CH_3}{\underset{\displaystyle }{C}}$—$CH_3$	tert-Butyl or t-butyl (tertiary butyl)

Classification of Carbon in Alkyl Groups

Primary (1°) Carbon	Secondary (2°) Carbon	Tertiary (3°) Carbon
CH_3—CH_2—CH_2—$\overset{\displaystyle H}{\underset{\displaystyle H}{C}}$—	CH_3—CH_2—$\overset{\displaystyle CH_3}{\underset{\displaystyle H}{C}}$—	H_3C—$\overset{\displaystyle CH_3}{\underset{\displaystyle CH_3}{C}}$—
Butyl group	sec-Butyl group	tert-Butyl group

Naming Branched-Chain Alkanes

In the IUPAC system of naming, the longest carbon chain is the main chain, which is numbered to give the location of substituents attached to it. Some common names are still used for compounds with one to four carbon atoms, but compounds with five or more carbon atoms follow the IUPAC rules. Let's take a look at the way the IUPAC system is used to name the following alkane.

$$CH_3—\overset{\displaystyle CH_3}{CH}—CH_2—CH_2—CH_3$$

Rule 1   Find the longest continuous chain of carbon atoms and name it as the main chain. In this example, the longest chain has five carbon atoms, which is *pentane*.

$$CH_3$$
$$|$$
$$\boxed{CH_3—CH—CH_2—CH_2—CH_3}$$ Pentane

Rule 2   **Number the carbon in the main chain starting from the end nearest a substitutent.** Once you start numbering the main chain, continue in that same direction.

$$CH_3$$
$$|$$
$$CH_3—CH—CH_2—CH_2—CH_3$$ Pentane
$$12345$$

Rule 3   **Give the location and name of each alkyl group in front of the name of the main chain.** Use a prefix (di-, tri-, tetra- and so on) to indicate a group that appears more than once. In the name, hyphens separate numbers from words and commas separate numbers. When multiple substituents allow numbering from both ends of the main chain, use the direction that gives the lowest series of numbers.

$$CH_3$$
$$|$$
$$CH_3—CH—CH_2—CH_2—CH_3$$ 2-Methylpentane
$$12345$$

$$CH_3CH_3$$
$$||$$
$$CH_3—CH_2—CH—CH—CH_3$$ 2,3-**Di**methylpentane
$$54321$$ *not* 3,4-dimethylpentane

$$CH_3CH_3$$
$$||$$
$$CH_3—CH—CH_2—C—CH_3$$ 2,2,4-**Tri**methylpentane
$$|$$ *not* 2,4,4-trimethylpentane
$$CH_3$$
$$54321$$

Rule 4   **List the substituents in alphabetical order.** The prefixes for repeated substituents are *not* used in deciding alphabetical order.

Examples of Alphabetical Order of Alkyl Substituents

<u>E</u>thyl, <u>m</u>ethyl, <u>p</u>ropyl   Tri<u>e</u>thyl, di<u>m</u>ethyl   <u>E</u>thyl, <u>i</u>sopropyl, di<u>m</u>ethyl

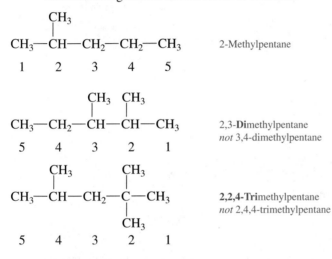

$$CH_3—CH_2CH_3$$
$$||$$
$$CH_3—CH_2—CH_2—CH—CH—CH_3$$
$$654321$$

3-Ethyl-2-methylhexane
not 2-methyl-3-ethylhexane
not 4-ethyl-5-methylhexane

$$CH_3$$
$$|$$
$$CH_3CH—CH_3$$
$$||$$
$$CH_3—CH—CH_2—C—CH_2—CH_2—CH_3$$
$$|$$
$$CH_2—CH_3$$

4-Ethyl-4-isopropyl-2-methylheptane
not 2-methyl-4-ethyl-4-isopropylheptane

Be sure that you always find the longest continuous carbon chain. It may not be the most obvious horizontal one. Name the following alkane.

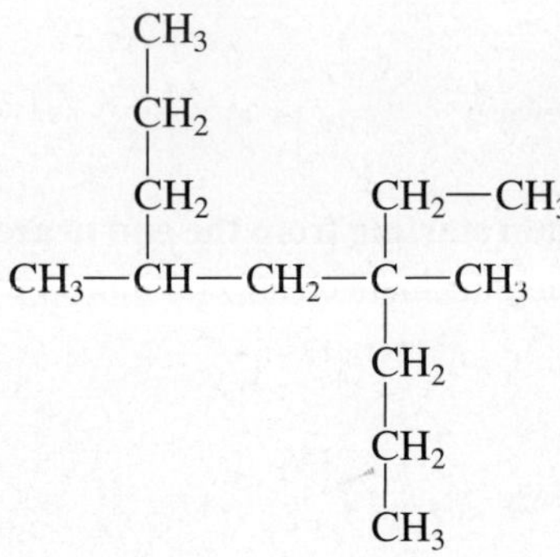

$$CH_3$$
$$|$$
$$CH_2$$
$$|$$
$$CH_2 \qquad\quad CH_2—CH_3$$
$$| \qquad\qquad\quad |$$
$$CH_3—CH—CH_2—C—CH_3$$
$$|$$
$$CH_2$$
$$|$$
$$CH_2$$
$$|$$
$$CH_3$$

How many carbon atoms are in the main chain? If you said five, look again. There is a chain of nine carbon atoms that goes around some corners!

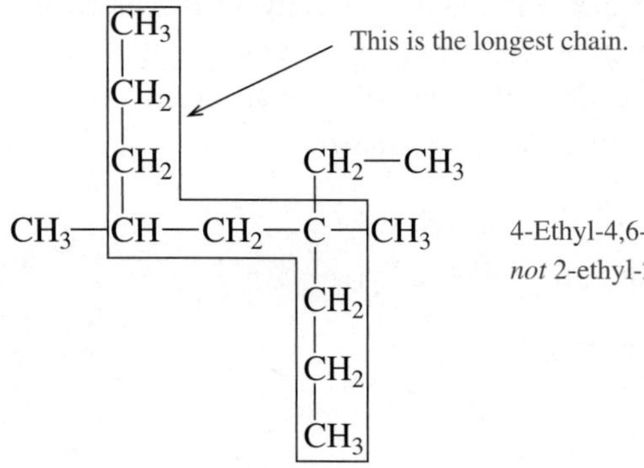

$$\boxed{CH_3}$$ This is the longest chain.
$$CH_2$$
$$CH_2 \qquad\quad CH_2—CH_3$$
$$CH_3—CH—CH_2—C—CH_3$$ 4-Ethyl-4,6-dimethylnonane
$$CH_2$$ *not* 2-ethyl-2,4-dipropylpentane
$$CH_2$$
$$CH_3$$

SAMPLE PROBLEM 12.3

Writing IUPAC names

Give the IUPAC name for each of the following alkanes:

a. 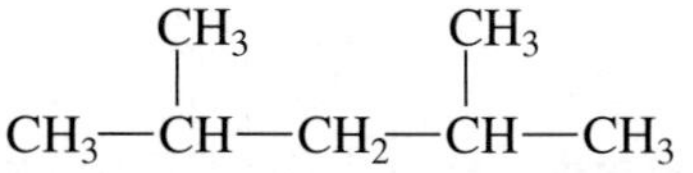
$$\qquad\quad CH_3 \qquad\quad CH_3$$
$$\qquad\quad | \qquad\qquad\quad |$$
$$CH_3—CH—CH_2—CH—CH_3$$

b. 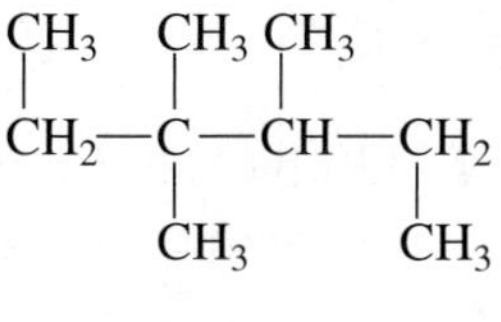
$$CH_3 \quad CH_3\ CH_3$$
$$| \qquad | \quad |$$
$$CH_2—C—CH—CH_2$$
$$\qquad | \qquad\qquad |$$
$$\qquad CH_3 \qquad\quad CH_3$$

c.
$$\qquad\qquad\qquad\qquad CH_3$$
$$\qquad\qquad\qquad\qquad |$$
$$\qquad CH_3—CH_2\ \ CH—CH_3 \quad CH_3$$
$$\qquad\qquad\quad | \qquad\qquad\qquad |$$
$$CH_3—CH_2—CH—CH—CH_2—CH—CH_3$$

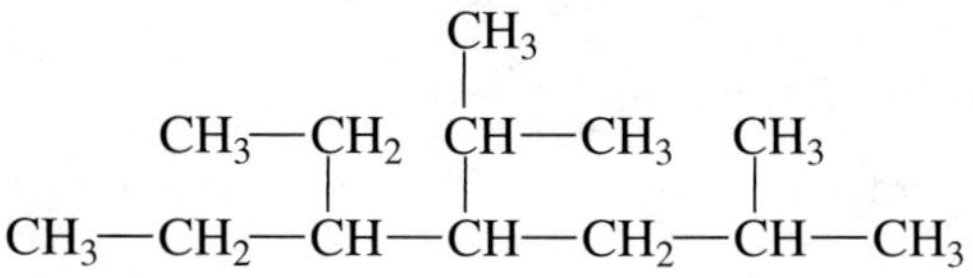

Solution

a. The main chain of five carbon atoms is *pentane*. The two methyl groups named *dimethyl* are on carbons 2 and 4. The IUPAC name is 2,4-dimethylpentane.

b. The longest chain of six carbons is *hexane*. There are three methyl groups indicated by the prefix *tri*, two on carbon 3 and one on carbon 4. The IUPAC name is 3,3,4-trimethylhexane.

c. The main chain with seven carbon atoms is *heptane*. Counting from *right to left*, there is a methyl group on carbon 2, an isopropyl on carbon 4, and an ethyl group on carbon 5. Placing the alkyl group names in alphabetical order gives the IUPAC name 5-ethyl-4-isopropyl-2-methylheptane.

Study Check

Give the IUPAC name of the following compound:

$$CH_3$$
$$|$$
$$CH_3 \quad\quad CH_2 \quad CH_3$$
$$| \quad\quad\quad | \quad\quad |$$
$$CH_3-CH_2-CH-CH_2-CH-CH_2$$

QUESTIONS AND PROBLEMS

IUPAC Naming System for Alkanes

12.7 Indicate whether each of the following pairs of structural formulas represent constitutional isomers or different conformations.

a.
$$CH_3 \quad\quad\quad\quad\quad CH_3$$
$$| \quad\quad\quad\quad\quad\quad\quad |$$
$$CH_3-CH-CH_3 \quad and \quad CH-CH_3$$
$$\quad\quad\quad\quad\quad\quad\quad\quad\quad\quad |$$
$$\quad\quad\quad\quad\quad\quad\quad\quad\quad\quad CH_3$$

b.
$$CH_3 \quad\quad\quad\quad\quad\quad CH_3 \quad\quad\quad CH_3$$
$$| \quad\quad\quad\quad\quad\quad\quad\quad | \quad\quad\quad\quad |$$
$$CH_3-CH-CH_2-CH_3 \quad and \quad CH_2-CH_2-CH_2$$

c.
$$CH_3 \quad CH_3 \quad\quad\quad\quad\quad\quad CH_3 \quad CH_3$$
$$| \quad\quad | \quad\quad\quad\quad\quad\quad\quad\quad | \quad\quad |$$
$$CH_2-CH-CH_2-CH_3 \quad and \quad CH_3-CH-CH-CH_3$$

12.8 Indicate whether each of the following pairs of structural formulas represent constitutional isomers or different conformations.

a.
$$CH_3 \quad\quad\quad\quad\quad\quad CH_3$$
$$| \quad\quad\quad\quad\quad\quad\quad\quad |$$
$$CH_3-C-CH_3 \quad and \quad CH-CH_2-CH_3$$
$$| \quad\quad\quad\quad\quad\quad\quad\quad |$$
$$CH_3 \quad\quad\quad\quad\quad\quad CH_3$$

b.
$$CH_3 \quad CH_3 \quad CH_3 \quad\quad\quad\quad\quad CH_3 \quad\quad\quad CH_3$$
$$| \quad\quad | \quad\quad | \quad\quad\quad\quad\quad\quad\quad | \quad\quad\quad\quad |$$
$$CH_3-CH-CH-CH_2 \quad and \quad CH_3-CH-CH_2-CH-CH_3$$

c.
$$CH_3 \quad\quad\quad\quad\quad\quad\quad\quad\quad CH_3$$
$$| \quad\quad\quad\quad\quad\quad\quad\quad\quad\quad |$$
$$CH_3-CH-CH_2-CH_3 \quad and \quad CH_3-CH_2-CH-CH_3$$

12.9 Name each of the following alkyl groups:

a. $CH_3-CH_2-CH_2-$

b.
$$CH_3$$
$$|$$
$$CH_3-CH$$

c. $CH_3-CH_2-CH_2-CH_2-$

d. CH_3-

12.10 Name each of the following alkyl groups:

a. CH_3-CH_2-

b.
$$CH_3$$
$$|$$
$$CH_3-CH-CH_2-$$

c.
$$CH_3$$
$$|$$
$$CH_3-CH_2-CH-$$

d.
$$CH_3$$
$$|$$
$$CH_3-C-$$
$$|$$
$$CH_3$$

12.11 Give the IUPAC name for each of the following alkanes:

a.
$$CH_3$$
$$|$$
$$CH_3-CH-CH_3$$

b.
$$CH_3 \quad\quad\quad CH_3$$
$$| \quad\quad\quad\quad |$$
$$CH_2-CH_2-CH-CH_3$$

c.
$$CH_3-CH_2 \quad\quad\quad CH_3$$
$$| \quad\quad\quad\quad\quad\quad |$$
$$CH_3-CH_2-CH-CH_2-CH-CH_3$$

d. 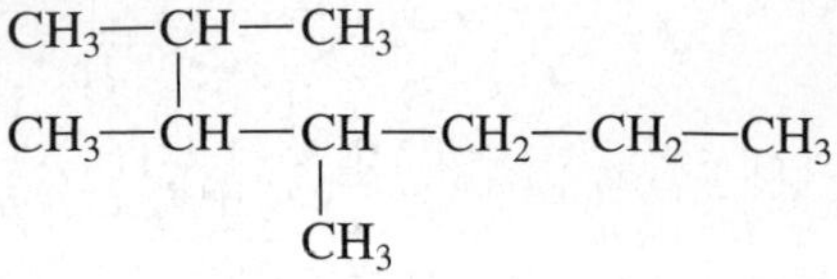

e. 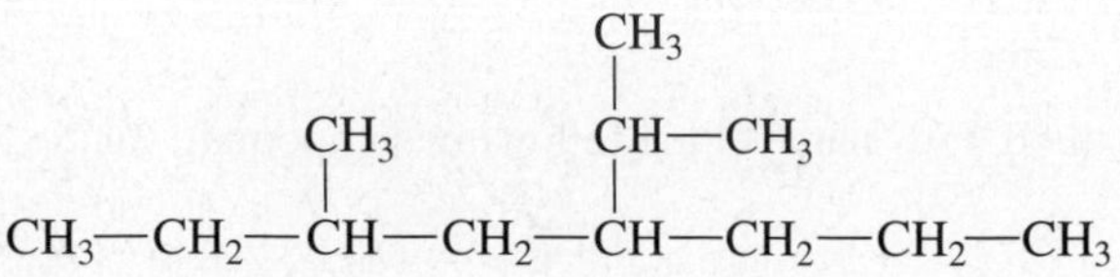

12.12 Give the IUPAC name for each of the following alkanes:

a. 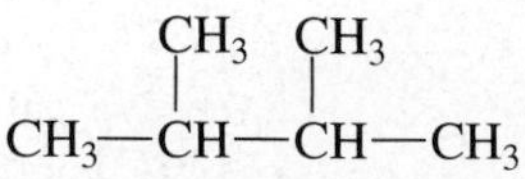

b. 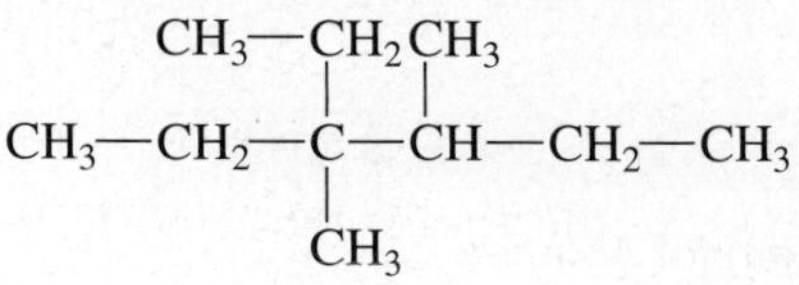

c. 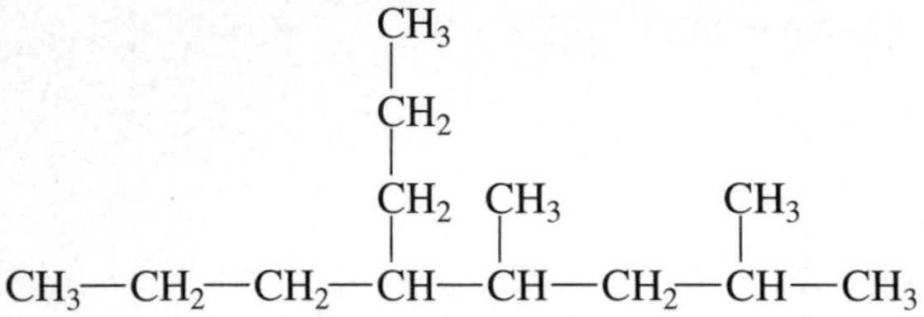

d. 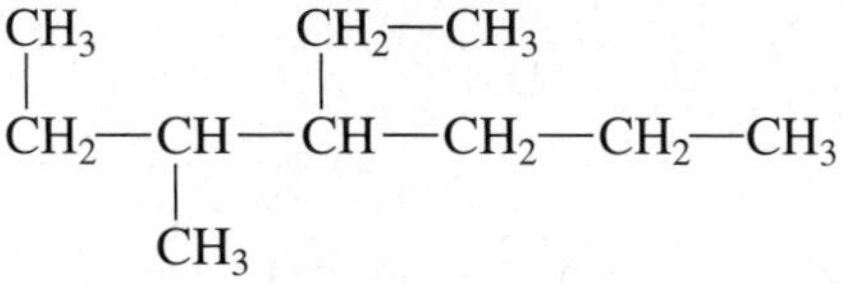

e. 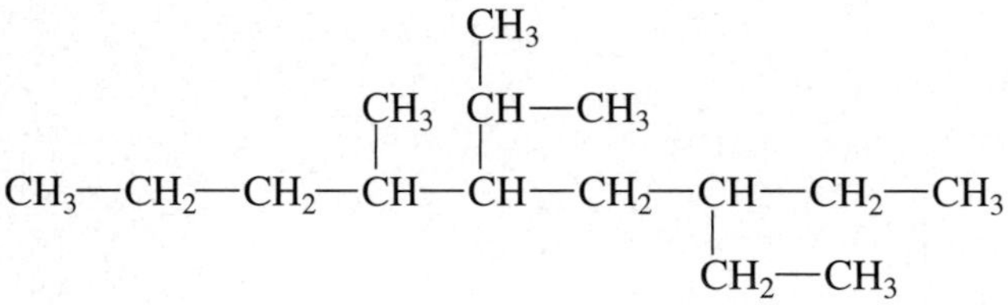

12.13 Give the IUPAC name for each of the following line-bond formulas:

a. 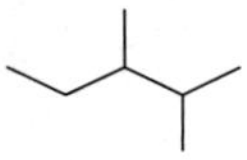

b. 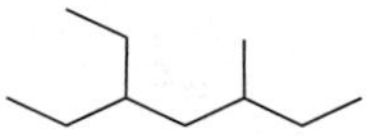

c. 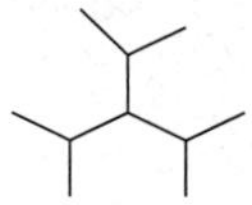

12.14 Give the IUPAC name for each of the following line-bond formulas:

a. 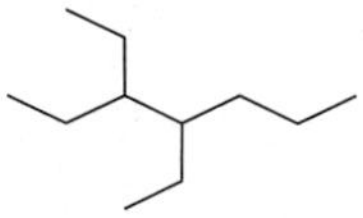

b. 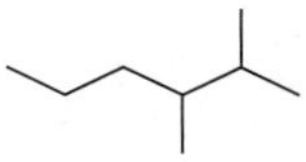

c. 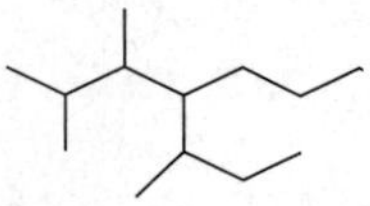

LEARNING GOAL

Draw the correct structural formulas of alkanes and their constitutional isomers.

12.3 Drawing Structural Formulas

Suppose you are asked to draw the condensed structural formula of 2,3-dimethylhexane. The IUPAC name gives all the information needed to draw the structure. The alkane name at the end gives the number of carbon atoms in the main chain. The first part of the IUPAC name gives the substituents and where they are attached. We can break down the name in the following way.

2,3-Dimethylhexane

2,3-	di	methyl	hex	ane
Substituents on carbons 2 and 3	Two identical groups	CH_3— alkyl group	6 Carbon atoms in main chain	Single C—C bonds

To draw the condensed structural formula, we proceed as follows:

Step 1   Draw the main chain of carbon atoms.

C—C—C—C—C—C   2,3-Dimethyl**hexane**

Step 2   Draw the substituents on the carbon atoms indicated by the location numbers.

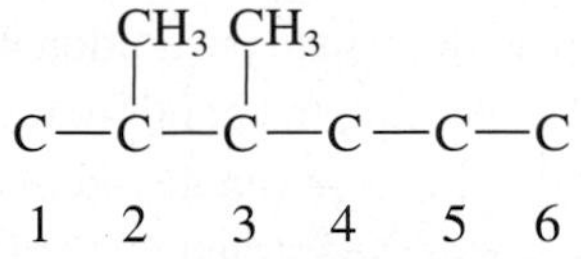

$$\begin{array}{c}\quad CH_3\ CH_3\\\quad\ \ |\quad\ |\\ C-C-C-C-C-C\\ 1\quad 2\quad 3\quad 4\quad 5\quad 6\end{array}$$
2,3-Dimethylhexane

Step 3   Add the correct number of hydrogen atoms to give four bonds to each carbon atom .

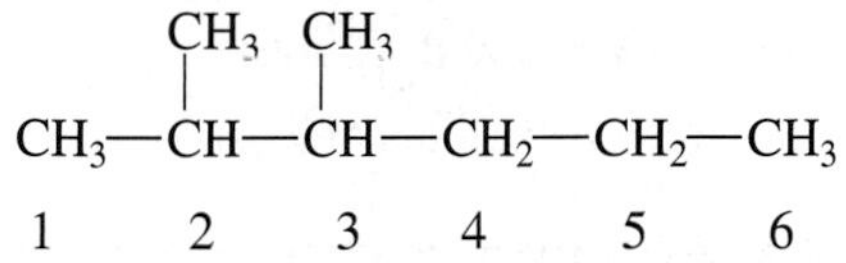

$$\begin{array}{c}\qquad CH_3\ CH_3\\\qquad\ \ |\quad\ |\\ CH_3-CH-CH-CH_2-CH_2-CH_3\\ \ 1\qquad 2\qquad 3\qquad 4\qquad 5\qquad 6\end{array}$$
2,3-Dimethylhexane

SAMPLE PROBLEM 12.4

Drawing Structural Formulas from IUPAC Names

Draw the structural formulas for each of the following alkanes:

a. 3-ethyl-5-methylheptane

b. 4-*tert*-butyloctane

Solution

a. Using the alkane name, heptane, at the end of the IUPAC name, draw a chain of seven carbon atoms and number it.

$$\begin{array}{c}C-C-C-C-C-C-C\\ 1\ \ \ 2\ \ \ 3\ \ \ 4\ \ \ 5\ \ \ 6\ \ \ 7\end{array}$$

The beginning of the name indicates an ethyl group on carbon 3 and a methyl group on carbon 5.

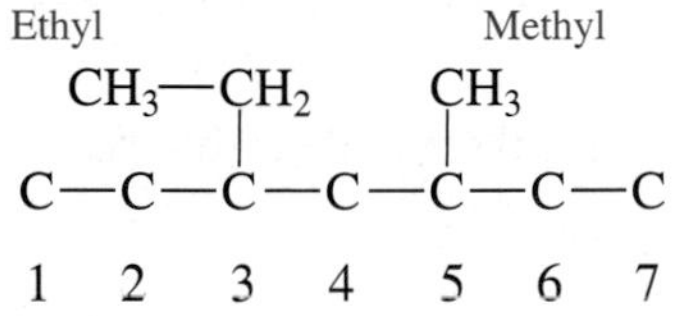

Ethyl   Methyl

$$\begin{array}{c}CH_3-CH_2\qquad CH_3\\\qquad\quad |\qquad\quad |\\ C-C-C-C-C-C-C\\ 1\ \ \ 2\ \ \ 3\ \ \ 4\ \ \ 5\ \ \ 6\ \ \ 7\end{array}$$

Complete the formula by adding the correct number of hydrogen atoms to each carbon.

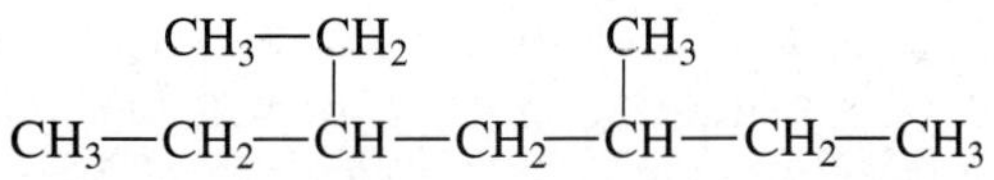

$$\begin{array}{c}\qquad\ CH_3-CH_2\qquad\quad CH_3\\\qquad\qquad\ \ |\qquad\qquad\ \ |\\ CH_3-CH_2-CH-CH_2-CH-CH_2-CH_3\end{array}$$

3-Ethyl-5-methylheptane

b. Draw a main chain of eight carbon atoms with a *tert*-butyl group on carbon 4.

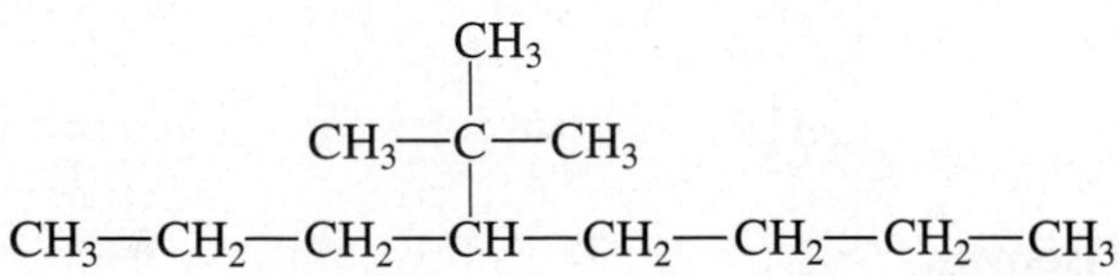

$$\begin{array}{c}\qquad\qquad\quad CH_3\\\qquad\qquad\quad\ |\\ \qquad\quad CH_3-C-CH_3\\\qquad\qquad\quad\ |\\ CH_3-CH_2-CH_2-CH-CH_2-CH_2-CH_2-CH_3\end{array}$$

4-*tert*-Butyloctane

Study Check

What is the structural formula for 3-ethyl-2,4-dimethylpentane?

Drawing Constitutional Isomers

A common problem in organic chemistry is how to draw the constitutional isomers of a particular molecular formula such as C_6H_{14}. Writing the continuous chain of six carbon atoms is obvious, but drawing the other isomers can get more difficult. Using this example, we can proceed to draw the condensed structural formulas and the line-bond formulas as follows.

A System for Writing Constitutional Isomers of an Alkane

Step 1: Draw the longest continuous (unbranched) chain.

In this example, we draw the structure formula of a carbon chain with six carbon atoms.

$CH_3-CH_2-CH_2-CH_2-CH_2-CH_3$ 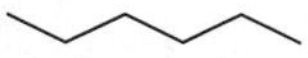 Hexane

Step 2: Remove one carbon from the chain and attach it as a methyl group to the shorter main chain in as many different locations as possible.

$$\begin{array}{l} \quad\;\; CH_3 \\ \quad\;\;\; | \\ CH_3-CH-CH_2-CH_2-CH_3 \end{array}$$ 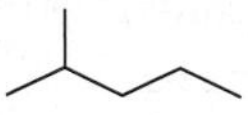 2-Methylpentane

$$\begin{array}{l} \qquad\qquad\; CH_3 \\ \qquad\qquad\;\; | \\ CH_3-CH_2-CH-CH_2-CH_3 \end{array}$$ 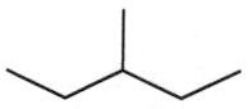 3-Methylpentane

Make sure that you do not repeat one of the isomers. Each different isomer has a different name. For example, moving the methyl group to the next carbon atom in pentane repeats the structure and name 2-methylpentane.

$$\begin{array}{l} \qquad\qquad\qquad\quad CH_3 \\ \qquad\qquad\qquad\quad\; | \\ CH_3-CH_2-CH_2-CH-CH_3 \end{array}$$ 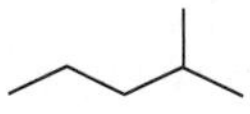 2-Methylpentane

Step 3: Remove another carbon atom from the main chain and attach as another alkyl group.

As we proceed, the substitution increases for the middle carbon atoms in the shortened main chain

$$\begin{array}{l} \quad\;\; CH_3\;\; CH_3 \\ \quad\;\;\; |\qquad\; | \\ CH_3-CH-CH-CH_3 \end{array}$$ 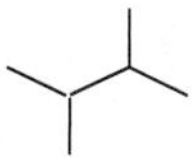 2,3-Dimethylbutane

$$\begin{array}{l} \quad\;\; CH_3 \\ \quad\;\;\; | \\ CH_3-C-CH_2-CH_3 \\ \quad\;\;\; | \\ \quad\;\; CH_3 \end{array}$$ 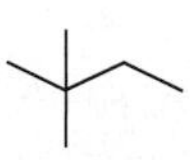 2,2-Dimethylbutane

The chain cannot be shortened again because one of the above structures would be repeated. Therefore, we have drawn five constitutional isomers for C_6H_{14}.

SAMPLE PROBLEM 12.5

Drawing Constitutional Isomers

There are three constitutional isomers with molecular formula C_5H_{12}. Draw and name each of their condensed structural formulas.

Solution

One isomer is a main chain of five carbon atoms.

$$CH_3 — CH_2 — CH_2 — CH_2 — CH_3 \qquad \text{Pentane}$$

Shorten the main chain by one carbon and attach a methyl group.

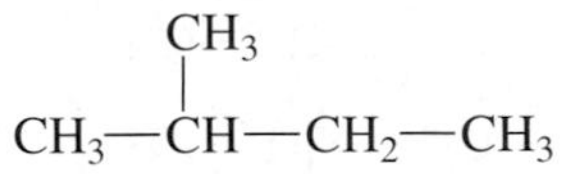

$$\begin{array}{c} CH_3 \\ | \\ CH_3 — CH — CH_2 — CH_3 \end{array} \qquad \text{2-Methylbutane}$$

Shorten the main chain again and attach two methyl groups.

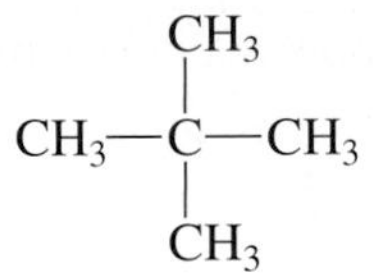

$$\begin{array}{c} CH_3 \\ | \\ CH_3 — C — CH_3 \\ | \\ CH_3 \end{array} \qquad \text{2,2-Dimethylpropane}$$

Study Check

What are the names of four constitutional isomers of dimethylpentane?

QUESTIONS AND PROBLEMS

Drawing Structural Formulas

12.15 Draw a condensed structural formula for each of the following alkanes:
   a. 2-methylbutane
   b. 3,3-dimethylpentane
   c. 2,3,5-trimethylhexane
   d. 3-ethyl-2,5-dimethyloctane
   e. 4-isopropyl-2-methylheptane
   f. 4-propylnonane

12.16 Draw a condensed structural formula for each of the following alkanes:
   a. 3-ethylpentane
   b. 3-ethyl-2-methylpentane
   c. 4-propylheptane
   d. 2,2,3,5-tetramethylhexane
   e. 4-ethyl-2,2-dimethyloctane
   f. 3-ethyl-4-isobutyl-2,3-dimethyldecane

12.17 Give the IUPAC names for the constitutional isomers of each of the following:
   a. methylheptane
   b. C_7H_{16} with a main chain of 5 carbon atoms

12.18 Give the IUPAC names for the constitutional isomers of each of the following:
   a. methyloctane
   b. dimethylhexane

12.4 Haloalkanes

A **haloalkane** is an alkane in which halogen atoms replace one or more hydrogen atoms in a carbon chain. In the IUPAC system, halogen atoms are named as substituents: fluorine is *fluoro*, chlorine is *chloro*, bromine is *bromo*, and iodine is *iodo*. The halo-substituents are numbered and arranged alphabetically just as we did with the alkyl groups. Simple haloalkanes are commonly named as alkyl halides; the carbon group is named as an alkyl group followed by the halide name.

LEARNING GOAL

Write the IUPAC and common name for a haloalkane; draw the condensed structural formula given the name.

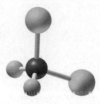

Methylene chloride

Chloroform

Carbon tetrachloride

	CH_3—Cl	CH_3—CH_2—Br	CH_3—$\overset{\overset{\text{F}}{\mid}}{CH}$—$CH_3$
IUPAC:	chloromethane	bromoethane	2-fluoropropane
Common:	methyl chloride	ethyl bromide	isopropyl fluoride

	Cl—CH_2—CH_2—Br	CH_3—CH_2—$\overset{\overset{\text{Cl}}{\mid}}{CH}$—$\overset{\overset{CH_3}{\mid}}{CH}$—$CH_3$
IUPAC:	1-bromo-2-chloroethane	3-chloro-2-methylpentane

Methane compounds with two or more halogens have another set of common names that do not indicate their structures.

Cl—CH_2—Cl	Cl—$\overset{\overset{\text{Cl}}{\mid}}{CH}$—Cl	Cl—$\overset{\overset{\text{Cl}}{\mid}}{\underset{\underset{\text{Cl}}{\mid}}{C}}$—Cl
CH_2Cl_2	$CHCl_3$	CCl_4
IUPAC: dichloromethane	trichloromethane	tetrachloromethane
Common: methylene chloride	chloroform	carbon tetrachloride

SAMPLE PROBLEM 12.6

Naming Haloalkanes

Freon 11 and Freon 12 are compounds known as chlorofluorocarbons (CFCs) and are widely used as refrigerants and aerosol propellants. What are their IUPAC names?

HEALTH NOTE

Common Uses of Halogenated Alkanes

Some common uses of halogenated alkanes include solvents and anesthetics. For many years, carbon tetrachloride was widely used by dry cleaners and in-home spot removers to take oils and grease out of clothes. However this use was discontinued when it was found to be toxic to the liver, where it can cause cancer. Today, dry cleaners use other halogenated compounds such as methylene chloride, 1,1,1-trichloroethane, and 1,1,2-trichloro-1,2,2-trifluoroethane.

CH_2Cl_2	Cl_3C—CH_3	FCl_2C—$CClF_2$
methylene chloride	1,1,1-trichloroethane	1,1,2-trichloro-1,2,2-trifluoroethane

General anesthetics are compounds that are inhaled or injected to cause a loss of sensation so that surgery or other procedures can be done without causing pain to the patient. As nonpolar compounds they are soluble in the nonpolar nerve membranes, where they decrease the ability of the nerve cells to conduct the sensation of pain. Chloroform $CHCl_3$ was once used as an anesthetic, but it is toxic and may be carcinogenic. One of the most widely used general anesthetics is halothane, also called Fluothane. It has a pleasant odor, is nonexplosive, has few side effects, undergoes few reactions within the body, and is eliminated quickly.

$$F—\overset{\overset{\text{F}}{\mid}}{\underset{\underset{\text{F}}{\mid}}{C}}—\overset{\overset{\text{Cl}}{\mid}}{\underset{\underset{\text{H}}{\mid}}{C}}—Br \qquad \text{Halothane (Fluothane)}$$

For minor surgeries, a local anesthetic such as chloroethane (ethyl chloride) CH_3—CH_2—Cl is applied to an area of the skin where it evaporates quickly, which cools the skin causing a loss of sensation.

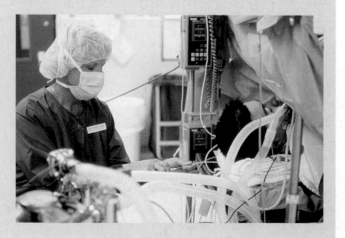

$$Cl-\underset{\underset{Cl}{|}}{\overset{\overset{Cl}{|}}{C}}-F \qquad F-\underset{\underset{Cl}{|}}{\overset{\overset{Cl}{|}}{C}}-F$$

Freon 11   Freon 12

Solution

Freon 11, trichlorofluoromethane; Freon 12, dichlorodifluoromethane.

Study Check

Halothane or Fluothane is the commercial name of a haloalkane widely used as an anesthetic. What is its IUPAC name?

ENVIRONMENTAL NOTE

CFCs and Ozone Depletion

The compounds called chlorofluorocarbons (CFCs) were used as propellants for hairs sprays, paints, and as refrigerants in home and car air conditioners. Two widely used CFCs, Freon 11 (CCl_3F) and Freon 12 (CCl_2F_2), were developed in the 1920s as nontoxic refrigerants, which were safer than the sulfur dioxide and ammonia used at the time.

$$Cl-\underset{\underset{Cl}{|}}{\overset{\overset{Cl}{|}}{C}}-F \qquad F-\underset{\underset{Cl}{|}}{\overset{\overset{Cl}{|}}{C}}-F$$

Freon 11   Freon 12

In the stratosphere, a layer of ozone (O_3) absorbs the ultraviolet (UV) radiation of the sun and acts as a protective shield for plants and animals on earth. Ozone is produced in the stratosphere when oxygen reacts with ultraviolet light and breaks into oxygen atoms that quickly combine with oxygen molecules to form ozone.

$$O_2 \xrightarrow{\text{UV}} O + O$$
$$O_2 + O \longrightarrow O_3$$

In the 1970s scientists became concerned that CFCs entering the atmosphere were accelerating the depletion of ozone and threatening the stability of the ozone layer. CFCs decompose in the upper atmosphere in the presence of UV light to produce highly reactive chlorine atoms.

$$CCl_3F \xrightarrow{\text{UV light}} CCl_2F + Cl$$

The reactive chlorine atoms catalyze the breakdown of ozone molecules.

$$Cl + O_3 \longrightarrow ClO + O_2$$
$$ClO + O_3 \longrightarrow Cl + 2O_2$$

It has been estimated that one chlorine atom can destroy as many as 100,000 ozone molecules. Normally, there is a balance between the ozone and oxygen in the atmosphere but the rapid destruction of ozone has upset that equilibrium.

Reports of polar ozone depletion over Antarctica in March 1985 prompted scientists to call for a freeze on the production of CFCs. In some areas as much as 50% of the ozone had been

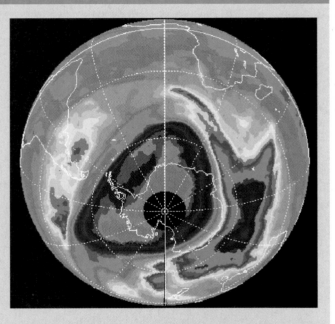

In the color image, the pink areas have the lowest levels of ozone.

depleted, and at certain times of the year an ozone hole appears. There is evidence of thinning in the ozone layer over the Arctic as well but to a somewhat lesser degree due to warmer temperatures. It is interesting that in the lower atmosphere, ozone is an automobile and industrial pollutant, but in the stratosphere ozone is a life-protecting compound.

Today the use of CFCs is being phased out. However, it is expected that ozone levels will remain low for several decades due to the stability of CFCs. Chemical companies are developing substitutes to CFCs that are not as damaging to the ozone. Replacement compounds such as hydrochlorofluorocarbons (HCFCs) contain chlorine atoms, but these compounds break down in the lower atmosphere reducing the amount of chlorine that reaches the stratosphere. Hydrofluorocarbons (HFCs), which contain no chlorine, are being considered as another replacement for CFCs. However, the potential effects of fluorine compounds on ozone destruction must be determined.

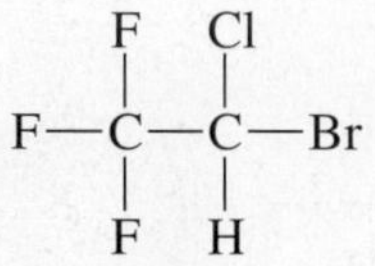

$$F-\underset{\underset{\displaystyle F}{|}}{\overset{\overset{\displaystyle F}{|}}{C}}-\underset{\underset{\displaystyle H}{|}}{\overset{\overset{\displaystyle Cl}{|}}{C}}-Br \qquad \text{Halothane}$$

QUESTIONS AND PROBLEMS

Haloalkanes

12.19 Give the IUPAC and common names for each of the following compounds:

a. CH_3-CH_2-Br **b.** $CH_3-CH_2-CH_2-F$

c. 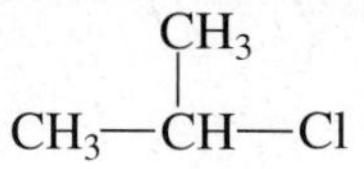 $CH_3-\underset{\underset{\displaystyle}{\overset{\overset{\displaystyle CH_3}{|}}{CH}}}-Cl$ **d.** $CHCl_3$

12.20 Give the IUPAC and common names for each of the following compounds:

a. 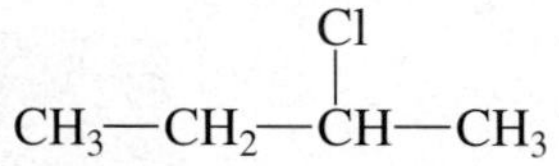 $CH_3-CH_2-\underset{\underset{\displaystyle}{\overset{\overset{\displaystyle Cl}{|}}{CH}}}-CH_3$ **b.** CCl_4

c. 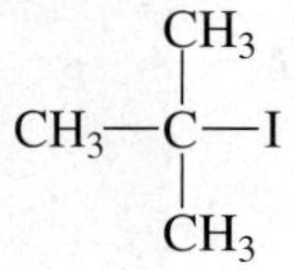 $CH_3-\underset{\underset{\displaystyle CH_3}{|}}{\overset{\overset{\displaystyle CH_3}{|}}{C}}-I$ **d.** CH_3F

12.21 Give the IUPAC name for each of the following compounds:

a. 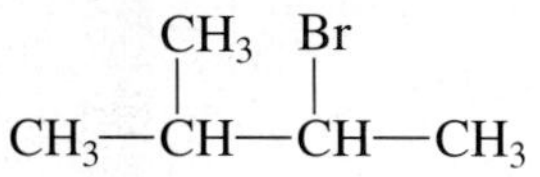 $CH_3-\underset{\underset{\displaystyle}{\overset{\overset{\displaystyle CH_3}{|}}{CH}}}-\underset{\underset{\displaystyle}{\overset{\overset{\displaystyle Br}{|}}{CH}}}-CH_3$ **b.**

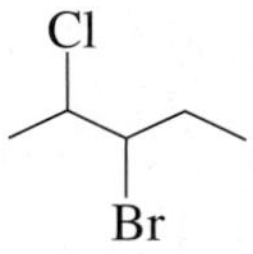

c. 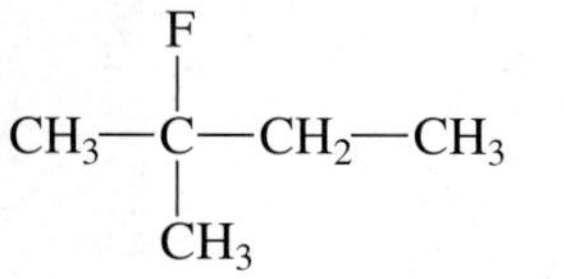 $CH_3-\underset{\underset{\displaystyle CH_3}{|}}{\overset{\overset{\displaystyle F}{|}}{C}}-CH_2-CH_3$

12.22 Give the IUPAC name for each of the following compounds:

a. 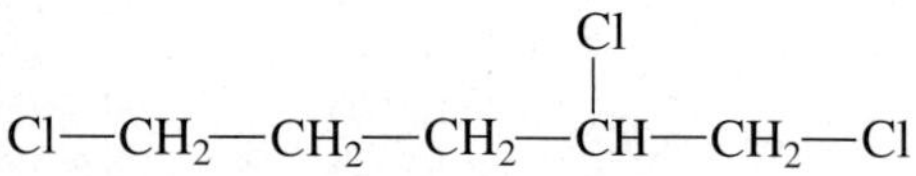 $Cl-CH_2-CH_2-CH_2-\underset{\underset{\displaystyle}{\overset{\overset{\displaystyle Cl}{|}}{CH}}}-CH_2-Cl$ **b.**

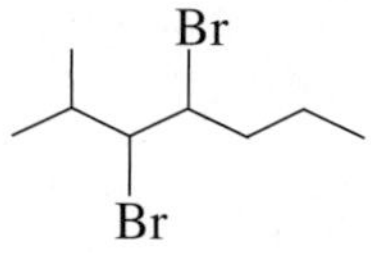

c. 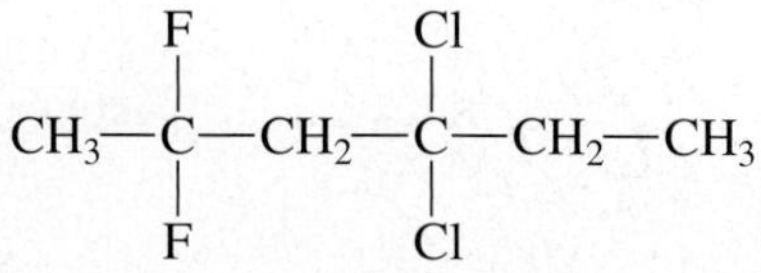 $CH_3-\underset{\underset{\displaystyle F}{|}}{\overset{\overset{\displaystyle F}{|}}{C}}-CH_2-\underset{\underset{\displaystyle Cl}{|}}{\overset{\overset{\displaystyle Cl}{|}}{C}}-CH_2-CH_3$

12.23 Write the condensed structural formula for each of the following compounds:

a. 2-chloropropane
b. 2-bromo-3-chlorobutane
c. methyl bromide
d. butyl chloride
e. 1,1-dibromo-2-chloro-4-fluoropentane
f. tetrabromomethane

12.24 Write the condensed structural formula for each of the following compounds:

 a. 1,1,2,2-tetrabromopropane   **b.** 2,2-dibromo-3,4-dichloropentane
 c. isopropyl bromide   **d.** methylene chloride
 e. 2,3-dichloro-2-methybutane   **f.** dibromodichloromethane

12.25 The topical anesthetics, methyl chloride and ethyl chloride, sprayed on the skin lower the temperature of the skin and numb nerve endings. What are the structural formulas of the anesthetics?

12.26 The isomers of $C_2H_4Cl_2$ are used as industrial solvents. Write the condensed structural formulas of two isomers and give their IUPAC names.

12.27 Write four isomers of C_4H_9Cl and give their IUPAC names.

12.28 Write four isomers of $C_3H_6F_2$ and give their IUPAC names.

12.5 Cycloalkanes

LEARNING GOAL

Give the IUPAC name for a cycloalkane; draw the structural formulas from the IUPAC name.

Up to now, we have looked at carbon atoms in continuous or branched chains. However, hydrocarbons can also form cyclic structures called **cycloalkanes.** Cycloalkanes, which have a general formula of C_nH_{2n}, have two fewer hydrogen atoms than the corresponding alkanes. Thus, the simplest cycloalkane, cyclopropane, C_3H_6, has a ring of three carbon atoms bonded to six hydrogen atoms. There are several ways to draw the structural formula of cyclopropane. The line-bond formula, which omits the hydrogen atoms and looks like a simple geometric figure, is a convenient way to show cyclic structures. Each corner of the triangle represents a carbon atom with four bonds to other carbon and hydrogen atoms.

Cyclopropane C_3H_6

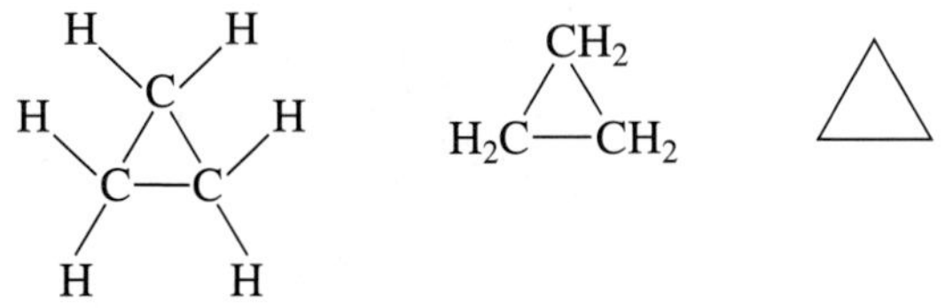

The ball-and-stick models and their structural formulas for several cycloalkanes are shown in Table 12.6.

IUPAC Names for Cycloalkanes

The names of cycloalkanes are similar to the names of the continuous-chain alkanes except that the prefix *cyclo* appears in front of the name of the alkane. When one substituent is attached to a carbon atom in the ring, the name of the substituent is placed in front of the cycloalkane name. No number is needed for a single alkyl group or halogen atom because the carbon atoms in the cycloalkane are equivalent. However, if two or more groups are attached, the ring is numbered to show the location of each group. The numbering starts by assigning carbon 1 to the substituent that gives the lowest numbers to the other substituents. Therefore, we may count clockwise or counterclockwise around a cycloalkane to give the lowest combination of numbers to the substituents.

methylcyclopentane

1,3-dimethylcyclopentane
not 2,4-dimethylcyclohexane

1-chloro-3-methylcyclohexane
not 1-methyl-3-chlorocyclohexane

2,4-dichloro-1-methylcyclohexane
not 1,3-dichloro-4-methylcyclohexane

Table 12.6 Formulas of Some Common Cycloalkanes

Condensed Structural Formula	Geometric Formula	Name	
CH$_2$ H$_2$C—CH$_2$	△	Cyclopropane	
H$_2$C—CH$_2$ H$_2$C—CH$_2$	□	Cyclobutane	
CH$_2$ H$_2$C CH$_2$ H$_2$C—CH$_2$	⬠	Cyclopentane	
CH$_2$ H$_2$C CH$_2$ H$_2$C CH$_2$ CH$_2$	⬡	Cyclohexane	

SAMPLE PROBLEM 12.7

Naming Cycloalkanes

Give the IUPAC name for each of the following cycloalkanes:

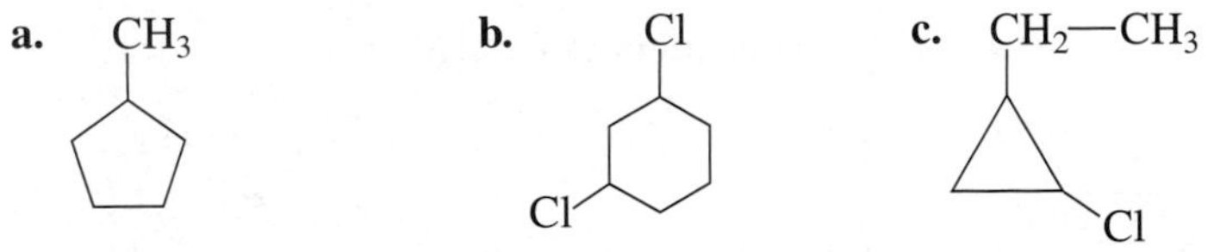

a. CH$_3$
b. Cl
c. CH$_2$—CH$_3$

Cl

Cl

Solution

a. The pentagon indicates a main chain that is a ring with five carbon atoms, cyclopentane. The single substituent is a methyl group. The compound is named methylcyclopentane.

b. The ring of six carbon atoms is numbered to give the lowest numbers starting with carbon 1. It is named 1,3-dichlorocyclohexane.

c. The groups are numbered starting from carbon 1 with the substituent that is first alphabetically. It is named 1-chloro-2-ethylcyclopropane.

Study Check

What is the IUPAC name of the following compound?

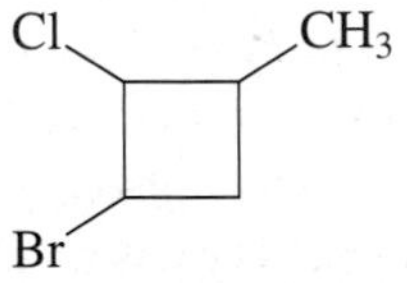

Cl   CH$_3$

Br

SAMPLE PROBLEM 12.8

Writing Structural Formulas for Cycloalkanes

Write the structural formula for 1-chloro-4-isopropyl cyclohexane.

Solution

The name cyclohexane tells us that the substituents are bonded to a ring of six carbon atoms. The first part indicates a chlorine atom on carbon 1 and an isopropyl group on carbon 4.

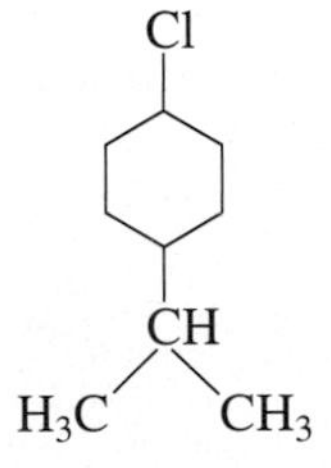

Cl

CH

H$_3$C   CH$_3$

Study Check

What is the structural formula for 1-bromo-3-ethylcyclopentane?

Constitutional Isomers of Cycloalkanes

Constitutional isomers are possible for cycloalkanes. For example, the molecular formula C_5H_{10} can be written with rings containing three, four, or five carbon atoms.

Constitutional Isomers of C_5H_{10}

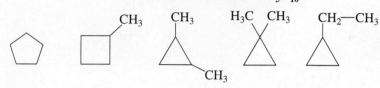

Cis–Trans Isomers

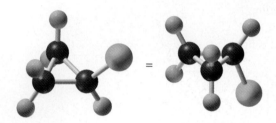

We saw that alkanes have rotation around the carbon–carbon single bonds. However, in cycloalkanes, there is no rotation of the carbon atoms in the ring, which gives the carbon ring two distinct sides. This structural characteristic produces stereoisomers called **cis–trans isomers** that differ only in the orientation of atoms in space. For example, chlorocyclopropane has just one isomer. Writing the chlorine atom on the "up" side of the ring is the same as writing it on the "down" side. By turning over the molecule, the chlorine atom will point down instead of up.

However, there are two stereoisomers of 1,2-dichlorocyclopropane. In the cis isomer, the two chlorine atoms are on the same side; both can be written "up" or both can be written "down." In the trans isomer, the chlorine atoms are on opposite sides of the carbon ring; if one is written "up," the other is "down." There is no way to change from one isomer into the other without breaking the bonds in the cyclic carbon structure. (See Figure 12.5.)

Chlorine atoms are
on the same side

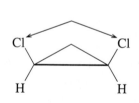

cis-1,2-dichlorocyclopropane

Chlorine atoms are
on opposite sides

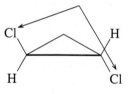

trans-1,2-dichlorocyclopropane

Figure 12.5 The cis–trans isomers of 1,2-dichlorocyclopropane are called *cis*-1,2-dichlorocyclopropane and *trans*-1,2-dichloropropane.
Q *What prevents the molecules from changing between the cis- and trans isomers?*

SAMPLE PROBLEM 12.9

Cis–Trans Isomers

Name each of the following as cis or trans isomers.

a.

b.

c.

Solution

a. Two methyl groups are on opposite sides of cyclobutane, which gives the name *trans*-1,3-dimethylcyclobutane.

b. Two bromine atoms groups are on the same side of cyclopentane; *cis*-1,3-dibromocyclopentane.

c. Two methyl groups are on opposite sides of a cyclohexane; *trans*-1,2-dimethylcyclohexane.

Study Check

Draw the structural formula of *cis*-1,2-dimethylcyclopropane.

QUESTIONS AND PROBLEMS

Cycloalkanes

12.29 Use the general formula for cycloalkanes to determine the following:
  a. The molecular formula of a cycloalkane with 5 carbon atoms.
  b. The number of hydrogen atoms in a cycloalkane with 4 carbon atoms.
  c. The number of carbon atoms in a cycloalkane bonded to 12 hydrogen atoms.

12.30 Use the general formula for cycloalkanes to determine the following:
  a. The molecular formula of a cycloalkane with 6 carbon atoms.
  b. The number of hydrogen atoms in a cycloalkane with 5 carbon atoms.
  c. The number of carbon atoms in a cycloalkane bonded to 16 hydrogen atoms.

12.31 Give the IUPAC name for each of the following cycloalkanes:

a.   **b.** 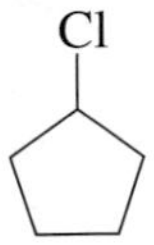   **c.**

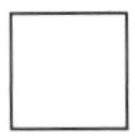

d.   **e.** 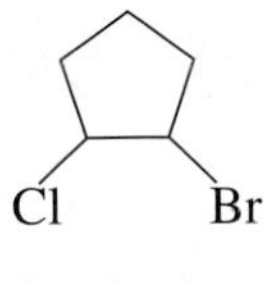   **f.**

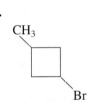   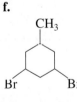

12.32 Give the IUPAC name for each of the following cycloalkanes.

a.   **b.** 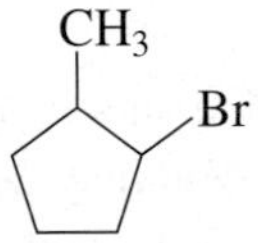   **c.**

   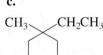

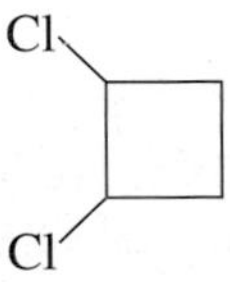

d.   **e.** 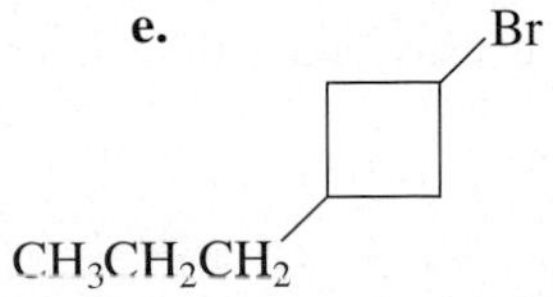   **f.**

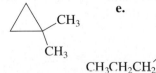

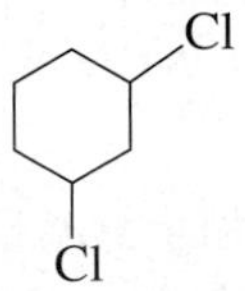

12.33 Draw the structural formulas for each of the following cycloalkanes.
  a. methylcyclopentane
  b. isopropylcyclohexane

c. 1,3-dimethylcyclobutane
d. 1-bromo-2,3-dimethylcyclopentane

12.34 Draw the structural formula for each of the following cycloalkanes.
a. bromocyclopropane
b. ethylcyclohexane
c. 1,2-dichlorocyclobutane
d. 1,3-dibromocyclopentane

12.35 Give the names of three cycloalkanes that are constitutional isomers of cyclopentane.

12.36 Give the names of four cycloalkanes that are constitutional isomers of cyclohexane.

12.37 Identify whether each of the following structures is a cis or trans isomer. If so, give its IUPAC name using the cis or trans prefix.

a.

b.

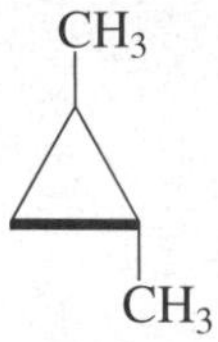

c.

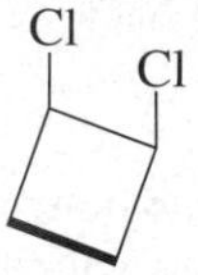

d.

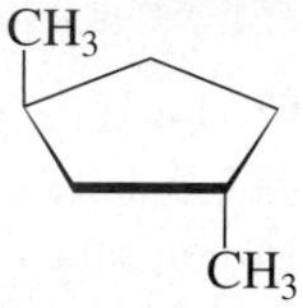

12.38 Identify whether each of the following structures is a cis or trans isomer. If so, give its IUPAC name using the cis or trans prefix.

a.

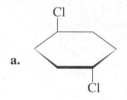

b.

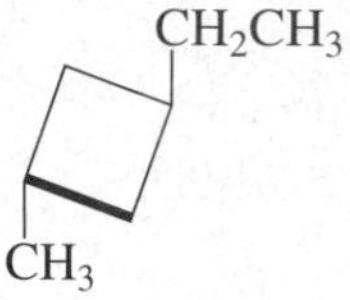

c.

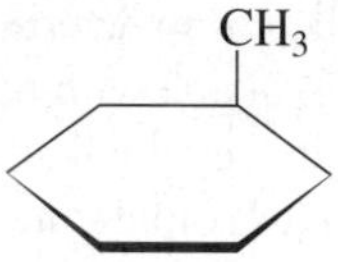

d.

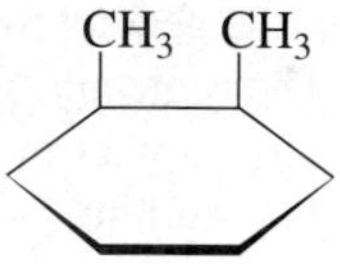

12.39 Draw structural formulas for each of the following compounds.
a. *cis*-1,2-dimethylcyclopentane
b. *trans*-1,3-dichlorocyclohexane
c. *cis*-1-bromo-2-methylcyclobutane

12.40 Draw structural formulas for each of the following compounds.
a. *trans*-1,2-dibromocyclopropane
b. *cis*-1,4-dichlorocyclohexane
c. *trans*-1-chloro-2-ethylcyclopentane

LEARNING GOAL

Identify the physical properties of alkanes and cycloalkanes.

12.6 Physical Properties of Alkanes and Cycloalkanes

Useful alkanes include the gasoline and diesel fuels that power our cars and the heating oils that heat our homes. You may have used a mixture of hydrocarbons such as mineral oil or petrolatum as a laxative or to soften your skin. The differences in uses of many of the alkanes and cycloalkanes result from their physical properties including solubility, density, and boiling point.

Solubility and Density

Alkanes and cycloalkanes are nonpolar, which makes them insoluble in water. However, they are soluble in nonpolar solvents such as other alkanes or cycloalkanes. Alkanes have densities from 0.65 g/mL to about 0.70 g/mL, which is less dense than the density of water (1.0 g/mL). If there is an oil spill in the ocean, the alkane mixtures in the crude oil remain on the surface and spread over a large area. In the *Exxon Valdez* oil spill in 1989, 40 million liters of oil covered over 25,000 square kilometers of water in Prince William Sound, Alaska. (See Figure 12.6.) If the crude oil reaches the beaches and inlets, there can be considerable damage to beaches, shellfish, fish, birds, and wildlife habitats. Even today, there is still oil on the surface, or just beneath the surface, in some areas of Prince William Sound. Cleanup includes both mechanical and chemical methods. In one method, a nonpolar compound that is "oil-attracting" is used to pick up oil, which is then scraped off into recovery tanks.

Melting and Boiling Points

Alkanes have the lowest melting and boiling points of all the organic compounds. The attractions between nonpolar alkanes in the solid and liquid states result from dispersion forces. In longer carbon chains, the greater number of electrons produces more attractions between molecules, which results in higher melting and boiling points.

CH_4
methane, bp −164°C

$CH_3—CH_3$
ethane, bp −89°C

$CH_3—CH_2—CH_3$
propane, bp −42°C

The boiling points of branched alkanes are generally lower than continuous alkanes with the same number of carbon atoms. The branched chain alkanes tend to be more compact, which reduces the amount of contact between the molecules. Cycloalkanes have higher boiling points than the continuous-chain alkanes. Because rotation of carbon bonds is restricted, cycloalkanes maintain a rigid structure. Those rigid structures are like a set of dishes that can be stacked closely together with many points of contact and therefore attractions to each other. We can compare the boiling points of alkanes and cycloalkanes with five carbon atoms as shown in Table 12.7.

Some Uses of Alkanes

The first four alkanes—methane, ethane, propane, and butane—are gases at room temperature and are widely used as heating fuels. Tanks of liquid propane and butane are used to provide fuels for heating homes and cooking on the barbecues. As chlorofluorocarbons (CFCs) are phased out, butane has replaced Freon as a propellant in aerosol containers.

Alkanes having 5–8 carbon atoms (pentane, hexane, heptane, and octane) are liquids at room temperature. They are highly volatile, which makes them useful in fuels such as gasoline. Liquid alkanes with 9–17 carbon atoms have higher boiling points and are found in kerosene, diesel, and jet fuels. Motor oil is a mixture of high-molecular-weight liquid hydrocarbons and is used to lubricate the internal components of engines. Mineral oil is a mixture of liquid hydrocarbons and is used as a laxative and a lubricant. Alkanes with 18 or more carbon atoms are waxy solids at room temperature. The high-molecular-weight alkanes, known as paraffins, are used in waxy coatings of fruits and vegetables to retain moisture, inhibit mold growth, and enhance appearance. (See Figure 12.7.) Petrolatum, or Vaseline, is a mixture of liquid hydrocarbons, which have low boiling points, that are encapsu-

Figure 12.6 In oil spills, large quantities of oil spread over the water.
Q *What physical properties cause oil to remain on the surface of water?*

Figure 12.7 The solid alkanes that make up waxy coatings on fruits and vegetables help retain moisture, inhibit mold, and enhance appearance.
Q *Why does the waxy coating help the fruits and vegetables retain moisture?*

Table 12.7   Comparison of Boiling Points of Alkanes and Cycloalkanes with Five Carbons

Formula	Name	Boiling Point (°C)
Cycloalkanes		
(cyclopentane structure)	cyclopentane	49
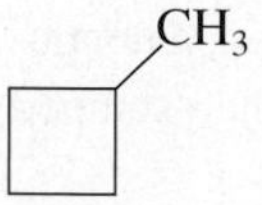 (methylcyclobutane structure) CH₃	methylcyclobutane	36.3
Continuous chain		
CH_3—CH_2—CH_2—CH_2—CH_3	pentane	36
Branched chains		
CH_3—$\overset{\displaystyle CH_3}{\overset{\displaystyle \|}{CH}}$—$CH_2$—$CH_3$	2-methylbutane	28
CH_3—$\overset{\displaystyle CH_3}{\underset{\displaystyle CH_3}{\overset{\displaystyle \|}{\underset{\displaystyle \|}{C}}}}$—$CH_3$	dimethylpropane	10

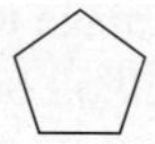

lated in solid hydrocarbons. It is used in ointments and cosmetics and as a lubricant and a solvent.

Crude Oil

Crude oil or petroleum contains a wide variety of hydrocarbons. At an oil refinery, the components in crude oil are separated by fractional distillation, a process that removes groups or fractions of hydrocarbons by continually heating the mixture to higher temperatures. (See Table 12.8.) Fractions containing alkanes with longer carbon chains require higher temperatures before they reach their boiling temperature and form gases. The gases are removed and passed through a distillation column where they cool and condense back to liquids. (See Figure 12.8.) The major use of crude oil is to obtain gasoline, but a barrel of crude oil is only about 35%

Table 12.8   Typical Alkane Mixtures Obtained by Distillation of Crude Oil

Distillation Temperatures (°C)	Number of Carbon Atoms	Product
Below 30	1–4	Natural gas
30–200	5–12	Gasoline
200–250	12–16	Kerosene, jet fuel
250–350	15–18	Diesel fuel, heating oil
350–450	18–25	Lubricating oil
Nonvolatile residue	Over 25	Asphalt, tar

Figure 12.8 At an oil refinery, crude oil is separated into hydrocarbon fractions by separating groups with different boiling point ranges.

Q *How does fractional distillation separate crude oil into different compounds?*

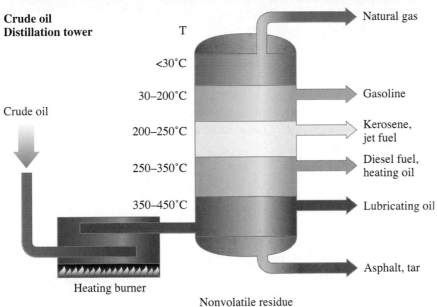

gasoline. To increase the production of gasoline, heating oils are broken down by "cracking" to give the lower weight alkanes.

$C_{14}H_{30}$ $+ H_2$ $\xrightarrow{\text{"cracking"}}$ C_7H_{16} C_7H_{16}

SAMPLE PROBLEM 12.10

Properties of Alkanes and Cycloalkanes

Indicate the compound in each pair that has the higher boiling point:

a. propane or pentane

b. 2-methylbutane or cyclopentane

Solution

a. Pentane, with the longer continuous chain, has the higher boiling point in the pair.

b. Cyclopentane.The rigid structure of a cycloalkane provides more contact between molecules than do the branched-chain molecules with the same number of carbons.

Study Check

Which of the following compounds would have the highest boiling point: 2-methylpentane, hexane, or cyclohexane?

QUESTIONS AND PROBLEMS

Physical Properties of Alkanes

12.41 Heptane, C_7H_{16}, has a density of 0.68 g/mL, and boils at 98°C.
   a. What is the structural formula of heptane?
   b. Is it a solid, liquid, or gas at room temperature?
   c. Is it soluble in water?
   d. Will it float or sink in water?
   e. Would you expect the branched isomers of heptane to have higher or lower boiling points?

12.42 Nonane, C_9H_{20}, has a density of 0.79 g/mL.
   a. What is the structural formula of nonane?
   b. Is it a solid, liquid, or gas at room temperature?
   c. Is it soluble in water?
   d. Will it float or sink in water?
   e. If heptane has a boiling point of 98°C, would you expect nonane to have a boiling point of 75°C or 151°C?

12.43 In each of the following pairs of hydrocarbons, which one would you expect to have the higher boiling point?
   a. pentane or heptane   b. propane or cyclopropane
   c. hexane or 2-methylpentane

12.44 In each of the following pairs of hydrocarbons, which one would you expect to have the higher boiling point?
   a. propane or butane   b. hexane or cyclohexane
   c. 2,2-dimethylpentane or heptane

LEARNING GOAL

Write balanced equations for the combustion and halogenation of alkanes and cycloalkanes.

12.7 Chemical Properties of Alkanes and Cycloalkanes

Carbon–carbon single bonds are difficult to break, which makes alkanes the least reactive family of organic compounds. However, alkanes burn readily in oxygen and will undergo substitution reactions with halogens.

Combustion

The growth of civilization was strongly influenced by the discovery of fire. Heat from burning wood was used to make pottery and glass, extract metals, make weapons, and forge tools. An alkane undergoes **combustion** when it reacts with oxygen to produce carbon dioxide, water, and energy.

$$Alkane + O_2 \longrightarrow CO_2 + H_2O + energy$$

Methane is the gas we use to cook our foods and heat our homes. Propane is the gas used in portable heaters and gas barbecues. (See Figure 12.9.) Gasoline, a mixture of liquid hydrocarbons, is the fuel that powers our cars, lawn mowers, and snow blowers. As alkanes, they all undergo combustion. The equations for the combustion of methane (CH_4) and propane (C_3H_8) follow:

$$CH_4 + 2O_2 \longrightarrow CO_2 + 2H_2O + energy$$
$$C_3H_8 + 5O_2 \longrightarrow 3CO_2 + 4H_2O + energy$$

In the cells of our bodies, energy is produced by the combustion of glucose. Although a series of reactions is involved, we can write the overall combustion of glucose in our cells as follows:

$$C_6H_{12}O_6 + 6O_2 \xrightarrow{\text{enzymes}} 6CO_2 + 6H_2O + energy$$

Figure 12.9 The propane fuel in the tank undergoes combustion, which provides energy.
Q *What is the balanced equation for the combustion of propane?*

SAMPLE PROBLEM 12.11

Combustion

Write a balanced equation for the complete combustion of butane.

Solution

The balanced equation for the complete combustion of butane can be written

$$2C_4H_{10} + 13O_2 \longrightarrow 8CO_2 + 10H_2O$$

Study Check

Write a balanced equation for the complete combustion of the following:

$$CH_3-\overset{\displaystyle CH_3}{\underset{\displaystyle |}{CH}}-CH_2-CH_3$$

HEALTH NOTE

Incomplete Combustion

You may already know that it is dangerous to burn natural gas, oil, or wood in a closed room where ventilation and fresh air are not adequate. A gas heater, fireplace, or wood stove must have proper ventilation. If the supply of oxygen is limited, incomplete combustion produces carbon monoxide. The incomplete combustion of methane in natural gas is written as:

$$2CH_4(g) + 3O_2(g) \longrightarrow 2CO(g) + 4H_2O(g) + heat$$

Limited oxygen supply   Carbon monoxide

Carbon monoxide (CO) is a colorless, odorless, poisonous gas. When inhaled, CO passes into the bloodstream, where it attaches to hemoglobin (as you saw in the Health Note on Oxygen, Hemoglobin and Carbon Monoxide Poisoning in Chapter 6). When CO binds to the hemoglobin, it reduces the amount of oxygen (O_2) reaching the organs and cells. As a result, a healthy person can experience a reduction in exercise capability, visual perception, and manual dexterity.

When the amount of hemoglobin bound to CO (COHb) is 10% or less, a person may experience shortness of breath, mild headache, and drowsiness, which are symptoms that may be mistaken for the flu. Heavy smokers can have as high as 9% COHb in their blood. When as much as 30% of the hemoglobin is bound to CO, a person may experience more severe symptoms including dizziness, mental confusion, severe headache, and nausea. If 50% or more of the hemoglobin is bound to CO, a person could become unconscious and die if not treated immediately with oxygen.

Combustion

In this exploration, we will look at the behavior of the products of combustion. You will need one or two candles, a Pyrex glass such as a measuring cup, and some matches or wooden splints.

Hold a Pyrex cup upside down, and place a burning match inside it. The match should continue to burn as long as oxygen is available. Light a candle and hold the inverted Pyrex cup above it for 15–20 seconds. Remove the cup from the candle and immediately place a burning match into it. The CO_2 accumulated from the combustion of the candle should extinguish the match.

Add some water and a lot of ice to the same Pyrex cup. It should become cold to the touch. Wipe the bottom of the cup and carefully hold the Pyrex cup over a burning candle. Look for condensation as water is formed from the combustion reaction. This may be more noticeable if two candles are used or the Pyrex cup is held for a few seconds over a gas flame on a cook top.

Questions

1. What are the products of combustion?
2. What was the evidence for the production of CO_2?
3. What observations gave evidence of the production of H_2O during combustion?

Halogenation of Alkanes (Substitution)

Alkanes react with halogen to produce a mixture of halogenated compounds. In a **halogenation reaction,** atoms of a halogen bond to a carbon atom. This type of reaction is also called a **substitution** reaction because halogen atoms replace one or more hydrogen atoms in an alkane. When halogenation uses chlorine Cl_2 (Cl—Cl), it is also called chlorination and with bromine Br_2 (Br—Br), it is called bromination. When a mixture of an alkane such as methane and chlorine are heated or exposed to light, halogenation can take place.

$$\text{Alkane} + \text{halogen} \xrightarrow{\text{light or heat}} \text{haloalkane and hydrogen halide}$$

The first halogenation of methane occurs when one chlorine atom replaces a hydrogen atom as follows:

methane *alkane*   chlorine *halogen*   →   chloromethane haloalkane   hydrogen chloride *hydrogen halide*

In the presence of heat or light and chlorine, the reaction continues. Chlorine replaces hydrogen atoms in chloromethane, which gives a mixture of halogenated products.

+ HCl   + HCl   + HCl

When halogenation occurs with larger alkanes, the halogen atoms may substitute for any of the hydrogen atoms. For example, the first substitution of propane (monohalogenation) produces both 1-bromopropane and 2-bromopropane. Because secondary hydrogens are more reactive than primary hydrogens, there is more 2-bromopropane produced.

When cycloalkanes undergo monohalogenation, a single product results.

SAMPLE PROBLEM 12.12

Halogenation of Alkanes

Write the structural formula of the monohalogenated product formed in the following reactions:

a. $CH_3CH_3 + Br_2 \xrightarrow{\text{light or heat}}$

b.  ⬡ $+ Cl_2 \xrightarrow{\text{light or heat}}$

Solution

a. CH_3CH_2Br

b. 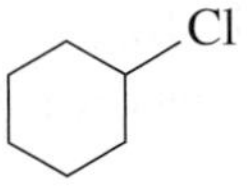 (cyclohexane ring with Cl substituent)

Study Check

Write the structural formulas of the organic products that form when two bromine atoms replace two hydrogen atoms in the halogenation of ethane.

QUESTIONS AND PROBLEMS

Chemical Properties of Alkanes and Cycloalkanes

12.45 Write a balanced equation for the complete combustion of each of the following compounds:

    a. ethane, C_2H_6 **b.** cyclopropane, C_3H_6

    c. octane   **d.** cyclohexane

12.46 Write a balanced equation for the complete combustion of each of the following compounds:

    a. hexane, C_6H_{14} **b.** cyclopentane, C_5H_{10}

    c. nonane   **d.** 2-methylbutane

12.47 Write the condensed structural formulas of the monohalogenated products from the chlorination of each of the following:

    a. ethane   **b.** cyclopentane

    c. 2-methylpropane

12.48 Write the condensed structural formulas of the monohalogenated products from the bromination of each of the following:

    a. butane   **b.** pentane

    c. cyclobutane

Chapter Review

12.1 Alkanes

Alkanes are hydrocarbons with the general formula C_nH_{2n+2} that have only C—C single bonds. In the expanded structural formula, a separate line is drawn for every bonded atom. A condensed structural formula depicts groups composed of each carbon atom and its attached hydrogen atoms. In a line-bond formula, only the carbon–carbon bonds are drawn. Conformational isomers exist because of free rotation of groups around the C—C single bonds.

12.2 IUPAC Naming System for Alkanes

The IUPAC system is used to name organic compounds in a systematic manner. The IUPAC name indicates the number of carbon atoms and family of the compounds. For example, in the name *pentane*, the prefix *pent* indicates a main chain of five carbon atoms and the ending *ane* indicates an alkane.

12.3 Drawing Structural Formulas

For a continuous alkane, the carbon atoms are connected in a chain and bonded to hydrogen atoms. Substituents such as alkyl groups and/or halogen atoms can replace hydrogen atoms on the main chain. Constitutional isomers have the same molecular formula but differ in the order of bonding. Conformational isomers, which are the different orientations of groups in a molecule, result as groups rotate freely about C—C bonds.

12.4 Haloalkanes

A haloalkane contains one or more F, Cl, Br, or I atoms. In the IUPAC system, halogen atoms are named as fluoro, chloro, bromo, or iodo substituents attached to the main chain. In the common name, alkyl halides, the name of the alkyl group precedes the halide, for example methyl chloride.

12.5 Cycloalkanes

In cycloalkanes, the carbon atoms form a ring or cyclic structure with the general formula C_nH_{2n}. The name is written by placing the prefix *cyclo* before the alkane name with the same number of carbon atoms. Substituents are named as alkyl and halo groups and numbered if two or more are bonded to the ring. Cyclic structures with two attached groups may have cis and trans isomers, which differ in the orientation of atoms in space. The two substituents are on the same side of the ring in the cis isomer, but on opposite sides in the trans isomer.

12.6 Physical Properties of Alkanes and Cycloalkanes

As nonpolar molecules, alkanes are not soluble in water. They are usually less dense than water. With only weak attractions, they have low melting and boiling points. For alkanes of similar mass, cycloalkanes have higher boiling points and branched alkanes have lower boiling points than the continuous, nonbranched chain alkanes.

12.7 Chemical Properties of Alkanes and Cycloalkanes

Although the C—C bonds in alkanes resist most reactions, alkanes undergo combustion and halogenation. In combustion or burning, alkanes react with oxygen to produce carbon dioxide and water. In the halogenation of alkanes, halogen atoms replace one or more hydrogen atoms. A source of energy such as UV radiation or heat is required for this substitution reaction.

Summary of Naming

Type	Example	Characteristic	Structure
Alkane	Propane	single C—C, C—H bonds	CH_3—CH_2—CH_3
	Methylpropane		CH_3—$\overset{\overset{\displaystyle CH_3}{\vert}}{CH}$—$CH_3$
Haloalkane	1-Chloropropane	halogen atom	CH_3—CH_2—CH_2—Cl
Cycloalkane	Cyclobutane	carbon ring	□

Summary of Reactions

Combustion

Alkane + O_2 ⟶ CO_2 + H_2O + energy

Halogenation (substitution by halogen)

Alkane + halogen $\xrightarrow{\text{Light or heat}}$ haloalkane + hydrogen halide

CH_4 + Cl_2 $\xrightarrow{\text{Light or heat}}$ CH_3Cl + HCl

 + Cl_2 $\xrightarrow{\text{Light or heat}}$ (cyclopentyl chloride) + HCl

Key Terms

alkane A hydrocarbon that has only carbon–carbon single bonds.

alkyl group An alkane minus one hydrogen atom that bonds to a main chain. Alkyl groups are named like the alkanes except a *yl* ending replaces *ane*.

branch A carbon group or halogen bonded to the main carbon chain.

branched-chain alkane A hydrocarbon containing a substituent bonded to the main chain.

cis–trans isomer In cycloalkanes, two substituents bonded to the ring can differ in the orientation in space. In cis isomers, the groups are on the same side of the ring but on opposite sides in the trans isomer.

combustion A chemical reaction in which an alkane reacts with oxygen to produce CO_2, H_2O, and energy.

condensed structural formula A structural formula that shows the arrangement of the carbon atoms in a molecule but groups each carbon atom with its bonded hydrogen atoms (CH_3, CH_2, or CH).

conformations The different orientations of groups in a molecule resulting from the free rotation of groups about C—C bonds.

constitutional isomers Molecules that have the same molecular formula but different arrangement of atoms.

continuous alkane An alkane in which the carbon atoms are connected in a row, one after the other.

cycloalkane An alkane that is a ring or cyclic structure.

expanded structural formula A type of structural formula that shows the arrangement of the atoms by drawing each bond in the hydrocarbon as C—H or C—C.

haloalkane A type of alkane that contains one or more halogen atoms.

halogenation reaction A substitution reaction that occurs in the presence of light or heat in which halogen atoms replace hydrogen atoms in an alkane.

IUPAC system An organization known as International Union of Pure and Applied Chemistry that determines the system for naming organic compounds.

line-bond formula A type of structural formula that shows only the bonds from carbon to carbon.

primary carbon A carbon bonded to one other carbon atom.

secondary carbon A carbon bonded to two other carbon atoms.

structural formula A type of formula that shows the order of bonding for the carbon atoms in the molecule.

substituent Groups of atoms such as an alkyl group or a halogen bonded to the main chain or ring of carbon atoms.

substitution A reaction such as halogenation in which halogen atoms replace one or more hydrogen atoms in a hydrocarbon.

tertiary carbon A carbon bonded to three other carbon atoms.

Additional Problems

12.49 Identify whether each of the following pairs of structural formulas represent constitutional isomers, the same molecule, or different molecules.

a. CH_2—CH_2—CH_3 (with CH_3 branch) and CH_3—CH—CH_3 (with CH_3 branch)

b. (methylcyclopentane) and (cyclohexane)

c. CH_3—CH—CH_3 (with CH_2—CH—CH_3 and CH_3 branches) and CH_3—CH—CH_2—CH (with CH_3 and CH_3 branches)

d. (line structures) and

12.50 Identify whether each of the following pairs of structural formulas represent constitutional isomers, the same molecule, or different molecules.

a. (line structure) and (line structure)

b. (cyclohexane) and (dimethylcyclobutane with CH_3 groups)

c. CH_3—C—CH_2—CH_3 (with Cl, Cl) and CH_2—C—CH_3 (with CH_3, Cl, Cl)

d. (cyclohexane with Br and Br) and (cyclohexane with Br Br)

12.51 Write the name of each of the following alkyl groups:

a. CH_3—

b. CH_3—CH_2—CH_2—

c. 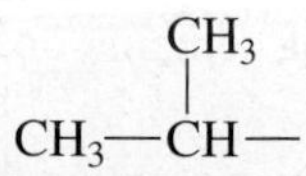 CH_3—$\overset{\displaystyle CH_3}{\overset{|}{CH}}$—

12.52 Write the name of each of the following alkyl groups:

a. CH_3—CH_2—

b. 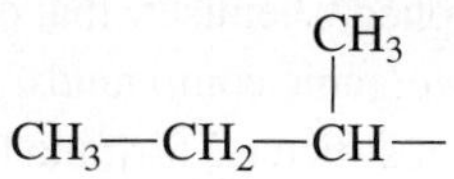 CH_3—CH_2—$\overset{\displaystyle CH_3}{\overset{|}{CH}}$—

c. CH_3—$\overset{\displaystyle CH_3}{\underset{\displaystyle CH_3}{\overset{|}{\underset{|}{C}}}}$—

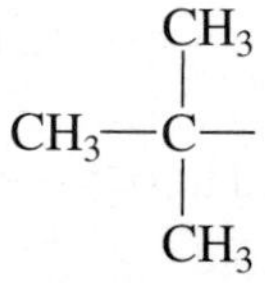

12.53 Give the IUPAC names for each of the following molecules:

a. 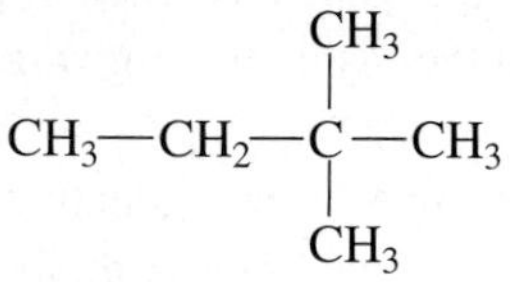 CH_3—CH_2—$\overset{\displaystyle CH_3}{\underset{\displaystyle CH_3}{\overset{|}{\underset{|}{C}}}}$—$CH_3$

b. CH_3—CH_2—Cl

c. 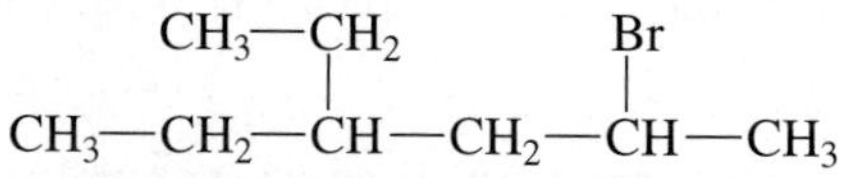 CH_3—CH_2—$\overset{\displaystyle CH_3-CH_2}{\overset{|}{CH}}$—$CH_2$—$\overset{\displaystyle Br}{\overset{|}{CH}}$—$CH_3$

d. 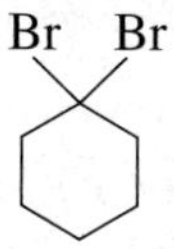 Br Br
(cyclohexane with two Br)

12.54 Give the IUPAC names for each of the following molecules:

a. 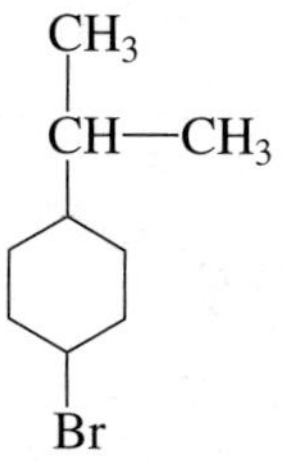
$\overset{\displaystyle CH_3}{\overset{|}{CH}}$—$CH_3$
(cyclohexane with Br)
Br

b. 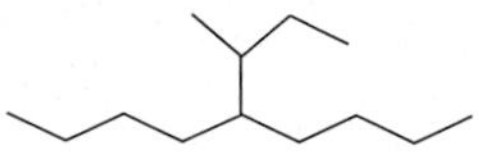
(line-bond structure)

c. 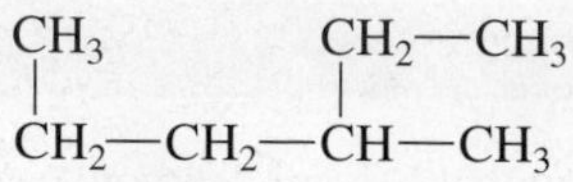 CH_3—CH_2—$\overset{\displaystyle CH_2-CH_3}{\overset{|}{CH}}$—$CH_3$

d. 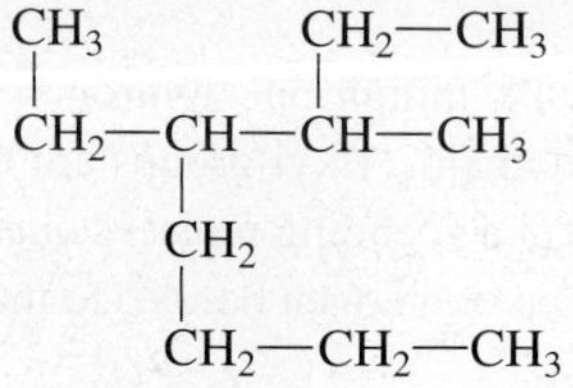 CH_3—CH—$\overset{\displaystyle CH_2-CH_3}{\overset{|}{\underset{\displaystyle CH_2}{\underset{|}{}}}}$—$CH_3$
CH_2—CH_2—CH_3

12.55 Write the condensed structural formula for each of the following molecules:

a. 3-ethylhexane

b. 2,3-dimethylpentane

c. 1,3-dichloro-3-methylheptane

d. bromocyclobutane

e. 1-bromo-4-isopropylcyclohexane

f. *sec*-butyl chloride

12.56 Write the condensed structural formula for each of the following molecules:

a. ethylcyclopropane

b. 1,3-dimethylcyclohexane

c. isopropylcyclopentane

d. 1,1-dimethylcyclopentane

e. 2-bromo-1,1-dichlorocyclopentane

f. *tert*-butyl chloride

12.57 Draw the line-bond structural formula for each of the following molecules:

a. pentane

b. 2,3-dimethylhexane

c. 2-bromo-4-isopropylheptane

12.58 Draw the line-bond structural formula for each of the following molecules:

a. butane

b. 3-ethyl-2-methylhexane

c. 3,4,5-trimethyloctane

12.59 Write structural formulas and IUPAC names for two constitutional isomers of each of the following:

a. hexane

b. 3-ethylpentane

c. 2,2-dimethylbutane

d. 1,1-dibromocyclohexane   (cyclohexanes only)

12.60 Write structural formulas and IUPAC names for two constitutional isomers of each of the following:

a. pentane

b. 1,1-dibromocyclobutane (cyclobutanes only)

c. 2,2-dichlorobutane

d. 1,2-dimethylcyclopentane (cyclopentanes only)

12.61 The following names are incorrect. Write the correct IUPAC name.
   a. 3-methylbutane
   b. 3,4-dimethylpentane
   c. 2,4-dibromocyclohexane
   d. 1,4-dimethylbutane

12.62 The following names are incorrect. Write the correct IUPAC name.
   a. 1-ethylbutane
   b. 4,5-dimethylpentane
   c. 2,3-dichlorocyclopentane
   d. 2-isopropylbutane

12.63 Draw at least six constitutional isomers of $C_5H_{11}Cl$.

12.64 Draw at least five constitutional isomers of $C_4H_6Br_2$.

12.65 Draw the structural formulas for each of the following:
   a. *sec*-butyl bromide
   b. *cis*-1-bromo-2-chlorocyclopropane
   c. *trans*-1,3-dimethylcyclopentane
   d. *cis*-1,2-dibromocyclobutane

12.66 Draw the structural formulas for each of the following:
   a. *cis*-1,2-dimethylcyclobutane
   b. *trans*-1-bromo-2-chlorocyclopentane
   c. isopropyl fluoride
   d. *trans*-1-chloro-2-methylcyclohexane

12.67 In an automobile engine, "knocking" occurs when the combustion of gasoline occurs too rapidly. The octane number of gasoline represents the ability of a gasoline mixture to reduce knocking. A sample of gasoline is compared with heptane, rated 0 because it reacts with severe knocking, and 2,2,4-trimethylpentane (isooctane), which has a rating of 100 because of its low knocking. Write the condensed structural formula, molecular formula, and equation for the complete combustion of isooctane.

12.68 Draw the structures of the following halogenated compounds, which are used as refrigerants.
   a. Freon 14, tetrafluoromethane
   b. Freon 114, 1,2-dichloro-1,1,2,2-tetrafluoroethane
   c. Freon C318, octafluorocyclobutane

12.69 Identify the compound in each pair that has the higher boiling point.
   a. pentane or heptane
   b. pentane or cyclopentane

   c. hexane or 2-methylpentane
   d. cyclobutane or cyclohexane

12.70 Identify the compound in each pair that has the higher boiling point.
   a. butane or octane
   b. cyclopentane or pentane
   c. 2,2-dimetylbutane or cyclohexane
   d. propane or pentane

12.71 Write a balanced equation for the complete combustion of each of the following:
   a. propane
   b. C_5H_{12}
   c. cyclobutane
   d. octane

12.72 Write a balanced equation for the complete combustion of each of the following:
   a. hexane
   b. methylcyclohexane
   c. cyclopentane
   d. 2-methylpropane

12.73 Draw the structural formulas of the monohalogenated organic products that result from the chlorination of each of the following:
   a. ethane
   b. propane
   c. cyclopentane

12.74 Draw the structural formulas of the monohalogenated organic products that result from the bromination of each of the following:
   a. pentane
   b. cyclohexane
   c. 2-methylbutane

12.75 The density of pentane, a component of gasoline, is 0.63 g/mL. The heat of combustion for pentane is 845 kcal per mole (3536 kJ/mole).
   a. Write an equation for the complete combustion of pentane.
   b. What is the molar mass?
   c. How much heat is produced when 1 gallon of pentane is burned (1 gallon = 3.78 liters)?
   d. How many liters of CO_2 at STP are produced from the complete combustion of 1 gallon of pentane?

13 Unsaturated Hydrocarbons

"When we have a hazardous materials spill, the first thing we do is isolate it," says Don Dornell, assistant fire chief, Burlingame Fire Station. "Then our technicians and a county chemist identify the product from its flammability and solubility in water so we can use the proper materials to clean up the spill. We use different methods for alcohol, which mixes with water, than for gasoline, which floats. Because hydrocarbons are volatile, we use foam to cover them and trap the vapors. At oil refineries, we will use foams, but many times we squirt water on the tanks to cool the contents below their boiling points. By knowing the boiling point of the product and its density and vapor density, we know if it floats or sinks in water and where its vapors will go."

LOOKING AHEAD

the Chemistry place

www.chemplace.com/college

Visit the URL above or use the CD-ROM in the book for extra quizzing, interactive tutorials, and career resources.

In Chapter 12, we looked at **saturated hydrocarbons**, which contain all single bonds between their carbon atoms. Now we will investigate unsaturated hydrocarbons, carbon compounds containing one or more carbon–carbon double bonds or triple bonds. When we cook with vegetable oils such as corn oil, safflower oil, or olive oil, we are using unsaturated lipids that have double bonds in their long carbon chains. Saturated fats from animal sources also have long chains of carbon atoms, but they are connected only by the single bonds of alkanes. If we compare the two types of fats, we find considerable differences in their physical and chemical properties. Vegetable oils are liquid at room temperature whereas animal fats are solid. Because double bonds are reactive, unsaturated fats can be oxidized by oxygen in the air, especially at warm temperatures, forming products that have rancid, unpleasant odors. The saturated fats are more resistant to reactions.

13.1 Alkenes and Alkynes

The **unsaturated hydrocarbons** have fewer hydrogen atoms than alkanes. **Alkenes** contain at least one double bond between carbons. If a straight-chain alkene contains just one double bond between carbons, it has the general formula C_nH_{2n}. Alkenes are useful compounds in industry and have important functions in plants and animals. Ethylene is an important industrial compound used to make a plastic called polyethylene. The unsaturated sites are very reactive, which distinguishes unsaturated compounds from the alkanes. **Alkynes** contain at least one triple bond between carbons. If a straight-chain alkyne contains just one triple bond between carbons, it has the general formula C_nH_{2n-2}. (See Figure 13.1.) Only a small number of alkynes are found in nature. However, ethyne, commonly called acetylene, is used in welding where it burns at a very high temperature.

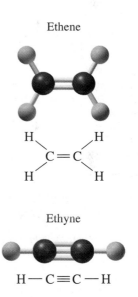

Ethene

Ethyne

Figure 13.1 Ball-and-stick models of ethene and ethyne show the functional groups of double or triple bonds.
Q *Why are these compounds called unsaturated hydrocarbons?*

$H_2C{=}CH_2$

Ethene (ethylene)
C_2H_4

$CH_3(CH_2)_7 \quad (CH_2)_{12}CH_3$

C=C

H   H

Cis-9-tricosene, muscalure, housefly sex attractant
$C_{23}H_{46}$

$HC{\equiv}CH$

Ethyne (acetylene)
C_2H_2

Identifying Unsaturated Compounds

Classify each of the following structural formulas as an alkane, alkene, or alkyne. Write their molecular formulas.

a. $CH_3—C\equiv C—CH_3$ **b.** $CH_3—CH_2—CH_3$

c. $CH_3—CH_2—\overset{\overset{\displaystyle CH_3}{|}}{C}=CH—CH_2—CH_3$

Solution

The structural formula with a double bond is an alkene, and the one with a triple bond is an alkyne.

a. alkyne: C_4H_6 **b.** alkane; C_3H_8 **c.** alkene; C_7H_{14}

Study Check

Is the following structure a cycloalkane or cycloalkene?

A σ bond between carbon and hydrogen

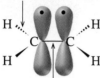

A σ bond between two carbons

Overlap of *p* orbitals forms a π bond

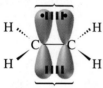

Figure 13.2 The π bond is formed by the overlapping of two *p* orbitals on adjacent carbon atoms.
Q *What are the two types of bonds in a double bond?*

Figure 13.3 In the flat ethene molecule, the bond angles are 120°. In ethyne, the bond angles are 180°.
Q *What accounts for the different bond angles in double and triple bonds?*

Alkene Structure

The simplest alkene C_2H_4 is called ethene, but it is more likely to be called by its common name ethylene. There are two CH_2 groups connected by a double bond. In the electron dot structure, the double bond is represented by two sets of electrons. We already know that carbon has four electrons and needs four bonds. In a double bond, each carbon atom is attached to three other atoms (one carbon and two hydrogens). The C—C and two C—H bonds each share a pair of electrons to form strong bonds called sigma (σ) bonds. To form a second bond between the carbon atoms, the single electrons in the *p* orbitals overlap to produce a weaker bond known as a pi (π) bond. Thus, a double bond is a combination of one sigma and one pi bond. (See Figure 13.2.)

According to VSEPR theory (Chapter 4), three groups bonded to each carbon in the double bond are planar and arranged at angles of 120°. (See Figure 13.3.) The pi (π) bond is weaker than a sigma bond, and because it is easier to break, it is the key to the chemical properties of alkenes.

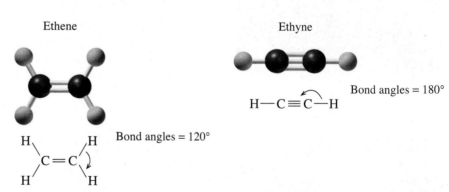

Ethene

Bond angles = 120°

Ethyne

$H—C\equiv C—H$ Bond angles = 180°

Alkyne Structure

The simplest alkyne is called ethyne, but it is commonly known as acetylene. In ethyne there are two CH groups connected by a triple bond. We already know that carbon has four electrons and needs four bonds. In a triple bond, each carbon atom is attached to two other atoms (one carbon and one hydrogen). The C—C and C—H bonds each share a pair of electrons to form strong bonds called sigma (σ) bonds. We have seen how a pi (π) bond forms in alkenes when the *p* orbitals overlap. In a triple bond, two sets of *p* orbitals overlap forming two pi (π) bonds. Thus, a triple bond is a combination of one sigma and two pi bonds. (See Figure 13.4.)

According to VSEPR theory, two groups bonded to each carbon in the triple bond are linear and arranged at angles of 180°.

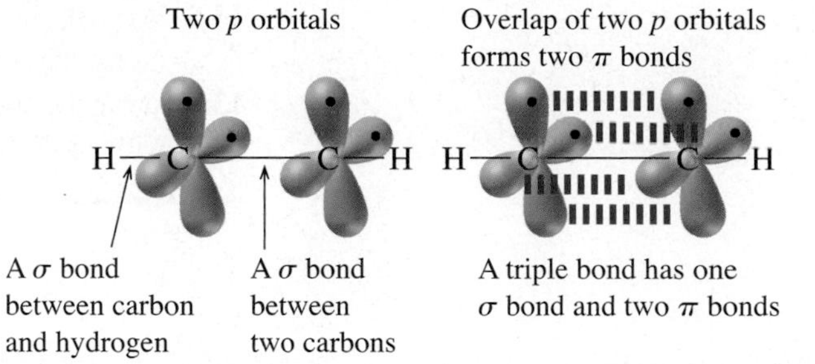

Figure 13.4 In ethyne, C_2H_2, the overlap of two *p* orbitals forms two π bonds required for the triple bond.

Q *How is the requirement of four bonds for a carbon atom maintained in ethyne?*

SAMPLE PROBLEM 13.2

Structure of Alkenes and Alkanes

What do double and triple bonds have in common and how do they differ?

Solution

In both types of bonds, a sigma bond connects two carbon atoms. The additional bonds are pi bonds, which are formed by overlapping one or two sets of *p* orbitals. A double bond has one sigma bond and one pi bond, whereas a triple bond has one sigma bond and two pi bonds.

Study Check

Why are the atoms arranged at angles of 120° in double bonds, but at 180° in triple bonds?

QUESTIONS AND PROBLEMS

Alkenes and Alkynes

13.1 Indicate whether each of the following statements describes a feature of an alkane, alkene, or alkyne:

a. Has only sigma (σ) bonds in the molecule.

b. The molecule contains one sigma (σ) and two pi (π) carbon–carbon bonds.

c. The groups on a carbon atom are arranged at 120°.

13.2 Indicate whether each of the following statements describes a feature of an alkane, alkene, or alkyne:

a. The molecule contains one sigma (σ) and one pi (π) carbon–carbon bond.

b. The groups on the carbon atoms are arranged at 109°.

c. The molecule is linear.

13.3 Identify the following as alkanes, alkenes, cycloalkenes, or alkynes:

a. C_6H_{14} **b.** $H-\underset{H}{\overset{H}{C}}-\overset{H}{C}=\overset{H}{C}-H$ **c.** $CH_3-CH_2-C\equiv C-H$ **d.** 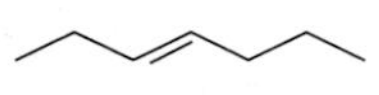 **e.** 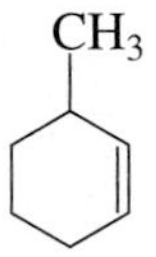

13.4 Identify the following as alkanes, alkenes, cycloalkenes, or alkynes:

a. 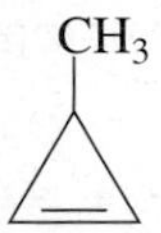

b. 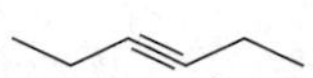

c. $CH_3-\overset{CH_3}{C}=\underset{CH_3}{C}-CH_3$ **d.** 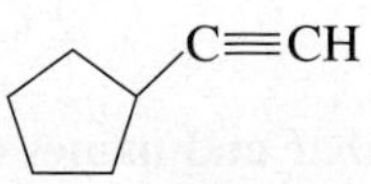

e. 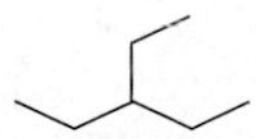

13.5 Write the condensed structural formulas of an alkene and cycloalkane with a molecular formula of C_3H_6.

13.6 Write the condensed structural formulas of two alkenes and a cycloalkane with a molecular formula of C_4H_8.

LEARNING GOAL

Write the IUPAC names for alkenes and alkynes; give common names for simple structures.

EXPLORE YOUR WORLD

Ripening Fruit

Obtain two unripe green bananas. Place one in a plastic bag and seal it. Leave them on the counter. Check the bananas twice a day to observe any difference in the ripening process.

Questions

1. What compound helps ripen the bananas?
2. What are some possible reasons for any difference in the ripening rate?
3. If you wish to ripen an avocado, what procedure might you use?

13.2 Naming Alkenes and Alkynes

The IUPAC names for alkenes and alkynes are similar to those of alkanes. The simplest alkene, ethene, often named by its common name, ethylene, is an important plant hormone involved in promoting the ripening of fruit. Commercially grown fruit, such as avocados, bananas, and tomatoes, are often picked before they are ripe. Before the fruit is brought to market, it is exposed to ethylene to accelerate the ripening process. Ethylene also accelerates the breakdown of cellulose in plants, which causes flowers to wilt and leaves to fall from trees.

The IUPAC name of the simplest alkyne is ethyne, although acetylene, its common name, is often used. See Table 13.1 for a comparison of the naming for alkanes, alkenes, and alkynes.

Table 13.1   Comparison of Names for Alkanes, Alkenes, and Alkynes

Alkane	Alkene	Alkyne
H_3C-CH_3 Ethane	$H_2C=CH_2$ Ethene (ethylene)	$HC\equiv CH$ Ethyne (acetylene)
$CH_3-CH_2-CH_3$ Propane	$CH_3-CH=CH_2$ Propene	$CH_3-C\equiv CH$ Propyne

IUPAC Rules for Naming Alkenes and Alkynes

For alkenes and alkynes, the longest carbon chain that contains the unsaturated site is the main chain, which is numbered to show the location of the unsaturated site.

1. **Name the longest carbon chain that contains the double or triple bond.** Replace the corresponding alkane ending with *–ene* for an alkene and *–yne* for an alkyne.

2. **Number the main chain from the end nearest the double or triple bond.** Place the lowest number of the carbon atom in the double or triple bond in front of the name of the main chain.

$$CH_3-CH_2-CH=CH_2 \qquad CH_3-CH=CH-CH_3 \qquad CH_3-C\equiv C-CH_3$$

| 4 | 3 | 2 | 1 | | 1 | 2 | 3 | 4 | | 1 | 2 | 3 | 4 |

1-butene   2-butene   2-butyne

3. **Place the number and names of substituents in front of the alkene or alkyne name.**

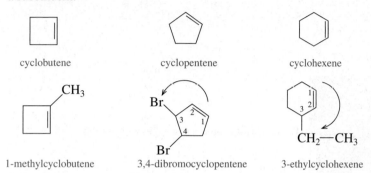

4-methyl-1-pentene   2,3-dimethyl-2-butene   3-methyl-1-butyne

4. **Cyclic alkenes are named as cycloalkenes with numbers for substituents.** With a substituent on the ring, the double bond is understood to be carbon 1 and 2 and the ring is numbered in the direction to give the lowest number to a substituent.

cyclobutene   cyclopentene   cyclohexene

1-methylcyclobutene   3,4-dibromocyclopentene   3-ethylcyclohexene

5. **Name a compound with two double bonds as a diene.** Use numbers to give the location of each double bond.

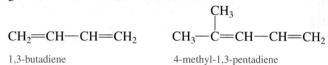

1,3-butadiene   4-methyl-1,3-pentadiene

SAMPLE PROBLEM 13.3

Naming Alkenes and Alkynes

Write the IUPAC name for each of the following unsaturated compounds:

a. $CH_3CH_2CH=CHCH_3$ b.   c.   d. $HC\equiv CCH_2CH_2CH_3$ e.

(c.) Cl / $CH_3C=CHCH_3$ (e.) CH_3

Solutions

a. 2-pentene   b. cyclohexene   c. 2-chloro-2-butene
d. 1-pentyne   e. 1-methylcyclohexene

Study Check

Draw the structural formulas for each of the following:
a. 2-pentyne   b. 3-chlorocyclohexene

QUESTIONS AND PROBLEMS

Naming Alkenes and Alkynes

13.7 Compare the structural formulas of the following:
  a. propene and propyne   b. cyclohexane and cyclohexene
13.8 Compare the structural formulas of the following:
  a. 1-butyne and 2-butyne
  b. 1-methylcyclohexene and 3-methylcyclohexene

Fragrant Alkenes

The odors you associate with lemons, oranges, roses, and lavender are due to volatile compounds that are synthesized by the plants. Often it is unsaturated compounds that are responsible for the pleasant flavors and fragrances of many fruits and flowers. They were some of the first kinds of compounds to be extracted from natural plant material. In ancient times, they were highly valued in their pure forms. Limonene and myrcene give the characteristic odors and flavors to lemons and oranges and bay leaves, respectively. Geraniol and citronellal give roses and lemon grass their distinct aromas. In the food and perfume industries, these compounds are extracted or synthesized and used as perfumes and flavorings.

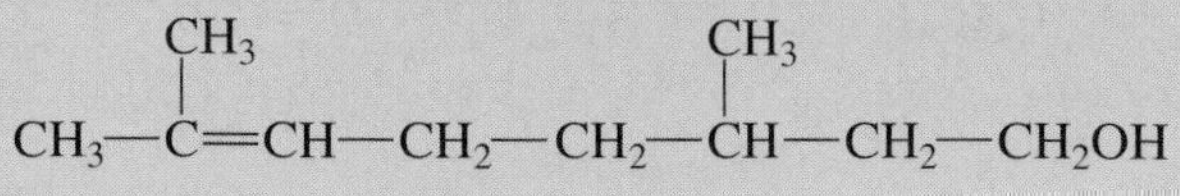

$$CH_3-C(CH_3)=CH-CH_2-CH_2-CH-CH_2-CH_2OH$$

Geraniol, roses

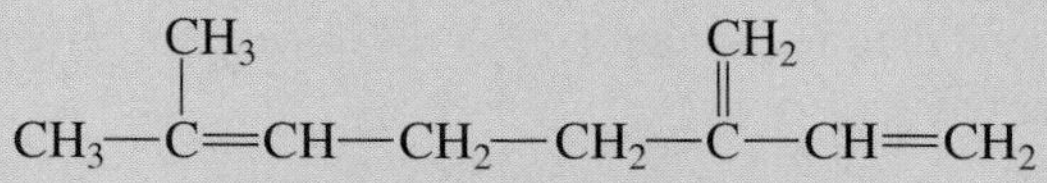

$$CH_3-C(CH_3)=CH-CH_2-CH_2-C(CH_2)-CH=CH_2$$

Myrcene, bay leaves

$$CH_3-C(CH_3)=CH-CH_2-CH_2-C(CH_3)=CH-CHO$$

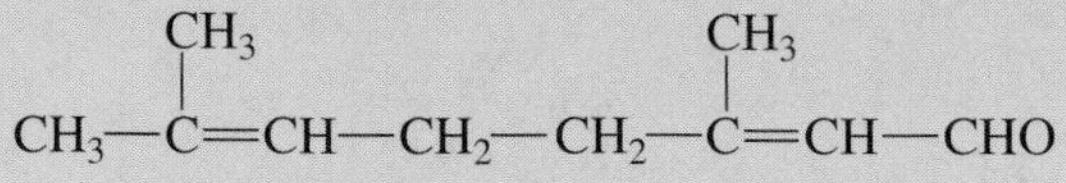

Citronellal, lemon grass

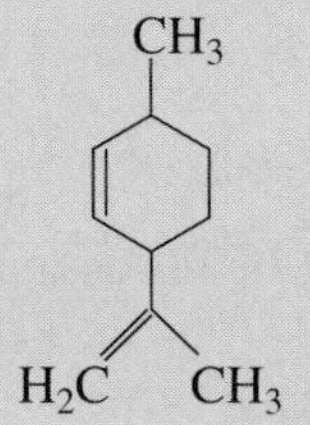

Limonene, lemons, and oranges

13.9 Give the IUPAC name for each of the following:

a. $CH_2=CH_2$　　**b.** 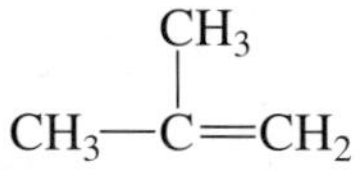 $CH_3-C(CH_3)=CH_2$　　**c.** $CH_3-CH(Br)-C\equiv C-CH_3$

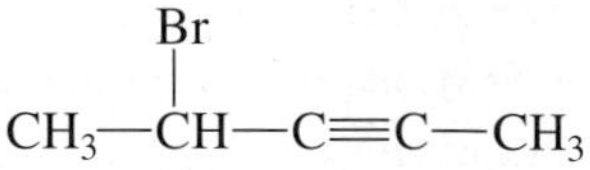

d.  □∥　　**e.** 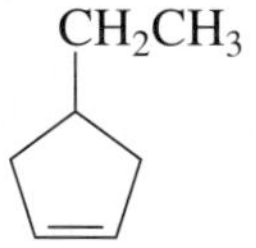 (cyclopentene with CH_2CH_3)　　**f.** 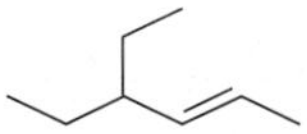 (structure)

13.10 Give the IUPAC name for each of the following:

a. $CH_2=CHCH_2CH_2CH_2CH_3$　　**b.**  $CH_3C\equiv CCH_2CH_2CH(CH_3)CH_3$　　**c.** 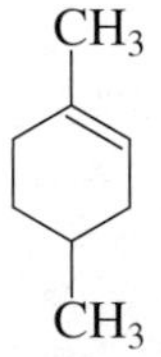 (methylcyclohexene with CH₃)　　**d.** 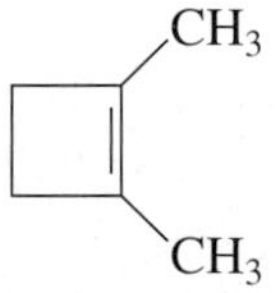 (cyclobutene with CH₃ groups)

e. 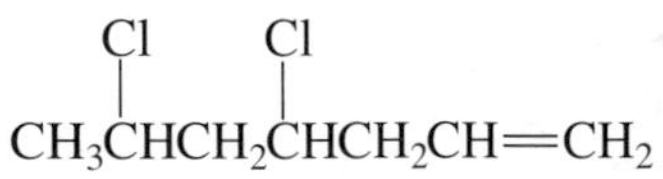 $CH_3CH(Cl)CH_2CH(Cl)CH_2CH=CH_2$　　**f.** 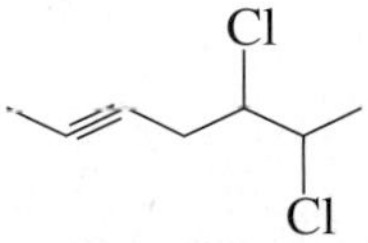 (structure with Cl, Cl)

13.11 Draw the structural formula for each of the following compounds:
a. propene　　　　　　　**b.** 1-pentene　　**c.** 2-methyl-1-butene
d. 3-methylcyclohexene　　**e.** 2-chloro-3-hexyne

13.12 Draw the structural formula for each of the following compounds:
a. 1-methylcyclopentene　　　　**b.** 3-methyl-1-butyne
c. 3,4-dimethyl-1-pentene　　　　**d.** 4-ethyl-1-methylcyclohexene
e. 1,2-dichlorocyclopentene

13.3 Cis–Trans Isomers

Several constitutional isomers can be written for alkenes similar to what we did with the structural formulas of alkanes, which gives the possibilities of a large number of constitutional isomers. Let's take a look at the structures for some constitutional isomers, including a cycloalkane that can be written for a molecular formula C_4H_8.

Constitutional Isomers of C_4H_8

$$H_2C=CH—CH_2—CH_3 \quad CH_3—CH=CH—CH_3 \quad H_2C=\overset{CH_3}{\underset{}{C}}—CH_3 \quad \square$$

1-butene   2-butene   2-methyl-1-propene   cyclobutane

Cis–Trans Isomers for Alkenes

In Chapter 12, we saw that cycloalkanes can have cis and trans isomers due to a lack of rotation of the carbon atoms in the ring. Alkenes can also have cis–trans isomers because there is no rotation around the carbons in the rigid double bond. As a result, any groups connected to the double bond remain fixed on one side or the other. In a **cis isomer,** two groups are on the same side of the double bond. In the **trans isomer,** the groups are on opposite sides of the double bond. For example, cis–trans isomers can be written for 2-butene. (See Figure 13.5.) In general, trans isomers are more stable than their cis counterparts because the large groups attached to the double bond are further apart. As with any pair of isomers, the cis–trans isomers of 2-butene are different compounds with different physical and chemical properties as shown in the following:

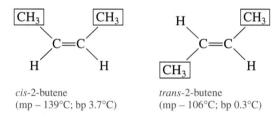

cis-2-butene
(mp – 139°C; bp 3.7°C)

trans-2-butene
(mp – 106°C; bp 0.3°C)

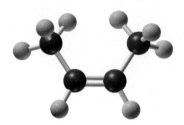

cis-2-butene

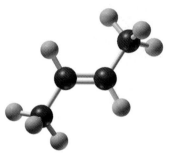

trans-2-butene

Figure 13.5 Ball-and-stick models of the cis and trans isomers of 2-butene.
Q *What feature in 2-butene accounts for the different cis and trans isomers?*

Not every alkene shows cis–trans isomerism. If one of the carbons in the double bond is attached to identical groups, the molecule does not have cis–trans isomers. This is the case of 1-butene and 2-methyl-1-propene, constitutional isomers of 2-butene. If the hydrogen atoms are interchanged on carbon 1, the same structure results. Alkynes do not have cis–trans isomers because the carbons in the triple bond are each attached to only one group.

Identical atoms   1-butene   2-methyl-1-propene   Identical groups

As long as the groups attached to the double bond are different, an alkene will show cis–trans isomers. Another example of cis–trans isomers is the following:

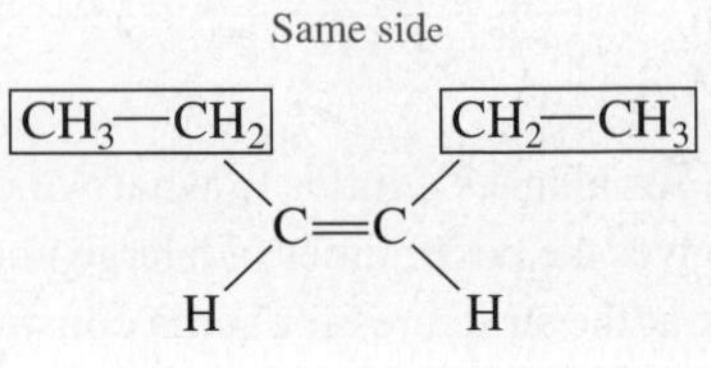

Same side   Opposite side

cis-3-hexene 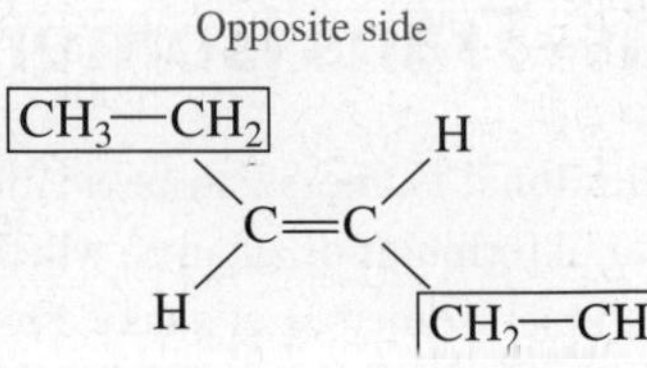   *trans*-3-hexene

---

SAMPLE PROBLEM 13.4

Writing and Naming Cis–Trans Isomers

Determine if each of the following shows cis–trans isomers. If so, write the names.

a. Br—CH=CH—Cl   **b.** CH_3—C=CH_2
                                              |
                                              Cl

Solution

Draw the double bond and attach the separate groups or atoms to each carbon using the following model:

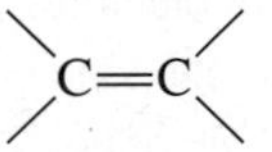

a. In a molecule of Br—CH=CH—Cl, the first carbon is bonded to a bromine atom and a hydrogen atom; the second carbon has a hydrogen atom and a chlorine atom. They can be drawn as follows:

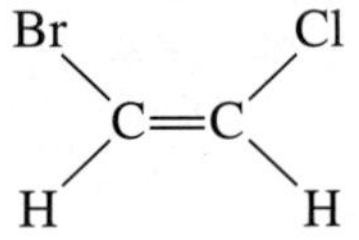

cis-1-bromo-2-chloroethene

In this drawing, the halogen atoms are on the same side and the hydrogen atoms are on the same side. It is the cis isomer.

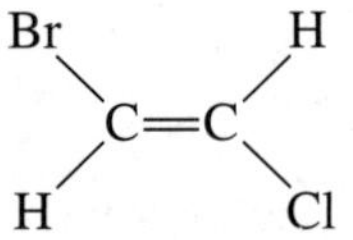

trans-1-bromo-2-chloroethene

In the trans isomer, the halogen atoms are drawn on the opposite sides of the double bond.

b.

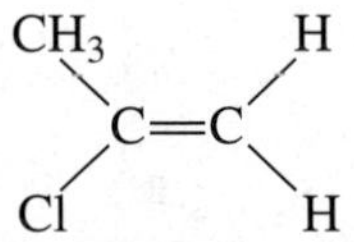

2-chloro-1-propene

In this drawing, there are two hydrogen atoms attached to one of the carbons in the double bond. This compound does not have cis–trans isomers.

Study Check

Draw the structural formula for *cis*-3-heptene.

EXPLORE YOUR WORLD

Modeling Cis–Trans Isomers

Because cis–trans isomerism is not easy to imagine, here are some things you can do to understand the difference in rotation around a single bond compared to a double bond and how it affects groups that are attached to the carbon atoms in the double bond.

Using Your Hands and Fingers as Single and Double Bonds

Put the fingertips of your index fingers together. This is a model of a single bond. Consider the index fingers as a pair of carbon atoms and think of your thumbs and other fingers as other parts of a carbon chain. While your index fingers are touching, twist your hands and change the position of the thumbs relative to each other. Notice how the relationship of your other fingers changes.

Now place the tips of your index fingers and middle fingers together in a model of a double bond. As you did before, twist your hands to move the thumbs apart. What happens? Can you change the location of your thumbs relative to each other without breaking the double bond? The difficulty of moving your hands with two fingers touching represents the lack of rotation about a double bond. You have made a model of a cis isomer when both thumbs are on the same side. If you

Cis-hands (*cis*-thumbs/fingers)

Trans-hands (*trans*-thumbs/fingers)

turn one hand over so one thumb points down and one points up, you have made a model of a trans isomer.

Using Toothpicks and Gumdrops to Form Single and Double Bonds

Use toothpicks for bonds and gumdrops for atoms. Place a toothpick between two black gumdrops (carbon atoms). To each carbon atom attach three toothpicks with a yellow gumdrop on the other end. Rotate one of the gumdrop carbon atoms to show the movement of the attached atoms.

Trans-gumdrop isomer

Remove a toothpick and yellow gumdrop from each carbon atom. Place a second toothpick between the carbon atoms. Each carbon atom should now be attached to two noncarbon atoms. Make those two atoms different colors, such as yellow and red. First place the red gumdrops on the same side of the double bond, which makes a model of a cis isomer. Try to twist the double toothpicks. Can you do it? If you turn the whole molecule upside down, are the red gumdrops still on the same side?

Now switch one of the red gumdrops with the yellow gumdrop on the same carbon. The red gumdrops should be opposite or trans to each other. Twist the double bond again. Can you move the red gumdrops to the same side of the double bond? It should not happen without breaking the double bond. How does this illustrate two isomers of a double bond molecule when the atoms in the double bond are attached to different groups of atoms? If you attach two yellow or two red gumdrops to the same carbon atom, can you make cis–trans isomers? Why or why not?

ENVIRONMENTAL NOTE

Pheromones in Insect Communication

Insects and many other organisms emit minute quantities of chemicals called pheromones. Insects use pheromones to send messages to individuals of the same species. Some pheromones warn of danger, others call for defense, mark a trail, or attract the opposite sex. In the last 40 years, the structures of many pheromones have been chemically determined. One of the most studied is bombykol, the sex pheromone produced by the female of the silkworm moth species. The bombykol molecule is a 16 carbon chain with one cis double bond, one trans double bond, and an alcohol group. A few molecules of synthetic bombykol will attract male silkworm moths from distances of over one kilometer. The effectiveness of many of these pheromones depends on the cis or trans configuration of the double bonds in the molecules. A certain species will respond to one isomer but not the other.

Scientists are interested in synthesizing pheromones for use as nontoxic alternatives to pesticides. When used in a trap, bombykol can be used to isolate male silkworm moths. When a synthetic pheromone is released in several areas of a field or crop, the males cannot locate the females, which disrupts the reproductive cycle. This technique has been successful with controlling the oriental fruit moth, the grapevine moth, and the pink bollworm.

Bombykol, sex attractant for the silkworm moth

HEALTH NOTE

Cis–Trans Isomers for Night Vision

The retinas of the eyes consist of two types of cells, rods and cones. The rods on the edge of the retina allow us to see in dim light, and the cones, in the center, produce our vision in bright light. In the rods, there is a substance called rhodopsin that absorbs light. Rhodopsin is composed of *cis*-11-retinal, an unsaturated compound, attached to a protein. When rhodopsin absorbs light, the *cis*-11-retinal isomer is converted to its trans isomer, which changes its shape. The trans form no longer fits the protein and it separates from the protein. The change from the cis to trans isomer and the separation from the protein generates an electrical signal that the brain converts into an image.

An enzyme (isomerase) converts the trans isomer back to the *cis*-11-retinal isomer and the rhodopsin re-forms. If there is a deficiency of rhodopsin in the rods of the retina, night blindness may occur. One common cause is a lack of vitamin A in the diet. In our diet, we obtain vitamin A from plant pigments containing *β*-carotene, which is found in foods such as carrots, squash, and spinach. In the small intestine, the *β*-carotene is converted to vitamin A, which can be converted to *cis*-11-retinal or stored in the liver for future use. Without a sufficient quantity of retinal, not enough rhodopsin is produced to enable us to see adequately in dim light.

Cis-trans isomers of retinal

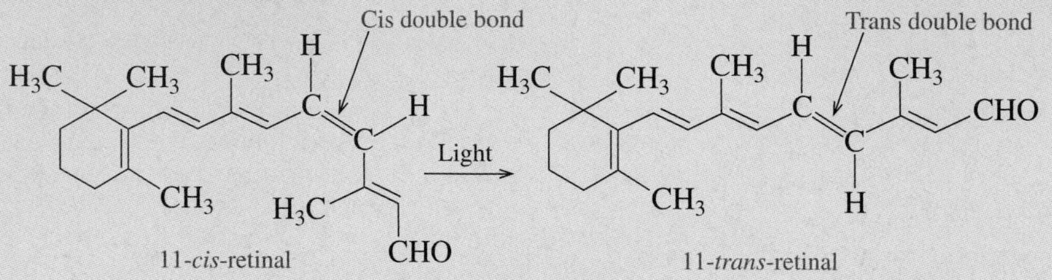

11-*cis*-retinal   →Light→   11-*trans*-retinal

QUESTIONS AND PROBLEMS

Cis-Trans Isomers

13.13 What is the difference between constitutional isomers and cis-trans isomers?

13.14 How does *cis*-2-butene differ from *trans*-2-butene?

13.15 Draw structural formulas for the constitutional isomers of molecular formula C_3H_5Cl.

13.16 Draw structural formulas for the constitutional isomers of molecular formula C_4H_6.

13.17 Which of the following can be written as cis–trans isomers?

a. $CH_2\!=\!CHCH_3$ b. $CH_3CH_2CH\!=\!CHCH_3$ c. (structure shown)

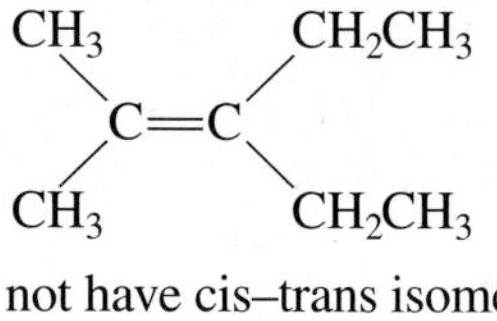

13.18 Which of the following do not have cis–trans isomers?

a. (structure shown)   b. $CH_3CH_2CH_2CH\!=\!CH_2$ c. (structure shown)

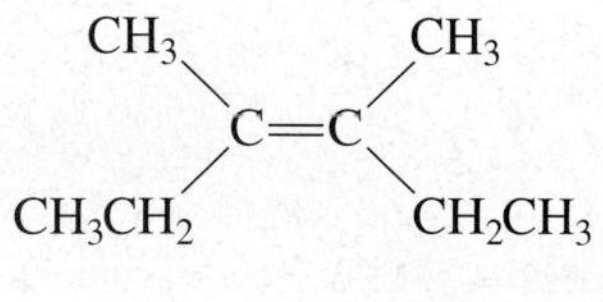

13.19 Write the IUPAC name of each of the following using cis or trans prefixes:

a. (structure shown)   b. (structure shown)   c. (structure shown)

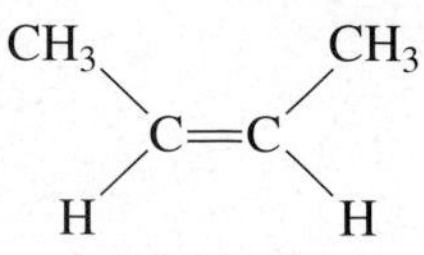

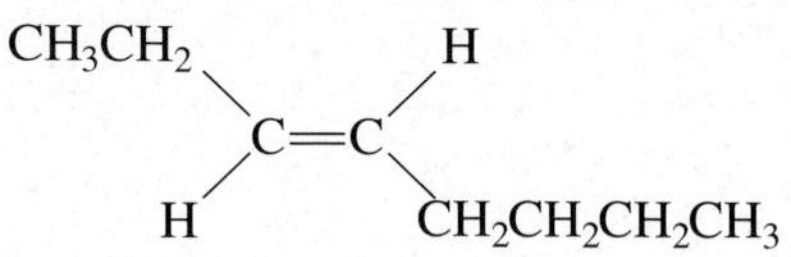

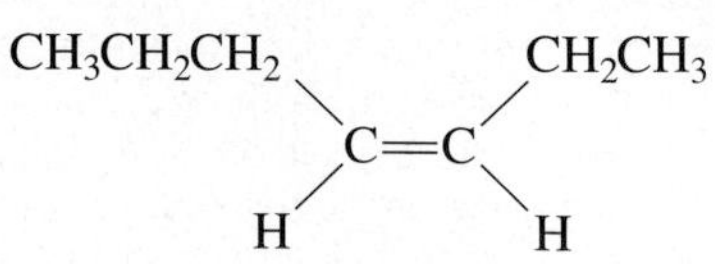

13.20 Write the IUPAC name of each of the following using cis or trans prefixes:

a.
$$CH_3 \quad\quad CH_2CH_3$$
$$\diagdown C = C \diagup$$
$$H \quad\quad\quad\quad H$$

b.
$$CH_3 \quad\quad\quad H$$
$$\diagdown C = C \diagup$$
$$H \quad\quad CH_2CH_2CH_2CH_3$$

c.
$$CH_3CH_2CH_2 \quad\quad H$$
$$\diagdown C = C \diagup$$
$$H \quad\quad\quad CH_2CH_3$$

13.21 Draw the structural formula for each of the following:
  a. *trans*-2-butene   b. *cis*-2-pentene   c. *trans*-3-heptene
13.22 Draw the structural formula for each of the following:
  a. *cis*-3-hexene   b. *trans*-2-pentene   c. *cis*-4-octene

13.4 Addition Reactions

LEARNING GOAL

Write the structural formulas and names for the organic products of addition reactions of alkenes and alkynes.

For alkenes and alkynes, the most characteristic reaction is the **addition** of atoms or groups of atoms to the carbons of the double or triple bond. Addition occurs because the weak pi bonds in double and triple bonds are easily broken, which provides electrons for new single bonds. Many different kinds of reactants can be added to double and triple bonds to form more stable products. Some addition reactions require a catalyst but others do not. The general equation for the addition of a reactant A—B to an alkene can be written as follows:

$$\diagup C = C \diagdown \quad + \quad A - B \quad \xrightarrow{\text{Addition}} \quad \begin{array}{cc} A & B \\ | & | \\ -C - C- \\ | & | \end{array}$$

Alkene

The addition reactions have different names that depend on the type of reactant we add to the alkene, as Table 13.2 shows.

Table 13.2 Reactants and Addition Reactions

Reactant Added	Name of Addition Reaction
H_2	Hydrogenation
Cl_2, Br_2	Halogenation
HCl, HBr, HI	Hydrohalogenation
HOH	Hydration

Hydrogenation

In a reaction called **hydrogenation**, atoms of hydrogen add to the carbons in a double or triple bond to form alkanes. A catalyst such as platinum (Pt), nickel (Ni), or palladium (Pd) is added to speed up the reaction. The general equation for hydrogenation can be written as follows:

$$\diagup C = C \diagdown \quad + \quad H - H \quad \xrightarrow{\text{Catalyst}} \quad \begin{array}{cc} H & H \\ | & | \\ -C - C- \\ | & | \end{array} \left\{ \begin{array}{l} \text{Electrons from the pi bond} \\ \text{are used to form these} \\ \text{single bonds} \end{array} \right.$$

Double bond (unsaturated)   Single bond (saturated)

Some examples of the hydrogenation of alkenes and alkynes follow:

$$CH_3-CH=CH-CH_3 + H-H \xrightarrow{Pt} CH_3-\overset{\overset{\displaystyle H}{|}}{C}H-\overset{\overset{\displaystyle H}{|}}{C}H-CH_3$$

2-Butene   Butane

Cyclohexene + H—H $\xrightarrow{Ni}$ Cyclohexane

EXPLORE YOUR WORLD

Unsaturation in Fats and Oils

Read the labels on some containers of vegetable oils, margarine, peanut butter, and shortenings.

Questions

1. What terms on the label tell you that the compounds contain double bonds?
2. A label on a bottle of canola oil lists saturated, polyunsaturated, and monounsaturated fats. What do these terms tell you about the type of bonding in the fats?
3. A peanut butter label states that it contains partially hydrogenated vegetable oils or completely hydrogenated vegetable oils. What does this tell you about the type of reaction that took place in preparing the peanut butter?

The hydrogenation of alkynes requires two molecules of hydrogen to form the alkane product.

$$CH_3-C\equiv C-CH_3 + 2\,H-H \xrightarrow{Pt} CH_3-\underset{\underset{H}{|}}{\overset{\overset{H}{|}}{C}}-\underset{\underset{H}{|}}{\overset{\overset{H}{|}}{C}}-CH_3$$

2-Butyne   Butane

SAMPLE PROBLEM 13.5

Writing Equations for Hydrogenation

Write the structural formula for the product of the following hydrogenation reactions:

a. $CH_3-CH=CH_2 + H_2 \xrightarrow{Pt}$ **b.** 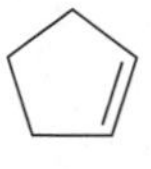 ⬠ $+ H_2 \xrightarrow{Pt}$ **c.** $HC\equiv CH + 2H_2 \xrightarrow{Ni}$

Solution

In an addition reaction, hydrogen adds to the double or triple bond to give an alkane.

a. $CH_3-CH_2-CH_3$ **b.** 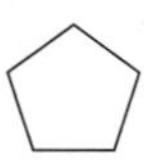 ⬠   **c.** H_3C-CH_3

Study Check

Draw the structural formula of the product of the hydrogenation of 2-methyl-1-butene using a platinum catalyst.

HEALTH NOTE

Hydrogenation of Unsaturated Fats

Vegetable oils such as corn oil or safflower oil are unsaturated fats composed of fatty acids that contain double bonds. The process of hydrogenation is used commercially to convert the double bonds in the unsaturated fats in vegetable oils to saturated fats such as margarine, which are more solid. Adjusting the amount of added hydrogen produces partially hydrogenated fats such as soft margarine, solid margarine in sticks, and shortenings, which are used in cooking. For example, oleic acid is a typical unsaturated fatty acid in olive oil and has a cis-double bond at carbon 9. When oleic acid is hydrogenated, it is converted to stearic acid, a saturated fatty acid.

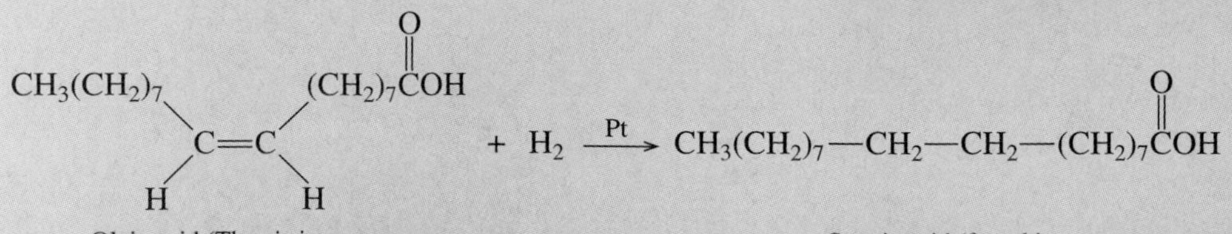

$$CH_3(CH_2)_7 \qquad (CH_2)_7\overset{\overset{\displaystyle O}{\|}}{C}OH$$
$$\underset{\underset{H}{|}}{C}=\underset{\underset{H}{|}}{C}$$

$$+\ H_2 \xrightarrow{Pt} CH_3(CH_2)_7-CH_2-CH_2-(CH_2)_7\overset{\overset{\displaystyle O}{\|}}{C}OH$$

Oleic acid (The cis isomer
is found in olive oil and other
unsaturated fats)

Stearic acid (found in
saturated fats)

Halogenation

In the **halogenation** reactions of alkenes or alkynes, halogen atoms such as chlorine or bromine are added to the double or triple bonds. The reaction occurs readily, without the use of any catalyst, and adds halogen atoms to yield a di- or tetrahaloalkane product. In the general equation for halogenation, the symbol X—X or X_2 is used for Cl_2 or Br_2.

$$\begin{array}{c} \quad\quad\quad\quad\quad \text{X} \quad \text{X} \\ \diagdown \text{C}=\text{C}\diagup \;+\; \text{X}-\text{X} \;\longrightarrow\; \begin{array}{c}||\\-\text{C}-\text{C}-\\||\end{array} \end{array}$$

Here are some examples of adding Cl_2 or Br_2 to alkenes:

$$\begin{array}{c} \quad\quad\quad\quad\quad\quad\quad\quad \text{Cl} \quad \text{Cl} \\ CH_2{=}CH_2 \;+\; Cl{-}Cl \;\longrightarrow\; \begin{array}{c}||\\ CH_2{-}CH_2 \end{array} \end{array}$$

Ethene   1,2-Dichloroethane

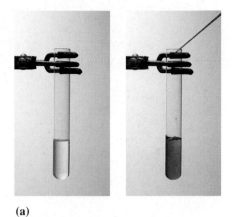

$$\text{Cyclohexene} \;+\; Br{-}Br \;\longrightarrow\; \text{1,2-Dibromocyclohexane}$$

Cyclohexene   1,2-Dibromocyclohexane

$$\begin{array}{c} \quad\quad\quad\quad\quad\quad\quad\quad\quad\quad \text{Cl} \quad \text{Cl} \\ CH_3{-}C{\equiv}CH \;+\; 2Cl{-}Cl \;\longrightarrow\; \begin{array}{c} || \\ CH_3{-}C{-}CH \\ || \\ \text{Cl} \quad \text{Cl} \end{array} \end{array}$$

Propyne   1,1,2,2-Tetrachloropropane

The addition reaction of bromine is sometimes used to test for the presence of double and triple bonds, as shown in Figure 13.6.

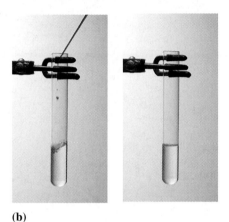

Figure 13.6 **(a)** When bromine is added to an alkane in the first test tube, the red color of bromine remains because the alkane does not react or reacts slowly. **(b)** When bromine is added to an alkene in the second test tube, the red color immediately disappears as bromine atoms add to the double bond.

Q *Will the red color disappear when bromine is added to cyclohexane or cyclohexene?*

SAMPLE PROBLEM 13.6

Writing Products of Halogenation

Write the condensed structural formula of the product of the following reaction:

$$\begin{array}{c} \quad\quad \text{CH}_3 \\ \quad\quad | \\ CH_3{-}C{=}CH_2 \;+\; Br_2 \;\longrightarrow\; \end{array}$$

Solution

The addition of bromine to an alkene places a bromine atom on each of the carbon atoms of the double bond.

$$\begin{array}{c} \quad\quad \text{CH}_3 \\ \quad\quad | \\ CH_3{-}C{-}CH_2 \\ \quad\quad | \quad\;\; | \\ \quad\;\; \text{Br} \;\; \text{Br} \end{array}$$

Study Check

What is the name of the product formed when chlorine is added to 1-butene?

Hydrohalogenation

In the reaction called **hydrohalogenation,** a hydrogen halide (HCl, HBr, or HI) adds to an alkene to yield a haloalkane. The hydrogen atom bonds to one carbon of the double bond, and the halogen atom adds to the other carbon. The general reaction, in which HX represents HCl, HBr, or HI, can be written as follows:

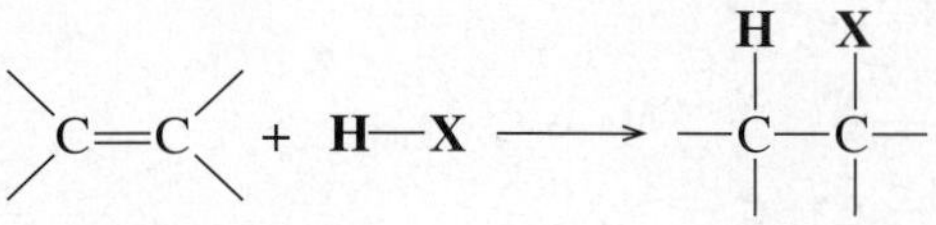

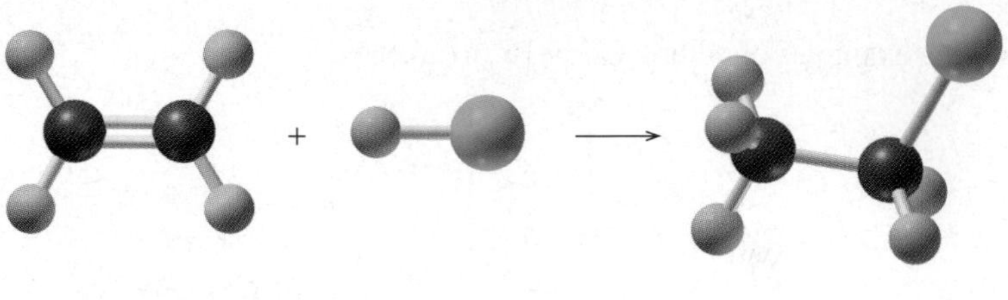

Two examples of hydrohalogenation follow:

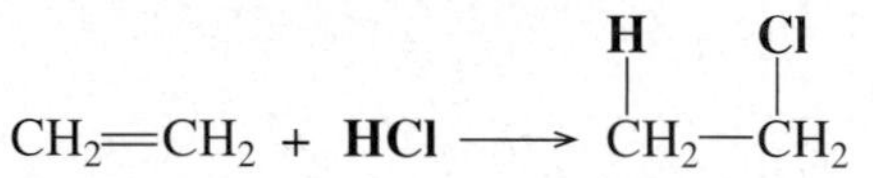

$$CH_2\!=\!CH_2 \;+\; HCl \longrightarrow \overset{\displaystyle H}{\underset{\displaystyle CH_2}{|}}\!-\!\overset{\displaystyle Cl}{\underset{\displaystyle CH_2}{|}}$$

Ethene (ethylene)   Chloroethane (ethyl chloride)

$$CH_3\!-\!CH\!=\!CH\!-\!CH_3 \;+\; HBr \longrightarrow CH_3\!-\!\overset{\displaystyle H}{\underset{}{CH}}\!-\!\overset{\displaystyle Br}{\underset{}{CH}}\!-\!CH_3$$

2-Butene   2-Bromobutane

Steps in Addition Reactions of Alkenes

We have seen that in the addition reaction of an alkene, two groups add to the carbons in the double bond to give a saturated compound. To understand how the addition of H—X or H—OH takes place, we need to look at how the electrons from the pi bond are used to form new bonds. We can consider the steps involved when HBr adds to ethene. Initially, a proton from the HBr reacts with one of the carbons in the double bond. This makes the other carbon into a **carbocation** (a carbon cation)**,** which has only three bonds and a positive charge. The carbocation reacts quickly with the bromide ion Br$^-$.

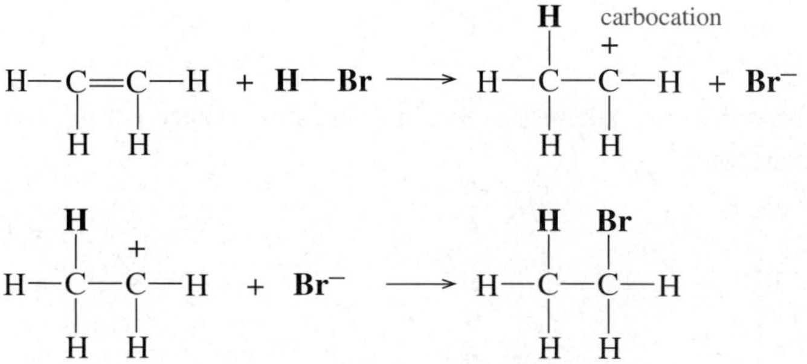

Single product

Markovnikov's Rule

When HBr adds to a symmetrical alkene, a single product results. However, HBr can also add to a double bond with alkyl substituents in unsymmetrical alkenes.

The most stable carbocation that forms is the one with the most alkyl groups. Therefore, in the initial step, the proton adds to the carbon that is less substituted, which is also the carbon in the double that has the greater number of protons.

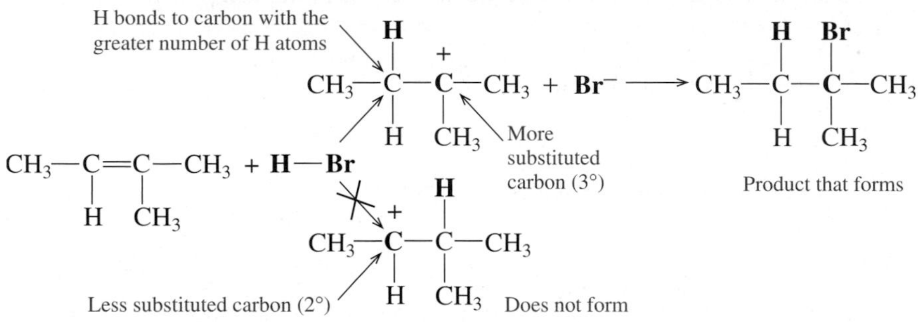

In 1870, Markovnikov, a Russian chemist, observed that the hydrohalogenation addition products of alkenes were limited to the more substituted halide product. His observation now called **Markovnikov's rule** states that when HX adds a double bond, the proton is bonded to the carbon atom that has the greater number of protons. Today, we know that this occurs because the proton adds to the position that produces the most stable carbocation, which also leads to the most substituted halide product.

SAMPLE PROBLEM 13.7

Addition to Alkenes

Predict the organic product for each of the following reactions:

a. $CH_3—CH=CH—CH_3$ + HBr $\longrightarrow$

b. 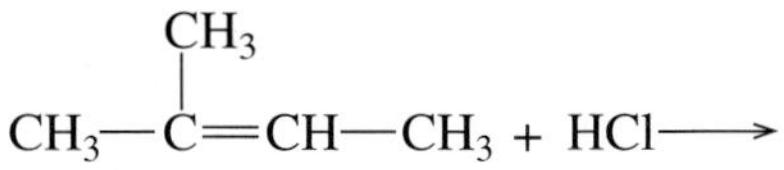
$$CH_3—\overset{\overset{\textstyle CH_3}{|}}{C}=CH—CH_3 + HCl \longrightarrow$$

Solution

a. This is a symmetrical alkene. Only one product forms when the H^+ and Br^- add to the carbons of the double bond

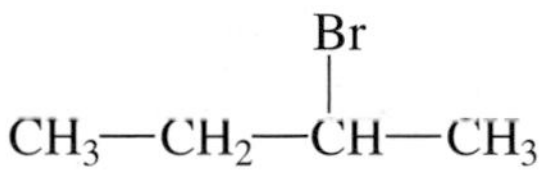

$$CH_3—CH_2—\overset{\overset{\textstyle Br}{|}}{C}H—CH_3$$

b. In the double bond of this unsymmetrical alkene, carbon 3 has the greater number of hydrogen atoms. Using Markovnikov's rule, the H from HCl adds to carbon 3 and the Cl adds to carbon 2. The product is the most substituted halide.

$$CH_3—\overset{\overset{\textstyle CH_3}{|}}{\underset{\underset{\textstyle Cl}{|}}{C}}—CH_2—CH_3$$

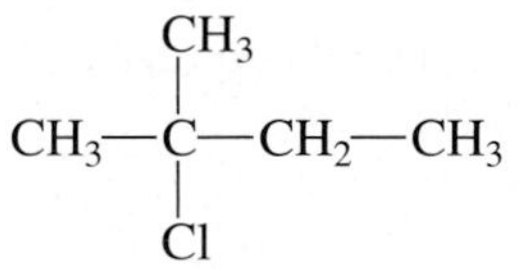

Study Check

Draw the structural formula of the organic product obtained when HBr adds to 1-methylcyclopentene.

SAMPLE PROBLEM 13.8

Writing a Synthesis by Using an Alkene

What alkene would you start with to prepare the following compound?

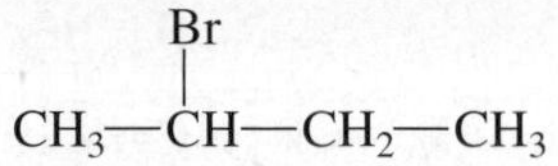

$$CH_3-\overset{\overset{\displaystyle Br}{|}}{CH}-CH_2-CH_3$$

Solution

The starting alkene could be 1-butene or 2-butene.

Study Check

Write an equation using an alkene and HCl to prepare 2-chloropentane.

Adding Water

Alkenes react with water (HOH) when the reaction is catalyzed by a strong acid such as H_2SO_4. In this reaction called **hydration,** H— attaches to one of the carbon atoms in the double bond, and —OH group to the other carbon. Hydration is used to prepare alcohols, which have the functional group —OH. In the general equation the acid is represented by H^+.

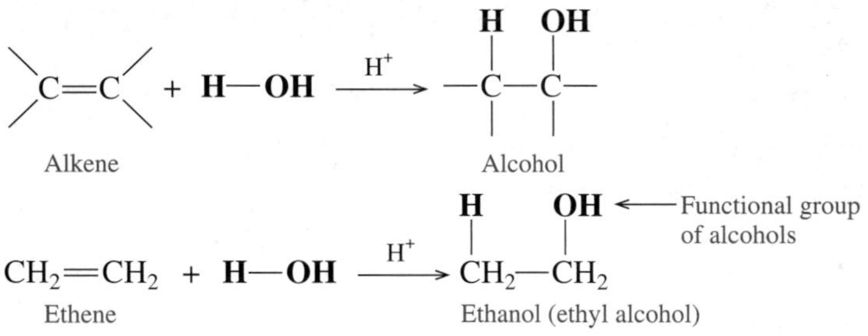

$$\overset{\displaystyle \diagdown}{\diagup}C=C\overset{\displaystyle \diagup}{\diagdown} + \text{H—OH} \xrightarrow{H^+} -\overset{\overset{\displaystyle H}{|}}{C}-\overset{\overset{\displaystyle OH}{|}}{C}-$$

Alkene   Alcohol

$$CH_2{=}CH_2 + \text{H—OH} \xrightarrow{H^+} \overset{\overset{\displaystyle H}{|}}{CH_2}-\overset{\overset{\displaystyle OH}{|}}{CH_2} \xleftarrow{\text{Functional group of alcohols}}$$

Ethene   Ethanol (ethyl alcohol)

The addition of water to a double bond in which the carbon atoms are attached to different groups follows Markovnikov's rule.

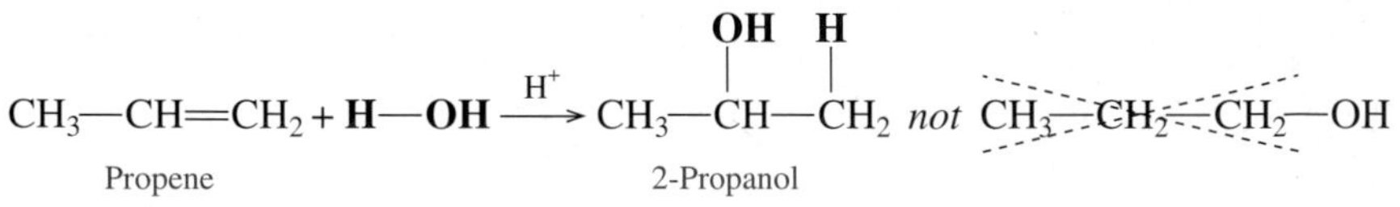

$$CH_3-CH{=}CH_2 + \text{H—OH} \xrightarrow{H^+} CH_3-\overset{\overset{\displaystyle OH}{|}}{CH}-\overset{\overset{\displaystyle H}{|}}{CH_2}\ not\ CH_3{=}CH_2{=}CH_2-OH$$

Propene   2-Propanol

SAMPLE PROBLEM 13.9

Writing Products of Hydration

Write the structural formulas for the products that form in the following hydration reactions:

a. $CH_3CH_2CH_2-CH{=}CH_2 + HOH \xrightarrow{H^+}$

b. 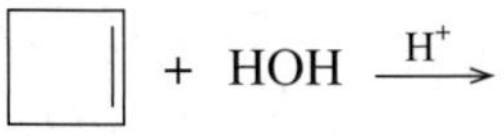 ▢ $+ HOH \xrightarrow{H^+}$

Solution

a. Water adds H— and —OH to the double bond. We use Markovnikov's rule to add the H— to the CH_2 in the double bond, and the —OH to the CH.

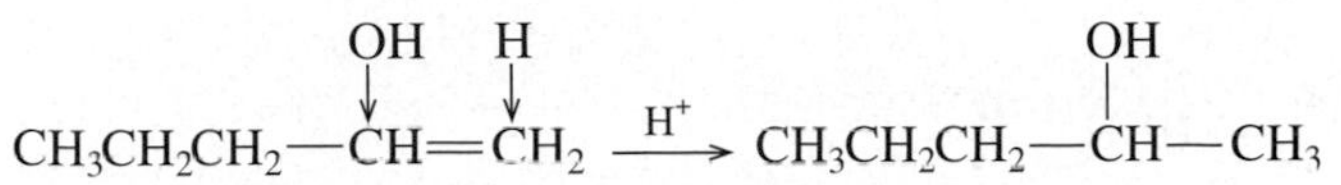

$$\text{CH}_3\text{CH}_2\text{CH}_2-\overset{\overset{\displaystyle \text{OH}}{|}}{\text{CH}}=\overset{\overset{\displaystyle \text{H}}{|}}{\text{CH}}_2 \xrightarrow{\text{H}^+} \text{CH}_3\text{CH}_2\text{CH}_2-\overset{\overset{\displaystyle \text{OH}}{|}}{\text{CH}}-\text{CH}_3$$

b. In cyclobutene, the H— adds to one side of the double bond, and the —OH adds to the other side.

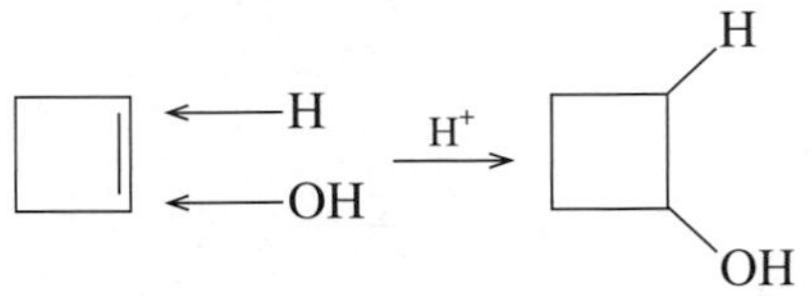

Study Check

Draw the structural formula for the alcohol obtained by the hydration of 2-methyl-2-butene.

QUESTIONS AND PROBLEMS

Addition Reactions

13.23 Give the condensed structural formulas and names of the products in each of the following reactions:

a. $\text{CH}_3-\text{CH}_2-\text{CH}_2-\text{CH}=\text{CH}_2 + \text{H}_2 \xrightarrow{\text{Pt}}$

b. $\text{CH}_2=\overset{\overset{\displaystyle \text{CH}_3}{|}}{\text{C}}-\text{CH}_2-\text{CH}_3 + \text{Cl}_2 \longrightarrow$

c. 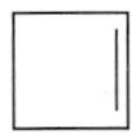 $+ \text{Br}_2 \longrightarrow$

d. cyclopentene $+ \text{H}_2 \xrightarrow{\text{Pt}}$
e. 2-methyl-2-butene $+ \text{Cl}_2 \longrightarrow$
f. 2-pentyne $+ 2\text{H}_2 \xrightarrow{\text{Pd}}$

13.24 Give the condensed structural formulas and names of the products in each of the following reactions:

a. $\text{CH}_3-\text{CH}_2-\text{CH}=\text{CH}_2 + \text{Br}_2 \longrightarrow$

b. cyclohexene $+ \text{H}_2 \xrightarrow{\text{Pt}}$ **c.** cis-2-butene $+ \text{H}_2 \xrightarrow{\text{Pt}}$

d. $\text{CH}_3-\overset{\overset{\displaystyle \text{CH}_3}{|}}{\text{C}}=\text{CH}-\text{CH}_2-\text{CH}_3 + \text{Cl}_2 \longrightarrow$

e. 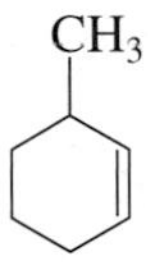 $+ \text{Br}_2 \longrightarrow$

f. $\text{CH}_3-\overset{\overset{\displaystyle \text{CH}_3}{|}}{\text{CH}}-\text{C}\equiv\text{CH} + 2\text{Cl}_2 \longrightarrow$

13.25 Give the condensed structural formulas of the products in each of the following reactions using Markovnikov's rule when necessary:
a. $\text{CH}_3-\text{CH}=\text{CH}-\text{CH}_3 + \text{HBr} \longrightarrow$

b. cyclopentene $+ \text{HOH} \xrightarrow{\text{H}^+}$ **c.** $\text{CH}_2=\text{CH}-\text{CH}_2-\text{CH}_3 + \text{HCl} \longrightarrow$

d. $\text{CH}_3-\overset{\overset{\displaystyle \text{CH}_3}{|}}{\text{C}}=\underset{\underset{\displaystyle \text{CH}_3}{|}}{\text{C}}-\text{CH}_3 + \text{HI} \longrightarrow$ **e.** $\text{CH}_3\text{CH}_2-\overset{\overset{\displaystyle \text{CH}_3}{|}}{\text{C}}=\text{CHCH}_3 + \text{HBr} \longrightarrow$

f.

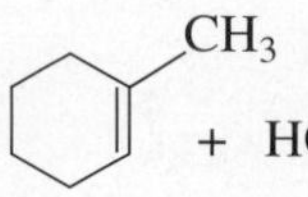

$$CH_3$$

(cyclohexene with CH₃) + HOH $\xrightarrow{H^+}$

13.26 Give the condensed structural formulas of the products in each of the following reactions using Markovnikov's rule when necessary:

a.
$$CH_3$$
$$|$$
$$CH_3—C=CH—CH_3 + HCl \longrightarrow$$

b. $CH_3CH_2—CH=CH—CH_2CH_3 + HOH \xrightarrow{H^+}$

c.
$$CH_3$$
$$|$$
$$CH_3—C=CH_2 + HBr \longrightarrow$$

d. 4-methylcyclopentene + HOH $\xrightarrow{H^+}$

e. 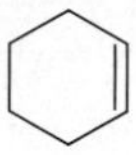 (cyclohexene) + HBr $\longrightarrow$

f. $CH_3—C\equiv C—CH_3 + 2HCl \longrightarrow$

13.27 Write an equation including any catalysts for the following reactions:
   a. hydrogenation of 2-methylpropene
   b. addition of hydrogen chloride to cyclopentene
   c. addition of bromine to 2-pentene
   d. hydration of propene
   e. addition of chlorine to 2-butyne

13.28 Write an equation including any catalysts for the following reactions:
   a. hydration of 1-methylcyclobutene
   b. hydrogenation of 3-hexene
   c. addition of hydrogen bromide to 2-methyl-2-butene
   d. addition of chlorine to 2,3-dimethyl-2-pentene
   e. addition of HCl to 1-methylcyclopentene

LEARNING GOAL

Draw structural formulas of monomers that form a polymer or a three-monomer section of a polymer.

13.5 Polymerization

Polymers have been a part of life since prehistoric times. Cellulose in wood and starches obtained from vegetables are carbohydrate polymers made from many thousands of glucose molecules. Silk and wool are polymers made from amino acids. DNA, the molecule that carries genetic information, is a polymer of nucleotides.

Polymers are large molecules that consist of small repeating units called **monomers**. In the past hundred years, the plastics industry has made synthetic polymers that are in many of the materials we use every day, such as carpeting, plastic wrap, nonstick pans, plastic cups, and rain gear. In medicine, synthetic polymers are used to replace diseased or damaged body parts such as hip joints, teeth, heart valves, and blood vessels. (See Figure 13.7.)

Addition Polymers

Many of the synthetic polymers are made by addition reactions of monomers that are small alkenes. The conditions for many polymerization reactions require high temperatures and very high pressure (over 1000 atm). In an addition reaction, the polymer grows as monomers are added to the end of the chain. Polyethylene, a polymer made from ethylene $CH_2=CH_2$, is used in plastic bottles, film, and plastic dinnerware. (See Figure 13.8.) In the polymerization, a series of addition reac-

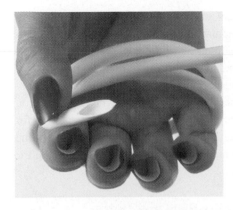

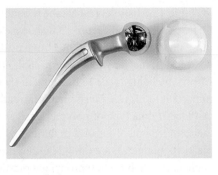

Figure 13.7 Synthetic polymers are used to replace diseased veins and arteries. A broken hip is repaired using a metal piece that fits into an artificial plastic cup socket.
Q *Why are the substances in these plastic devices called polymers?*

Polyethylene

Vinyl chloride

Polypropylene

Polytetrafluoroethylene (Teflon)

Polydichloroethylene (Saran)

Polystyrene

Figure 13.8 Synthetic polymers provide a wide variety of items that we use every day.
Q *What are some alkenes used to make the polymers in these plastic items?*

tions joins one monomer to the next until a long carbon chain forms that contains as many as 1000 monomers.

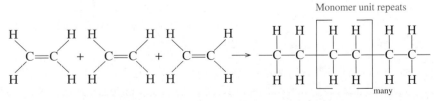

Ethene (ethylene) monomers

Polyethylene section

Table 13.3 lists several alkene monomers that are used to produce common synthetic polymers and Figure 13.8 shows examples of each. The alkane-like nature of these plastic synthetic polymers makes them unreactive. Thus, they do not decompose easily (they are nonbiodegradable) and have become contributors to pollution. Efforts are being made to make them more degradable. It is becoming increasingly important to recycle plastic material, rather than add to our growing landfills. You can identify the type of polymer used to manufacture a plastic item by looking for the recycle symbol (arrows in a triangle) found on the label or on the bottom of the plastic container. For example, either the number 5 or the letters PP inside the triangle is a code for a polypropylene plastic.

Table 13.3   Some Alkenes and Their Polymers

Monomer	Polymer Section	Common Uses
$CH_2{=}CH_2$ Ethene (ethylene)	Polyethylene	Plastic bottles, film, insulation materials
$CH_2{=}CH$ (Cl) Chloroethene (vinyl chloride)	Polyvinyl chloride (PVC)	Plastic pipes and tubing, garden hoses, garbage bags
$CH_2{=}CH$ (CH_3) Propene (propylene)	Polypropylene	Ski and hiking clothing, carpets, artificial joints
$F{-}C{=}C{-}F$ (F, F) Tetrafluoroethene	Polytetrafluoroethylene (Teflon)	Nonstick coatings
$CH_2{=}C{-}Cl$ (Cl) 1,1-Dichloroethene	Polydichloroethylene (Saran)	Plastic film and wrap
$H_2C{=}CH$ Phenylethene (styrene)	$-CH_2-CH-CH_2-CH-CH_2-CH-$ Polystyrene	Plastic coffee cups and cartons, insulation

1	2	3	4	5	6
PETE	HDPE	PV	LDPE	PP	PS
Polyethylene terephthalate	High-density polyethylene	Polyvinyl chloride	Low-density polyethylene	Polypropylene	Polystyrene

SAMPLE PROBLEM 13.10

Polymers

What are the starting monomers for the following polymers?

a. polypropylene

b. Saran

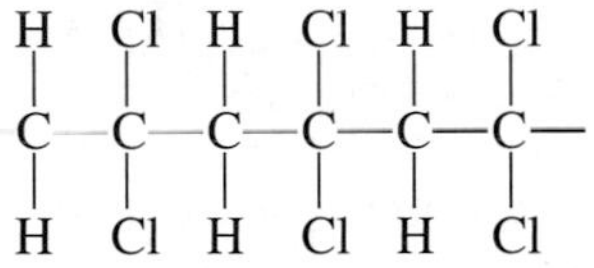

$$\begin{array}{cccccc} H & Cl & H & Cl & H & Cl \\ | & | & | & | & | & | \\ C & C & C & C & C & C- \\ | & | & | & | & | & | \\ H & Cl & H & Cl & H & Cl \end{array}$$

Solution

a. propene (propylene) $CH_2{=}\overset{\overset{\displaystyle CH_3}{|}}{CH}$

b. 1,1-dichloroethene, $CH_2{=}\overset{\overset{\displaystyle Cl}{|}}{C}{-}Cl$

Study Check

What is the monomer for PVC?

QUESTIONS AND PROBLEMS

Addition Polymers of Alkenes

13.29 What is a polymer?

13.30 What is a monomer?

13.31 Write an equation that represents the formation of a part of the Teflon polymer from three of the monomer units.

13.32 Write an equation that represents the formation of a part of the polystyrene polymer from three of the monomer units.

13.33 A plastic called polyvinylidene difluoride, PVDF, is made from monomers of 1,1-difluoroethene. Write the structure of the polymer formed from the addition of three monomers of the 1,1-difluoroethene.

13.34 An alkene called acrylonitrile is the monomer used to form the polymer used in the fabric material called Orlon. Write an equation that represents the formation of a part of the polyacrylonitrile polymer from three of the monomer units.

Acrylonitrile $CH_2{=}\overset{\overset{\displaystyle CN}{|}}{CH}$

13.6 Aromatic Compounds

In 1825, Michael Faraday isolated a hydrocarbon called benzene, which had the molecular formula C_6H_6. Because many compounds containing benzene had fragrant odors, the family of benzene compounds became known as **aromatic compounds**. A molecule of **benzene** consists of a ring of six carbon atoms with one

LEARNING GOAL

Describe the bonding in benzene; name aromatic compounds, and write their structural formulas.

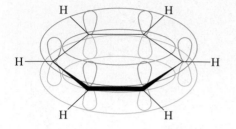

hydrogen atom attached to each carbon. Each carbon atom uses 3 valence electrons to bond to the hydrogen atom and two adjacent carbons. That leaves 1 valence electron to share in a double bond with an adjacent carbon. In 1865, August Kekulé dreamed that the carbon atoms were arranged in a flat ring with alternating single and double bonds between the carbon atoms. This idea led to two ways of writing the benzene structure, as follows:

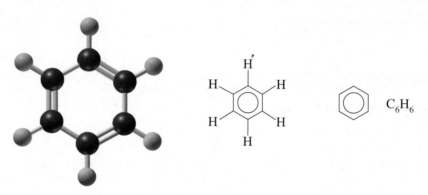

Structures for benzene

However, there is only one structure of benzene. Today we know that all the bonds in benzene are identical. In benzene, each carbon bonds to two other carbons and one hydrogen by sigma bonds. That leaves six p orbitals that overlap to form a continuous ring of orbitals above and below the carbon atoms. In this circle of p orbitals, the pi electrons are shared equally, a unique feature that makes aromatic compounds especially stable. The benzene structure is represented as a hexagon with a circle in the center.

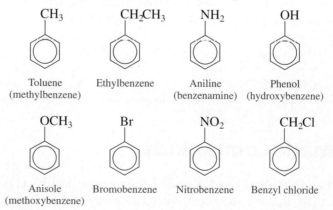

Naming Aromatic Compounds

Aromatic compounds that contain a benzene ring with a single substituent are usually named as benzene derivatives. However, many of these compounds have been important in chemistry for many years and still use their common names. Some common names such as toluene are now accepted as IUPAC names.

CH_3 CH_2CH_3 NH_2 OH

Toluene   Ethylbenzene   Aniline   Phenol
(methylbenzene)   (benzenamine)   (hydroxybenzene)

OCH_3 Br NO_2 CH_2Cl

Anisole   Bromobenzene   Nitrobenzene   Benzyl chloride
(methoxybenzene)

Aromatic Compounds in Health and Medicine

Aromatic compounds are common in nature and in medicine. Toluene is used as a starting reactant to make drugs, dyes, and explosives such as TNT (trinitrotoluene). The benzene ring is found in some amino acids (the building blocks of proteins), in pain relievers such as aspirin, acetaminophen, and ibuprofen, and flavorings such as vanillin.

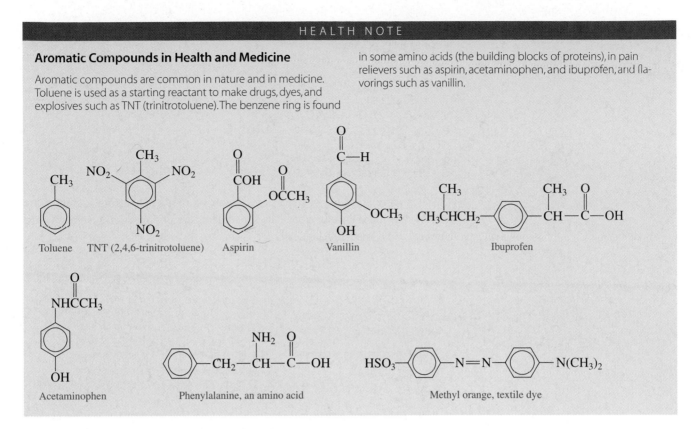

Toluene   TNT (2,4,6-trinitrotoluene)   Aspirin   Vanillin   Ibuprofen

Acetaminophen   Phenylalanine, an amino acid   Methyl orange, textile dye

When a benzene ring is a substituent C_6H_5—, it is named as a phenyl group. A benzyl group is a benzene ring and a CH_2— group.

phenyl group   benzyl group

benzyl chloride   3-phenyl-1-butene

When there are two substituents on benzene, the ring is numbered to give the lowest numbers to the substituents. However, common names use the prefixes **ortho, meta,** and **para** to show the substituent arrangement. The prefix *ortho (o)* indicates a 1,2 arrangement, *meta (m)* is a 1,3 arrangement, and *para (p)* is used for 1,4 arrangements.

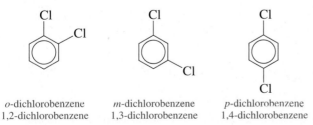

o-dichlorobenzene   *m*-dichlorobenzene   *p*-dichlorobenzene
1,2-dichlorobenzene   1,3-dichlorobenzene   1,4-dichlorobenzene

Common names are used for many disubstituted benzenes such as the constitutional isomers of dimethylbenzene.

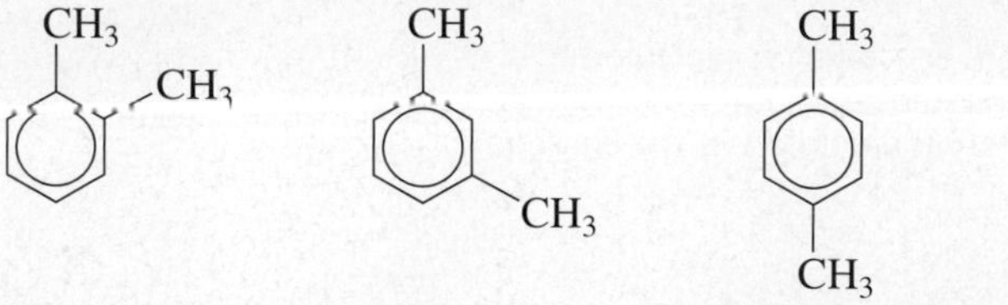

o-xylene
1,2-dimethylbenzene

m-xylene
1,3-dimethylbenzene

p-xylene
1,4-dimethylbenzene

When there are three or more substituents on the benzene ring, numbers are used to show their arrangement. The substituents are numbered to give the lowest numbers and named alphabetically.

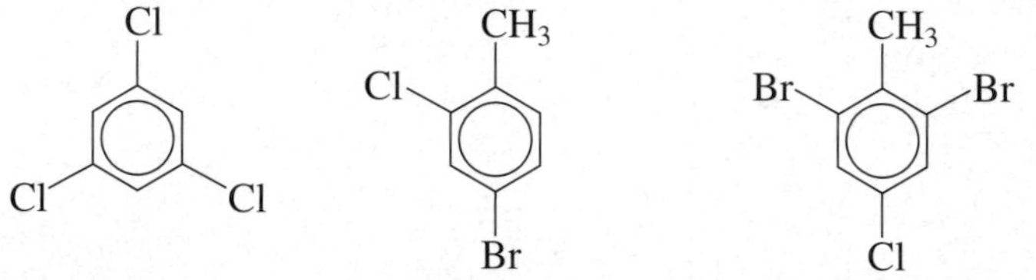

1,3,5-trichlorobenzene   4-bromo-2-chlorotoluene   2,6-dibromo-4-chlorotoluene

---

SAMPLE PROBLEM 13.11

Naming Aromatic Compounds

Give IUPAC and common names for each of the following aromatic compounds:

a.

b.

c.

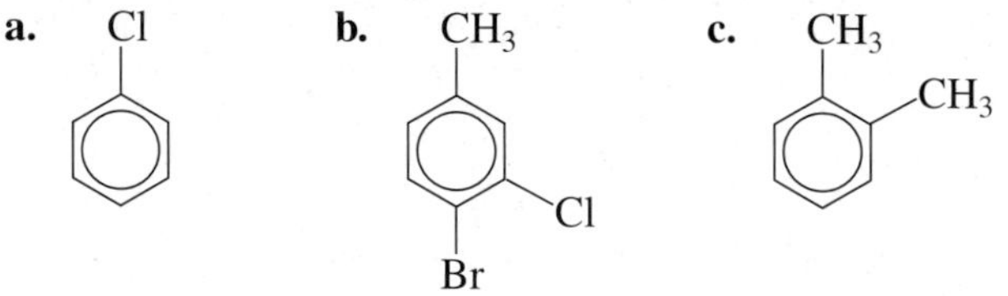

Solution

a. chlorobenzene; phenyl chloride
b. 4-bromo-3-chlorotoluene
c. 1,2-dimethylbenzene; *o*-xylene

Study Check

Name the following compound.

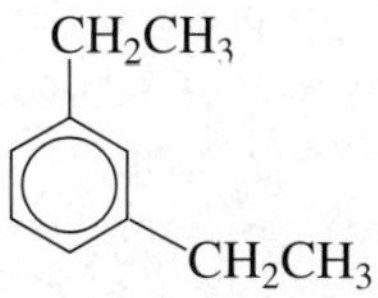

QUESTIONS AND PROBLEMS

Aromatic Compounds

13.35 Cyclohexane and benzene each have six carbon atoms. How are they different?

13.36 In the Health Note "Aromatic Compounds in Health and Medicine," what part of each molecule is the aromatic portion?

Polycyclic Aromatic Hydrocarbons (PAHs)

Large aromatic compounds known as polycyclic aromatic hydrocarbons are formed by fusing together two or more benzene rings edge-to-edge. In a fused ring compound, neighboring benzene rings share two or more carbon atoms. Naphthalene with two benzene rings is well known for its use in mothballs. Anthracene with three rings is used in the manufacture of dyes.

 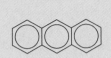

Napthalene   Anthracene 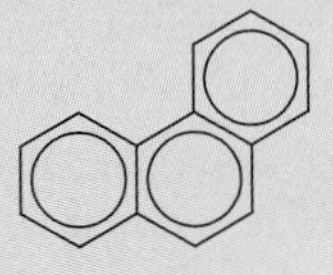   Phenanthrene

uct of combustion, has been identified in coal tar, tobacco smoke, barbecued meats, and automobile exhaust.

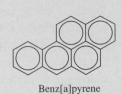

Benz[a]pyrene

When a polycyclic compound contains phenanthrene, it may act as a carcinogen, a substance known to cause cancer. For example, some aromatic compounds in cigarette smoke cause cancer, as seen in the lung tissue of a heavy smoker. Benz[a]pyrene, a prod-

Compounds containing five or more fused benzene rings such as benz[a]pyrene are potent carcinogens. The molecules interact with the DNA in the cells, causing abnormal cell growth and cancer. Increased exposure to carcinogens increases the chance of DNA alterations in the cells.

13.37 Give the IUPAC and any common names for each of the following:

a.   CH₃ ... Cl
b. CH₂CH₃
c. Cl ... Cl ... Cl

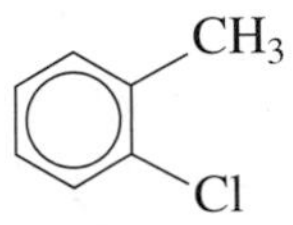

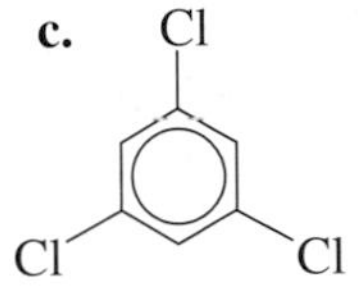

d. CH₃ CH₃
e. CH₃ Br ... Cl
f. CH₃—CH—CH₃

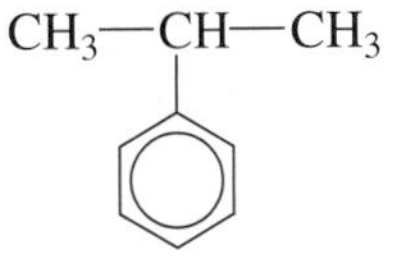

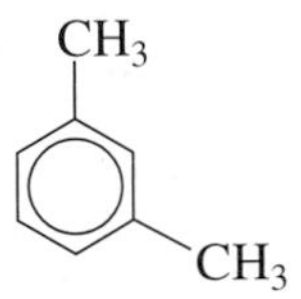

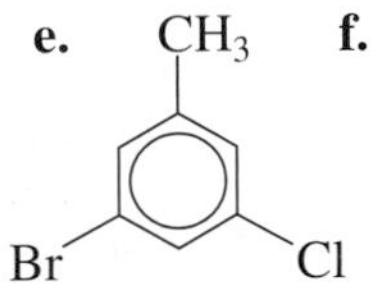

13.38 Give the IUPAC and any common names for each of the following:

a.
b. CH₃
c. Cl Cl

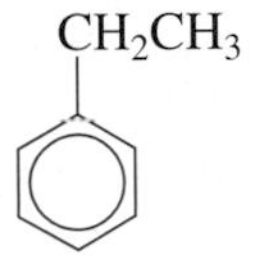

d. CH₃ 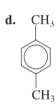 CH₃
e. CH₂Br
f. Br 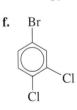 Cl ... Cl

13.39 Draw the structural formulas for each of the following compounds:
  a. methylbenzene
  b. *m*-dichlorobenzene
  c. 1-ethyl-4-methylbenzene
  d. *p*-chlorotoluene

13.40 Draw the structural formulas for each of the following compounds:
  a. benzene
  b. *o*-chloromethylbenzene
  c. propylbenzene
  d. 1,2,4-trichlorobenzene

LEARNING GOAL

Describe the physical and chemical properties of aromatic compounds; draw structural formulas produced by substitution of benzene.

13.7 Properties of Aromatic Compounds

The symmetrical structure of benzene allows the cyclic structures to stack close together, which contributes to higher melting points and boiling points of benzene and its derivatives. For example, hexane melts at $-95°C$, while benzene melts at $6°C$. Among the disubstituted benzene compounds, the para consitutional isomers are more symmetrical and have higher melting points than the ortho and meta isomers: o-xylene melts at $-26°C$ and m-xylene melts at $-48°C$, while p-xylene melts at $13°C$.

Aromatic compounds are less dense than water, although they are somewhat denser than other hydrocarbons. Halogenated benzene compounds are denser than water. Aromatic hydrocarbons are insoluble in water and are used as solvents for other organic compounds. Only those containing strongly polar functional groups such as —OH or —COOH will be more soluble. Benzene and other aromatic compounds are resistant to reactions that break up the aromatic system, although they are flammable, as are other hydrocarbon compounds.

Chemical Properties

The most important type of reaction for benzene and aromatic compounds is **substitution,** in which an atom or group of atoms replaces a hydrogen atom on a benzene ring. A substitution reaction, rather than addition, retains the stability of the aromatic bonding system. Substitution reactions of benzene include halogenation, nitration, and sulfonation.

Halogenation

In the chlorination or bromination of benzene, a chlorine or bromine atom replaces a hydrogen atom on the benzene ring. A catalyst such as $FeCl_3$ is required for chlorination; $FeBr_3$ is a catalyst in bromination.

Chlorobenzene

When toluene (methylbenzene) undergoes halogenation, a mixture of constitutional isomers is obtained as products. However, the presence of a methyl group in toluene has the effect of producing mostly ortho and para isomers. In most substitution reactions of toluene, the meta isomer is produced in very low amounts.

Toluene   o-Chlorotoluene   m-Chlorotoluene   p-Chlorotoluene
(very little)

Nitration

When benzene is heated with nitric acid, nitrobenzene is produced. Sulfuric acid (H_2SO_4) is required as a catalyst for the nitration.

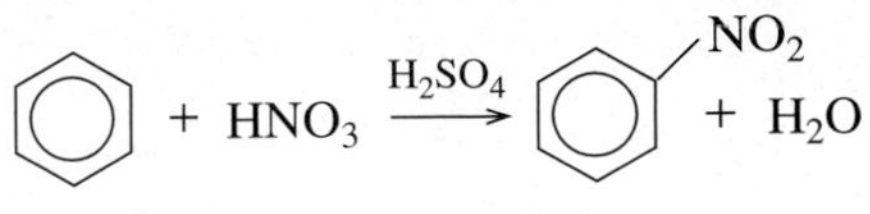

Nitrobenzene

Sulfonation

When benzene reacts with a mixture of SO_3 + H_2SO_4, known as "fuming sulfuric acid," the product is benzenesulfonic acid.

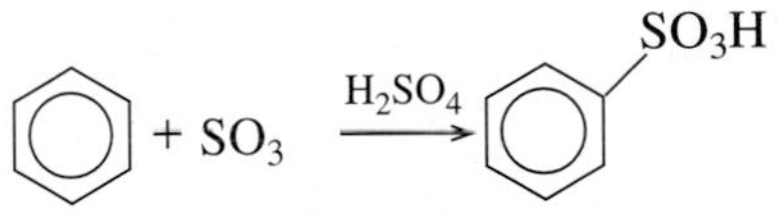

Benzenesulfonic acid

The sulfonation of aromatic compounds is one way to produce sulfa drugs.

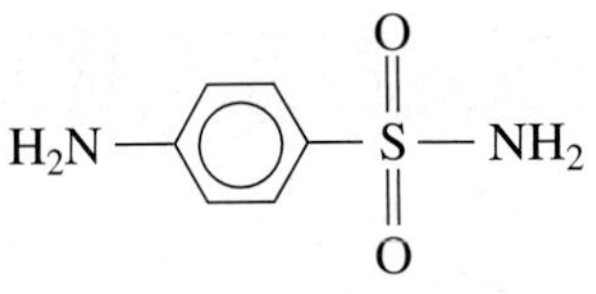

Sulfanilamide, a sulfa drug

SAMPLE PROBLEM 13.12

Reactions of Benzene

Write the structure of the organic product when benzene reacts with the following:
a. Br_2 and $FeBr_3$ **b.** HNO_3 and H_2SO_4

Solution

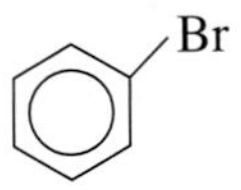

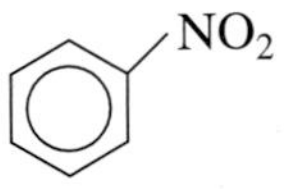

Study Check

A chemist needs to synthesize chlorobenzene. If benzene is available in the lab, how could she prepare this compound?

QUESTIONS AND PROBLEMS

Reactions of Aromatic Compounds

13.41 Alkenes undergo addition reactions, but benzene does not. How does benzene react and why?

13.42 When toluene is reacted with chlorine in light, the product is

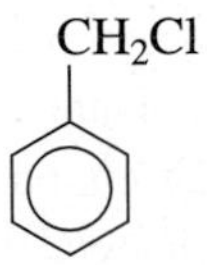

How would you explain this result?

13.43 Draw the structures of the organic product(s), if any, for the following reactants:

a. Benzene + Cl_2 $\xrightarrow{\text{no catalyst}}$

b. Benzene + Cl_2 $\xrightarrow{\text{FeCl}_3}$

c. Benzene + HNO_3 $\xrightarrow{\text{H}_2\text{SO}_4}$

13.44 Draw the structures of the organic product(s), if any, for the following reactants:

a. Toluene + Br_2 $\xrightarrow{\text{FeBr}_3}$

b. Benzene + SO_3 $\xrightarrow{\text{H}_2\text{SO}_4}$

c. Benzene + HNO_3 $\xrightarrow{\text{H}_2\text{SO}_4}$

Chapter Review

13.1 Alkenes and Alkynes
Alkenes are unsaturated hydrocarbons that contain carbon–carbon double bonds ($C=C$) made of one sigma and one pi bond. Alkynes contain a triple bond ($C\equiv C$) composed of one sigma and two pi bonds.

13.2 Naming Alkenes and Alkynes
The IUPAC names of alkenes end with *ene,* while alkyne names end with *yne*. The main chain is numbered from the end nearest the double or triple bond. In a cycloalkene, the double bond is carbon 1 and 2, and the ring is numbered to give the lowest numbers to any substituents, which are named alphabetically.

13.3 Cis–Trans Isomers
Isomers of alkenes occur when the carbon atoms in the double bond are connected to different atoms or groups. In the cis isomer the attached groups are on the same side of the double bond, whereas in the trans isomer they are connected on the opposite sides of the double bond.

13.4 Addition Reactions
The addition of small molecules to the double bond is a characteristic reaction of alkenes. Hydrogenation adds hydrogen atoms to the double bond of an alkene to yield an alkane. Halogenation adds bromine or chlorine atoms to produce dihaloalkanes. Hydrogen halides and water can also add to a double bond. When there are a different number of groups attached to the carbons in the double bond, the H from the reactant (HX or H—OH) adds to the carbon with the greater number of hydrogen atoms.

13.5 Polymerization
Polymers are long-chain molecules that consist of many repeating units of smaller carbon molecules called monomers. In nature, cellulose and starch are polymers of glucose and proteins are polymers of amino acids. Many materials that we use every day are synthetic polymers, including carpeting, plastic wrap, nonstick pans, and nylon. These synthetic materials are often made by addition reactions in which a catalyst links the carbon atoms from various kinds of alkene molecules.

13.6 Aromatic Compounds
Most aromatic compounds contain benzene, a cyclic structure containing six CH units. The structure of benzene is represented as a hexagon with a circle in the center. Many aromatic compounds use the parent name benzene, although common names such as toluene are retained. The benzene ring is numbered and the branches are listed in alphabetical order. For two branches, the positions are often shown by the prefixes *ortho* (1,2-), *meta* (1,3-), and *para* (1,4-).

13.7 Properties of Aromatic Compounds
Aromatic compounds undergo substitution reactions such as halogenation, nitration, and sulfonation. They do not undergo addition reactions, which would disrupt their stable aromatic bonding system.

Summary of Naming

Type	Example	Characteristic	Structure
Alkene	Propene (propylene)	double bond	$CH_3-CH=CH_2$
Cycloalkene	Cyclopropene	double bond in a carbon ring	
Alkyne	Propyne	triple bond	$CH_3C\equiv CH$
Aromatic	Benzene	Aromatic ring of six carbons	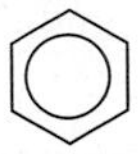
	Methylbenzene or toluene		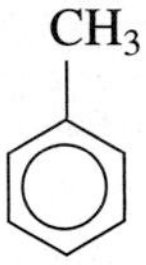
	1,4-dichlorobenzene or *para*-dichlorobenzene		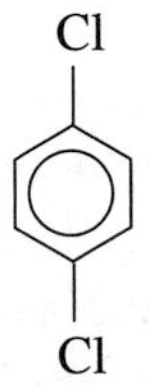

Summary of Reactions

Hydrogenation

$$Alkene + H_2 \xrightarrow{Pt} alkane$$

$$CH_2=CH-CH_3 + H_2 \xrightarrow{Pt} CH_3-CH_2-CH_3$$

$$Alkyne + 2H_2 \xrightarrow{Pt} Alkane$$

$$CH_3-C\equiv CH + 2H_2 \xrightarrow{Pt} CH_3-CH_2-CH_3$$

Halogenation

$$Alkene + Cl_2 \text{ (or } Br_2) \longrightarrow dihaloalkane$$

$$CH_2=CH-CH_3 + Cl_2 \longrightarrow \underset{\begin{array}{c}|\\Cl\end{array}}{CH_2}-\underset{\begin{array}{c}|\\Cl\end{array}}{CH}-CH_3$$

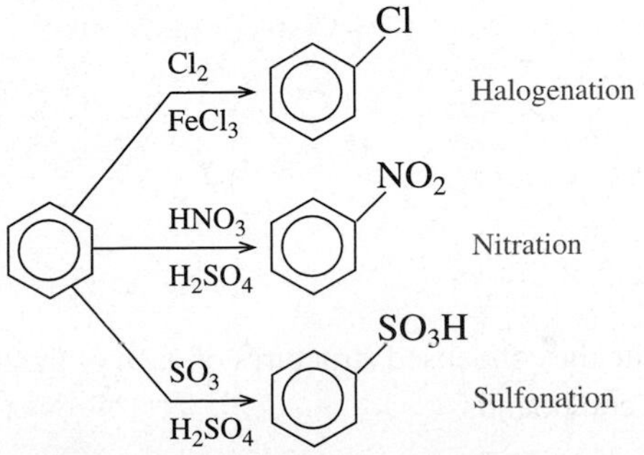

Hydrohalogenation

$$Alkene + HX \longrightarrow haloalkane$$

$$CH_2=CH-CH_3 + HCl \longrightarrow CH_3-\underset{\begin{array}{c}|\\Cl\end{array}}{CH}-CH_3$$

Markovnikov's rule

Hydration of Alkenes

$$Alkene + H-OH \xrightarrow{H^+} alcohol$$

$$CH_2=CH-CH_3 + H-OH \xrightarrow{H^+} CH_3-\underset{\begin{array}{c}|\\OH\end{array}}{CH}-CH_3$$

Markovnikov's rule

Substitution Reactions of Benzene

Cl — Halogenation (Cl₂, FeCl₃)

NO₂ — Nitration (HNO₃, H₂SO₄)

SO₃H — Sulfonation (SO₃, H₂SO₄)

Key Terms

addition A reaction in which atoms or groups of atoms bond to a double bond. Addition reactions include the addition of hydrogen (hydrogenation), halogens (halogenation), hydrogen halides (hydrohalogenation), or water (hydration).

alkene An unsaturated hydrocarbon containing a carbon–carbon double bond.

alkyne An unsaturated hydrocarbon containing a carbon–carbon triple bond.

aromatic compounds Compounds that contain the ring structure of benzene.

benzene A ring of six carbon atoms each of which is attached to a hydrogen atom, C_6H_6.

carbocation A carbon cation that has only three bonds and a positive charge and is formed during the addition reactions of hydration and hydrohalogenation.

cis isomer A geometric isomer in which similar groups are connected on the same side of the double bond.

cycloalkene A cyclic hydrocarbon that contains a double bond in the ring.

halogenation The addition of Cl_2 or Br_2 to an alkene to form halogen-containing compounds.

hydration An addition reaction in which the components of water, H— and —OH, bond to the carbon–carbon double bond to form an alcohol.

hydrogenation The addition of hydrogen (H_2) to the double bond of alkenes to yield alkanes.

hydrohalogenation The addition of a hydrogen halide such as HCl or HBr to a double bond.

Markovnikov's rule When adding HX or HOH to alkenes with different numbers of groups attached to the double bonds, the H adds to the carbon that has the greater number of hydrogen atoms.

meta A method of naming that indicates two substituents at carbons 1 and 3 of benzene.

monomer The small organic molecule that is repeated many times in a polymer.

nitration The addition of a nitro group (—NO_2) to benzene.

ortho A method of naming that indicates two substituents at carbons 1 and 2 of a benzene ring.

para A method of naming that indicates two substituents at carbons 1 and 4 of a benzene ring.

polymer A very large molecule that is composed of many small, repeating structural units that are identical.

saturated hydrocarbons A compound of carbon and hydrogen in which the carbon chain consists of only single carbon–carbon bonds (H_3C—CH_3).

substitution The reactions of benzene and other aromatic compounds in which an atom or group of atoms replaces a hydrogen on a benzene ring .

sulfonation The reaction of benzene with SO_3 and H_2SO_4 to give benzenesulfonic acid.

trans isomer A geometric isomer in which similar groups are connected to opposite sides of the double bond in an alkene.

unsaturated hydrocarbons A compound of carbon and hydrogen in which the carbon chain contains at least one double (alkene) or triple carbon–carbon bond (alkyne). An unsaturated compound is capable of an addition reaction with hydrogen, which converts the double or triple bonds to single carbon–carbon bonds.

Additional Problems

13.45 Compare the formulas and bonding in propane, cyclopropane, propene, and propyne.

13.46 Compare the formulas and bonding in butane, cyclobutane, cyclobutene, and 2-butyne.

13.47 Give the IUPAC name for each of the following compounds:

a.   Cl

b. $CH_3CHCH_2CHCH_3$ (Cl on C2, CH_3 on C4)

c. CH_2=$CCH_2CH_2CH_3$ (CH_3 substituent)

d. $CH_3CH_2C \equiv CCH_3$

e.   Cl (on cyclopentene)

f. CH_3, H / C=C / H, CH_2CH_3

g.   Cl (on cyclohexene ring with Cl)

13.48 Write the condensed structures of each of the following compounds:

a. 1,2-dibromocyclopentane
b. 2-pentyne
c. *cis*-2-heptene
d. 3,3-dichloro-2-methylpentane
e. *trans*-3-hexene
f. 2-bromo-3-chlorocyclohexene
g. 2,3-dichloro-1-butene

13.49 Indicate if the following pairs of structures represent constitutional isomers, cis–trans isomers, or identical compounds.

a.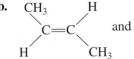

and

Cl

b.

CH₃   H   CH₃   CH₃
  \C=C/   and   \C=C/
H/   \CH₃   H/   \H

c. CH₂=CH   and   CH₃CH₂CH₂CH=CH₂
       |
       CH₂CH₂
       |
       CH₃

d.

CH₃   CH₃   CH₃
 | | |
CH₃CHCH₂CHCH₃ and CH₃CH₂CHCH₂CH₂CH₃

13.50 Draw the condensed structures and give the names for all the isomers of C₄H₈ including cyclic and cis–trans isomers.

13.51 Methylcyclopentane is formed by four different alkenes that react with hydrogen (H₂) in the presence of a Ni catalyst. Draw the condensed structures of each of these alkenes.

13.52 What role do cis–trans isomers play in night vision?

13.53 Write the cis and trans isomers for each of the following:
a. 2-pentene   b. 3-hexene
c. 2-butene   d. 2-hexene

13.54 Write the structures of the products, if any, for the following.

a. CH₃CH=CHCH₃ + H₂ $\xrightarrow{\text{Ni}}$

b. $\bigcirc$ + H₂ $\xrightarrow{\text{Ni}}$

c. CH₃CH=CHCH₃ + HBr $\longrightarrow$

d. $\bigcirc$ + HBr $\longrightarrow$

e. CH₃CH=CHCH₃ + Cl₂ $\longrightarrow$

f. CH₃CH₂CH₃ + H₂ $\xrightarrow{\text{Ni}}$

g. CH₃CH=CHCH₃ + HOH $\xrightarrow{\text{H}^+}$

h. $\bigcirc$ + HOH $\xrightarrow{\text{H}^+}$

i.

         CH₃
          |
CH₃C=CCH₃ + HCl $\longrightarrow$
          |
         CH₃

13.55 What is the condensed structural formula of the organic compound needed to prepare each of the following products?

a. ? + H₂ $\xrightarrow{\text{Ni}}$ $\bigcirc$

b. ? + Br₂ $\longrightarrow$

          Br Br
           | |
          CH₃CHCHCH₂CH₃

c. ? + HCl $\longrightarrow$

          Cl
          |
          CH₃CHCH₃

d. ? + HOH $\xrightarrow{\text{H}^+}$

          OH
          |
          $\bigcirc$

13.56 Write the combustion reaction for the acetylene used in a welder's torch.

13.57 Copolymers contain more than one type of monomer. One copolymer used in medicine is made of alternating units of styrene and acrylonitrile. Write a section of the copolymer that would have these alternating units. (For structure of the styrene, see Table 13.3.)

          CN
           |
H₂C=CH
acrylonitrile

13.58 Lucite or Plexiglas is a polymer of methylmethacrylate. Write the part of the polymer that is made from the addition of three of these monomers.

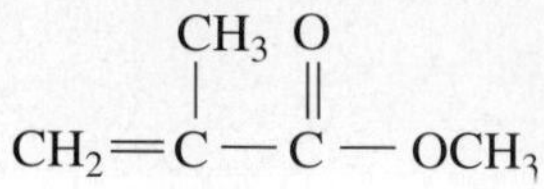

$$CH_2 = \overset{\overset{\displaystyle CH_3}{|}}{C} - \overset{\overset{\displaystyle O}{||}}{C} - OCH_3$$

Methylmethacrylate

13.59 Name each of the following aromatic compounds:

a.  CH₃

b. 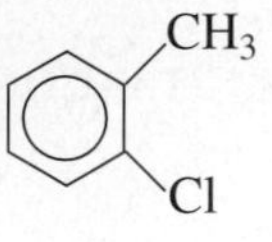 CH₃ Cl

c.  CH₃ CH₂CH₃

d. 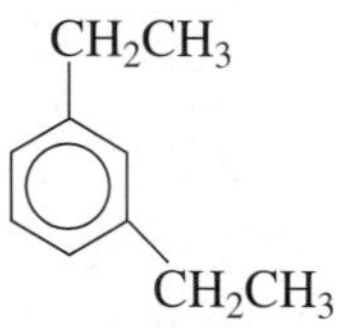 CH₂CH₃ CH₂CH₃

13.60 Write the structural formulas for each of the following:

a. ethylbenzene

b. *m*-dichlorobenzene

c. 1,2,4-trimethylbenzene

d. *p*-dimethylbenzene

13.61 Name the organic product(s) produced, if any, in each of the following reactions.

a. benzene and Cl_2 $\xrightarrow{\text{FeCl}_3}$

b. toluene and Br_2 $\xrightarrow{\text{FeBr}_3}$

c. benzene and SO_3 $\xrightarrow{\text{H}_2\text{SO}_4}$

d. benzene and Br_2 $\xrightarrow{\text{light}}$

13.62 What reactants and catalysts are needed to synthesize the following products?

a. nitrobenzene

b. benzenesulfonic acid

c. bromobenzene

14 Alcohols, Phenols, Ethers, and Thiols

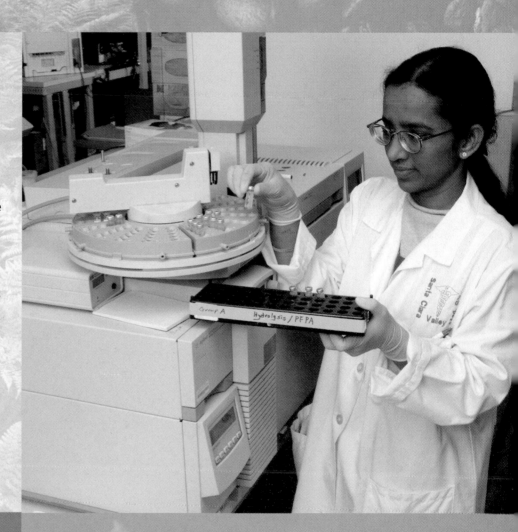

"We use mass spectrometry to analyze and confirm the presence of drugs," says Valli Vairavan, clinical lab technologist—Mass Spectrometry, Santa Clara Valley Medical Center. "A mass spectrometer separates and identifies compounds including drugs by mass. When we screen a urine sample, we look for metabolites, which are the products of drugs that have metabolized in the body. If the presence of one or more drugs such as heroin and cocaine is indicated, we confirm it by using mass spectrometry."

Drugs or their metabolites are detected in urine 24–48 hours after use. Cocaine metabolizes to benzoylecgonine and hydroxycocaine, morphine to morphine-3-glucuronide, and heroin to acetylmorphine. Amphetamines and methamphetamines are detected unchanged.

LOOKING AHEAD

the Chemistry place

www.chemplace.com/college

Visit the URL above or use the CD-ROM in the book for extra quizzing, interactive tutorials, and career resources.

In this chapter, we will look at organic compounds that contain single bonds to oxygen atoms or sulfur atoms. Alcohols, which contain the hydroxyl group (—OH) are commonly found in nature and used in industry and at home. For centuries, grains, vegetables, and fruits have been fermented to produce the ethanol present in alcoholic beverages. The hydroxyl group is important in biomolecules such as sugars and starches as well as in steroids such as cholesterol and estradiol. Menthol is a cyclic alcohol with a minty odor and flavor that is used in cough drops, shaving creams, and ointments. The ethers are compounds that contain an oxygen atom connected to two carbon atoms (—O—). Ethers are important solvents in chemistry and medical laboratories. Beginning in 1842, diethyl ether was used for about 100 years as a general anesthesia. Today less flammable and more easily tolerated anesthetics are used. Thiols, which contain an —SH group, give the strong odors we associate with garlic and onions.

14.1 Structure and Classification of Alcohols

In an **alcohol,** a **hydroxyl group (—OH)** replaces a hydrogen atom in an alkane. In a **phenol,** the hydroxyl group is attached to an aromatic ring. Both types of compounds have bent structures similar to water with an alkyl or aromatic group replacing one hydrogen atom (see Figure 14.1).

Figure 14.1 A hydrogen atom in water is replaced by an alkyl group in methanol and by an aromatic ring in phenol.

Q *Why are the structures of alcohols and phenols similar to water?*

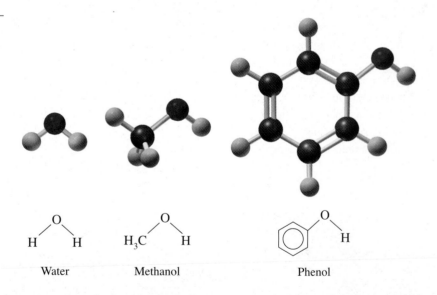

Water   Methanol   Phenol

Classification of Alcohols

When we studied alkanes in Chapter 12, we saw that carbon atoms can be classified as primary, secondary, or tertiary. Alcohols are classified in a similar way with an emphasis on the type of carbon atom bonded to the hydroxyl (—OH) group. In a **primary (1°) alcohol,** there is one alkyl group attached to the carbon bonded to the —OH group. Methanol is considered a primary alcohol. In a **secondary (2°) alcohol,** there are two alkyl groups attached. When there are three alkyl groups, the alcohol is classified as a **tertiary (3°) alcohol.**

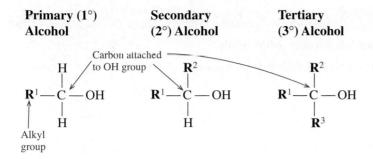

Primary (1°)
Alcohol

Secondary
(2°) Alcohol

Tertiary
(3°) Alcohol

Examples:

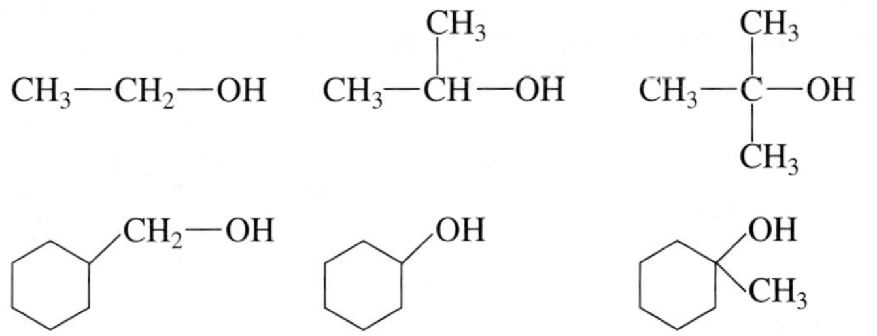

$$CH_3{-}CH_2{-}OH \qquad CH_3{-}\overset{\overset{\textstyle CH_3}{|}}{CH}{-}OH \qquad CH_3{-}\overset{\overset{\textstyle CH_3}{|}}{\underset{\underset{\textstyle CH_3}{|}}{C}}{-}OH$$

(One alkyl group)   (Two alkyl groups)   (Three alkyl groups)

SAMPLE PROBLEM 14.1

Classifying Alcohols

Classify each of the following alcohols as primary, secondary, or tertiary.

a. $CH_3{-}CH_2{-}CH_2{-}OH$

b. $CH_3{-}CH_2{-}\overset{\overset{\textstyle OH}{|}}{\underset{\underset{\textstyle CH_3}{|}}{C}}{-}CH_3$

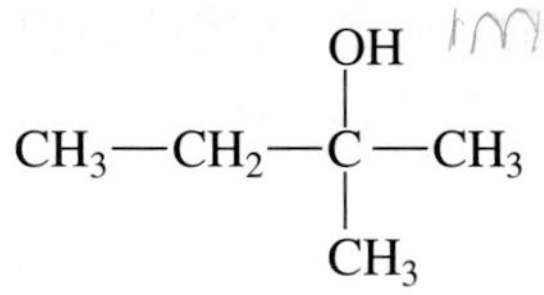

IMGN S.AT

c. OH

Solution

a. One alkyl group attached to the carbon atom bonded to the —OH makes this a primary alcohol.

b. Three alkyl groups attached to the carbon atom bonded to the —OH makes this a tertiary alcohol.

c. Two alkyl groups attached to the carbon atom bonded to the —OH makes this a secondary alcohol.

Study Check

Classify the following as primary, secondary, or tertiary:

H_3C OH

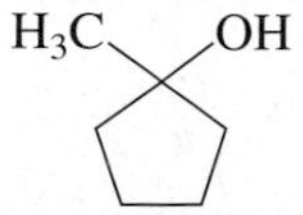

QUESTIONS AND PROBLEMS

Structure and Classification of Alcohols

14.1 Classify each of the following as a primary, secondary, or tertiary alcohol:

a. $CH_3-\overset{\overset{\displaystyle CH_3}{|}}{CH}-CH_2-CH_2-OH$

b. $CH_3-CH_2-CH_2-CH_2-OH$

c. $CH_3-\overset{\overset{\displaystyle OH}{|}}{\underset{\underset{\displaystyle CH_3}{|}}{C}}-CH_2-CH_3$

d.

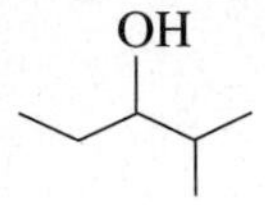

e.

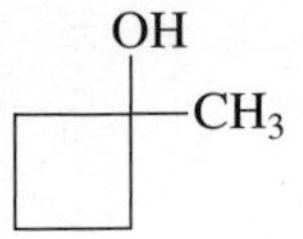

14.2 Classify each of the following as a primary, secondary, or tertiary alcohol:

a.

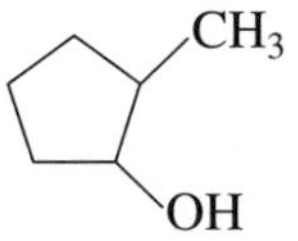

b. $CH_3-\overset{\overset{\displaystyle CH_3}{|}}{CH}-CH_2-OH$

c. 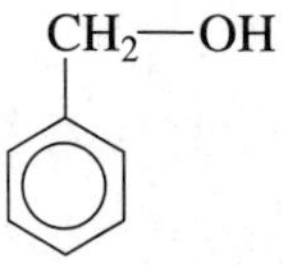 CH_2-OH

d. $CH_3-CH_2-CH_2-\overset{\overset{\displaystyle CH_3}{|}}{\underset{\underset{\displaystyle CH_3}{|}}{C}}-OH$

e.

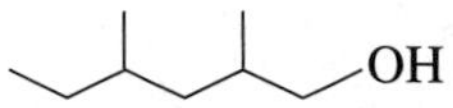

14.2 Naming Alcohols, Phenols, and Thiols

The rules for IUPAC names of alcohols are similar to those we used to name other families of organic compounds. The alcohol family is indicated in the name by an *–ol* ending, which is numbered to show the location of the hydroxyl group on the main chain.

Step 1 Name the longest carbon chain containing the —OH group. Replace the *–e* in the alkane name by *–ol*. The common names of simple alcohols give the alkyl group followed by *alcohol*.

Step 2 Number the main chain starting at the end closest to the —OH group. Alcohols with two —OH groups are named *diols;* three —OH groups, *triols.*

CH_3OH
Methanol
(methyl alcohol)

CH_3OH_2OH
Ethanol
(ethyl alcohol)

$CH_3CH_2CH_2OH$
1-Propanol
(propyl alcohol)

$CH_3\overset{\overset{\displaystyle OH}{|}}{CH}CH_3$
2-Propanol
(isopropyl alcohol)

$HO-CH_2-CH_2-OH$
1,2-Ethanediol
(ethylene glycol)

$HOCH_2\overset{\overset{\displaystyle OH}{|}}{CH}CH_2OH$
1,2,3-Propanetriol
(glycerol)

Step 3 Name and number other substituents relative to the —OH group. The —OH takes precedence over double and triple bonds.

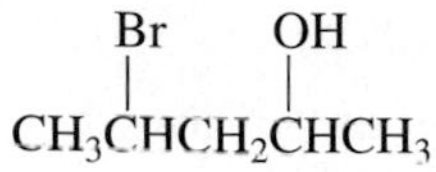

Br   OH
| |
$CH_3CHCH_2CHCH_3$

4-Bromo-2-pentanol

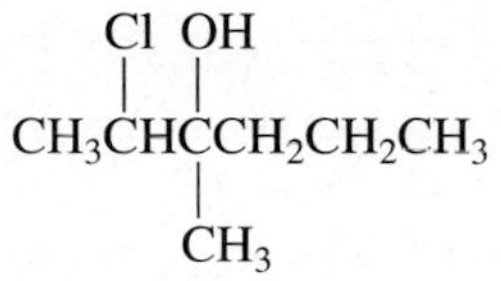

Cl OH
| |
$CH_3CHCCH_2CH_2CH_3$
        |
        CH_3

2-Chloro-3-methyl-3- hexanol

$CH_3-CH=CH-CH_2-OH$

2-Butene-1-ol

Step 4   Name a cyclic alcohol as a *cycloalkanol*. For other substituents, the ring is numbered with the —OH group on carbon 1.

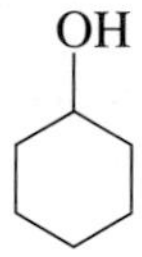

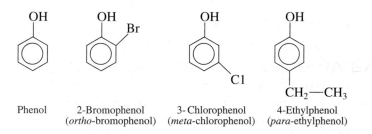

Cyclohexanol 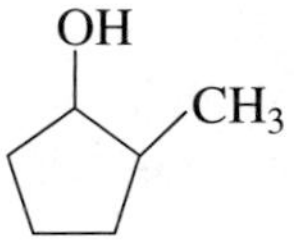   2-Methylcyclopentanol 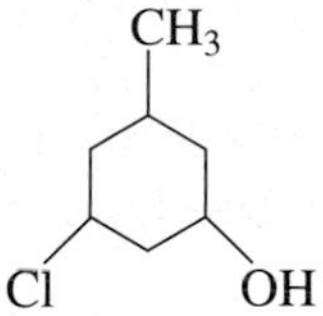   3-Chloro-5-methylcyclohexanol

Naming Phenols

The term *phenol* is the IUPAC name for a benzene ring bonded to a hydroxyl group (—OH) and is used in the name of the family of organic compounds derived from phenol. When there is a second substituent, the benzene ring is numbered starting from the carbon 1, which is bonded to the —OH group. The terms *ortho, meta,* and *para* are used for the common names of simple phenols.

Phenol

2-Bromophenol
(*ortho*-bromophenol)

3- Chlorophenol
(*meta*-chlorophenol)

4-Ethylphenol
(*para*-ethylphenol)

Certain disubstituted phenols have common names based on historical uses. The methylphenols are commonly named as *cresols*, while benzenediols have a variety of common names.

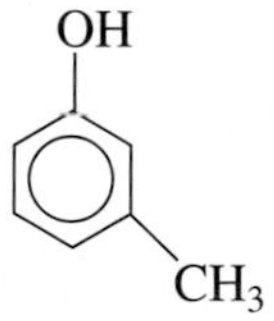

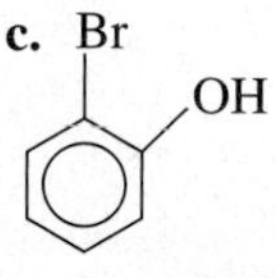

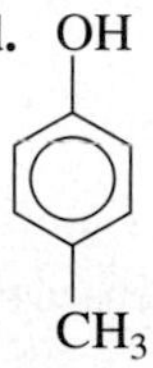

3-Methylphenol
(*meta*-cresol)

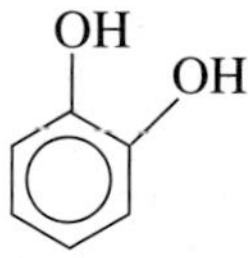

1,2-Benzenediol
(catechol)

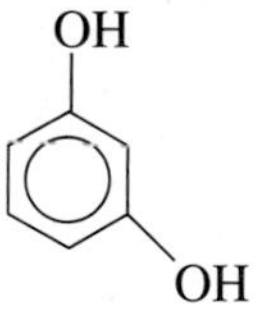

1,3-Benzenediol
(resorcinol)

1,4-Benzenediol
(hydroquinone)

SAMPLE PROBLEM 14.2

Naming Alcohols and Phenols

Give the IUPAC name for each of the following:

a. 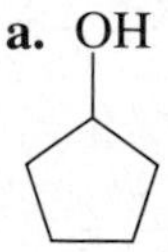 OH

b. 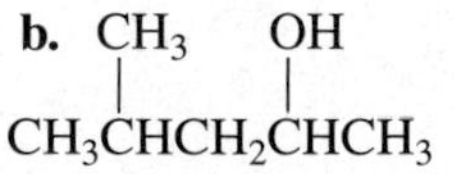 CH_3 OH
     | |
$CH_3CHCH_2CHCH_3$

c. Br
     OH

d. OH

     CH_3

Solution

a. The —OH is attached to a cyclopentane ring. Changing the *-e* to *-ol* gives the IUPAC name of *cyclopentanol.*

b. The parent chain is pentane; the alcohol is named *pentanol.* The carbon chain is numbered to give the position of the —OH group on carbon 2 and the methyl group on carbon 4. The compound is named *4-methyl-2-pentanol.*

c. The compound is a *phenol* because the —OH is attached to a benzene ring. The ring is numbered with carbon 1 attached to the —OH in the direction that gives the bromine the lower number. The compound is named *2-bromophenol* or *o-bromophenol.*

d. The —OH takes priority in naming this compound as 4-methylphenol. The common name is based on an isomer of cresol, which is *para-cresol.*

Study Check

Give the IUPAC name for the following:

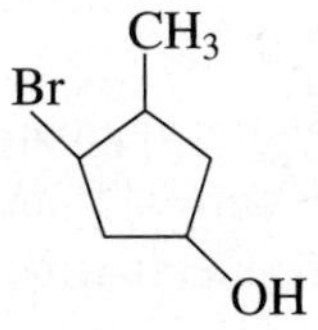

SAMPLE PROBLEM 14.3

Writing Condensed Structural Formulas of Alcohols and Phenols

a. The alcohol 3,5,5-trimethyl-1-hexanol is used as a plasticizer, a substance added to plastics to keep them pliable. Draw its condensed structural formula.

b. Thymol, used as an antiseptic in mouthwashes, has an IUPAC name of 5-methyl-2-isopropylphenol. Draw its condensed structural formula.

Solution

a. The parent chain of six carbon atoms has an —OH group on carbon 1. Methyl groups are bonded to carbon 3 and to carbon 5.

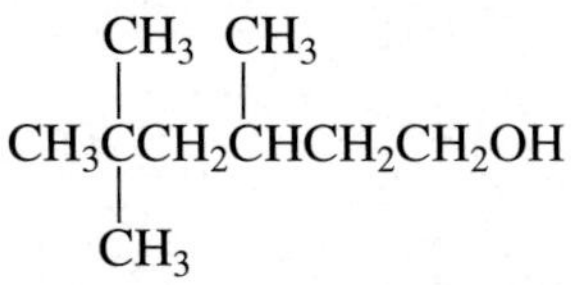

b. The phenol part of the name tells us to draw a benzene ring with an —OH group on carbon 1. Then the alkyl groups are attached, methyl to carbon 5 and isopropyl to carbon 2.

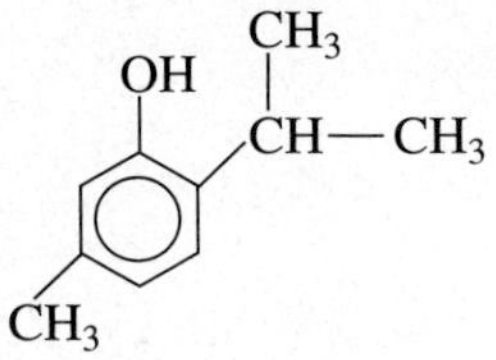

Study Check

Draw the condensed structural formula of 5-chloro-4-methyl-3-heptanol.

Thiols

Thiols are a family of sulfur-containing organic compounds that have a **sulfhydryl** (—SH) **group.** They have structures similar to alcohols except that a —SH group takes the place of an —OH group.

Naming Thiols

In the IUPAC system, thiols are named by adding *thiol* to the alkane name of the longest carbon chain bonded to the —SH group. As we did for alcohols, the location of the —SH group is indicated by numbering the main chain from the closest end.

$$CH_3—OH \qquad CH_3—SH \qquad \overset{\displaystyle SH}{\underset{\displaystyle}{CH_3—\overset{|}{CH}—CH_2—CH_3}}$$

Methanol   Methanethiol   2-Butanethiol

An important property of thiols is a strong, usually disagreeable, odor. Methanethiol is the characteristic odor of oysters and cheddar cheese. (See Figure 14.2.) To help us detect natural gas (methane) leaks, a small amount of ethanethiol

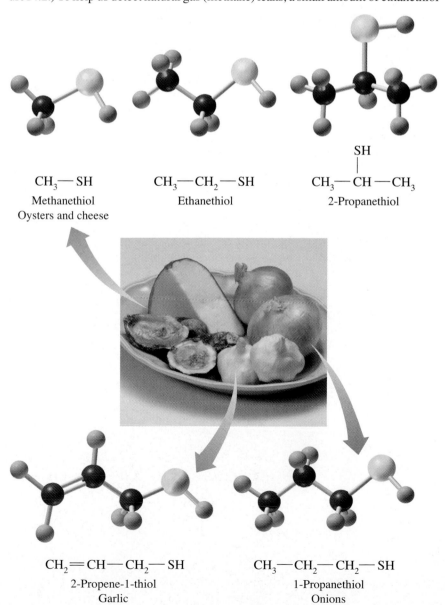

Figure 14.2 Thiols are sulfur-containing compounds with an —SH group.
Q *Why do thiols have structures similar to alcohols?*

$CH_3—SH$
Methanethiol
Oysters and cheese

$CH_3—CH_2—SH$
Ethanethiol

$\overset{\displaystyle SH}{\underset{\displaystyle}{CH_3—\overset{|}{CH}—CH_3}}$
2-Propanethiol

$CH_2=CH—CH_2—SH$
2-Propene-1-thiol
Garlic

$CH_3—CH_2—CH_2—SH$
1-Propanethiol
Onions

is added to the gas supply. There are thiols in the spray emitted when a skunk senses danger. The odor of onions is due to 1-propanethiol, which is also a lachrymator, a substance that makes eyes tear. Garlic contains thiols such as 2-propene-1-thiol. We can break this name down as follows.

2-	prop	ene	-1-	thiol
Carbon 2 has C=C	3 carbons in chain	Alkene	On carbon 1	—SH group

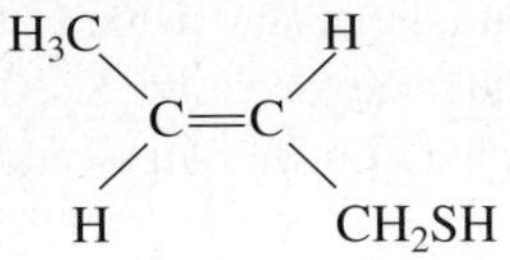

H₃C   H
     \   /
      C=C
     /   \
H   CH₂SH

CH₃CH₂CH₂—SH   CH₂=CH—CH₂—SH

trans-2-Butene-1-thiol
(in skunk spray)

1-Propanethiol
(in onions)

2-Propene-1-thiol
(in garlic)

SAMPLE PROBLEM 14.4

Thiols

Draw the condensed structural formula of the following:

a. 1-butanethiol   **b.** cyclohexanethiol

Solution

a. This compound has a —SH group on the first carbon of a butane chain.

CH₃CH₂CH₂CH₂—SH

b. This compound has a —SH group on cyclohexane.

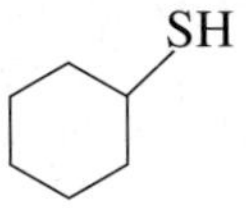

Study Check

What is the condensed structural formula of ethanethiol?

QUESTIONS AND PROBLEMS

Naming Alcohols, Phenols, and Thiols

14.3 Give the IUPAC name for each of the following alcohols:

a. CH₃CH₂OH

b. 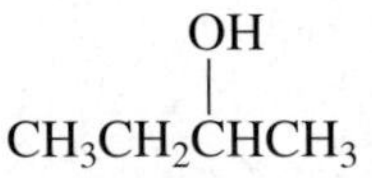
            OH
            |
   CH₃CH₂CHCH₃

c. 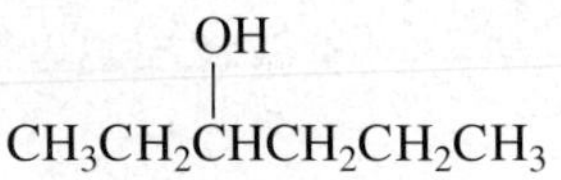
              OH
              |
   CH₃CH₂CHCH₂CH₂CH₃

d. 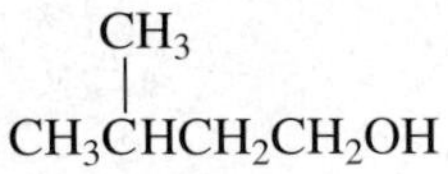
       CH₃
       |
   CH₃CHCH₂CH₂OH

e. 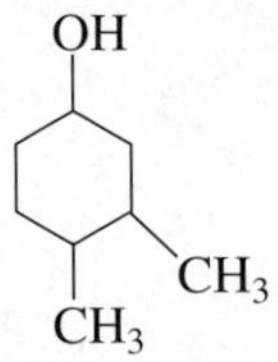
   OH
   CH₃

f. 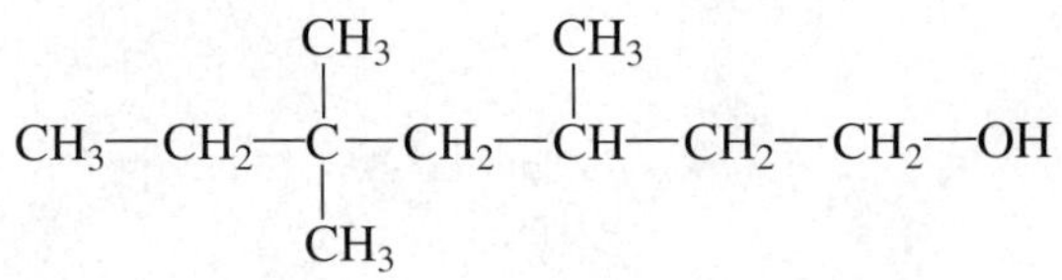
                 CH₃   CH₃
                 | |
   CH₃—CH₂—C—CH₂—CH—CH₂—CH₂—OH
                 |
                 CH₃

14.4 Give the IUPAC name for each of the following alcohols:

a.

b.
$$CH_3-\overset{\overset{\textstyle Cl}{|}}{CH}-\overset{\overset{\textstyle CH_3}{|}}{CH}-CH_2-OH$$

c. $CH_3CH_2\overset{\overset{\textstyle CH_3}{|}}{CH}\overset{\underset{\textstyle CH_3}{|}}{CH}CH_2OH$

d. $Cl-\overset{\overset{\textstyle Cl}{|}}{CH}-CH_2-CH_2-\overset{\overset{\textstyle OH}{|}}{CH}-CH_3$

e.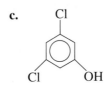

f.
$$CH_3-CH_2-\overset{\overset{\textstyle OH}{\overset{\textstyle |}{\overset{\textstyle CH_2}{\overset{\textstyle |}{CH_2}}}}}{CH}-CH_2-CH_3$$

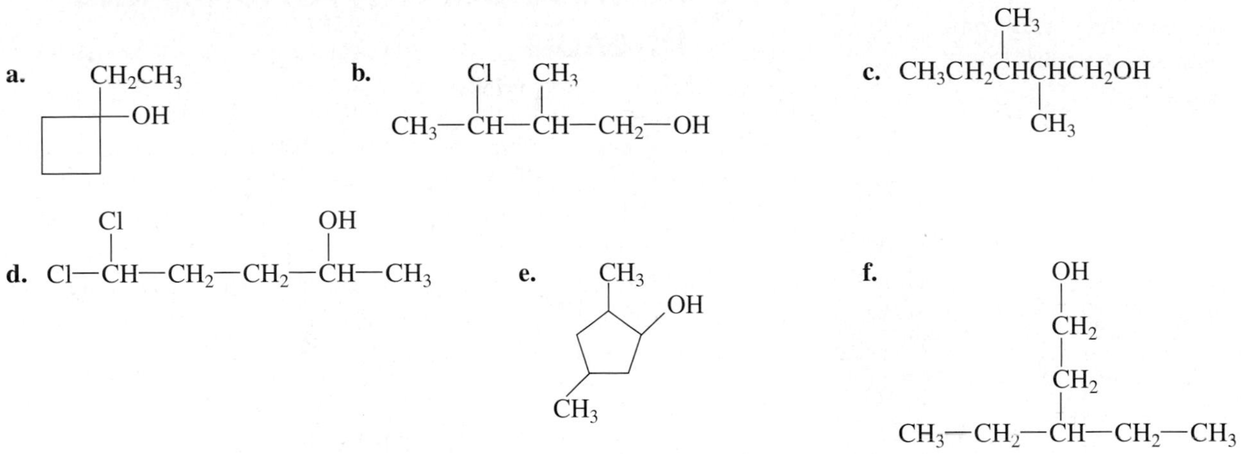

14.5 Write the condensed structural formula of each of the following alcohols:
  a. 1-propanol   **b.** methyl alcohol   **c.** 3-pentanol
  d. 2-methyl-2-butanol   **e.** cyclohexanol   **f.** 1,4-butanediol

14.6 Write the condensed structural formula of each of the following alcohols:
  a. ethyl alcohol   **b.** 3-methyl-1-butanol
  c. 2,4-dichlorocyclohexanol   **d.** propyl alcohol
  e. 1,3-cyclopentanediol   **f.** 2,2,4-trimethyl-3-hexanol

14.7 Name each of the following phenols:

a. OH

b. OH / Br

c. Cl / Cl OH

d. OH / Br

14.8 Name each of the following phenols:

a. OH / CH₂CH₃

b. Br OH / Br

c. OH / Cl

d. OH Cl

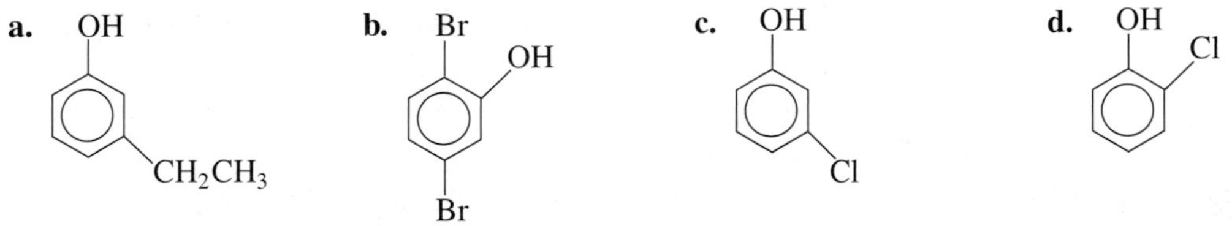

14.9 Write the condensed structural formula of each of the following phenols:
  a. *m*-bromophenol   **b.** *p*-chlorophenol   **c.** 2,5-dichlorophenol
  d. *o*-phenylphenol   **e.** 4-ethylphenol

14.10 Write the condensed structural formula of each of the following phenols:
  a. *o*-ethylphenol   **b.** 2,4-dichlorophenol
  c. 2,4-dimethylphenol   **d.** 2-ethyl-5-methylphenol
  e. *m*-phenylphenol

14.11 Give the IUPAC name for each of the following thiols:

a. CH_3-SH

b. $CH_3-\overset{\overset{\textstyle SH}{|}}{CH}-CH_3$

c. $CH_3-\overset{\overset{\textstyle CH_3}{|}}{CH}-\overset{\overset{\textstyle CH_3}{|}}{CH}-CH_2-SH$

d.

14.12 Give the IUPAC name for each of the following thiols:

a. $CH_3CH_2CH_2-SH$

b. $CH_3-CH_2-CH_2-\overset{\overset{\textstyle SH}{|}}{CH}-CH_3$

c. $CH_3-\overset{\overset{\textstyle CH_3}{|}}{\underset{\underset{\textstyle CH_3}{|}}{C}}-CH_2-SH$

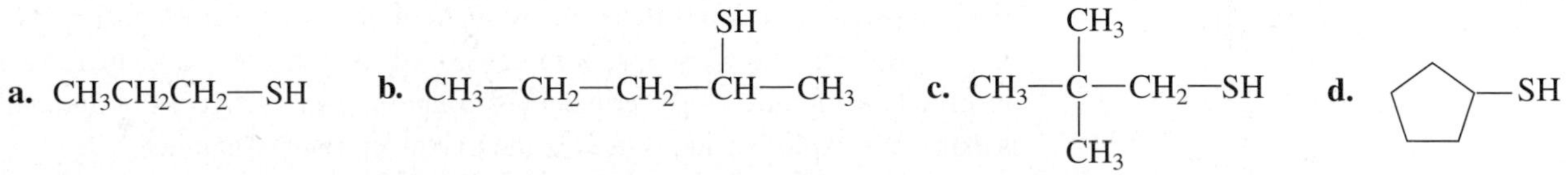

d.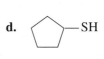

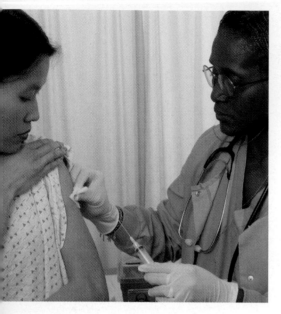

14.3 Some Important Alcohols and Phenols

Methanol (methyl alcohol), the simplest alcohol, is found in many solvents and paint removers. It is sometimes called *wood alcohol* because methanol is produced when wood is heated to high temperatures in the absence of air. If ingested, methanol is oxidized to formaldehyde, which can cause headaches, blindness, and death. Methanol is used to make plastics, medicines, and fuels. In car racing, it is used as a fuel because it is less flammable and has a higher octane rating than gasoline. Today, methanol is synthesized by reacting carbon monoxide with hydrogen gas at high temperatures and pressures.

$$CO + 2H_2 \xrightarrow[\text{300–400°C, 200 atm}]{\text{ZnO/Cr}_2\text{O}_3 \text{ catalyst,}} CH_3-OH$$

Ethanol (ethyl alcohol) has been known since prehistoric times as an intoxicating product formed by the fermentation of grains and starches.

$$C_6H_{12}O_6 \xrightarrow{\text{fermentation}} 2CH_3-CH_2-OH + 2CO_2$$

Today, ethanol for commercial uses is produced from ethene and water reacted at high temperatures and pressures. It is used as a solvent for perfumes, varnishes, and some medicines, such as tincture of iodine. "Gasohol" is a mixture of ethanol and gasoline used as a fuel.

$$H_2C=CH_2 + H_2O \xrightarrow{\text{300°C, 200 atm, catalyst}} CH_3-CH_2-OH$$

2-Propanol (isopropyl alcohol), commonly referred to as *rubbing alcohol,* is used as an astringent because it evaporates rapidly and cools the skin, reducing the size of blood vessels near the surface. An antiseptic solution of isopropyl alcohol is used to clean the skin before an injection or taking a blood sample. It is also used to sterilize equipment because it destroys bacteria by coagulating protein. The toxicity of 2-propanol is similar to that of methanol.

1,2-Ethanediol (ethylene glycol) is used as antifreeze in heating and cooling systems. It is also a solvent for paints, inks, and plastics, and is used in the production of synthetic fibers such as Dacron. If ingested, it is extremely toxic. In the body, it is oxidized to oxalic acid, which forms insoluble salts in the kidneys that cause renal damage, convulsions, and death. Because its sweet taste is attractive to pets and children, ethylene glycol solutions must be carefully stored.

$$HO-CH_2-CH_2-OH \xrightarrow{[O]} HO-\overset{\overset{\displaystyle O}{\|}}{C}-\overset{\overset{\displaystyle O}{\|}}{C}-OH$$

1,2-Ethanediol (ethylene glycol) · Oxalic acid

1,2,3-Propanetriol (glycerol or glycerin), a trihydroxy alcohol, is a viscous liquid obtained from oils and fats during the production of soaps. The presence of several polar —OH groups makes it strongly attracted to water, a feature that makes glycerin useful as a skin softener in products such as skin lotions, cosmetics, shaving creams, and liquid soaps. It is also used as antifreeze and in fluid shock absorbers. A mixture of glycerin and a strong oxidizer such as potassium chlorate is extremely explosive and is used in the manufacture of dynamite.

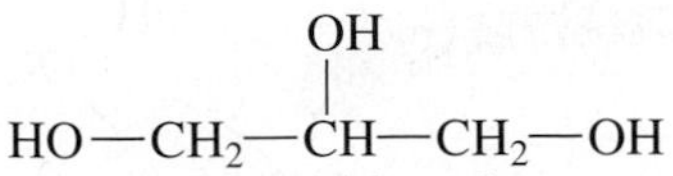

$$\underset{\text{1,2,3-Propanetriol (glycerol)}}{\text{HO}-\text{CH}_2-\overset{\overset{\displaystyle \text{OH}}{|}}{\text{CH}}-\text{CH}_2-\text{OH}}$$

Menthol is a cyclohexanol with a peppermint taste and odor that is used in candy, throat lozenges, and nasal inhalers. It causes the mucous membranes to increase their secretions and soothes the respiratory tract.

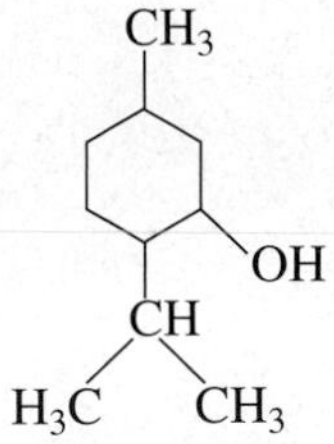

Menthol

Derivatives of Phenol

Phenol and its derivatives are used as antiseptics in throat lozenges and mouthwashes. In household disinfectant sprays such as Lysol, the active ingredient is *ortho*-phenylphenol. Poison ivy contains a substance called urushiol that causes itching and blistering of the skin.

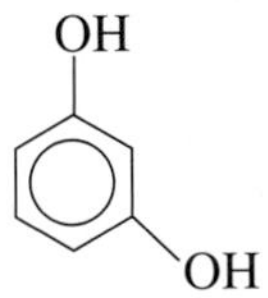

Resorcinol, antiseptic

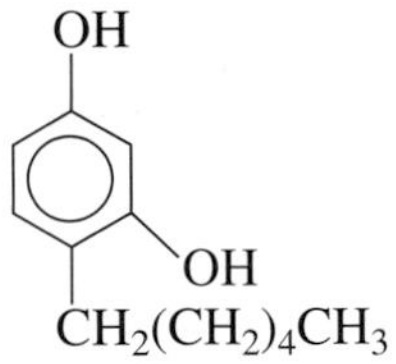

4-Hexylresorcinol, antiseptic
$CH_2(CH_2)_4CH_3$

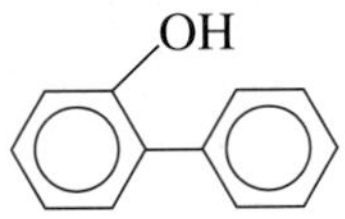

ortho-phenylphenol

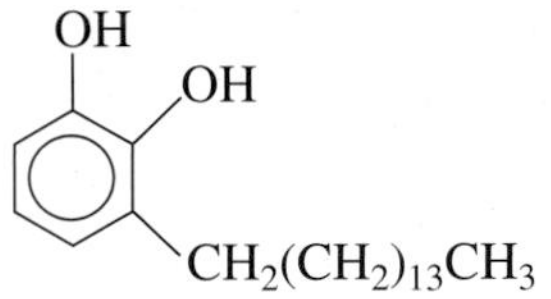

Urushiol
$CH_2(CH_2)_{13}CH_3$

Several of the essential oils of plants, which produce the odor or flavor of the plant, are derivatives of phenol. Eugenol is found in cloves, vanillin in vanilla bean, isoeugenol in nutmeg, and thymol in thyme and mint. Thymol has a pleasant, minty taste and is used in mouthwashes and by dentists to disinfect a cavity before adding a filling compound. (See Figure 14.3.)

Foods such as cereals and oils spoil when they react with the oxygen in the air. One way to prolong their shelf life is to add an antioxidant such as BHA or BHT. The phenol portion of these additives reacts readily with oxygen and keeps the food from spoiling.

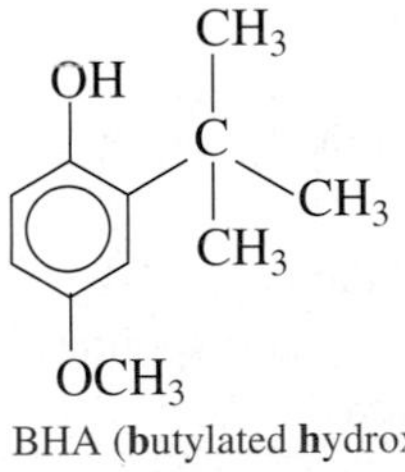

BHA (butylated hydroxyanisole)   BHT (butylated hydroxytoluene)

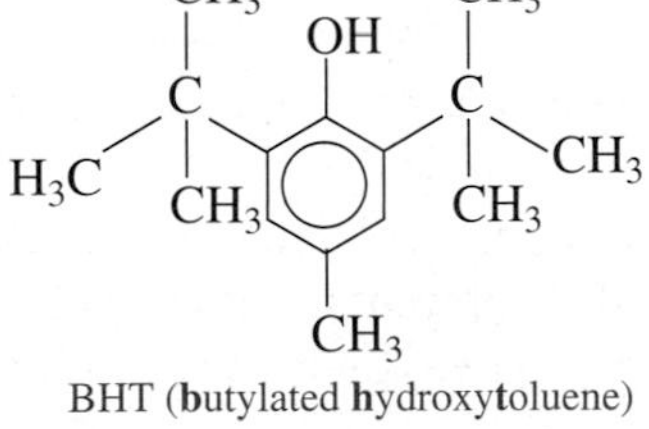

SAMPLE PROBLEM 14.5

Some Uses of Alcohols and Phenols

Identify the alcohol or phenol that best matches each of the following descriptions:

a. used to clean the skin prior to giving an injection

b. used in antifreeze

c. provides the odor and taste of cloves

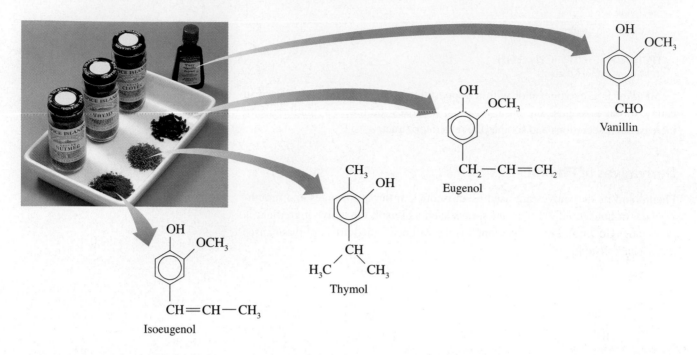

Figure 14.3 Derivatives of phenol are active ingredients found in the essential oils of cloves, vanilla, nutmeg, and mint.
Q *If phenol is a structural feature common to these spices, what accounts for their different odors and tastes?*

Solution

a. 2-propanol; isopropyl alcohol
b. 1,2-ethanediol; ethylene glycol
c. eugenol

Study Check

Is methanol or ethanol found in alcoholic beverages?

QUESTIONS AND PROBLEMS

Some Important Alcohols and Phenols

14.13 Identify the alcohol or phenol that best matches each of the following descriptions:
   a. found in alcoholic beverages
   b. gives a peppermint taste to candy and throat lozenges
   c. used as household disinfect in Lysol

14.14 Identify the alcohol or phenol that best matches each of the following descriptions:
   a. if ingested, it forms insoluble salts that cause renal damage
   b. added to foods to prevent reactions with oxygen that cause spoilage
   c. used to soften skin in skin lotions and shaving creams

LEARNING GOAL

Give the IUPAC and common names of ethers; draw the condensed structural formula.

14.4 Ethers

An **ether** contains an oxygen atom that is attached by single bonds to two carbon groups that are alkyls or aromatic rings. Ethers have a bent structure like water and alcohols except both hydrogen atoms are replaced by alkyl groups. (See Figure 14.4.)

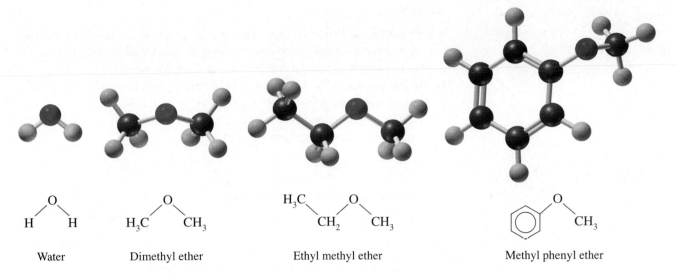

| Water | Dimethyl ether | Ethyl methyl ether | Methyl phenyl ether |

Naming Ethers

Most ethers are named by their common names. The IUPAC names are used only when the ether is more complex.

Step 1   For a simple ether, use the common name. Write the name of each alkyl or aryl (aromatic) group attached to the oxygen atom in alphabetical order followed by the word *ether.*

Step 2   For complex ethers, use the IUPAC name. Use the alkane name of the larger alkyl group as the main chain. Name the oxygen and smaller alkyl group as a substituent called an **alkoxy (alkyl + oxygen) group.**

Step 3   Number the main chain beginning at the end nearest the alkoxy group and give the location of the alkoxy group on the main chain.

Alkoxy group

$$\overbrace{CH_3-O}-\boxed{CH_2-CH_2-CH_3}\;\;\text{main chain = propane}$$
$$\quad\quad\quad\quad\quad\;\; 1\quad\;\;\; 2\quad\;\;\; 3$$

IUPAC name: 1-methoxypropane
Common name: methyl propyl ether

Some examples of naming ethers follow:

$$\overset{\displaystyle OCH_3}{\underset{}{}}$$

$$CH_3-O-CH_3 \quad\quad CH_3CH_2-O-CH_2CH_3 \quad\quad CH_3-\overset{\displaystyle |}{CH}-CH_2-CH_3$$

Methoxymethane   Ethoxyethane   2-Methoxybutane
(dimethyl ether)   (diethyl ether)

$$CH_3CH_2-O-\text{⬡}$$ $$\text{⬡}-O-\text{⬡}$$

Ethoxybenzene   Phenoxybenzene
(ethyl phenyl ether)   (diphenyl ether)

SAMPLE PROBLEM 14.6

Ethers

Assign a common name and the IUPAC name to the following ethers:

a. $CH_3CH_2-O-CH_2CH_2CH_3$

b.

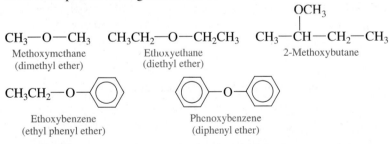

OCH₃

Figure 14.4 The structures of ethers are similar to that of water.
Q *What part of the structure of water is found in ethers?*

Solution

a. The groups attached to the oxygen are an ethyl group and a propyl group. The common name is *ethyl propyl ether.* Naming the shorter alkyl group and the oxygen as ethoxy gives the IUPAC name of *1-ethoxypropane.*

b. The groups attached to the oxygen are a methyl group and a cyclobutyl group. The common name is *cyclobutyl methyl ether.* In the IUPAC name, the CH_3O— is named methoxy and the cyclobutane is the longer carbon chain, which gives the name *methoxycyclobutane.*

Study Check

What is the common name of ethoxybenzene?

Constitutional Isomers of Alcohols and Ethers

The general formula $C_nH_{2n+2}O$ represents the molecular formula of both alcohols and ethers formed from alkanes. For example, we can write structural formulas for constitutional isomers with the molecular formula C_2H_6O as follows:

$$CH_3—CH_2—OH \qquad\qquad CH_3—O—CH_3$$
   Ethyl alcohol   Dimethyl ether

SAMPLE PROBLEM 14.7

Constitutional Isomers

Draw the structural formulas and give the common names of two alcohols and one ether with a molecular formula C_3H_8O.

Solution

To draw the structural formulas for alcohols, the hydroxyl group is bonded to two different atoms in a chain of three carbon atoms. For the ether, two alkyl groups are bonded to an oxygen atom.

$$\overset{\textstyle OH}{\underset{\textstyle }{}}$$

$$CH_3—CH_2—CH_2—OH \qquad CH_3—\overset{|}{C}H—CH_3 \qquad CH_3—CH_2—O—CH_3$$
   Propyl alcohol   Isopropyl alcohol   Ethyl methyl ether

Study Check

Write the IUPAC names of the unbranched isomers of $C_4H_{10}O$.

Cyclic Ethers

Cyclic ethers contain an oxygen atom in a carbon ring. They are **heterocyclic compounds** because there is a ring with one or more atoms that are not carbon. The cyclic ethers are usually given common names. The five-atom rings with an oxygen atom use common names derived from the aromatic ring **furan.** The four-atom cyclic ethers are not common. The rings are numbered from the oxygen atom as 1.

HEALTH NOTE

Ethers as Anesthetics

Anesthesia is the loss of all sensation and consciousness. A general anesthetic is a substance that blocks signals to the awareness centers in the brain so the person has a loss of memory, a loss of feeling pain, and an artificial sleep. The term *ether* has been associated with anesthesia because diethyl ether was the most widely used anesthetic for more than a hundred years. Although it is easy to administer, ether is very volatile and highly flammable. A small spark in the operating room could cause an explosion. Since the 1950s, anesthetics such as Forane (isoflurane), Ethrane (enflurane), and Penthrane (methoxyflurane) have been developed that are not as flammable and do not cause nausea. Most of these anesthetics retain the ether group, but the addition of many halogen atoms reduces the volatility and flammability of the ethers. More recently, they have been replaced by halothane (1-bromo-1-chloro-2,2,2-trifluoroethane), discussed in Chapter 12, because of the side effects of the ether-type inhalation anesthetics.

Forane
(isoflurane)

Ethrane
(enflurane)

Penthrane
(methoxyflurane)

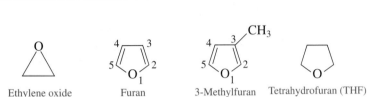

Ethylene oxide   Furan   3-Methylfuran   Tetrahydrofuran (THF)

A cyclic ether with six atoms is named *pyran,* which is the common name for an unsaturated ether ring of six atoms.

Pyran   Tetrahydropyran (THP)   4-Methylpyran

Cyclic ethers containing two oxygen atoms in a ring of six atoms are called *dioxanes.* The oxygen atoms are numbered because they can take different positions in the ring.

1,4-Dioxane   1,3-Dioxane

Dioxin is a term used for a group of highly toxic compounds composed of dioxanes bonded to aromatic rings. One of the most toxic is 2,3,7,8-tetrachlorodibenzodioxin (TCDD), now considered carcinogenic (cancer causing) because its structure interferes with DNA. Dioxin is formed during forest fires and as a by-product of many industrial processes involving chlorine such as chemical and pesticide manufacturing and pulp and paper bleaching. The herbicide Agent Orange used in Vietnam was contaminated by highly toxic dioxin, which formed during the synthesis of Agent Orange.

2,4,5-Trichlorophenoxyacetic acid
(2,4,5-T; Agent Orange)

2,3,7,8-Tetrachlorodibenzodioxin
(TCDD, "dioxin")

SAMPLE PROBLEM 14.8

Cyclic Ethers

Identify the following as a cyclic alcohol, ether, or cyclic ether.

a. 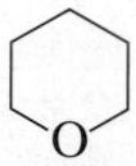   **b.**   **c.**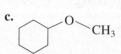

Solution

a. A cyclic ether has an oxygen atom in the ring.
b. A cyclic alcohol has a hydroxyl group bonded to a cyclic alkane.
c. An ether has an oxygen atom with single bonds to two alkyl or aryl groups.

Study Check

What is the difference between furan and pyran?

QUESTIONS AND PROBLEMS

Ethers

14.15 Give the IUPAC name and a common name for each of the following ethers:

a. CH_3—O—CH_2CH_3 b. 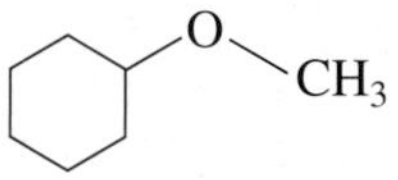 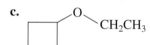   c.   d. CH_3—O—$CH_2CH_2CH_3$

14.16 Give the IUPAC name and a common name for each of the following ethers:

a. CH_3CH_2—O—$CH_2CH_2CH_3$ b. 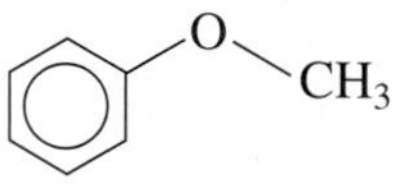   c. 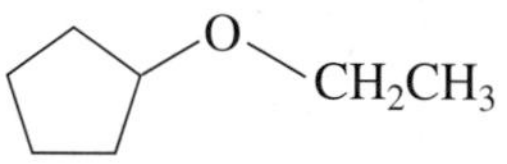   d. CH_3—O—CH_3

14.17 Write the condensed structural formula for each of the following ethers:
   a. ethyl propyl ether   b. ethyl cyclopropyl ether
   c. methoxycyclopentane   d. 1-ethoxy-2-methylbutane
   e. 2,3-dimethoxypentane

14.18 Write the condensed structural formula for each of the following ethers:
   a. diethyl ether   b. diphenyl ether
   c. ethoxycyclohexane   d. 2-methoxy-2,3-dimethylbutane
   e. 1,2-dimethoxybenzene

14.19 Indicate whether each of the following pairs represent constitutional isomers, the same compound, or different compounds:
   a. 2-pentanol and 2-methoxybutane
   b. 2-butanol and cyclobutanol
   c. ethyl propyl ether and 2-methyl-1-butanol

14.20 Indicate whether each of the following pairs represent constitutional isomers, the same compound, or different compounds:
   a. 2-methoxybutane and 3-methyl-2-butanol
   b. 1-hexanol and dipropyl ether
   c. *tert*-butyl alcohol and diethyl ether

14.21 Give the name for each of the following cyclic ethers:

a.    b. 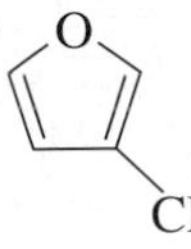   c.

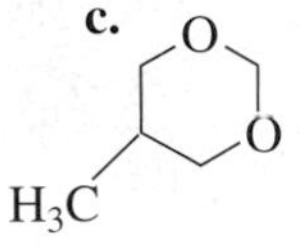

14.22 Give the name for each of the following cyclic ethers:

a.    b. 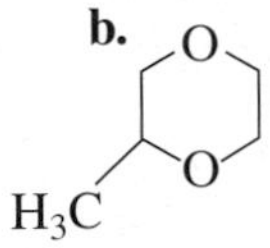   c.

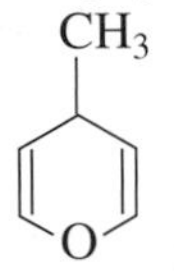

14.5 Physical Properties of Alcohols, Phenols, and Ethers

LEARNING GOAL

Describe some physical properties of alcohols, phenols, and ethers.

In Chapters 11 and 12, we learned that hydrocarbons, which are composed of only carbon and hydrogen, are nonpolar. In this chapter, we looked at compounds containing the element oxygen, which is strongly electronegative. As a result the oxygen and hydrogen atoms in the hydroxyl group —OH form hydrogen bonds. Although ethers also contain an oxygen atom, there is no hydrogen atom attached. Thus, ethers do not hydrogen bond with each other, but they do hydrogen bond with water.

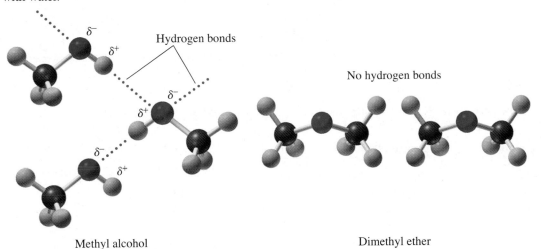

Methyl alcohol   Dimethyl ether

Boiling Points

Higher temperatures are required to provide the energy needed to break the hydrogen bonds between alcohol molecules. Thus, alcohols have higher boiling points than alkanes and ethers of similar mass. The boiling points of ethers are close to those of alkanes of similar mass because hydrogen bonding does not occur. The boiling points of some typical alkanes, alcohols, and ethers are listed in Table 14.1.

Solubility in Water

The oxygen atom in alcohols and ethers influences their solubility in water. In alcohols, the polar —OH group can hydrogen bond with water, which makes alcohols with one to four carbon atoms very soluble in water.

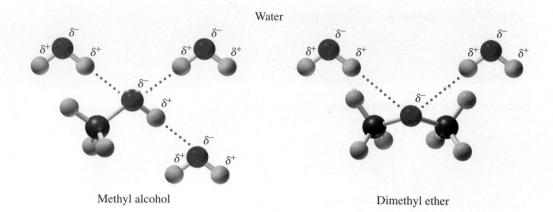

Methyl alcohol   Dimethyl ether

When there are more carbon atoms in the alkyl portion of the alcohol, the effect of the —OH group is diminished. The alkane portion does not participate in hydrogen bonding with water. Thus alcohols with five or more carbon atoms are not very soluble in water.

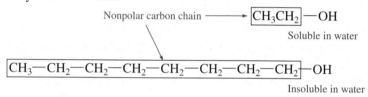

Ethers are more soluble in water than alkanes of similar mass because the oxygen can hydrogen bond with water, but ethers are not as soluble as alcohols. Ethers such as diethyl ether are very useful as solvents for hydrocarbons. However, ether vapors are highly flammable and react with oxygen to form explosive compounds. The utmost care must be taken when working with ethers.

Table 14.1 also compares the solubility of some alkanes, alcohols, and ethers by mass.

Table 14.1   Solubility and Boiling Points of Some Typical Alkanes, Alcohols, and Ethers

Compound	Structural Formula	Molar Mass (g/mole)	Boiling Point (°C)	Solubility in Water
Methane	CH_4	16	−161	No
Ethane	CH_3CH_3	30	−88	No
Methanol	CH_3—OH	32	65	Yes
Propane	$CH_3CH_2CH_3$	44	−42	No
Dimethyl ether	CH_3—O—CH_3	46	−23	Yes
Ethanol	CH_3CH_2—OH	46	78	Yes
Butane	$CH_3CH_2CH_2CH_3$	58	0	No
Ethyl methyl ether	CH_3—O—CH_2CH_3	60	8	Yes
1-Propanol	$CH_3CH_2CH_2$—OH	60	97	Yes
2-Propanol	$\overset{\displaystyle OH}{\underset{\displaystyle CH_3CHCH_3}{|}}$	60	83	Yes
Pentane	$CH_3CH_2CH_2CH_2CH_3$	72	36	No
Diethyl ether	CH_3CH_2—O—CH_2CH_3	74	35	Slightly
1-Butanol	$CH_3CH_2CH_2CH_2$—OH	74	118	Yes
2-Butanol	$\overset{\displaystyle OH}{\underset{\displaystyle CH_3CHCH_2CH_3}{|}}$	74	100	Yes

Solubility of Phenols

Phenol is soluble in water because the hydroxyl group ionizes slightly as a weak acid. In fact, an early name for phenol was *carbolic acid*. A concentrated solution of phenol is very corrosive and highly irritating to the skin; it can cause severe burns and ingestion can be fatal. Dilute solutions of phenol were previously used in hospitals as antiseptics, but they have generally been replaced.

$$\text{Phenol} + H_2O \rightleftharpoons \text{Phenoxide ion} + H_3O^+$$

Phenol   Phenoxide ion

SAMPLE PROBLEM 14.9

Physical Properties of Alcohols, Ethers, and Phenols

Predict which compounds in each pair will be more soluble in water.

a. propane or ethanol   **b.** 1-propanol or 1-hexanol

Solution

a. Ethanol is more soluble because it can form hydrogen bonds with water.
b. The 1-propanol is more soluble because it has a shorter carbon chain.

Study Check

Dimethyl ether and ethanol both have molar masses of 46. However, ethanol has a much higher boiling point than dimethyl ether. How would you explain this difference in boiling points?

QUESTIONS AND PROBLEMS

Physical Properties of Alcohols, Phenols, and Ethers

14.23 Predict the compound with the higher boiling point in the following pairs:
　　　a. ethane or methanol   **b.** diethyl ether or 1-butanol
　　　c. 1-butanol or pentane

14.24 Glycerol (1,2,3-propanetriol) has a boiling point of 290°C. 1-Pentanol, which has about the same molar mass as glycerol, boils at 138°C. Why is the boiling point of glycerol so much higher?

14.25 Are each of the following soluble in water? Explain.
　　　a. CH_3CH_2OH **b.** $CH_3—O—CH_3$
　　　c. $CH_3CH_2CH_2CH_2CH_2CH_2—OH$
　　　d. $CH_3CH_2CH_3$ **e.**　OH

14.26 Give an explanation for the following observations:
　　　a. Ethanol is soluble in water, but propane is not.
　　　b. Dimethyl ether is soluble in water, but pentane is not.
　　　c. 1-Propanol is soluble in water, but 1-hexanol is not.

Figure 14.5 A flaming dessert is prepared using a liquor that undergoes combustion.
Q *What is the equation for the combustion of the ethanol in the liquor?*

EXPLORE YOUR WORLD

Combustion of Alcohol

Find a recipe for a dessert that is flamed, such as cherries jubilee, bananas Foster, or bananas flambé. Get a group of chemistry friends together and prepare your dessert. Explain to your friends how this dramatic presentation works.

Questions

1. What reaction does an alcohol undergo when you have crêpes suzette or bananas flambéd?
2. In Sterno, also called "canned heat" and used to heat fondue dishes, methanol is prepared as a gel or a liquid. Methanol is a highly toxic alcohol and can cause blindness if just 4 mL of methanol is ingested. What is the combustion reaction for methanol?

14.6 Reactions of Alcohols

In Chapter 12, we learned that hydrocarbons undergo combustion in the presence of oxygen. Alcohols burn with oxygen too. For example, in a restaurant, a dessert may be prepared by pouring a liquor on fruit or ice cream and lighting it. (See Figure 14.5.) The combustion of the ethanol in the liquor proceeds as follows:

$$CH_3-CH_2-OH + 3O_2 \longrightarrow 2CO_2 + 3H_2O + \text{energy}$$

Dehydration of Alcohols to Form Alkenes

Earlier we saw that alkenes can add water to yield alcohols. In a reverse reaction alcohols lose a water molecule when they are heated with an acid catalyst such as H_2SO_4. During the **dehydration** of an alcohol, a H— and —OH are removed from *adjacent carbon atoms of the same alcohol* to produce a water molecule. A double bond forms between the same two carbon atoms to produce an alkene product.

$$-\underset{\substack{|}}{C}-\underset{\substack{| \\ H}}{C}-\underset{\substack{| \\ OH}}{C}- \xrightarrow[\text{Heat}]{H^+} \; C=C \; + \; H_2O$$

Alcohol   Alkene   Water

Examples

$$H-\underset{\substack{| \\ H}}{\overset{\substack{H}}{C}}-\underset{\substack{| \\ H}}{\overset{\substack{OH}}{C}}-H \xrightarrow[\text{Heat}]{H^+} H-\underset{\substack{| \\ H}}{\overset{\substack{}}{C}}=\underset{\substack{| \\ H}}{\overset{\substack{}}{C}}-H + H_2O$$

Ethanol   Ethene

Cyclopentanol $\xrightarrow[\text{Heat}]{H^+}$ Cyclopentene $+ H_2O$

The dehydration of a secondary alcohol can result in the formation of either of two products. **Saytzeff's rule** states that the major product is the one that results when the hydrogen (H—) is removed from the carbon atom with the smallest number of hydrogen atoms. This occurs because a hydrogen atom on a secondary carbon atom is easier to remove than a hydrogen from a primary carbon atom.

Adjacent carbon with the smallest number of H atoms

$$CH_3-\underset{\substack{| \\ H}}{\overset{\substack{H}}{C}}-\underset{\substack{| \\ OH}}{\overset{\substack{}}{C}}-CH_3 \xrightarrow[\text{Heat}]{H^+}$$

2-Butanol

$CH_3-CH=CH-CH_3 + H_2O$
2-Butene (major product: 90%)

$CH_3-CH_2-CH=CH_2 + H_2O$
1-Butene (minor product: 10%)

SAMPLE PROBLEM 14.10

Dehydration of Alcohols

Draw the condensed structural formulas for the alkenes produced by the dehydration of the following alcohols:

a.

$$\underset{\text{Heat}}{\overset{\text{H}^+}{\longrightarrow}}$$

OH
|
$CH_3-CH_2-CH-CH_2-CH_3$

b.

OH

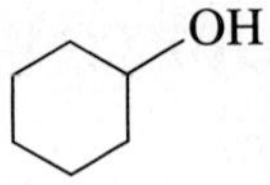

$$\underset{\text{Heat}}{\overset{\text{H}^+}{\longrightarrow}}$$

Solution

a. $CH_3-CH_2-CH=CH-CH_3 + H_2O$

b. The —OH of this alcohol is removed along with an H from an adjacent carbon. Remember that the hydrogens are not drawn in this type of geometric figure.

Study Check

What is the name of the alkene produced by the dehydration of cyclopentanol?

SAMPLE PROBLEM 14.11

Predicting Reactants

Draw the structural formula of the alcohol that is needed to produce each of the following products.

a.

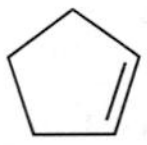

b. $CH_3-\underset{\underset{CH_3}{|}}{C}=CH-CH_3$

Solution

a.

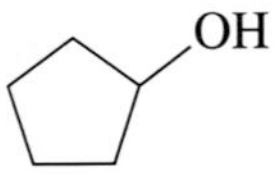

OH

b. $CH_3-\underset{\underset{CH_3}{|}}{\overset{\overset{OH}{|}}{C}}-CH_2-CH_3$ or $CH_3-\underset{\underset{CH_3}{|}}{CH}-\overset{\overset{OH}{|}}{CH}-CH_3$

Study Check

What is the name of an alcohol that forms 2-methylpropene?

Formation of Ethers

Ethers form when the dehydration of alcohols occurs at lower temperatures in the presence of an acid catalyst. Then the components of water are removed from two molecules: an H— from one alcohol and the —OH from another. When the remaining portions of the two alcohols join, an ether is produced.

$$R-O\boxed{H + HO}-R \underset{\text{Heat}}{\overset{\text{H}^+}{\longrightarrow}} R-O-R + H_2O$$

Example

$$CH_3-OH + HO-CH_3 \underset{\text{Heat}}{\overset{\text{H}^+}{\longrightarrow}} CH_3-O-CH_3 + H_2O$$

Methanol   Methanol   Dimethyl ether

Ether Formation

Write the structural formula and name of the ether produced by the following reaction:

$$2CH_3CH_2OH \xrightarrow[\text{Heat}]{H^+}$$

Solution

The structural formula of the ether product is written by attaching the alkyl groups of the reacting alcohols to an oxygen atom.

$$CH_3CH_2-O-CH_2CH_3 \qquad \text{Diethyl ether}$$

Study Check

What alcohol is needed to form the ether named dicyclohexyl ether?

The reactions of alcohols have a central role in organic chemistry because alcohols can be converted to many of the other functional groups.

Oxidation of Alcohols

In Chapter 6, we learned that an oxidation was a loss of electrons and a reduction was a gain of electrons. However, in organic chemistry, it is convenient to think of **oxidation** as a loss of hydrogen atoms or the addition of oxygen. By counting the number of bonds between carbon and oxygen, we can decide if an oxidation reaction occurred. An increase in the number of carbon–oxygen bonds is the result of a loss of electrons. In a reduction, hydrogen is added, which gives a compound with fewer bonds between carbon and oxygen. (See Figure 14.6.)

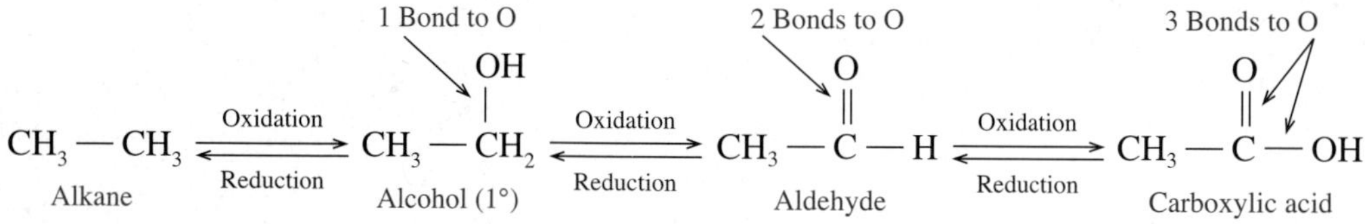

Figure 14.6 A compound is oxidized when the number of bonds increases between carbon and oxygen.

Q *Is an aldehyde more or less oxidized than the corresponding primary alcohol?*

Oxidation of Primary Alcohols

The oxidation of a primary alcohol produces an aldehyde, which contains a double bond between carbon and oxygen. The oxidation occurs by removing two hydrogen atoms, one from the —OH group and another from the carbon that is bonded to the —OH. In the laboratory, the oxidation requires some type of oxidizing agent such as O_2, $KMnO_4$, H_2CrO_4, or $K_2Cr_2O_7$. To indicate the presence of an oxidizing agent, reactions are often written with the symbol [O].

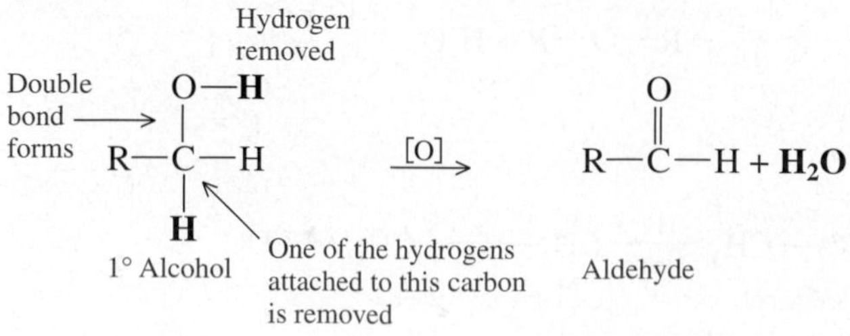

Examples

$$OH$$
$$H-\overset{\displaystyle |}{\underset{\displaystyle |}{C}}-H \quad \xrightarrow{[O]} \quad H-\overset{\displaystyle O}{\overset{\|}{C}}-H + H_2O$$
$$H$$

Methanol
(methyl alcohol)

Methanal
(formaldehyde)

$$OH$$
$$CH_3-\overset{\displaystyle |}{CH_2} \quad \xrightarrow{[O]} \quad CH_3-\overset{\displaystyle O}{\overset{\|}{C}}-H + H_2O$$

Ethanol
(ethyl alcohol)

Ethanal
(acetaldehyde)

Aldehydes oxidize further, this time by the addition of another oxygen to form a carboxylic acid. This step occurs so readily that it is often difficult to isolate the aldehyde product during oxidation. We will learn more about carboxylic acids in Chapter 17.

$$R-\overset{\displaystyle O}{\overset{\|}{C}}-H \quad \xrightarrow{[O]} \quad R-\overset{\displaystyle O}{\overset{\|}{C}}-OH$$

Aldehyde

Carboxylic acid

Example

$$H-\overset{\displaystyle O}{\overset{\|}{C}}-H \quad \xrightarrow{[O]} \quad H-\overset{\displaystyle O}{\overset{\|}{C}}-OH$$

Methanal
(formaldehyde)

Methanoic acid
(formic acid)

Oxidation of Methanol

Methanol, "wood alcohol," is a highly toxic alcohol present in products such as windshield washer fluid, Sterno, and paint strippers. Methanol is rapidly absorbed in the gastrointestinal tract. In the liver, it is metabolized to formaldehyde and then formic acid, a substance that causes nausea, severe abdominal pain, and blurred vision. Blindness can occur because the intermediate products destroy the retina of the eye. As little as 4 mL of methanol can produce blindness. The formic acid, which is not readily eliminated from the body, lowers blood pH so severely that just 30 mL of methanol can lead to coma and death.

$$CH_3OH \quad \xrightarrow[\text{[O]}]{\text{Liver enzymes}} \quad H-\overset{\displaystyle O}{\overset{\|}{C}}-H \quad \longrightarrow \quad H-\overset{\displaystyle O}{\overset{\|}{C}}-OH$$

Methanol
(methyl alcohol)

Methanal
(formaldehyde)

Methanoic acid
(formic acid)

The treatment for methanol poisoning involves giving sodium bicarbonate to neutralize the formic acid in the blood. In some cases, ethanol is given intravenously to the patient. The enzymes in the liver pick up ethanol molecules to oxidize instead of methanol molecules, which gives more time for the methanol to be eliminated via the lungs without the formation of its dangerous oxidation products.

Oxidation of Secondary Alcohols

The oxidation of secondary alcohols is similar to that of primary alcohols except the products are ketones. One hydrogen is removed from the —OH and another from the carbon bonded to the —OH group. The result is a ketone that has the carbon–oxygen double bond attached to alkyl groups on both sides.

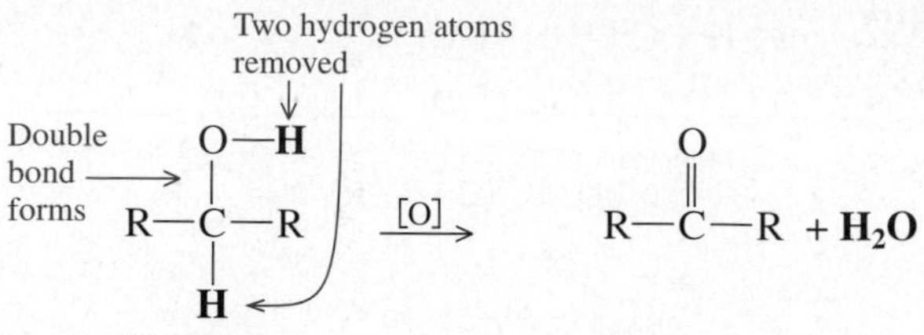

Two hydrogen atoms
removed

$$
\underset{\text{2° Alcohol}}{\text{Double bond forms} \rightarrow \underset{\underset{\text{H}}{|}}{\overset{\overset{\text{O—H}}{|}}{R\text{—}C\text{—}R}}} \xrightarrow{[O]} \underset{\text{Ketone}}{R\text{—}\overset{\overset{\text{O}}{||}}{C}\text{—}R} + H_2O
$$

Examples

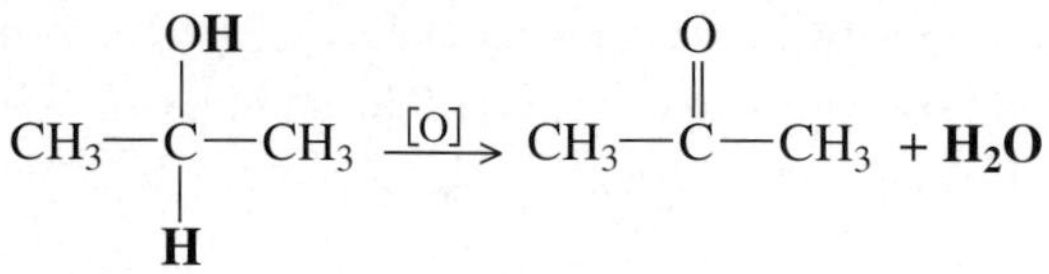

$$
\underset{\substack{\text{2-Propanol}\\\text{(isopropyl alcohol)}}}{CH_3\text{—}\underset{\underset{\text{H}}{|}}{\overset{\overset{\text{OH}}{|}}{C}}\text{—}CH_3} \xrightarrow{[O]} \underset{\substack{\text{Propanone}\\\text{(dimethyl ketone: acetone)}}}{CH_3\text{—}\overset{\overset{\text{O}}{||}}{C}\text{—}CH_3} + H_2O
$$

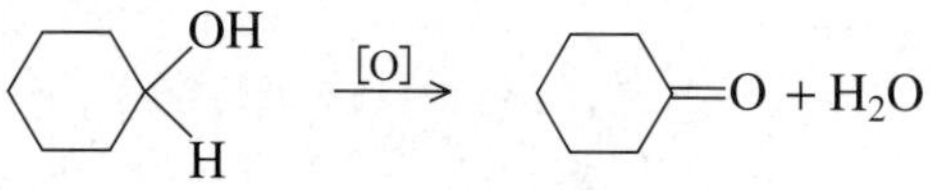

Cyclohexanol $\xrightarrow{[O]}$ Cyclohexanone $=O + H_2O$

During vigorous exercise, lactic acid accumulates in the muscles and causes fatigue. When the activity level is decreased, oxygen enters the muscles and oxidizes the secondary —OH group in lactic acid to a ketone group in pyruvic acid. Pyruvic acid is metabolized further until it is completely oxidized to CO_2 and H_2O. The muscles in highly trained athletes are capable of taking up greater quantities of oxygen so that vigorous exercise can be maintained for longer periods of time.

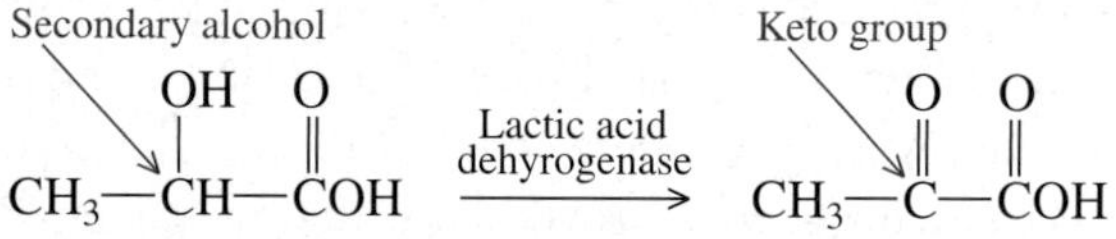

Secondary alcohol   Keto group

$$
\underset{\text{Lactic acid}}{CH_3\text{—}\underset{\underset{\text{OH}}{|}}{CH}\text{—}\overset{\overset{\text{O}}{||}}{C}OH} \xrightarrow{\substack{\text{Lactic acid}\\\text{dehyrogenase}}} \underset{\text{Pyruvic acid}}{CH_3\text{—}\overset{\overset{\text{O}}{||}}{C}\text{—}\overset{\overset{\text{O}}{||}}{C}OH}
$$

Tertiary alcohols do not oxidize readily because there are no hydrogen atoms on the alcohol carbon. Because C—C bonds are usually too strong to oxidize, tertiary alcohols resist oxidation.

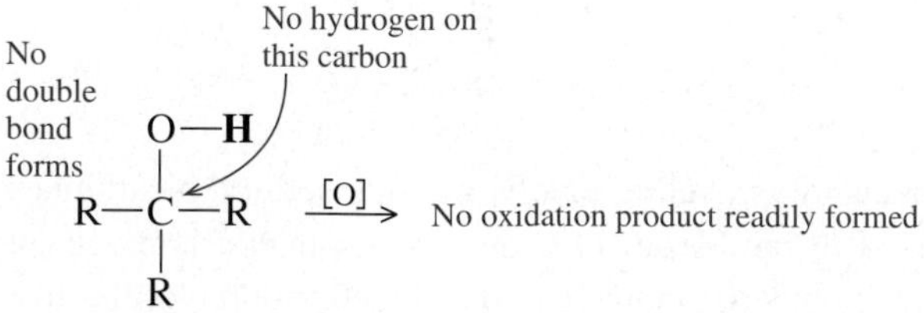

No hydrogen on
this carbon

$$
\underset{\text{3° Alcohol}}{\text{No double bond forms} \underset{\underset{\text{R}}{|}}{\overset{\overset{\text{O—H}}{|}}{R\text{—}C\text{—}R}}} \xrightarrow{[O]} \text{No oxidation product readily formed}
$$

SAMPLE PROBLEM 14.13

Oxidation of Alcohols

Draw the structural formulas of the organic products that form when the following alcohols undergo oxidation.

OH
|
a. $CH_3CH_2CHCH_3$ —[O]→   **b.** $CH_3CH_2CH_2OH$ —[O]→

Solution

a. Because the reactant is a secondary alcohol, oxidation produces a ketone and water.

O
‖
$CH_3CH_2CCH_3$

b. This primary alcohol can oxidize to an aldehyde and further oxidize to a carboxylic acid.

O   O
‖   ‖
CH_3CH_2C—H CH_3CH_2C—OH

Study Check

When propene is hydrated in the presence of an acid cataylst, the product forms a ketone when it is oxidized. Explain using equations.

HEALTH NOTE

Oxidation of Alcohol in the Body

Ethanol is the most commonly abused drug in the United States. When ingested in small amounts, ethanol may produce a feeling of euphoria in the body although it is a depressant. In the liver, enzymes such as alcohol dehydrogenases oxidize ethanol to acetaldehyde, a substance that impairs mental and physical coordination. If the blood alcohol concentration exceeds 0.4%, coma or death may occur. Table 14.2 gives some of the typical behaviors exhibited at various levels of blood alcohol.

O
‖
CH_3CH_2OH —[O]→ CH_3CH —[O]→ $2CO_2 + H_2O$
Ethanol   Ethanal
(ethyl alcohol)   (acetaldehyde)

The acetaldehyde produced from ethanol in the liver is further oxidized to acetic acid, which is converted to carbon dioxide and water in the citric acid (Krebs) cycle. Thus, the enzymes in the liver

Table 14.2 Typical Behaviors Exhibited by a 150-lb Person Consuming Alcohol

Number of Beers (12 oz) or Glasses of Wine (5 oz)	Blood Alcohol Level (w/v %)	Typical Behavior
1	0.025	Slightly dizzy, talkative
2	0.05	Euphoria, loud talking, and laughing
4	0.10	Loss of inhibition, loss of coordination, drowsiness, legally drunk in most states
8	0.20	Intoxicated, quick to anger, exaggerated emotions
12	0.30	Unconscious
16–20	0.40–0.50	Coma and death

can eventually break down ethanol, but the aldehyde and carboxylic acid intermediates can cause considerable damage while they are present within the cells of the liver.

A person weighing 150 lb requires about one hour to completely metabolize 10 ounces of beer. However, the rate of metabolism of ethanol varies between nondrinkers and drinkers. Typically, nondrinkers and social drinkers can metabolize 12–15 mg of ethanol/dL of blood in one hour, but an alcoholic can metabolize as much as 30 mg of ethanol/dL in one hour. Some effects of alcohol metabolism include an increase in liver lipids (fatty liver), an increase in serum triglycerides, gastritis, pancreatitis, ketoacidosis, alcoholic hepatitis, and psychological disturbances.

When the Breathalyzer test is used for suspected drunk drivers, the driver exhales a volume of breath into a solution containing the orange Cr^{6+} ion. If there is ethyl alcohol present in the exhaled air, the alcohol is oxidized, and the Cr^{6+} is reduced to give a green solution of Cr^{3+}.

O
‖
$CH_3CH_2OH + Cr^{6+}$ —[O]→ $CH_3COH + Cr^{3+}$
Ethanol   Orange   Acetic acid   Green

Sometimes alcoholics are treated with a drug called Antabuse (disulfiram), which prevents the oxidation of acetaldehyde to acetic acid. As a result, acetaldehyde accumulates in the blood, which causes nausea, profuse sweating, headache, dizziness, vomiting, and respiratory difficulties. Because of these unpleasant side effects, the patient is less likely to use alcohol.

Oxidation of Thiols

Thiols also undergo oxidation by a loss of hydrogen atoms from the —SH groups. The oxidized product is called a **disulfide.**

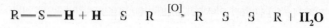

Example

$$CH_3—S—H + H—S—CH_3 \xrightarrow{[O]} CH_3—S—S—CH_3 + H_2O$$

Methanethiol   Dimethyl disulfide

Much of the protein in the hair is cross-linked by disulfide bonds, which occur mostly between the thiol groups of the amino acid cysteine.

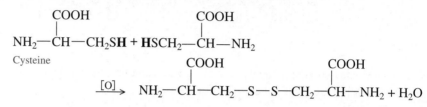

When a person is given a "perm," a reducing substance is used to break the disulfide bonds. While the hair is still wrapped around the curlers, an oxidizing substance is applied that causes new disulfide bonds to form between different parts of the protein hair strands, which gives the hair a new shape.

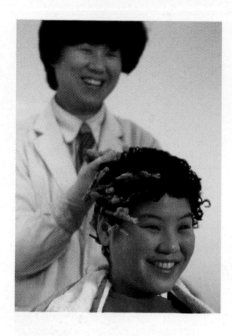

QUESTIONS AND PROBLEMS

Reactions of Alcohols

14.27 Draw the condensed structural formula of the alkene produced by each of the following dehydration reactions:

a. $CH_3—CH_2—CH_2—CH_2—OH \xrightarrow[\text{Heat}]{H^+}$

b.
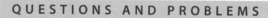

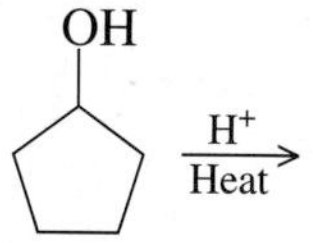

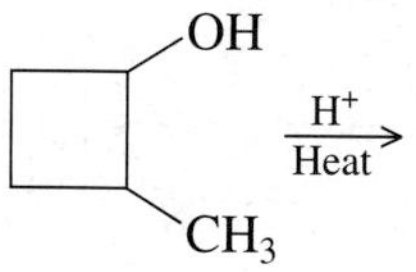

d.

$$CH_3—CH_2—CH_2—\overset{\displaystyle OH}{\underset{|}{CH}}—CH_3 \xrightarrow[\text{Heat}]{H^+}$$

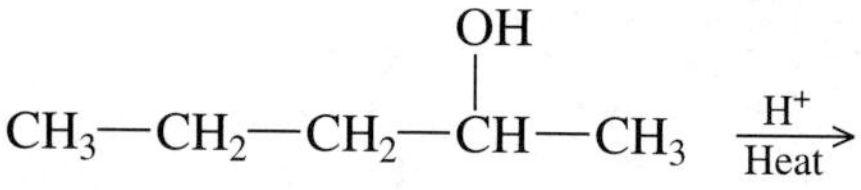

14.28 Draw the condensed structural formula of the alkene produced by each of the following dehydration reactions:

a.

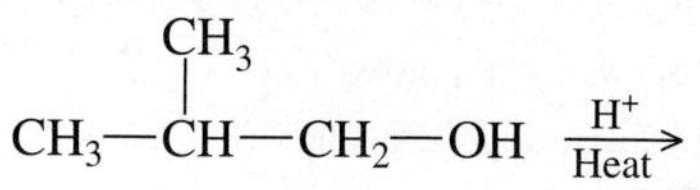

$$CH_3—\overset{\displaystyle CH_3}{\underset{|}{CH}}—CH_2—OH \xrightarrow[\text{Heat}]{H^+}$$

b.

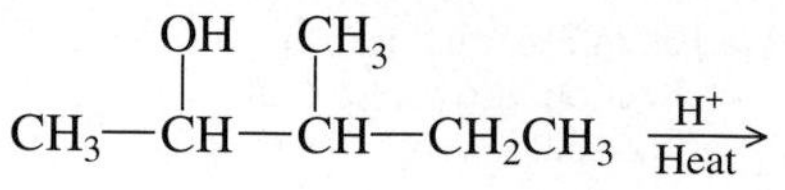

$$CH_3—\overset{\displaystyle OH}{\underset{|}{CH}}—\overset{\displaystyle CH_3}{\underset{|}{CH}}—CH_2CH_3 \xrightarrow[\text{Heat}]{H^+}$$

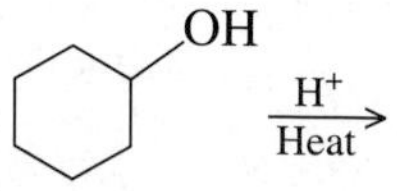

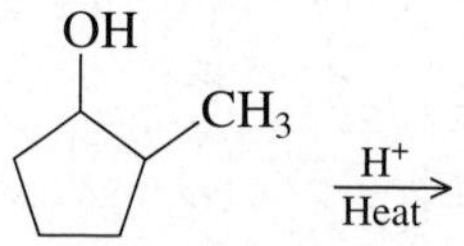

14.29 Write the ether product from the reaction of each of the following:

a. $2CH_3OH \xrightarrow[\text{Heat}]{H^+}$

b. $2CH_3CH_2CH_2OH \xrightarrow[\text{Heat}]{H^+}$

14.30 Write the ether product from the reaction of each of the following:

a. $2CH_3CH_2OH \xrightarrow[\text{Heat}]{H^+}$

b.
$$\begin{array}{c} CH_3 \\ | \\ 2CH_3CHCH_2OH \end{array} \xrightarrow[\text{Heat}]{H^+}$$

14.31 What alcohol(s) could be used to produce each of the following compounds?

a. $CH_2{=}CH_2$

b. $CH_3{-}O{-}CH_2CH_3$

c.

14.32 What alcohol(s) could be used to produce each of the following compounds?

a. $CH_3CH_2{-}O{-}CH_2CH_3$

b.
$$\begin{array}{c} CH_3 \\ | \\ CH_3{-}C{=}CHCH_3 \end{array}$$

c.

14.33 Draw the condensed structural formula of the organic product when each of the following alcohols is oxidized [O] (if no reaction, write *none*):

a. $CH_3CH_2CH_2CH_2CH_2OH$

b.
$$\begin{array}{c} OH \\ | \\ CH_3CH_2CHCH_3 \end{array}$$

c.

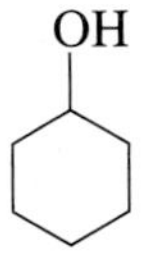

d.
$$\begin{array}{c} OH \qquad\quad CH_3 \\ | \qquad\qquad | \\ CH_3{-}CH{-}CH_2{-}CH{-}CH_3 \end{array}$$

e.
$$\begin{array}{c} CH_3 \\ | \\ CH_3{-}CH{-}CH_2{-}CH_2{-}OH \end{array}$$

14.34 Draw the condensed structural formula of the organic product when each of the following alcohols is oxidized [O] (if no reaction, write *none*):

a.

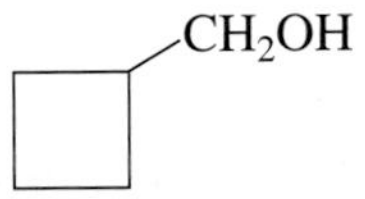

b.
$$\begin{array}{c} CH_3 \\ | \\ CH_3{-}CH{-}CH_2{-}CH{-}OH \\ | \\ CH_3 \end{array}$$

c.
$$\begin{array}{c} OH \\ | \\ CH_3{-}CH_2{-}C{-}CH_3 \\ | \\ CH_3 \end{array}$$

d.
$$\begin{array}{c} OH \\ | \\ CH_3CHCHCH_2CH_3 \\ | \\ OH \end{array}$$

e.

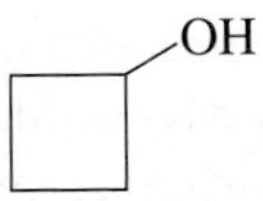

14.35 Draw the condensed structural formula of the alcohol needed to give each of the following oxidation products:

a.
$$\begin{array}{c} O \\ \| \\ H{-}C{-}H \end{array}$$

b.

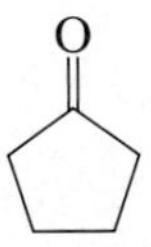

c.
$$\begin{array}{c} O \\ \| \\ CH_3{-}C{-}CH_2{-}CH_3 \end{array}$$

d. (benzaldehyde structure)   e. (3-methylcyclohexanone structure)

14.36 Draw the condensed structural formula of the alcohol needed to give each of the following oxidation products:

$$
\text{a.} \quad CH_3-\overset{\overset{\displaystyle O}{\|}}{C}-H
\qquad
\text{b.} \quad CH_3-\overset{\overset{\displaystyle O}{\|}}{C}-\overset{\overset{\displaystyle CH_3}{|}}{CH}-CH_3
\qquad
\text{c.} \quad \text{(cyclohexanone structure)}
$$

$$
\text{d.} \quad CH_3-CH_2-\overset{\overset{\displaystyle O}{\|}}{C}-H
\qquad
\text{e.} \quad CH_3-\overset{\overset{\displaystyle CH_3}{|}}{CH}-CH_2-\overset{\overset{\displaystyle O}{\|}}{C}-H
$$

Chapter Review

14.1 Structure and Classification of Alcohols
The functional group of an alcohol is the hydroxyl group —OH bonded to a carbon chain. In a phenol, the hydroxyl group is bonded to an aromatic ring. Alcohols are classified according to the number of alkyl or aromatic groups bonded to the carbon that holds the —OH group. In a primary (1°) alcohol one alkyl group is attached to the hydroxyl carbon. In a secondary (2°) alcohol two alkyl groups are attached, and in a tertiary (3°) alcohol there are three alkyl groups bonded to the hydroxyl carbon. In thiols, the functional group is —SH, which is analogous to the —OH group of alcohols.

14.2 Naming Alcohols, Phenols, and Thiols
In the IUPAC system, the names of alcohols have *ol* endings, and the location of the —OH group is given by numbering the carbon chain. Simple alcohols are generally named by their common names with the alkyl name preceding the term *alcohol*. A cyclic alcohol is named as a cycloalkanol. An aromatic alcohol is named as a phenol.

14.3 Some Important Alcohols and Phenols
Methanol, the simplest alcohol, is used as a solvent and as a fuel in race cars. It is toxic if ingested. Ethanol is found in alcoholic beverages and used as a solvent for perfumes and varnishes. Isopropyl alcohol or rubbing alcohol is used as an astringent and to sterilize medical equipment. Menthol is a cyclohexanol with a peppermint taste that is used in candy and throat lozenges. Phenol is used in antiseptics and disinfectants. Many plants produce oils containing derivatives of phenol such as eugenol in cloves and thymol in thyme.

14.4 Ethers
In an ether, an oxygen atom is connected by single bonds to two alkyl or aromatic groups, R—O—R. In the common names of ethers, the alkyl groups are listed alphabetically followed by the name *ether*. In the IUPAC name, the smaller alkyl group with the oxygen is named as an *alkoxy group* and is attached to the longer alkane chain, which is numbered to give the location of the alkoxy group.

14.5 Physical Properties of Alcohols, Phenols, and Ethers
The —OH group allows alcohols to hydrogen bond, which causes alcohols to have higher boiling points than alkanes and ethers of similar mass. Short-chain alcohols and ethers can hydrogen bond with water, which makes them soluble in water.

14.6 Reactions of Alcohols
At high temperatures, alcohols dehydrate in the presence of an acid to yield alkenes. At lower temperatures, two molecules of alcohol lose H and OH to produce an ether. Primary alcohols are oxidized to aldehydes, which can oxidize further to carboxylic acids. Secondary alcohols are oxidized to ketones. Tertiary alcohols do not oxidize.

Summary of Naming

Structure	Family	IUPAC Name	Common Name
CH_3—**OH**	Alcohol	Methanol	Methyl alcohol
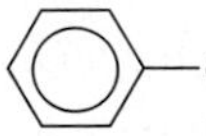 ⬡—**OH**	Phenol	Phenol	Phenol
CH_3—**SH**	Thiol	Methanethiol	
CH_3—**O**—CH_3	Ether	Methoxymethane	Dimethyl ether
 (furan ring)	Cyclic ether	Furan	
CH_3—**S**—**S**—CH_3	Disulfide	Dimethyldisulfide	

Summary of Reactions

Combustion of Alcohols

$$CH_3—CH_2—OH + 3O_2 \longrightarrow 2CO_2 + 3H_2O$$

Ethanol   Oxygen   Carbon dioxide   Water

Intramolecular Dehydration of Alcohols to Form Alkenes

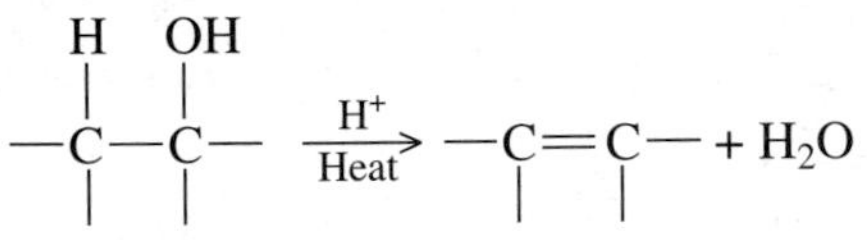

Alcohol   Alkene

$$CH_3—CH_2—CH_2—OH \xrightarrow[\text{Heat}]{H^+}$$

1-Propanol

$$CH_3—CH{=}CH_2 + H_2O$$

Propene

Intermolecular Dehydration of Alcohols to Form Ethers

$$R—OH + HO—R \xrightarrow[\text{Heat}]{H^+} R—O—R + H_2O$$

Two alcohols   Ether

$$CH_3—OH + HO—CH_3 \xrightarrow[\text{Heat}]{H^+}$$

Methanol

$$CH_3—O—CH_3 + H_2O$$

Dimethyl ether

Oxidation of Primary Alcohols to Form Aldehydes

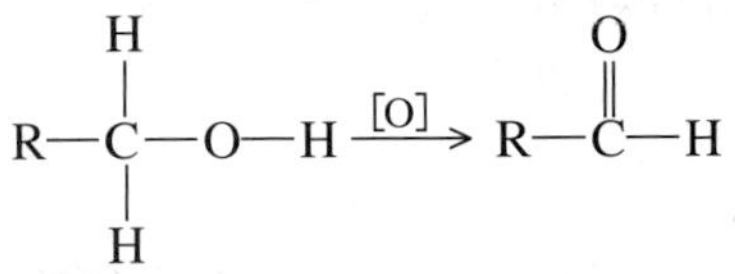

Primary alcohol   Aldehyde

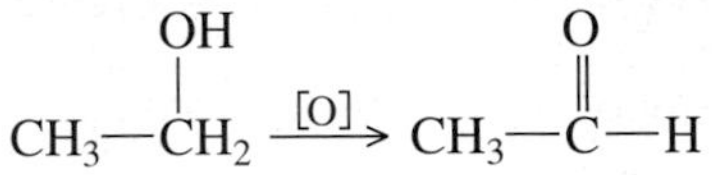

Ethanol   Acetaldehyde

Oxidation of Secondary Alcohols to Form Ketones

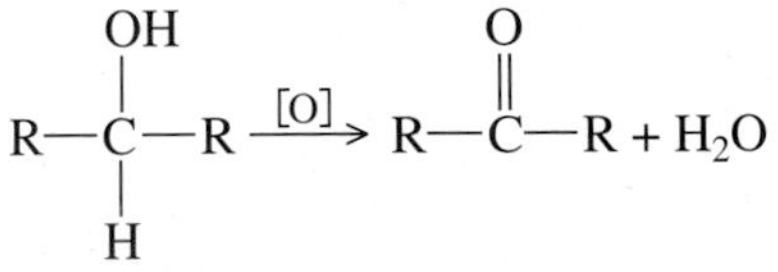

Secondary alcohol   Ketone

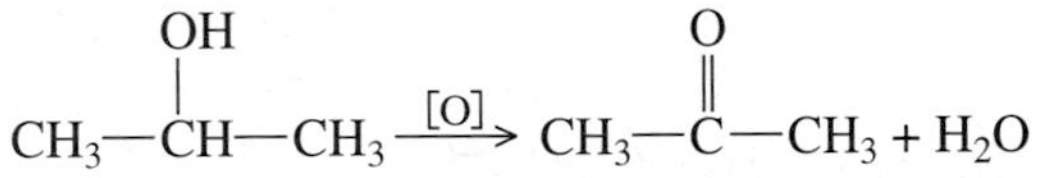

2-Propanol   Propanone

Oxidation of Thiols to Form Disulfides

$$R—S—H + H—S—R \xrightarrow{[O]} R—S—S—R + H_2O$$

Two thiols   Disulfide

$$CH_3—S—H + H—S—CH_3 \xrightarrow{[O]} CH_3—S—S—CH_3 + H_2O$$

Methanethiol   Dimethyl disulfide

Key Terms

alcohols   Organic compounds that contain the hydroxyl (—OH) functional group, attached to a carbon chain, R—OH.

alkoxy group   A group that contains oxygen bonded to an alkyl group.

cyclic ethers   Compounds that contain an oxygen atom in a carbon ring.

dehydration   A reaction that removes water from an alcohol in the presence of an acid to form alkenes at high temperature, or ethers at lower temperatures.

disulfides   Compounds formed from thiols, disulfides contain the —S—S— functional group.

ether   An organic compound, R—O—R, in which an oxygen atom is bonded to two alkyl or two aromatic groups, or a mix of the two.

furan   An unsaturated cyclic ether that is a five-atom ring with an oxygen atom.

heterocyclic compounds   Compounds composed of a ring that contains one or more atoms other than carbon.

hydroxyl group   The —OH functional group.

oxidation   The loss of two hydrogen atoms from a reactant to give a more oxidized compound, e.g., primary alcohols oxidize to aldehydes, secondary alcohols oxidize to ketones. An oxidation can also be the addition of an oxygen atom as in the oxidation of aldehydes to carboxylic acids.

phenol   An organic compound that has an —OH group attached to a benzene ring.

primary (1°) alcohol   An alcohol that has one alkyl group bonded to the alcohol carbon atom, R—CH$_2$OH.

Saytzeff's rule   In the dehydration of an alcohol, hydrogen is removed from the carbon that already has the smallest number of hydrogen atoms to form an alkene.

secondary (2°) alcohol   An alcohol that has two alkyl groups bonded to the carbon atom with the —OH group.

sulfhydryl group   The —SH functional group.

tertiary (3°) alcohol   An alcohol that has three alkyl groups bonded to the carbon atom with the —OH.

thiols   Organic compounds that contain a thiol group (—SH) in place of the —OH group of an alcohol.

Additional Problems

14.37 Classify each of the following as primary, secondary, or tertiary alcohols:

a. (cyclohexane with OH)

b. (cyclohexane with CH$_2$—OH)

c. CH$_3$—CH(CH$_3$)—CH$_2$—OH

d. CH$_3$—C(CH$_3$)(CH$_3$)—CH$_2$—CH(OH)—CH$_3$

e. HO—CH$_2$—CH$_2$—CH$_3$

f. (cyclopentane with C(CH$_3$)(CH$_3$)—OH, C—CH$_3$)

14.38 Classify each of the following as primary, secondary, or tertiary alcohols:

a. (cyclohexane with OH and CH$_3$)

b. (cyclohexane with OH and CH$_2$—CH$_3$)

c. CH$_3$—CH(CH$_2$—OH)—CH$_2$—CH$_3$

d. CH$_3$—C(OH)(CH$_3$)—CH$_2$—CH(CH$_3$)—CH$_3$

e. CH$_3$—CH$_2$—CH$_2$—CH$_2$—OH

f. (cyclopentane with CH(CH$_3$)—OH)

14.39 Identify each of the following as an alcohol, a phenol, an ether, a cyclic ether, or a thiol:

a.

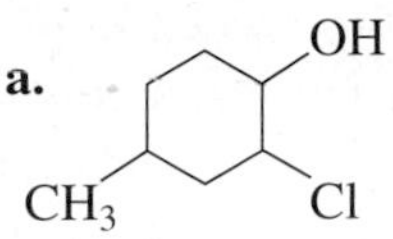

b.

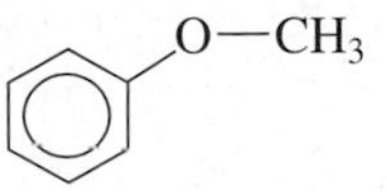

c. 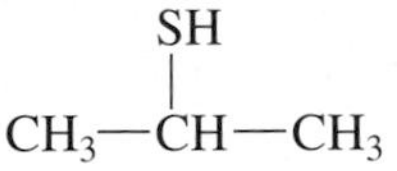 CH₃—CH—CH₃ (with SH)

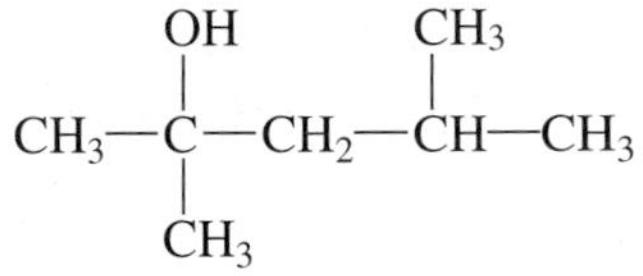

$$\text{SH}$$
$$\text{c. } CH_3-\overset{|}{CH}-CH_3$$

d.
$$\text{d. } CH_3-\overset{\overset{OH}{|}}{\underset{\underset{CH_3}{|}}{C}}-CH_2-\overset{\overset{CH_3}{|}}{CH}-CH_3$$

e. CH₃—CH₂—CH₂—O—CH₃

f.

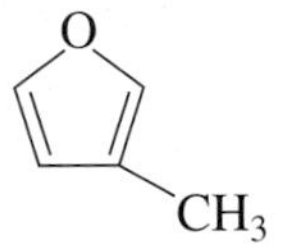

g. 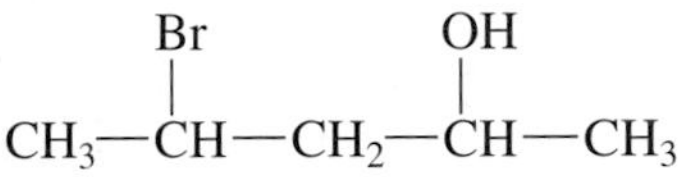
$$\text{g. } CH_3-\overset{\overset{Br}{|}}{CH}-CH_2-\overset{\overset{OH}{|}}{CH}-CH_3$$

h.

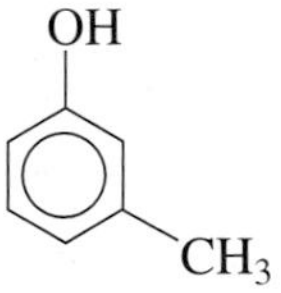

14.40 Identify each of the following as an alcohol, a phenol, an ether, a cyclic ether, or a thiol:

a.

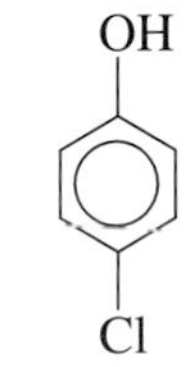

b. CH₃—CH₂—CH₂—SH

c.

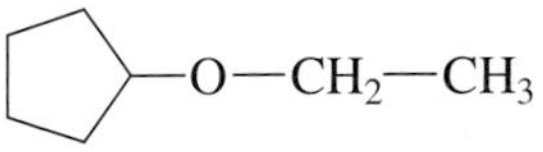

d.
$$\text{d. } CH_3-\overset{\overset{OH}{|}}{\underset{\underset{CH_3}{|}}{C}}-CH_2-\overset{\overset{CH_3}{|}}{CH}-CH_3$$

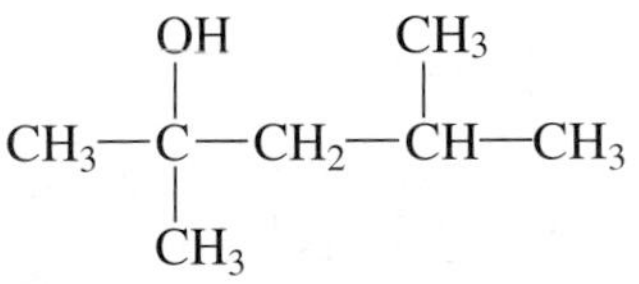

e.
$$\text{e. } CH_3-CH_2-\overset{\overset{O-CH_3}{|}}{CH}-CH_2-CH_3$$

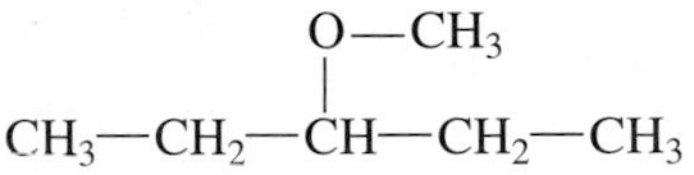

f.

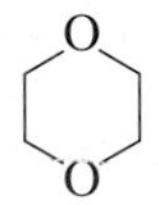

g.

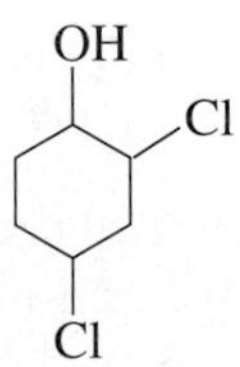

h.

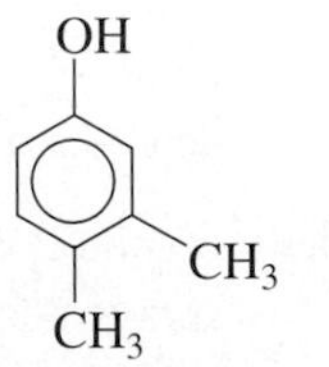

14.41 Give the IUPAC and common names (if any) for each of the compounds in problem 14.39.

14.42 Give the IUPAC and common names (if any) for each of the compounds in problem 14.40.

14.43 Draw the condensed structural formula of each of the following compounds:
  a. 3-methylcyclopentanol   **b.** *p*-chlorophenol
  c. 2-methyl-3-pentanol   **d.** phenyl ethyl ether
  e. 3-pentanethiol   **f.** *ortho*-cresol
  g. 2,4-dibromophenol

14.44 Draw the condensed structural formula of each of the following compounds:
  a. 3-methoxypentane   **b.** *meta*-chlorophenol
  c. 2,3-pentanediol   **d.** methyl propyl ether
  e. methanethiol   **f.** 3-methyl-2-butanol
  g. 3,4-dichlorocyclohexanol

14.45 Identify the alcohol described by each of the following:
  a. a syrupy liquid used in skin lotions
  b. used in antifreeze
  c. produced by fermentation of grains and sugars

14.46 Identify the alcohol described by each of the following:
  a. produced by the reaction of carbon monoxide and hydrogen at high temperatures
  b. used to clean the skin before an injection
  c. the "alcohol" in alcoholic drinks

14.47 Draw the structural formulas of all the alcohols with a molecular formula $C_4H_{10}O$.

14.48 Draw the structural formulas of all the ethers with a molecular formula $C_5H_{12}O$.

14.49 Which compound in each pair would you expect to have the higher boiling point?
  a. butane or 1-propanol
  b. 1-propanol or methyl ethyl ether
  c. ethanol or 1-butanol

14.50 Which compound in each pair would you expect to have the higher boiling point?
  a. propane or ethyl alcohol
  b. 2-propanol or 2-pentanol
  c. diethyl ether or 1-butanol

14.51 Explain why each of the following compounds would be soluble or insoluble in water.
  a. 2-propanol
  b. dimethyl ether
  c. 1-hexanol

14.52 Explain why each of the following compounds would be soluble or insoluble in water.
a. glycerol
b. butane
c. 1,3-hexanediol

14.53 Draw the condensed structural formula for the major product of each of the following reactions:

a. $CH_3-CH_2-CH_2-OH \xrightarrow{H^+, \text{ heat}}$

b. $CH_3-CH_2-CH_2-OH \xrightarrow{[O]}$

c. $CH_3-CH_2-\overset{\overset{\displaystyle OH}{|}}{CH}-CH_3 \xrightarrow{H^+, \text{ heat}}$

d. $CH_3-CH_2-\overset{\overset{\displaystyle OH}{|}}{CH}-CH_3 \xrightarrow{[O]}$

e. $2CH_3-CH_2-CH_2-OH \xrightarrow{H^+}$

f.

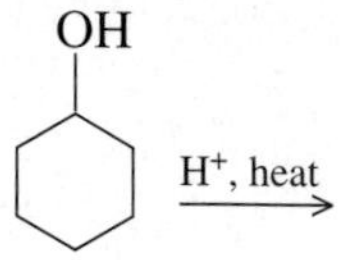

![cyclohexanol] $\xrightarrow{H^+, \text{ heat}}$

g. ![cyclohexanol] $\xrightarrow{[O]}$

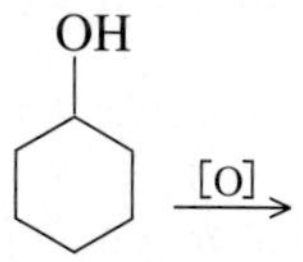

14.54 Draw the condensed structural formula for the major product of each of the following reactions:

a. $CH_3-\overset{\overset{\displaystyle CH_3}{|}}{CH}-CH_2-OH \xrightarrow{H^+, \text{ heat}}$

b. $CH_3-\overset{\overset{\displaystyle CH_3}{|}}{CH}-\overset{\overset{\displaystyle OH}{|}}{CH}-CH_3 \xrightarrow{H^+, \text{ heat}}$

c. $CH_3-\overset{\overset{\displaystyle CH_3}{|}}{CH}-\overset{\overset{\displaystyle OH}{|}}{CH}-CH_3 \xrightarrow{[O]}$

d. ![2-methylcyclopentanol with OH and CH3] $\xrightarrow{H^+, \text{ heat}}$

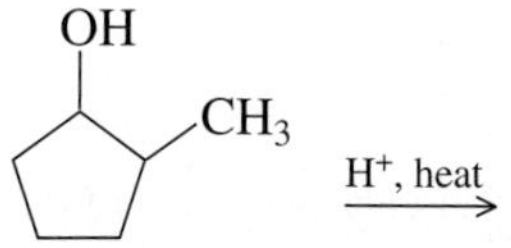

e. ![2-methylcyclopentanol with OH and CH3] $\xrightarrow{[O]}$

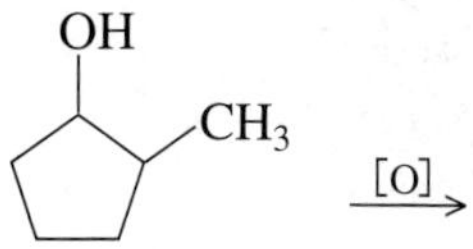

f. $2CH_3-CH_2-OH \xrightarrow{H^+, \text{ heat}}$

g. $CH_3-CH_2-CH_2-\overset{\overset{\displaystyle OH}{|}}{CH}-CH_3 \xrightarrow{H^+, \text{ heat}}$

14.55 Sometimes several steps are needed to prepare a compound. Using a combination of the reactions we have studied, indicate how you might prepare the following from the starting substance given. For example, 2-propanol could be prepared from 1-propanol by first dehydrating the alcohol to give propene and then hydrating it again to give 2-propanol according to Markovnikov's rule.

$$CH_3-CH_2-CH_2-OH \xrightarrow{H^+, \text{ heat}} CH_3-CH=CH_2 + H_2O \xrightarrow{H^+}$$

$$\underset{\text{2-Propanol}}{CH_3-\overset{\overset{\displaystyle OH}{|}}{CH}-CH_3}$$

a. Prepare 2-chloropropane from 1-propanol.
b. Prepare 2-methylpropane from 2-methyl-2-propanol.
c. Prepare $CH_3-\overset{\overset{\displaystyle O}{\|}}{C}-CH_3$ from 1-propanol.

14.56 As in problem 14.55, indicate how you might prepare the following from the starting substance given:
a. Prepare 1-pentene from 1-pentanol.
b. Prepare chlorocyclohexane from cyclohexanol.
c. Prepare 1,2-dibromobutane from 1-butanol.

14.57 Identify the functional groups in the following molecule:

![Testosterone structure]

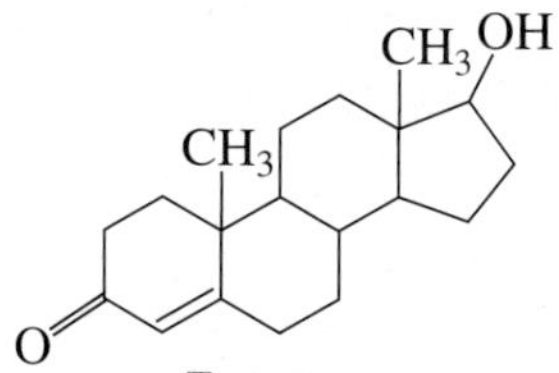

Testosterone

14.58 Identify the functional groups in the following molecule:

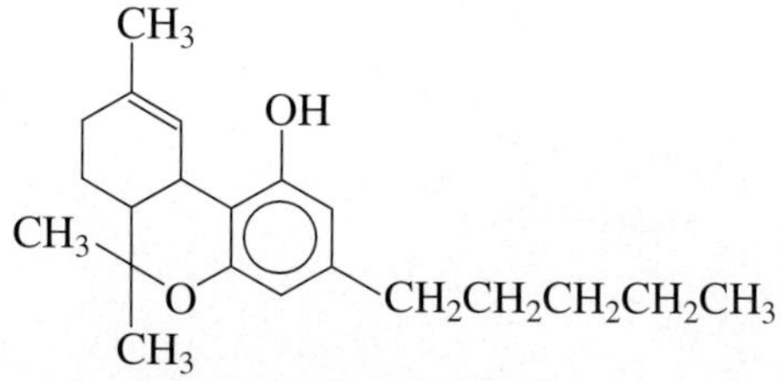

![Tetrahydrocannabinol structure]

Tetrahydrocannabinol (THC)

14.59 Hexylresorcinol, an antiseptic ingredient used in mouthwashes and throat lozenges, has the IUPAC name of 4-hexyl-1,3-benzenediol. Draw its condensed structural formula.

14.60 Menthol, which has a minty flavor, is used in throat sprays and lozenges. Thymol is used as a topical antiseptic to destroy mold. Give each of their IUPAC names. What is similar and what is different about their structures?

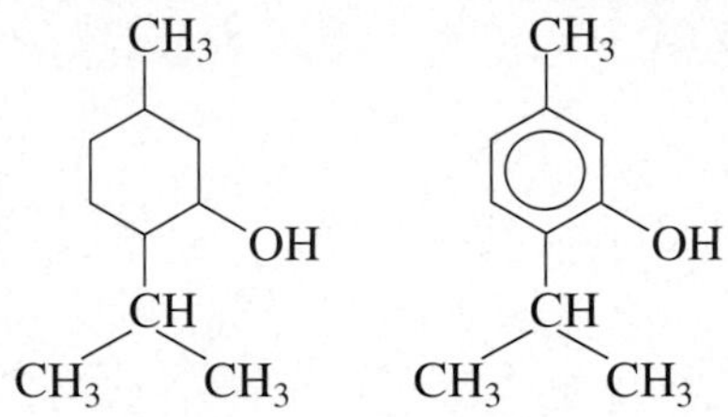

Menthol   Thymol

14.61 Write the condensed structural formulas for each of the following naturally occurring compounds:

a. 2,5-dichlorophenol, a defense pheromone of a grasshopper.

b. Skunk scent, a mixture of 3-methyl-1-butanethiol and *trans*-2-butene-1-thiol.

c. Pentachlorophenol, which is a wood preservative.

14.62 Dimethyl ether and ethyl alcohol both have the molecular formula C_2H_6O. One has a boiling point of $-24°C$, and the other, $79°C$. Draw the condensed structural formulas of each compound. Decide which boiling point goes with which compound and explain. Check the boiling points in a chemistry handbook.

15 Aldehydes, Ketones, and Chiral Molecules

"Dentures replace natural teeth that are extracted due to cavities, bad gums, or trauma," says Dr. Irene Hilton, dentist, La Clinica De La Raza. "I make an impression of teeth using alginate, which is a polysaccharide extracted from seaweed. I mix the compound with water and place the gel-like material in the patient's mouth, where it becomes a hard, cement-like substance. I fill this mold with gypsum ($CaSO_4$) and water, which form a solid to which I add teeth made of plastic or porcelain. When I get a good match to the patient's own teeth, I prepare a preliminary wax denture. This is placed in the patient's mouth to check the bite and adjust the position of the replacement teeth. Then a permanent denture is made using a hard plastic polymer (methyl methacrylate)."

LOOKING AHEAD

15.1 **Structure and Bonding**

15.2 **Naming Aldehydes and Ketones**

15.3 **Some Important Aldehydes and Ketones**

15.4 **Physical Properties**

15.5 **Chiral Molecules**

15.6 **Oxidation and Reduction**

15.7 **Addition Reactions**

Many of the odors and flavors that you associate with solvents, flavorings, paint removers, and perfumes are from organic molecules called aldehydes and ketones. In biology, you may have seen specimens preserved in a solution of formaldehyde. You may have noticed the odor of acetone if you used paint or polish remover. Foods and perfumes with the odors and flavors of vanilla, almond, and cinnamon contain aldehydes that occur in nature.

The feature responsible for this wide variety of flavors and odors is a carbon–oxygen double bond called a **carbonyl group** (C=O). In this chapter we will study two families with carbonyl groups: aldehydes and ketones. They are prevalent in nature and play an important role in biochemical pathways. In later chapters, we will see how the carbonyl group influences the structures of carbohydrates, proteins, and nucleic acids. Aldehydes and ketones are also important compounds in industry, providing the solvents and reactants that make up many common materials we use in our lives.

15.1 Structure and Bonding

In an **aldehyde,** the carbon of the carbonyl group is bonded to at least one hydrogen atom. That carbon may also be bonded to another hydrogen, a carbon of an alkyl group, or an aromatic ring. (See Figure 15.1.) In a **ketone,** the carbonyl group is bonded to two alkyl groups or aromatic rings.

Writing Aldehyde and Ketone Structures

There are several ways to write the structural formulas of aldehydes and ketones. In the condensed structural formula, the aldehyde group may be drawn as separate atoms or it may be written as —CHO, with the double bond understood. An aldehyde would not be written as —COH, which looks like a hydroxy group. The keto

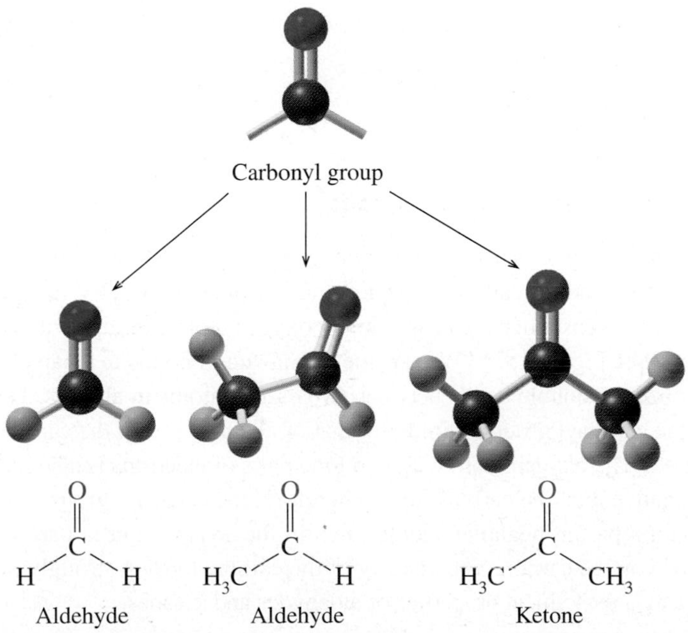

Carbonyl group

Aldehyde   Aldehyde   Ketone

Figure 15.1 The carbonyl group in aldehydes and ketones.
Q *If aldehydes and ketones both contain a carbonyl group, how can you differentiate between compounds from each family?*

group (C=O) is sometimes written as CO. For convenience, the line-bond formulas are also used with the functional group atoms written separately. Constitutional isomers, which have the same molecular formula, can be written for aldehydes and ketones as follows:

Formulas for Constitutional Isomers of C₃H₆O

Aldehyde

$$\text{CH}_3\text{—CH}_2\text{—}\overset{\overset{\displaystyle O}{\|}}{\text{C}}\text{—H} \quad = \quad \text{CH}_3\text{—CH}_2\text{—CHO} \quad =$$

Ketone

$$\text{CH}_3\text{—}\overset{\overset{\displaystyle O}{\|}}{\text{C}}\text{—CH}_3 \quad = \quad \text{CH}_3\text{—CO—CH}_3 \quad =$$

SAMPLE PROBLEM 15.1

Identifying Aldehydes and Ketones

Identify each of the following compounds as an aldehyde or ketone:

a.
$$\text{CH}_3\text{—}\overset{\overset{\displaystyle \text{CH}_3}{|}}{\underset{\underset{\displaystyle \text{CH}_3}{|}}{\text{C}}}\text{—CH}_2\text{—}\overset{\overset{\displaystyle O}{\|}}{\text{C}}\text{—H}$$

b.
(cyclopentyl)—C(=O)—CH₃

c.
(benzene ring)—CHO

d.
(line structure ketone)

Solution

a. aldehyde   **b.** ketone   **c.** aldehyde   **d.** ketone

Study Check

Draw the condensed structural formula of a ketone that has a carbonyl group bonded to two ethyl groups.

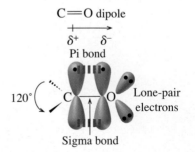

C=O dipole

δ⁺   δ⁻

Pi bond

120°   C   O   Lone-pair electrons

Sigma bond

Figure 15.2 The sigma and pi bond in the carbon–oxygen double bond of the carbonyl group.
Q *Why does the carbonyl bond have a dipole?*

Structure of the Carbonyl Group

The carbonyl group consists of a carbon–oxygen double bond. Sigma bonds at angles of 120° bond the carbon to oxygen and two other atoms. The second bond is a pi bond that forms when a *p* orbital in the oxygen atom overlaps with a *p* orbital in carbon. (See Figure 15.2.) We saw the sigma and pi bonds in Chapter 13 when we described the double bonds between two carbon atoms in alkenes. The double bond in the carbonyl group is similar to that of alkenes, except the carbonyl group has a dipole. The oxygen atom with two lone pairs of electrons is also much more electronegative than the carbon atom. Therefore, the carbonyl group has a strong dipole with a partial negative charge (δ^-) on the oxygen, and a partial positive charge (δ^+) on the carbon. The polarity of the carbonyl group strongly influences the physical and chemical properties of aldehydes and ketones.

SAMPLE PROBLEM 15.2

Carbonyl Group

Using the dipole symbol, $+\!\!\longrightarrow$, illustrate the polarity of the carbonyl group in each of the following compounds:

a. 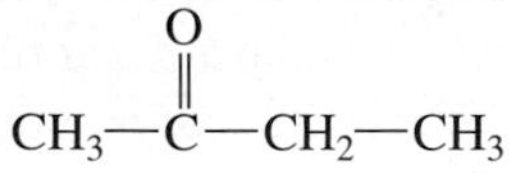 CH₃—C(=O)—CH₂—CH₃   b. 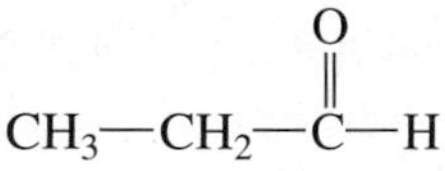 CH₃—CH₂—C(=O)—H

Solution

a. 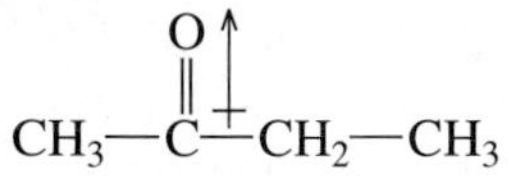 CH₃—C(=O↑)—CH₂—CH₃   b. 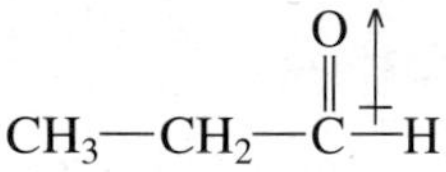 CH₃—CH₂—C(=O↑)—H

Study Check

Why does the carbonyl group have a dipole?

QUESTIONS AND PROBLEMS

Structure and Bonding

15.1 Identify the following compounds as aldehydes or ketones:

a. 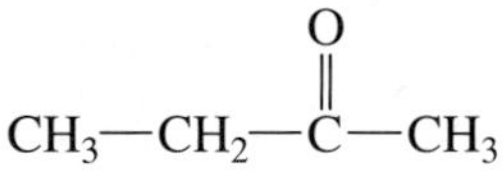 CH₃—CH₂—C(=O)—CH₃   b. [structure with CHO]   c. [cyclopentanone with CH₃]   d. 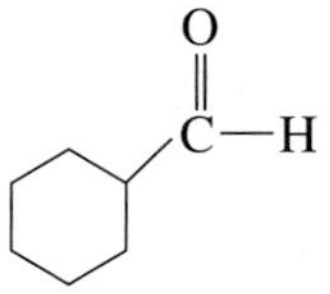 [cyclohexane with C—H carbonyl]

15.2 Identify the following compounds as aldehydes or ketones:

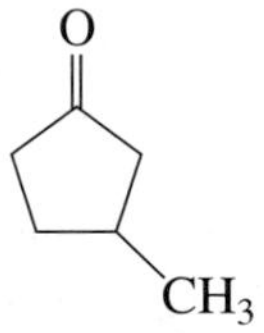

a. 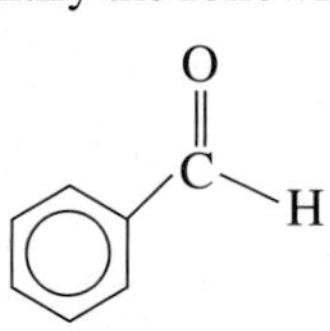 [benzaldehyde structure C—H]   b. CH₃—CH(CH₃)—C(=O)—H   c. [branched ketone structure]   d. 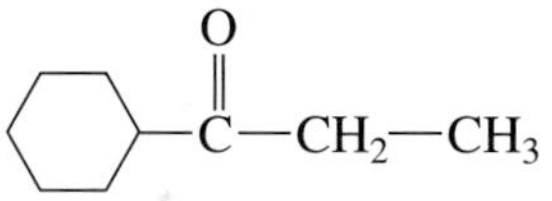 [cyclohexane-C(=O)—CH₂—CH₃]

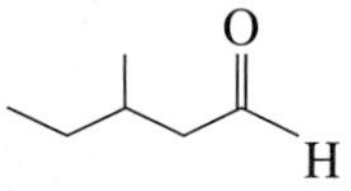

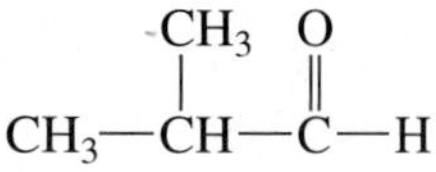

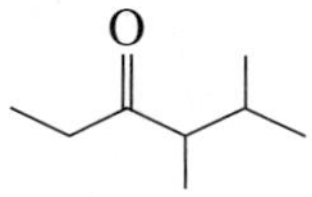

15.3 Indicate if each of the following pairs of structural formulas represent (1) constitutional isomers, (2) the same compound, or (3) different compounds.

a. 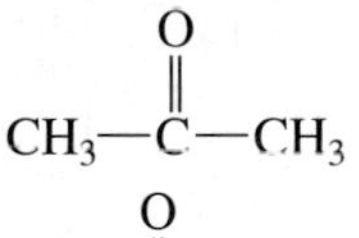 CH₃—C(=O)—CH₃ and CH₃—CH₂—C(=O)—H   b. [structure] and [structure]

c. CH₃—C(=O)—CH₂—CH₃ and CH₃—CH₂—C(=O)—CH₃

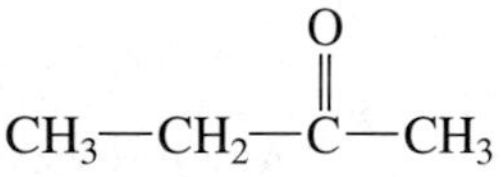

15.4 Indicate if each of the following pairs of structural formulas represent (1) constitutional isomers, (2) the same compound, or (3) different compounds.

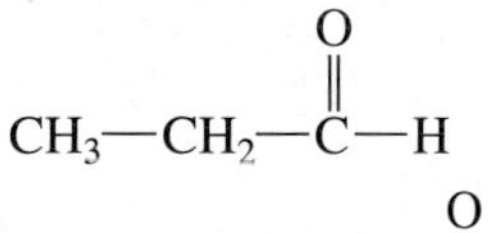

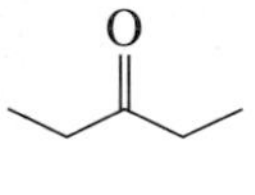

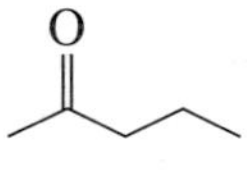

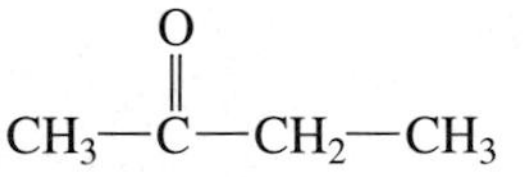

a. CH₃—CH₂—CH₂—C(=O)—H and [cyclobutane-C—H structure]   b. [ether structure] ～O～ and [structure]

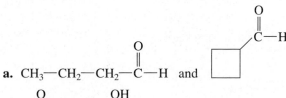

c. [cyclohexanone] and [cyclohexenol with OH]

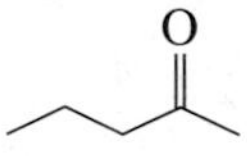

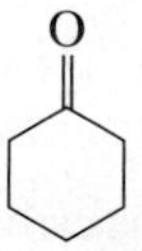

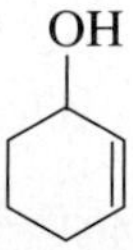

15.2 Naming Aldehydes and Ketones

Aldehydes and ketones are some of the most important functional groups in organic chemistry. Because they have played a major role in organic chemistry for more than a century, the common names for unbranched ketones are still in use. In the common names the alkyl groups bonded to the carbonyl group are named as substituents and listed alphabetically followed by *ketone*. The name acetone, which is propanone, has been retained by the IUPAC system.

IUPAC Names for Ketones

In the IUPAC system, the name of a ketone is obtained by replacing the -*e* in the corresponding alkane name with -*one*.

Step 1   Name the longest carbon chain containing the carbonyl group by replacing the -*e* in the corresponding alkane name by -*one*.

Step 2   Number the main chain starting from the end nearest the carbonyl group. Place the number of the carbonyl carbon in front of the ketone name. (Propanone and butanone do not require numbers.)

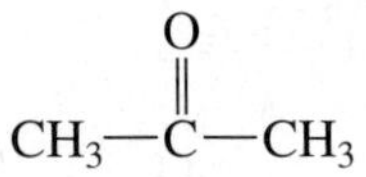

$$CH_3-\overset{\overset{\displaystyle O}{\|}}{C}-CH_3 \qquad CH_3-CH_2-\overset{\overset{\displaystyle O}{\|}}{C}-CH_3 \qquad CH_3-CH_2-\overset{\overset{\displaystyle O}{\|}}{C}-CH_2-CH_3$$

Propanone   Butanone   3-Pentanone
(dimethyl ketone; acetone)   (ethyl methyl ketone)   (diethyl ketone)

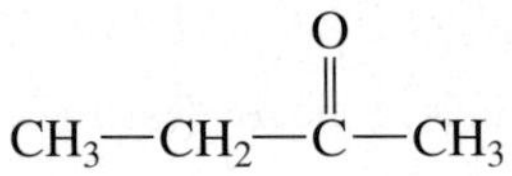

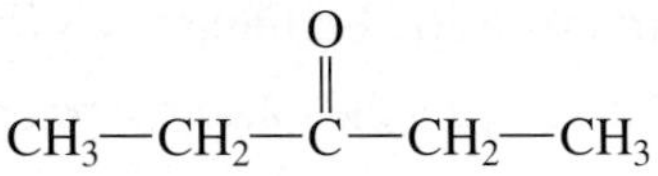

Step 3   Name and number any substituents on the carbon chain.

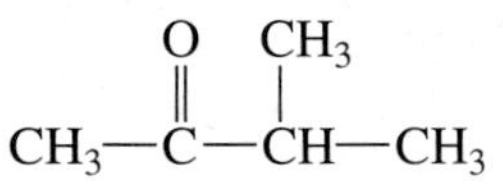

$$CH_3-\overset{\overset{\displaystyle O}{\|}}{C}-\overset{\overset{\displaystyle CH_3}{|}}{CH}-CH_3 \qquad CH_3-\overset{\overset{\displaystyle Br}{|}}{CH}-\overset{\overset{\displaystyle O}{\|}}{C}-CH_2-CH_3$$

3-Methylbutanone 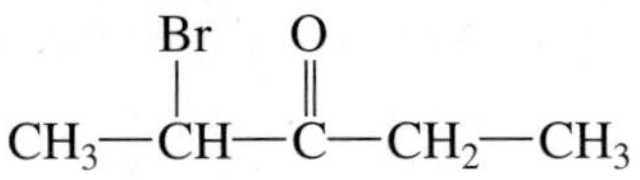   2-Bromo-3-pentanone 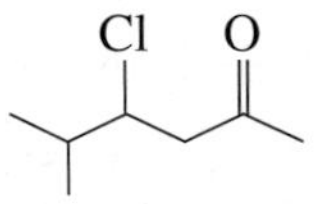   4-Chloro-5-methyl-2-hexanone

Step 4   For cyclic ketones, the prefix *cyclo* is used in front of the ketone name. Any substituents are located by numbering the ring starting with the carbonyl carbon as carbon 1.

Cyclopentanone 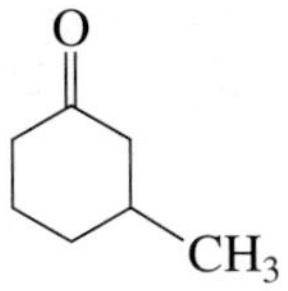   3-Methylcyclohexanone 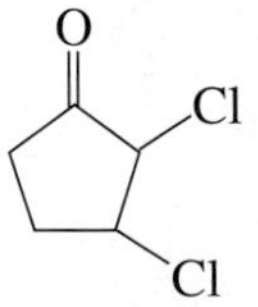   2,3-Dichlorocyclopentanone

---

SAMPLE PROBLEM 15.3

Names of Ketones

Give the IUPAC name for the following ketone:

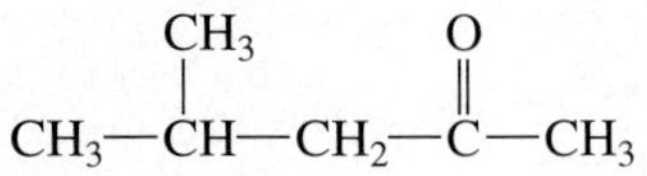

$$CH_3-\overset{\overset{\displaystyle CH_3}{|}}{CH}-CH_2-\overset{\overset{\displaystyle O}{\|}}{C}-CH_3$$

Solution

The longest chain is five carbon atoms. Counting from the right, the carbonyl group is on carbon 2 and a methyl group is on carbon 4. The IUPAC name is *4-methyl-2-pentanone.*

Study Check

What is the common name of 3-hexanone?

Naming Aldehydes

In the IUPAC names of aldehydes, the *-e* of the alkane name is replaced with *-al.*

Step 1   Name the longest carbon chain containing the carbonyl group by replacing the *-e* in the corresponding alkane name by *-al.* No number is needed for the aldehyde group because it always appears at the end of the chain. The IUPAC system names the aldehyde of benzene as benzaldehyde.

Benzaldehyde

The first four unbranched aldehydes are often referred to by their common names, which end in *aldehyde.* (See Figure 15.3.) The roots of these common names are derived from Latin or Greek words that indicate the source of the corresponding carboxylic acid. We will study carboxylic acids in the next chapter.

The carbonyl carbon is at the end of the chain

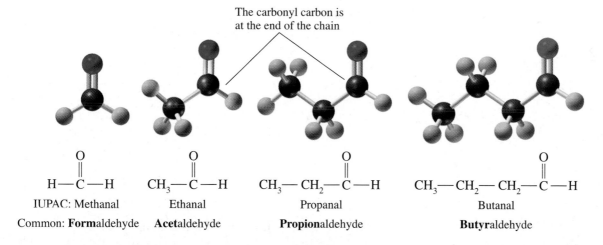

H—C—H	CH₃—C—H	CH₃— CH₂—C —H	CH₃—CH₂— CH₂—C —H
IUPAC: Methanal	Ethanal	Propanal	Butanal
Common: **Form**aldehyde	**Acet**aldehyde	**Propion**aldehyde	**Butyr**aldehyde

Step 2   Name and number any substituents on the carbon chain by counting the carbonyl carbon as carbon 1. An aldehyde group takes precedence in numbering over a ketone or an alcohol. When present with an aldehyde functional group, a ketone group is numbered and named oxo. When present with an aldehyde, an alcohol is named hydroxy.

Figure 15.3 In the structures of aldehydes, the carbonyl group is always the end carbon.

Q *Why is the carbon in the carbonyl group in aldehydes always at the end of the chain?*

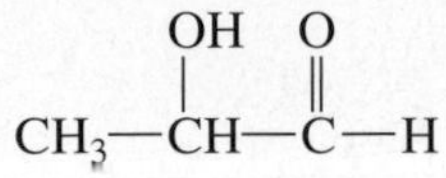

$$CH_3-\underset{\underset{OH}{|}}{CH}-\underset{\underset{\overset{\|}{O}}{}}{C}-H$$
2-Hydroxypropanal

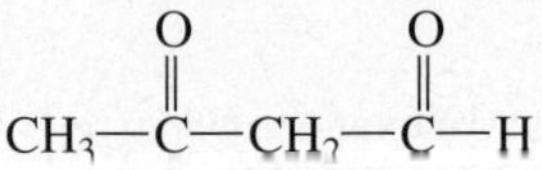

$$CH_3-\underset{\overset{\|}{O}}{C}-CH_2-\underset{\overset{\|}{O}}{C}-H$$
3-Oxobutanal

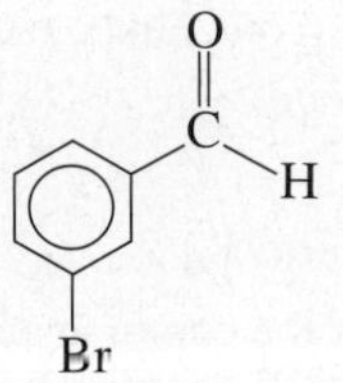

3-Bromobenzaldehyde

SAMPLE PROBLEM 15.4

Naming Aldehydes

Give the IUPAC names for the following aldehydes:

a. 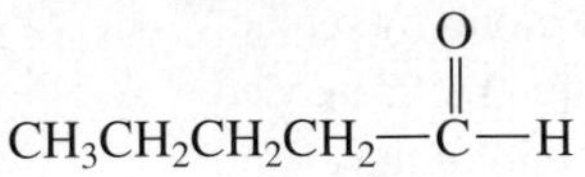 $CH_3CH_2CH_2CH_2-\underset{\overset{\|}{O}}{C}-H$

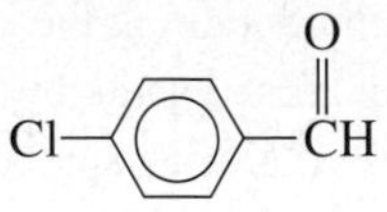

b. $Cl-\bigcirc-\underset{\overset{\|}{O}}{CH}$

c. $CH_3-\underset{\underset{CH_3}{|}}{CH}-CH_2-\underset{\overset{\|}{O}}{C}-H$

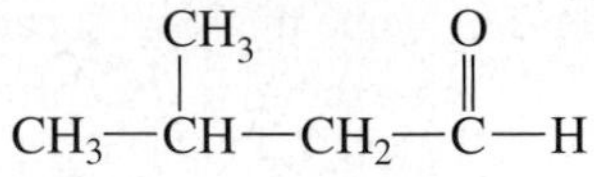

Solution

a. pentanal

b. 4-chlorobenzaldehyde

c. The longest unbranched chain has four atoms with a methyl group on the third carbon. The IUPAC name is *3-methylbutanal*.

Study Check

What are the IUPAC and common names of the aldehyde with three carbon atoms?

QUESTIONS AND PROBLEMS

Naming Aldehydes and Ketones

15.5 Give the IUPAC name for each of the following compounds:

a. 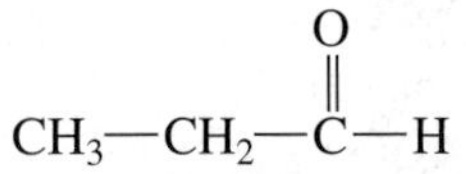 $CH_3-CH_2-\underset{\overset{\|}{O}}{C}-H$

b. $CH_3-CH_2-\underset{\overset{\|}{O}}{C}-\underset{\underset{CH_3}{|}}{CH}-CH_3$

c. $CH_3-\underset{\underset{OH}{|}}{CH}-CH_2-\underset{\overset{\|}{O}}{C}-H$

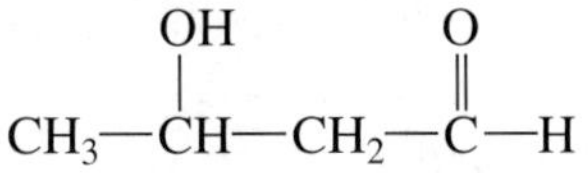

d.

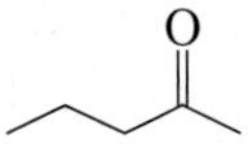

e.

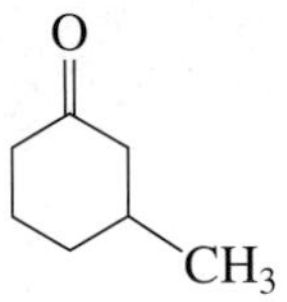

f.

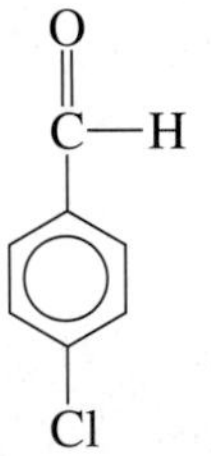

15.6 Give the IUPAC name for each of the following compounds:

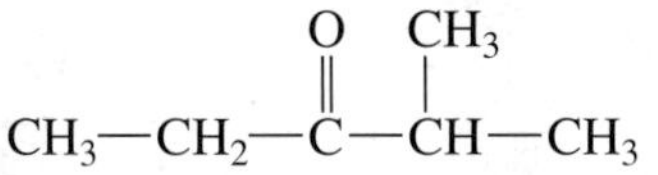

a. 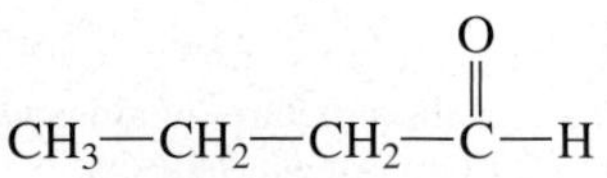 $CH_3-CH_2-CH_2-\underset{\overset{\|}{O}}{C}-H$

b. 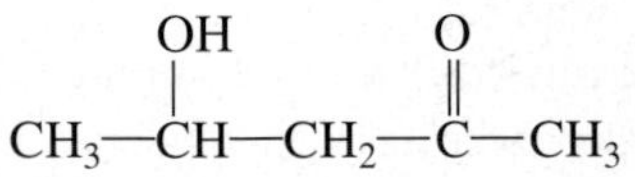 $CH_3-\underset{\underset{OH}{|}}{CH}-CH_2-\underset{\overset{\|}{O}}{C}-CH_3$

c.

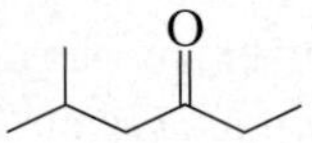

d.

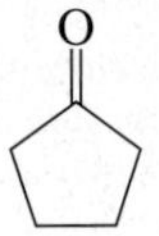

e.

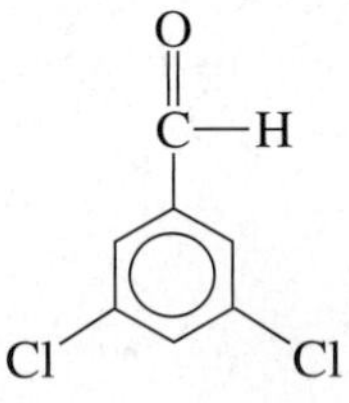

f. 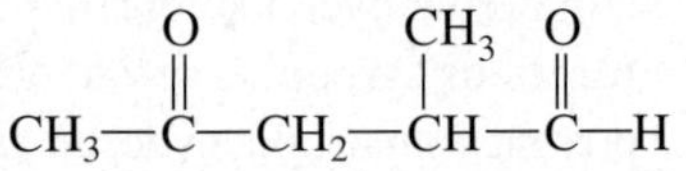 $CH_3-\underset{\overset{\|}{O}}{C}-CH_2-\underset{\underset{CH_3}{|}}{CH}-\underset{\overset{\|}{O}}{C}-H$

15.7 Give a common name for each of the following compounds:

a. $CH_3-\overset{\displaystyle O}{\overset{\displaystyle \|}{C}}-H$ **b.** $CH_3-\overset{\displaystyle O}{\overset{\displaystyle \|}{C}}-CH_2-CH_2-CH_3$ **c.** $H-\overset{\displaystyle O}{\overset{\displaystyle \|}{C}}-H$

15.8 Give the common name for each of the following compounds:

a. $CH_3-\overset{\displaystyle O}{\overset{\displaystyle \|}{C}}-CH_2-CH_3$ **b.** $CH_3-CH_2-\overset{\displaystyle O}{\overset{\displaystyle \|}{C}}-CH_2-CH_3$ **c.** $CH_3-CH_2-\overset{\displaystyle O}{\overset{\displaystyle \|}{C}}-H$

15.9 Write the condensed structural formula for each of the following compounds:
  a. acetaldehyde   **b.** 4-hydroxy-2-pentanone
  c. 2,3-dibromobutanal   **d.** methyl butyl ketone
  e. 3-methylpentanal   **f.** 3-oxopentanal

15.10 Write the condensed structural formula for each of the following compounds:
  a. propionaldehyde   **b.** butanal
  c. 4-bromo-3-hydroxyhexanal   **d.** 4-bromobutanone
  e. 2-oxobutanal   **f.** acetone

15.11 The IUPAC name of anisaldehyde used in perfumes is 4-methoxybenzaldehyde. What is the structural formula of anisaldehdye?

15.12 The IUPAC name of vanillin, a naturally occurring compound in vanilla beans, is 4-hydroxy-3-methoxybenzaldehyde. What is the structural formula of vanillin?

15.3 Some Important Aldehydes and Ketones

Formaldehyde, the simplest aldehyde, is a colorless gas with a pungent odor. Industrially, it is a reactant in the synthesis of polymers used to make fabrics, insulation materials, carpeting, pressed wood products such as plywood, and plastics for kitchen counters. An aqueous solution called formalin, which contains 40% formaldehyde, is used as a germicide and to preserve biological specimens. Exposure to formaldehyde fumes can irritate eyes, nose, upper respiratory tract, and cause skin rashes, headaches, dizziness, and general fatigue.

Acetone, or propanone (dimethyl ketone), which is the simplest ketone, is a colorless liquid with a mild odor that has wide use as a solvent in cleaning fluids, paint and nail-polish removers, and rubber cement. (See Figure 15.4.) It is extremely flammable and care must be taken when using acetone. In the body, acetone may

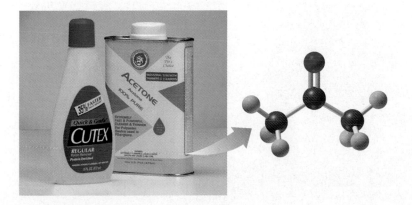

Figure 15.4 Acetone is used as a solvent in paint and nail-polish removers.
Q *What is the IUPAC name of acetone?*

be produced in uncontrolled diabetes, fasting, and high-protein diets when large amounts of fats are metabolized for energy.

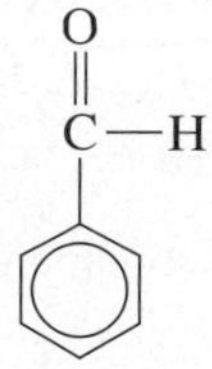

Benzaldehyde
(almond)

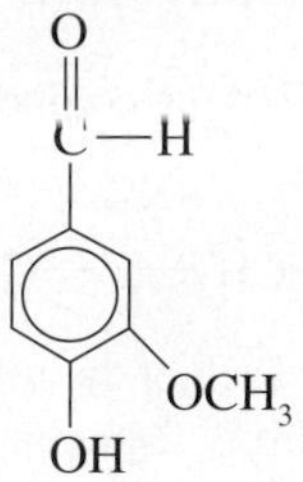

Vanillin
(vanilla)

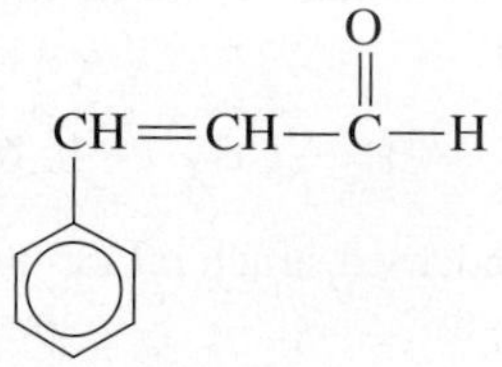

Cinnamaldehyde
(cinnamon)

Several naturally occurring aromatic aldehydes are used to flavor food and as fragrances in perfumes. Benzaldehyde is found in almonds, vanillin in vanilla beans, and cinnamaldehyde in cinnamon.

SAMPLE PROBLEM 15.5

Glyceraldehyde is an intermediate in carbohydrate metabolism. What are the functional groups in glyceraldehyde?

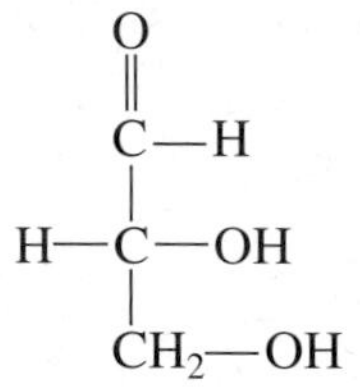

Glyceraldehyde

ENVIRONMENTAL NOTE

Vanilla

Vanilla has been used as a flavoring for over a thousand years. After drinking a beverage made from powdered vanilla and cocoa beans with Emperor Montezuma in Mexico, Cortez took vanilla back to Europe where it became popular for flavoring and for scenting perfumes and tobacco. Thomas Jefferson introduced vanilla to the United States in the late 1700s. Today much of the vanilla we use in the world is grown in Mexico, Madagascar, Réunion, Seychelles, Tahiti, Ceylon, Java, the Philippines, and Africa.

The vanilla plant is a member of the orchid family. There are many species of *Vanilla*, but *Vanilla planifolia* (or *V. fragrans*) is considered to produce the best flavor. The vanilla plant grows like a vine, and can grow to 100 feet in length. Its flowers are

hand-pollinated to produce a green fruit that is picked in 8 or 9 months. It is sun-dried to form a long, dark brown pod, which is called "vanilla bean" because it looks like a string bean.

The flavor and fragrance of the vanilla bean comes from the black seeds found inside the dried bean; the seeds and pod are used to flavor desserts such as custards and ice cream. The extract of vanilla is made by chopping up vanilla beans and mixing them with a 35% alcohol–water mixture. The liquid, which contains the aldehyde vanillin, is drained from the bean residue and used for flavoring.

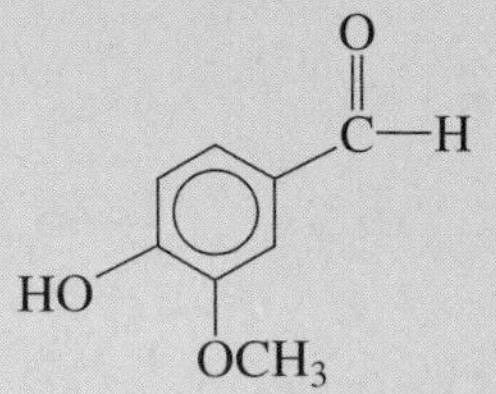

Vanillin

Solution

Glyceraldehyde has aldehyde and alcohol functional groups.

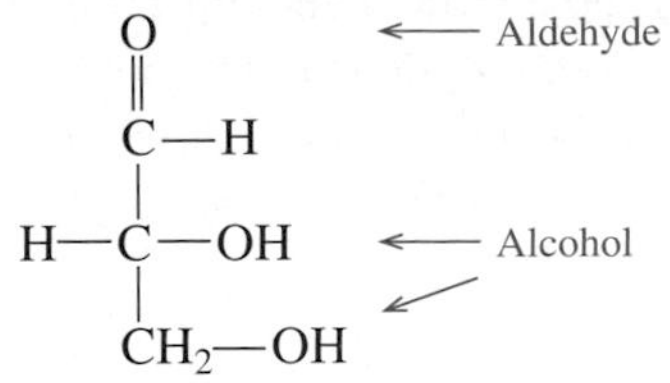

$$
\begin{array}{l}
\quad\; \overset{\textstyle O}{\underset{\textstyle \|}{}} \\
\quad\; C\!-\!H \qquad \longleftarrow \text{ Aldehyde} \\
H\!-\!C\!-\!OH \qquad \longleftarrow \text{ Alcohol} \\
\quad\; CH_2\!-\!OH
\end{array}
$$

Study Check

The compound dihydroxyacetone (DHA) used in "sunless" tanning lotions darkens the skin without sun. What is its structural formula?

The flavor of butter or margarine is from butanedione; muscone is used to make musk perfumes; and oil of spearmint contains carvone.

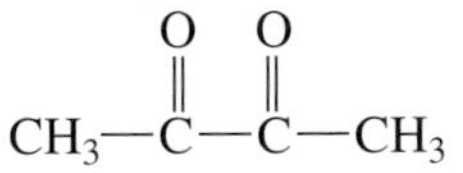

$$
CH_3\!-\!\overset{O}{\overset{\|}{C}}\!-\!\overset{O}{\overset{\|}{C}}\!-\!CH_3
$$

Butanedione
(butter flavor)

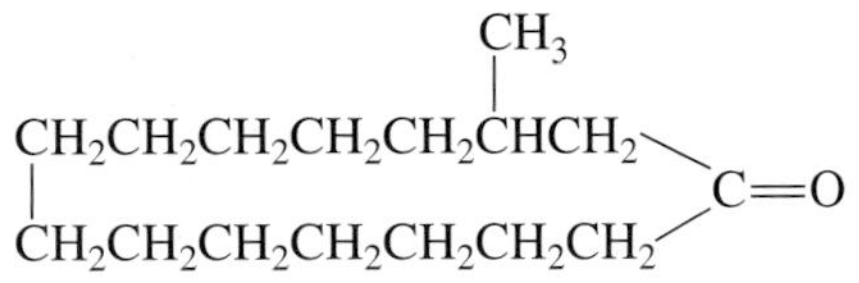

Muscone
(musk)

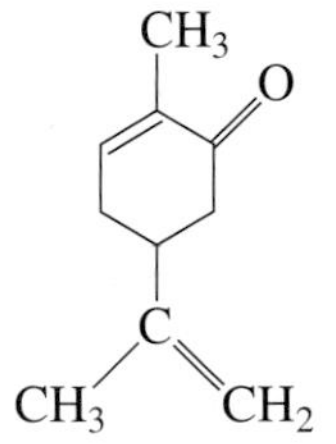

Carvone
(spearmint oil)

Important hormones of the steroid family, such as cortisone, testosterone, and progesterone, contain one or more carbonyl groups.

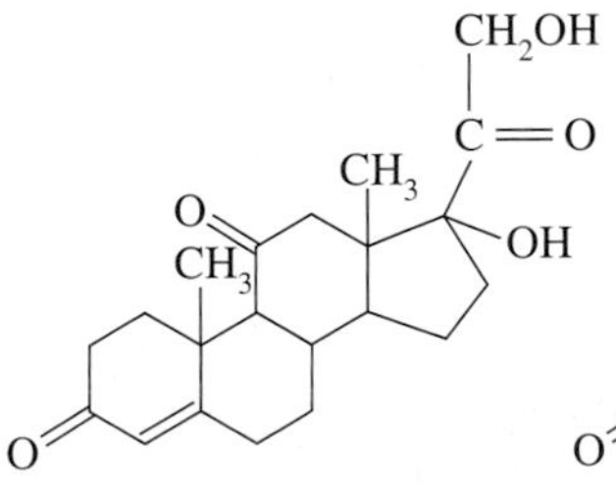

Cortisone
(protein metabolism;
inflammation reducer)

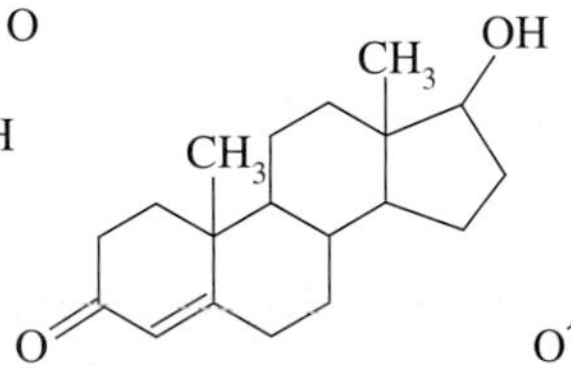

Testosterone
(male sex hormone)

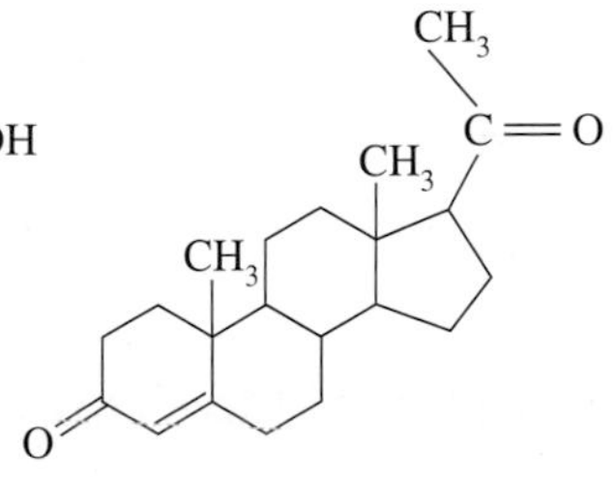

Progesterone
(female sex hormone)

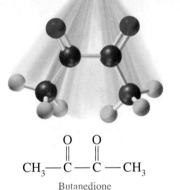

$$
CH_3\!-\!\overset{O}{\overset{\|}{C}}\!-\!\overset{O}{\overset{\|}{C}}\!-\!CH_3
$$

Butanedione

QUESTIONS AND PROBLEMS

Some Important Aldehydes and Ketones

15.13 Identify the aldehyde or ketone with each of the following uses:
 a. almond flavoring   **b.** fingernail-polish remover
 c. making plywood

15.14 Identify the aldehyde or ketone with each of the following uses:
 a. butter flavoring   **b.** produced in uncontrolled diabetes
 c. preserve tissues

15.4 Physical Properties

At room temperature, formaldehyde (bp −21°C) and acetaldehyde (bp 21°C) are gases. Aldehydes containing from 3 to 10 carbon atoms are liquids. The polar carbonyl group with a partially negative oxygen atom and a partially positive carbon atom has an influence on the boiling points and the solubility of aldehydes and ketones in water.

Boiling Points

The polar carbonyl group gives aldehydes and ketones higher boiling points than alkanes and ethers of similar mass. The increase in boiling points is due to dipole–dipole interactions.

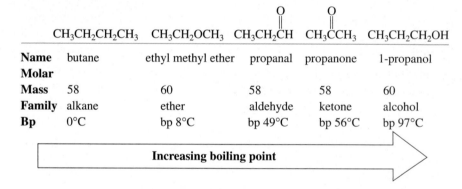

Dipole-dipole interaction

However, because there is no hydrogen on the oxygen atom, aldehydes and ketones cannot form hydrogen bonds with each other. Thus they have boiling points that are lower than alcohols.

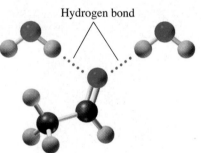

Hydrogen bond

Acetaldehyde

Hydrogen bond

Acetone

	$CH_3CH_2CH_2CH_3$	$CH_3CH_2OCH_3$	$CH_3CH_2\overset{O}{\overset{\|}{C}}H$	$CH_3\overset{O}{\overset{\|}{C}}CH_3$	$CH_3CH_2CH_2OH$
Name	butane	ethyl methyl ether	propanal	propanone	1-propanol
Molar Mass	58	60	58	58	60
Family	alkane	ether	aldehyde	ketone	alcohol
Bp	0°C	bp 8°C	bp 49°C	bp 56°C	bp 97°C

Increasing boiling point →

Solubility of Aldehydes and Ketones in Water

Although aldehydes and ketones do not hydrogen bond with each other, the electronegative oxygen atom does hydrogen bond with water molecules. Carbonyl compounds with one to four carbons are very soluble in water. However, those with five carbon atoms or more are not very soluble because the alkyl portions diminish the effect of the polar carbonyl group. (See Figure 15.5.)

Table 15.1 compares the boiling points of some carbonyl compounds, as well as their solubilities in water.

Figure 15.5 Hydrogen bonding of acetaldehyde and acetone with water.
Q *Would you expect propanal to be soluble in water?*

SAMPLE PROBLEM 15.6

Boiling Point and Solubility

Would you expect ethanol CH_3CH_2OH to have a higher or lower boiling point than ethanal CH_3CHO? Explain.

Solution

Ethanol would have a higher boiling point because its molecules can hydrogen bond with each other, while molecules of ethanal cannot.

Study Check

If acetone molecules cannot hydrogen bond with each other, why is acetone soluble in water?

QUESTIONS AND PROBLEMS

Physical Properties

15.15 Which compound in each of the following pairs would have the higher boiling point? Explain.

a. CH_3—CH_2—CH_3 or CH_3—CH_2—$\overset{\displaystyle O}{\overset{\|}{C}}$—H

b. propanal or pentanal   **c.** butanal or 1-butanol

15.16 Which compound in each of the following pairs would have the higher boiling point? Explain.

a. 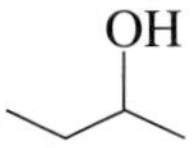 (structure with OH) or (structure with O)

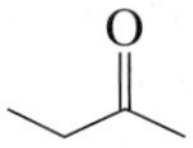

b. butane or butanone   **c.** propanone or pentanone

15.17 Which compound in each of the following pairs would be more soluble in water? Explain.

a. CH_3—$\overset{\displaystyle O}{\overset{\|}{C}}$—$CH_2$—$CH_3$ or CH_3—$\overset{\displaystyle O}{\overset{\|}{C}}$—$\overset{\displaystyle O}{\overset{\|}{C}}$—$CH_3$

b. dimethyl ether or acetaldehyde   **c.** acetone or 2-pentanone

15.18 Which compound in each of the following pairs would be more soluble in water? Explain.

a. CH_3—CH_2—CH_3 or CH_3—CH_2—CHO

b. propanone or 3-hexanone   **c.** propane or propanone

15.19 Would you expect an aldehyde with a formula of $C_8H_{16}O$ to be soluble in water? Explain.

15.20 Would you expect an aldehyde with a formula of C_3H_6O to be soluble in water? Explain.

Table 15.1   Comparison of Physical Properties of Some Selected Compounds

Compound	Boiling Point (°C)	Solubility in Water
Methanal (formaldehyde)	−21	Very soluble
Ethanal (acetaldehyde)	21	Very soluble
Propanal (propionaldehyde)	49	Soluble
Propanone (acetone)	56	Soluble
Butanal (butyraldehyde)	75	Soluble
Butanone	80	Soluble
Pentanal	103	Slightly soluble
2-Pentanone	102	Slightly soluble
3-Pentanone	102	Slightly soluble
Hexanal	129	Not soluble
2-Hexanone	127	Not soluble
3-Hexanone	124	Not soluble
Acetophenone	202	Not soluble

LEARNING GOAL

Identify chiral and achiral carbon atoms in an organic molecule.

15.5 Chiral Molecules

In the preceding chapters, we have looked at some types of isomers. Let's review those now. Molecules can be constitutional (also called structural) isomers when they have the same molecular formula, but different bonding arrangements. In the last chapter, we saw that alcohols and ethers can be constitutional isomers. In this chapter, we wrote constitutional isomers for aldehydes and ketones.

Constitutional Isomers

C_2H_6O CH_3-CH_2-OH CH_3-O-CH_3

                 Ethanol   Dimethyl ether

C_3H_6O $CH_3-CH_2-\overset{\overset{\displaystyle O}{\|}}{C}-H$ $CH_3-\overset{\overset{\displaystyle O}{\|}}{C}-CH_3$

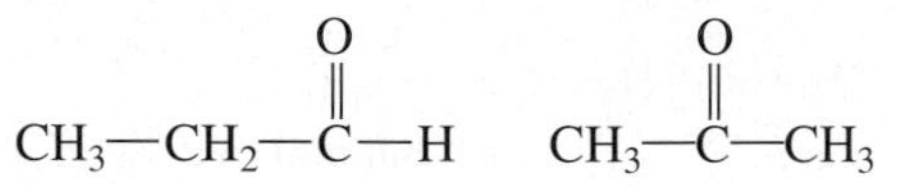

                 Propanal   Propanone

Another group of isomers called **stereoisomers** have identical molecular formulas too, but they are not constitutional isomers. In stereoisomers the atoms are bonded in the same sequence, but differ in the way they are arranged in space. Both the cis-trans isomers of cycloalkanes (Chapter 12), and the cis-trans isomers of alkenes (Chapter 13) are stereoisomers.

In this chapter, we take at look at stereoisomers called enantiomers and diastereoisomers. Enantiomers are stereoisomers that are mirror images, but cannot be superimposed on each other. Diastereoisomers are stereoisomers that are not mirror images. Many of the cis-trans stereoisomers such as *cis-* and *trans-*dimethyl-cyclopropane and *cis-* and *trans-*2-butene shown below are diastereoisomers, because the structures are not mirror images. We can summarize the types of isomers as follows:

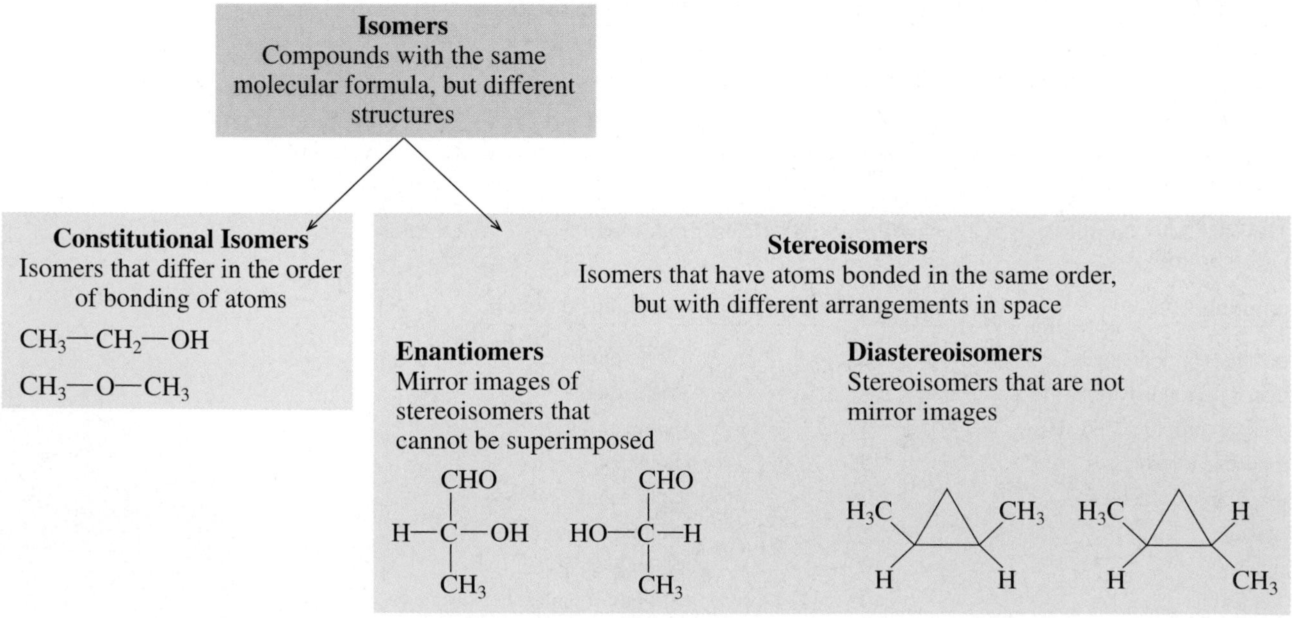

Isomers
Compounds with the same molecular formula, but different structures

Constitutional Isomers
Isomers that differ in the order of bonding of atoms

CH_3-CH_2-OH

CH_3-O-CH_3

Stereoisomers
Isomers that have atoms bonded in the same order, but with different arrangements in space

Enantiomers
Mirror images of stereoisomers that cannot be superimposed

        CHO   CHO
         | |
$H-\overset{|}{\underset{|}{C}}-OH$ $HO-\overset{|}{\underset{|}{C}}-H$
        CH_3   CH_3

Diastereoisomers
Stereoisomers that are not mirror images

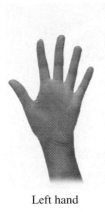

Left hand   Right hand

Mirror image of right hand

Figure 15.6 A pair of hands are chiral because they have mirror images that cannot be superimposed on each other.
Q *Why are your shoes chiral objects?*

Chirality

When stereoisomers have mirror images that are different, they are said to have "handedness." If you look at the palms of your hands, your thumbs are on opposite sides. If you turn your palms toward each other, you have mirror images. (See Figure 15.6.) The left hand is the mirror image of the right hand. Your hands are not superimposable. Objects such as hands that have nonsuperimposable mirror images are **chiral.** Left and right shoes are chiral; left- and right-handed golf clubs are chiral. When one mirror image can be superimposed on the other, the object is **achiral.** The object has no "handedness."

Molecules in nature also have mirror images, and often the "left-handed" stereoisomer has a different biological effect than does the "right-handed" one. For some compounds, one isomer has a certain odor, and the mirror image has a completely different odor. For example, one enantiomer of limonene smells like oranges, while its mirror image has the odor of lemons. When we think of how difficult it is to put a left-hand glove on our right hand, or a right shoe on our left foot, or use left-handed scissors if we are right handed, we begin to realize that certain properties of mirror images can be very different. (See Figure 15.7.)

Limonene

CH_3

CH_3 CH_2

SAMPLE PROBLEM 15.7

Chiral Objects

Classify each of the following objects as chiral or achiral:
a. left ear   **b.** clear drinking glass   **c.** a glove

Solution

a. Chiral; the left ear cannot be superimposed on the right ear.
b. Achiral; mirror images of a clear drinking glass can be superimposed on each other.
c. Chiral; there is a right glove for the right hand and a left glove for the left hand.

Study Check

Would a bowling pin be chiral or achiral?

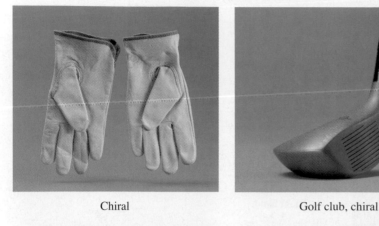

Chiral   Golf club, chiral    Achiral

Achiral   Chiral    Right-handed scissors, chiral

Figure 15.7 Everyday objects can be chiral or achiral.

Q *Why are some of the above objects chiral and others are achiral?*

Chiral Carbon Atoms

A carbon compound is chiral if it has at least one carbon atom bonded to four different atoms or groups. This type of carbon atom is called a **chiral carbon** because there are two different ways that it can bond to four atoms or groups of atoms. The resulting structures are mirror images of each other. Let's look at the mirror images of a carbon bonded to four different atoms. (See Figure 15.8.) If we line up the hydrogen and iodine atoms in the mirror images, the bromine and chorine atoms appear on opposite sides. No matter how we turn the models, we cannot align all four atoms at the same time. When stereoisomers cannot be superimposed, they are called **enantiomers.** If two or more atoms are the same, the atoms can be aligned (superimposed) and the mirror images represent the same structure. (See Figure 15.9.)

Figure 15.8 (a) The enantiomers of a chiral molecule are mirror images. (b) The enantiomers of a chiral molecule cannot be superimposed on each other.

Q *Why is the carbon atom in this compound a chiral carbon?*

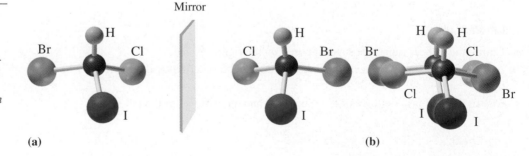

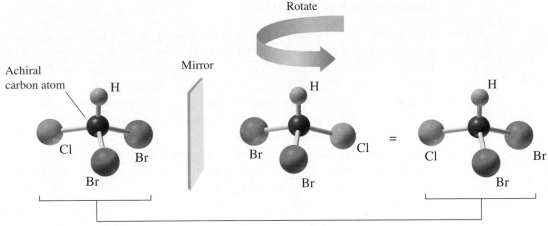

Figure 15.9 The mirror images of an achiral compound can be superimposed on each other.

Q *Why can the mirror images of the compound be superimposed?*

SAMPLE PROBLEM 15.8

Chiral Carbons

Indicate whether the carbon in red is chiral or not chiral.

a. $Cl-\underset{\underset{H}{|}}{\overset{\overset{Cl}{|}}{C}}-CH_3$
b. $CH_3-\underset{\underset{H}{|}}{\overset{\overset{OH}{|}}{C}}-CH_2-CH_3$
c. $CH_3-\overset{\overset{O}{\|}}{C}-CH_2-CH_3$
d. $CH_3-CH_2-\underset{\underset{H}{|}}{\overset{\overset{CH_3}{|}}{C}}-\overset{\overset{O}{\|}}{C}-H$

Solution

a. Not chiral. Two of the substituents on the carbon are the same (Cl). A chiral carbon must be bonded to four different groups or atoms.

b. Chiral. Carbon 2 is bonded to four different groups: one OH, one CH_3, one CH_2-CH_3, and one H.

c. Not chiral. Carbon 2 is bonded to only three groups, not four.

d. Chiral. Carbon 2 is bonded to four different groups: one H, one CH_3, one CH_2-CH_3, and one CHO.

Study Check

Circle the two chiral carbons in the structural formula of the carbohydrate erythrose.

$HO-CH_2-\underset{}{\overset{\overset{OH}{|}}{CH}}-\underset{}{\overset{\overset{OH}{|}}{CH}}-\overset{\overset{O}{\|}}{C}-H$

Erythrose

Drawing Fischer Projections

Emil Fischer devised a simplified system for drawing stereoisomers that shows the arrangements of the atoms. Fischer received the Nobel Prize in 1902 for his contributions to carbohydrate and protein chemistry. Using his method called a **Fischer projection,** the bonds to a chiral atom are drawn as intersecting lines with the chiral carbon being at the center where the lines cross. The horizontal lines represent the bonds that come forward in the three-dimensional structure, and the vertical lines represent the bonds that point away.

By convention the carbon chain in the Fischer projection is written vertically with the most highly oxidized carbon at the top. For glyceraldehyde, the carbonyl group, which is the most highly oxidized group in the molecule, is written at the top. The letter L is assigned to the left-handed stereoisomer, which has the —OH group on the left of the chiral carbon. The letter D is assigned to the right-handed structure where the —OH is on the right of the chiral carbon. Let's look at how glyceraldehyde, the simplest sugar, is converted from a three-dimensional view to a Fischer projection. (See Figure 15.10.)

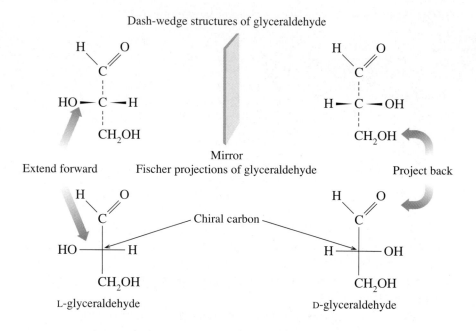

Figure 15.10 In a Fischer projection, the chiral carbon atom is at the center with horizontal lines for bonds that extend toward the viewer and vertical lines for bonds that point away. **Q** *Why does glyceraldehyde have only one chiral carbon atom?*

Fischer projections can also be written for larger compounds that have two or more chiral carbons. For example, the mirror images below are different. Each chiral atom is bonded to four different groups. To draw the mirror image of a Fischer projection, the positions of the substituents on the horizontal lines are reversed, while the groups on the vertical line are left unchanged.

We can also draw the mirror image of the carbohydrate erythrose, which has two chiral carbons.

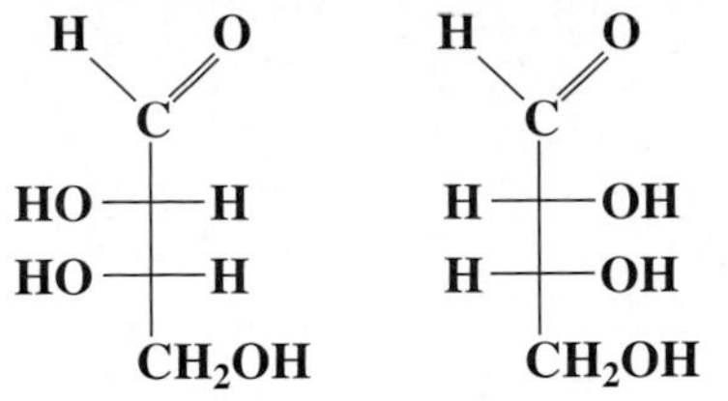

L-Erythrose   D-Erythrose

SAMPLE PROBLEM 15.9

Fischer Projections

Determine if each Fischer projection is a chiral compound. If so, identify it as the D or L isomer and draw the mirror image.

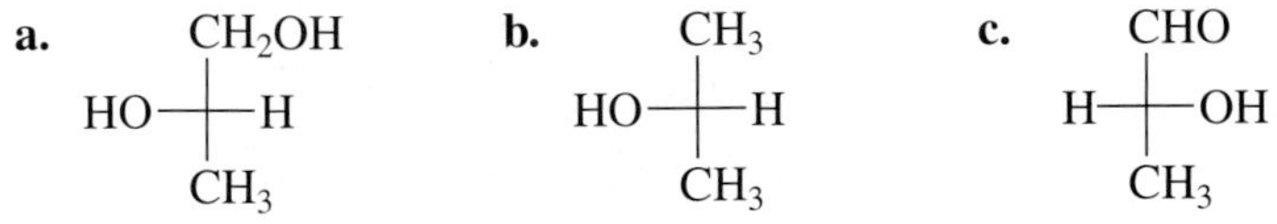

a.
$$CH_2OH$$
$$HO \!-\!\!\!|\!-\! H$$
$$CH_3$$

b.
$$CH_3$$
$$HO \!-\!\!\!|\!-\! H$$
$$CH_3$$

c.
$$CHO$$
$$H \!-\!\!\!|\!-\! OH$$
$$CH_3$$

Solution

a. This chiral compound with four different substituents is the L isomer. The mirror image is written by reversing the H and OH on the horizontal lines.

$$CH_2OH$$
$$H \!-\!\!\!|\!-\! OH$$
$$CH_3$$

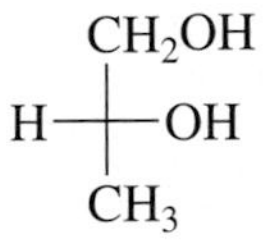

b. The compound is achiral because it has two identical groups (CH_3).

c. This chiral compound with four different substituents is the D isomer. The mirror image is written by reversing the H and OH on the horizontal lines.

$$CHO$$
$$HO \!-\!\!\!|\!-\! H$$
$$CH_3$$

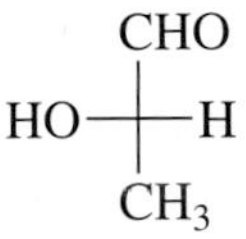

Study Check

Draw the Fischer projections for the D and L stereoisomers of 2-hydroxypropanal.

QUESTIONS AND PROBLEMS

Chiral Molecules

15.21 Identify each of the following structures as chiral or achiral. If chiral, indicate the chiral carbon.

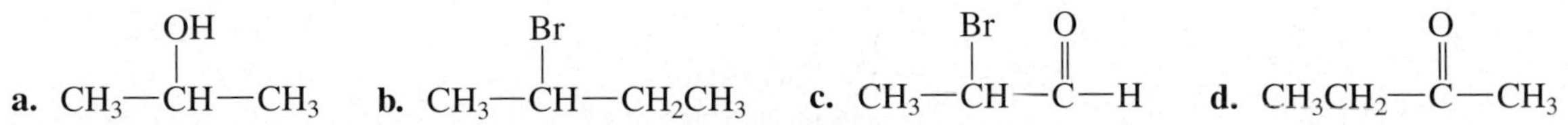

a.
$$OH$$
$$CH_3\!-\!CH\!-\!CH_3$$

b.
$$Br$$
$$CH_3\!-\!CH\!-\!CH_2CH_3$$

c.
$$Br \quad O$$
$$CH_3\!-\!CH\!-\!\overset{\|}{C}\!-\!H$$

d.
$$O$$
$$CH_3CH_2\!-\!\overset{\|}{C}\!-\!CH_3$$

Enantiomers in Biological Systems

Most stereoisomers that are active in biological systems consist of only one enantiomer. Rarely are both enantiomers of biological molecules active. This happens because the enzymes and cell surface receptors on which metabolic reactions take place also have "handedness." Thus, only one of the enantiomers of a reactant or a drug interacts with its enzymes or receptors; the other is inactive. At the target site, the chiral receptor fits the arrangement of the substituents in only one enantiomer. Its mirror image, which is inactive, does not fit properly. This is similar to the idea that your right hand will only fit into a right-handed glove. (See Figure 15.11.)

A substance called carvone exists as two enantiomers. One enantiomer gives the odor of spearmint oil, while the other enantiomer produces the flavor of caraway seeds. Such differences in odor are due to the chiral receptor sites in the nose that fit the shape of one enantiomer, but not the other. Thus, our senses of smell and also taste are sensitive to the chirality of molecules.

Enantiomers of Carvone

From spearmint oil   From caraway seeds

In the brain, one stereoisomer of LSD causes hallucinations because it affects the production of serotonin, a chemical that is important in sensory perception. However, its enantiomer LSD produces little effect in the brain. The behavior of nicotine and epinephrine (adrenaline) also depends upon one of their enantiomers. One enantiomer of nicotine is more toxic than the other. In epinephrine (adrenaline), one enantiomer is responsible for the constriction of blood vessels.

Nicotine   Adrenalin (epinephrine)

A substance used to treat Parkinson's disease is L-dopa, which is converted to dopamine in the brain, where it raises the serotonin level. However, the D-dopa enantiomer has no biological effect.

L-Dopa, anti-Parkinsonian drug   D-Dopa has no biological effect

For many drugs, only one of the enantiomers is biologically active. However, for many years, drugs have been produced that were mixtures of their enantiomers. Today, drug researchers are using *chiral technology* to produce the active enantiomers of chiral drugs. Chiral catalysts are being designed that direct the formation of just one enantiomer rather than both. The benefits of producing only the active enantiomer include using a lower dose, enhancing activity, reducing interactions with other drugs, and eliminating possible harmful side effects from the nonactive enantiomer. Several active enantiomers are now being produced such as L-dopa and the active enantiomer of the popular analgesic ibuprofen used in Advil, Motrin, and Nuprin.

Ibuprofen

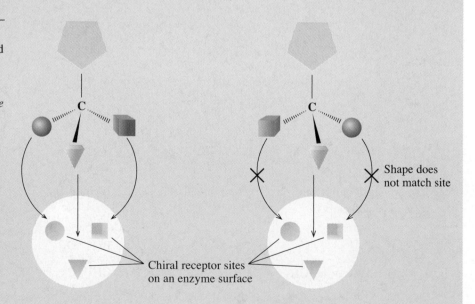

Figure 15.11 (a) The substituents on the biologically active enantiomer bind to all the sites on a chiral receptor; **(b)** its enantiomer does not bind properly and is not active biologically.

Q *Why don't all the substituents of the mirror image of the active enantiomer fit into a chiral receptor site?*

Shape does not match site

Chiral receptor sites on an enzyme surface

15.22 Identify each of the following structures as chiral or achiral. If chiral, indicate the chiral carbon.

a. 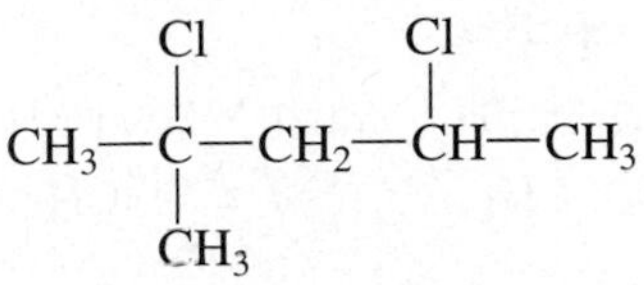 $CH_3-\overset{\overset{\displaystyle Cl}{|}}{\underset{\underset{\displaystyle CH_3}{|}}{C}}-CH_2-\overset{\overset{\displaystyle Cl}{|}}{CH}-CH_3$

b. $CH_3-\overset{\overset{\displaystyle Br}{|}}{C}=CH-CH_3$

c. $CH_3-\overset{\overset{\displaystyle OH}{|}}{\underset{\underset{\displaystyle CH_3}{|}}{C}}-\overset{\overset{\displaystyle OH}{|}}{CH}-CH_3$

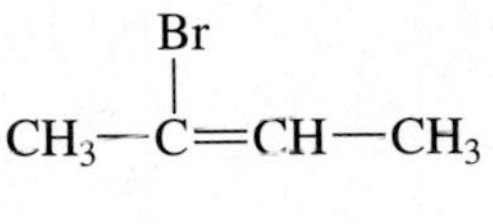

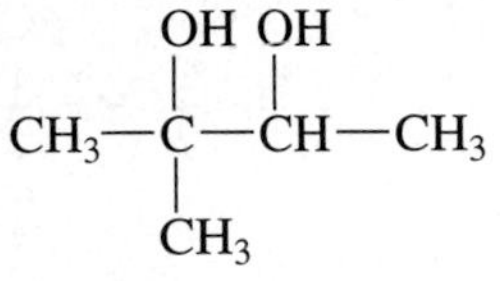

d. 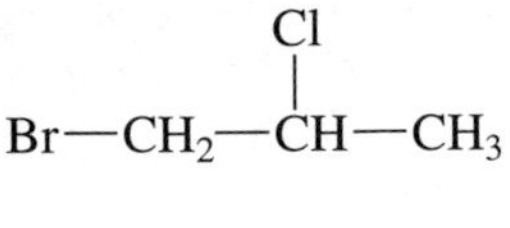 $Br-CH_2-\overset{\overset{\displaystyle Cl}{|}}{CH}-CH_3$

15.23 Identify the chiral carbon in each of the following naturally occurring compounds.

a. citronellol; one enantiomer has the geranium odor.

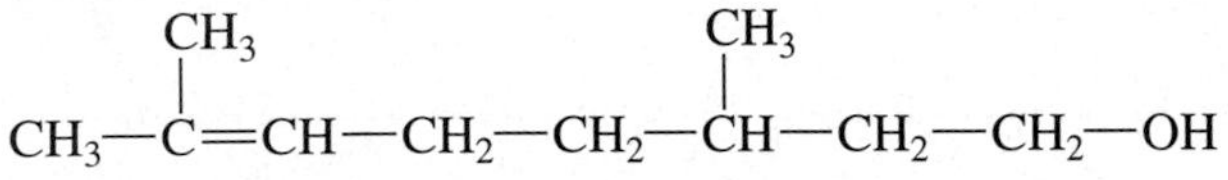

$$CH_3-\overset{\overset{\displaystyle CH_3}{|}}{C}=CH-CH_2-CH_2-\overset{\overset{\displaystyle CH_3}{|}}{CH}-CH_2-CH_2-OH$$

b. alanine, amino acid

$$H_2N-\overset{\overset{\displaystyle CH_3}{|}}{CH}-\overset{\overset{\displaystyle O}{||}}{C}-OH$$

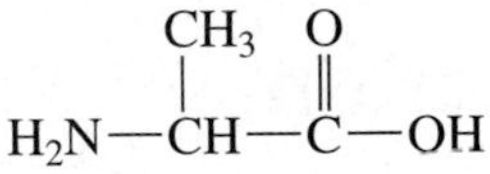

15.24 Identify the chiral carbon in each of the following naturally occurring compounds.

a. amphetamine (Benzedrine), stimulant, treatment of hyperactivity

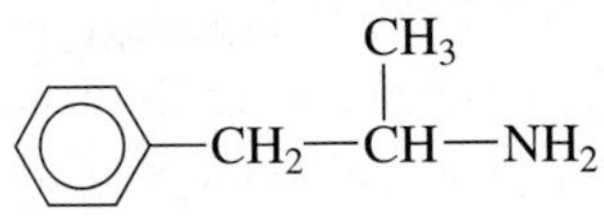

$$\text{(benzene ring)}-CH_2-\overset{\overset{\displaystyle CH_3}{|}}{CH}-NH_2$$

b. Norepinephrine, increases blood pressure and nerve transmission

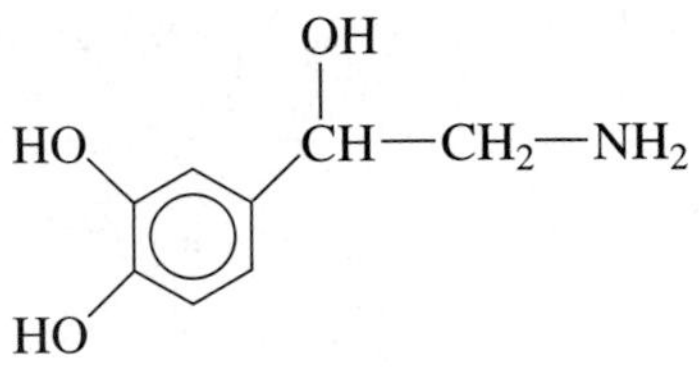

$$\text{HO, HO (dihydroxybenzene ring)}-\overset{\overset{\displaystyle OH}{|}}{CH}-CH_2-NH_2$$

15.25 Draw Fischer projections for each of the following dash-wedge structures.

a. 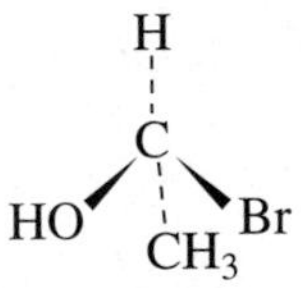   H — C (HO, CH₃, Br)

b.   CH₃ — C (Cl, OH, Br)

c.   CHO — C (HO, H, CH₂CH₃)

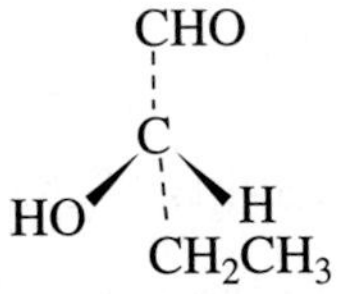

15.26 Draw Fischer projections for each of the following dash-wedge structures.

a. 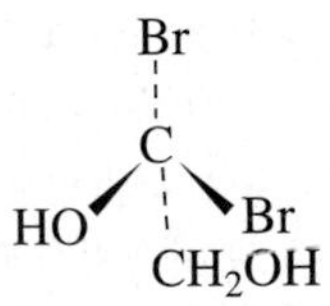   Br — C (HO, Br, CH₂OH)

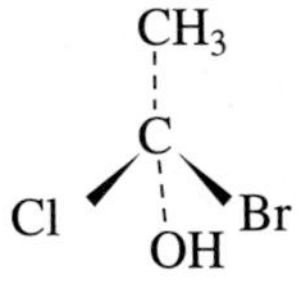

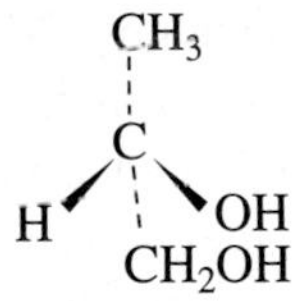

b.   CH₃ — C (H, OH, CH₂OH)

c.   CHO — C (HO, H, CH₂OH)

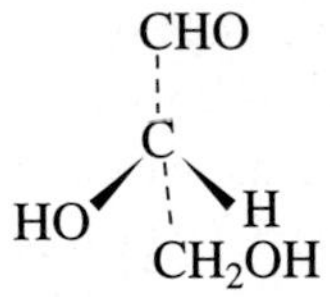

15.27 Indicate whether each pair of Fischer projections represent enantiomers or identical structures.

a. 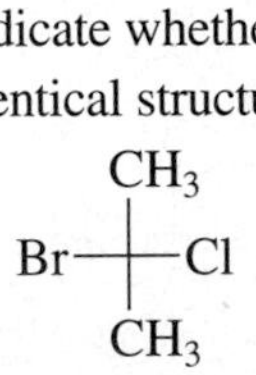 $Br-\!\!\!\underset{CH_3}{\overset{CH_3}{|}}\!\!\!-Cl$ and $Cl-\!\!\!\underset{CH_3}{\overset{CH_3}{|}}\!\!\!-Br$

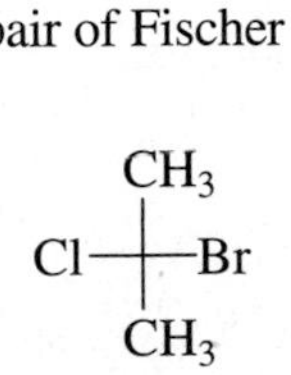

b. 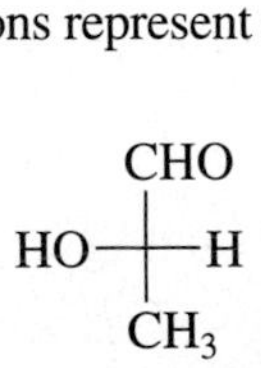 $HO-\!\!\!\underset{CH_3}{\overset{CHO}{|}}\!\!\!-H$ and $H-\!\!\!\underset{CH_3}{\overset{CHO}{|}}\!\!\!-OH$

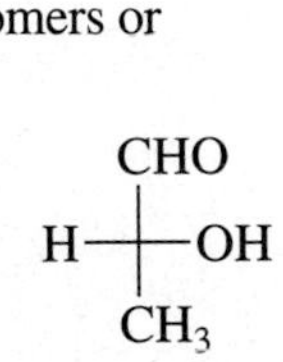

c. 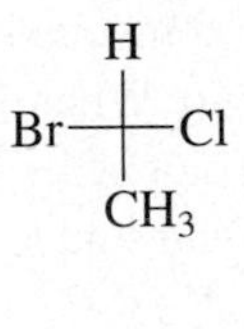 $Br-\!\!\!\underset{CH_3}{\overset{H}{|}}\!\!\!-Cl$ and $Br-\!\!\!\underset{H}{\overset{CH_3}{|}}\!\!\!-Cl$

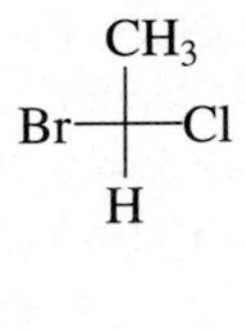

d. 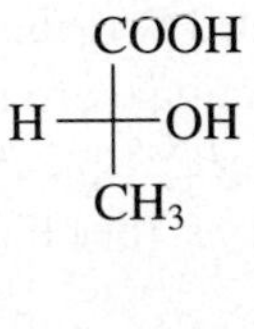 $H-\!\!\!\underset{CH_3}{\overset{COOH}{|}}\!\!\!-OH$ and $HO-\!\!\!\underset{CH_3}{\overset{COOH}{|}}\!\!\!-H$

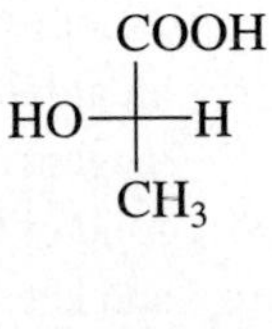

15.28 Indicate whether each pair of Fischer projections represent enantiomers or identical structures.

a.
$$\begin{array}{c} CH_2OH \\ Br-\!\!+\!\!-Cl \\ CH_3 \end{array} \quad \text{and} \quad \begin{array}{c} CH_2OH \\ Cl-\!\!+\!\!-Br \\ CH_3 \end{array}$$

b.
$$\begin{array}{c} CHO \\ H-\!\!+\!\!-H \\ CH_3 \end{array} \quad \text{and} \quad \begin{array}{c} CHO \\ H-\!\!+\!\!-H \\ CH_3 \end{array}$$

c.
$$\begin{array}{c} CH_3 \\ H-\!\!+\!\!-OH \\ CH_2CH_3 \end{array} \quad \text{and} \quad \begin{array}{c} CH_3 \\ HO-\!\!+\!\!-H \\ CH_2CH_3 \end{array}$$

d.
$$\begin{array}{c} COOH \\ H-\!\!+\!\!-NH_2 \\ CH_3 \end{array} \quad \text{and} \quad \begin{array}{c} COOH \\ H_2N-\!\!+\!\!-H \\ CH_3 \end{array}$$

15.6 Oxidation and Reduction

LEARNING GOAL

Draw the structural formulas of reactants and products for the oxidation or reduction of aldehydes and ketones.

In Chapter 14, we saw that aldehydes produced by the oxidation of primary alcohols oxidize readily to carboxylic acids. In fact they oxidize so easily that even the aldehydes exposed to the air in the laboratory quickly form carboxylic acids. In contrast, ketones produced by the oxidation of secondary alcohols do not undergo further oxidation. Let's review examples of the oxidation reactions of primary and secondary alcohols that form aldehydes and ketones.

$$CH_3-CH_2-OH \xrightarrow{\text{Oxidation}} CH_3-\overset{\overset{\displaystyle O}{\|}}{C}-H \xrightarrow[\text{oxidation}]{\text{Further}} CH_3-\overset{\overset{\displaystyle O}{\|}}{C}-OH$$

Ethanol (1°)   Acetaldehyde   Acetic acid

$$CH_3-\overset{\overset{\displaystyle OH}{|}}{CH}-CH_3 \xrightarrow{\text{Oxidation}} CH_3-\overset{\overset{\displaystyle O}{\|}}{C}-CH_3 \xrightarrow[\text{oxidation}]{\text{Further}} \text{no reaction}$$

2-Propanol (2°)   Propanone

Tollens' Text

The ease of oxidation of aldehydes allows certain mild oxidizing agents to oxidize the aldehyde functional group without oxidizing other functional groups such as alcohols or ethers. In the laboratory, **Tollens' test** may be used to distinguish between an aldehyde and ketone. Tollens' reagent, a solution of Ag^+ ($AgNO_3$) and ammonia, oxidizes aldehydes, but not ketones. The silver ion is reduced to metallic silver, which forms a layer called a "silver mirror" on the inside of the container. Commercially, a similar process is used to make mirrors by applying a mixture of $AgNO_3$ and ammonia on glass with a spray gun. (See Figure 15.12.)

$$CH_3-\overset{\overset{\displaystyle O}{\|}}{C}-H + 2Ag^+ \xrightarrow{[O]} 2Ag(s) + CH_3-\overset{\overset{\displaystyle O}{\|}}{C}-OH$$

Acetaldehyde   Tollens reagent   Silver mirror   Acetic acid

Another test, called **Benedict's test,** gives a positive test with compounds that have an aldehyde functional group and an adjacent hydroxyl group. When Benedict's reagent containing Cu^{2+} ($CuSO_4$) ions is added to this type of aldehyde, a brick-red solid of Cu_2O forms. (See Figure 15.13.) The test is negative with simple aldehydes and ketones.

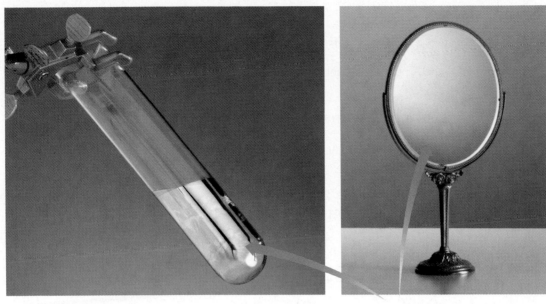

Figure 15.12 In Tollens' test, a silver mirror forms when the oxidation of an aldehyde reduces silver ion to metallic silver. The silvery surface of a mirror is formed in a similar way.

Q *What is the product of the oxidation of an aldehyde?*

$$Ag^+ + 1e^- \longrightarrow Ag(s)$$

$$\underset{\text{2-Hydroxypropanal}}{CH_3-\overset{OH}{\underset{|}{CH}}-\overset{O}{\overset{\|}{C}}-H} + \underset{\substack{\text{Benedict's} \\ \text{reagent}}}{Cu^{2+}} \longrightarrow \underset{\text{Brick-red solid}}{Cu_2O(s)} + \underset{\text{2-Hydroxypropanoic acid}}{CH_3-\overset{OH}{\underset{|}{CH}}-\overset{O}{\overset{\|}{C}}-OH}$$

Because many sugars such as glucose contain this type of aldehyde grouping, Benedict's reagent can be used to determine the presence of glucose in blood or urine.

$$\begin{array}{c}
\overset{O}{\overset{\|}{C}}-H \\
H-C-OH \\
HO-C-H \\
H-C-OH \\
H-C-OH \\
CH_2OH \\
\text{D-Glucose}
\end{array} + \underset{\substack{\text{Benedict's} \\ \text{(blue)}}}{Cu^{2+}} \longrightarrow
\begin{array}{c}
\overset{O}{\overset{\|}{C}}-OH \\
H-C-OH \\
HO-C-H \\
H-C-OH \\
H-C-OH \\
CH_2OH \\
\text{D-Gluconic acid}
\end{array} + \underset{\substack{\text{(brick-red)}}}{Cu_2O(s)}$$

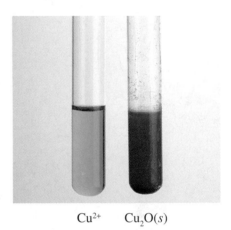

Cu^{2+} $Cu_2O(s)$

Figure 15.13 The blue Cu^{2+} in Benedict's solution forms a brick-red solid of Cu_2O in a positive test for many sugars and aldehydes with adjacent hydroxyl groups.

Q *Which test tube contains an aldehyde with an adjacent hydroxyl group?*

SAMPLE PROBLEM 15.10

Alcohol Oxidation

Draw the condensed structural formula of the alcohol needed to give each of the following oxidation products:

a. $CH_3-\overset{O}{\overset{\|}{C}}-CH_2-CH_3$ **b.** $CH_3-\overset{CH_3}{\underset{|}{CH}}-\overset{O}{\overset{\|}{C}}-H$ **c.** $CH_3-\overset{O}{\overset{\|}{C}}-OH$

Solution

a. A secondary alcohol oxidizes to a ketone.

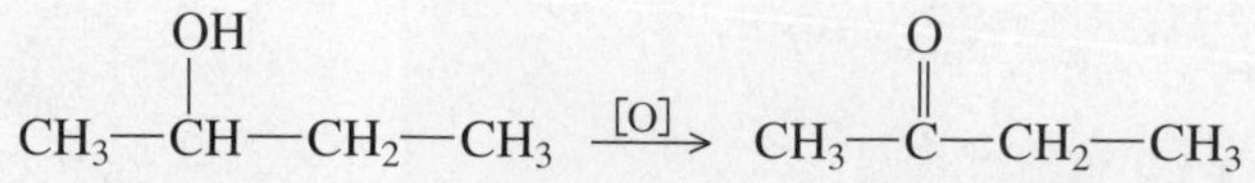

$$\underset{\underset{CH_3-CH-CH_2-CH_3}{\overset{OH}{|}}}{} \xrightarrow{[O]} \underset{\underset{CH_3-C-CH_2-CH_3}{\overset{O}{\parallel}}}{}$$

b. A primary alcohol oxidizes to an aldehyde with a mild oxidizing agent.

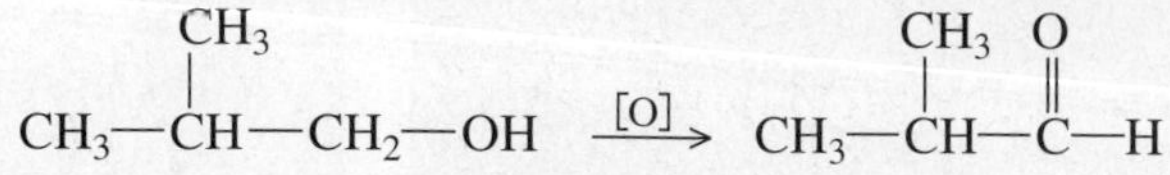

$$\underset{CH_3-\underset{\overset{|}{CH}}{\overset{CH_3}{|}}-CH_2-OH}{} \xrightarrow{[O]} CH_3-\underset{\overset{|}{CH}}{\overset{CH_3}{|}}-\underset{\overset{\parallel}{C}}{\overset{O}{\parallel}}-H$$

c. A primary alcohol oxidizes to an aldehyde, which oxidizes further to a carboxylic acid.

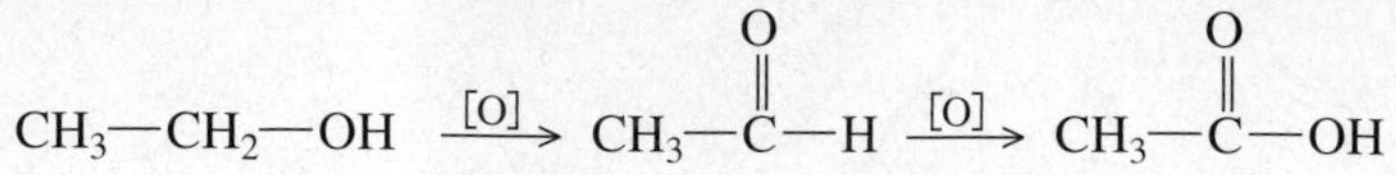

$$CH_3-CH_2-OH \xrightarrow{[O]} CH_3-\overset{O}{\underset{\parallel}{C}}-H \xrightarrow{[O]} CH_3-\overset{O}{\underset{\parallel}{C}}-OH$$

Study Check

What is the IUPAC name of the alcohol that oxidized to cyclohexanone?

SAMPLE PROBLEM 15.11

Tollens' Test

Draw the condensed structural formula of the product of oxidation, if any, when Tollens' reagent is added to each of the following compounds:

a. propanal   **b.** propanone   **c.** 2-methylbutanal

Solution

Tollens' reagent will oxidize aldehydes, but not ketones.

a. 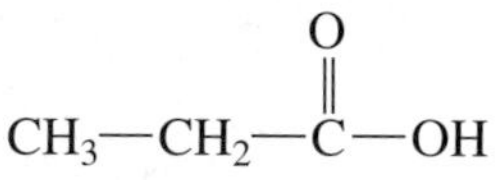 $CH_3-CH_2-\overset{O}{\underset{\parallel}{C}}-OH$ **b.** no reaction   **c.** 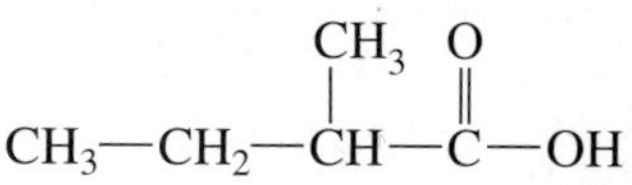 $CH_3-CH_2-\underset{\overset{|}{CH}}{\overset{CH_3}{|}}-\overset{O}{\underset{\parallel}{C}}-OH$

Study Check

Why does a silver mirror form when Tollens' reagent is added to a test tube containing benzaldehyde?

Reduction of Aldehydes and Ketones

Aldehydes and ketones are reduced by sodium borohydride ($NaBH_4$) or hydrogen (H_2). **Reduction** decreases the number of carbon–oxygen bonds by the addition of hydrogen or the loss of oxygen. Aldehydes are reduced to primary alcohols, and ketones to secondary alcohols. A catalyst such as nickel, platinum, or palladium is used with hydrogenation.

Aldehydes Reduce to Primary Alcohols

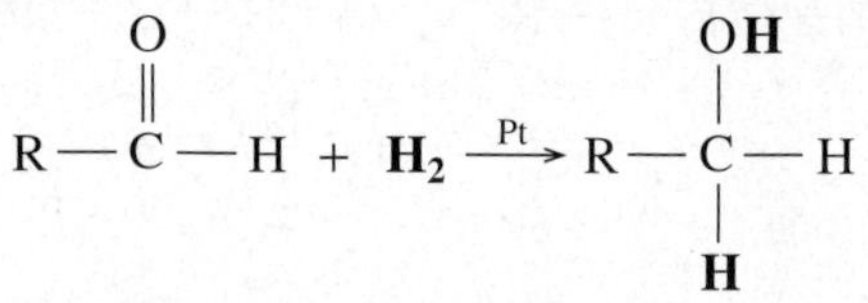

$$R-\overset{O}{\underset{\parallel}{C}}-H + \mathbf{H_2} \xrightarrow{Pt} R-\underset{\underset{\mathbf{H}}{\overset{|}{|}}}{\overset{OH}{\overset{|}{C}}}-H$$

Aldehyde   1° alcohol

Ketones Reduce to Secondary Alcohols

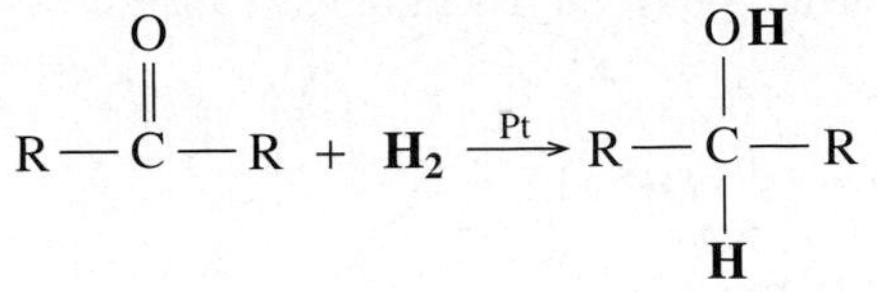

$$R-\underset{\parallel O}{C}-R + H_2 \xrightarrow{Pt} R-\underset{\underset{H}{|}}{\overset{\overset{OH}{|}}{C}}-R$$

Ketone   2° alcohol

Examples

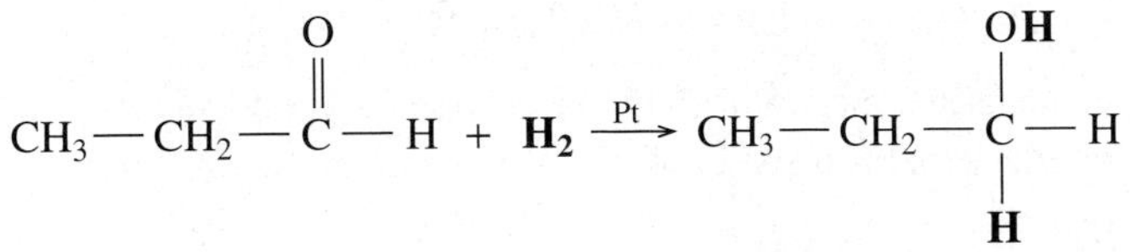

$$CH_3-CH_2-\underset{\parallel O}{C}-H + H_2 \xrightarrow{Pt} CH_3-CH_2-\underset{\underset{H}{|}}{\overset{\overset{OH}{|}}{C}}-H$$

Propionaldehyde   1-Propanol (1° alcohol)

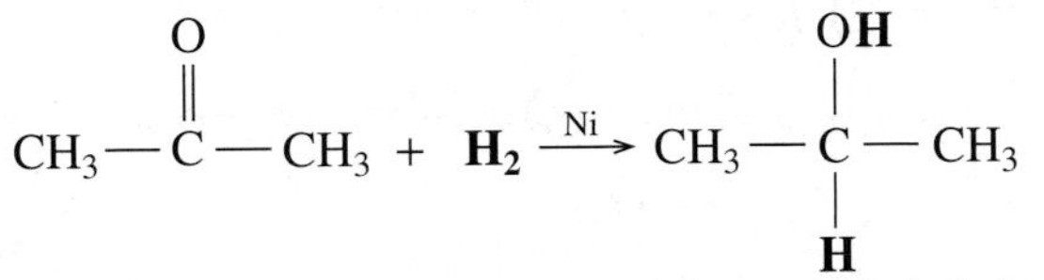

$$CH_3-\underset{\parallel O}{C}-CH_3 + H_2 \xrightarrow{Ni} CH_3-\underset{\underset{H}{|}}{\overset{\overset{OH}{|}}{C}}-CH_3$$

Dimethyl ketone   2-Propanol (2° alcohol)

SAMPLE PROBLEM 15.12

Reduction of Carbonyl Groups

Write an equation for the reduction of cyclopentanone in the presence of a nickel catalyst.

Solution

The reacting molecule is a cyclic ketone that has five carbon atoms. Hydrogen atoms will add to the carbon and oxygen in the carbonyl group to form the corresponding secondary alcohol.

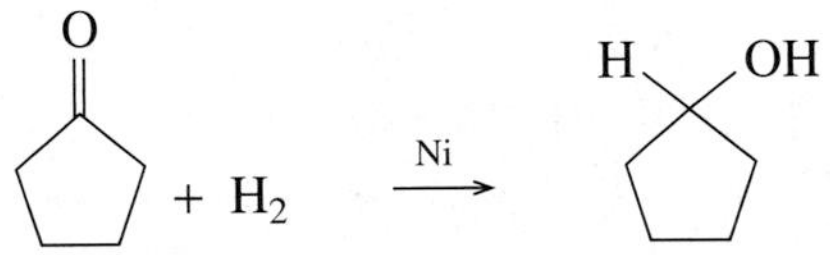

Cyclopentanone + H₂ →(Ni) Cyclopentanol

Study Check

What is the name of the product obtained from the hydrogenation of propionaldehyde?

QUESTIONS AND PROBLEMS

Oxidation and Reduction

15.29 Draw the condensed structural formula of the alcohol needed to give each of the following oxidation products:
a. formaldehyde **b.** cyclopentanone **c.** 2-butanone
d. benzaldehyde **e.** 3-methylcyclohexanone

15.30 Draw the condensed structural formula of the alcohol needed to give each of the following oxidation products:
a. acetaldehyde **b.** 3-methylbutanone **c.** cyclohexanone
d. propionaldehyde **e.** 3-methylbutanal

15.31 Draw the condensed structural formula of the organic product when each of the following alcohols is oxidized [O] (if no reaction, write *none*):

a. $CH_3CH_2CH_2CH_2CH_2OH$ b.   OH
                                      |
                            $CH_3CH_2CHCH_3$

c.   OH
       |
      (cyclohexanol)

     OH   CH₃   CH₃
     | | |

d. $CH_3 — CH — CH_2 — CH — CH_3$ e. $CH_3 — CH — CH_2 — CH_2 — OH$

15.32 Draw the condensed structural formula of the organic product when each of the following alcohols is oxidized [O] (if no reaction, write *none*):

a. ⬜—CH_2OH b. CH_3
                         |
                $CH_3 — CH — CH_2 — CH — OH$
                                     |
                                    CH_3

c.   OH
       |
$CH_3 — CH_2 — C — CH_3$
       |
      CH_3

d. $CH_3CHCHCH_2CH_3$ with OH on C2 and OH below

e. ⬜—OH

15.33 Give the condensed structural formula of the organic product formed when each of the following is reduced by hydrogen in the presence of a nickel catalyst:

a. butyraldehyde   **b.** acetone
c. 3-bromohexanal   **d.** 2-methyl-3-pentanone

15.34 Give the condensed structural formula of the organic product formed when each of the following is reduced by hydrogen in the presence of a nickel catalyst:

a. ethyl propyl ketone   **b.** formaldehyde
c. 3-chlorocyclopentanone **d.** 2-pentanone

LEARNING GOAL

Write the products of the addition of alcohols to aldehydes and ketones.

15.7 Addition Reactions

One of the most common reactions of aldehydes and ketones is the addition of polar molecules to the carbonyl group. The carbonyl group is reactive because of the polarity of the $C=O$ double bond. In addition reactions, the negative part of the adding molecule bonds with the partially positively charged carbonyl carbon. The positive part, usually a proton, combines with the partially negatively charged carbonyl oxygen. This type of addition to the carbonyl group can be illustrated as follows:

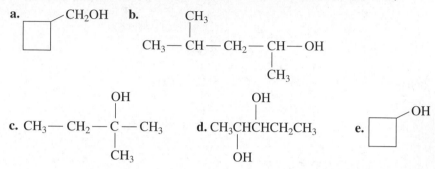

Carbonyl group
of aldehyde or
ketone   Adding molecule

In general, aldehydes are more reactive than ketones because the carbonyl carbon is more positive in aldehydes. Also, the presence of two alkyl groups makes it more difficult for a molecule to form bonds with the carbon in the carbonyl group.

Addition of Water

The components of water add to aldehydes and ketones to give carbonyl hydrates in the presence of acid or base. The negative — OH group bonds with the carbonyl carbon, while the — H bonds to the negative oxygen. In water, the simplest aldehyde, formaldehyde, forms its hydrate called formalin, which is used to preserve tissues. Other aldehydes form hydrates in water as well, but not with as high a percentage as formaldehyde. The carbonyl group in ketones also reacts with water, but their hydrates are not very stable.

$$
\begin{array}{c}
\text{H} \\
\diagdown \\
\quad \text{C}=\text{O} \\
\diagup \\
\text{H}
\end{array}
\;+\; \text{H}_2\text{O} \; \underset{}{\overset{\text{H}^+}{\rightleftharpoons}} \;
\begin{array}{c}
\text{H} \qquad \text{OH} \\
\diagdown \;\; \diagup \\
\text{C} \\
\diagup \;\; \diagdown \\
\text{H} \qquad \text{OH}
\end{array}
$$

Formaldehyde   Formalin

$$
\begin{array}{c}
\text{H}_3\text{C} \\
\diagdown \\
\quad \text{C}=\text{O} \\
\diagup \\
\text{H}
\end{array}
\;+\; \text{H}_2\text{O} \; \underset{}{\overset{\text{H}^+}{\rightleftharpoons}} \;
\begin{array}{c}
\text{H}_3\text{C} \qquad \text{OH} \\
\diagdown \;\; \diagup \\
\text{C} \\
\diagup \;\; \diagdown \\
\text{H} \qquad \text{OH}
\end{array}
$$

Acetaldehyde   Hydrate

Chloral, which is an aldehyde with chlorine atoms, forms a hydrate known as chloral hydrate, the substance in "knock out" drops.

$$
\begin{array}{c}
\text{Cl}_3\text{C} \\
\diagdown \\
\quad \text{C}=\text{O} \\
\diagup \\
\text{H}
\end{array}
\;+\; \text{H}_2\text{O} \; \rightleftharpoons \;
\begin{array}{c}
\text{Cl}_3\text{C} \qquad \text{OH} \\
\diagdown \;\; \diagup \\
\text{C} \\
\diagup \;\; \diagdown \\
\text{H} \qquad \text{OH}
\end{array}
$$

Chloral   Chloral hydrate

Acetal Formation

Similar to the addition of water to form hydrates, aldehydes and ketones react with alcohols in the presence of an acid catalyst to form **acetals.** (Ketal is an older term previously used for acetals from ketones.) In the acetal product, the two — OR groups are added to the carbonyl carbon and a molecule of water is eliminated. Recall that the symbol R represents an alkyl group. The general reaction is written as follows:

$$
\begin{array}{c}
\text{O} \\
\parallel \\
\text{R}-\text{C}-\text{H}
\end{array}
\;+\; 2\text{R}-\text{OH} \; \underset{}{\overset{\text{H}^+}{\rightleftharpoons}} \;
\begin{array}{c}
\text{OR} \\
| \\
\text{R}-\text{C}-\text{H} \\
| \\
\text{OR}
\end{array}
\;+\; \text{H}_2\text{O}
$$

Aldehyde   Two alcohols   Acetal

$$
\begin{array}{c}
\text{O} \\
\parallel \\
\text{R}-\text{C}-\text{R}
\end{array}
\;+\; 2\text{R}-\text{OH} \; \underset{}{\overset{\text{H}^+}{\rightleftharpoons}} \;
\begin{array}{c}
\text{OR} \\
| \\
\text{R}-\text{C}-\text{R} \\
| \\
\text{OR}
\end{array}
\;+\; \text{H}_2\text{O}
$$

Ketone   Two alcohols   Acetal

Examples of acetal formation

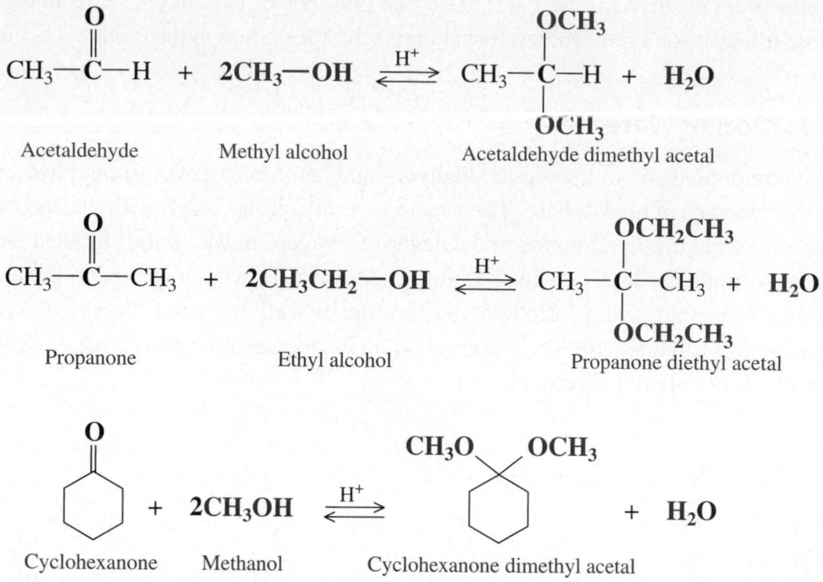

$$CH_3-\overset{\overset{\displaystyle O}{\|}}{C}-H \; + \; 2CH_3-OH \; \underset{\longleftarrow}{\overset{H^+}{\longrightarrow}} \; CH_3-\overset{\overset{\displaystyle OCH_3}{|}}{\underset{\underset{\displaystyle OCH_3}{|}}{C}}-H \; + \; H_2O$$

Acetaldehyde   Methyl alcohol   Acetaldehyde dimethyl acetal

$$CH_3-\overset{\overset{\displaystyle O}{\|}}{C}-CH_3 \; + \; 2CH_3CH_2-OH \; \underset{\longleftarrow}{\overset{H^+}{\longrightarrow}} \; CH_3-\overset{\overset{\displaystyle OCH_2CH_3}{|}}{\underset{\underset{\displaystyle OCH_2CH_3}{|}}{C}}-CH_3 \; + \; H_2O$$

Propanone   Ethyl alcohol   Propanone diethyl acetal

Cyclohexanone   Methanol   Cyclohexanone dimethyl acetal

Hemiacetal Intermediate

In the process of forming acetals, an intermediate called a **hemiacetal** forms when one of the two alcohol molecules adds to the carbonyl carbon. The term *hemi* indicates that the hemiacetal is halfway to an acetal. Most of the hemiacetal intermediates are unstable and difficult to isolate from the reaction mixture. In the next step, the second alcohol is added to produce the more stable acetal. Acetals are stable and can be isolated from the reaction mixture.

Examples of forming a hemiacetal intermediate

R—C—R + R—OH ⇌ { R—C—H } + R—OH ⇌ R—C—H + H₂O

Aldehyde   Alcohol   Hemiacetal   Acetal
                                intermediate

CH₃—C—H + CH₃OH ⇌ { CH₃—C—H } + CH₃OH ⇌ CH₃—C—H + H₂O

Acetaldehyde   Methyl   Hemiacetal   Acetaldehyde dimethyl acetal
               alcohol   intermediate

Cyclohexanone + CH₃OH ⇌ { Hemiacetal intermediate } + CH₃OH ⇌ + H₂O   Cyclohexanone dimethyl acetal

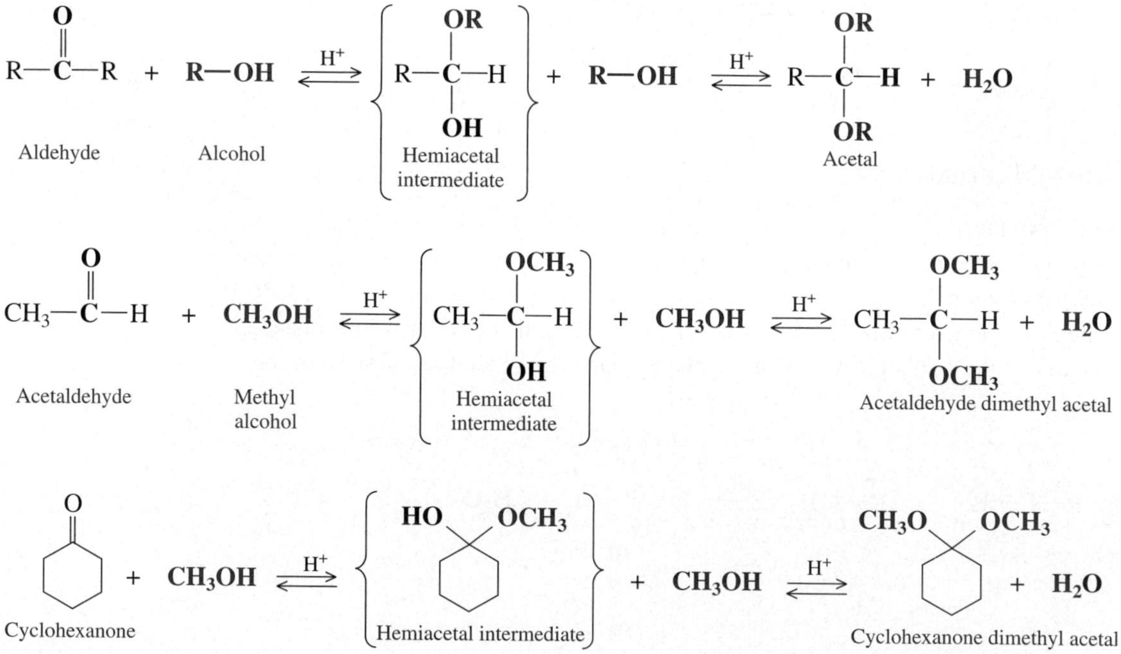

Both the step to the hemiacetal and the step to the acetal are reversible. The forward reaction to form the acetal is favored by removing water from the reaction mixture. The reverse reaction, which is the hydrolysis of an acetal, is favored by adding water to drive the equilibrium back to the ketone or aldehyde.

SAMPLE PROBLEM 15.13

Identifying Addition Products

Identify each of the following structural formulas as a hemiacetal or acetal. Write the structural formulas for the carbonyl compounds and alcohols that are the reactants.

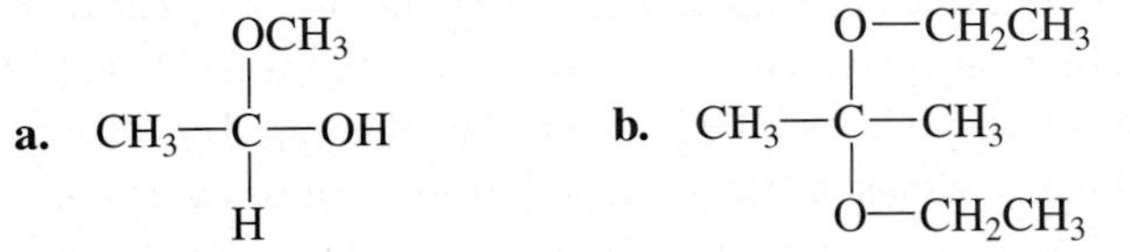

$$
\text{a.} \quad CH_3-\underset{\underset{H}{|}}{\overset{\overset{OCH_3}{|}}{C}}-OH
\qquad\qquad
\text{b.} \quad CH_3-\underset{\underset{O-CH_2CH_3}{|}}{\overset{\overset{O-CH_2CH_3}{|}}{C}}-CH_3
$$

Solution

a. Hemiacetal. The reactants are CH_3—CHO and CH_3—OH.
b. Acetal. The reactants are CH_3—CO—CH_3 and CH_3CH_2—OH.

Study Check

Identify the compound below as a hemiacetal or an acetal.

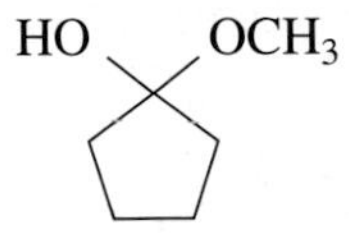

SAMPLE PROBLEM 15.14

Acetals

Write the structural formula of the hemiacetal and acetal products when methanol adds to propionaldehyde.

Solution

To form the hemiacetal, the hydrogen from the alcohol adds to the oxygen of the carbonyl group to form a new hydroxyl group and the remaining part of the alcohol adds to the carbon atom in the carbonyl group. The acetal forms when a second molecule of methanol is added to the carbonyl carbon atom.

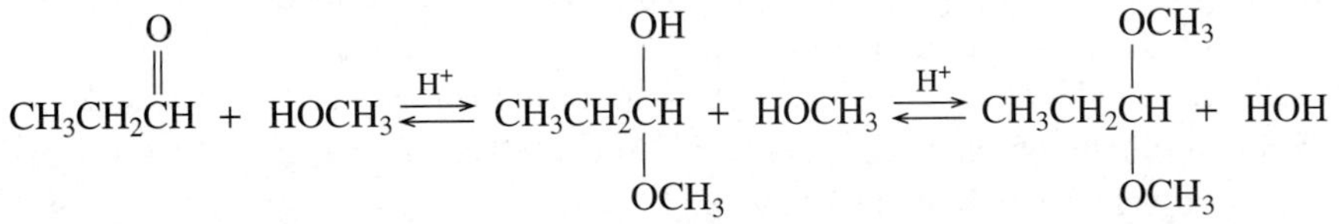

$$
\underset{\text{Aldehyde}}{CH_3CH_2\overset{\overset{O}{\|}}{C}H} + \underset{\text{Methanol}}{HOCH_3} \underset{}{\overset{H^+}{\rightleftarrows}} \underset{\text{Hemiacetal}}{CH_3CH_2\underset{\underset{OCH_3}{|}}{\overset{\overset{OH}{|}}{C}H}} + HOCH_3 \overset{H^+}{\rightleftarrows} \underset{\text{Acetal}}{CH_3CH_2\underset{\underset{OCH_3}{|}}{\overset{\overset{OCH_3}{|}}{C}H}} + HOH
$$

Study Check

What is the structural formula of the acetal produced when methanol adds to propanone?

Cyclic Hemiacetals

One very important type of hemiacetal that can be isolated is a cyclic hemiacetal that forms when the carbonyl group and the —OH group are in the *same* molecule.

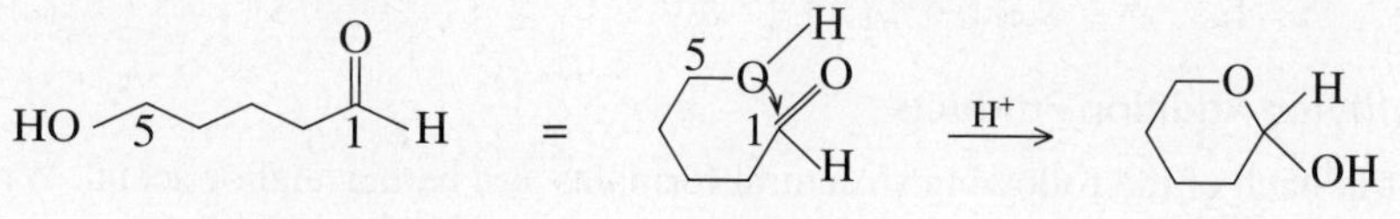

Open chain   Cyclic hemiacetal

The five- and six-atom cyclic hemiacetals and acetals are more stable than their open-chain structures. For example, glucose, a simple sugar, forms a hemiacetal when the hydroxyl group on carbon 5 bonds with the carbonyl group. The hemiacetal of glucose is so stable that almost all the glucose (99%) exists as the hemiacetal in aqueous solution. We will discuss carbohydrates and their structures in Chapter 16.

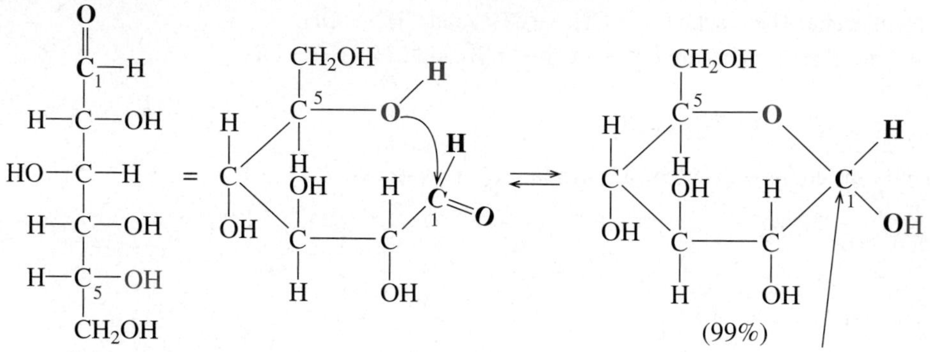

Glucose   Formation of cyclic hemiacetal   New chiral carbon

An alcohol can add to the cyclic hemiacetal to form a cyclic acetal. This reaction is also very important in carbohydrate chemistry. It is the linkage used by glucose molecules to bond to other glucose molecules to form long chains.

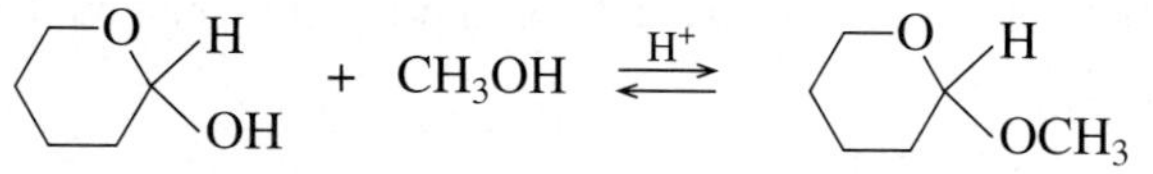

Cyclic hemiacetal   Cyclic acetal

QUESTIONS AND PROBLEMS

Addition Reactions

15.35 Write the structural formula of the organic product formed by the addition of water to each of the following:
       a. acetaldehyde   **b.** formaldehyde

15.36 Write the structural formula of the organic product formed by the addition of water to each of the following:
       a. propanal   **b.** propanone

15.37 Indicate whether each of the following structural formulas is a hemiacetal, acetal, or neither.

a. $CH_3{-}CH_2{-}O{-}CH_2{-}OH$ **b.** 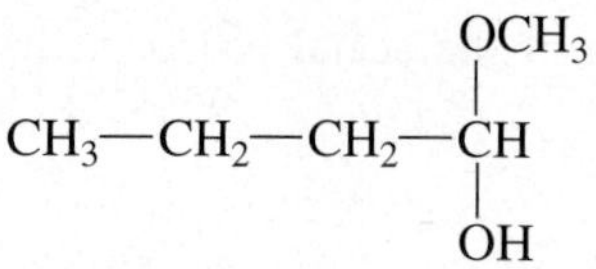   **c.**

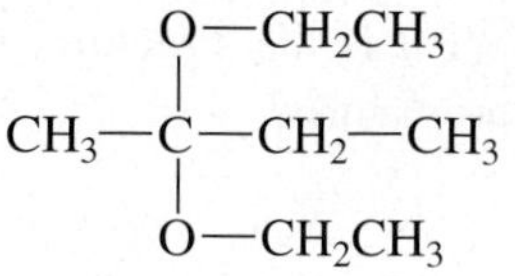

d. 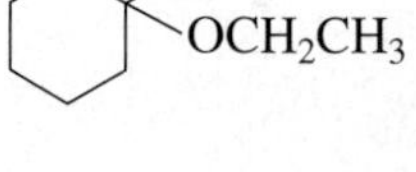   **e.**

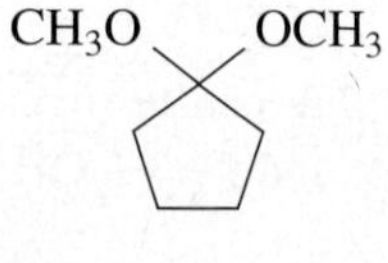

15.38 Indicate whether each of the following structural formulas is a hemiacetal, acetal, or neither.

a. $CH_3—CH_2—O—CH_2—CH_3$ **b.** $HO—CH_2—CH_2—O—CH_2—CH_2—O—CH_3$

c. 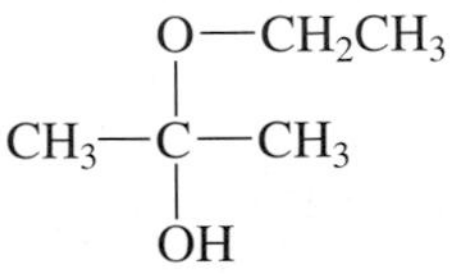

d. 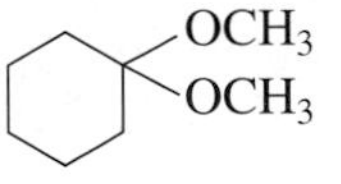

e. 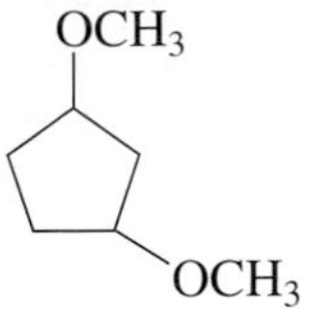

15.39 Draw the structural formula of the hemiacetal formed by adding methanol to each of the following compounds.

a. ethanal **b.** propanone
c. cyclopentanone **d.** butanal

15.40 Draw the structural formula of the hemiacetal formed by adding ethanol to each of the following compounds.

a. propanal **b.** 2-butanone
c. cyclohexanone **d.** formaldehyde

15.41 Draw the structural formulas of the acetal formed by adding a second methanol to the compounds in problem 15.39.

15.42 Draw the structural formulas of the acetal formed by adding a second ethanol to the compounds in problem 15.40.

Chapter Review

15.1 Structure and Bonding

Aldehydes and ketones contain a carbonyl group (C=O), which consists of a double bond between a carbon and an oxygen atom. Similar to the double bond in alkenes, the second bond is a pi bond, which forms when *p* orbitals overlap. However, in contrast to the C=C double bond, the C=O is strongly polar. In aldehydes, the carbonyl group appears at the end of carbon chains. In ketones, the carbonyl group occurs between two alkyl groups.

15.2 Naming Aldehydes and Ketones

In the IUPAC system, the *e* in the corresponding alkane is replaced with *al* for aldehydes, and *one* for ketones. For ketones with more than four carbon atoms in the main chain, the carbonyl group is numbered to show its location. Many of the simple aldehydes and ketones use common names.

15.3 Some Important Aldehydes and Ketones

Formaldehyde is the simplest aldehyde used in solution as formalin to preserve tissues as well as in the manufacture of many commercial products. Acetone is used as a solvent in paint and nail-polish removers. Many aldehydes and ketones are found in biological systems, flavorings, and drugs.

15.4 Physical Properties

Because they contain a polar carbonyl group, aldehydes and ketones have higher boiling points than alkanes and ethers. However, their boiling points are lower than alcohols because aldehydes and ketones cannot hydrogen bond with each other. Aldehydes and ketones can hydrogen bond with water molecules, which makes carbonyl compounds with one to four carbon atoms soluble in water.

15.5 Chiral Molecules

Chiral molecules are molecules with mirror images that cannot be superimposed on each other. These types of stereoisomers are called enantiomers. A chiral molecule must have at least one chiral carbon, which is a carbon bonded to four different atoms or groups of atoms. The Fischer projection is a simplified way to draw the arrangements of atoms by placing the chiral carbons at the center of crossed lines. The names of the mirror images are labeled D or L to differentiate between the enantiomers.

15.6 Oxidation and Reduction

Aldehydes are easily oxidized to carboxylic acids, but ketones do not oxidize further. Aldehydes, but not ketones, react with Tollens' reagent to give silver mirrors. In Benedict's test, aldehydes with adjacent hydroxyl groups reduce blue Cu^{2+} to give a brick-red Cu_2O solid. The reduction of aldehydes with hydrogen produces primary alcohols, while ketones are reduced to secondary alcohols.

15.7 Addition Reactions

Water and alcohols can add to the carbonyl group of aldehydes and ketones. The addition of one alcohol forms a hemiacetal, while the addition of two alcohols forms an acetal. Hemiacetals are not usually stable, except for cyclic hemiacetals, which are the most common form of simple sugars such as glucose.

Summary of Naming

Structure	Family	IUPAC Name	Common Name
$H-\overset{\displaystyle O}{\overset{\|}{C}}-H$	Aldehyde	Methanal	Formaldehyde
$CH_3-\overset{\displaystyle O}{\overset{\|}{C}}-CH_3$	Ketone	Propanone	Acetone; Dimethyl ketone

Summary of Reactions

Oxidation of Aldehydes to Carboxylic Acids

$$R-\overset{\displaystyle O}{\overset{\|}{C}}-H \xrightarrow{[O]} CH_3-\overset{\displaystyle O}{\overset{\|}{C}}-OH$$

Aldehyde   Carboxylic acid

$$CH_3-\overset{\displaystyle O}{\overset{\|}{C}}-H \xrightarrow{[O]} CH_3-\overset{\displaystyle O}{\overset{\|}{C}}-OH$$

Acetaldehyde   Acetic acid

Reduction of Aldehydes to Primary Alcohols

$$R-\overset{\displaystyle O}{\overset{\|}{C}}-H + H_2 \xrightarrow{Ni} R-\overset{\displaystyle OH}{\overset{\|}{\underset{\underset{\displaystyle H}{\|}}{C}}}-H$$

Aldehyde   Primary alcohol

$$CH_3-\overset{\displaystyle O}{\overset{\|}{C}}-H + H_2 \xrightarrow{Ni} CH_3-\overset{\displaystyle OH}{\overset{\|}{CH_2}}$$

Acetaldehyde   Ethanol

Reduction of Ketones to Secondary Alcohols

$$R-\overset{\displaystyle O}{\overset{\|}{C}}-R + H_2 \xrightarrow{Ni} R-\overset{\displaystyle OH}{\overset{\|}{\underset{\underset{\displaystyle H}{\|}}{C}}}-R$$

Ketone   Secondary alcohol

$$CH_3-\overset{\displaystyle O}{\overset{\|}{C}}-CH_3 + H_2 \xrightarrow{Ni} CH_3-\overset{\displaystyle OH}{\overset{\|}{CH}}-CH_3$$

Acetone   2-Propanol

Addition of Water to Aldehydes

$$\overset{\displaystyle H}{\underset{\displaystyle H}{>}}C=O + H_2O \underset{}{\overset{H^+}{\rightleftharpoons}} \overset{\displaystyle H}{\underset{\displaystyle H}{>}}C\overset{\displaystyle OH}{\underset{\displaystyle OH}{<}}$$

Formaldehyde   Formalin

$$\overset{\displaystyle H_3C}{\underset{\displaystyle H}{>}}C=O + H_2O \underset{}{\overset{H^+}{\rightleftharpoons}} \overset{\displaystyle H_3C}{\underset{\displaystyle H}{>}}C\overset{\displaystyle OH}{\underset{\displaystyle OH}{<}}$$

Acetaldehyde   Hydrate

Addition of Alcohols to Form Hemiacetals and Acetals

From aldehydes

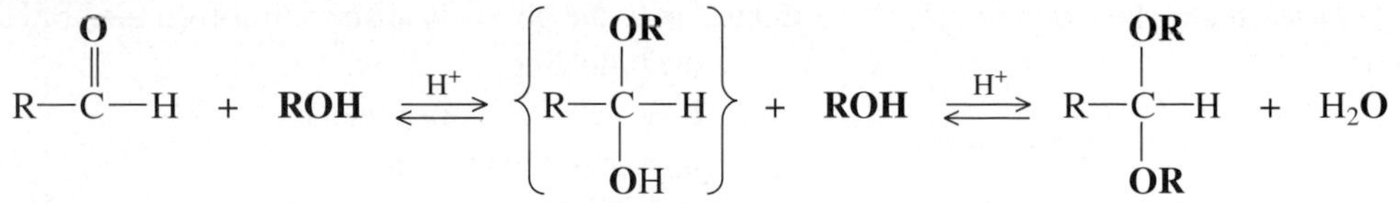

Aldehyde   Alcohol   Hemiacetal   Acetal

Example

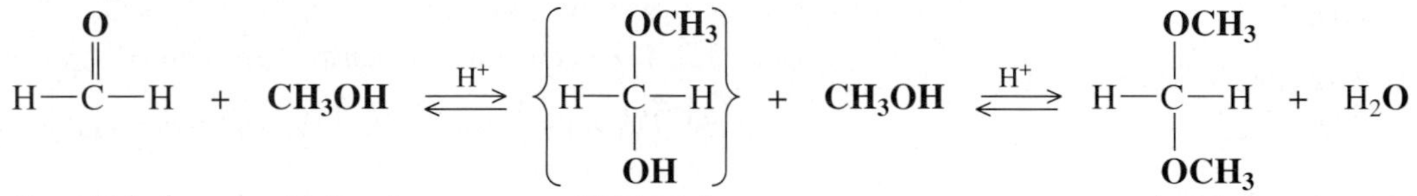

Formaldehyde   Methanol   Hemiacetal   Acetal

From ketones

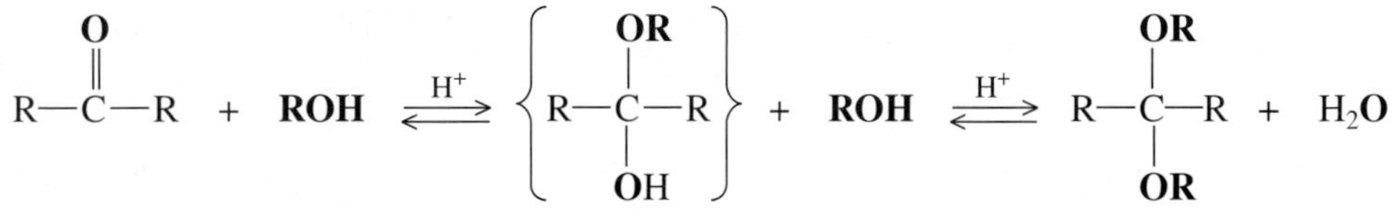

Ketone   Alcohol   Hemiacetal   Acetal

Example

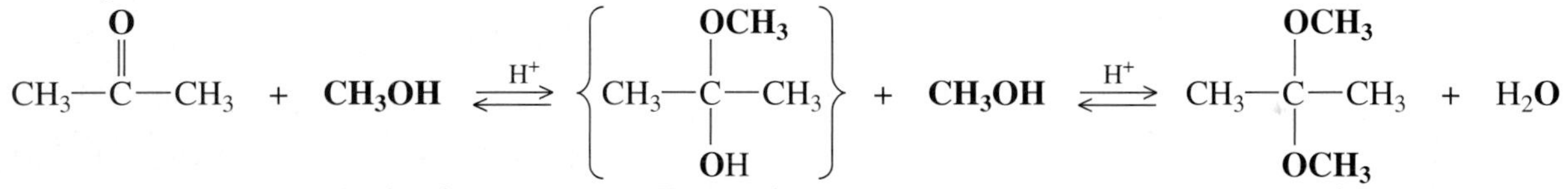

Acetone   Methanol   Hemiacetal   Acetal

Key Terms

acetal The product of the addition of two alcohols to an aldehyde or ketone.

achiral Molecules with mirror images that are superimposable.

aldehyde An organic compound with a carbonyl functional group and at least one hydrogen.

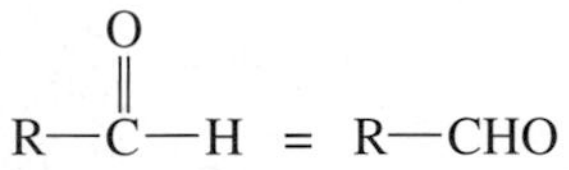

$$R-\overset{\overset{\displaystyle O}{\|}}{C}-H \; = \; R-CHO$$

Benedict's test A test for aldehydes with adjacent hydroxyl groups in which Cu^{2+} ($CuSO_4$) ions in Benedict's reagent are reduced to a brick-red solid of Cu_2O.

carbonyl group A functional group that contains a carbon–oxygen double bond ($C=O$).

chiral Objects or molecules that have mirror images that cannot be superimposed on each other.

chiral carbon A carbon atom that is bonded to four different atoms or groups of atoms.

enantiomers Stereoisomers that are mirror images that cannot be superimposed on each other.

Fischer projection A system for drawing stereoisomers that shows horizontal lines for bonds coming forward, and vertical lines for bonds going back with the chiral atom at the center.

hemiacetal The product of the addition of one alcohol to the double bond of the carbonyl group in aldehydes and ketones.

ketone An organic compound in which the carbonyl functional group is bonded to two alkyl groups.

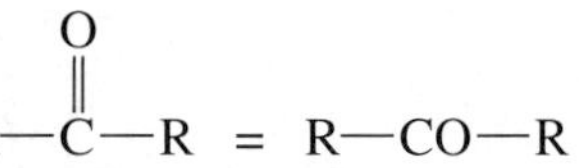

$$R-\overset{\overset{\displaystyle O}{\|}}{C}-R \; = \; R-CO-R$$

reduction A decrease in the number of carbon–oxygen bonds by the addition of hydrogen to a carbonyl bond. Aldehydes are reduced to primary alcohols; ketones to secondary alcohols.

stereoisomers Isomers that have atoms bonded in the same order, but with different arrangements in space.

Tollens' test A test for aldehydes in which Ag^+ in Tollens' reagent is reduced to metallic silver, which forms a "silver mirror" on the walls of the container.

Additional Problems

15.43 Describe the bonds between the carbon and oxygen in the carbonyl group.

15.44 Why does the $C=O$ double bond have a dipole, while the $C=C$ does not?

15.45 Write the constitutional isomers for the carbonyl compounds of C_4H_8O.

15.46 Write the constitutional isomers for the carbonyl compounds of $C_5H_{10}O$.

15.47 Give the IUPAC and common names (if any) for each of the following compounds:

a.

b.

c. $Cl-CH_2-CH_2-\overset{\displaystyle O}{\overset{\|}{C}}-H$

d. $CH_3-CH_2-\overset{\displaystyle O}{\overset{\|}{C}}-CH_2-\overset{\displaystyle OH}{\overset{|}{CH}}-CH_3$

e. $CH_3-\overset{\displaystyle Cl}{\overset{|}{CH}}-\overset{\displaystyle O}{\overset{\|}{C}}-CH_2-CH_3$

f.

15.48 Give the IUPAC and common names (if any) for each of the following compounds:

a. $CH_3-CH_2-\overset{\displaystyle O}{\overset{\|}{C}}-CH_3$

b.

c.

d. $CH_3-\overset{\displaystyle CH_3}{\overset{|}{CH}}-\overset{\displaystyle OH}{\overset{|}{CH}}-CH_2-\overset{\displaystyle O}{\overset{\|}{C}}-H$

e.

f. $CH_3-\overset{\displaystyle O}{\overset{\|}{C}}-CH_2-\overset{\displaystyle O}{\overset{\|}{C}}-H$

15.49 Draw the condensed structural formulas of each of the following:
a. 3-methylcyclopentanone
b. 4-chlorobenzaldehyde
c. 3-chloropropionaldehyde
d. ethyl methyl ketone
e. 3-methylhexanal
f. 2-heptanone, an alarm pheromone of bees

15.50 Draw the condensed structural formulas of each of the following:
a. propionaldehyde
b. 2-chlorobutanal
c. 2-methylcyclohexanone
d. 3,5-dimethylhexanal
e. 3-bromocyclopentanone
f. *trans*-2-hexenal, an alarm pheromone of an ant

15.51 Which of the following compounds are soluble in water?
a. $CH_3-CH_2-CH_2-CH_3$

b. $CH_3-CH_2-\overset{\displaystyle O}{\overset{\|}{C}}-H$　　**c.** $CH_3-\overset{\displaystyle O}{\overset{\|}{C}}-CH_3$

d. $CH_3-CH_2-CH_2-OH$

e. $CH_3-CH_2-\overset{\displaystyle O}{\overset{\|}{C}}-CH_2-CH_2-CH_3$

15.52 Which of the following compounds are soluble in water?

a. $CH_3-CH_2-\overset{\displaystyle O}{\overset{\|}{C}}-CH_3$　　**b.** $H-\overset{\displaystyle O}{\overset{\|}{C}}-H$

c. $CH_3-\overset{\displaystyle O}{\overset{\|}{C}}-H$　　　　　**d.** $CH_3-CH_2-CH_3$

e. $CH_3-CH_2-\overset{\displaystyle CH_3}{\overset{|}{CH}}-CH_2-CH_2-\overset{\displaystyle O}{\overset{\|}{C}}-H$

15.53 In each of the following pairs of compounds, select the compound with the higher boiling point:

a. CH_3-CH_2-OH or $CH_3-\overset{\displaystyle O}{\overset{\|}{C}}-H$

b. $CH_3-CH_2-CH_2-CH_3$ or $CH_3-CH_2-\overset{\displaystyle O}{\overset{\|}{C}}-H$

c. $CH_3-CH_2-CH_2-OH$ or $CH_3-\overset{\displaystyle O}{\overset{\|}{C}}-CH_3$

15.54 In each of the following pairs of compounds, select the compound with the higher boiling point:

a. 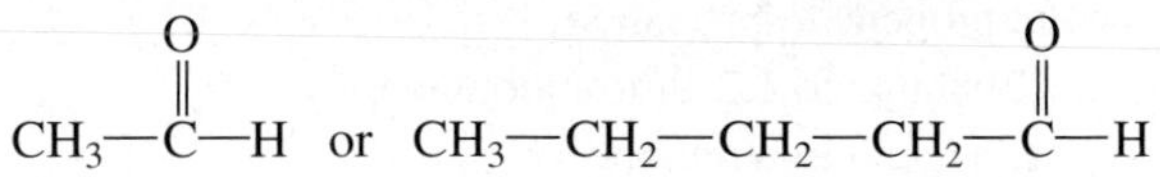 CH₃—C(=O)—H or CH₃—CH₂—CH₂—CH₂—C(=O)—H

b. CH₃—CH₂—CH₂—CH₃ or CH₃—C(=O)—CH₃

c. CH₃—CH₂—C(=O)—H or CH₃—CH(OH)—CH₃

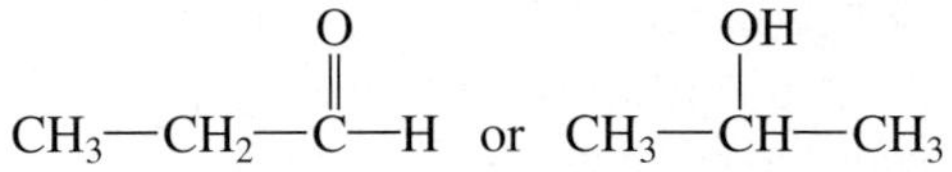

15.55 Identify the chiral carbons, if any, in each of the following compounds.

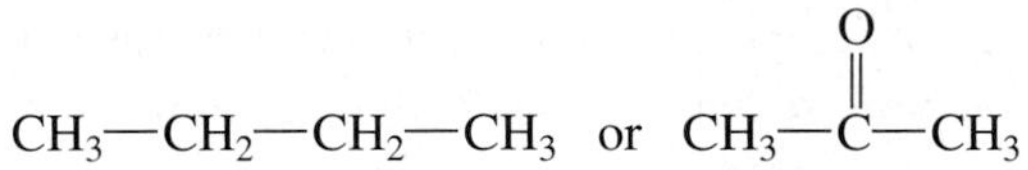

a. 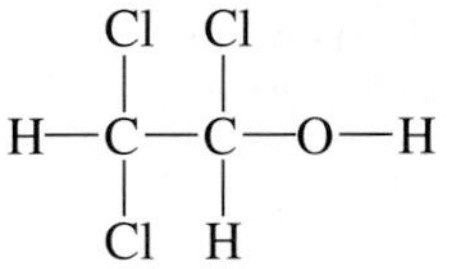 H—C(Cl)(Cl)—C(Cl)(H)—O—H

b. CH₃—C(H)=C(CH₃)—CH₃

c. HO—CH₂—CH(OH)—CH₂—OH

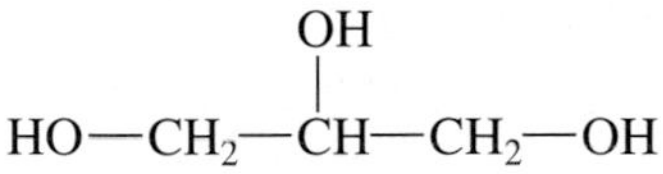

d. 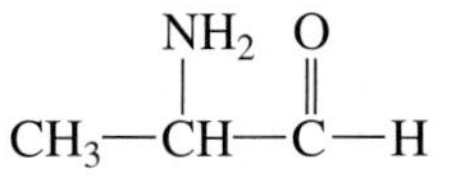 CH₃—CH(NH₂)—C(=O)—H

e. 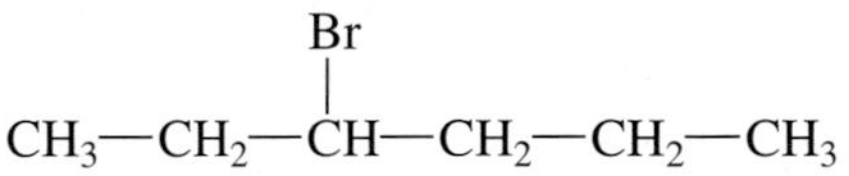 CH₃—CH₂—CH(Br)—CH₂—CH₂—CH₃

f. 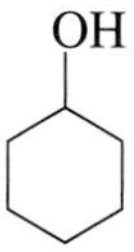 cyclohexanol (OH on ring)

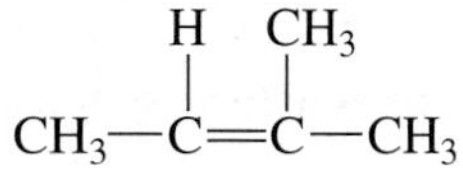

15.56 Identify the chiral carbons, if any, in each of the following compounds.

a. 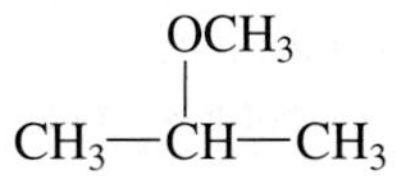 CH₃—CH(OCH₃)—CH₃

b. 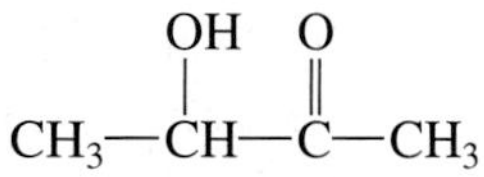 CH₃—CH(OH)—C(=O)—CH₃

c. CH₃—C(OH)(OH)—CH₃

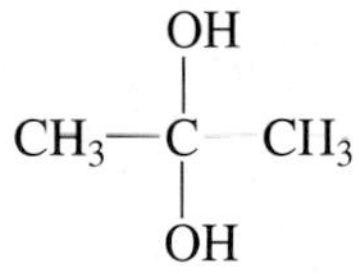

d. 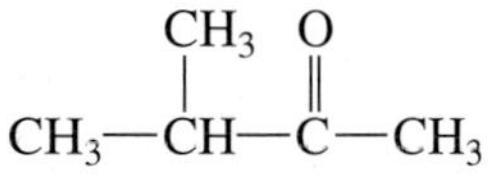 CH₃—CH(CH₃)—C(=O)—CH₃

e. 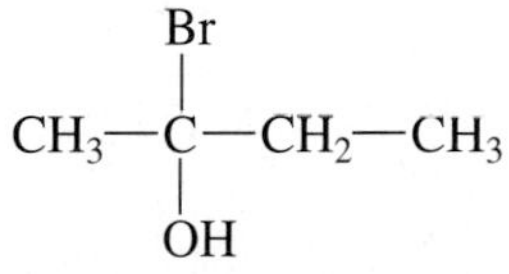 CH₃—C(Br)(OH)—CH₂—CH₃

f. 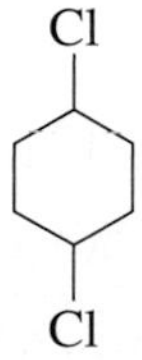 1,4-dichlorocyclohexane (Cl on ring)

15.57 Identify each of the following pairs of Fischer projections as enantiomers or identical compounds:

a. 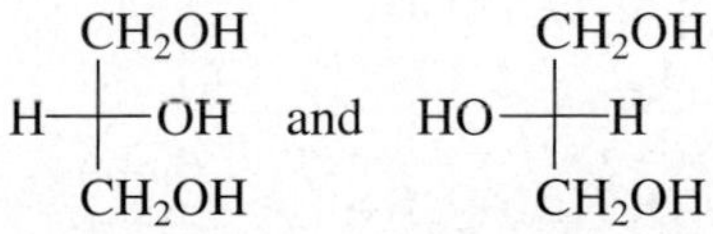
CH₂OH
H——OH
CH₂OH

and

CH₂OH
HO——H
CH₂OH

b. 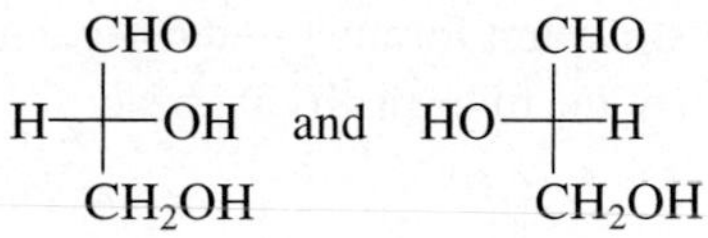
CHO
H——OH
CH₂OH

and

CHO
HO——H
CH₂OH

c. 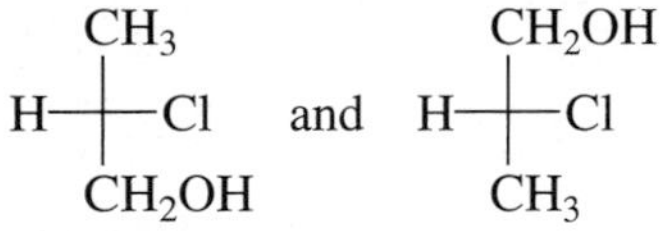
CH₃
H——Cl
CH₂OH

and

CH₂OH
H——Cl
CH₃

d. 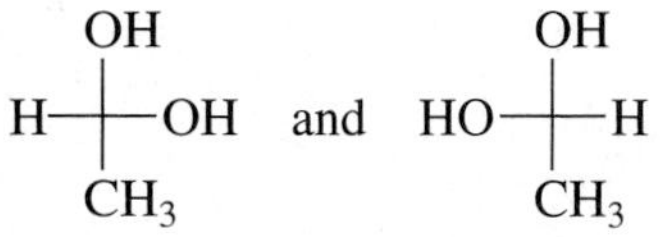
OH
H——OH
CH₃

and

OH
HO——H
CH₃

15.58 Identify each of the following pairs of Fischer projections as enantiomers or identical compounds:

a. 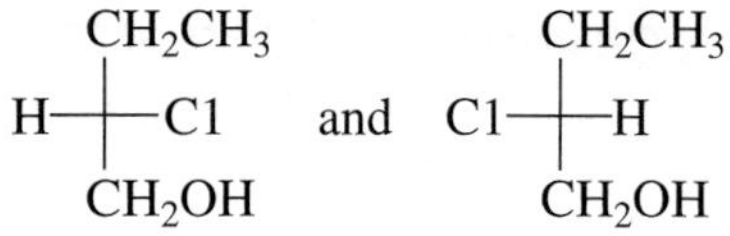
CH₂CH₃
H——Cl
CH₂OH

and

CH₂CH₃
Cl——H
CH₂OH

b. 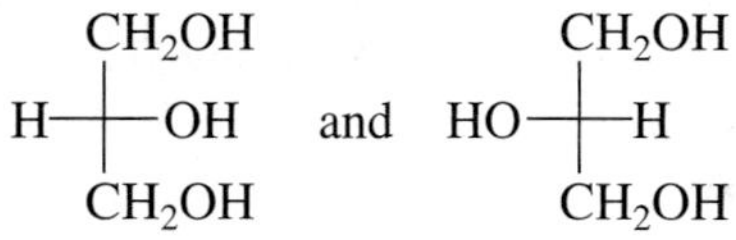
CH₂OH
H——OH
CH₂OH

and

CH₂OH
HO——H
CH₂OH

c. 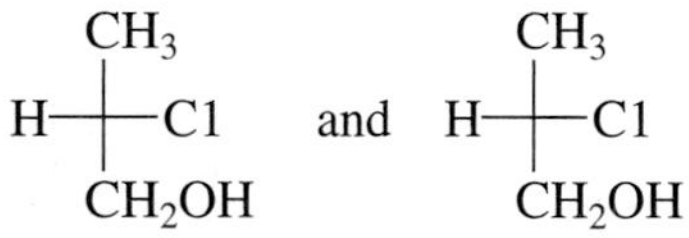
CH₃
H——Cl
CH₂OH

and

CH₃
H——Cl
CH₂OH

d. 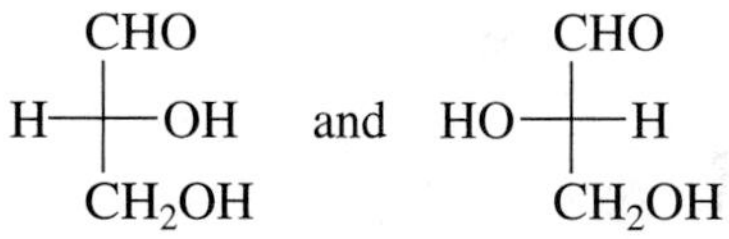
CHO
H——OH
CH₂OH

and

CHO
HO——H
CH₂OH

15.59 Draw the structural formula of the organic product when each of the following is oxidized:

a. CH₃—CH₂—CH₂—OH

b. 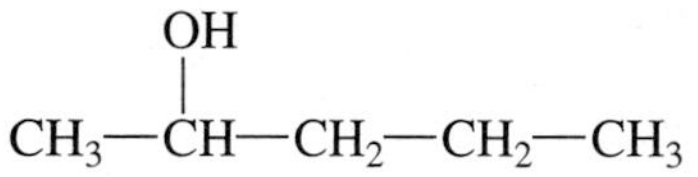 CH₃—CH(OH)—CH₂—CH₂—CH₃

c. 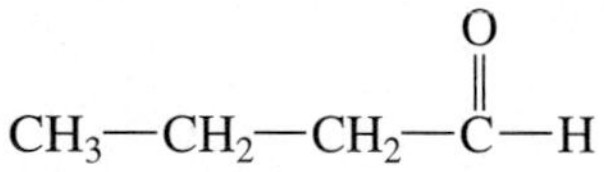 CH₃—CH₂—CH₂—C(=O)—H

d. 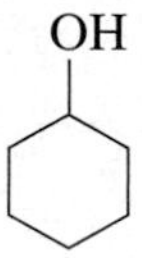 cyclohexanol (OH on ring)

15.60 Draw the structural formula of the organic product when each of the following is oxidized:

OH
|
a. $CH_3-CH_2-CH-CH_2OH$

OH
|
b. $CH_3-CH_2-CH-CH_3$

$\quad\quad\;\; CH_3 \quad\quad\; O$
$\quad\quad\;\; | \quad\quad\quad\; \|$
c. $CH_3-CH-CH_2-C-H$

OH
|
CH—CH₃

d. (cyclohexane ring)

15.61 Draw the structural formula of the organic product when hydrogen and a nickel catalyst reduce each of the following:

$\quad\quad\quad O$
$\quad\quad\quad \|$
a. CH_3-C-CH_3

$\quad\quad\quad\quad\quad O$
$\quad\quad\quad\quad\quad \|$
b. (benzene ring)CH_2-C-H

$\quad\quad CH_3 \quad\quad\; O$
$\quad\quad | \quad\quad\quad\; \|$
c. $CH_3-CH-CH_2-C-CH_3$

15.62 Draw the structural formula of the organic product when hydrogen and a nickel catalyst reduce each of the following:

$\quad\quad\quad O$
$\quad\quad\quad \|$
a. CH_3-C-H

b. O (cyclopentane ring with CH₃)
CH₃

$\quad\quad\quad O$
$\quad\quad\quad \|$
c. $H-C-H$

15.63 Using reactions such as dehydration, hydrogenation, oxidation, reduction, and hydration, indicate how

you might prepare the following from the starting substance given:

a. propene to propanone
b. butanal to 1,2-dibromobutane
c. butanal to butanone

15.64 Using reactions such as dehydration, hydrogenation, oxidation, reduction, and hydration, indicate how you might prepare the following from the starting substance given:

a. pentanal to 1-pentene
b. 1-butanol to butanone
c. cyclohexene to cyclohexanone

15.65 Identify the following as hemiacetals or acetals. Give the names of the carbonyl compounds and alcohols used in their synthesis.

$\quad\quad\quad\quad\quad OCH_3$
$\quad\quad\quad\quad\quad |$
a. CH_3-CH_2-CH
$\quad\quad\quad\quad\quad OCH_3$

$\quad\quad\quad\quad\quad OCH_2CH_3$
$\quad\quad\quad\quad\quad |$
b. $CH_3-CH_2-C-CH_3$
$\quad\quad\quad\quad\quad OH$

c. $CH_3CH_2O \quad\quad OCH_2CH_3$ (cyclohexane ring)

15.66 Identify the following as hemiacetals or acetals, and give the names of the carbonyl compounds and alcohols used in their synthesis.

$\quad\quad\quad\quad\quad OCH_3$
$\quad\quad\quad\quad\quad |$
a. CH_3-CH_2-CH
$\quad\quad\quad\quad\quad OH$

$\quad\quad\quad\quad CH_3$
$\quad\quad\quad\quad |$
b. $HO \quad\quad OCHCH_3$ (cyclohexane ring)

$\quad\quad\quad\quad\quad OCH_2CH_2CH_3$
$\quad\quad\quad\quad\quad |$
c. CH_3-CH
$\quad\quad\quad\quad\quad OCH_2CH_2CH_3$

16 Carbohydrates

"We use a refractometer to measure sugar content in a small sample of juices from the grapes in different areas of the vineyard," says Leslie Bucher, laboratory director at Bouchaine Winery. "We also measure the alcohol content during fermentation and run tests for sulfur, pH, and total acid."

As grapes ripen, there is an increase in the sugars, which are the monosaccharides fructose and glucose. The sugar content is affected by soil conditions and the amount of sun and water. When the grapes are ripe and sugar content is at a desirable level, they are harvested. During fermentation, enzymes from yeast convert about half the sugar to ethanol, and half to carbon dioxide. Grapes harvested with 22.5% sugar will ferment to give a wine with 12.5–13.5% alcohol content.

LOOKING AHEAD

the Chemistry place

www.chemplace.com/college

Visit the URL above or use the CD-ROM in the book for extra quizzing, interactive tutorials, and career resources.

Of all the organic compounds in nature, carbohydrates are the most abundant. In plants, energy from the sun converts carbon dioxide and water into the carbohydrate glucose. Many of the glucose molecules are made into long-chain polymers of starch that store energy or into cellulose to build the structural framework of the plant. About 65% of the foods in our diet consist of carbohydrates. Each day you enjoy carbohydrates such as bread, pasta, potatoes, and rice. During digestion and cellular metabolism, the starches are broken down into glucose, which is oxidized further in our cells to provide our bodies with energy and to provide the cells with carbon atoms for building molecules of protein, lipids, and nucleic acids. Cellulose has other important uses, too. The wood in our furniture, the pages in this book, and the cotton in our clothing are made of cellulose.

LEARNING GOAL

Classify carbohydrates as monosaccharides, disaccharides, and polysaccharides.

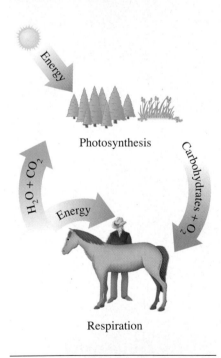

Figure 16.1 During photosynthesis, energy from the sun combines CO_2 and H_2O to form glucose $C_6H_{12}O_6$ and O_2. During respiration in the body, carbohydrates are oxidized to CO_2 and H_2O, while energy is produced.
Q *What are the reactants and products of respiration?*

16.1 Types of Carbohydrates

Carbohydrates such as table sugar, lactose in milk, and cellulose are all made of carbon, hydrogen, and oxygen. A carbohydrate is also called a **saccharide**, a word that comes from the Latin term *saccharum,* "sugar." Simple sugars, which have formulas of $C_n(H_2O)_n$, were once thought to be hydrates of carbon, thus the name *carbohydrate.* In a series of reactions called photosynthesis, energy from the sun is used to combine the carbon atoms from carbon dioxide (CO_2) and the hydrogen and oxygen atoms of water into the carbohydrate glucose.

$$6CO_2 + 6H_2O + energy \underset{\text{Respiration}}{\overset{\text{Photosynthesis}}{\rightleftharpoons}} \underset{\text{Glucose}}{C_6H_{12}O_6} + 6O_2$$

In our body tissues, glucose is oxidized in a series of metabolic reactions known as respiration, which releases chemical energy to do work in the cells. Carbon dioxide and water are produced and returned to the atmosphere. The combination of photosynthesis and respiration is called the carbon cycle, in which energy from the sun is stored in plants by photosynthesis and made available to us when the carbohydrates in our diets are metabolized. (See Figure 16.1.)

Types of Carbohydrates

The simplest carbohydrates are the **monosaccharides.** A monosaccharide cannot be split or hydrolyzed into smaller carbohydrates. One of the most common carbohydrates, glucose, $C_6H_{12}O_6$, is a monosaccharide. **Disaccharides** consist of two monosaccharide units joined together. A disaccharide can be split into two monosaccharide units. For example, ordinary table sugar, sucrose, $C_{12}H_{22}O_{11}$, is a disaccharide that can be hydrolyzed in the presence of an acid or an enzyme to give one molecule of glucose and one molecule of another monosaccharide, fructose.

$$\underset{\text{Sucrose}}{C_{12}H_{22}O_{11}} + H_2O \xrightarrow{\text{H}^+ \text{ or enzyme}} \underset{\text{Glucose}}{C_6H_{12}O_6} + \underset{\text{Fructose}}{C_6H_{12}O_6}$$

Polysaccharides are carbohydrates that are naturally occurring polymers containing many monosaccharide units. In the presence of an acid or an enzyme, a polysaccharide can be completely hydrolyzed to yield many molecules of monosaccharide.

EXPLORE YOUR WORLD

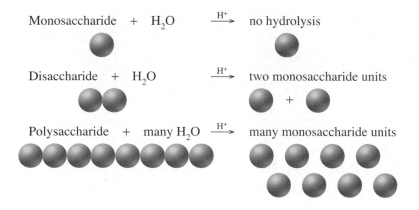

Photosynthesis

Carefully wrap four or five leaves of a houseplant with aluminum foil and masking tape. Place the plant in the sun for 1 week. Be sure to water. After 1 week unwrap the leaves.

Questions

1. What differences do you observe between the leaves that were wrapped and the unwrapped leaves?
2. What explanation would you give?
3. What are the requirements for photosynthesis?

SAMPLE PROBLEM 16.1

Types of Carbohydrates

Classify the following carbohydrates as mono-, di-, or polysaccharides:

a. When lactose, milk sugar, is hydrolyzed, two monosaccharide units are produced.

b. Cellulose, a carbohydrate in cotton, yields thousands of monosaccharide units when completely hydrolyzed.

Solution

a. A disaccharide contains two monosaccharide units.

b. A polysaccharide contains many monosaccharide units.

Study Check

Fructose found in fruits does not undergo hydrolysis. What type of carbohydrate is it?

QUESTIONS AND PROBLEMS

Classification of Carbohydrates

16.1 What reactants are needed for photosynthesis and respiration?

16.2 What is the relationship between photosynthesis and respiration?

16.3 What is a monosaccharide? A disaccharide?

16.4 What is a polysaccharide?

16.2 Classification of Monosaccharides

LEARNING GOAL

Classify a monosaccharide as an aldose or ketose and indicate the number of carbon atoms.

Monosaccharides are simple sugars that have an unbranched chain of three to eight carbon atoms, one of them in a carbonyl group and the rest attached to hydroxyl groups. There are two types of monosaccharide structures. In an **aldose,** the carbonyl group is on the first carbon (—CHO); a **ketose** contains the carbonyl group on the second carbon atom as a ketone (C═O).

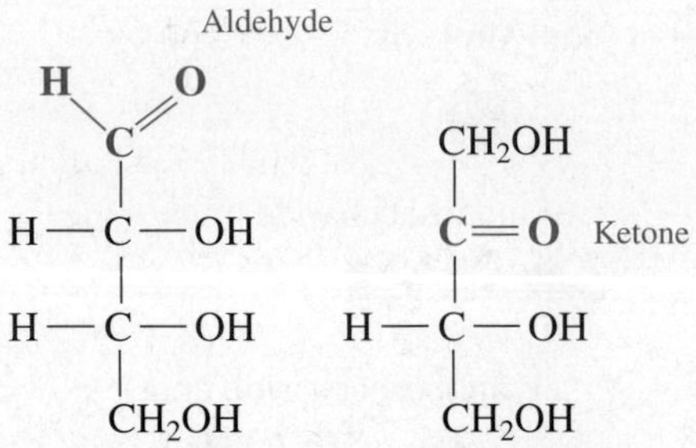

Aldehyde

Erythrose, a
polyhydroxy
aldehyde

Erythrulose, a
polyhydroxy
ketone

A monosaccharide with three carbon atoms is a *triose*, one with four carbon atoms is a *tetrose*, a *pentose* has five carbons, and a *hexose* contains six carbons. We can use both classification systems to indicate the type of carbonyl group and the number of carbon atoms. An aldopentose is a five-carbon monosaccharide that is an aldehyde; a ketohexose would be a six-carbon monosaccharide that is a ketone. Some examples:

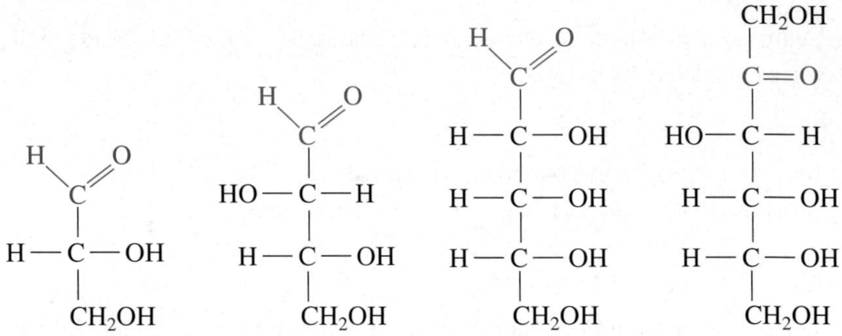

Glyceraldehyde
(aldotriose)

Threose
(aldotetrose)

Ribose
(aldopentose)

Fructose
(ketohexose)

SAMPLE PROBLEM 16.2

Monosaccharides

Classify each of the following monosaccharides to indicate their carbonyl group and number of carbon atoms:

a.

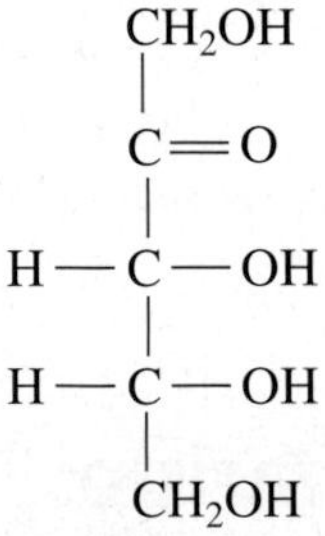

Ribulose

b.

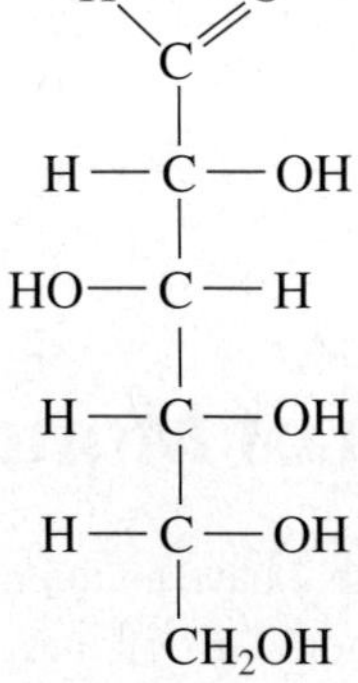

Glucose

Solution

a. The structural formula has a ketone group; ribulose is a ketose. Because there are five carbon atoms, it is a pentose. Combining these classifications makes it a ketopentose.

b. The structural formula has an aldehyde group; glucose is an aldose. Because there are six carbon atoms, it is an aldohexose.

Study Check

The simplest ketose is a triose named dihydroxyacetone. Draw its structural formula.

QUESTIONS AND PROBLEMS

Classification of Monosaccharides

16.5 What functional groups are found in all monosaccharides?

16.6 What is the difference between an aldose and a ketose?

16.7 What are the functional groups and number of carbons in a ketopentose?

16.8 What are the functional groups and number of carbons in an aldohexose?

16.9 Classify each of the following monosaccharides as an aldose or ketose.

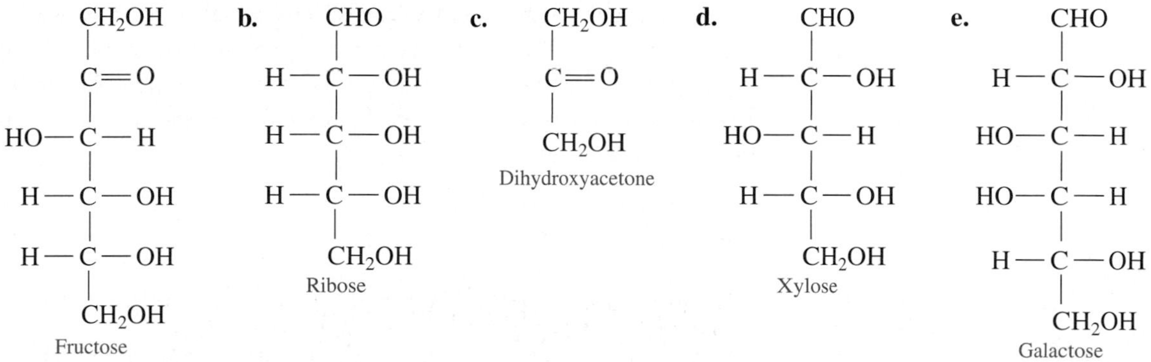

a.
$$CH_2OH$$
$$|$$
$$C=O$$
$$|$$
$$HO-C-H$$
$$|$$
$$H-C-OH$$
$$|$$
$$H-C-OH$$
$$|$$
$$CH_2OH$$
Fructose

b.
$$CHO$$
$$|$$
$$H-C-OH$$
$$|$$
$$H-C-OH$$
$$|$$
$$H-C-OH$$
$$|$$
$$CH_2OH$$
Ribose

c.
$$CH_2OH$$
$$|$$
$$C=O$$
$$|$$
$$CH_2OH$$
Dihydroxyacetone

d.
$$CHO$$
$$|$$
$$H-C-OH$$
$$|$$
$$HO-C-H$$
$$|$$
$$H-C-OH$$
$$|$$
$$CH_2OH$$
Xylose

e.
$$CHO$$
$$|$$
$$H-C-OH$$
$$|$$
$$HO-C-H$$
$$|$$
$$HO-C-H$$
$$|$$
$$H-C-OH$$
$$|$$
$$CH_2OH$$
Galactose

16.10 Classify each of the monosaccharides in problem 16.9 according to the number of carbon atoms in the chain.

16.3 D and L Notations from Fischer Projections

In Chapter 15 we learned that compounds with chiral carbons can exist in forms that are mirror images of each other and cannot be superimposed. The monosaccharides are chiral because they contain one or more chiral carbons in their carbon chains. Thus a monosaccharide can exist as either of two molecular forms that are mirror images of each other. The Fischer projections that represent enantiomers are also used for sugars, particularly monosaccharides.

Let's take a look again at the Fischer projection for the simplest sugar, glyceraldehyde. By convention the carbon chain is written vertically with the aldehyde group (most oxidized carbon) at the top and the —CH₂OH group at the bottom. The center carbon is the chiral carbon to which four different groups are attached. The letter L is assigned to the stereoisomer if the —OH group is on the left of the chiral carbon. In D-glyceraldehyde, the —OH is on the right.

> **LEARNING GOAL**
>
> Identify a Fischer projection of a sugar as the D or L configuration.

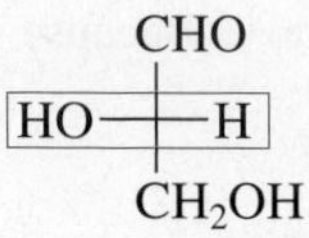

CHO
HO———H
CH₂OH
L-Glyceraldehyde

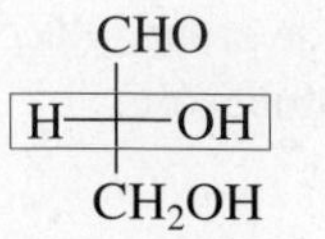

CHO
H———OH
CH₂OH
D-Glyceraldehyde

In an aldotetrose, there are two chiral carbon atoms, carbon 2 and 3. Then the bottom chiral atom, which is the chiral atom farthest from the carbonyl group, is used to assign the D or L configuration. The very bottom carbon atom in the Fischer projection of a carbohydrate —CH₂OH, is not chiral because it does not have four different groups bonded to it.

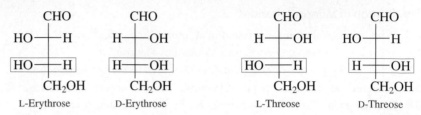

L-Erythrose	D-Erythrose	L-Threose	D-Threose

Most of the carbohydrates we will study have carbon chains with five or six carbon atoms, which means that they have several chiral carbons. However, we still use the chiral carbon furthest from the carbonyl group to determine the D or L isomer. Most of the naturally occurring sugars are the D isomers. The following are the isomers of ribose, which is a five-carbon monosaccharide, and glucose, a six-carbon monosaccharide.

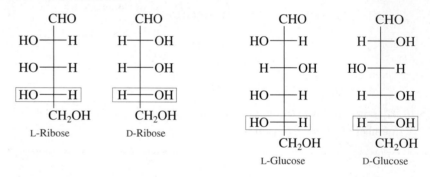

SAMPLE PROBLEM 16.3

Identifying D and L Isomers of Sugars

Is the following structure the D or L enantiomer of ribose?

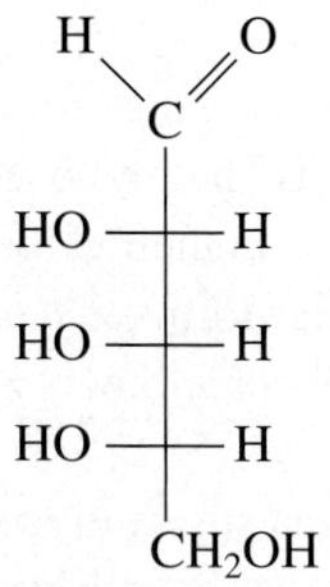

H   O
 \   ⫽
   C
HO———H
HO———H
HO———H
CH₂OH

Solution

In ribose, carbon 4 is the chiral atom furthest from the carbonyl group. Because the hydroxyl group on carbon 4 is on the left, this enantiomer is L-ribose.

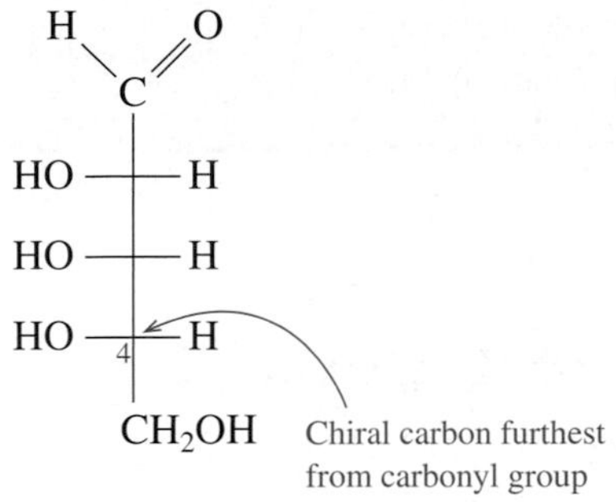

H
C=O
HO —— H
HO —— H
HO —4— H
CH$_2$OH   Chiral carbon furthest
from carbonyl group

Study Check

Draw the Fischer projection for D-ribose.

QUESTIONS AND PROBLEMS

D and L Notations from Fischer Projections

16.11 What is a Fischer projection?

16.12 Write the Fischer projection formula for D-glyceraldehyde and L-glycer-aldehyde.

16.13 State whether each of the following sugars is the D or L isomer:

a. 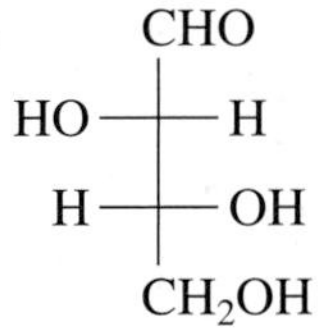
CHO
HO —— H
H —— OH
CH$_2$OH
Threose

b. 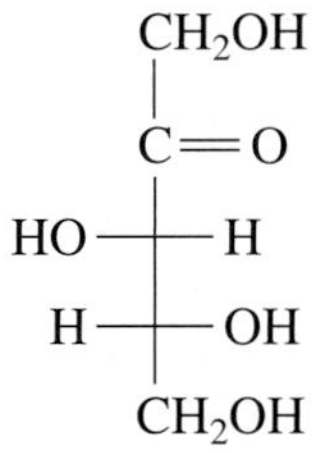
CH$_2$OH
C=O
HO —— H
H —— OH
CH$_2$OH
Xylulose

c. 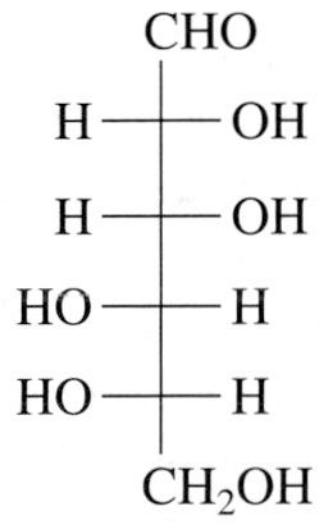
CHO
H —— OH
H —— OH
HO —— H
HO —— H
CH$_2$OH
Mannose

d. 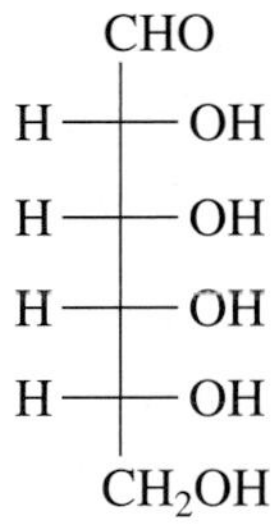
CHO
H —— OH
H —— OH
H —— OH
H —— OH
CH$_2$OH
Allose

16.14 State whether each of the following sugars is the D or L isomer:

a. 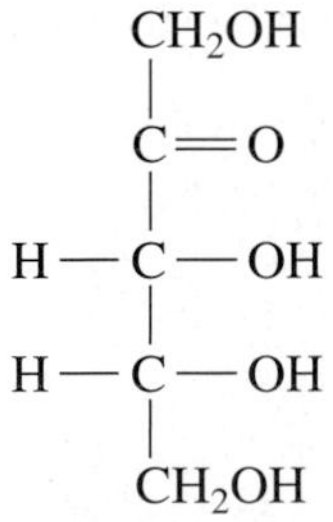
CH$_2$OH
C=O
H — C — OH
H — C — OH
CH$_2$OH
Ribulose

b. 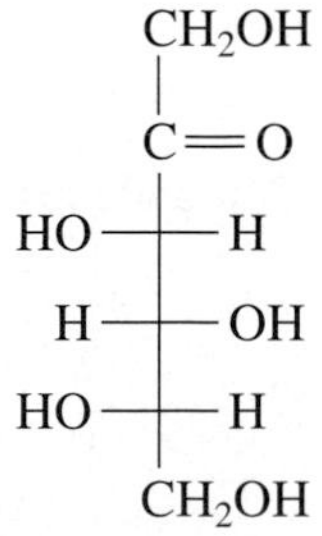
CH$_2$OH
C=O
HO —— H
H —— OH
HO —— H
CH$_2$OH
Sorbose

c. 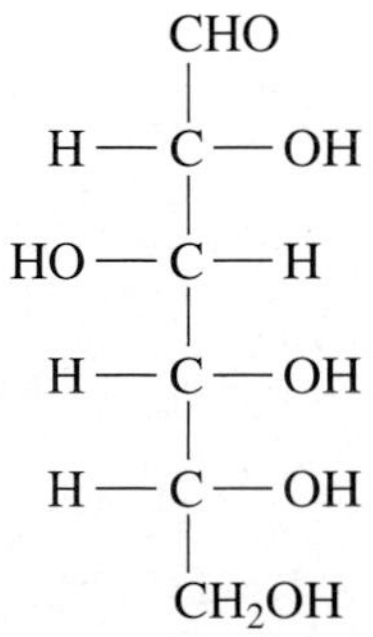
CHO
H — C — OH
HO — C — H
H — C — OH
H — C — OH
CH$_2$OH
Glucose

d. 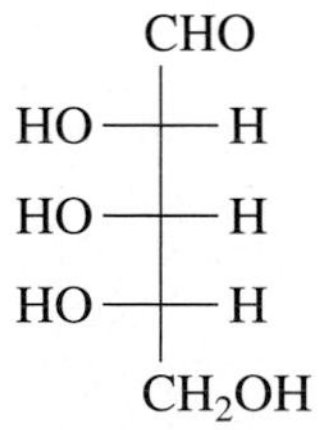
CHO
HO —— H
HO —— H
HO —— H
CH$_2$OH
Ribose

16.15   Write the enantiomers for a–d in problem 16.13.
16.16   Write the enantiomers for a–d in problem 16.14.

LEARNING GOAL

Draw the open-chain structures for D-glucose, D-galactose, and D-fructose.

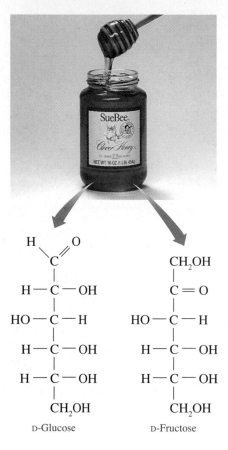

Figure 16.2 The sweet taste of honey is due to the monosaccharides of glucose and fructose.

Q *What are some differences in the structures of glucose and fructose?*

16.4 Structures of Some Important Monosaccharides

The hexoses glucose, galactose, and fructose are important monosaccharides. Although structural formulas for both the D and L isomers can be written for these hexoses, the D isomers are the more common form of carbohydrates found in nature. (See Figure 16.2.) We can write their open-chain structures as follows:

D-Glucose   D-Galactose   D-Fructose

D-Glucose

The most common hexose is D-glucose, which has the molecular formula $C_6H_{12}O_6$. Also known as dextrose, **glucose** is found in fruits, vegetables, corn syrup, and honey. It is a building block of the disaccharides sucrose, lactose, and maltose and is the monosaccharide unit in polysaccharides such as starch, cellulose, and glycogen.

In the body, glucose normally occurs at a concentration of 70–90 mg/dL (1 dL = 100 mL) of blood. However, the amount of glucose depends on the time that has passed since eating. In the first hour after a meal, the level of glucose rises to about 130 mg/dL of blood, and then decreases over the next 2–3 hours as it is used in the tissues. Some glucose is converted to glycogen and stored in the liver and muscle. When the amount of glucose exceeds what is needed for energy or glycogen, the excess glucose is converted to fat, which can be stored in unlimited amounts.

Glycogen (liver and muscle)

Fat ←—Excess— Glucose

Excess

Urine   Metabolism CO_2 + H_2O + energy

Hyperglycemia and Hypoglycemia

A doctor may order a glucose tolerance test to evaluate the body's ability to return to normal glucose concentration in response to the ingestion of a specified amount of glucose. The patient fasts for 12 hours and then drinks a solution containing glucose. A blood sample is taken immediately, followed by more blood samples each half-hour for 2 hours, and then every hour for a total of 5 hours. If the blood glucose exceeds 140 mg/dL in plasma and remains high, hyperglycemia may be indicated. The term *glyc* or *gluco* refers to "sugar". The prefix *hyper* means above or over, and *hypo* is below or under. Thus the blood sugar level in *hyperglycemia* is above normal and below normal in *hypoglycemia*.

An example of a disease that can cause hyperglycemia is diabetes mellitus, which occurs when the pancreas is unable to produce sufficient quantities of insulin. As a result, glucose levels in the body fluids can rise as high as 350 mg/dL plasma. Symptoms of diabetes in people under the age of 40 include thirst, excessive urination,

increased appetite, and weight loss. In older persons, diabetes is sometimes a consequence of excessive weight gain.

When a person is hypoglycemic, the blood glucose level rises and then decreases rapidly to levels as low as 40 mg/dL plasma. In some cases, hypoglycemia is caused by overproduction of insulin by the pancreas. Low blood glucose can cause dizziness, general weakness, and muscle tremors. A diet may be prescribed that consists of several small meals high in protein and low in carbohydrate. Some hypoglycemic patients are finding success with diets that include more complex carbohydrates rather than simple sugars.

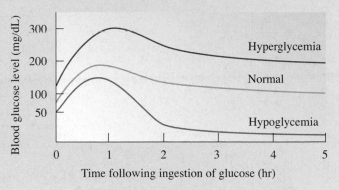

D-Galactose

Galactose is an aldohexose that does not occur in the free form in nature. It is obtained as a hydrolysis product of the disaccharide lactose, a sugar found in milk and milk products. Galactose is important in the cellular membranes of the brain and nervous system. The only difference in the structures of D-glucose and D-galactose is the arrangement of the —OH group on carbon 4.

D-Glucose   D-Galactose

In a condition called *galactosemia,* an enzyme needed to convert galactose to glucose is missing. The accumulation of galactose in the blood and tissues can lead to cataracts, mental retardation, and cirrhosis. The treatment for galactosemia is the removal of all galactose-containing foods, mainly milk and milk products, from the diet. If this is done for an infant immediately after birth, the damaging effects of galactose accumulation can be avoided.

D-**Fructose**

In contrast to glucose and galactose, **fructose** is a ketohexose. The structure of fructose differs from glucose at carbons 1 and 2 by the location of the carbonyl group.

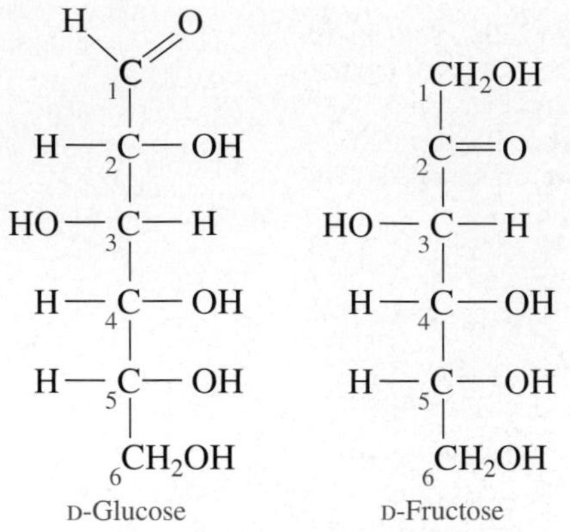

D-Glucose   D-Fructose

Fructose is the sweetest of the carbohydrates, twice as sweet as sucrose (table sugar). This makes fructose popular with dieters because less fructose, and therefore fewer calories, are needed to provide a pleasant taste. After fructose enters the bloodstream, it is converted to its isomer, glucose. Fructose is found in fruit juices and honey; it is also called levulose and fruit sugar. Fructose is also obtained as one of the hydrolysis products of sucrose, the disaccharide known as table sugar.

SAMPLE PROBLEM 16.4

Monosaccharides

Ribulose has the following open-chain structure.

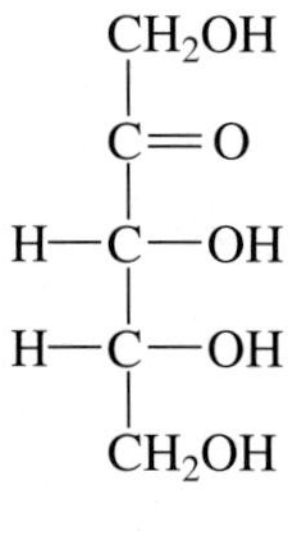

a. Identify the above compound D- or L-ribulose.
b. Write the open-chain structural formula of its enantiomer.

Solution

a. D-ribulose
b. The enantiomer, L-ribulose, has the following structural formula:

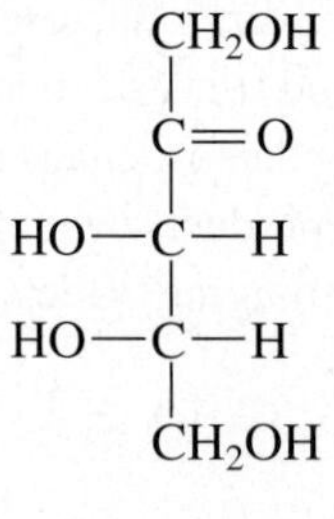

Study Check

What type of carbohydrate is ribulose?

QUESTIONS AND PROBLEMS

Structures of Some Important Monosaccharides

16.17 Draw the open-chain structure of D-glucose and L-glucose.

16.18 Draw the open-chain structure of D-fructose and L-fructose.

16.19 How does the open-chain structure of D-galactose differ from D-glucose?

16.20 How does the open-chain structure of D-fructose differ from D-glucose?

16.21 Identify a monosaccharide that fits each of the following descriptions:
  a. also called blood sugar
  b. not metabolized in galactosemia
  c. also called fruit sugar

16.22 Identify a monosaccharide that fits each of the following descriptions:
  a. high blood levels in diabetes
  b. obtained as a hydrolysis product of lactose
  c. the sweetest of the monosaccharides

16.5 Cyclic Structures of Monosaccharides

LEARNING GOAL

Draw and identify the cyclic structures of monosaccharides.

In Chapter 15, we saw that an aldehyde group reacts with one alcohol molecule to form a hemiacetal. For example, acetaldehyde reacts with methanol to form the following hemiacetal. In the product, the carbonyl carbon is bonded by an ether link to the alkyl group and to a new —OH group.

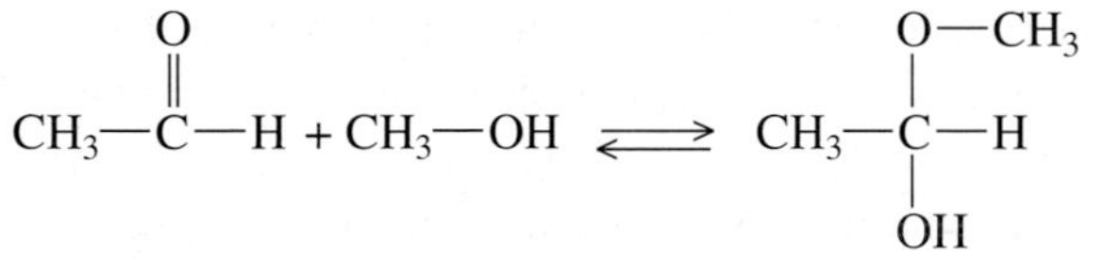

$$CH_3-\overset{\overset{\textstyle O}{\|}}{C}-H + CH_3-OH \rightleftharpoons CH_3-\overset{\overset{\textstyle O-CH_3}{|}}{\underset{\underset{\textstyle OII}{|}}{C}}-H$$

This same reaction occurs when a carbonyl group and —OH group are in the *same* molecule. The product, called a *cyclic hemiacetal,* forms a ring structure that is the most stable form of aldopentoses and aldohexoses. In the following general diagram, the hydroxyl group on carbon 5 bonds with the carbonyl carbon 1 to produce a heterocyclic six-atom ring containing an oxygen atom and a new —OH group on carbon 1.

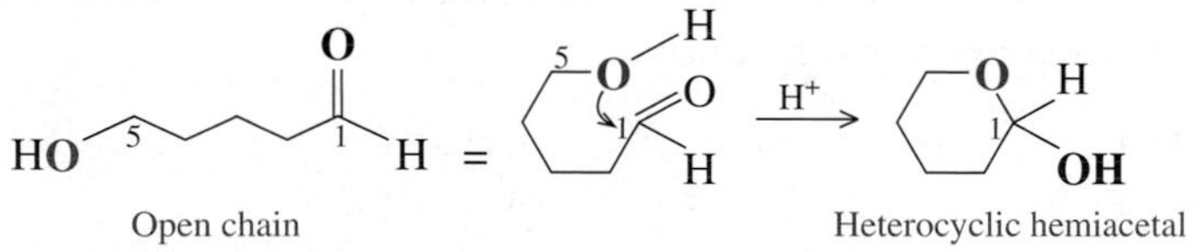

Open chain   Heterocyclic hemiacetal

Drawing Haworth Structures for Cyclic Forms

While the carbonyl group in an aldohexose could react with several of the —OH groups, the equilibrium for aldohexoses favors the formation of six-atom rings. Let's look at how we draw the cyclic hemiacetal for D-glucose starting with the Fischer projection.

Step 1   Think of turning the open chain of glucose clockwise to the right. Then the —OH groups written on the right, other than the one on carbon 5, are drawn down, and the —OH groups on the left are up.

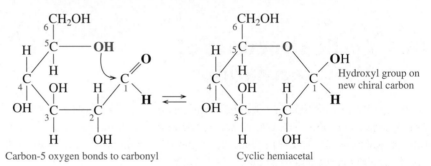

D-Glucose (open chain)

Step 2   Rotate the groups around carbon 5 placing the —CH₂OH up, and the —OH group close to the carbonyl carbon 1. Form the cyclic hemiacetal by bonding the oxygen in the —OH group to the carbonyl carbon. This way of depicting a cyclic hemiacetal structure is known as a **Haworth structure**.

Carbon-5 oxygen bonds to carbonyl   Cyclic hemiacetal

Step 3   In the cyclic hemiacetal, carbon 1 is now a chiral carbon bonded to a new —OH group. There are two ways to place the —OH, either up or down, which gives two stereoisomers called **anomers**. The —OH group on carbon 1 is down in the α (alpha) anomer and up in the β (beta) anomer.

Such differences in structural forms may seem trivial. However, we can digest starch products such as pasta to obtain glucose because the polysaccharide contains the α isomers of glucose. We cannot digest paper or wood because cellulose consists of only β-D-glucose units. Humans have an α-amylase, an enzyme needed for the digestion of starches, but not a β-amylase for the digestion of cellulose.

Sometimes, the cyclic structure is simplified to show only the position of the hydroxyl groups on the six-atom ring structure.

α-D-Glucose (simplified structure)   β-D-Glucose

Mutarotation

In solution, the α-D-glucose is in equilibrium with β-D-glucose. In a process called **mutarotation,** each isomer converts from the closed ring to the open chain and back again. As the ring opens and closes, the bond between carbons 1 and 2 can rotate, which allows the hydroxyl (—OH) group on carbon 1 to shift between the α and the β position. Although the open chain is an essential part of mutarotation, only a small amount of open chain is present at any given time.

Haworth Structures for α- and β-D-Glucose

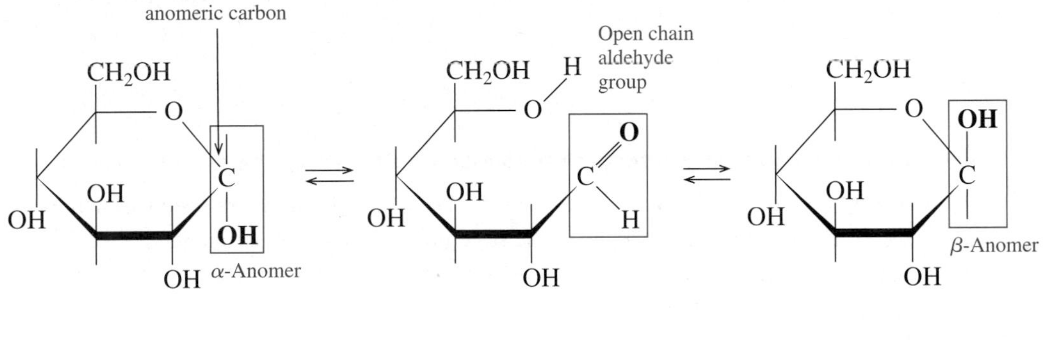

α-D-Glucose
(36% in equilibrium mixture)

D-Glucose
open-chain (trace)

β-D-Glucose
(64% in equilibrium mixture)

Cyclic Structures of Galactose

Galactose is an aldohexose like glucose, differing only in the arrangement of the —OH group on carbon 4. Thus, its cyclic structure is also similar to glucose, except that in galactose the —OH on carbon 4 is up. With the formation of a new hydroxyl group on carbon 1, galactose also exists as α and β anomers and undergoes mutarotation via the open-chain form in solution.

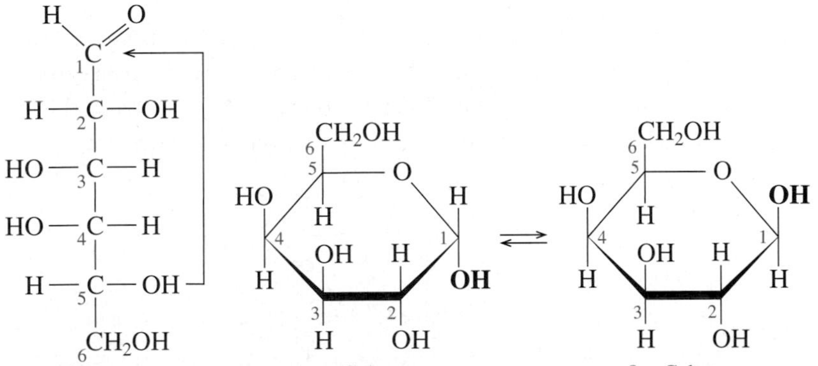

D-Galactose

α-D-Galactose

β-D-Galactose

Cyclic Structures of Fructose

In contrast to glucose and galactose, fructose is a ketohexose. It forms a hemiacetal when a hydroxyl group on carbon 5 reacts with the ketone group. The cyclic structure for fructose is a five-atom ring with carbon 2 at the right corner. A new hydroxyl group is on carbon 2 in addition to the carbon 1 from the CH_2OH group. There are also α and β anomers of fructose that undergo mutarotation in solution.

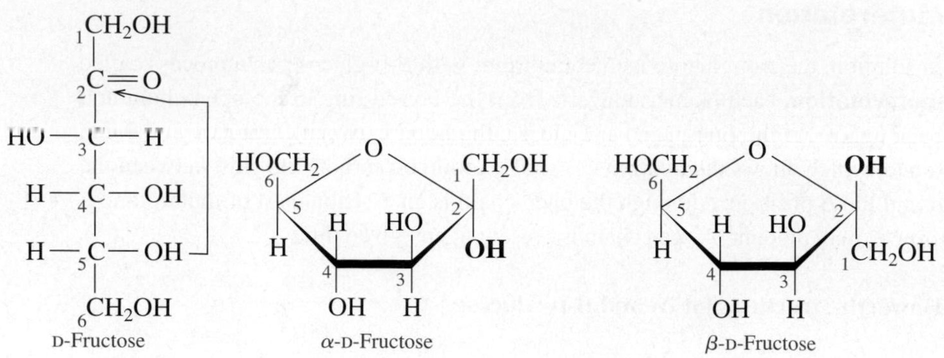

$$\begin{array}{c}
{}^{1}CH_2OH \\
{}_{2}C{=}O \\
HO{-}{}_{3}C{-}H \\
H{-}{}_{4}C{-}OH \\
H{-}{}_{5}C{-}OH \\
{}_{6}CH_2OH
\end{array}$$

D-Fructose

α-D-Fructose

β-D-Fructose

SAMPLE PROBLEM 16.5

Drawing Cyclic Structures for Sugars

D-Mannose, a carbohydrate found in immunoglobulins, has the following open-chain structure. Draw the cyclic structure for β-D-mannose anomer.

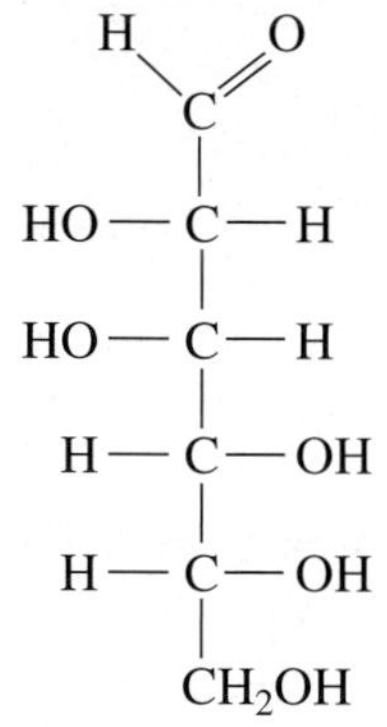

$$\begin{array}{c}
H \quad O \\
\diagdown C \diagup \\
HO{-}C{-}H \\
HO{-}C{-}H \\
H{-}C{-}OH \\
H{-}C{-}OH \\
CH_2OH
\end{array}$$

D-Mannose

Solution

Number the carbon atoms in the open chain starting at the aldehyde group. Turn the chain on its side and bend it into a hexagon so that the —OH group on carbon 5 is close to the carbon 1 carbonyl group. Draw the other —OH groups on the left of the open chain above the ring, and the —OH groups on the right below.

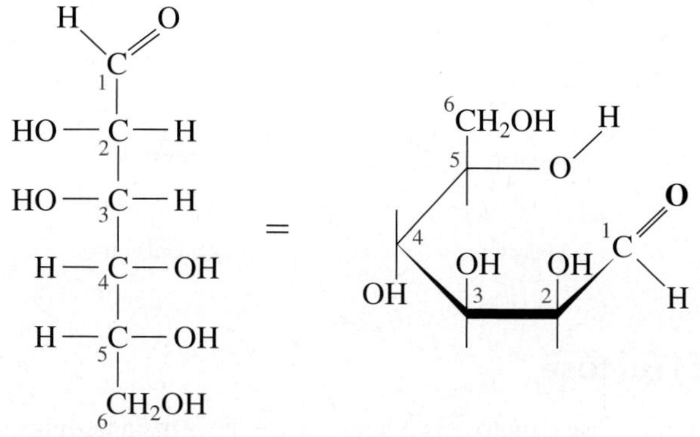

$$\begin{array}{c}
H \quad O \\
\diagdown C \diagup \\
{}_{1} \\
HO{-}{}_{2}C{-}H \\
HO{-}{}_{3}C{-}H \\
H{-}{}_{4}C{-}OH \\
H{-}{}_{5}C{-}OH \\
{}_{6}CH_2OH
\end{array}$$

D-Mannose

Form the cyclic hemiacetal by bonding the oxygen in —OH to carbon 1. Write the new hydroxyl group upward to make the β-D-mannose anomer.

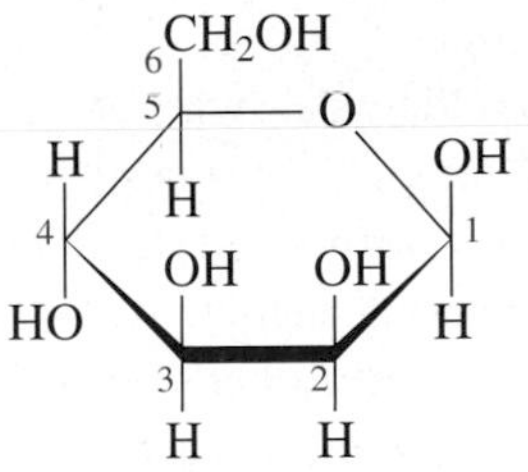

β-D-Mannose

Study Check

Draw the cyclic structure for α-D-glucose.

QUESTIONS AND PROBLEMS

Cyclic Structures of Monosaccharides

16.23 What are the kind and number of atoms in the ring portion of the cyclic structure of glucose?

16.24 What are the kind and number of atoms in the ring portion of the cyclic structure of fructose?

16.25 Draw the cyclic structures for the α and β anomers of D-glucose.

16.26 Draw the cyclic structures for the α and β anomers of D-fructose.

16.27 Identify each of the following cyclic structures as the α or β anomer:

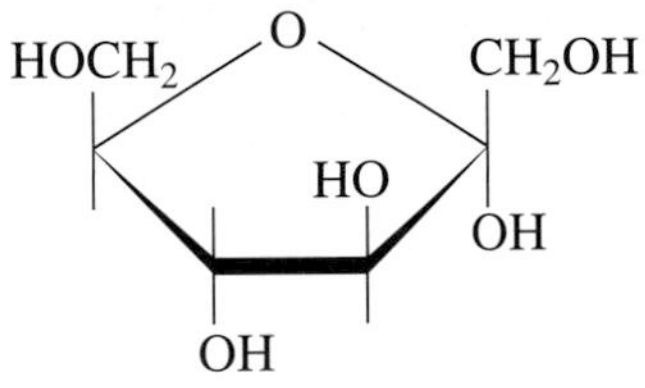

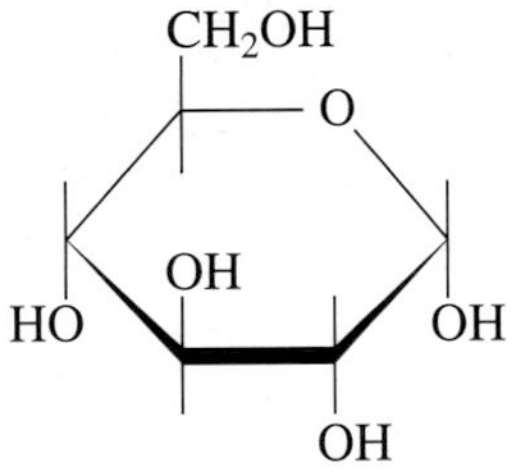

16.28 Identify each of the following cyclic structures as the α or β anomer.

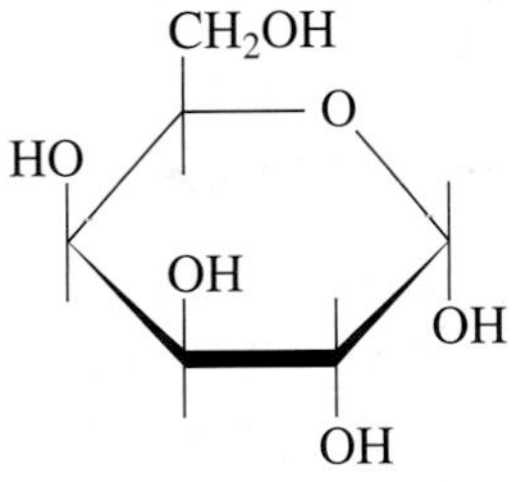

16.6 Chemical Properties of Monosaccharides

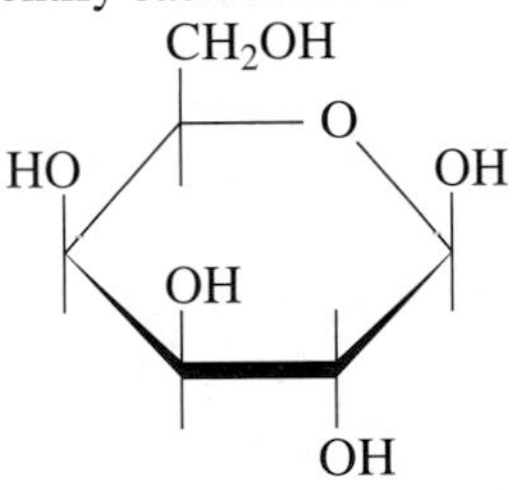

LEARNING GOAL

Identify the products of oxidation or reduction of monosaccharides; determine whether a carbohydrate is a reducing sugar.

Monosaccharides contain several functional groups that can undergo chemical reactions. In an aldose, the aldehyde group can be oxidized to a carboxylic acid. The carbonyl group in both an aldose and ketose can be reduced to give a hydroxyl group. The hydroxyl groups can react with other compounds to form a variety of derivatives. Several of these derivatives have important roles in metabolism and biological structures.

Oxidation of Monosaccharides

In Chapter 15, we looked at the Benedict's test for aldehydes with adjacent hydroxyl groups, which we find in monosaccharides. Although monosaccharides exist mostly in their cyclic forms, the aldehyde group of the open-chain structure in the equilibrium mixture does oxidize easily. When the monosaccharide oxidizes, a carboxylic group is produced. At the same time the Cu^{2+} in Benedict's reagent is reduced to Cu^+, forming a brick-red precipitate of Cu_2O. Monosaccharides that reduce another substance such as Benedict's reagent are called **reducing sugars.**

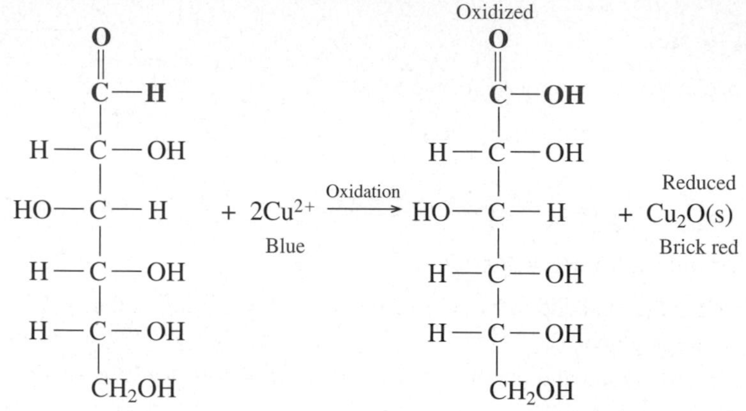

$$\text{Open chain of D-glucose, a reducing sugar} \quad + \; 2Cu^{2+} \xrightarrow{\text{Oxidation}} \quad \text{D-Gluconic acid} \quad + \; Cu_2O(s)$$

Open chain of D-glucose, a reducing sugar

D-Gluconic acid

Fructose is also a reducing sugar. In the open-chain form, a rearrangement between the hydroxyl group on carbon 1 and the ketone group provides an aldehyde group that can be oxidized.

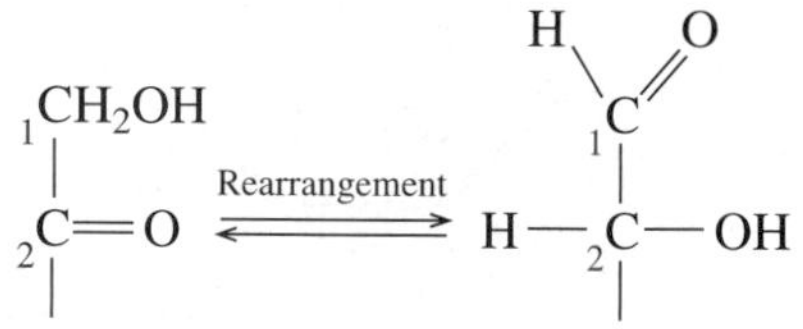

$$\text{D-Fructose (ketose)} \xrightleftharpoons{\text{Rearrangement}} \text{D-Glucose (aldose)}$$

D-Fructose
(ketose)

D-Glucose
(aldose)

Reduction of Monosaccharides

The reduction of the carbonyl group in monosaccharides produces sugar alcohols, which are also called *alditols*. D-glucose is reduced to D-glucitol, better known as sorbitol. D-Mannose is reduced to give D-mannitol.

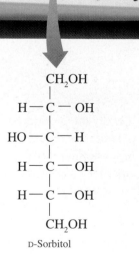

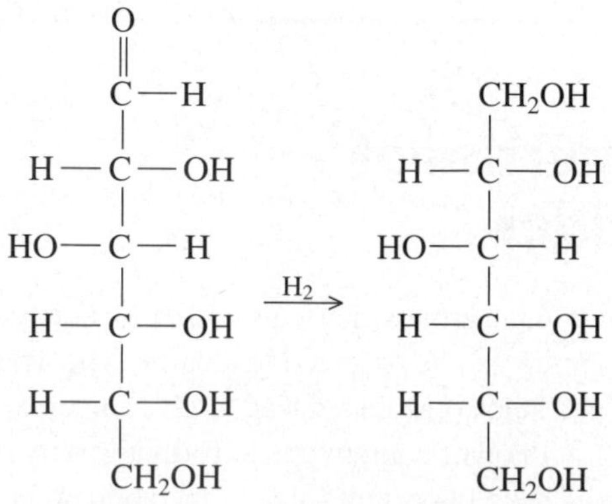

D-Sorbitol

D-Glucose

$\xrightarrow{H_2}$

D-Glucitol or D-Sorbitol

Testing for Glucose in Urine

Normally, blood glucose flows through the kidneys and is reabsorbed into the bloodstream. However, if the blood level exceeds about 160 mg of glucose/dL of blood, the kidneys cannot reabsorb it all, and glucose spills over into the urine, a condition known as glucosuria. A symptom of diabetes mellitus is a high level of glucose in the urine.

Benedict's test can be used to determine the presence of glucose in urine. The amount of cuprous oxide (Cu_2O) formed is proportional to the amount of reducing sugar present in the urine. Low to moderate levels of reducing sugar turn the solution green; solutions with high glucose levels turn Benedict's yellow or brick-red. Table 16.1 lists some colors associated with the concentration of glucose in the urine.

In another clinical test that is more specific for glucose, the enzyme glucose oxidase is used. The oxidase enzyme converts glucose to gluconic acid and hydrogen peroxide, H_2O_2. The peroxide produced reacts with a dye in the test strip to give different colors. The level of glucose present in the urine is found by matching the color produced to a color chart on the container.

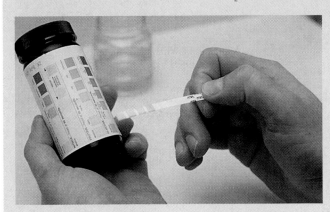

Table 16.1   Glucose Test Results

	Glucose Present in Urine	
Color	%	mg/dL
Blue	0	0
Blue-green	0.25	250
Green	0.50	500
Yellow	1.00	1000
Brick-red	2.00	2000

Sugar alcohols such as sorbitol, xylitol from xylose, and mannitol from mannose are used as sweeteners in many sugar-free products such as diet drinks and sugarless gum as well as products for people with diabetes. However, there are some side effects of these sugar substitutes. Some people experience some discomfort such as gas and diarrhea from the ingestion of sugar alcohols. The development of cataracts in diabetics is attributed to the accumulation of sorbitol in the lens of the eye.

Formation of Acetals (Glycosides)

In Chapter 15, we saw that hemiacetals react with other alcohol molecules to form acetals.

$$CH_3-\overset{\displaystyle O}{\overset{\displaystyle \|}{C}}-H \;+\; CH_3OH \;\underset{}{\overset{H^+}{\rightleftharpoons}}\; CH_3-\overset{\displaystyle OCH_3}{\underset{\displaystyle OH}{C}}-H \;+\; CH_3OH \;\underset{}{\overset{H^+}{\rightleftharpoons}}\; CH_3-\overset{\displaystyle OCH_3}{\underset{\displaystyle OCH_3}{C}}-H \;+\; H_2O$$

Aldehyde   Alcohol   Hemiacetal   Alcohol   Acetal

A similar reaction occurs when an alcohol reacts with a cyclic hemiacetal of a monosaccharide to give an acetal. These acetal products are called **glycosides**, and the acetal linkage in sugars is called a **glycosidic bond.** In cyclic acetals, the glycosidic bond cannot open to give the open-chain form. Thus, the acetals of monosaccharides cannot undergo mutarotation or oxidation, which means that acetals are nonreducing sugars. This property is indicated by the ending *–oside.* The glycosidic bond will remain in the α or β position, depending on the anomer that reacted. This is an important reaction in carbohydrate chemistry because it is the way that monosaccharides bond to form disaccharides and polysaccharides.

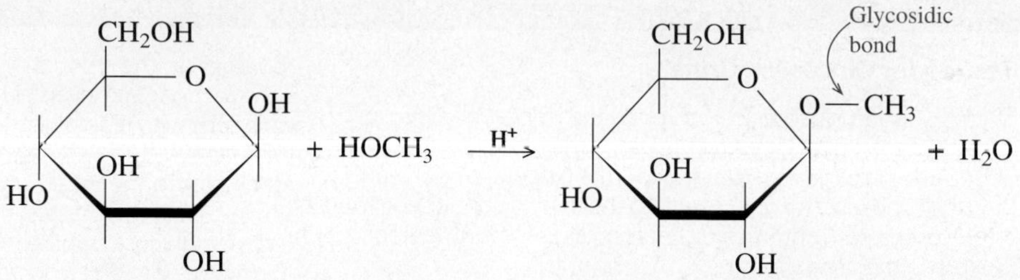

β-D-Glucose   Methanol   Methyl β-D-glucoside

SAMPLE PROBLEM 16.6

Reducing Sugars

Why is D-glucose called a *reducing sugar*?

Solution

D-glucose is easily oxidized by Benedict's reagent. A carbohydrate that reduces Cu^{2+} to Cu^+ is called a reducing sugar.

Study Check

A test using Benedict's reagent turns brick-red with a urine sample. According to Table 16.1, what might this result indicate?

QUESTIONS AND PROBLEMS

Chemical Properties of Monosaccharides

16.29 Draw the product xylitol produced from the reduction of D-xylose.

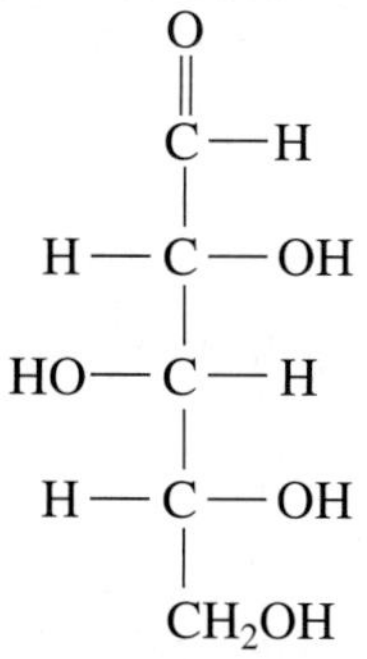

D-Xylose

16.30 Draw the product mannitol produced from the reduction of D-mannose.

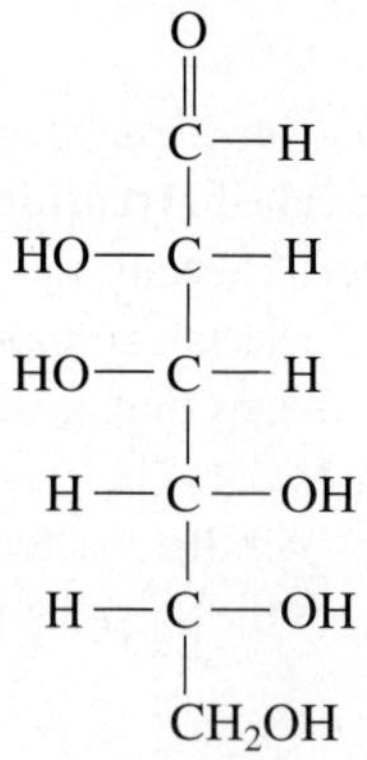

D-Mannose

16.31 Write the oxidation and reduction products of D-arabinose. What is the name of the sugar alcohol produced?

$$
\begin{array}{c}
\text{O} \\
\parallel \\
\text{C}\!-\!\text{H} \\
| \\
\text{HO}\!-\!\text{C}\!-\!\text{H} \\
| \\
\text{H}\!-\!\text{C}\!-\!\text{OH} \\
| \\
\text{H}\!-\!\text{C}\!-\!\text{OH} \\
| \\
\text{CH}_2\text{OH}
\end{array}
$$

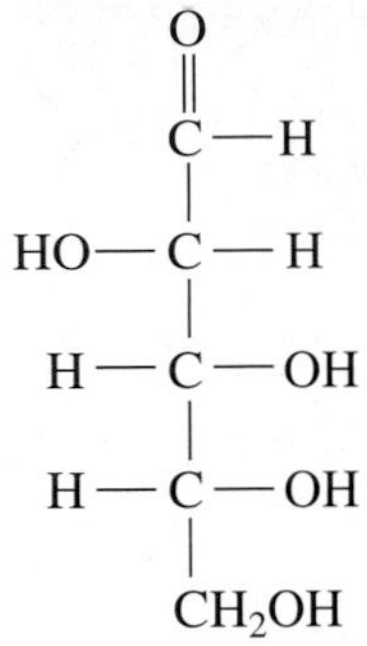

D Arabinose

16.32 Write the oxidation and reduction products of D-ribose. What is the name of the sugar alcohol produced?

$$
\begin{array}{c}
\text{O} \\
\parallel \\
\text{C}\!-\!\text{H} \\
| \\
\text{H}\!-\!\text{C}\!-\!\text{OH} \\
| \\
\text{H}\!-\!\text{C}\!-\!\text{OH} \\
| \\
\text{H}\!-\!\text{C}\!-\!\text{OH} \\
| \\
\text{CH}_2\text{OH}
\end{array}
$$

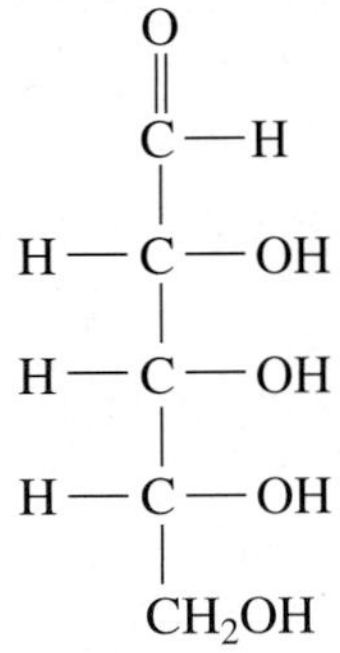

D-Ribose

16.33 Write the acetal anomers (α and β) of D-galactose and methyl alcohol.

16.34 Write the acetal anomers (α and β) of D-glucose and ethyl alcohol.

16.7 Disaccharides

A disaccharide is composed of two monosaccharides linked together. The most common disaccharides are maltose, lactose, and sucrose. Their hydrolysis, by an acid or an enzyme, gives the following monosaccharides.

$$\text{Maltose} + \text{H}_2\text{O} \xrightarrow{\text{H}^+} \text{glucose} + \text{glucose}$$

$$\text{Lactose} + \text{H}_2\text{O} \xrightarrow{\text{H}^+} \text{glucose} + \text{galactose}$$

$$\text{Sucrose} + \text{H}_2\text{O} \xrightarrow{\text{H}^+} \text{glucose} + \text{fructose}$$

Maltose

Maltose, or malt sugar, is a disaccharide obtained from starch. When maltose in barley and other grains is hydrolyzed by yeast enzymes, glucose is obtained that can undergo fermentation to give ethanol. Maltose is used in cereals, candies, and the brewing of beverages.

How Sweet Is My Sweetener?

Although many of the monosaccharides and disaccharides taste sweet, they differ considerably in their degree of sweetness. Dietetic foods contain sweeteners that are noncarbohydrate or carbohydrates that are sweeter. Some examples of sweeteners compared with sucrose are shown in Table 16.2.

Sucralose is made from sucrose by replacing some of the hydroxyl groups with chlorine atoms.

Sucralose

Aspartame, which is marketed as Nutra-Sweet, is used in a large number of sugar-free products. It is a noncarbohydrate sweetener made of aspartic acid and a methyl ester of phenylalanine. It does have some caloric value, but it is so sweet that a very small quantity is needed. However, one of the breakdown products phenylalanine, poses a danger to anyone who cannot metabolize it properly, a condition called phenylketonuria (PKU).

From aspartic acid   From phenylalanine   Methyl ester
Aspartame (Nutra-Sweet)

Table 16.2   Relative Sweetness of Sugars and Artificial Sweeteners

	Sweetness Relative to Sucrose (= 100)
Monosaccharides	
Galactose	30
Sorbitol	36
Glucose	75
Fructose	175
Disaccharides	
Lactose	16
Maltose	33
Sucrose	100 ← reference standard
Artificial Sweeteners (Noncarbohydrate)	
Sucralose	600
Aspartame	18,000
Saccharin	45,000

Saccharin has been used as a noncarbohydrate artificial sweetener for the past 25 years. The use of saccharin has been banned in Canada because studies indicate that it may cause bladder tumors. However, it is still approved for use by the FDA in the United States.

Saccharin

As we saw in the previous section, the hemiacetal carbon of a monosaccharide reacts with the hydroxyl group of an alcohol. If the hydroxyl group is another monosaccharide, the glycoside product is a disaccharide.

To make maltose, the hydroxyl group on carbon 1 of one glucose molecule bonds with the hydroxyl group on carbon 4 of the second glucose molecule. A glycosidic bond joins the two glucose molecules with a loss of a molecule of water. The glycosidic bond is designated as an α-1,4 linkage to show that —OH on carbon 1 of the α anomer is joined to carbon 4 of the second glucose. Because the second glucose molecule has a free —OH on the anomeric carbon, there are α and β anomers of maltose. This anomeric carbon also opens up to give a free aldehyde group to oxidize, which makes maltose a reducing sugar.

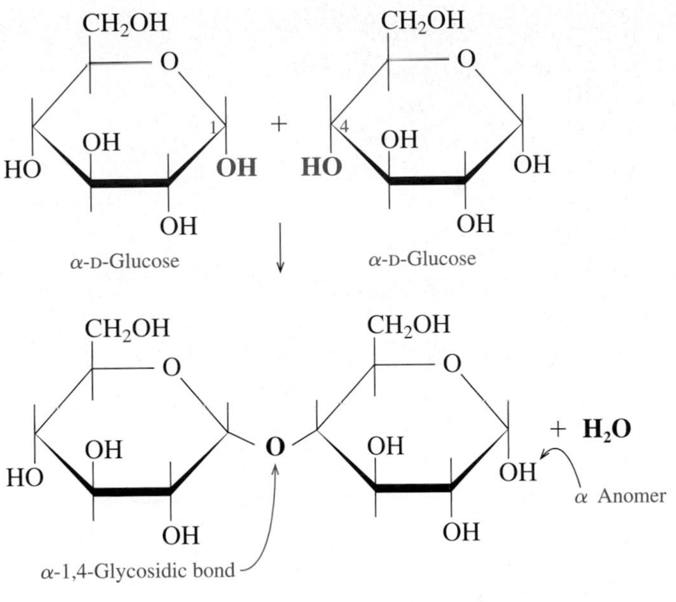

α-D-Glucose + α-D-Glucose

α-1,4-Glycosidic bond

+ H$_2$O

α Anomer

α-Maltose, a disaccharide

Lactose

Lactose, milk sugar, is a disaccharide found in milk and milk products. (See Figure 16.3.) It makes up 6–8% of human milk and about 4–5% of cow's milk and is used in products that attempt to duplicate mother's milk. Some people do not produce sufficient quantities of the enzyme needed to hydrolyze lactose, and the sugar remains undigested, causing abdominal cramps and diarrhea. In some commercial milk products, an enzyme called lactase is added to break down lactose.

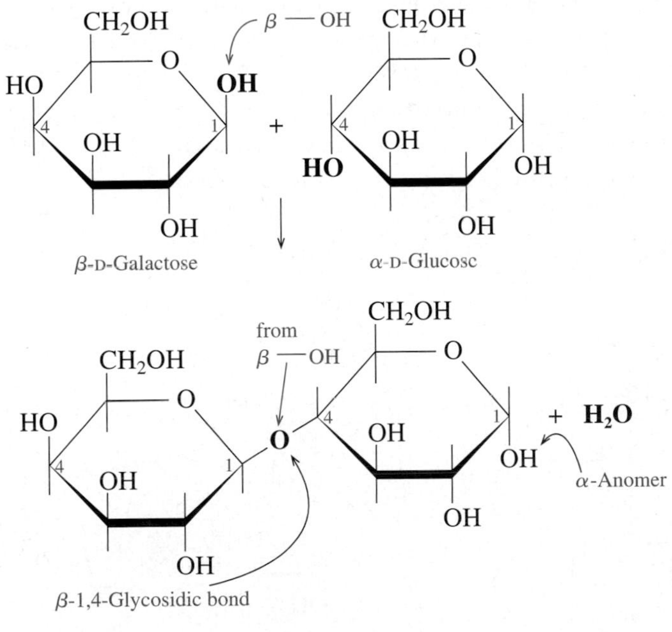

β-D-Galactose + α-D-Glucose

β-1,4-Glycosidic bond

from β—OH

+ H$_2$O

α-Anomer

α-Lactose, a disaccharide

The bond in lactose is a β-1,4-glycosidic bond because the β anomer of galactose forms an acetal with a hydroxyl group on carbon 4 of glucose. In the lactose molecule, the acetal of galactose cannot open, however the hemiacetal carbon in glucose

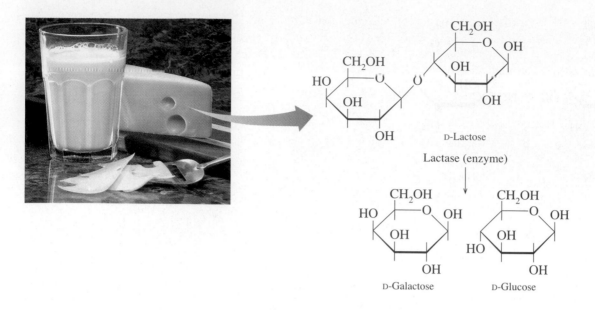

Figure 16.3 Lactose, a disaccharide found in milk and milk products undergoes digestion to give galactose and glucose.
Q *What type of glycosidic bond links galactose and glucose in lactose?*

undergoes mutarotation to give both α- and β-lactose. Because the open chain has an aldehyde group that can be oxidized, lactose is a reducing sugar.

Sucrose

You already know that ordinary table sugar is sucrose, a disaccharide that is the most abundant carbohydrate in the world. Most of the sucrose for table sugar comes from sugar cane (20% by mass) or sugar beets (15% by mass). (See Figure 16.4.) Both the raw and refined forms of sugar are sucrose. Some estimates indicate that each person in the United States consumes an average of 45 kg (100 lb) of sucrose every year either by itself or in a variety of food products.

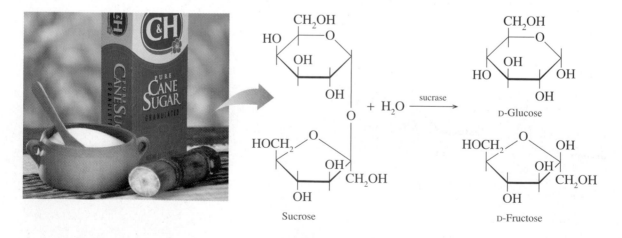

Figure 16.4 Sucrose, a disaccharide obtained from sugar beets and sugar cane, undergoes digestion to give glucose and fructose.
Q *Why is sucrose a nonreducing sugar?*

Sucrose consists of α-D-glucose and β-D-fructose molecules joined by an α, β-1,2-glycosidic bond. The structure of sucrose differs from the other disaccharides because the glycosidic bond ties up the anomeric carbons of both monosaccharide units. There are no isomers for sucrose. Without a free aldehyde group, there is no mutarotation. Sucrose cannot react with Benedict's reagent; sucrose is not a reducing sugar.

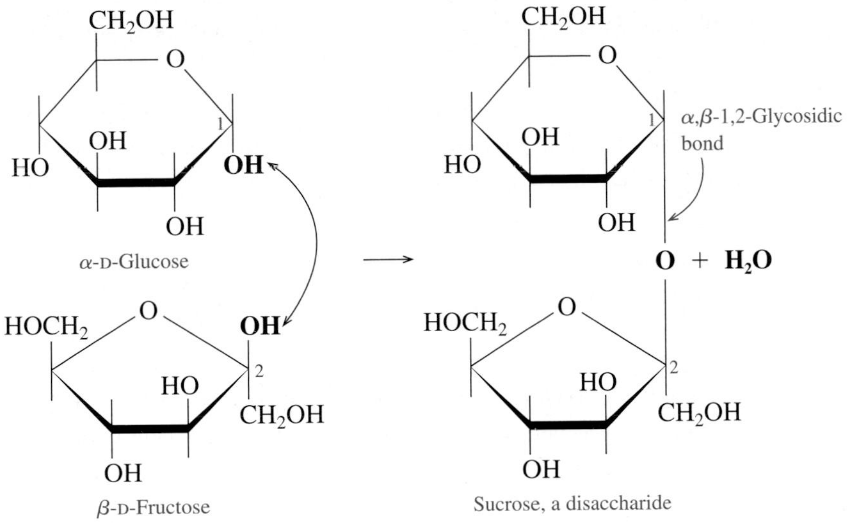

α-D-Glucose

β-D-Fructose

Sucrose, a disaccharide

α,β-1,2-Glycosidic bond

Fermentation

When yeast is added, the monosaccharides glucose and fructose (but not galactose) will undergo **fermentation** to produce ethanol and carbon dioxide gas.

$$C_6H_{12}O_6 \xrightarrow{\text{Yeast enzymes}} 2C_2H_5OH + 2CO_2(g)$$

Glucose or fructose → Ethanol

The disaccharides maltose and sucrose can also undergo fermentation because yeast contains enzymes for their hydrolysis to give glucose and fructose. However, lactose will not ferment because the enzyme lactase required for its hydrolysis is not present in yeast. Fermentation does not occur with polysaccharides.

SAMPLE PROBLEM 16.7

Glycosidic Bonds in Disaccharides

Melebiose is a disaccharide that has a sweetness of about 30 compared with sucrose (= 100).

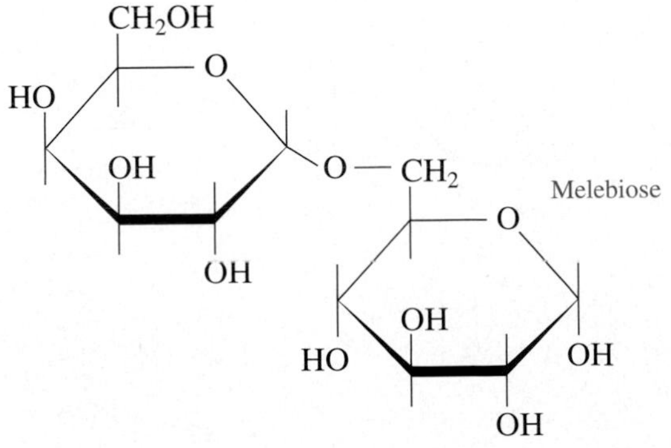

Melebiose

a. What are the monosaccharide units in melebiose?

b. What type of glycosidic bond links the monosaccharides?

c. Is the compound drawn as α- or β-melebiose?

Solution

a. The monosaccharide on the left side is α-D-galactose; on the right is α-D-glucose.

b. The monosaccharide units are linked by an α-1,6-glycosidic bond.

c. The downward position of the anomeric OH makes it α-melebiose.

Study Check

Cellobiose is a disaccharide composed of two β-D-glucose molecules linked by a β-1,4-glycosidic linkage. Draw a structural formula for β-cellobiose.

QUESTIONS AND PROBLEMS

Disaccharides

16.35 For each of the following disaccharides, give the monosaccharide units produced by hydrolysis, the type of glycosidic bond, and the identity of the disaccharide including the α or β anomer:

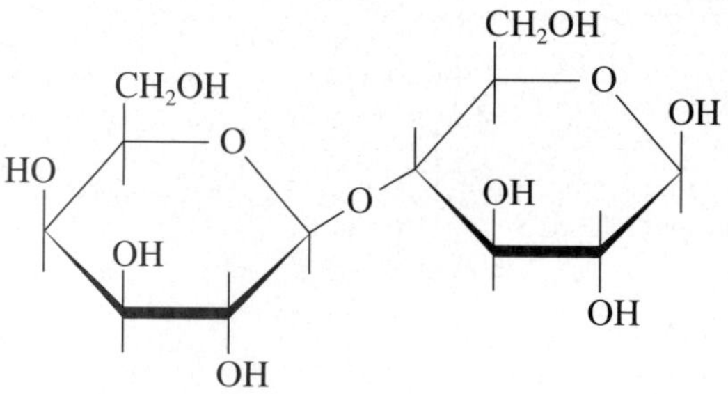

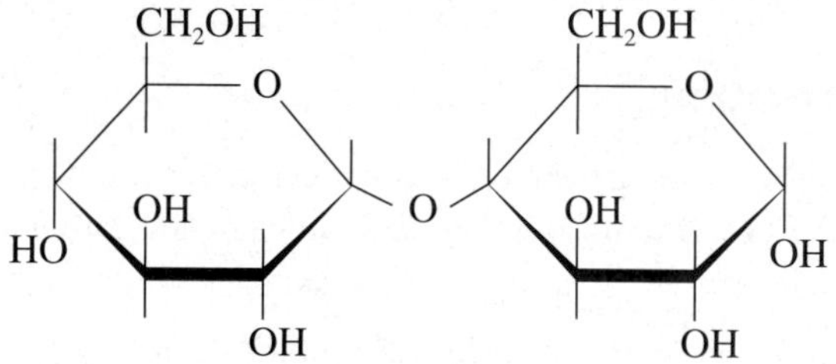

16.36 For each of the following disaccharides, give the monosaccharide units produced by hydrolysis, the type of glycosidic bond, and the identity of the disaccharide including the α or β anomer:

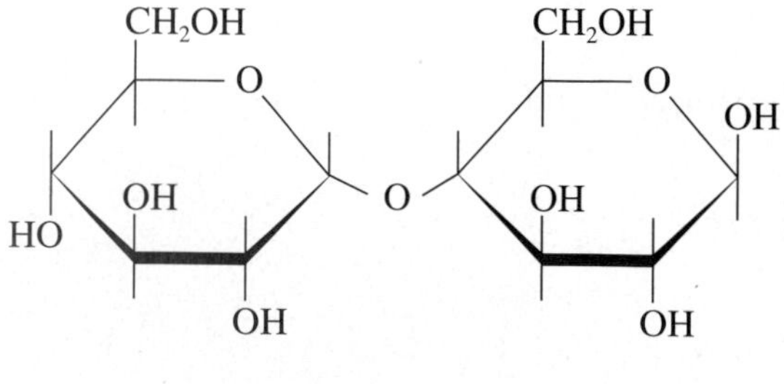

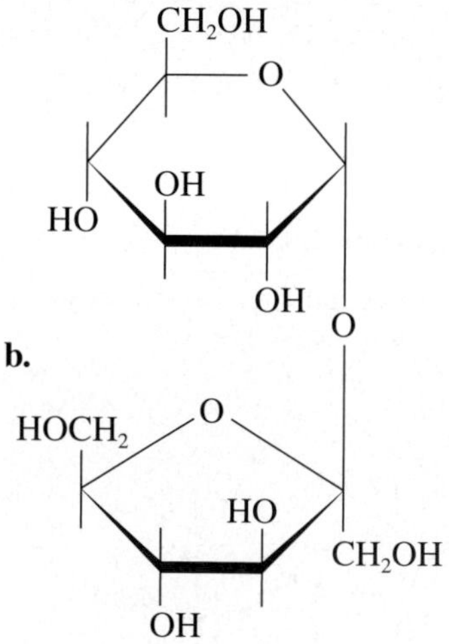

16.37 Indicate whether the sugars in problem 16.35 will undergo mutarotation and oxidation.

16.38 Indicate whether the sugars in problem 16.36 will undergo mutarotation and oxidation.

16.39 Identify disaccharides that fit each of the following descriptions:
a. Ordinary table sugar. **b.** Found in milk and milk products.
c. Also called *malt* sugar. **d.** Hydrolysis gives galactose and glucose.

16.40 Identify disaccharides that fit each of the following descriptions:
a. Not a reducing sugar. **b.** Composed of two glucose units.
c. Also called *milk* sugar. **d.** Hydrolysis gives glucose and fructose.

16.8 Polysaccharides

LEARNING GOAL

Describe the structural features of amylose, amylopectin, glycogen, and cellulose.

A polysaccharide is a polymer of many monosaccharides joined together. Three biologically important polysaccharides—starch, cellulose, and glycogen—are all polymers of D-glucose, which differ only in the type of glycosidic bonds and the amount of branching in the molecule.

Plant Starch: Amylose and Amylopectin

Starch, a storage form of glucose in plants, is found as insoluble granules in rice, wheat, potatoes, beans, and cereals. Starch is composed of two kinds of polysaccharides, amylose and amylopectin. **Amylose,** which makes up about 20% of starch, consists of α-D-glucose molecules connected by α-1,4-glycosidic bonds in a continuous chain. A typical polymer of amylose may contain from 250 to 4000 glucose units. Sometimes called a straight-chain polymer, polymers of amylose are actually coiled in helical fashion.

Amylopectin, which makes up as much as 80% of plant starch, is a branched-chain polysaccharide. Like amylose, the glucose molecules are connected by α-1,4-glycosidic bonds. However, at about every 25 glucose units, there is a branch of glucose molecules attached by an α-1,6-glycosidic bond between carbon 1 of the branch and carbon 6 in the main chain. (See Figure 16.5.)

Starches hydrolyze easily in water and acid to give smaller saccharides called dextrins, which then hydrolyze to maltose and finally glucose. In our bodies, these complex carbohydrates are digested by the enzymes amylase (in saliva) and maltase. The glucose obtained provides about 50% of our nutritional calories.

$$\text{Amylose, amylopectin} \xrightarrow{\text{H}^+ \text{ or amylase}} \text{dextrins} \xrightarrow{\text{H}^+ \text{ or amylase}} \text{maltose} \xrightarrow{\text{H}^+ \text{ or maltase}} \text{many D-glucose units}$$

Animal Starch: Glycogen

Glycogen, or animal starch, is a polymer of glucose that is stored in the liver and muscle of animals. It is hydrolyzed in our cells at a rate that maintains the blood level of glucose and provides energy between meals. The structure of glycogen is very similar to that of amylopectin found in plants except that glycogen is more highly branched. In glycogen, the glucose units are joined by α-1,4-glycosidic bonds, and branches occurring about every 10–15 glucose units are attached by α-1,6-glycosidic bonds.

Figure 16.5 The structure of **(a)** amylose is a straight-chain polysaccharide of glucose units, and **(b)** amylopectin is a branched chain of glucose.

Q *What are the two types of glycosidic bonds that link glucose molecules in amylopectin?*

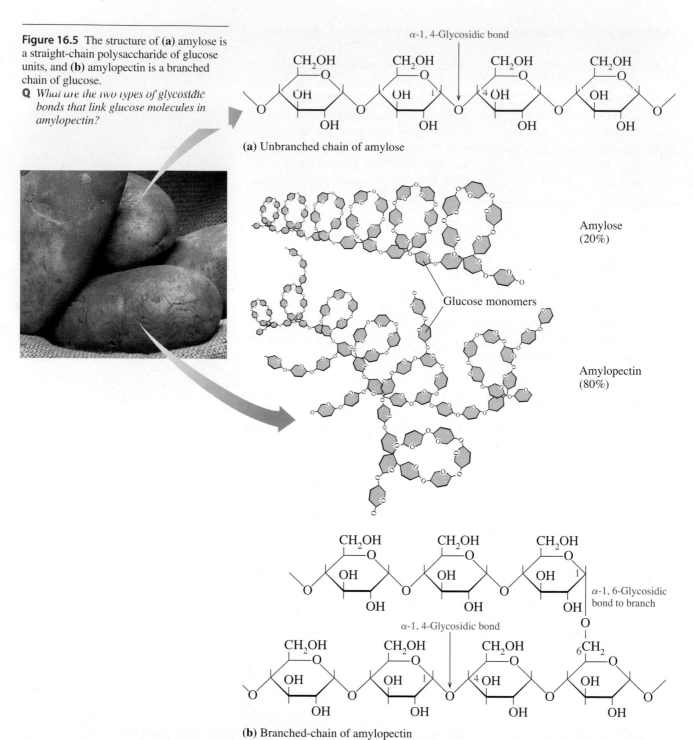

(a) Unbranched chain of amylose

Amylose (20%)

Glucose monomers

Amylopectin (80%)

(b) Branched-chain of amylopectin

Structural Polysaccharide: Cellulose

Cellulose is the major structural material of wood and plants. Cotton is almost pure cellulose. In **cellulose,** glucose molecules form a long unbranched chain similar to that of amylose. However, the glucose units in cellulose are linked by β-1,4-glycosidic bonds. The β isomers do not form coils like the α isomers but are aligned in parallel rows that are held in place by hydrogen bonds between hydroxyl groups in adjacent chains making cellulose insoluble in water. This gives a rigid structure to the cell walls in wood and fiber that is more resistant to hydrolysis than the starches. (See Figure 16.6.)

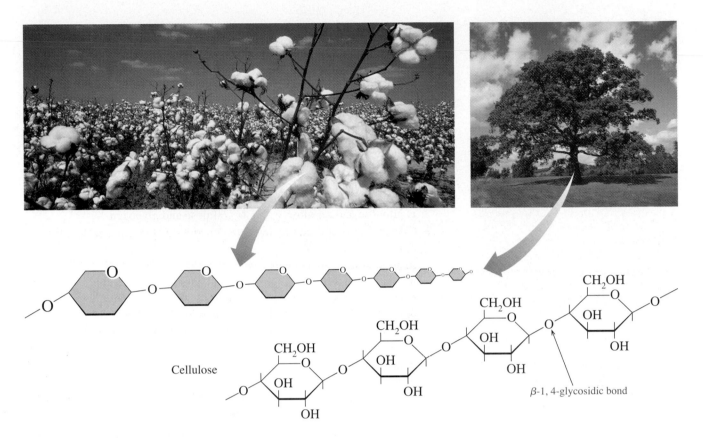

Cellulose

β-1, 4-glycosidic bond

Enzymes in our saliva and pancreatic juices hydrolyze the α-1,4-glycosidic bonds of the starches. However, there are no enzymes in humans that are able to hydrolyze the β-1,4-glycosidic bonds of cellulose; we cannot digest cellulose. Some animals such as goats and cows and insects like termites are able to obtain glucose from cellulose. Their digestive systems contain bacteria and protozoa with enzymes such as cellulase that can hydrolyze β-1,4-glycosidic bonds.

Figure 16.6 The polysaccharide cellulose is composed of β-1,4-glycosidic bonds.
Q *Why are humans unable to digest cellulose?*

Iodine Test

In the **iodine test**, iodine (I_2) is used to test for the presence of starch. The unbranched helical shape of the polysaccharide amylose in starch reacts strongly with iodine to form a deep blue-black complex. Amylopectin, cellulose, and glycogen produce reddish-purple and brown colors. Such colors do not develop when iodine is added to samples of mono- or disaccharides.

EXPLORE YOUR WORLD

Polysaccharides

Read the nutrition label on a box of crackers, cereal, bread, chips, or pasta. The major ingredient in crackers is flour, a starch. Chew on a single cracker for 4–5 minutes. An enzyme (amylase) in your saliva breaks apart the bonds in starch.

Questions

1. How are carbohydrates listed?
2. What other carbohydrates are listed?
3. How does the taste of the cracker change during the time that you chewed it?
4. What happened to the starches in the cracker as the amylase enzyme in your saliva reacted with the amylose or amylopectin?

SAMPLE PROBLEM 16.8

Structures of Polysaccharides

Identify the polysaccharide described by each of the following statements:

a. A polysaccharide that is stored in the liver and muscle tissues
b. An unbranched polysaccharide containing β-1,4-glycosidic bonds
c. A starch containing α-1,4- and α-1,6-glycosidic bonds

Solution

a. glycogen   **b.** cellulose   **c.** amylopectin, glycogen

Iodine Test for Starch

If you have some iodine solution for cuts, you can use it to test for starch. Collect samples of food such as crackers, bread, pasta, cereals, and candy. You may also test aspirin tablets, antacids, and the glue on an envelope. Place a few drops of iodine on each of the items. If starch is present, a deep blue-black color will develop. After recording your observations, throw all your samples away! Iodine is toxic.

Questions

1. Which of the foods test positive for starch?
2. What other items contain starch?
3. What type of carbohydrate is starch?
4. Why do you think starch is used in preparing tablets such as aspirin?
5. Why is starch present in the glue of the envelope flap?

Study Check

Cellulose and amylose are both unbranched glucose polymers. How do they differ?

QUESTIONS AND PROBLEMS

Polysaccharides

16.41 Describe the similarities and differences in the following polysaccharides:
  a. amylose and amylopectin
  b. amylopectin and glycogen

16.42 Describe the similarities and differences in the following polysaccharides:
  a. amylose and cellulose
  b. cellulose and glycogen

16.43 Give the name of one or more polysaccharides that matches each of the following descriptions:
  a. not digestible by humans
  b. the storage form of carbohydrates in plants
  c. contains only α-1,4-glycosidic bonds
  d. the most highly branched polysaccharide

16.44 Give the name of one or more polysaccharides that matches each of the following descriptions:
  a. the storage form of carbohydrates in animals
  b. contains only β-1,4-glycosidic bonds
  c. contains both α-1,4- and α-1,6-glycosidic bonds
  d. produces maltose during digestion

Chapter Review

16.1 Types of Carbohydrates

Carbohydrates are classified as monosaccharides (simple sugars), disaccharides (two monosaccharide units), and polysaccharides (many monosaccharide units).

16.2 Classification of Monosaccharides

Monosaccharides are polyhydroxy aldehydes (aldoses) or ketones (ketoses). Monosaccharides are also classified by their number of carbon atoms: *triose*, *tetrose*, *pentose*, or *hexose*.

16.3 D and L Notations from Fischer Projections

Chiral molecules can exist in two different forms which are mirror images of each other. These two different molecules are called enantiomers. In a Fischer projection (straight chain), the prefixes D- and L- are used to distinguish between the mirror images. In D-monosaccharides, the —OH is on the right of the chiral carbon farthest from the carbonyl carbon; it is on the left in L-monosaccharides.

16.4 Structures of Some Important Monosaccharides

Important monosaccharides are the aldohexoses glucose and galactose and the ketohexose fructose.

16.5 Cyclic Structures of Monosaccharides

The predominant form of monosaccharides is the cyclic form of five or six atoms. The cyclic structure forms by a reaction between an OH (usually the one on carbon 5 in hexoses) with the carbonyl group of the same molecule. The formation of a new hydroxyl group on carbon 1 (or 2 in fructose) gives α and β anomers of the cyclic monosaccharide. Because the molecule opens and closes continuously (mutarotation) while in solution, both anomers are present. Monosaccharides are reducing sugars because the open-chain aldehyde group (also available in ketoses) can be oxidized by a metal ion such as Cu^{2+}.

16.6 Chemical Properties of Monosaccharides

The aldehyde group in an aldose can be oxidized to a carboxylic acid, while the carbonyl group in an aldose or a ketose can be reduced to give a hydroxyl group. The anomeric hydroxyl groups can react with hydroxyl groups of an alcohol to form an acetal, which makes the compound a nonreducing sugar.

16.7 Disaccharides

Disaccharides are glycosides of two monosaccharide units joined together by a glycosidic bond. In the most common disaccharides, maltose, lactose, and sucrose, there is at least one glucose unit. In the glycoside, the acetal carbon cannot mutarotate or oxidize.

16.8 Polysaccharides

Polysaccharides are polymers of monosaccharide units. Starches consist of amylose, an unbranched chain of glucose, and amylopectin, a branched polymer of glucose. Glycogen, the storage form of glucose in animals, is similar to amylopectin with more branching. Cellulose is also a polymer of glucose, but in cellulose the glycosidic bonds are β bonds rather than α bonds as in the starches. Humans can digest starches, but not cellulose, to obtain energy. However, cellulose is important as a source of fiber in our diets.

Summary of Carbohydrates

Carbohydrate	Food Sources	Monosaccharides
Monosaccharides		
Glucose	Fruit juices, honey, corn syrup	Glucose
Galactose	Lactose hydrolysis	Galactose
Fructose	Fruit juices, honey, sucrose hydrolysis	Fructose
Disaccharides		
Maltose	Germinating grains, starch hydrolysis	Glucose + glucose
Lactose	Milk, yogurt, ice cream	Glucose + galactose
Sucrose	Sugar cane, sugar beets	Glucose + fructose
Polysaccharides		
Amylose	Rice, wheat, grains, cereals	Unbranched polymer of glucose joined by α-1,4-glycosidic bonds
Amylopectin	Rice, wheat, grains, cereals	Branched polymer of glucose joined by α-1,4- and α-1,6-glycosidic bonds
Glycogen	Liver, muscles	Highly branched polymer of glucose joined by α-1,4- and α-1,6-glycosidic bonds
Cellulose	Plant fiber, bran, beans, celery	Unbranched polymer of glucose joined by β-1,4-glycosidic bonds

Summary of Reactions

Glycoside (Acetal) Formation

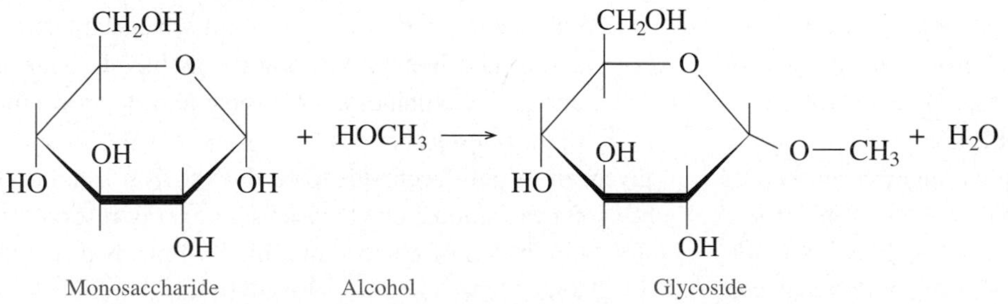

Monosaccharide   Alcohol   Glycoside

Monosaccharide   +   Monosaccharide   →   Disaccharide, a glycoside   + H_2O

Glycosidic bond

Oxidation and Reduction of Monosaccharides

D-Glucitol   ← reduction   D-Glucose   oxidation →   D-Gluconic acid

Hydrolysis of Disaccharides

Sucrose + H_2O ⟶ glucose + fructose

Lactose + H_2O ⟶ glucose + galactose

Maltose + H_2O ⟶ glucose + glucose

Key Terms

aldose Monosaccharides that contain an aldehyde group.

amylopectin A branched-chain polymer of starch composed of glucose units joined by α-1,4- and α-1,6-glycosidic bonds.

amylose An unbranched polymer of starch composed of glucose units joined by α-1,4-glycosidic bonds.

anomers The isomers of cyclic hemiacetals of monosaccharides that have a hydroxyl group on carbon 1 (or carbon 2). In the α anomer, the OH is drawn downward; in the β isomer the OH is up.

carbohydrate A simple or complex sugar composed of carbon, hydrogen, and oxygen.

cellulose An unbranched polysaccharide composed of glucose units linked by β-1,4-glycosidic bonds that cannot be hydrolyzed by the human digestive system.

disaccharides Carbohydrates composed of two monosaccharides joined by a glycosidic bond.

fermentation A reaction of glucose, fructose, maltose, or sucrose in which the sugar reacts with enzymes in yeast to give ethanol and carbon dioxide gas.

fructose A monosaccharide found in honey and fruit juices; it is combined with glucose in sucrose. Also called levulose and fruit sugar.

galactose A monosaccharide that occurs combined with glucose in lactose.

glucose The most prevalent monosaccharide in the diet. An aldohexose that is found in fruits, vegetables, corn syrup, and honey. Also known as blood sugar and dextrose. Combines in glycosidic bonds to form most of the polysaccharides.

glycogen A polysaccharide formed in the liver and muscles for the storage of glucose as an energy reserve. It is composed of glucose in a highly branched polymer joined by α-1,4- and α-1,6-glycosidic bonds.

glycosides Acetal products of a monosaccharide reacting with an alcohol or another sugar.

glycosidic bond The acetal bond that forms when an alcohol or a hydroxyl group of a monosaccharide adds to a hemiacetal. It is the type of bond that links monosaccharide units in di- or polysaccharides.

Haworth structure The cyclic structure that represents the closed chain of a monosaccharide.

iodine test A test for amylose that forms a blue-black color after iodine is added to the sample.

ketose A monosaccharide that contains a ketone group.

lactose A disaccharide consisting of glucose and galactose found in milk and milk products.

maltose A disaccharide consisting of two glucose units; it is obtained from the hydrolysis of starch and in germinating grains.

monosaccharide A polyhydroxy compound that contains an aldehyde or ketone group.

mutarotation The conversion between α and β anomers.

polysaccharides Polymers of many monosaccharide units, usually glucose. Polysaccharides differ in the types of glycosidic bonds and the amount of branching in the polymer.

reducing sugar A carbohydrate with a free aldehyde group capable of reducing the Cu^{2+} in Benedict's reagent.

saccharide A term from the Latin word *saccharum*, meaning "sugar"; it is used to describe the carbohydrate family.

sucrose A disaccharide composed of glucose and fructose; a nonreducing sugar, commonly called table sugar or "sugar."

Additional Problems

16.45 What are the structural differences in D-glucose and D-galactose?

16.46 What are the structural differences in D-glucose and D-fructose?

16.47 How do D-galactose and L-galactose differ?

16.48 How do α-D-glucose and β-D-glucose differ?

16.49 Consider the sugar D-gulose.

D-Gulose

a. What is the Fisher projection for L-gulose?
b. Draw the Haworth structure for α- and β-D-gulose.

16.50 Consider the structures for D-gulose in question 16.49.
a. What is the structure and name of the product formed by the reduction of D-gulose?
b. Write the structure and name of the product formed by the oxidation of D-gulose.

16.51 D-Sorbitol, a sweetener found in seaweed and berries, contains only hydroxyl functional groups. When D-sorbitol is oxidized, it forms D-glucose. What is the structural formula of D-sorbitol?

16.52 Raffinose is a trisaccharide found in Australian manna and in cottonseed meal. It is composed of three different monosaccharides. Identify the monosaccharides in raffinose.

16.53 If α-galactose is dissolved in water, β-galactose is eventually present. Explain how this occurs.

16.54 Why are lactose and maltose reducing sugars, but sucrose is not?

16.55 β-Cellobiose is a disaccharide obtained from the hydrolysis of cellulose. It is quite similar to maltose except it has a β-1,4-glycosidic bond. What is the structure of β-cellobiose?

16.56 The disaccharide trehalose found in mushrooms is composed of two α-D-glucose molecules joined by an α-1,1-glycosidic bond. Draw the structure of trehalose.

16.57 Gentiobiose is found in saffron.
a. Gentiobiose contains two glucose molecules linked by a β-1,6-glycoside bond. Draw the structure of α-gentiobiose.
b. Would gentiobiose be a reducing sugar? Why or why not?

16.58 From the compounds shown, select those that meet
the following statements:

   a. is the L-enantiomer of mannose.

   b. a ketopentose,

   c. an aldopentose.

   d. a ketohexose.

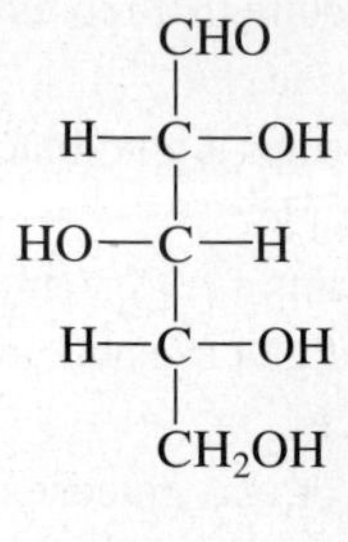

A

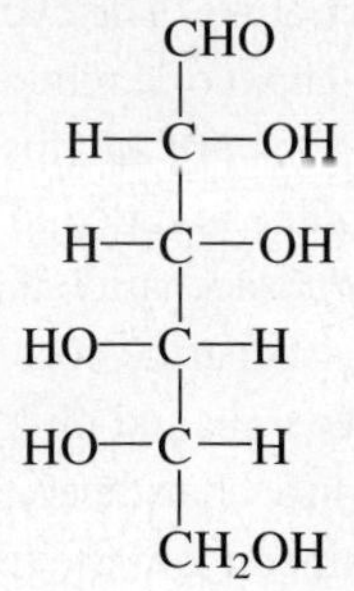

B

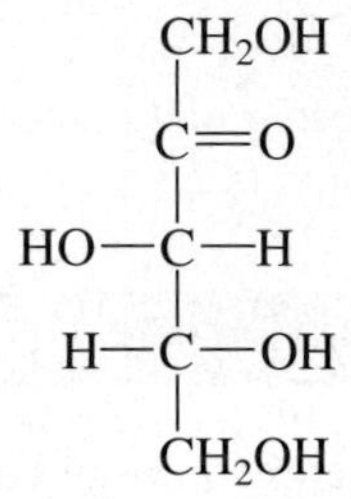

C

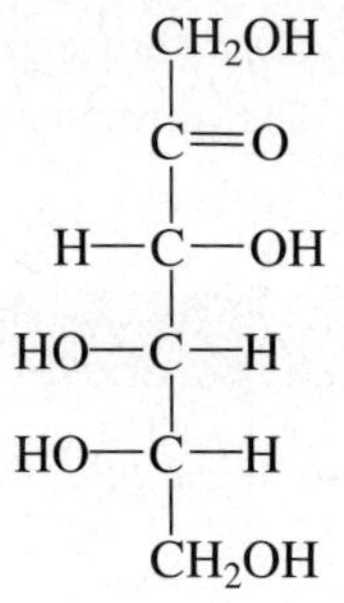

D

17 Carboxylic Acids and Esters

"There are many carboxylic acids, including the alpha hydroxy acids, that are found today in skin products," says Dr. Ken Peterson, pharmacist and cosmetic chemist, Oakland. "When you take a carboxylic acid called a fatty acid and react it with a strong base, you get a salt called soap. Soap has a high pH because the weak fatty acid and the strong base won't have a neutral pH of 7. If you take soap and drop its pH down to 7, you will convert the soap to the fatty acid. When I create fragrances, I use my nose and my chemistry background to identify and break down the reactions that produce good scents. Many fragrances are esters, which form when an alcohol reacts with a carboxylic acid. For example, the ester that smells like pineapple is made from ethanol and butyric acid."

LOOKING AHEAD

the Chemistry place

www.chemplace.com/college

Visit the URL above or use the CD-ROM in the book for extra quizzing, interactive tutorials, and career resources.

Carboxylic acids are similar to the weak acids we studied in Chapter 10. They have a sour or tart taste, produce hydronium ions in water, and neutralize bases. You encounter carboxylic acids when you use a vinegar salad dressing, which is a solution of acetic acid and water, or experience the sour taste of citric acid in a grapefruit or lemon. When a carboxylic acid combines with an alcohol, an ester is produced. Aspirin is an ester as well as a carboxylic acid. Fats known as trigycerides are esters of glycerol and fatty acids, which are long-chain carboxylic acids. Many fruits and flavorings including bananas, oranges, and strawberries contain esters, which produce their pleasant aromas and flavors.

17.1 Carboxylic Acids

In Chapter 15, we described the carbonyl group (C=O) as the functional group in aldehydes and ketones. In a **carboxylic acid,** a hydroxyl group is attached to the carbonyl group, forming a **carboxyl group.** The carboxyl functional group may be attached to an alkyl (R) group or an aromatic (Ar) group.

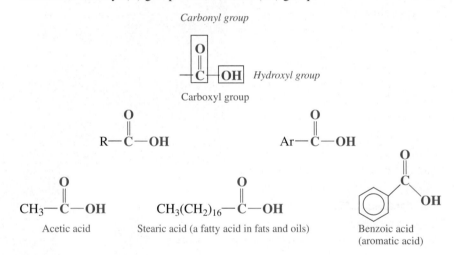

The carboxyl group can be written in several different ways. For example, the condensed structural formula for propanoic acid can be written as follows:

$$CH_3-CH_2-\overset{\overset{\displaystyle O}{\|}}{C}-OH \quad CH_3-CH_2-COOH \quad CH_3-CH_2-CO_2H$$

Some condensed structural formulas for propanoic acid

Naming Carboxylic Acids

The IUPAC names of carboxylic acids use the alkane names of the corresponding carbon chains.

Step 1   Identify the longest carbon chain containing the carboxyl group and replace the *e* of the alkane name by *oic acid*.

Step 2   Number the carbon chain beginning with the carboxyl carbon as carbon 1.

Step 3   Give the location and names of substituents on the main chain. The carboxyl function group takes priority over all the functional groups we have discussed.

Figure 17.1 Red ants inject formic acid under the skin which causes burning and irritation.
Q *What is the IUPAC name of formic acid?*

$$H-\overset{\overset{O}{\|}}{C}-OH \qquad CH_3-\overset{\overset{CH_3}{|}}{CH}-\overset{\overset{O}{\|}}{C}-OH \qquad CH_3-\overset{\overset{OH}{|}}{CH}-CH_2-\overset{\overset{O}{\|}}{C}-OH$$

Methanoic acid   2-Methylpropanoic acid   3-Hydroxybutanoic acid

Many carboxylic acids are still named by their common names, which are derived from their natural sources. In Chapter 15, we named aldehydes using the prefixes that represent the typical sources of carboxylic acids.

Formic acid is injected under the skin from bee or red ant stings and other insect bites. (See Figure 17.1.) Acetic acid is the oxidation product of the ethanol in wines and apple cider. The resulting solution of acetic acid and water is known as vinegar. Butyric acid gives the foul odor to rancid butter. (See Table 17.1.) Some ball-and-stick models of carboxylic acids are shown in Figure 17.2.

When using the common names, the Greek letters alpha (α), beta (β), and gamma (γ) are assigned to the carbons adjacent to the carboxyl carbon.

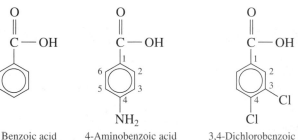

IUPAC	4	3	2	1
Common		γ	β	α

The aromatic carboxylic acid is called benzoic acid. With the carboxyl carbon as carbon 1, the ring is numbered in the direction that gives any substituents the smallest possible numbers. As we did with other aromatic compounds (Chapter 13), the prefixes *ortho, meta, and para* may be used to show the position of one other substituent.

Benzoic acid   4-Aminobenzoic acid
(*p*-aminobenzoic acid)   3,4-Dichlorobenzoic acid
(*not* 4,5-dichlorobenzoic acid)

Methanoic acid
(formic acid)

Ethanoic acid
(acetic acid)

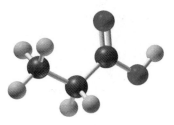

Propanoic acid
(propionic acid)

Figure 17.2 In carboxylic acids, a carbonyl group is bonded to a hydroxyl group.
Q *What is the IUPAC and common name of a carboxylic acid with a chain of four carbons?*

Table 17.1 Names and Natural Sources of Carboxylic Acids

Condensed Structural Formulas	IUPAC Name	Common Name	Occurs In
$H-\overset{\overset{O}{\|}}{C}-OH$	Methanoic acid	Formic acid	Ant and bee stings (Latin *formica,* "ant")
$CH_3-\overset{\overset{O}{\|}}{C}-OH$	Ethanoic acid	Acetic acid	Vinegar (Latin *acetum,* "sour")
$CH_3-CH_2-\overset{\overset{O}{\|}}{C}-OH$	Propanoic acid	Propionic acid	Dairy products (Greek, *pro,* "first," *pion,* "fat")
$CH_3-CH_2-CH_2-\overset{\overset{O}{\|}}{C}-OH$	Butanoic acid	Butyric acid	Rancid butter (Latin *butyrum,* "butter")

Naming Carboxylic Acids

Give the IUPAC and common name, if any, for each of the following carboxylic acids:

a. $CH_3-CH_2-\overset{\displaystyle O}{\overset{\displaystyle \|}{C}}-OH$

b. (structure with CH_3 branch) $\overset{\displaystyle O}{\overset{\displaystyle \|}{C}}-OH$

c. aromatic ring with $\overset{\displaystyle O}{\overset{\displaystyle \|}{C}}-OH$ and Cl

Solution

a. This carboxylic acid has 3 carbon atoms. In the IUPAC system, the *e* in propane is replaced by *oic acid,* to give the name, *propanoic acid.* Its common name is *propionic acid.*

b. This carboxylic acid has a methyl group on the second carbon. It has the IUPAC name *2-methylbutanoic acid.* In the common name, the Greek letter α specifies the carbon atom next to the carboxyl carbon, *α-methylbutyric acid.*

c. An aromatic carboxylic acid is named as benzoic acid. Counting from the carboxyl carbon places the Cl on carbon 3, or the *meta* carbon. The name is 3-chlorobenzoic acid or *meta*-chlorobenzoic acid.

Study Check

Write the condensed structural formula of 3-phenylpropanoic acid.

Alpha Hydroxy Acids

Alpha hydroxy acids (AHAs) are naturally occurring carboxylic acids found in fruits, milk, and sugarcane. Cleopatra reportedly bathed in sour milk to smooth her skin. Dermatologists have been using products with a high concentration of AHAs to remove acne scars and reduce irregular pigmentation and age spots. Now lower concentrations (8–10%) of AHAs have been added to skin care products for the purpose of smoothing fine lines, improving skin texture, and cleansing pores. Several alpha hydroxy acids may be found in skin care products singly or in combination. Glycolic acid and lactic acid are most frequently used.

Recent studies indicate that products with AHAs increase sensitivity of the skin to sun and UV radiation. It is recommended that a sunscreen with a sun protection factor (SPF) of at least 15 be used when treating the skin with products that include AHAs. Products containing AHAs at concentrations under 10% and pH values greater than 3.5 are generally considered safe. However, the Food and Drug Administration has reports of AHAs causing skin irritation including blisters, rashes, and discoloration of the skin. The FDA does not require product safety reports from cosmetic manufacturers, although they are responsible

for marketing safe products. The FDA advises that you test any product containing AHAs on a small area of skin before you use it on a large area.

Alpha Hydroxy Acid (Source)	Structure
Glycolic acid (Sugarcane, sugar beet)	$HO-CH_2-\overset{O}{\overset{\|}{C}}OH$
Lactic acid (Sour milk)	$CH_3-\overset{OH}{\overset{\|}{C}H}-\overset{O}{\overset{\|}{C}}OH$
Tartaric acid (Grapes)	$HO\overset{O}{\overset{\|}{C}}-\overset{OH}{\overset{\|}{C}H}-\overset{OH}{\overset{\|}{C}H}-\overset{O}{\overset{\|}{C}}OH$
Malic acid (Apples, grapes)	$HO\overset{O}{\overset{\|}{C}}-CH_2-\overset{OH}{\overset{\|}{C}H}-\overset{O}{\overset{\|}{C}}OH$
Citric acid (Citrus fruits: lemons, oranges, grapefruit)	CH_2-COOH $HO-\overset{\|}{C}-COOH$ CH_2-COOH

Preparation of Carboxylic Acids

Carboxylic acids can be prepared from primary alcohols or aldehydes. As we saw in Chapter 15, there is an increase in carbon–oxygen bonds as a primary alcohol is oxidized to an aldehyde. Oxidation continues easily as another oxygen is added to yield a carboxylic acid. For example, when ethyl alcohol in wine comes in contact with the oxygen in the air, vinegar is produced. The oxidation process converts the ethyl alcohol (primary alcohol) to acetaldehyde, and then to acetic acid, the carboxylic acid in vinegar. (See Figure 17.3.)

$$CH_3-CH_2 \xrightarrow[\text{Ethyl alcohol}]{\overset{OH}{|}} \overset{[O]}{\longrightarrow} CH_3-\overset{\overset{O}{\|}}{C}-H \xrightarrow{[O]} CH_3-\overset{\overset{O}{\|}}{C}-OH$$

Ethyl alcohol   Acetaldehyde   Acetic acid

Figure 17.3 Vinegar is a 5% solution of acetic acid and water.
Q *What is the IUPAC name for acetic acid?*

SAMPLE PROBLEM 17.2

Preparation of Carboxylic Acids

Write an equation for the oxidation of 1-propanol and name each product.

Solution

A primary alcohol will oxidize to an aldehyde, which can oxidize further to a carboxylic acid.

$$CH_3-CH_2-CH_2-OH \xrightarrow{[O]} CH_3-CH_2-\overset{\overset{O}{\|}}{C}-H \xrightarrow{[O]} CH_3-CH_2-\overset{\overset{O}{\|}}{C}-OH$$

1-Propanol   Propanal   Propanoic acid
(propyl alcohol)   (propionaldehyde)   (propionic acid)

Study Check

Write the condensed structural formula of the carboxylic acid produced by the oxidation of 1-butanol.

QUESTIONS AND PROBLEMS

Carboxylic Acids

17.1 What carboxylic acid is responsible for the pain of an ant sting?

17.2 What carboxylic acid is found in a solution of vinegar?

17.3 Explain the differences in the condensed structural formulas of propanal and propanoic acid.

17.4 Explain the differences in the condensed structural formulas of benzaldehyde and benzoic acid.

17.5 Give the IUPAC and common names (if any) for the following carboxylic acids:

a. $CH_3-\overset{\overset{O}{\|}}{C}-OH$

b. (structure) $\overset{\overset{O}{\|}}{}$ OH

c. $CH_3-\overset{\overset{Cl}{|}}{CH}-\overset{\overset{O}{\|}}{C}-OH$

d. (structure) $\overset{CH_3}{|}$ $\overset{O}{\|}$ OH

e. (structure) $\overset{\overset{O}{\|}}{C}-OH$ with HO and OH on ring

f. $CH_3-\overset{\overset{Br}{|}}{CH}-CH_2-CH_2-\overset{\overset{O}{\|}}{C}-OH$

17.6 Give the IUPAC and common names (if any) for the following carboxylic acids:

a. $H-\overset{\displaystyle O}{\overset{\|}{C}}-OH$ b. [structure with Br] OH   c. [benzene ring with $\overset{\displaystyle O}{\overset{\|}{C}}-OH$]   d. [benzene ring with $\overset{\displaystyle O}{\overset{\|}{C}}-OH$ and Cl]

e. $CH_3-\overset{\displaystyle CH_3}{\overset{|}{CH}}-CH_2-\overset{\displaystyle O}{\overset{\|}{C}}-OH$ f. $Cl-CH_2-\overset{\displaystyle O}{\overset{\|}{C}}-OH$

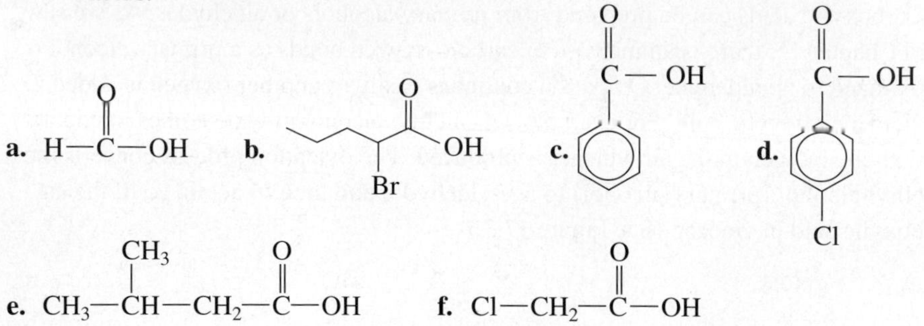

17.7 Draw the condensed structural formulas of each of the following carboxylic acids:
  a. propionic acid   **b.** benzoic acid
  c. 2-chloroethanoic acid   **d.** 3-hydroxypropanoic acid
  e. α-methylbutyric acid   **f.** 3,5-dibromoheptanoic acid

17.8 Draw the condensed structural formulas of each of the following carboxylic acids:
  a. butyric acid   **b.** 3-ethylbenzoic acid
  c. α-hydroxyacetic acid   **d.** 2,4-dibromobutanoic acid
  e. m-methylbenzoic acid   **f.** 4,4-dibromohexanoic acid

17.9 Draw the condensed structural formulas of the carboxylic acids formed by the oxidation of each of the following:

a. CH_3-OH b. $CH_3-\overset{\displaystyle O}{\overset{\|}{C}}-H$ c. $CH_3-\overset{\displaystyle CH_3}{\overset{|}{CH}}-CH_2-CH_2-OH$ d. [cyclopentane ring with CH_2-CH_2-OH]

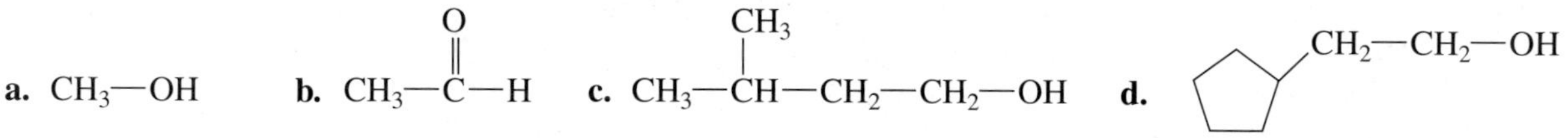

17.10 Draw the condensed structural formulas of the carboxylic acids formed by the oxidation of each of the following:

a. $CH_3-CH_2-CH_2-CH_2-CH_2-CH_2-OH$ b. $CH_3-CH_2-CH_2-CH_2-\overset{\displaystyle O}{\overset{\|}{C}}-H$

c. $CH_3-\overset{\displaystyle CH_3}{\overset{|}{CH}}-CH_2-\overset{\displaystyle O}{\overset{\|}{C}}-H$ d. [benzene ring with CH_2-CH_2-OH]

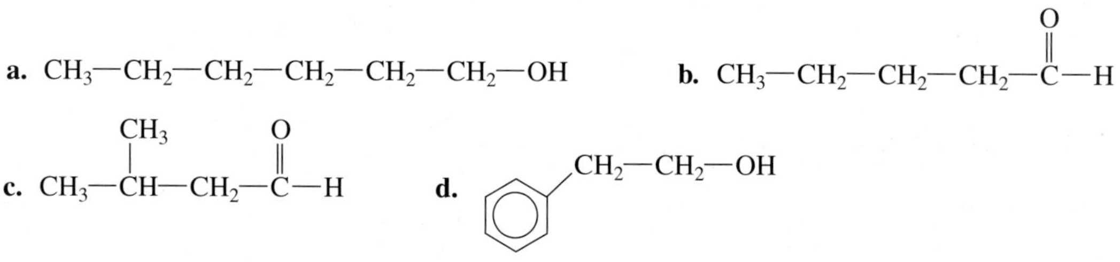

LEARNING GOAL

Describe the boiling points and solubility of carboxylic acids in water.

17.2 Physical Properties of Carboxylic Acids

Carboxylic acids are among the most polar organic compounds because the functional group consists of two polar groups: a hydroxyl (—OH) group and a carbonyl (C=O) group. This C=O double bond is similar to that of the aldehydes and ketones.

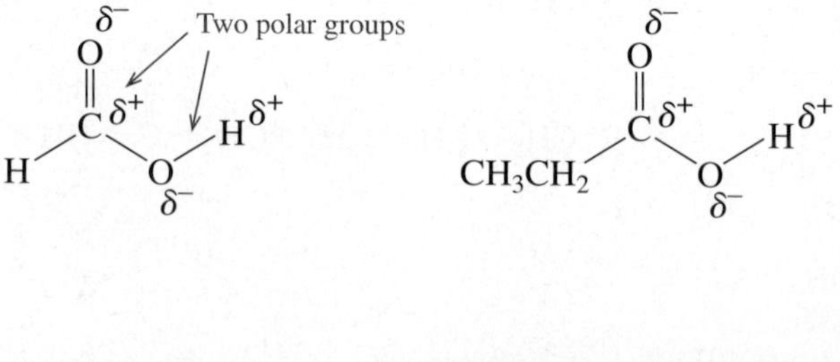

[structure diagrams showing two polar groups with δ⁻, δ⁺ charges labeled; formic acid (H) and propanoic acid (CH₃CH₂) carboxylic acid structures]

Therefore, carboxylic acids form hydrogen bonds with other carboxylic acid molecules and water. This ability to form hydrogen bonds has a major influence on both their boiling points and solubility in water.

Boiling Points

Carboxylic acids have higher boiling points than alcohols, ketones, and aldehydes of similar mass.

$$CH_3-CH_2-\overset{\overset{\displaystyle O}{\|}}{C}-H \qquad CH_3-CH_2-CH_2OH \qquad CH_3-\overset{\overset{\displaystyle O}{\|}}{C}-OH$$

Compound	Propanal	1-Propanol	Acetic acid
Molar mass	58	60	60
bp	49°C	97°C	118°C

One reason for the high boiling points of carboxylic acids is the formation of hydrogen bonds between two carboxylic acids to give a dimer. Because the dimers are stable as gases, the mass of the carboxylic acid is effectively doubled, which means that higher temperatures are required to reach the boiling point and form gases.

Two hydrogen bonds

$$CH_3-C\overset{\displaystyle O\cdots H-O}{\underset{\displaystyle O-H\cdots O}{}}C-CH_3$$

A dimer of two acetic acid molecules

Solubility in Water

Carboxylic acids with one to four carbons are very soluble in water because the carboxyl group forms hydrogen bonds with several water molecules. (See Figure 17.4.) However, as the length of the carbon chain increases, the nonpolar portion reduces solubility. Carboxylic acids having five or more carbons are not very soluble in water. Table 17.2 lists boiling points and solubilities for some selected carboxylic acids.

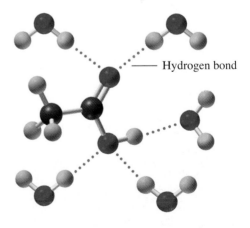

—— Hydrogen bond

Figure 17.4 Acetic acid forms hydrogen bonds with water molecules.
Q *Why do the atoms in the carboxyl group hydrogen bond with water molecules?*

Table 17.2   Physical Properties of Selected Carboxylic Acids

IUPAC name	bp (°C)	Solubility in Water
Methanoic acid	101	Very soluble
Ethanoic acid	118	Very soluble
Propanoic acid	141	Very soluble
Butanoic acid	164	Very soluble
Pentanoic acid	187	Slightly soluble
Hexanoic acid	205	Slightly soluble
Benzoic acid	250	Slightly soluble

SAMPLE PROBLEM 17.3

Physical Properties of Carboxylic Acids

Put the following organic compounds in order of increasing boiling points: butanoic acid, butanone, pentane, and 2-butanol.

Solution

The boiling point increases when the molecules of a compound can form hydrogen bonds or dipole-dipole interactions. The alkane has the lowest boiling point because alkanes cannot hydrogen bond. Ketones have low boiling points because they can only have dipole-dipole interactions via the $C=O$ group. Alcohols and carboxylic acids have higher boiling points because they hydrogen bond via the — OH. The higher boiling points of carboxylic acids are due to the formation of stable dimers, which increase the effective mass and therefore the boiling point.

Pentane < butanone < 2-butanol < butanoic acid

Study Check

At its boiling point, acetic acid forms gas molecules that have a mass of 120 rather than the molar mass of 60 g/mole for CH_3COOH. Explain.

QUESTIONS AND PROBLEMS

Physical Properties of Carboxylic Acids

17.11 Identify the compound in each pair that has the higher boiling point.
   a. acetic acid or butanoic acid   **b.** 1-propanol or propanoic acid
   c. butanone or butanoic acid

17.12 Identify the compound in each pair that has the higher boiling point.
   a. acetone or propanoic acid   **b.** propanoic acid or hexanoic acid
   c. ethanol or acetic acid

17.13 Place the compounds in each group in order of increasing solubility in water:
   a. propanoic acid, propanol, acetone
   b. butanoic acid, propanoic acid, acetic acid
   c. propane, ethanol, acetic acid

17.14 Place the compounds in each group in order of increasing solubility in water:
   a. butanone, butanoic acid, diethyl ether
   b. propanoic acid, acetic acid, formic acid
   c. butanone, 1-butanol, butanoic acid

LEARNING GOAL

Describe the ionization of carboxylic acids as weak acids.

17.3 Acidity of Carboxylic Acids

One of the most important properties of carboxylic acids is their ionization in water, which makes them weak acids (Chapter 10). In the ionization, a carboxylic acid donates a proton to a water molecule to produce an anion called a **carboxylate ion** and a hydronium ion.

$$\underset{\substack{\text{Carboxylic acid}\\ \text{(weak acid)}}}{R-\overset{\overset{\displaystyle O}{\|}}{C}-OH} + H_2O \rightleftharpoons \underset{\text{Carboxylate ion}}{R-\overset{\overset{\displaystyle O}{\|}}{C}-O^-} + \underset{\substack{\text{Hydronium}\\ \text{ion}}}{H_3O^+}$$

$$\underset{\substack{\text{Ethanoic acid}\\ \text{(acetic acid)}}}{CH_3-\overset{\overset{\displaystyle O}{\|}}{C}-OH} + H_2O \rightleftharpoons \underset{\substack{\text{Ethanoate ion}\\ \text{(acetate ion)}}}{CH_3-\overset{\overset{\displaystyle O}{\|}}{C}-O^-} + \underset{\substack{\text{Hydronium}\\ \text{ion}}}{H_3O^+}$$

SAMPLE PROBLEM 17.4

Ionization of Carboxylic Acids in Water

Write the equation for the ionization of propionic acid in water.

Solution

The ionization of propionic acid produces a carboxylate ion and a hydronium ion.

$$CH_3-CH_2-\overset{\overset{\displaystyle O}{\|}}{C}-OH + H_2O \rightleftharpoons CH_3-CH_2-\overset{\overset{\displaystyle O}{\|}}{C}-O^- + H_3O^+$$

Study Check

Write an equation for the ionization of formic acid in water.

Carboxylic acids are more acidic than other organic compounds including phenols, but only a small percentage (~1%) of the carboxylic acid molecules in a dilute solution are ionized, which means that most of the acid is not ionized. The acid dissociation constants of carboxylic acid are between 10^{-4} to 10^{-5} as seen in Table 17.3.

Salts of Carboxylic Acids

Although carboxylic acids are weak acids, they are completely neutralized by strong bases such as NaOH and KOH. The products are water and a **carboxylic acid salt,** which is a carboxylate ion and the metal ion from the base. The carboxylate ion is named by replacing the *–oic acid* ending of the acid name with *ate*.

Table 17.3   Acid Dissociation Constants K_a for Selected Carboxylic Acids

Name	K_a (25°C)
Methanoic acid	1.8×10^{-4}
Ethanoic acid	1.8×10^{-5}
Propanoic acid	1.3×10^{-5}
Butanoic acid	1.5×10^{-5}
Pentanoic acid	1.5×10^{-5}
Benzoic acid	6.5×10^{-5}

$$\underset{\text{Carboxylic acid}}{R-\overset{\overset{\displaystyle O}{\|}}{C}-OH} + \underset{\text{Strong base}}{MOH} \longrightarrow \underset{\text{Carboxylic acid salt}}{R-\overset{\overset{\displaystyle O}{\|}}{C}-O^-\ M^+} + \underset{\text{Water}}{H_2O}$$

$$\underset{\text{Formic acid}}{H-\overset{\overset{\displaystyle O}{\|}}{C}-OH} + NaOH \longrightarrow \underset{\text{Sodium formate}}{H-\overset{\overset{\displaystyle O}{\|}}{C}-O^-\ Na^+} + H_2O$$

$$\underset{\text{Benzoic acid}}{C_6H_5-\overset{\overset{\displaystyle O}{\|}}{C}-OH} + KOH \longrightarrow \underset{\text{Potassium benzoate}}{C_6H_5-\overset{\overset{\displaystyle O}{\|}}{C}-O^-\ K^+} + H_2O$$

Carboxylic Acids in Metabolism

There are several carboxylic acids that are part of the metabolic processes within our cells. For example, during glycolysis, a molecule of glucose is broken down into two molecules of pyruvic acid or actually its carboxylate ion pyruvate. During strenuous exercise when oxygen levels are low (anaerobic), pyruvic acid is reduced to give lactic acid or the lactate ion. The buildup of lactate ion in the muscle leads to fatigue and pain.

$$CH_3-\overset{\overset{O}{\|}}{C}-\overset{\overset{O}{\|}}{C}-OH \; +2H \xrightarrow{\text{Reduction}}$$

Pyruvic acid

$$CH_3-\overset{\overset{OH}{|}}{C}H-\overset{\overset{O}{\|}}{C}-OH$$

Lactic acid

In the citric acid cycle or Krebs cycle, several dicarboxylic acids are oxidized and decarboxylated (loss of CO_2) in order to produce energy for the cell. These carboxylic acids are normally referred to by their common names. At the start of the citric acid cycle, citric acid with six carbons is converted to five-carbon α-ketoglutaric acid. Citric acid is also the acid that gives the sour tastes to citrus fruits such as lemons and grapefruits.

$$
\begin{array}{c}
COOH \\
| \\
CH_2 \\
| \\
HO-C-COOH \\
| \\
CH_2 \\
| \\
COOH
\end{array}
\xrightarrow{[O]}
\begin{array}{c}
COOH \\
| \\
CH_2 \\
| \\
CH_2 \\
| \\
C=O \\
| \\
COOH
\end{array}
+ CO_2
$$

Citric acid α-Ketoglutaric acid

The citric acid cycle continues as α-ketoglutaric acid loses CO_2 to give a four-carbon succinic acid. Then a series of reactions converts succinic acid to oxaloacetic acid. We see that some of the functional groups we have studied along with reactions such as hydration and oxidation are part of the metabolic processes that take place in our cells.

$$
\begin{array}{c}
COOH \\
| \\
CH_2 \\
| \\
CH_2 \\
| \\
COOH
\end{array}
\xrightarrow{[O]}
\begin{array}{c}
COOH \\
| \\
C-H \\
\| \\
H-C \\
| \\
COOH
\end{array}
\xrightarrow{H_2O}
$$

Succinic acid   Fumaric acid

$$
\begin{array}{c}
COOH \\
| \\
HO-C-H \\
| \\
CH_2 \\
| \\
COOH
\end{array}
\xrightarrow{[O]}
\begin{array}{c}
COOH \\
| \\
C=O \\
| \\
CH_2 \\
| \\
COOH
\end{array}
$$

Malic acid   Oxaloacetic acid

At the pH of the aqueous environment in the cells, the carboxylic acids are ionized, which means it is actually the carboxylate ions that take part in the reactions of citric acid cycle. For example, in water, succinic acid is in equilibrium with its carboxylate ion succinate.

$$
\begin{array}{c}
COOH \\
| \\
CH_2 \\
| \\
CH_2 \\
| \\
COOH
\end{array}
+ 2H_2O \rightleftharpoons
\begin{array}{c}
COO^- \\
| \\
CH_2 \\
| \\
CH_2 \\
| \\
COO^-
\end{array}
+ 2H_3O^+
$$

Succinic acid   Succinate ion

In Chapters 23 and 24, we will study glycolysis and the citric acid cycle in more detail.

Sodium propionate, a preservative, is added to bread, cheeses, and bakery items to inhibit the spoilage of the food by microorganisms. Sodium benzoate, an inhibitor of mold and bacteria, is added to juices, margarine, relishes, salads, and jams. Monosodium glutamate (MSG) is added to meats, fish, vegetables, and bakery items to enhance flavor, although it causes headaches in some people. (See Figure 17.5.)

$$CH_3-CH_2-\underset{\displaystyle\|}{\overset{\displaystyle O}{C}}-O^-Na^+$$

Sodium propionate

Sodium benzoate

$$HO-\underset{\displaystyle\|}{\overset{\displaystyle O}{C}}-\underset{\displaystyle\underset{\displaystyle NH_2}{|}}{CH}-CH_2-CH_2-\underset{\displaystyle\|}{\overset{\displaystyle O}{C}}-O^-Na^+$$

Monosodium glutamate

The carboxylic acid salts are solids at room temperature and have high melting points. Because they are ionic compounds, carboxylic acid salts of the alkali metals (Li^+, Na^+, and K^+) and NH_4^+ are usually soluble in water.

SAMPLE PROBLEM 17.5

Neutralization of a Carboxylic Acid

Write the equation for the neutralization of propionic acid with sodium hydroxide.

Solution

The neutralization of an acid with a base produces the salt of the acid and water.

$$CH_3-CH_2-\underset{\displaystyle\|}{\overset{\displaystyle O}{C}}-OH + NaOH \longrightarrow CH_3-CH_2-\underset{\displaystyle\|}{\overset{\displaystyle O}{C}}-O^-Na^+ + H_2O$$

Propionic acid   Sodium propionate

Study Check

What carboxylic acid will give potassium butyrate when it is neutralized by KOH?

Figure 17.5 Preservatives and flavor enhancers in soups and seasonings are often carboxylic acids or their salts.
Q *What is the carboxylate salt produced by the neutralization of butanoic acid and lithium hydroxide?*

QUESTIONS AND PROBLEMS

Acidity of Carboxylic Acids

17.15 Write equations for the ionization of each of the following carboxylic acids in water:

a. $H-\underset{\displaystyle\|}{\overset{\displaystyle O}{C}}-OH$ b. $CH_3-CH_2-\underset{\displaystyle\|}{\overset{\displaystyle O}{C}}-OH$ c. acetic acid

17.16 Write equations for the ionization of each of the following carboxylic acids in water:

a. $CH_3-\underset{\displaystyle\underset{\displaystyle CH_3}{|}}{CH}-\underset{\displaystyle\|}{\overset{\displaystyle O}{C}}-OH$ b. α-hydroxyacetic acid   c. butanoic acid

17.17 Write equations for the reaction of each of the following carboxylic acids with NaOH:
a. formic acid   b. propanoic acid   c. benzoic acid

17.18 Write equations for the reaction of each of the following carboxylic acids with KOH:
  a. acetic acid   **b.** 2-methylbutanoic acid   **c.** *p*-chlorobenzoic acid
17.19 Give the IUPAC and common names, if any, of the carboxylic acid salts in problem 17.17.
17.20 Give the IUPAC and common names, if any, of the carboxylic acid salts in problem 17.18.

LEARNING GOAL

Write the products of the esterification reaction of an alcohol and carboxylic acid.

17.4 Esters of Carboxylic Acids

Esters are the derivatives of carboxylic acid in which an alkoxy group (—OR) replaces the hydroxyl (—OH) group in the carboxylic acid.

$$
\underset{\text{Carboxylic acid}}{R-\overset{\displaystyle O}{\overset{\|}{C}}-O-H} \qquad\qquad \underset{\text{Ester}}{R-\overset{\displaystyle O}{\overset{\|}{C}}-O-R}
$$

Esterification

In a reaction called **esterification,** a carboxylic acid reacts with an alcohol when heated in the presence of an acid catalyst (usually H_2SO_4). In the reaction, water is produced from the —OH removed from the carboxylic acid and an —H lost by the alcohol.

$$
\underset{\text{Acid}}{R-\overset{\displaystyle O}{\overset{\|}{C}}-O-H} \; + \; \underset{\text{Alcohol}}{H-O-R} \; \underset{\text{Heat}}{\overset{H^+}{\rightleftharpoons}} \; \underset{\text{Ester}}{R-\overset{\displaystyle O}{\overset{\|}{C}}-O-R} \; + \; H-O-H
$$

Example:

$$
\underset{\text{Acetic acid}}{CH_3-\overset{\displaystyle O}{\overset{\|}{C}}-O-H} \; + \; \underset{\text{Methyl alcohol}}{H-O-CH_3} \; \underset{\text{Heat}}{\overset{H^+}{\rightleftharpoons}} \; \underset{\text{Methyl acetate}}{CH_3-\overset{\displaystyle O}{\overset{\|}{C}}-O-CH_3} \; + \; H-O-H
$$

If we use acetic acid and 1-propanol, we can write an equation for the formation of the ester that is responsible for the flavor and odor of pears.

$$
\underset{\text{Acetic acid}}{CH_3-\overset{\displaystyle O}{\overset{\|}{C}}-OH} \; + \; \underset{\text{1-Propanol}}{H-O-CH_2CH_2CH_3} \; \underset{\text{Heat}}{\overset{H^+}{\rightleftharpoons}} \; \underset{\substack{\text{Propyl acetate}\\ \text{(pears)}}}{CH_3-\overset{\displaystyle O}{\overset{\|}{C}}-O-CH_2CH_2CH_3} \; + \; H_2O
$$

SAMPLE PROBLEM 17.6

Writing Esterification Equations

The ester that gives the flavor and odor of apples can be synthesized from butyric acid and methyl alcohol. What is the equation for the formation of the ester in apples?

Solution

$$CH_3-CH_2-CH_2-\overset{\overset{\displaystyle O}{\|}}{C}-OH \;+\; H-O-CH_3 \underset{Heat}{\overset{H^+}{\rightleftharpoons}}$$

Butyric acid · Methyl alcohol

$$CH_3-CH_2-CH_2-\overset{\overset{\displaystyle O}{\|}}{C}-O-CH_3 \;+\; H_2O$$

Methyl butyrate

Study Check

What carboxylic acid and alcohol are needed to form the following ester, which gives the flavor and odor to apricots?

$$CH_3-CH_2-\overset{\overset{\displaystyle O}{\|}}{C}-O-CH_2-CH_2-CH_2-CH_2-CH_3$$

HEALTH NOTE

Salicylic Acid and Aspirin

Chewing on a piece of willow bark was a way of relieving pain for many centuries. By the 1800s, chemists discovered that salicylic acid was the agent in the bark responsible for the relief of pain. However, salicylic acid, which has both a carboxylic group and a hydroxyl group, irritates the stomach lining. An ester of salicylic acid and acetic acid called acetylsalicylic acid or "aspirin" that is less irritating was prepared in 1899 by the Bayer chemical company in Germany. In some aspirin preparations, a buffer is added to neutralize the carboxylic acid group and lessen its irritation of the stomach. Aspirin is used as an analgesic (pain reliever), antipyretic (fever reducer), and anti-inflammatory agent.

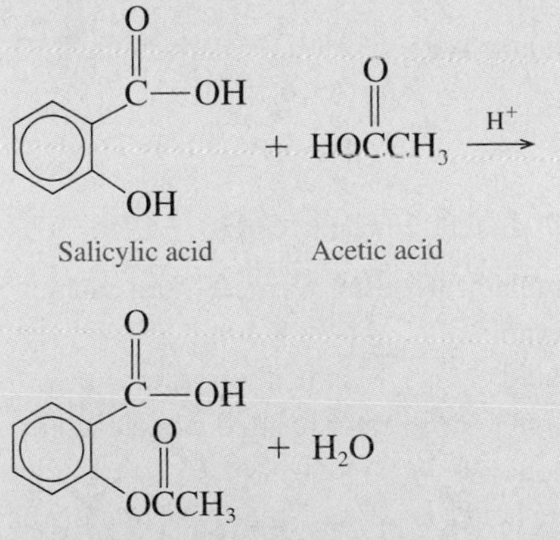

Salicylic acid · Acetic acid

Acetylsalicylic acid, "aspirin" · + H$_2$O

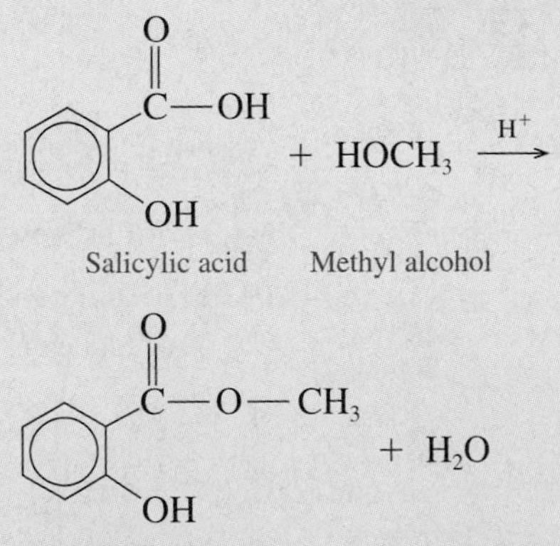

Oil of wintergreen, or methyl salicylate, has a spearmint odor and flavor. Because it can pass through the skin, methyl salicylate is used in skin ointments where it acts as a counterirritant, producing heat to soothe sore muscles.

Salicylic acid · Methyl alcohol · + HOCH$_3$ $\xrightarrow{H^+}$

Methyl salicylate (oil of wintergreen) · + H$_2$O

ENVIRONMENTAL NOTE

Plastics

Terephthalic acid (an acid with two carboxyl groups) is produced in large quantities for the manufacture of polyesters such as Dacron, and plastics. When terephthalic acid reacts with ethylene glycol, ester bonds can form on both ends of the molecules, allowing many molecules to combine until they have formed a long polymer known as a *polyester.*

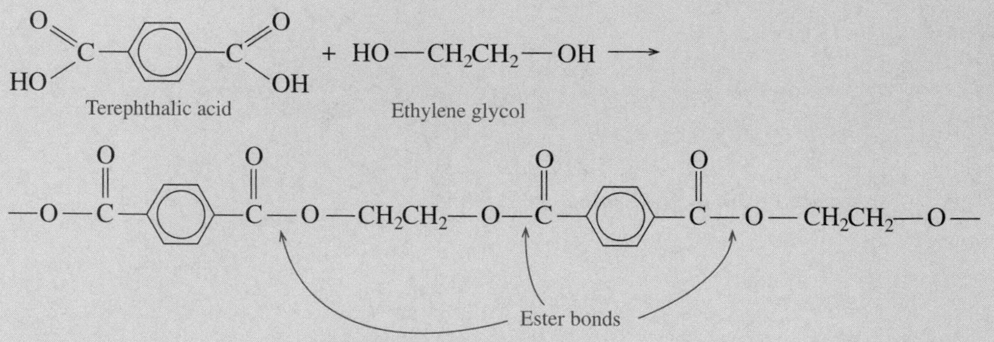

A section of the polyester Dacron

Dacron polyester is used to make permanent press fabrics, carpets, and clothes. In medicine, artificial blood vessels and valves are made of Dacron, which is biologically inert and does not clot the blood. The polyester can also be made as a film called Mylar and as a plastic known as PETE (**p**oly**e**thylene**te**rphthalate). PETE is used for plastic soft drink bottles as well as for containers of salad dressings, shampoos, and dishwashing liquids.

Today PETE is the most widely recycled of all the plastics. In 1992, 365 million pounds (166 million kilograms) of PETE were recycled. After it is separated from other plastics, PETE can be changed into other useful items including polyester fabric for T-shirts and coats, fill for sleeping bags, door mats, and tennis ball containers.

QUESTIONS AND PROBLEMS

Esters of Carboxylic Acids

17.21 Identify each of the following as an aldehyde, a ketone, a carboxylic acid, or an ester:

a. $CH_3-\overset{\overset{O}{\|}}{C}-H$ **b.** $CH_3-\overset{\overset{O}{\|}}{C}-O-CH_3$ **c.** $CH_3-CH_2-\overset{\overset{O}{\|}}{C}-CH_3$ **d.** $CH_3-CH_2-\overset{\overset{O}{\|}}{C}-O-H$

17.22 Identify each of the following as an aldehyde, a ketone, a carboxylic acid, or an ester:

a. $CH_3-\overset{\overset{O}{\|}}{C}-OH$ **b.** $CH_3-\overset{\overset{O}{\|}}{C}-O-CH_2-CH_3$ **c.** $CH_3-CH_2-\overset{\overset{O}{\|}}{C}-H$ **d.** $CH_3-\overset{\overset{CH_3}{|}}{CH}-\overset{\overset{O}{\|}}{C}-O-CH_2-CH_3$

17.23 Write the condensed structural formula of the ester formed when each of the following react with methyl alcohol:

a. acetic acid **b.** butyric acid **c.** benzoic acid

17.24 Write the condensed structural formula of the ester formed when each of the following react with methyl alcohol:

a. formic acid **b.** propionic acid **c.** 2-methylpentanoic acid

17.25 Draw the condensed structural formulas of the ester formed when each of the following carboxylic acids and alcohols react:

a. $CH_3-CH_2-\overset{\overset{O}{\|}}{C}-OH + HO-CH_2-CH_2-CH_3 \overset{H^+}{\rightleftarrows}$

b. $CH_3-CH_2-CH_2-CH_2-\overset{\overset{O}{\|}}{C}-OH + HO-\overset{\overset{CH_3}{|}}{CH}-CH_3 \overset{H^+}{\rightleftarrows}$

17.26 Draw the condensed structural formula of the ester formed when each of the following carboxylic acids and alcohols react:

a. CH₃—CH₂—C(=O)—OH + HO—CH₃ $\xrightarrow{H^+}$

b. C(=O)—OH (with benzene ring) + HO—CH₂—CH₂—CH₂—CH₃ $\xrightarrow{H^+}$

17.5 Naming Esters

LEARNING GOAL

Write the IUPAC and common names for esters; draw condensed structural formulas.

The name of an ester consists of two words taken from the names of the alcohol and the acid. The first word indicates the *alkyl* part of the alcohol. The second word is the *carboxylate* name of the carboxylic acid. The IUPAC names of esters use the IUPAC names for the alkyl group and the carboxylate ion, while the common names of esters use the common names of each.

Ester

R—O—C(=O)—R

From alcohol (alkyl)   From carboxylic acid (carboxylate)

Let's take a look at the following ester and break it into two parts, one from the alcohol and one from the acid. By writing and naming the alcohol and carboxylic acid that produced the ester, we can determine the name of the ester (Figure 17.6).

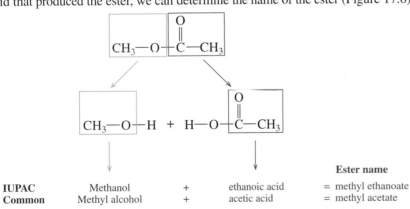

CH₃—O—C(=O)—CH₃

CH₃—O—H + H—O—C(=O)—CH₃

Ester name

IUPAC	Methanol	+ ethanoic acid	= methyl ethanoate
Common	Methyl alcohol	+ acetic acid	= methyl acetate

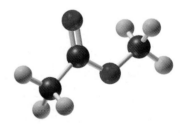

Figure 17.6 The ester methyl ethanoate (methyl acetate) is made from methyl alcohol and ethanoic acid (acetic acid).
Q *What change is made in the name of the carboxylic acid used to make the ester?*

The following examples of some typical esters show the IUPAC as well as the common names of esters.

CH₃—CH₃—O—C(=O)—CH₃

Ethyl ethanoate
(ethyl acetate)

CH₃—O—C(=O)—CH₂—CH₃

Methyl propanoate
(methyl propionate)

CH₃—CH₂—O—C(=O)— (benzene ring)

Ethyl benzoate

EXPLORE YOUR WORLD

Carboxylic Acids and Esters in Foods and Cosmetics

Smell some fruits such as a banana, strawberry, pear, and orange, and taste them if you like.

Look for the names of carboxylic acids and esters on the labels of food products, flavorings, and cosmetics you may use. List their names and write the formulas as you can. For complex names, use a *Merck Index* to look up the formulas of the substances whose names you read on the product labels.

Questions

1. For each fruit that you smell or taste, write the formula of the ester that is responsible for its characteristic smell and flavor.
2. What are the carboxylic acid or ester formulas found in some food products, flavorings, and cosmetics?
3. What are the formulas of some of the carboxylic acids and alcohols used in the esters?

SAMPLE PROBLEM 17.7

Naming Esters

Write the IUPAC and common names of the following ester:

$$CH_3-CH_2-\overset{\overset{\displaystyle O}{\|}}{C}-O-CH_2-CH_2-CH_3$$

Solution

The alcohol part of the ester is propyl, and the carboxylic acid part is propanoic (propionic) acid.

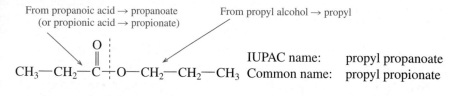

From propanoic acid → propanoate
(or propionic acid → propionate)

From propyl alcohol → propyl

$$CH_3-CH_2-\overset{\overset{\displaystyle O}{\|}}{C}\overset{\vdots}{-}O-CH_2-CH_2-CH_3$$

IUPAC name: propyl propanoate
Common name: propyl propionate

Study Check

Draw the condensed structural formula of pentyl acetate.

Table 17.4   Some Naturally Occurring Esters in Fruits and Flavorings

Condensed Structural Formula	Name	Flavor/Odor
$HC\overset{\overset{\displaystyle O}{\|}}{-}O-CH_2CH_3$	Ethyl methanoate (ethyl formate)	Rum
$HC\overset{\overset{\displaystyle O}{\|}}{-}O-CH_2\overset{\overset{\displaystyle CH_3}{\|}}{C}HCH_3$	Isobutyl methanoate (isobutyl formate)	Raspberries
$CH_3C\overset{\overset{\displaystyle O}{\|}}{-}O-CH_2CH_2CH_3$	Propyl ethanoate (propyl acetate)	Pears
$CH_3C\overset{\overset{\displaystyle O}{\|}}{-}O-CH_2CH_2CH_2CH_2CH_3$	Pentyl ethanoate (pentyl acetate)	Bananas
$CH_3C\overset{\overset{\displaystyle O}{\|}}{-}O-CH_2CH_2CH_2CH_2CH_2CH_2CH_2CH_3$	Octyl ethanoate (octyl acetate)	Oranges
$CH_3CH_2CH_2C\overset{\overset{\displaystyle O}{\|}}{-}O-CH_2CH_3$	Ethyl butanoate (ethyl butyrate)	Pineapples
$CH_3CH_2CH_2C\overset{\overset{\displaystyle O}{\|}}{-}O-CH_2CH_2CH_2CH_2CH_3$	Pentyl butanoate (pentyl butyrate)	Apricots
$CH_3CH_2CH_2C\overset{\overset{\displaystyle O}{\|}}{-}SCH_3$	Methyl thiobutanoate (methyl thiobutyrate)	Strawberries

Esters in Plants

Many of the fragrances of perfumes and flowers and the flavors of fruits are due to esters. Small esters are volatile so we can smell them and soluble in water so we can taste them. (See Figure 17.7.) Several of these are listed in Table 17.4.

Figure 17.7 Esters are responsible for part of the odor and flavor of oranges, bananas, pears, pineapples, and strawberries.
Q *What is the ester found in pineapple?*

QUESTIONS AND PROBLEMS

Naming Esters

17.27 Give the names of the carboxylic acid and alcohol needed to produce each of the following esters:

a. $H-\overset{\overset{\displaystyle O}{\|}}{C}-O-CH_3$

b. $CH_3-\overset{\overset{\displaystyle O}{\|}}{C}-O-CH_3$

c. $CH_3-CH_2-CH_2-\overset{\overset{\displaystyle O}{\|}}{C}-O-CH_3$

d. $CH_3-\overset{\overset{\displaystyle CH_3}{|}}{CH}-CH_2-\overset{\overset{\displaystyle O}{\|}}{C}-O-CH_2-CH_3$

17.28 Give the names of the carboxylic acid and alcohol needed to produce each of the following esters:

a. $CH_3-CH_2-\overset{\overset{\displaystyle O}{\|}}{C}-O-CH_2-CH_3$

b. $CH_3-CH_2-CH_2-CH_2-CH_2-\overset{\overset{\displaystyle O}{\|}}{C}-O-CH_3$

c. $CH_3-CH_2-\overset{\overset{\displaystyle O}{\|}}{\underset{\underset{\displaystyle CH_3}{|}}{CH}}-C-O-CH_3$

d. $CH_3-CH_2-\overset{\overset{\displaystyle O}{\|}}{C}-O-CH_2-CH_2-CH_2-CH_3$

17.29 Name each of the following esters:

a. $CH_3-O-\overset{\overset{\displaystyle O}{\|}}{C}-H$

b. $CH_3-O-\overset{\overset{\displaystyle O}{\|}}{C}-CH_3$

c. $CH_3-O-\overset{\overset{\displaystyle O}{\|}}{C}-CH_2-CH_2-CH_3$

d. $CH_3-\overset{\overset{\displaystyle CH_3}{|}}{CH}-CH_2-\overset{\overset{\displaystyle O}{\|}}{C}-O-CH_2-CH_3$

17.30 Name each of the following esters:

a. $CH_3-CH_2-O-\overset{\overset{\displaystyle O}{\|}}{C}-CH_2-CH_2-CH_3$

b. $CH_3-O-\overset{\overset{\displaystyle O}{\|}}{C}-CH_2-CH_2-CH_2-CH_2-CH_3$

c. $CH_3-O-\overset{\overset{\displaystyle O}{\|}}{C}-CH_2-\overset{\overset{\displaystyle CH_3}{|}}{CH}-CH_3$

d. $CH_3-CH_2-\overset{\overset{\displaystyle O}{\|}}{C}-O-CH_2-CH_2-CH_2-CH_3$

17.31 Draw the condensed structural formulas of each of the following esters:
a. methyl acetate
b. butyl formate
c. ethyl pentanoate
d. 2-bromopropyl propanoate

17.32 Draw the condensed structural formulas of each of the following esters:
a. hexyl acetate
b. propyl propionate
c. ethyl-2-hydroxybutanoate
d. methyl benzoate

17.33 What is the ester responsible for the flavor and odor of the following fruit?
  a. banana   **b.** orange   **c.** apricot   **d.** raspberry
17.34 What flavor would you notice if you smelled or tasted the following?
  a. ethyl butanoate   **b.** propyl acetate
  c. methyl thiobutanoate   **d.** ethyl formate

LEARNING GOAL

Describe the boiling points and solubility of esters. Draw the condensed structural formulas of the hydrolysis products.

17.6 Properties of Esters

Esters have higher boiling points than alkanes, but lower than alcohols and carboxylic acids of similar mass. Because ester molecules do not have hydroxyl groups, they cannot hydrogen bond to each other.

$$CH_3-CH_2-CH_2-CH_3 \qquad CH_3-O-CH_2-CH_3 \qquad CH_3-O-\overset{\overset{\displaystyle O}{\|}}{C}-H \qquad CH_3-CH_2-CH_2-OH \qquad CH_3-\overset{\overset{\displaystyle O}{\|}}{C}-OH$$

	Butane	Ethyl methyl ether	Methyl formate	1-Propanol	Acetic acid
Type	Alkane	Ether	Ester	Alcohol	Carboxylic acid
Bp	0°C	11°C	32°C	97°C	118°C
Mass	58	60	60	60	60

Increasing boiling points →

Solubility in Water

It is possible for esters to use the lone pairs of electrons on the oxygen atom to hydrogen bond with water. Small esters having only a few carbon atoms are soluble in water. However, the solubility of esters decreases as the number of carbons increases.

Acid Hydrolysis of Esters

In **hydrolysis,** esters are split apart with water when heated in the presence of a strong acid, usually H_2SO_4 or HCl. The products of acid hydrolysis are the carboxylic acid and alcohol. Therefore, hydrolysis is the reverse of the esterification reaction. When hydrolysis of biological compounds occurs in the cells, an enzyme replaces the acid as the catalyst. In the hydrolysis reaction, the —OH from a water molecule replaces the —OR part of the ester to form the carboxylic acid.

$$\underset{\text{Ester}}{R-\overset{\overset{\displaystyle O}{\|}}{C}-O-R} \quad + \quad \underset{\text{Water}}{H-OH} \quad \overset{H^+}{\rightleftharpoons} \quad \underset{\text{Carboxylic acid}}{R-\overset{\overset{\displaystyle O}{\|}}{C}-O-H} \quad + \quad \underset{\text{Alcohol}}{R-OH}$$

Example

$$\underset{\text{Methyl acetate}}{CH_3-\overset{\overset{\displaystyle O}{\|}}{C}-OCH_3} \quad + \quad H-OH \quad \overset{H^+}{\rightleftharpoons} \quad \underset{\text{Acetic acid}}{CH_3-\overset{\overset{\displaystyle O}{\|}}{C}-OH} \quad + \quad \underset{\text{Methyl alcohol}}{H-OCH_3}$$

SAMPLE PROBLEM 17.8

Acid Hydrolysis of Esters

Aspirin that has been stored for a long time may undergo hydrolysis in the presence of water and heat. What are the hydrolysis products of aspirin? Why does a bottle of old aspirin smell like vinegar?

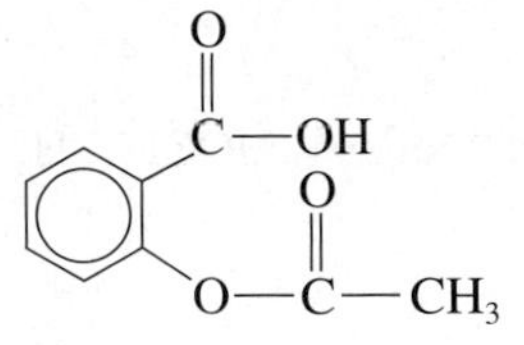

Aspirin, Acetylsalicylic acid

Solution

To write the hydrolysis products, separate the compound at the ester bond. Complete the formula of the carboxylic acid by adding —OH (from water) to the carbonyl group and the —H to complete the alcohol. The acetic acid in the products gives the vinegar odor to a sample of aspirin that has hydrolyzed.

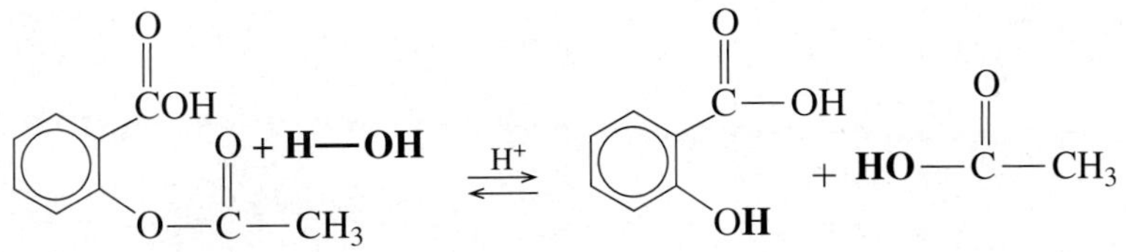

Aspirin — Salicylic acid — Acetic acid

Study Check

What are the names of the products from the acid hydrolysis of ethyl propionate?

Base Hydrolysis of Esters

When an ester undergoes hydrolysis with a strong base such as NaOH or KOH, the products are the carboxylic acid salt and the corresponding alcohol. The base hydrolysis reaction is also called **saponification,** which refers to the reaction of a long-chain fatty acid with NaOH to make soap. The carboxylic acid, which is produced in acid hydrolysis, is converted in an irreversible reaction to its carboxylate ion by the strong base.

$$\underset{\text{Ester}}{R{-}\overset{\overset{\displaystyle O}{\|}}{C}{-}O{-}R} + \underset{\text{Sodium hydroxide}}{NaOH} \longrightarrow \underset{\text{Carboxylic acid salt}}{R{-}\overset{\overset{\displaystyle O}{\|}}{C}{-}O^-\,Na^+} + \underset{\text{Alcohol}}{R{-}OH}$$

$$\underset{\substack{\text{Methyl ethanoate}\\ \text{(methyl acetate)}}}{CH_3{-}\overset{\overset{\displaystyle O}{\|}}{C}{-}O{-}CH_3} + \underset{\text{Sodium hydroxide}}{NaOH} \longrightarrow \underset{\substack{\text{Sodium acetate}\\ \text{(sodium ethanoate)}}}{CH_3{-}\overset{\overset{\displaystyle O}{\|}}{C}{-}O^-\,Na^+} + \underset{\substack{\text{Methanol}\\ \text{(methyl alcohol)}}}{CH_3{-}OH}$$

SAMPLE PROBLEM 17.9

Base Hydrolysis of Esters

Ethyl acetate is a solvent widely used for fingernail polish, plastics, and lacquers. Write the equation of the hydrolysis of ethyl acetate by NaOH.

Solution

The hydrolysis of ethyl acetate by NaOH gives the salt of acetic acid and ethyl alcohol.

$$CH_3-\overset{\overset{\displaystyle O}{\|}}{C}-OCH_2CH_3 \; + \; NaOH \; \xrightarrow{\text{Heat}} \; CH_3-\overset{\overset{\displaystyle O}{\|}}{C}-O^-Na^+ \; + \; HOCH_2CH_3$$

Ethyl acetate   Sodium acetate   Ethyl alcohol

Study Check

Write the condensed structural formulas of the products from the hydrolysis of methyl benzoate by KOH.

Cleaning Action of Soaps

For many centuries, soaps were made by heating a mixture of animal fats (tallow) with lye, a basic solution obtained from wood ashes. In the soap-making process, fatty acids, which are long-chain carboxylic acids, undergo saponification with the strong base in lye.

Fatty acid

$$\boxed{CH_3CH_2CH_2CH_2CH_2CH_2CH_2CH_2CH_2CH_2CH_2CH_2CH_2CH_2CH_2CH_2CH_2}-\overset{\overset{\displaystyle O}{\|}}{C}-OH \; + \; NaOH \; \longrightarrow$$

Carboxylic acid salt, "soap"

$$\boxed{CH_3CH_2CH_2CH_2CH_2CH_2CH_2CH_2CH_2CH_2CH_2CH_2CH_2CH_2CH_2CH_2CH_2}-\overset{\overset{\displaystyle O}{\|}}{C}-O^-Na^+$$

Nonpolar tail   Polar head
(hydrophobic)   (hydrophilic)

Today soaps are also prepared from fats such as coconut oil. Perfumes are added to give a pleasant-smelling soap. Because a soap is the salt of a long-chain fatty acid, the two ends of a soap molecule have different polarities. The long carbon chain end is nonpolar and *hydrophobic* (water-fearing). It is soluble in nonpolar substances such as oil or grease; but it is not soluble in water. The carboxylate salt end is ionic and *hydrophilic* (water-loving). It is very soluble in water but not in oils or grease.

When a soap is used to clean grease or oil, the nonpolar ends of the soap molecules dissolve in the nonpolar fats and oils that accompany dirt. The water-loving salt ends of the soap molecules extend outside where they can dissolve in water. The soap molecules coat the oil or grease, forming clusters called *micelles*. The ionic ends of the soap molecules provide polarity to the micelles, which makes them soluble in water. As a result, small globules of oil and fat coated with soap molecules are pulled into the water and rinsed away.

One of the problems of using soaps is that the carboxylate end reacts with ions in water such as Ca^{2+} and Mg^{2+} and forms insoluble substances.

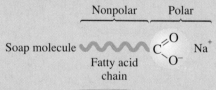

Nonpolar   Polar

Soap molecule ~~~~~ Na⁺

Fatty acid
chain

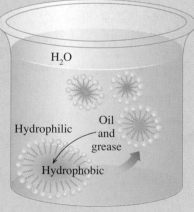

H_2O

Hydrophilic

Oil
and
grease

Hydrophobic

$$2CH_3(CH_2)_{16}COO^- + Mg^{2+} \longrightarrow \quad [CH_3(CH_2)_{16}COO^-]_2Mg^{2+}$$

Stearate ion   Magnesium ion   Magnesium stearate
(insoluble)

QUESTIONS AND PROBLEMS

Properties of Esters

17.35 For each of the following pairs of compounds, select the compound that has the higher boiling point:

a. $CH_3-\overset{\overset{O}{\|}}{C}-O-CH_3$ or $CH_3-\overset{\overset{O}{\|}}{C}-OH$

b. $CH_3-\overset{\overset{O}{\|}}{C}-O-CH_3$ or $CH_3-CH_2-CH_2-CH_2-OH$

c. $CH_3-CH_2-CH_2-CH_3$ or $CH_3-O-\overset{\overset{O}{\|}}{C}-CH_3$

17.36 For each of the following pairs of compounds, select the compound that has the higher boiling point:

a. $H-\overset{\overset{O}{\|}}{C}-O-CH_3$ or $CH_3-CH_2-CH_2-OH$ b. $CH_3-\overset{\overset{O}{\|}}{C}-O-CH_3$ or $CH_3-\overset{\overset{O}{\|}}{C}-CH_2-CH_3$

c. $CH_3-O-CH_2-CH_3$ or $CH_3-O-\overset{\overset{O}{\|}}{C}-H$

17.37 What are the products of the acid hydrolysis of an ester?

17.38 What are the products of the base hydrolysis of an ester?

17.39 Draw the condensed structural formulas of the products from the acid- or base-catalyzed hydrolysis of each of the following compounds:

a. $CH_3CH_2-\overset{\overset{O}{\|}}{C}-O-CH_3 + NaOH \longrightarrow$

b. $CH_3-\overset{\overset{O}{\|}}{C}-O-CH_2CH_2CH_3 + H_2O \xrightarrow{H^+}$

c. $CH_3CH_2CH_2-\overset{\overset{O}{\|}}{C}-O-CH_2CH_3 + H_2O \xrightarrow{H^+}$

d. $\langle\bigcirc\rangle-\overset{\overset{O}{\|}}{C}-O-CH_2CH_3 + H_2O \xrightarrow{H^+}$

e. $\langle\bigcirc\rangle-\overset{\overset{O}{\|}}{C}-O-CH_2CH_3 + NaOH \longrightarrow$

17.40 Draw the condensed structural formulas of the products from the acid- or base-catalyzed hydrolysis of each of the following compounds:

a. $CH_3CH_2-\overset{\overset{O}{\|}}{C}-O-CH_2CH_2CH_2CH_3 + H_2O \xrightarrow{H^+}$ b. $H-\overset{\overset{O}{\|}}{C}-O-CH_2CH_3 + NaOH \longrightarrow$

c. $CH_3CH_2-\overset{\overset{O}{\|}}{C}-O-CH_3 + H_2O \xrightarrow{H^+}$ d. $CH_3-CH_2-\overset{\overset{O}{\|}}{C}-O-\langle\bigcirc\rangle + H_2O \xrightarrow{H^+}$

e. $\langle\bigcirc\rangle-CH_2-\overset{\overset{O}{\|}}{C}-OCH_2CH_3 + NaOH \longrightarrow$

Chapter Review

17.1 Carboxylic Acids

A carboxylic acid contains the carboxyl functional group, which is a hydroxyl group connected to the carbonyl group.

17.2 Physical Properties of Carboxylic Acids

The carboxyl group contains polar bonds of O—H and C=O, which makes a carboxylic acid with one to four carbon atoms very soluble in water.

17.3 Acidity of Carboxylic Acids

As weak acids, carboxylic acids ionize slightly by donating a proton to water to form carboxylate and hydronium ions. Carboxylic acids are neutralized by base, producing the carboxylate salt and water.

17.4 Esters of Carboxylic Acids

In an ester, an alkyl or aromatic group has replaced the H of the hydroxyl group of a carboxylic acid. In the presence of a strong acid, a carboxylic acid reacts with an alcohol to produce an ester. A molecule of water is removed: —OH from the carboxylic acid and —H from the alcohol molecule.

17.5 Naming Esters

The names of esters consist of two words, one from the alcohol and the other from the carboxylic acid with the *ic* ending replaced by *ate*.

17.6 Properties of Esters

Esters undergo acid hydrolysis by adding water to yield the carboxylic acid and alcohol (or phenol). Base hydrolysis or saponification of an ester produces the carboxylate salt and an alcohol.

Summary of Naming

Family	Condensed Structural Formula	IUPAC Name	Common Name
Carboxylic acid	$CH_3-\overset{\overset{O}{\|\|}}{C}-OH$	Ethanoic acid	Acetic acid
Carboxylic acid salt	$CH_3-\overset{\overset{O}{\|\|}}{C}-O^-Na^+$	Sodium ethanoate	Sodium acetate
Ester	$CH_3-\overset{\overset{O}{\|\|}}{C}-OCH_3$	Methyl ethanoate	Methyl acetate

Summary of Reactions

Ionization of a Carboxylic Acid in Water

$$R-\overset{\overset{O}{\|\|}}{C}-OH \ + \ H_2O \ \rightleftharpoons \ R-\overset{\overset{O}{\|\|}}{C}-O^- \ + \ H_3O^+$$

Carboxylic acid (weak acid)   Water (weak base)   Carboxylate ion   Hydronium ion

$$CH_3-\overset{\overset{O}{\|\|}}{C}-OH \ + \ H_2O \ \rightleftharpoons \ CH_3-\overset{\overset{O}{\|\|}}{C}-O^- \ + \ H_3O^+$$

Ethanoic acid (acetic acid)   Ethanoate ion (acetate ion)   Hydronium ion

Neutralization of a Carboxylic Acid

$$R-\overset{\overset{\displaystyle O}{\|}}{C}-OH + NaOH \longrightarrow R-\overset{\overset{\displaystyle O}{\|}}{C}-O^-Na^+ + H_2O$$

Carboxylic acid · Base · Carboxylic acid salt

$$CH_3CH_2-\overset{\overset{\displaystyle O}{\|}}{C}-OH + NaOH \longrightarrow CH_3CH_2-\overset{\overset{\displaystyle O}{\|}}{C}-O^-Na^+ + H_2O$$

Propanoic acid (propionic acid) · Sodium hydroxide · Sodium propanoate (sodium propionate)

Esterification: Carboxylic Acid and an Alcohol

$$R-\overset{\overset{\displaystyle O}{\|}}{C}-OH + HO-R \underset{}{\overset{H^+}{\rightleftharpoons}} R-\overset{\overset{\displaystyle O}{\|}}{C}-O-R + H_2O$$

Carboxylic acid · Alcohol · Ester

$$CH_3-\overset{\overset{\displaystyle O}{\|}}{C}-OH + HO-CH_3 \underset{}{\overset{H^+}{\rightleftharpoons}} CH_3-\overset{\overset{\displaystyle O}{\|}}{C}-O-CH_3 + H_2O$$

Ethanoic acid (acetic acid) · Methanol (methyl alcohol) · Methyl ethanoate (methyl acetate)

Acid Hydrolysis of an Ester

$$R-\overset{\overset{\displaystyle O}{\|}}{C}-O-R + H-OH \underset{}{\overset{H^+}{\rightleftharpoons}} R-\overset{\overset{\displaystyle O}{\|}}{C}-OH + R-OH$$

Ester · Carboxylic acid · Alcohol

$$CH_3-\overset{\overset{\displaystyle O}{\|}}{C}-OCH_3 + H-OH \underset{}{\overset{H^+}{\rightleftharpoons}} CH_3-\overset{\overset{\displaystyle O}{\|}}{C}-OH + H-OCH_3$$

Methyl ethanoate (methyl acetate) · Ethanoic acid (acetic acid) · Methanol (methyl alcohol)

Base Hydrolysis of an Ester

$$R-\overset{\overset{\displaystyle O}{\|}}{C}-O-R + NaOH \overset{Heat}{\longrightarrow} R-\overset{\overset{\displaystyle O}{\|}}{C}-O^-Na^+ + R-OH$$

Ester · Base · Carboxylate salt · Alcohol

$$CH_3CH_2-\overset{\overset{\displaystyle O}{\|}}{C}-OCH_3 + NaOH \overset{Heat}{\longrightarrow} CH_3CH_2C-O^-Na^+ + HOCH_3$$

Methyl propanoate (methyl propionate) · Sodium hydroxide · Sodium propanoate (sodium propionate) · Methanol (methyl alcohol)

Key Terms

carboxyl group A functional group found in carboxylic acids composed of carbonyl and hydroxyl groups.

$$-\overset{\overset{\displaystyle O}{\|}}{C}-OH \quad \text{Carboxyl group}$$

carboxylate ion The anion produced when a carboxylic acid donates a proton to water.

carboxylic acids A family of organic compounds containing the carboxyl group.

$$R-\overset{\overset{\displaystyle O}{\|}}{C}-OH \quad \text{Carboxylic acid}$$

carboxylic acid salt The product of neutralization of a carboxylic acid; a carboxylate ion and the metal ion from the base.

esterification The formation of an ester from a carboxylic acid and an alcohol with the elimination of a molecule of water in the presence of an acid catalyst.

esters A family of organic compounds in which an alkyl group replaces the hydrogen atom in a carboxylic acid.

$$R-\overset{\overset{\displaystyle O}{\|}}{C}-O-R \quad \text{Ester}$$

hydrolysis The splitting of a molecule by the addition of water. Esters hydrolyze to produce a carboxylic acid and an alcohol.

saponification The hydrolysis of an ester with a strong base to produce a salt of the carboxylic acid and an alcohol.

Additional Problems

17.41 Give the IUPAC and common names (if any) for each of the following compounds:

a. $CH_3-\overset{\overset{\displaystyle CH_3}{|}}{CH}-CH_2-\overset{\overset{\displaystyle O}{\|}}{C}-OH$

b. (benzene ring)$-\overset{\overset{\displaystyle O}{\|}}{C}-O-CH_2-CH_3$

c. $CH_3-CH_2-O-\overset{\overset{\displaystyle O}{\|}}{C}-CH_2-CH_3$

d. (benzene ring with $COOH$ and Cl ortho)

e. $CH_3-\overset{\overset{\displaystyle OH}{|}}{CH}-CH_2-CH_2-\overset{\overset{\displaystyle O}{\|}}{C}-OH$

f. $CH_3-\overset{\overset{\displaystyle O}{\|}}{C}-O-\overset{\overset{\displaystyle CH_3}{|}}{CH}-CH_3$

17.42 Give the IUPAC and common names (if any) for each of the following compounds:

a. $CH_3-\overset{\overset{\displaystyle CH_3}{|}}{CH}-CH_2-CH_2-\overset{\overset{\displaystyle O}{\|}}{C}-OH$

b. (benzene ring with $\overset{\overset{\displaystyle O}{\|}}{C}-OH$ and two Cl)

c. (benzene ring)$-\overset{\overset{\displaystyle O}{\|}}{C}-O-CH_3$

d. $CH_3-CH_2-CH_2-\overset{\overset{\displaystyle O}{\|}}{C}-O-CH_3$

e. $CH_3-CH_2-O-\overset{\overset{\displaystyle O}{\|}}{C}-CH_2-\overset{\overset{\displaystyle CH_3}{|}}{CH}-CH_3$

f. $CH_3-\overset{\overset{\displaystyle CH_3}{|}}{CH}-CH_2-\overset{\overset{\displaystyle OH}{|}}{CH}-\overset{\overset{\displaystyle O}{\|}}{C}-OH$

17.43 Draw the structural formulas of at least three carboxylic acids with the molecular formula $C_5H_{10}O_2$.

17.44 Draw the structural formulas of at least three esters with the formula $C_4H_8O_2$.

17.45 Draw the condensed structural formulas of each of the following:
a. methyl acetate   b. *p*-chlorobenzoic acid
c. *β*-chloropropionic acid   d. ethyl butanoate
e. 3-methylpentanoic acid   f. ethyl benzoate

17.46 Draw the condensed structural formulas of each of the following:
a. *α*-bromobutyric acid
b. ethyl butyrate
c. 2-methylcyclohexanoic acid
d. 3,5-dimethylhexanoic acid
e. propyl acetate   f. 3,4-dibromobenzoic acid

17.47 In each of the following pairs of compounds, select the compound that would have the higher boiling point:

a. $CH_3-CH_2-CH_2-OH$ or $CH_3-\overset{\overset{\displaystyle O}{\|}}{C}-OH$

b. $CH_3-CH_2-CH_2-CH_3$ or $CH_3-CH_2-\overset{\overset{\displaystyle O}{\|}}{C}-OH$

c. $CH_3-\overset{\overset{\displaystyle O}{\|}}{C}-OH$ or $CH_3-CH_2-CH_2-\overset{\overset{\displaystyle O}{\|}}{C}-OH$

17.48 In each of the following pairs of compounds, select the compound that would have the higher boiling point:

a. $CH_3-CH_2-CH_2-OH$ or $CH_3-\overset{\overset{\displaystyle O}{\|}}{C}-O-CH_3$

b. $CH_3-O-\overset{\overset{\displaystyle O}{\|}}{C}-CH_3$ or $CH_3-CH_2-\overset{\overset{\displaystyle O}{\|}}{C}-OH$

c. $CH_3-\overset{\overset{\displaystyle O}{\|}}{C}-O-CH_3$ or $CH_3-CH_2-CH_2-CH_3$

17.49 Why does acetic acid have a higher boiling point than either 1-propanol or methyl formate when they all have the same molar mass?

17.50 Propionic acid, 1-butanol, and butanal all have the same molar mass. The possible boiling points are 76°C, 118°C, and 141°C. Match the compounds with the boiling points and explain your choice.

17.51 Which of the following compounds are soluble in water?

a. $CH_3-CH_2-CH_2-CH_2-CH_3$

b. $CH_3-CH_2-\overset{\overset{\displaystyle O}{\|}}{C}-O^-Na^+$

c. $CH_3-\overset{\overset{\displaystyle O}{\|}}{C}-O-CH_3$

d. $CH_3-CH_2-CH_2-OH$

e. $CH_3-CH_2-\overset{\overset{\displaystyle O}{\|}}{C}-OH$

17.52 Which of the following compounds are soluble in water?

a. $CH_3-CH_2-CH_2-\overset{\overset{\displaystyle O}{\|}}{C}-OH$

b. $CH_3-CH_2-\overset{\overset{\displaystyle O}{\|}}{C}-O-CH_2-CH_2-CH_3$

c. $CH_3-CH_2-CH_2-CH_3$

d. $CH_3-(CH_2)_8-CH_2-OH$

e. $CH_3-CH_2-CH_2-O-CH_2-CH_2-CH_3$

17.53 Methyl benzoate is not soluble in water; but when it is heated with KOH, it dissolves. Write an equation for the reaction and explain what happens. When HCl is added to the solution, a white solid forms. What is the solid?

17.54 Hexanoic acid is soluble in NaOH solution, but hexanal is not. Explain.

17.55 Salicylic acid could be named *o*-hydroxybenzoic acid.
   a. What two reactive functional groups are present?
   b. Draw the structure of the ester product that forms when the hydroxyl group of salicylic acid reacts with acetic acid.
   c. Draw the structure of methyl salicylate, oil of wintergreen, formed when salicylic acid forms an ester with methyl alcohol.

17.56 What volume of 0.100 M NaOH is needed to neutralize 3.00 g of benzoic acid?

17.57 Write the products of the following reactions:

a. $CH_3-CH_2-\overset{\overset{\displaystyle O}{\|}}{C}-OH + H_2O \rightleftharpoons$

b. $CH_3-CH_2-\overset{\overset{\displaystyle O}{\|}}{C}-OH + KOH \longrightarrow$

c. $CH_3-CH_2-\overset{\overset{\displaystyle O}{\|}}{C}-OH + CH_3OH \overset{H^+}{\rightleftharpoons}$

d. 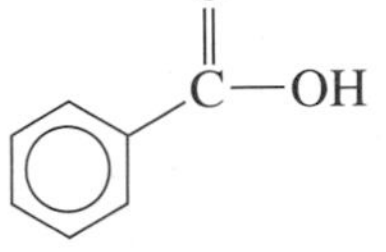 $\underset{}{\bigcirc}-\overset{\overset{\displaystyle O}{\|}}{C}-OH$ $+ CH_3-CH_2-OH \overset{H^+}{\rightleftharpoons}$

17.58 Write the products of the following reactions:

a. $CH_3-\overset{\overset{\displaystyle O}{\|}}{C}-OH + NaOH \longrightarrow$

b. $CH_3-\overset{\overset{\displaystyle O}{\|}}{C}-OH + H_2O \rightleftharpoons$

c. $CH_3-\overset{\overset{\displaystyle CH_3}{|}}{CH}-\overset{\overset{\displaystyle O}{\|}}{C}-OH$ + KOH $\longrightarrow$

d. $CH_3-\overset{\overset{\displaystyle CH_3}{|}}{CH}-\overset{\overset{\displaystyle O}{\|}}{C}-OH$ + CH_3OH $\overset{H^+}{\rightleftharpoons}$

17.59 Give the IUPAC names of the carboxylic acid and alcohol needed to prepare each of the following esters:

a. $CH_3-\overset{\overset{\displaystyle CH_3}{|}}{CH}-CH_2-\overset{\overset{\displaystyle O}{\|}}{C}-O-CH_3$

b.

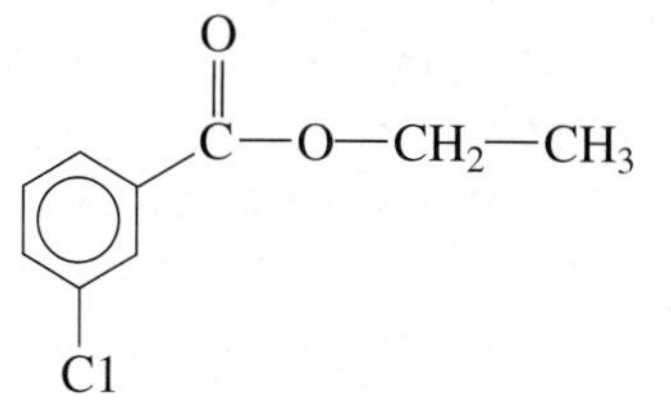

c. $CH_3-(CH_2)_4-\overset{\overset{\displaystyle O}{\|}}{C}-O-CH_3$

17.60 Give the IUPAC names of the carboxylic acid and alcohol needed to prepare each of the following esters:

a. $CH_3-CH_2-CH_2-\overset{\overset{\displaystyle O}{\|}}{C}-O-CH_2-CH_3$

b.

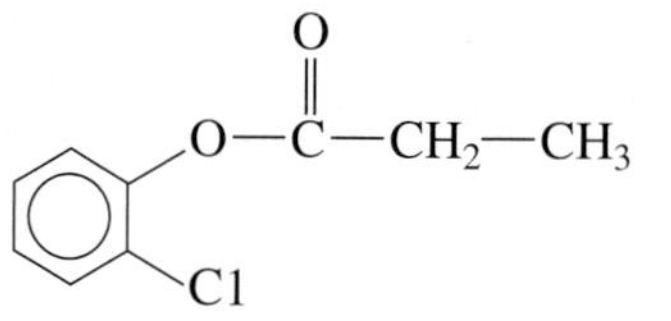

c. $CH_3-\overset{\overset{\displaystyle CH_3}{|}}{CH}-\overset{\overset{\displaystyle CH_3}{|}}{CH}-\overset{\overset{\displaystyle O}{\|}}{C}-O-CH_3$

17.61 Write the products of the following reactions:

a. $CH_3-CH_2-\overset{\overset{\displaystyle O}{\|}}{C}-O-\overset{\overset{\displaystyle CH_3}{|}}{CH}-CH_3$ + H_2O $\overset{H^+}{\underset{\longleftarrow}{\rightharpoonup}}$

b. $CH_3-\overset{\overset{\displaystyle CH_3}{|}}{CH}-\overset{\overset{\displaystyle O}{\|}}{C}-O-CH_2-CH_2-CH_3$ + NaOH $\longrightarrow$

17.62 Write the products of the following reactions:

a. $CH_3-CH_2-\overset{\overset{\displaystyle O}{\|}}{C}-O-\overset{\overset{\displaystyle CH_3}{|}}{CH}-CH_3$ + NaOH $\longrightarrow$

b. $CH_3-\overset{\overset{\displaystyle CH_3}{|}}{CH}-\overset{\overset{\displaystyle O}{\|}}{C}-O-CH_2-CH_2-CH_3$ + NaOH $\longrightarrow$

17.63 Using the reactions we have studied, indicate how you might prepare the following from the starting substance given:
   a. acetic acid from ethene
   b. butyric acid from 1-butanol

17.64 Using the reactions we have studied, indicate how you might prepare the following from the starting substance given:
   a. pentanoic acid from 1-pentanol
   b. ethyl acetate from 2 molecules of ethanol

18 Lipids

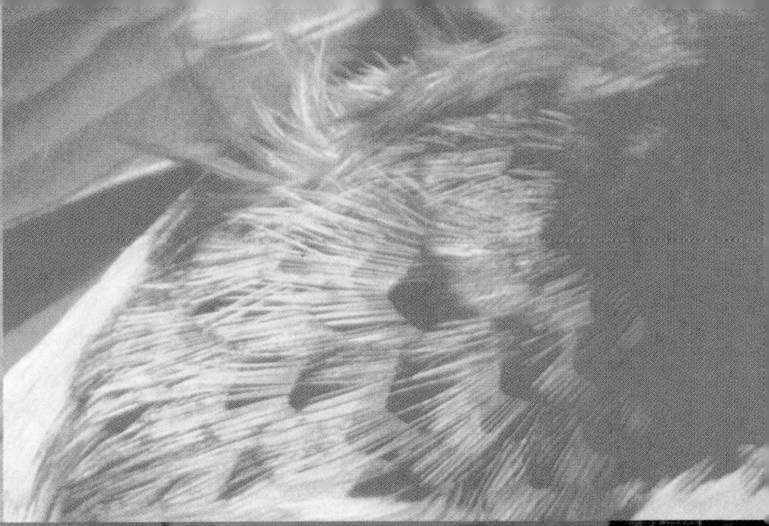

"In our toxicology lab, we measure the drugs in samples of urine or blood," says Penny Peng, assistant supervisor of chemistry, Toxicology Lab, Santa Clara Valley Medical Center. "But first we extract the drugs from the fluid and concentrate them so they can be detected in the machine we use. We extract the drugs by using different organic solvents such as methanol, ethyl acetate, or methylene chloride, and by changing the pH. We evaporate most of the organic solvent to concentrate any drugs it may contain. A small sample of the concentrate is placed into a machine called a gas chromatograph. As the gas moves over a column, the drugs in it are separated. From the results, we can identify as many as 10 to 15 different drugs from one urine sample."

LOOKING AHEAD

the Chemistry place

www.chemplace.com/college

Visit the URL above or use the CD-ROM in the book for extra quizzing, interactive tutorials, and career resources.

Solubility of Fats and Oils

Place some water in a small bowl. Add a drop of a vegetable oil. Then add a few more drops of the oil. Now add a few drops of liquid soap and mix. Record your observations.

Place a small amount of fat such as margarine, butter, shortening, or vegetable oil on a dish or plate. Run water over it. Record your observations. Mix some soap with the fat substance and run water over it again. Record your observations.

Questions

1. Do the drops of oil in the water separate or do they come together? Explain.
2. How does the soap affect the oil layer?
3. Why don't the fats on the dish or plate wash off with water?
4. In general what is the solubility of lipids in water?
5. Why does soap help to wash the fats off the plate?

Describe the classes of lipids.

When we talk of fats and oils, waxes, steroids, cholesterol, and fat-soluble vitamins, we are discussing lipids. All the lipids are naturally occurring compounds that vary considerably in structure but share a common feature of being soluble in nonpolar solvents, but not in water. Fats, which are one family of lipids, have many functions in the body; they store energy and protect and insulate internal organs. Other types of lipids are found in nerve fibers and in hormones, which act as chemical messengers. Because they are not soluble in water, a major function of lipids is to build the cell membranes that separate the internal contents of cells from the surrounding aqueous environment.

Many people are concerned about the amounts of saturated fats and cholesterol in our diets. Researchers suggest that saturated fats and cholesterol are associated with diseases such as diabetes, cancers of the breast, pancreas, and colon, and artherosclerosis, a condition in which deposits of lipid materials accumulate in the coronary blood vessels. These plaques restrict the flow of blood to the tissue, causing necrosis (death) of the tissue. In the heart, this could result in a *myocardial infarction* (heart attack).

The American Institute for Cancer Research has recommended that our diet contain more fiber and starch by adding more vegetables, fruits, and whole grains with moderate amounts of foods with low levels of fat and cholesterol such as fish, poultry, lean meats, and low-fat dairy products. They also suggest that we limit our intake of foods high in fat and cholesterol such as eggs, nuts, fatty or organ meats, cheeses, butter, and coconut and palm oil.

18.1 Lipids

Lipids are a family of biomolecules that have the common property of being soluble in organic solvents but not in water. The word "lipid" comes from the Greek word *lipos,* meaning "fat" or "lard." Typically, the lipid content of a cell can be extracted using an organic solvent such as ether or chloroform. Lipids are an important feature in cell membranes, fat-soluble vitamins, and steroid hormones.

Types of Lipids

Within the lipid family, there are distinct structures that distinguish the different types of lipids. Lipids such as waxes, fats, oils, and glycerophospholipids are esters that can be hydrolyzed to give fatty acids along with other products including an alcohol. Sphingolipids contain an alcohol called sphingosine, and glycosphingolipids contain a carbohydrate. Steroids are characterized by the steroid nucleus of four fused carbon rings. They do not contain fatty acids and cannot be hydrolyzed. Figure 18.1 illustrates the general structure of lipids we will discuss in this chapter.

SAMPLE PROBLEM 18.1

Classes of Lipids

What type of lipid does not contain fatty acids?

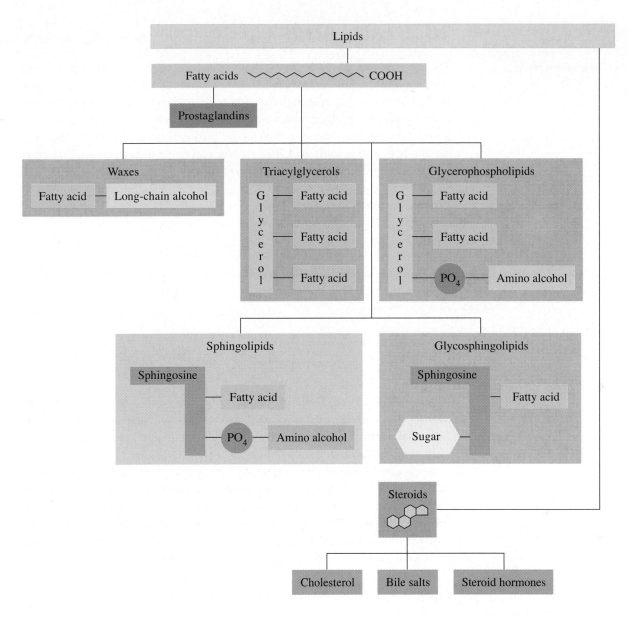

Figure 18.1 Structures for some classes of lipids that are naturally occurring compounds in cells and tissues.

Q *What chemical property do waxes, triacyclglycerols, and steroids have in common?*

Solution

The steroids are a group of lipids with no fatty acids.

Study Check

What type of lipid contains a carbohydrate?

QUESTIONS AND PROBLEMS

Lipids

18.1 What are some functions of lipids in the body?

18.2 What are some of the different kinds of lipids?

18.3 Lipids are not soluble in water. Are lipids polar or nonpolar molecules?

18.4 Which of the following solvents might be used to dissolve an oil stain?
   a. water   **b.** CCl_4 **c.** diethyl ether
   d. benzene   **e.** NaCl solution

18.2 Fatty Acids

The fatty acids are the simplest type of lipids and are found as components in more complex lipids. A **fatty acid** contains a long carbon chain attached to a carboxylic acid group at one end. Although the carboxylic acid part is hydrophilic, the long hydrophobic carbon chain makes long-chain fatty acids insoluble in water. Fatty acids have an even number of carbon atoms, usually between 10 and 20. An example of a fatty acid is lauric acid, a 12-carbon acid found in coconut oil. The structural formula of lauric acid can be written in several forms as follows.

Writing Formulas for Lauric Acid

$$CH_3-(CH_2)_{10}-\overset{\overset{\displaystyle O}{\|}}{C}-OH \qquad CH_3-(CH_2)_{10}-COOH$$

$$CH_3-CH_2-CH_2-CH_2-CH_2-CH_2-CH_2-CH_2-CH_2-CH_2-CH_2-\overset{\overset{\displaystyle O}{\|}}{C}\diagdown_{OH}$$
Condensed structural formula

Line-bond structural formula

Saturated fatty acids such as lauric acid contain only single bonds between carbons. **Monounsaturated fatty acids** have one double bond in the carbon chain, and **polyunsaturated fatty acids** have two or more double bonds. Table 18.1 lists some of the typical fatty acids in lipids.

In Chapter 13 we saw that compounds with double bonds can have cis and trans stereoisomers. This is also true of unsaturated fatty acids. For example, oleic acid, a monounsaturated fatty acid found in olives and corn, has one double bond at carbon 9. We can show its cis and trans structural formulas using the line-bond notation. The cis configuration is most prevalent in naturally occurring unsaturated fatty acids. In the cis isomer, the geometry of the carbon chain is not linear, but has a "kink" at the double bond site. As we will see, the cis bond has a major impact on the physical properties and uses of unsaturated fatty acids.

cis-Oleic acid
Cis double bond

trans-Oleic acid
Trans double bond

The human body is capable of synthesizing most fatty acids from carbohydrates or other fatty acids. However, humans do not synthesize sufficient amounts of fatty acids that have more than one double bond, such as linoleic acid, linolenic acid, and arachidonic acid. These fatty acids are called *essential* fatty acids because they must be provided by the diet. A deficiency of essential fatty acids can cause skin dermatitis in infants. However, the role of fatty acids in adult nutrition is not well understood. Adults do not usually have a deficiency of essential fatty acids.

Table 18.1 Structures and Melting Points of Common Fatty Acids

Name	Carbon Atoms	Structure	Melting Point (°C)	Source
Saturated Fatty Acids				
Capric acid	10	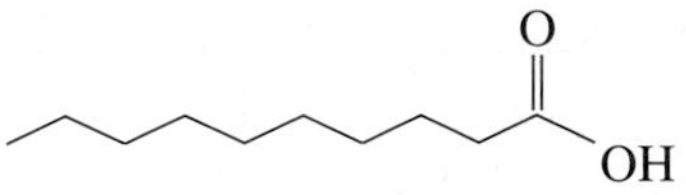	32	Saw palmetto
Lauric acid	12	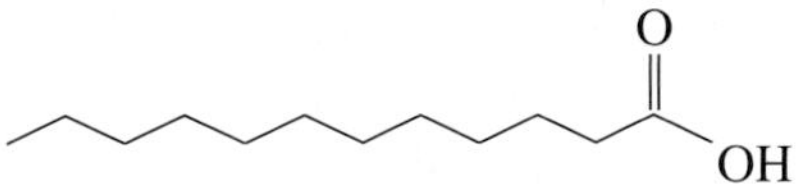	43	Coconut
Myristic acid	14	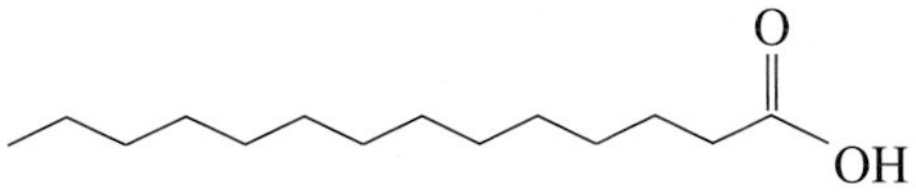	54	Nutmeg
Palmitic acid	16	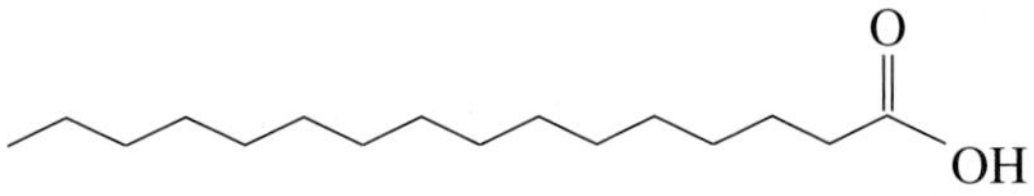	62	Palm
Stearic acid	18	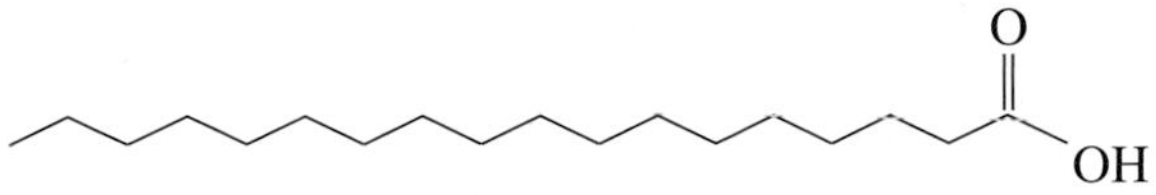	69	Animal fat
Arachidic acid	20	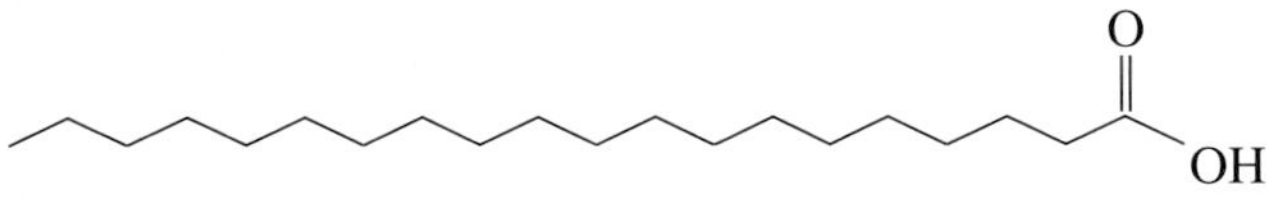	76	Peanut oil, vegetable and fish oils
Monounsaturated Fatty Acids				
Palmitoleic acid	16	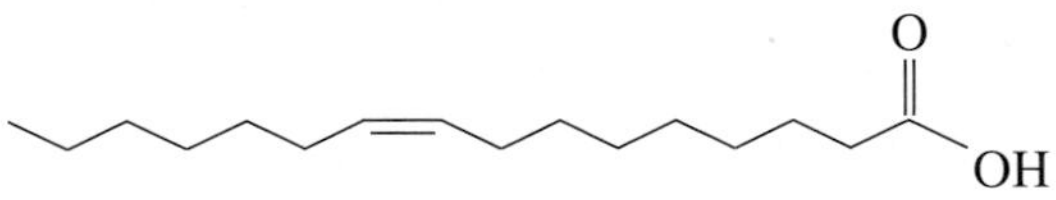	0	Butter
Oleic acid	18	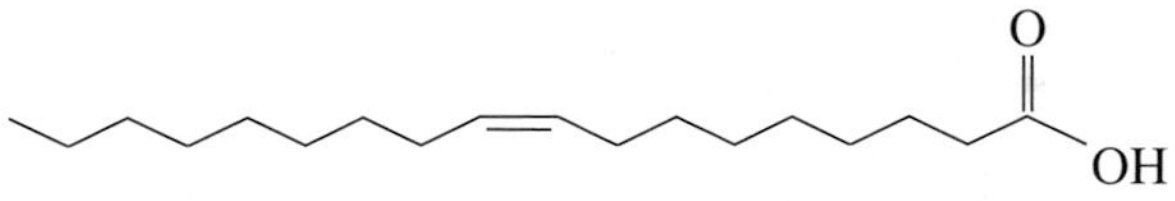	13	Olives, corn
Polyunsaturated Fatty Acids				
Linoleic acid	18	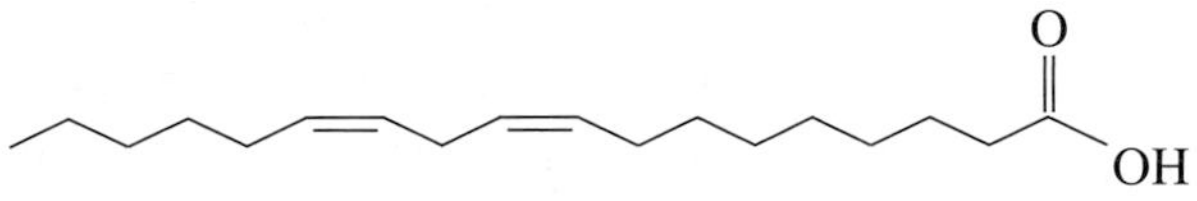	−9	Soybean, safflower, sunflower
Linolenic acid	18	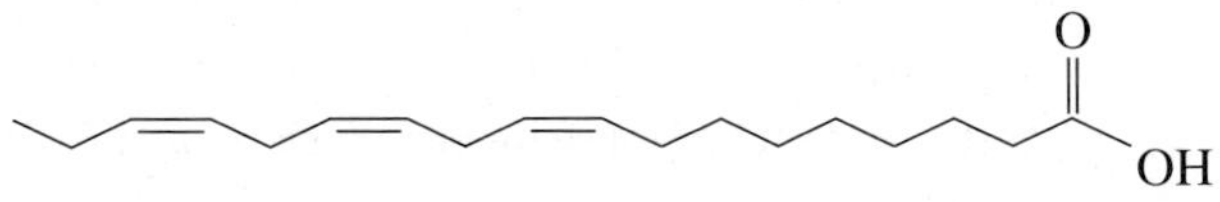	−17	Corn
Arachidonic acid	20	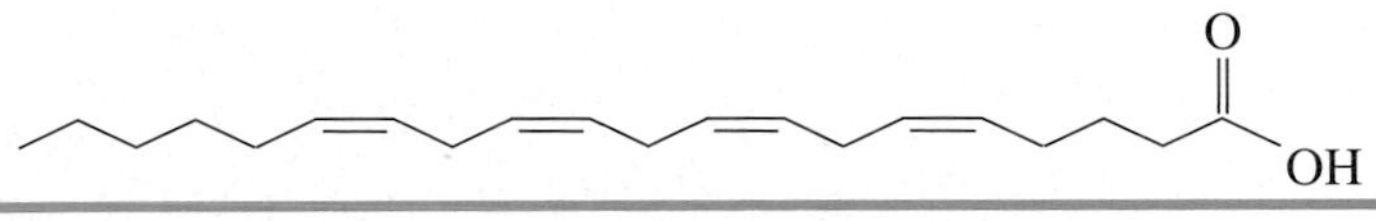	−50	Prostaglandins

Physical Properties of Fatty Acids

The uniform structure of the carbon chain in saturated fatty acids allows the molecules of saturated fatty acids to fit close together in a regular pattern. The close pattern of the saturated fatty acids allows strong attractions to occur between the car-

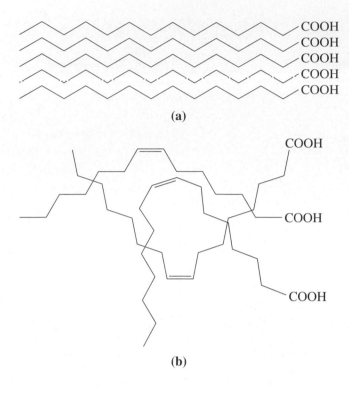

(a)

(b)

Figure 18.2 **(a)** In saturated fatty acids, the molecules fit closely together to give high melting points. **(b)** In unsaturated fatty acids, molecules cannot pack closely together resulting in lower melting points.
Q *Why does the cis double bond affect the melting points of unsaturated fatty acids?*

bon chains. (See Figure 18.2.) As a result, a significant amount of energy and high temperatures are required to separate the fatty acids and melt the fat. As the length of the carbon chain increases, more interactions occur between the carbon chains, requiring higher melting points. Saturated fatty acids are usually solids at room temperature.

In unsaturated fatty acids, the cis double bonds cause the carbon chain to bend, which gives the molecules an irregular shape. As a result, the molecules of an unsaturated fatty acid cannot stack together in a regular pattern, so fewer interactions occur between carbon chains. Consequently, the melting points of unsaturated fats are lower than those of saturated fats. (See Figure 18.3.) Most unsaturated fats are liquid oils at room temperature.

Prostaglandins

Prostaglandins are hormone-like substances produced in low amounts in most cells of the body. The variety of prostaglandins is formed from the unsaturated fatty acid arachidonic acid with 20 carbons. Prostaglandins are sometimes referred to as eicosanoids (eicos is the Greek word for 20). Most prostaglandins have a hydroxyl group on carbon 11 and carbon 15, and a trans double bond at carbon 13. Those with a ketone group on carbon 9 are designated as PGE, and as PGF, when there is an hydroxyl group on carbon 9. (See Figure 18.4.)

Although prostaglandins are broken down quickly, they have potent physiological effects. Some prostaglandins increase blood pressure, and others lower blood pressure. Other prostaglandins stimulate contraction and relaxation in the smooth muscle of the uterus. When tissues are injured, arachidonic acid is converted to prostaglandins such PGE and PGF that produce inflammation and pain in the area. (See Figure 18.5, p. 583.)

The treatment of pain, fever, and inflammation is based on inhibiting the enzymes that convert arachidonic acid to prostaglandins. Several nonsteroidal anti-inflammatory drugs (NSAIDs), such as aspirin, block the production of prostaglandins, and in doing so decrease pain and inflammation and reduce fever (antipyretics). Ibuprofen has similar anti-inflammatory and analgesic effects. Other NSAIDs include naproxen (Aleve and Naprosyn), ketoprofen (Actron), and nabumetone (Relafen). Long-term use of such products can result in liver, kidney, and gastrointestinal damage. Some forms of PGE are being tested as inhibitors of gastric secretion for use in the treatment of stomach ulcers.

Aspirin (acetylsalicyclic acid)   Ibuprofen (Advil, Motrin)   Naproxen (Aleve, Naprosyn)

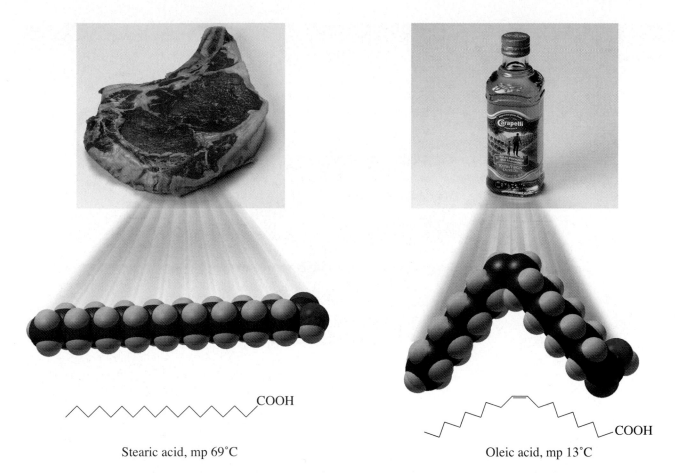

Stearic acid, mp 69°C

Oleic acid, mp 13°C

Figure 18.3 The cis double bond in oleic acid contributes to a lower melting point compared to stearic acid.

Q *If both stearic acid and oleic acid have 18 carbons, why is there such as big difference in their respective melting points?*

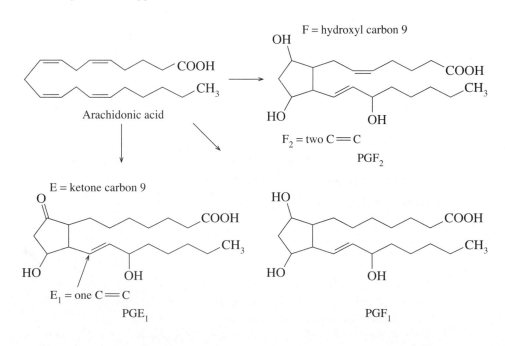

Figure 18.4 The structures of prostaglandins PGE have a ketone at carbon 9, while PGF has a hydroxyl group at carbon 9.

Q *How is the number of carbon–carbon double bonds indicated in the abbreviation for prostaglandins?*

HEALTH NOTE

Omega-3 Fatty Acids in Fish Oils

Over the past several decades, Americans have been changing their diets to include more polyunsaturated fats and fewer saturated fats. This change is a response to research that indicates that atherosclerosis and heart disease are associated with high levels of saturated fats in the diet. However, this association does not seem to be correct for the Inuit people of Alaska, who have a high-fat diet and high levels of blood cholesterol, but a very low occurrence of atherosclerosis and heart attacks. The fats in the Inuit diet are primarily from fish rather than from land animals, as in many other people's diets.

Both fish and vegetable oils have high levels of polyunsaturated fats. The fatty acids in vegetable oils are omega-6 acids, which means that the first double bond occurs at carbon 6 counting from the methyl group. Two common omega-6 acids are linoleic acid and arachidonic acid. However, the fatty acids in the fish

oils are mostly the omega-3 type, in which the first double bond occurs at the third carbon counting from the methyl group. Three common omega-3 fatty acids in fish are linolenic acid, eicosapentaenoic acid (EPA), and docosahexaenoic acid (DHA).

In atherosclerosis and heart disease, cholesterol forms plaque that adhere to the walls of the blood vessels. Blood pressure rises as blood has to squeeze through a smaller opening in the blood vessel. As more plaque forms, there is also a possibility of blood clots blocking the blood vessels and causing a heart attack. Omega-3 fatty acids lower the tendency of blood platelets to stick together, thereby reducing the possibility of blood clots. However, high levels of omega-3 fatty acids can increase bleeding if the ability of the platelets to form blood clots is reduced too much. It does seem that a diet that includes fish such as salmon, tuna, and herring can provide higher amounts of the omega-3 fatty acids, which help lessen the possibility of developing heart disease.

Linoleic acid — Omega-6 fatty acid

Linolenic acid — Omega-3 fatty acid

Eicosapentaenoic acid (EPA)

Docosahexaenoic acid (DHA)

SAMPLE PROBLEM 18.2

Structures and Properties of Fatty Acids

Consider the structural formula of oleic acid.

$$CH_3(CH_2)_7CH{=}CH(CH_2)_7\overset{\displaystyle O}{\overset{\|}{C}}OH$$

a. Why is the substance called an acid?

b. How many carbon atoms are in oleic acid?

c. Is it a saturated or unsaturated fatty acid?

d. Is it most likely to be solid or liquid at room temperature?

e. Would it be soluble in water?

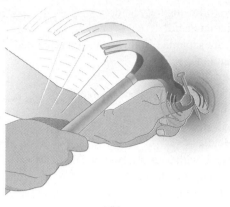

Figure 18.5 An injury to the body tissues releases arachidonic acid, which is converted to prostaglandins that cause pain, fever, and inflammation. Analgesics reduce the effects of prostaglandins by inhibiting the enzyme required for their synthesis.
Q *How do analgesics reduce pain and inflammation of an injury?*

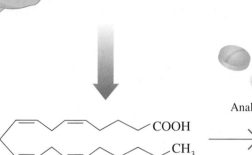

Analgesics

Arachidonic acid

PGF_2

Pain, fever, inflammation

Solution

a Oleic acid contains a carboxylic acid group.
b. It contains 18 carbon atoms.
c. It is an unsaturated fatty acid.
d. It is liquid at room temperature.
e. No, its long hydrocarbon chain makes it insoluble in water.

Study Check

Palmitoleic acid is a fatty acid with the following formula:

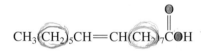

$CH_3(CH_2)_5CH\!=\!CH(CH_2)_7COH$

a. How many carbon atoms are in palmitoleic acid?
b. Is it a saturated or unsaturated fatty acid?
c. Is it most likely to be solid or liquid at room temperature?

QUESTIONS AND PROBLEMS

Fatty Acids

18.5 Describe some similarities and differences in the structures of a saturated fatty acid and an unsaturated fatty acid.

18.6 Stearic acid and linoleic acid both have 18 carbon atoms. Why does stearic acid melt at 69°C, but linoleic acid melts at −9°C?

18.7 Write the line-bond structure of the following fatty acids:
  a. palmitic acid   **b.** oleic acid

18.8 Write the line-bond structure of the following fatty acids:
    a. stearic acid   **b.** linoleic acid

18.9 Which of the following fatty acids are saturated, and which are unsaturated?
    a. lauric acid   **b.** linolenic acid
    c. palmitoleic acid   **d.** stearic acid

18.10 Which of the following fatty acids are saturated, and which are unsaturated?
    a. linoleic acid   **b.** palmitic acid
    c. myristic acid   **d.** oleic acid

18.11 How does the structure of a fatty acid with a cis double bond differ from the structure of a fatty acid with a trans double bond?

18.12 In each pair, identify the fatty acid with the lower melting point. Explain.
    a. myristic acid and stearic acid
    b. stearic acid and linoleic acid
    c. oleic acid and linolenic acid

18.13 Describe the position of the first double bond in an omega-3 and an omega-6 fatty acid.

18.14 **a.** What are some sources of omega-3 and omega-6 fatty acids?
    b. How may omega-3 fatty acids help in lowering the risk of heart disease?

18.15 What are some structural and functional differences in arachidonic acid and prostaglandins such as PGE_2?

18.16 What is the structural difference in PGE and PGF?

18.17 What are some functions of prostaglandins in the body?

18.18 How does an anti-inflammatory drug reduce inflammation?

LEARNING GOAL

Write the structural formula of a wax, fat, or oil produced by the reaction of a fatty acid and an alcohol or glycerol.

Figure 18.6 Honeycomb is made of beeswax, which is composed of esters of long-chain fatty acids and long-chain alcohols.
Q *Write the structural formulas of the alcohol and carboxylic acid for beeswax.*

18.3 Waxes, Fats, and Oils

Waxes are found in many plants and animals. Coatings of carnauba wax on fruits and the leaves and stems of plants help to prevent loss of water and damage from pests. Waxes on the skin, fur, and feathers of animals and birds provide a waterproof coating. A **wax** is an ester of a saturated fatty acid and a long-chain alcohol, each containing from 14 to 30 carbon atoms.

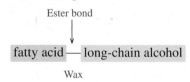

    The formulas of some common waxes are given in Table 18.2. Beeswax obtained from honeycombs and carnauba wax obtained from palm trees are used to give a protective coating to furniture, cars, and floors. (See Figure 18.6.) Jojoba wax is used in making candles and cosmetics such as lipstick. Lanolin, a mixture of waxes obtained from wool, is used in hand and facial lotions to aid retention of water, which softens the skin.

Fats and Oils: Triacylglycerols

In the body, fatty acids are stored as fats and oils known as **triacylglycerols**. These substances, also called *triglycerides*, are triesters of glycerol (a trihydroxy alcohol) and fatty acids. The general formula of a triacylglycerol follows.

Table 18.2   Some Typical Waxes

Type	Structural Formula	Source	Uses
Beeswax	$CH_3(CH_2)_{14}-\overset{\displaystyle O}{\overset{\|}{C}}-O-(CH_2)_{29}CH_3$	Honeycomb	Candles, shoe polish, wax paper
Carnauba wax	$CH_3(CH_2)_{24}-\overset{\displaystyle O}{\overset{\|}{C}}-O-(CH_2)_{29}CH_3$	Brazilian palm tree	Waxes for furniture, cars, floors, shoes
Jojoba wax	$CH_3(CH_2)_{18}-\overset{\displaystyle O}{\overset{\|}{C}}-O-(CH_2)_{19}CH_3$	Jojoba	Candles, soaps, cosmetics

A triacylglycerol is produced by **esterification,** a reaction in which the hydroxyl groups of glycerol form ester bonds with the carboxyl groups of fatty acids. For example, glycerol and three molecules of stearic acid form tristearin (glyceryl tristearate).

Triacylglycerols are the major form of energy storage for animals. Animals that hibernate eat large quantities of plants, seeds, and nuts that contain large amounts of fats and oils. They gain as much as 14 kilograms a week. As the external temperature drops, the animal goes into hibernation. The body temperature drops to nearly freezing, and there is a dramatic reduction in cellular activity, respiration, and heart rate. Animals who live in extremely cold climates will hibernate for 4–7 months. During this time, stored fat is the only source of energy.

However, most fats and oils are mixed triacylglycerols that contain two or three different fatty acids. For example, a mixed triacylglycerol might be made from lauric acid, myristic acid, and palmitic acid. One possible structure for the mixed triacylglycerol follows:

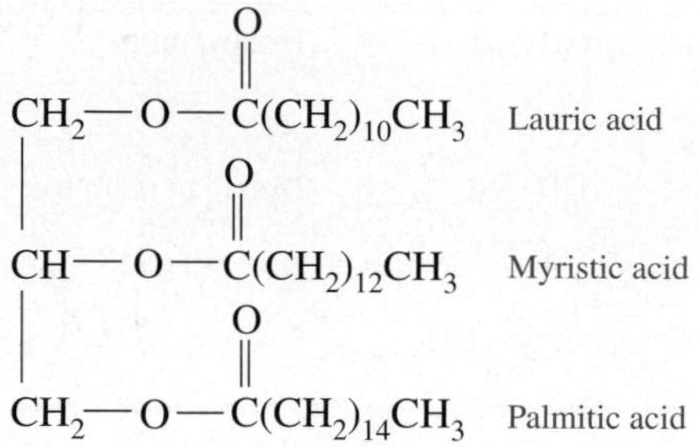

$$
\begin{array}{l}
CH_2-O-\overset{\overset{O}{\|}}{C}(CH_2)_{10}CH_3 \quad \text{Lauric acid} \\
CH-O-\overset{\overset{O}{\|}}{C}(CH_2)_{12}CH_3 \quad \text{Myristic acid} \\
CH_2-O-\overset{\overset{O}{\|}}{C}(CH_2)_{14}CH_3 \quad \text{Palmitic acid}
\end{array}
$$

A mixed triacylglycerol

SAMPLE PROBLEM 18.3

Writing Structures for a Triacylglycerol

Draw the structural formula of triolein, a simple triacylglycerol that uses oleic acid.

Solution

Triolein is the triacylglycerol of glycerol and three oleic acid molecules. Each fatty acid is attached by an ester bond to one of the hydroxyl groups in glycerol.

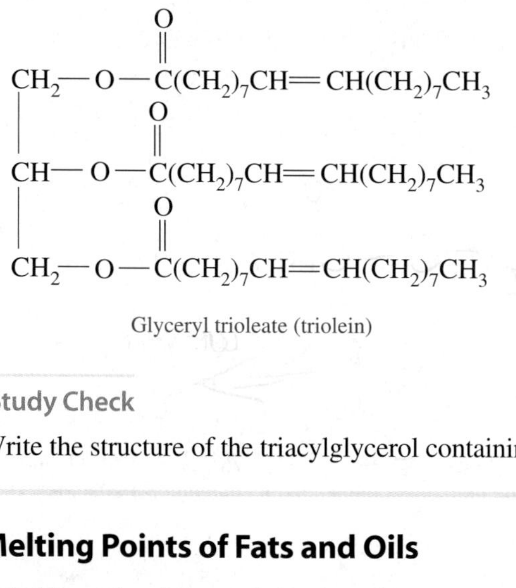

$$
\begin{array}{l}
CH_2-O-\overset{\overset{O}{\|}}{C}(CH_2)_7CH=CH(CH_2)_7CH_3 \\
CH-O-\overset{\overset{O}{\|}}{C}(CH_2)_7CH=CH(CH_2)_7CH_3 \\
CH_2-O-\overset{\overset{O}{\|}}{C}(CH_2)_7CH=CH(CH_2)_7CH_3
\end{array}
$$

Glyceryl trioleate (triolein)

Study Check

Write the structure of the triacylglycerol containing 3 molecules of myristic acid.

Melting Points of Fats and Oils

A **fat** is a triacylglycerol that is solid at room temperature, such as fats in meat, whole milk, butter, and cheese. Most fats come from animal sources.

An **oil** is a triacylglycerol that is usually liquid at room temperature. The most commonly used oils come from plant sources. Olive oil and peanut oil are monounsaturated because they contain large amounts of oleic acid. Oils from corn, cottonseed, safflower, and sunflower are polyunsaturated because they contain large amounts of fatty acids with two or more double bonds. (See Figure 18.7.) A few oils such as palm oil and coconut oil are solid at room temperature because they consist mostly of saturated fatty acids.

The amounts of saturated, monounsaturated, and polyunsaturated fatty acids in some typical fats and oils are shown in Figure 18.8. Saturated fatty acids have higher melting points than unsaturated fatty acids because they pack together more tightly.

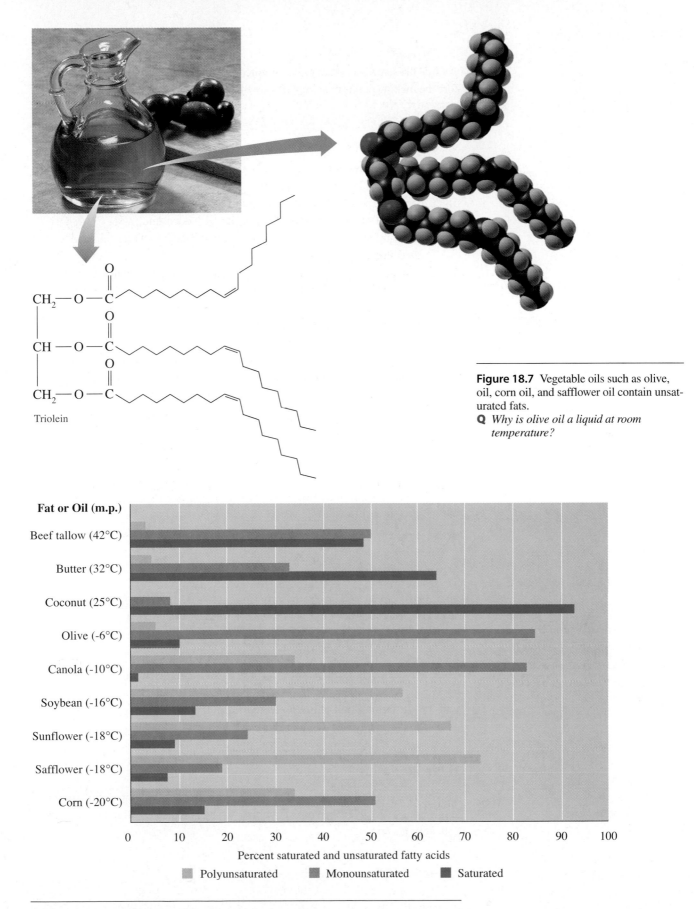

CH₂—O—C (Triolein structure)

Triolein

Figure 18.7 Vegetable oils such as olive, oil, corn oil, and safflower oil contain unsaturated fats.
Q *Why is olive oil a liquid at room temperature?*

Fat or Oil (m.p.)

Beef tallow (42°C)

Butter (32°C)

Coconut (25°C)

Olive (-6°C)

Canola (-10°C)

Soybean (-16°C)

Sunflower (-18°C)

Safflower (-18°C)

Corn (-20°C)

0   10   20   30   40   50   60   70   80   90   100

Percent saturated and unsaturated fatty acids

Polyunsaturated   Monounsaturated   Saturated

Figure 18.8 Fats and oils have low melting points because they have a higher percentage of unsaturated fatty acids than do animal fats.
Q *Why is the melting point of butter higher than olive or canola oil?*

Animal fats usually contain more saturated fatty acids than do vegetable oils. There-fore the melting points of animal fats are higher than those of vegetable oils.

QUESTIONS AND PROBLEMS

Waxes, Fats, and Oils

18.19 Draw the structure of a component of beeswax that contains myricyl alco-hol, $CH_3(CH_2)_{29}OH$, and palmitic acid.

18.20 Draw the structure of a component of jojoba wax that contains arachidic acid, a 20-carbon saturated fatty acid, and 1-docosanol, $CH_3(CH_2)_{21}OH$.

18.21 Draw the structure of a triacylglycerol that contains stearic acid and glycerol.

18.22 A mixed triacylglycerol contains two palmitic acid molecules to every one oleic acid molecule. Write two possible structures (isomers) for the compound.

18.23 Draw the structure of tripalmitin.

18.24 Draw the structure of triolein.

18.25 Safflower oil is called a polyunsaturated oil, whereas olive oil is a mono-unsaturated oil. Explain.

18.26 Why does olive oil have a lower melting point than butter fat?

18.27 Why does coconut oil, a vegetable oil, have a melting point similar to fats from animal sources?

18.28 A label on a bottle of 100% sunflower seed oil states that it is lower in saturated fats than all the leading oils.
   a. How does the percentage of saturated fats in sunflower seed oil com-pare to that of safflower, corn, and canola oil? (See Figure 18.8.)
   b. Is the claim valid?

LEARNING GOAL

Draw the structure of the product from the reaction of a triacylglycerol with hydrogen, an acid or base, or an oxidizing agent.

18.4 Chemical Properties of Triacylglycerols

The chemical reactions of the triacylglycerols (fats and oils) are the same as we dis-cussed for alkenes (Chapter 13), and carboxylic acids and esters (Chapter 17). We will look at the addition of hydrogen to the double bonds, the oxidation of double bonds, and the acid and base hydrolysis of the ester bonds of fats and oils.

Hydrogenation

The **hydrogenation** of unsaturated fats converts carbon–carbon double bonds to single bonds. The hydrogen gas is bubbled through the heated oil in the presence of a nickel catalyst.

$$-CH\!=\!CH- \; + \; H_2 \xrightarrow{\text{Ni}} \begin{array}{c} H \;\; H \\ | \;\;\; | \\ -C-C- \\ | \;\;\; | \\ H \;\; H \end{array}$$

For example, when hydrogen adds to all of the double bonds of triolein using a nickel catalyst, the product is the saturated fat tristearin.

Olestra: A Fat Substitute

In 1968, food scientists designed an artificial fat called *olestra* as a source of nutrition for premature babies. However, olestra could not be digested and was never used for that purpose. Then scientists realized that olestra had the flavor and texture of a fat without the calories.

Olestra is manufactured by obtaining the fatty acids from the fats in cottonseed or soybean oils and bonding the fatty acids with the hydroxyl groups on sucrose. Chemically, olestra is composed of six to eight long-chain fatty acids attached by ester links to a sugar (sucrose) rather than to a glycerol molecule found in fats. This makes olestra a very large molecule, which cannot be absorbed through the intestinal walls. The enzymes and bacteria in the intestinal tract are unable to break down the olestra molecule and it travels through the intestinal tract undigested.

In 1996 the Food and Drug Administration approved olestra for use in potato chips, tortilla chips, crackers and fried snacks. In 1996, olestra snack products were test marketed in parts of Iowa, Wisconsin, Indiana, and Ohio. By 1997, there were reports of some adverse reactions, including diarrhea, abdominal cramps, and anal leakage, indicating that olestra may act as a laxative in some people. However, the manufacturers contend there is no direct proof that olestra is the cause of those effects.

The large molecule of olestra also combines with fat-soluble vitamins (A, D, E, and K) as well as the carotenoids from the foods we eat before they can be absorbed through the intestinal wall. Carotenoids are plant pigments in fruits and vegetables that protect against cancer, heart disease, and macular degeneration, a form of blindness in the elderly. The FDA now requires manufacturers to add the four vitamins, but not the carotenoids, to olestra products. The label on an olestra product must state the following: "This product contains olestra. Olestra may cause abdominal cramping and loose stools. Olestra inhibits the absorption of some vitamins and other nutrients. Vitamins A, D, E, and K have been added." Some snack foods made with olestra are now in supermarkets nationwide. Since there are already low-fat snacks on the market, it remains to be seen whether olestra will have any significant effect on reducing the problem of obesity.

Fatty acids

$CH_3(CH_2)_6COOH$

$CH_3(CH_2)_8COOH$

Olestra

Glyceryl trioleate
(triolein)

Glyceryl tristearate
(tristearin)

In commercial hydrogenation, the addition of hydrogen is stopped before all the double bonds in an oil become completely saturated. Complete hydrogenation gives a very brittle product, whereas the partial hydrogenation of a liquid vegetable oil changes it to a soft, semisolid fat. As the oil becomes more saturated, the melting point increases and the fat becomes more solid at room temperature. Control of the degree of hydrogenation gives the various types of partially hydrogenated vegetable oil products on the market today—soft margarines, solid stick margarines, and solid shortenings. (See Figure 18.9.) Although these products now contain more saturated fatty acids than the original oils, they contain no cholesterol, unlike similar products from animal sources, such as butter and lard.

Trans Fatty Acids and Hydrogenation

In the early 1900s, margarine became a popular replacement for the highly saturated fats such as butter and lard. Margarine is produced by partially hydrogenating the unsaturated fats in vegetable oils such as safflower oil, corn oil, canola oil, cottonseed oil, and sunflower oil. Fats that are more saturated are more resistant to oxidation.

In vegetable oils, the unsaturated fats usually contain cis double bonds. As hydrogenation occurs, double bonds are converted to single bonds. However, some of the cis double bonds are converted to trans double bonds, which causes a change in the overall structure of the fatty acids. If the label on a product states that the oils have been "partially" or "fully hydrogenated," that product will also contain trans fatty acids. In the United States, it is estimated that 2–4% of our total calories come from trans fatty acids.

The concern about trans fatty acids is that their altered structure may make them behave like saturated fatty acids in the body. In the 1980s, research indicated that trans fatty acids have an effect on blood cholesterol similar to that of saturated fats, although study results vary. Several studies reported that trans fatty acids raise the levels of LDL-cholesterol, low-density lipoproteins containing cholesterol that can accumulate in the arteries. (LDLs and HDLs are described in the section on lipoproteins later in the chapter.) Some studies also report that trans fatty acids lower HDL-cholesterol, high-density lipoproteins that carry cholesterol to the liver to be excreted. But other studies did not report any decrease in HDL-cholesterol. In some American and European studies, an increased risk of breast cancer was associated with increased intake of trans fatty acids. However, these studies are not conclusive and not all studies have supported such findings. Current evidence does not yet indicate that the intake of trans fatty acids is a significant risk factor for heart disease. The trans fatty acids controversy will continue to be debated as more research is done.

Foods containing trans fatty acids include milk, bread, fried foods, ground beef, baked goods, stick margarine, butter, soft margarine, cookies, crackers, and vegetable shortening. The American Heart Association recommends that margarine should have no more than 2 grams of saturated fat per tablespoon and a liquid vegetable oil should be the first ingredient. They also recommend the use of soft margarine, which is lower in trans fatty acids because soft margarine is only slightly hydrogenated, and diet margarine because it has less fat and therefore fewer trans fatty acids.

Many health organizations agree that fat should account for less than 30% of daily calories (the current average for Americans is 34%) and saturated fat should be less than 10% of total calories. Lowering the overall fat intake would also decrease the amount of trans fatty acids. The Food and Drug Administration and the U.S. Department of Agriculture are encouraging the use of new food labels to inform consumers of the fat content of food. The best advice may be to reduce total fat in the diet by using fats and oils sparingly, cooking with little or no fat, substituting olive oil or canola oil for other oils, and limiting the use of coconut oil and palm oil, which are high in saturated fatty acids.

There are several products including peanut butter and butter-like spreads on the market that have 0% trans fatty acids. On the labels, they state that their products are nonhydrogenated, which avoids the production of the undesirable trans fatty acid. However, in the list of natural vegetable oils, such as soy and canola oil, there is also palm oil. Because palm oil has a melting point of 30°C, palm oil increases the overall melting point of the spread and gives a product that is solid at room temperature. However, palm oil contains high amounts of saturated fatty acids and has a similar effect in the body as does stearic acid (18 carbons) and fats derived from animal sources. Health experts recommend that we limit the amount of saturated fats, including palm oil, in our diets.

Cis-oleic acid

H_2/Ni

Double bond opens

Ni catalyst

H_2 Isomerization

Addition of H_2

Undesired side product (trans-oleic acid)

Desired saturated product (stearic acid)

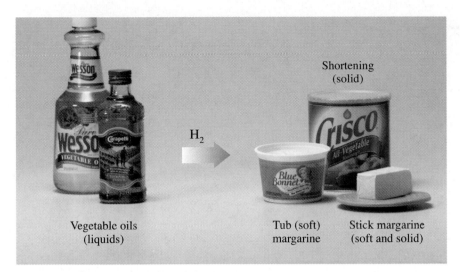

Figure 18.9 Many soft margarines, stick margarines, and solid shortenings are produced by the partial hydrogenation of vegetable oils.
Q *How does hydrogenation change the structure of the fatty acids in the vegetable oils?*

Oxidation of Unsaturated Fats

A fat or oil becomes rancid when its double bonds are oxidized in the presence of oxygen and microorganisms. The products are short-chain fatty acids and aldehydes that have disagreeable odors.

$$-CH=CH- \xrightarrow{[O]} \underset{\substack{\text{Short-chain}\\\text{aldehydes}}}{-\overset{O}{\overset{\|}{C}}-H + H-\overset{O}{\overset{\|}{C}}-} \xrightarrow{[O]} \underset{\substack{\text{Short-chain}\\\text{carboxylic acids}}}{-\overset{O}{\overset{\|}{C}}-OH + HO-\overset{O}{\overset{\|}{C}}-}$$

Unsaturated fatty acids

If a vegetable oil does not contain an antioxidant, it will oxidize rather easily. You can detect an oil that has become rancid by its unpleasant odor. If an oil is covered tightly and stored in a refrigerator, the process of oxidation can be slowed and the oil will last longer.

Oxidation also occurs in the oils that accumulate on the surface of the skin during heavy exercise. (See Figure 18.10.) At body temperature, microorganisms on the skin promote rapid oxidation of the oils as they are exposed to oxygen and water. The resulting short-chain aldehydes and fatty acids account for the body odor associated with a workout and heavy perspiration.

Hydrolysis

In Chapter 17, we saw that esters in the presence of a strong acid undergo hydrolysis with water to give an alcohol and a carboxylic acid.

$$CH_3-O-\overset{O}{\overset{\|}{C}}-CH_3 + H_2O \xrightarrow{H^+} CH_3-OH + HO-\overset{O}{\overset{\|}{C}}-CH_3$$

Ester   + H₂O   →   alcohol   + carboxylic acid

Triacylglycerols are also hydrolyzed (split by water) in the presence of strong acids or digestive enzymes called *lipases*. The products of hydrolysis of the ester bonds are glycerol and three fatty acids. The polar glycerol is soluble in water, but the fatty acids with their long hydrocarbon chains are not.

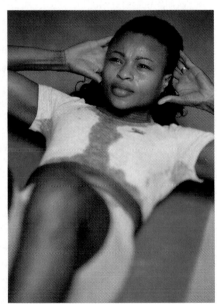

Figure 18.10 After strenuous exercise, increased body temperature and perspiration promote the oxidation of oils that accumulate on the surface of the skin.
Q *What types of organic products result from the oxidation of oils on the skin?*

Water adds to ester bonds

$$
\begin{array}{c}
\text{CH}_2-\text{O}-\overset{\overset{\displaystyle O}{\|}}{\text{C}}-(\text{CH}_2)_{14}\text{CH}_3 \\[2mm]
\text{CH}-\text{O}-\overset{\overset{\displaystyle O}{\|}}{\text{C}}-(\text{CH}_2)_{14}\text{CH}_3 + \mathbf{3H_2O} \\[2mm]
\text{CH}_2-\text{O}-\overset{\overset{\displaystyle O}{\|}}{\text{C}}-(\text{CH}_2)_{14}\text{CH}_3
\end{array}
\xrightarrow[\text{lipase}]{\substack{\text{H}^+ \\ \text{or}}}
\begin{array}{c}
\text{CH}_2-\text{OH} \\[2mm]
\text{CH}-\text{OH} + \mathbf{3HO}\overset{\overset{\displaystyle O}{\|}}{\text{C}}-(\text{CH}_2)_{14}\text{CH}_3 \\[2mm]
\text{CH}_2-\text{OH}
\end{array}
$$

Glyceryl tripalmitate   Glycerol   3 Palmitic acid
(tripalmitin)   molecules

Saponification

When a strong base is used in the hydrolysis of an ester, the products are an alcohol and the carboxylic acid salt.

$$
\text{CH}_3-\text{O}-\overset{\overset{\displaystyle O}{\|}}{\text{C}}-\text{CH}_3 + \text{NaOH} \longrightarrow \text{CH}_3-\text{OH} + \text{Na}^+\text{O}-\overset{\overset{\displaystyle O}{\|}}{\text{C}}-\text{CH}_3
$$

Ester   + NaOH $\longrightarrow$ alcohol   +   carboxylic acid salt

When a fat is heated with a strong base such as sodium hydroxide, saponification of the fat gives glycerol and the sodium salts of the fatty acids, which are soaps. When NaOH is used, a solid soap is produced that can be molded into a desired shape; KOH produces a softer, liquid soap. Oils that are polyunsaturated produce softer soaps. Names like "coconut" or "avocado shampoo" tell you the sources of the oil used in the hydrolysis reaction.

Fat or oil + strong base $\longrightarrow$ glycerol + salts of fatty acids (soaps)

$$
\begin{array}{c}
\text{CH}_2-\text{O}-\overset{\overset{\displaystyle O}{\|}}{\text{C}}-(\text{CH}_2)_{14}\text{CH}_3 \\[2mm]
\text{CH}-\text{O}-\overset{\overset{\displaystyle O}{\|}}{\text{C}}-(\text{CH}_2)_{14}\text{CH}_3 + \mathbf{3NaOH} \\[2mm]
\text{CH}_2-\text{O}-\overset{\overset{\displaystyle O}{\|}}{\text{C}}-(\text{CH}_2)_{14}\text{CH}_3
\end{array}
\longrightarrow
\begin{array}{c}
\text{CH}_2-\text{OH} \\[2mm]
\text{CH}-\text{OH} + \mathbf{3Na}^+\text{O}-\overset{\overset{\displaystyle O}{\|}}{\text{C}}-(\text{CH}_2)_{14}\text{CH}_3 \\[2mm]
\text{CH}_2-\text{OH}
\end{array}
$$

Glyceryl tripalmitate   Glycerol   3 sodium palmitate
(tripalmitin)   (soap)

SAMPLE PROBLEM 18.4

Reactions of Lipids

Write the equation for the reaction catalyzed by the enzyme lipase that hydrolyzes trilaurin during the digestion process.

Solution

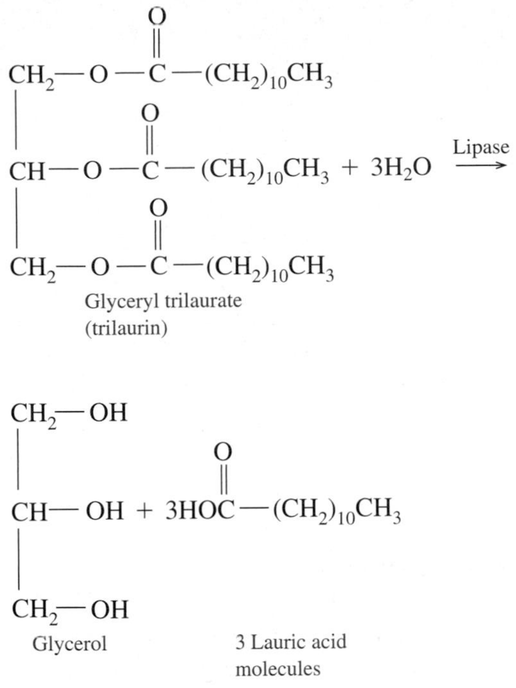

$$CH_2-O-\overset{\overset{\textstyle O}{\|}}{C}-(CH_2)_{10}CH_3$$

$$CH-O-\overset{\overset{\textstyle O}{\|}}{C}-(CH_2)_{10}CH_3 + 3H_2O \xrightarrow{\text{Lipase}}$$

$$CH_2-O-\overset{\overset{\textstyle O}{\|}}{C}-(CH_2)_{10}CH_3$$

Glyceryl trilaurate
(trilaurin)

$$CH_2-OH$$

$$CH-OH + 3HO\overset{\overset{\textstyle O}{\|}}{C}-(CH_2)_{10}CH_3$$

$$CH_2-OH$$

Glycerol   3 Lauric acid molecules

Study Check

What is the name of the product formed when a triacylglycerol containing oleic acid and linoleic acid is completely hydrogenated?

QUESTIONS AND PROBLEMS

Chemical Properties of Triacylglycerols

18.29 Write an equation for the hydrogenation of glyceryl trioleate, a fat containing glycerol and three oleic acid units.

18.30 Write an equation for the hydrogenation of glyceryl trilinolenate, a fat containing glycerol and three linolenic acid units.

18.31 A label on a container of margarine states that it contains partially hydrogenated corn oil.
a. How has the liquid corn oil been changed?
b. Why is the margarine product solid?

18.32 Why should a bottle of vegetable oil that has no preservatives be tightly covered and refrigerated?

18.33 a. Write an equation for the acid hydrolysis of glyceryl trimyristate (trimyristin).
b. Write an equation for the NaOH saponification of glyceryl trimyristate (trimyristin).

18.34 a. Write an equation for the acid hydrolysis of glyceryl trioleate (triolein).
b. Write an equation for the NaOH saponification of glyceryl trioleate (triolein).

18.35 Compare the structure of a triacylglycerol to the structure of olestra.

18.36 An oil is partially hydrogenated.
a. Are all or just some of the double bonds converted to single bonds?
b. What happens to many of the cis double bonds during hydrogenation?
c. How can you reduce the amount of trans fatty acids in your diet?

18.37 Write the product of the hydrogenation of the following triacylglycerol.

$$
\begin{array}{l}
\underset{|}{CH_2}-O-\overset{\overset{\displaystyle O}{\|}}{C}-(CH_2)_{16}CH_3 \\
\underset{|}{CH}-O-\overset{\overset{\displaystyle O}{\|}}{C}-(CH_2)_7CH=CH(CH_2)_7CH_3 \\
CH_2-O-\overset{\overset{\displaystyle O}{\|}}{C}-(CH_2)_{16}CH_3
\end{array}
$$

18.38 Write all the products that would be obtained when the triacylglycerol in problem 18.37 undergoes complete hydrolysis.

18.5 Glycerophospholipids

The **glycerophospholipids** are a family of lipids similar to triacylglycerols except that one hydroxyl group of glycerol is replaced by the ester of phosphoric acid and an amino alcohol, bonded through a phosphodiester bond. We can compare the general structures of a triacylglycerol and a glycerophospholipid as follows:

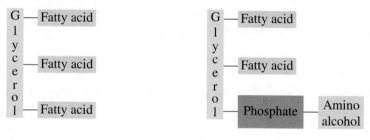

Triacylglycerol (triglyceride)   Glycerophospholipid

Phosphate Esters

In this group of compounds, ester bonds form between a hydroxyl group of an alcohol and phosphoric acid to give ester products similar to those formed with carboxylic acids.

$$
HO-\overset{\overset{\displaystyle O}{\|}}{\underset{\underset{\displaystyle OH}{|}}{P}}-OH \;+\; HO-CH_3 \;\longrightarrow\; HO-\overset{\overset{\displaystyle O}{\|}}{\underset{\underset{\displaystyle OH}{|}}{P}}-O-CH_3 \;+\; H_2O
$$

Phosphoric acid +   alcohol $\longrightarrow$ phosphate ester   + water

The phosphate ester forms a diester by reaction with another alcohol.

$$
HO-\overset{\overset{\displaystyle O}{\|}}{\underset{\underset{\displaystyle OH}{|}}{P}}-O-CH_3 \;+\; HO-CH_3 \;\longrightarrow\; CH_3-O-\overset{\overset{\displaystyle O}{\|}}{\underset{\underset{\displaystyle OH}{|}}{P}}-O-CH_3 \;+\; H_2O
$$

Phosphate ester   +   alcohol $\longrightarrow$ phosphate diester   + water

Three amino alcohols found in glycerophospholipids are choline, serine, and ethanolamine. Because these compounds exist in the body at physiological pH of 7.4, they are ionized. The nitrogen atom in the amino alcohol has a plus one charge while the —OH group from phosphoric acid gives up a proton.

$$HO-CH_2CH_2-\overset{+}{\underset{\underset{CH_3}{|}}{\overset{\overset{CH_3}{|}}{N}}}-CH_3 \qquad HO-CH_2-\overset{\overset{\overset{+}{N}H_3}{|}}{CH}-COO^- \qquad HO-CH_2CH_2-\overset{+}{N}H_3$$

Choline   Serine   Ethanolamine

Lecithins and **cephalins** are two types of glycerophospholipids that are particularly abundant in brain and nerve tissues as well as in egg yolks, wheat germ, and yeast. Lecithins contain choline, and cephalins contain ethanolamine and sometimes serine. In the following structural formulas, palmitic acid is used as an example of a fatty acid.

$$\begin{array}{l} CH_2-O-\overset{\overset{O}{\|}}{C}-(CH_2)_{14}CH_3 \\ CH-O-\overset{\overset{O}{\|}}{C}-(CH_2)_{14}CH_3 \\ CH_2-O-\overset{\overset{}{\underset{O^-}{|}}}{P}-O-CH_2CH_2\overset{+}{N}(CH_3)_3 \end{array}$$

Nonpolar fatty acids / Polar — Choline

A lecithin

$$\begin{array}{l} CH_2-O-\overset{\overset{O}{\|}}{C}-(CH_2)_{14}CH_3 \\ CH-O-\overset{\overset{O}{\|}}{C}-(CH_2)_{14}CH_3 \\ CH_2-O-\overset{\overset{}{\underset{O^-}{|}}}{P}-O-CH_2CH_2\overset{+}{N}H_3 \end{array}$$

Ethanolamine

A cephalin

Glycerophospholipids contain both polar and nonpolar regions, which allow them to interact with both polar and nonpolar substances. The ionized alcohol and phosphate portion called "the head" is polar and can hydrogen bond with water. (See Figure 18.11.) The two fatty acids connected to the glycerol molecule repre-

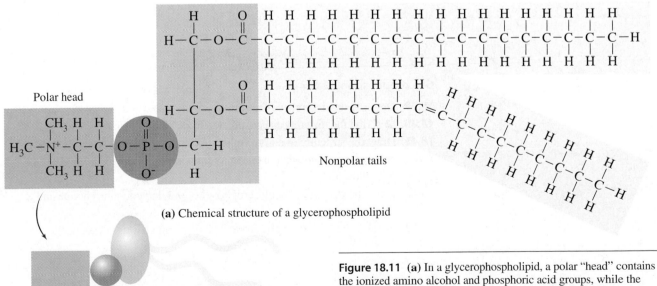

(a) Chemical structure of a glycerophospholipid

Polar head   Nonpolar tails

(b) Simplified way to draw a glycerophospholipid

Figure 18.11 (a) In a glycerophospholipid, a polar "head" contains the ionized amino alcohol and phosphoric acid groups, while the two fatty acids make up the nonpolar "tails." (b) A simplified drawing indicates the polar region and the nonpolar region.
Q *Why are glycerophospholipids polar?*

sent the nonpolar "tails" of the phospholipid. The hydrocarbon chains that make up the "tails" are only soluble in other nonpolar substances, mostly lipids.

Glycerophospholipids are the most abundant lipids in cell membranes, where they play an important role in cellular permeability. They make up much of the myelin sheath that protects nerve cells. In the body fluids, they combine with the less polar triglycerides and cholesterol to make them more soluble as they are transported in the body.

SAMPLE PROBLEM 18.5

Drawing Glycerophospholipid Structures

Draw the structure of cephalin, using stearic acid for the fatty acids and serine for the amino alcohol. Describe each of the components in the glycerophospholipid.

Solution

In general, glycerophospholipids are composed of a glycerol molecule in which two carbon atoms are attached to fatty acids such as stearic acid. The third carbon atom is attached in an ester bond to phosphate linked to an amino alcohol. In this example, the amino alcohol is serine.

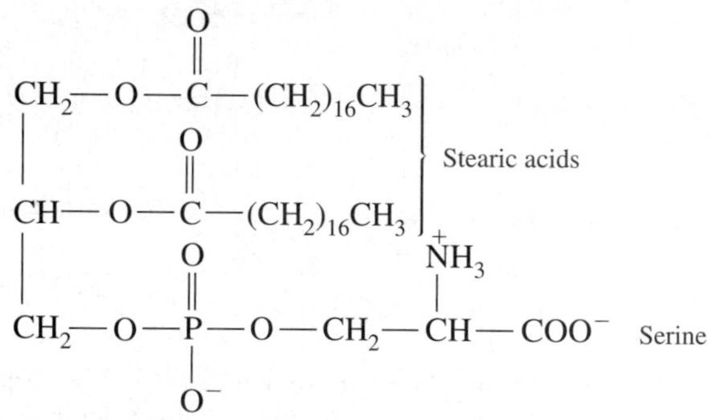

$$
\begin{array}{l}
\overset{\displaystyle O}{\overset{\displaystyle \|}{CH_2-O-C-(CH_2)_{16}CH_3}} \\[4pt]
\left.\begin{array}{l}
\overset{\displaystyle O}{\overset{\displaystyle \|}{CH-O-C-(CH_2)_{16}CH_3}} \\[4pt]
\overset{\displaystyle O}{\overset{\displaystyle \|}{CH_2-O-P-O-CH_2-CH-COO^-}} \\[4pt]
\quad\quad\quad\overset{\displaystyle |}{O^-}
\end{array}\right.
\end{array}
$$

Stearic acids

$\overset{+}{N}H_3$

Serine

Study Check

How do glycerophospholipids differ structurally from triacylgylcerols?

QUESTIONS AND PROBLEMS

Glycerophospholipids

18.39 Describe the differences between triacylglycerols and glycerophospholipids.

18.40 Describe the differences between lecithin and cephalin.

18.41 Draw the structure of a glycerophospholipid containing two molecules of palmitic acid, and ethanolamine. What is another name for this type of phospholipid?

18.42 Draw the structure of a glycerophospholipid that contains choline and palmitic acids.

18.43 Identify the following glycerophospholipid and list its components:

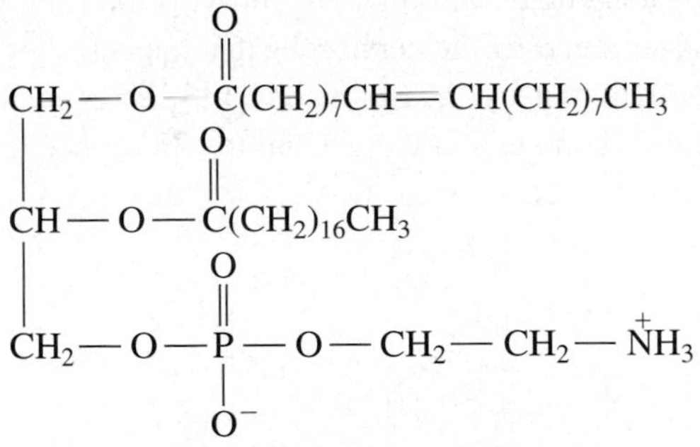

$$CH_2-O-\overset{\overset{\textstyle O}{\|}}{C}(CH_2)_7CH=CH(CH_2)_7CH_3$$

$$CH-O-\overset{\overset{\textstyle O}{\|}}{C}(CH_2)_{16}CH_3$$

$$CH_2-O-\overset{\overset{\textstyle O}{\|}}{\underset{\underset{\textstyle O^-}{|}}{P}}-O-CH_2-CH_2-\overset{+}{N}H_3$$

18.44 Identify the following glycerophospholipid and list its components:

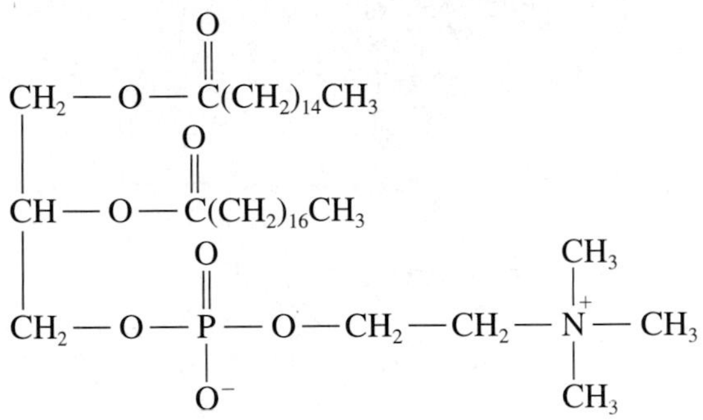

$$CH_2-O-\overset{\overset{\textstyle O}{\|}}{C}(CH_2)_{14}CH_3$$

$$CH-O-\overset{\overset{\textstyle O}{\|}}{C}(CH_2)_{16}CH_3$$

$$CH_2-O-\overset{\overset{\textstyle O}{\|}}{\underset{\underset{\textstyle O^-}{|}}{P}}-O-CH_2-CH_2-\overset{\overset{\textstyle CH_3}{|}}{\underset{\underset{\textstyle CH_3}{|}}{\overset{+}{N}}}-CH_3$$

18.6 Sphingolipids

LEARNING GOAL

Describe the types of lipids that contain sphingosine.

When the —NH$_2$ group in sphingosine bonds to a fatty acid by an *amide* link, a *ceramide* is obtained. The lipids known as sphingolipids form when an ester bond replaces the —OH group of the *ceramide* with a phosphate ester of an amino alcohol.

The *sphingolipids* are another group of phospholipids that are abundant in the biological membranes of the brain and nerve tissues. The **sphingolipids** are esters of an 18-carbon alcohol called sphingosine instead of glycerol.

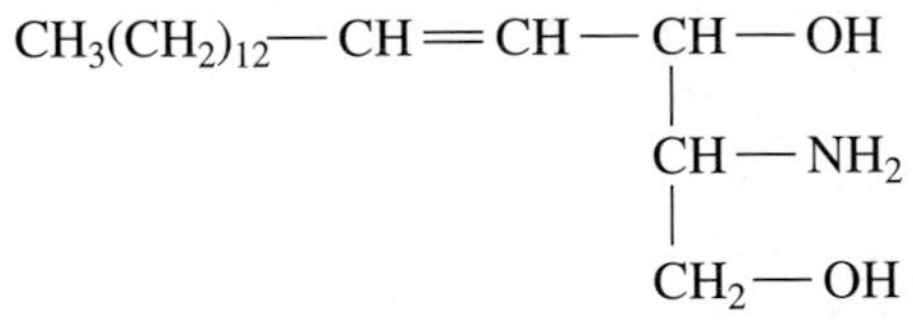

$$CH_3(CH_2)_{12}-CH=CH-\overset{\overset{\textstyle |}{CH}-OH}{\underset{\underset{\textstyle CH_2-OH}{|}}{\underset{\textstyle CH-NH_2}{|}}}$$

Sphingosine

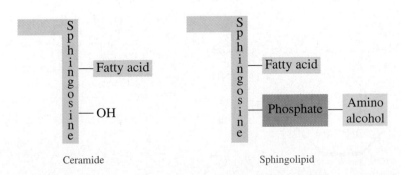

Ceramide   Sphingolipid

One of the most abundant sphingolipids is sphingomyelin. It is the white matter of the myelin sheath, a coating surrounding the nerve cells that increases the speed of nerve impulses and insulates and protects the nerve cells. In sphingomyelin, the sphingosine is linked by an amide bond to a fatty acid and to a phosphate ester of choline, an amino alcohol.

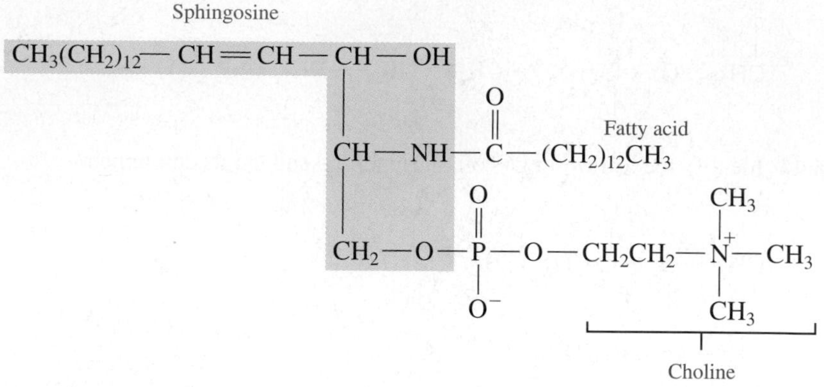

Sphingosine

$$CH_3(CH_2)_{12}-CH=CH-CH-OH$$

$$CH-NH-\overset{\displaystyle O}{\overset{\displaystyle \|}{C}}-(CH_2)_{12}CH_3 \quad \text{Fatty acid}$$

$$CH_2-O-\overset{\displaystyle O}{\overset{\displaystyle \|}{P}}-O-CH_2CH_2-\overset{\displaystyle CH_3}{\overset{\displaystyle |}{\overset{+}{N}}}-CH_3$$

$$\overset{\displaystyle |}{O^-} \qquad \overset{\displaystyle |}{CH_3}$$

Choline

Sphingomyelin, a sphingolipid

Glycosphingolipids

Glycosphingolipids are sphingolipids that contain a carbohydrate such as galactose or glucose. They are important components of nerve membranes and muscle. In a glycosphingolipid, one or more monosaccharides, bonded by a glycosidic bond, replace the —OH of the ceramide.

Cerebrosides

Although they are only present in membranes in small amounts, certain types of glycosphingolipids on the cell surface are important to cellular recognition and tissue immunity. In glycosphingolipids such as **cerebrosides,** one monosaccharide (usually galactose, though sometimes glucose), replaces the —OH of the ceramide.

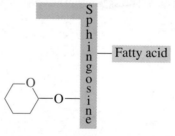

Cerebroside

Sphingosine

$$CH_3(CH_2)_{12}CH=CH-CH-OH \qquad O$$

$$CH_2OH \qquad\qquad CH-NH-\overset{\displaystyle O}{\overset{\displaystyle \|}{C}}-(CH_2)_{12}CH_3$$

$$\text{Fatty acid}$$

$$HO \qquad\qquad O-CH_2$$

$$OH$$

Galactose   OH

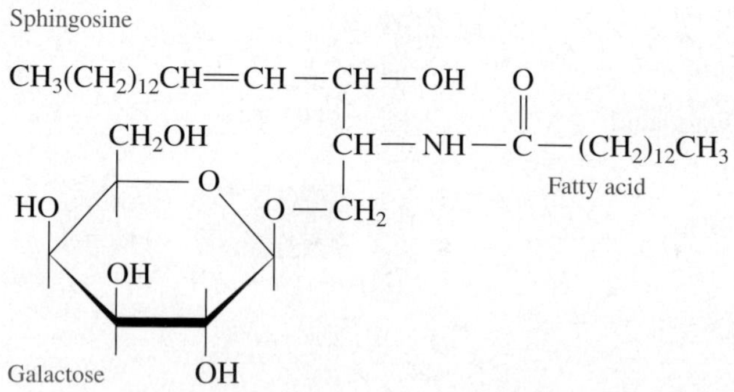

Galactocerebroside, a glycosphingolipid

Gangliosides

Gangliosides are similar to cerebrosides but contain two or more monosaccharides, such as glucose and galactose. Gangliosides are important in the membranes of neurons and act as receptors for hormones, viruses, and certain drugs. In Tay–Sachs disease, the ganglioside known as GM_2 accumulates because of a genetic defect in hexosaminidase A, an enzyme needed for the removal of the N-acetyl-D-galactosamine.

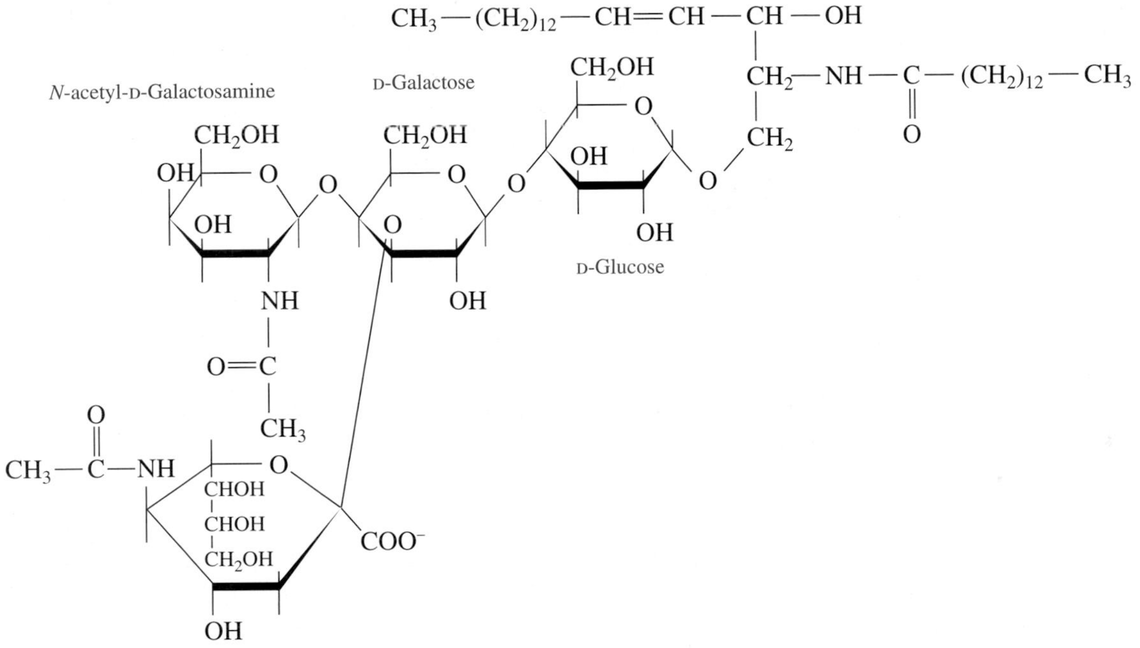

Glycosphingolipid GM_2 or Tay–Sachs ganglioside

Lipid Diseases

Many lipid diseases **(lipidoses)** involve the excessive accumulation of a sphingolipid or glycolipid because an enzyme needed for its breakdown is deficient or absent. The accumulation of these glycolipids may enlarge the spleen, liver, and bone marrow cells (Gaucher's disease) and cause mental retardation, seizures, blindness, and death in early infancy. Some lipid storage diseases are listed in Table 18.3.

In multiple sclerosis, sphingomyelins are lost from the myelin sheath, which is the protective membrane surrounding the neurons in the brain and spinal cord. As the disease progresses, the myelin sheath deteriorates. Scars form on the neurons and impair the transmission of nerve signals. The symptoms of multiple sclerosis include various levels of muscle weakness and loss of coordination and vision depending on the amount of damage. The cause of multiple sclerosis is not yet known, although some researchers suggest that a virus is involved.

Table 18.3 Lipid Diseases

Names of Disease	Lipid Stored	Type	Enzyme Absent
Fabry's	Gal-gal-glucosylceramide	Ganglioside	α-Galactosidase
Gaucher's	Glucosylceramide	Cerebroside	β-Glucosidase
Niemann–Pick	Sphingomyelin	Sphingolipid	Sphingomyelinase
Tay–Sachs	GM_2 ganglioside	Ganglioside	Hexosaminidase A

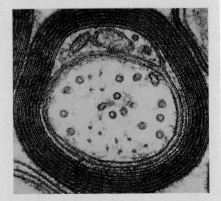

Myelin sheath

SAMPLE PROBLEM 18.6

Glycosphingolipid

In Fabry's disease, the ganglioside shown here accumulates due to a deficiency of α-galactosidase. Identify the components A–E in this glycolipid.

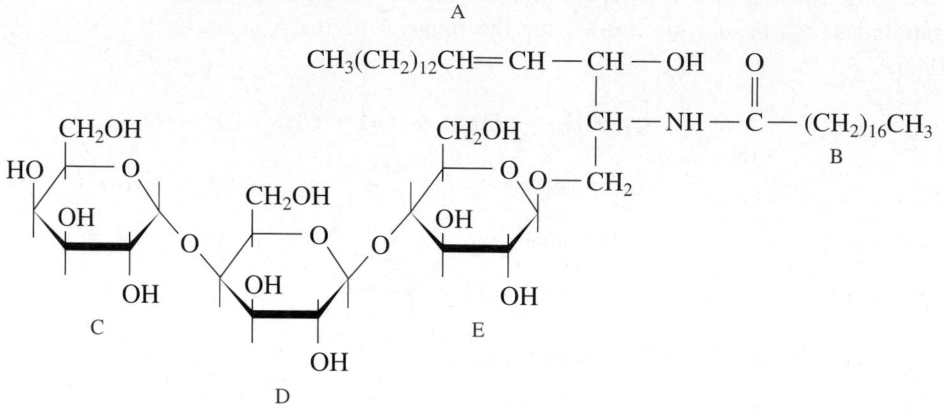

A

$$CH_3(CH_2)_{12}CH = CH - CH - OH$$

Solution

In this glycolipid, the components are sphingosine (A); stearic acid (B), an 18-carbon fatty acid; two galactose units (C, D); and one glucose (E).

Study Check

How do we know that this glycolipid is a ganglioside rather than a cerebroside?

QUESTIONS AND PROBLEMS

Sphingolipids

18.45 Describe the difference between glycerophospholipids and sphingolipids.
18.46 Describe the differences between a cerebroside and a ganglioside.
18.47 Draw the structure of a cerebroside containing palmitic acid and galactose.
18.48 What amino alcohol is found in sphingomyelin? Draw the structure of a sphingomyelin containing palmitic acid.

LEARNING GOAL

Describe the structures of steroids.

18.7 Steroids: Cholesterol, Bile Salts, and Steroid Hormones

Steroids are compounds containing the steroid nucleus, which consists of three cyclohexane rings and one cyclopentane ring fused together. Although they are large molecules, steroids do not hydrolyze to give fatty acids and alcohols. The four rings in the steroid nucleus are designated A, B, C, and D. The carbon atoms are numbered beginning with the carbons in ring A and ending with the two methyl groups.

Steroid

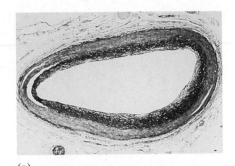

(a)

Cholesterol

Attaching other atoms and groups of atoms to the steroid structure forms a wide variety of steroid compounds. **Cholesterol**, which is one of the most important and abundant steroids in the body, is a *sterol* because it contains an oxygen atom as a hydroxyl (—OH) group on carbon 3. Like many steroids, cholesterol has methyl groups at carbon 10 and carbon 13 and a carbon chain at carbon 17 with a double bond between carbon 5 and carbon 6. In other steroids, the oxygen atom typically at carbon 3 forms a carbonyl (C=O) group.

Cholesterol

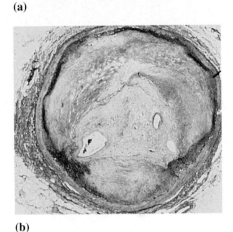

(b)

Figure 18.12 Excess cholesterol forms plaque that can block an artery, resulting in a heart attack. **(a)** A normal, open artery shows no buildup of plaque. **(b)** An artery that is almost completely clogged by atherosclerotic plaque.

Q *What property of cholesterol would cause it to form deposits along the coronary arteries?*

Cholesterol in the Body

Cholesterol is a component of cellular membranes, myelin sheath, and brain and nerve tissue. It is also found in the liver, bile salts, and skin, where it forms vitamin D. In the adrenal gland, it is used to synthesize steroid hormones. Cholesterol in the body is obtained from eating meats, milk, and eggs, and it is also synthesized by the liver from fats, carbohydrates, and proteins. There is no cholesterol in vegetable and plant products.

If a diet is high in cholesterol, the liver produces less. A typical daily American diet includes 400–500 mg of cholesterol, one of the highest in the world. The American Heart Association has recommended that we consume no more than 300 mg of cholesterol a day. The cholesterol contents of some typical foods are listed in Table 18.4.

When cholesterol exceeds its saturation level in the bile, gallstones may form. Gallstones are composed of almost 100% cholesterol with some calcium salts, fatty acids, and phospholipids. High levels of cholesterol are also associated with the accumulation of lipid deposits (plaque) that line and narrow the coronary arteries. (See Figure 18.12.) Clinically, cholesterol levels are considered elevated if the total plasma cholesterol level exceeds 200–220 mg/dL.

Table 18.4 Cholesterol Content of Some Foods

Food	Serving Size	Cholesterol (mg)
Liver (beef)	3 oz	370
Egg	1	250
Lobster	3 oz	175
Fried chicken	3½ oz	130
Hamburger	3 oz	85
Chicken (no skin)	3 oz	75
Fish (salmon)	3 oz	40
Butter	1 tablespoon	30
Whole milk	1 cup	35
Skim milk	1 cup	5
Margarine	1 tablespoon	0

Some research indicates that saturated fats in the diet may stimulate the production of cholesterol by the liver. A diet that is low in foods containing cholesterol and saturated fats appears to be helpful in reducing the serum cholesterol level. Other factors that may also increase the risk of heart disease are family history, lack of exercise, smoking, obesity, diabetes, gender, and age.

SAMPLE PROBLEM 18.7

Cholesterol

Observe the structure of cholesterol for the following questions:
a. What part of cholesterol is the steroid nucleus?
b. What features have been added to the steroid nucleus in cholesterol?
c. What classifies cholesterol as a sterol?

Solution

a. The four fused rings form the steroid nucleus.
b. The cholesterol molecule contains an alcohol group (— OH) on the first ring, one double bond in the second ring, and a branched carbon chain.
c. The alcohol group determines the sterol classification.

Study Check

Why is cholesterol in the lipid family?

Lipoproteins: Transporting Lipids

In the body, lipids must be transported through the bloodstream to tissues where they are stored, used for energy, or to make hormones. However, most lipids are nonpolar and insoluble in the aqueous environment of blood. They are made more soluble by combining them with phospholipids and proteins to form water-soluble complexes called **lipoproteins.** In general, lipoproteins are spherical particles with an outer surface of polar proteins and phospholipids that surround hundreds of nonpolar molecules of triacylglycerols and cholesteryl esters. (See Figure 18.13.) Cholesteryl esters are the prevalent form of cholesterol in the blood. They are formed by the esterification of the hydroxyl group in cholesterol with a fatty acid.

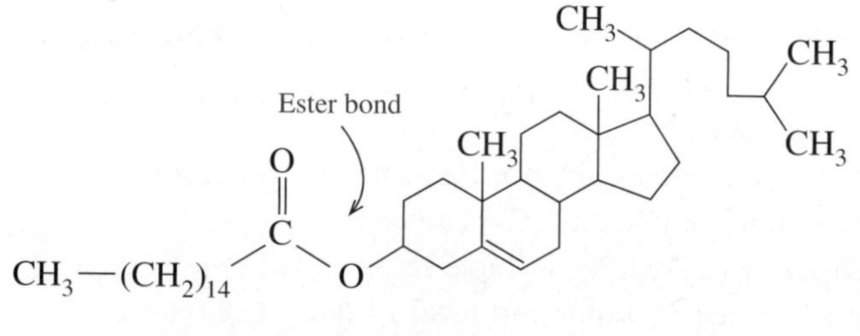

Cholesteryl ester

There are several types of lipoproteins that differ in density, lipid composition, and function. They include chylomicrons, very-low-density lipoprotein (VLDL), low-density lipoprotein (LDL), and high-density lipoprotein (HDL). (See Table 18.5.)

The chylomicrons formed in the mucosal cells of the small intestine and the VLDLs formed in the liver transport triacylglycerols, phospholipids, and choles-

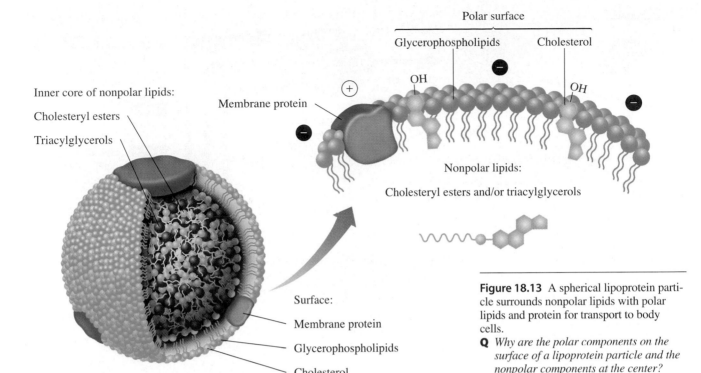

Figure 18.13 A spherical lipoprotein particle surrounds nonpolar lipids with polar lipids and protein for transport to body cells.

Q *Why are the polar components on the surface of a lipoprotein particle and the nonpolar components at the center?*

terol to the tissues for storage or to the muscles for energy. (See Figure 18.14.) The LDLs transport cholesterol to tissues to be used for the synthesis of cell membranes, steroid hormones, and bile salts. When the level of LDL exceeds the amount of cholesterol needed by the tissues, the LDLs deposit cholesterol in the arteries, which can restrict blood flow and increase the risk of developing heart disease and/or myocardial infarctions (heart attacks). This is why LDL cholesterol is called "bad" cholesterol.

Table 18.5   Composition and Properties of Plasma Lipoproteins

	Chylomicron	VLDL	LDL	HDL
Density (g/mL)	0.94	0.950–1.006	1.006–1.063	1.063–1.210
Composition (% by mass)				
Triacylglycerol	86	55	6	4
Phospholipids	7	18	22	24
Cholesterol	2	7	8	2
Cholesteryl esters	3	12	42	15
Protein	2	8	22	55

The HDLs remove excess cholesterol from the tissues and carry it to the liver where it is converted to bile salts and eliminated. When HDL levels are high, cholesterol that is not needed by the tissues is carried to the liver for elimination rather than deposited in the arteries, which gives the HDLs the name of "good" cholesterol. Most of the cholesterol in the body is synthesized in the liver, although some comes from the diet. However, a person on a high-fat diet reabsorbs cholesterol from the bile salts causing less cholesterol to be eliminated. In addition, higher levels of saturated fats stimulate the synthesis of cholesterol by the liver.

Because high cholesterol levels are associated with the onset of arteriosclerosis and heart disease, the serum levels of LDL and HDL are generally determined in a medical examination. For adults, recommended levels for total cholesterol are less than 200 mg/dL with LDL less than 130 mg/dL and HDL over 40 mg/dL. A lower level of serum cholesterol decreases the risk of heart disease. Higher HDL levels are found in people who exercise regularly and eat less saturated fat.

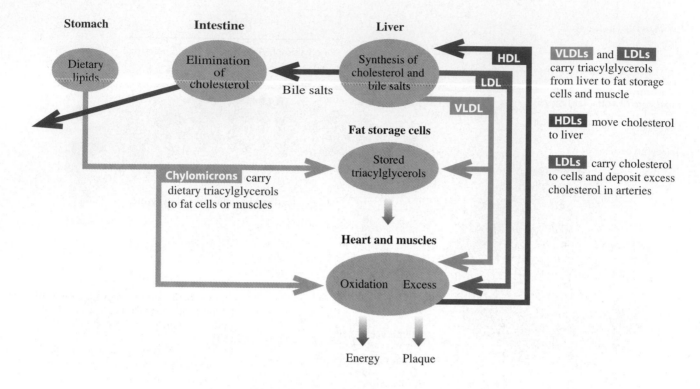

Figure 18.14 Lipoproteins such as HDLs and LDLs transport nonpolar lipids and cholesterol to cells and the liver.
Q *What type of lipoprotein transports cholesterol to the liver?*

Bile Salts

The *bile* salts are synthesized in the liver from cholesterol and stored in the gallbladder. When bile is secreted into the small intestine, the bile salts mix with the water-insoluble fats and oils in our diets. The bile salts with their nonpolar and polar regions act much like soaps, breaking apart and emulsifying large globules of fat. The emulsions that form have a larger surface area for the lipases, enzymes that digest fat. The bile salts also help in the absorption of cholesterol into the intestinal mucosa.

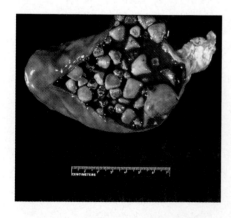

Figure 18.15 Gallstones form in the gallbladder when cholesterol levels are high.
Q *What type of steroid is stored in the gallbladder?*

If large amounts of cholesterol accumulate in the gallbladder, cholesterol can precipitate out and form gallstones. (See Figure 18.15.) If a gallstone passes into the bile duct, the pain can be severe. If the gallstone obstructs the duct, bile cannot be excreted. Then bile pigments known as bilirubin enter the blood where they cause jaundice, which gives a yellow color to the skin and eyes.

Steroid Hormones

The word *hormone* comes from the Greek "to arouse" or "to excite." Hormones are chemical messengers that serve as a kind of communication system from one part of the body to another. The *steroid* hormones, which include the sex hormones and the adrenocortical hormones, are closely related in structure to cholesterol and depend on cholesterol for their synthesis.

Two important male sex hormones, *testosterone* and *androsterone*, promote the growth of muscle and of facial hair and the maturation of the male sex organs and of sperm.

The *estrogens*, a group of female sex hormones, direct the development of female sexual characteristics: the uterus increases in size, fat is deposited in the breasts, and the pelvis broadens. *Progesterone* prepares the uterus for the implantation of a fertilized egg. If an egg is not fertilized, the levels of progesterone and estrogen drop sharply, and menstruation follows. Synthetic forms of the female sex hormones are used in birth-control pills. As with other kinds of steroids, side effects include weight gain and a greater risk of forming blood clots. The structures of some steroid hormones follow:

Hormone	Biological Effects
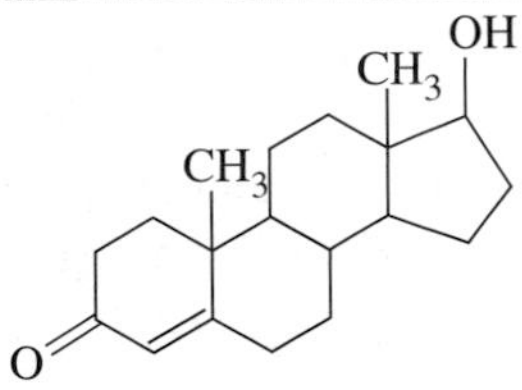 Testosterone (androgen) (produced in testes)	Development of male organs; male sexual characteristics including muscles and facial hair; sperm formation
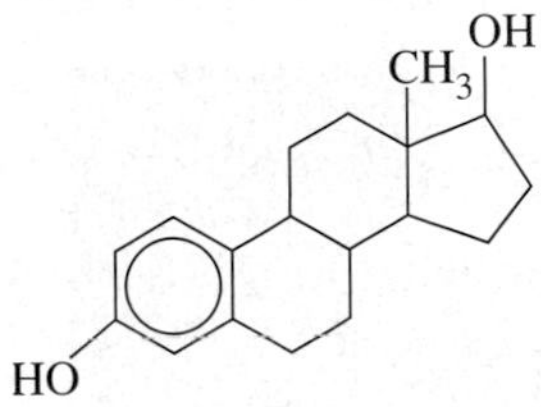 Estradiol (estrogen) (produced in ovaries)	Development of female sexual characteristics; ovulation
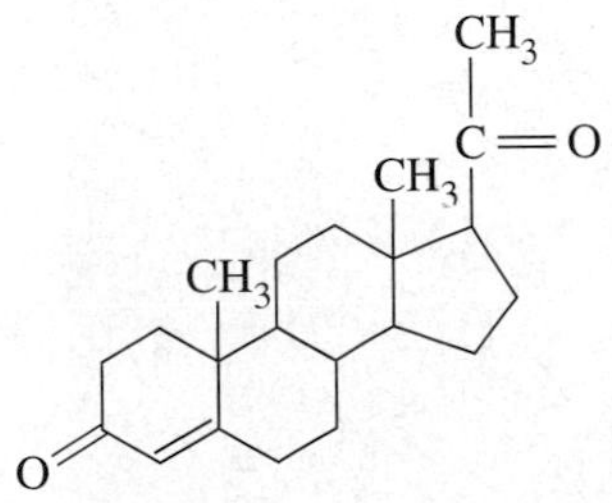 Progesterone (produced in ovaries)	Prepares uterus for fertilized egg
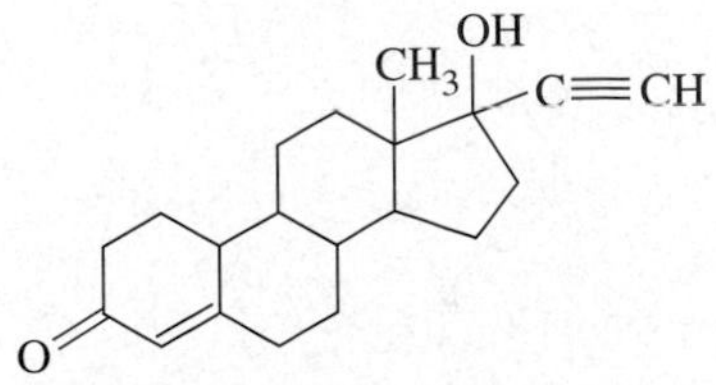 Norethindrone (synthetic progestin)	Contraceptive (birth control) pill

Adrenal Corticosteroids

The adrenal glands, located on the top of each kidney, produce the corticosteroids. *Aldosterone*, a mineralocorticoid, is responsible for electrolyte and water balance by the kidneys. *Cortisone*, a glucocorticoid, increases the blood glucose level and stimulates the synthesis of glycogen in the liver from amino acids. Synthetic corticoids such as *prednisone* are derived from cortisone and used medically for reducing inflammation and treating asthma and rheumatoid arthritis, although precautions are given for long-term use.

Corticosteroids

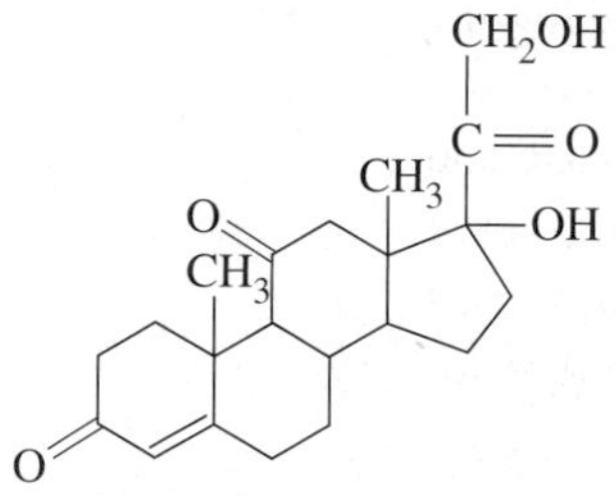

Cortisone
(produced in adrenal gland)

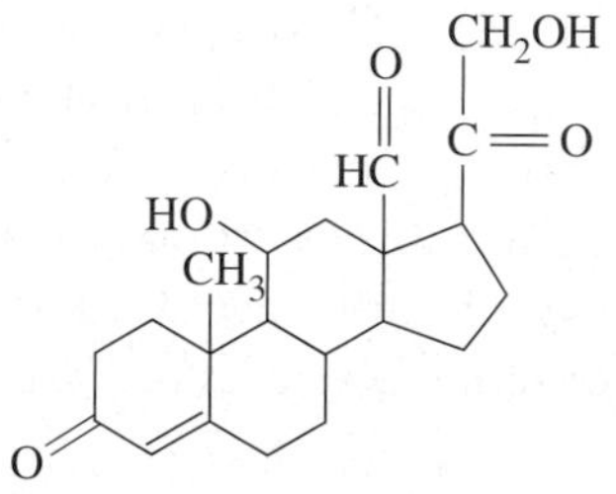

Aldosterone (mineralocorticoid)
(produced in adrenal gland)

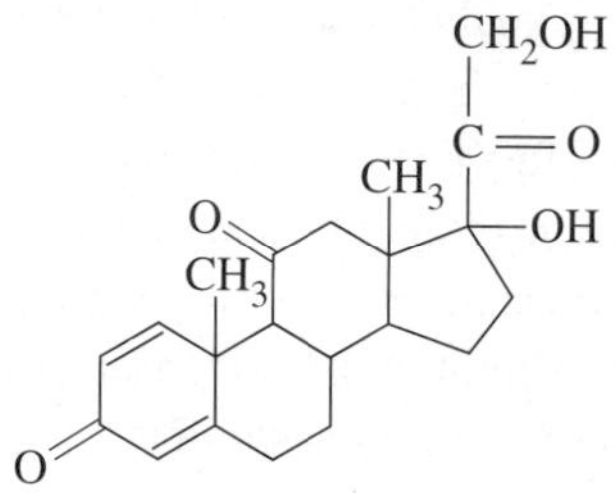

Prednisone
(synthetic corticoid)

Biological Effects

Increases the blood glucose and glycogen levels from fatty acids and amino acids

Increases the reabsorption of Na$^+$ in kidneys; retention of water

Reduces inflammation; treatment of asthma and rheumatoid arthritis

---

HEALTH NOTE

Anabolic Steroids

Some of the physiological effects of testosterone are to increase muscle mass and decrease body fat. Derivatives of testosterone called *anabolic steroids* that enhance these effects have been synthesized. Although they have some medical uses, anabolic steroids have been used in rather high dosages by some athletes in an effort to increase muscle mass. Such use is illegal.

Use of anabolic steroids in attempting to improve athletic strength can cause side effects including hypertension, fluid retention, increased hair growth, sleep disturbances, and acne. Over a long period of time, their use can be devastating and may cause irreversible liver damage and decreased sperm production.

Some Anabolic Steroids

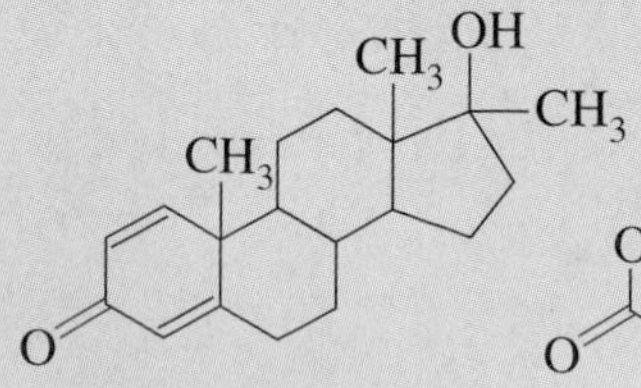

Methandienone

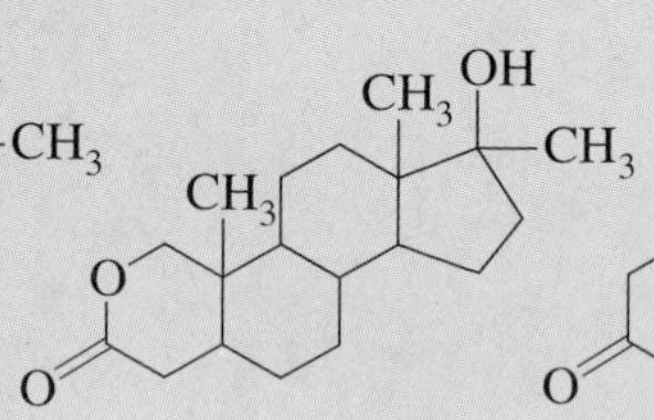

Oxandrolone

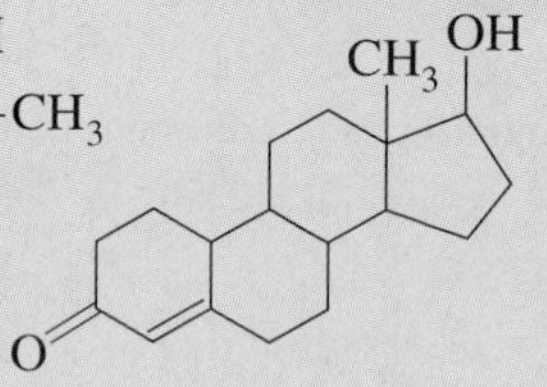

Nandrolone

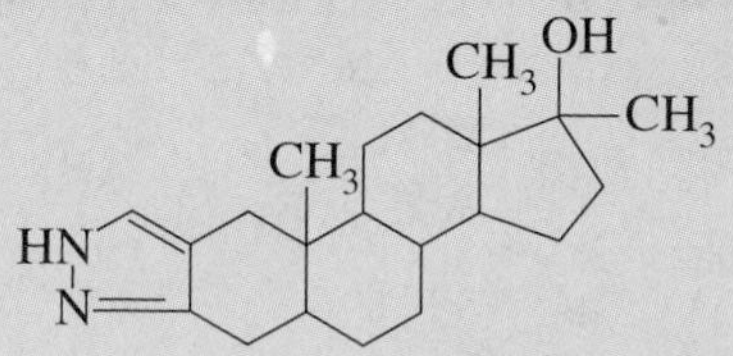

Stanozolol

Steroid Hormones

What are the functional groups on the steroid nucleus in the sex hormones estradiol and testosterone?

Solution

Estradiol contains a benzene ring and two alcohol groups. Testosterone contains a ketone group, a double bond, and an alcohol group.

Study Check

What are the similarities and differences in the structures of testosterone and the anabolic steroid nandrolone?

QUESTIONS AND PROBLEMS

Steroids: Cholesterol, Bile Salts, and Steroid Hormones

18.49 Draw the structure for the steroid nucleus.

18.50 Which of the following compounds are derived from cholesterol?
   a. glyceryl tristearate
   b. cortisone
   c. bile salts
   d. testosterone
   e. estradiol

18.51 What is the function of bile salts in digestion?

18.52 Why are gallstones composed of cholesterol?

18.53 What is the general structure of lipoproteins?

18.54 Why are lipoproteins needed to transport lipids in the bloodstream?

18.55 How do chylomicrons differ from very-low-density lipoproteins?

18.56 How do LDLs differ from HDLs?

18.57 Why are LDLs called "bad" cholesterol?

18.58 Why are HDLs called "good" cholesterol?

18.59 What are the similarities and differences between the sex hormones estradiol and testosterone?

18.60 What are the similarities and differences between the adrenal hormone cortisone and the synthetic corticoid prednisone?

18.61 Which of the following are male sex hormones?
   a. cholesterol
   b. aldosterone
   c. estrogen
   d. testosterone
   e. choline

18.62 Which of the following are adrenal steroids?
   a. cholesterol
   b. aldosterone
   c. estrogen
   d. testosterone
   e. choline

LEARNING GOAL

Describe the composition and function of the lipid bilayer in cell membranes.

18.8 Cell Membranes

The membrane of a cell separates the contents of a cell from the external fluids. It is semipermeable so that nutrients can enter the cell and waste products can leave. The main components of a cell membrane are the glycerophospholipids and sphingolipids. Earlier in this chapter we saw that the structures of such phospholipids consisted of a nonpolar region or tail with long-chain fatty acids and a polar region or head from phosphoric acid and amino alcohols that ionize at physiological pH. The lipid composition of the membranes of human red blood cells and bacterial cells is given in Table 18.6.

In a cell membrane, two rows of phospholipids are arranged like a sandwich. Their nonpolar tails, which are hydrophobic ("water-fearing"), move to the center, while their polar heads, which are hydrophilic ("water-loving"), align on the outer edges of the membrane. This double row arrangement of phospholipids is called a **lipid bilayer.** (See Figure 18.16.) One row of the phospholipids forms the outside surface of the membrane, which is in contact with the external fluids, and the other row forms the inside surface or edge of the membrane, which is in contact with the internal contents of the cell.

Most of the phospholipids in the lipid bilayer contain unsaturated fatty acids. Due to the kinks in the carbon chains at the cis double bonds, the phospholipids do not fit closely together. As a result, the lipid bilayer is not a rigid, fixed structure, but one that is dynamic and fluid-like. In this liquid-like bilayer, there are also proteins, carbohydrates, and cholesterol molecules. For this reason, the model of biological membranes is referred to as the **fluid mosaic model** of membranes.

In the fluid mosaic model, proteins known as peripheral proteins emerge on just one of the surfaces, outer or inner. The integral proteins extend through the entire lipid bilayer and appear on both surfaces of membrane. Some proteins and some lipids on the outer surface of the cell membrane are attached to carbohydrates. The saccharide chains of these glycoproteins and glycolipids project into the surrounding fluid environment where they are responsible for cell recognition and communication with chemical messengers such as hormones and neurotransmitters. In animals, cholesterol makes up 20–25% of the lipid bilayer. Embedded among the nonpolar tails of the fatty acids, cholesterol molecules add strength and rigidity to the cell membrane.

Table 18.6   Lipid Composition of Cell Membranes

Type of Lipid	Human Red Blood Cells	Bacterial Cells
Glycerophospholipids		
–choline	19	0
–ethanolamine	18	65
–serine	8	0
Triacylglycerol	0	18
Sphinogmylelin	18	0
Glycosphingolipids	10	0
Cholesterol	25	0
Others	2	17

Data adapted from Mathews, C. K.; Van Holde, K. K.; Ahern, K. G. *Biochemistry*; Addison Wesley/Longman/Benjamin Cummings: New York, 2000, p. 322.

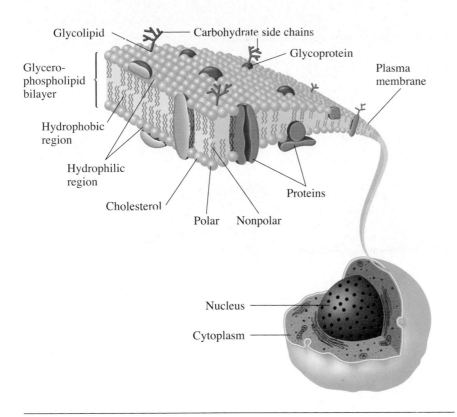

Figure 18.16 The fluid mosaic model of a cell membrane. Proteins and cholesterol are embedded in a lipid bilayer of phospholipids. The bilayer forms a membrane-type barrier with polar heads at the membrane surfaces and the nonpolar tails in the center away from the water.

Q *What types of fatty acids are found in the phospholipids of the lipid bilayer?*

Transport Through Cell Membranes

Ions and molecules flow in and out of the cell in several ways. In the simplest transport mechanism called diffusion or passive transport, ions and small molecules migrate from a higher concentration to a lower concentration. For example, some ions as well as small molecules such as O_2, urea, and water diffuse through cell membranes. If their concentration is greater outside the cell than inside, they diffuse into the cell. If water has a high concentration in the cell, it diffuses out of the cell.

Another type of transport called facilitated transport increases the rate of diffusion for substances that diffuse too slowly by passive diffusion to meet cell needs. This process utilizes the integral proteins that extend from one edge of the cell membrane to the other. Groups of integral proteins provide channels to transport chloride ion (Cl^-), bicarbonate ion (HCO_3^-), and glucose molecules in and out of the cell more rapidly.

Certain ions such as K^+, Na^+, and Ca^{2+}, move across a cell membrane against a concentration gradient. For example, the K^+ concentration is greater inside a cell, and the Na^+ concentration is greater outside. However, in the conduction of nerve impulses and contraction of muscles, K^+ moves into the cell, and Na^+ moves out. To move an ion from a lower to a higher concentration requires energy, which is accomplished by a process known as active transport. In active transport, a protein complex called a Na^+/K^+ pump breaks down ATP to ADP (Chapter 23), which

Figure 18.17 Substances are transported across a cell membrane by either simple diffusion, facilitated transport, or by active transport.
Q *What is the difference between simple diffusion and facilitated transport?*

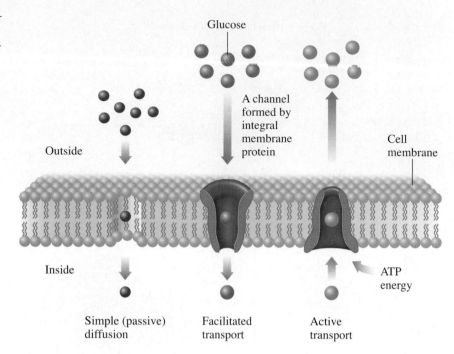

Glucose

Outside

A channel formed by integral membrane protein

Cell membrane

Inside

ATP energy

Simple (passive) diffusion

Facilitated transport

Active transport

releases energy to move Na^+ and K^+ against their concentration gradients. (See Figure 18.17.)

SAMPLE PROBLEM 18.9

Lipid Bilayer in the Cell Membranes

Describe the role of phospholipids in the lipid bilayer.

Solution

Phospholipids consist of polar and nonpolar parts. In a cell membrane, an alignment of the nonpolar sections toward the center with the polar sections on the outside produces a barrier that prevents the contents of a cell from mixing with the fluids on the outside of the cell.

Study Check

Why are protein channels needed in the lipid bilayer?

QUESTIONS AND PROBLEMS

Cell Membranes

18.63 What types of lipids are found in cell membranes?

18.64 Describe the structure of a lipid bilayer.

18.65 What is the function of the lipid bilayer in a cell membrane?

18.66 How do the unsaturated fatty acids in the phospholipids affect the structure of cell membranes?

18.67 Where are proteins located in cell membranes?

18.68 How are carbohydrates attached to the lipid bilayer?

18.69 What is the function of the carbohydrates on a cell membrane?

18.70 Why is a cell membrane semipermeable?

18.71 What are some ways that substances move in and out of cells?

18.72 Identify the type of transport described by each of the following:
   a. A molecule moves through a protein channel.
   b. O_2 moves into the cell from a higher concentration outside the cell.
   c. An ion moves from low to high concentration in the cell.

Chapter Review

18.1 Lipids

Lipids are nonpolar compounds that are not soluble in water. Classes of lipids include waxes, fats and oils, glycerophospholipids, and steroids.

18.2 Fatty Acids

Fatty acids are unbranched carboxylic acids that typically contain an even number (12–18) of carbon atoms. Fatty acids may be saturated, monounsaturated with one double bond, or polyunsaturated with two or more double bonds. The double bonds in unsaturated fatty acids are almost always cis.

18.3 Waxes, Fats, and Oils

A wax is an ester of a long-chain fatty acid and a long-chain alcohol. The triacylglycerols of fats and oils are esters of glycerol with three long-chain fatty acids. Fats contain more saturated fatty acids and have higher melting points than most vegetable oils.

18.4 Chemical Properties of Triacylglycerols

The hydrogenation of unsaturated fatty acids converts double bonds to single bonds. The oxidation of unsaturated fatty acids produces short-chain fatty acids with disagreeable odors. The hydrolysis of the ester bonds in fats or oils produces glycerol and fatty acids. In saponification, a fat heated with a strong base produces glycerol and the salts of the fatty acids or soaps.

18.5 Glycerophospholipids

Glycerophospholipids are esters of glycerol with two fatty acids and a phosphate group attached to an amino alcohol.

18.6 Sphingolipids

In sphingolipids, the alcohol sphingosine forms a bond with a fatty acid and a phosphate–amino alcohol group. In glycosphingolipids, sphingosine is bonded to a fatty acid and one or more monosaccharides.

18.7 Steroids: Cholesterol, Bile Salts, and Steroid Hormones

Steroids are lipids containing the steroid nucleus, which is a fused structure of four rings. Steroids include cholesterol, bile salts, and vitamin D. Lipids, which are nonpolar, are transported through the aqueous environment of the blood by forming lipoproteins. Lipoproteins such as chylomicrons and LDL transport triacylglycerols from the intestines and the liver to fat cells for storage and muscles for energy. HDLs transport cholesterol from the tissues to the liver for elimination. The steroid hormones are closely related in structure to cholesterol and depend on cholesterol for their synthesis. The sex hormones, such as estrogen and testosterone, are responsible for sexual characteristics and reproduction. The adrenal corticosteroids such as aldosterone and cortisone regulate water balance and glucose levels in the cells.

18.8 Cell Membranes

All animal cells are surrounded by a semipermeable membrane that separates the cellular contents from the external fluids. The membrane is composed of two rows of glycerophospholipids in a lipid bilayer. Nutrients and waste products move through the cell membrane using simple transport (diffusion), facilitated transport, or active transport.

Key Terms

cephalins Phospholipids found in brain and nerve tissues that incorporate the amino alcohol serine or ethanolamine.

cerebroside A glycolipid consisting of sphingosine, a fatty acid, and a monosaccharide (usually galactose).

cholesterol The most prevalent of the steroid compounds found in cellular membranes; needed for the synthesis of vitamin D, hormones, and bile acids.

esterification The reaction of an alcohol such as glycerol with acids to form ester bonds.

fat Another term for solid triacylglycerols.

fatty acids Long-chain carboxylic acids found in fats.

fluid mosaic model The concept that cell membranes are lipid bilayer structures that contain an assortment of polar lipids and proteins in a dynamic, fluid arrangement.

ganglioside A glycolipid consisting of sphingosine, a fatty acid, and two or more monosaccharides.

glycerophospholipids Polar lipids of glycerol attached to two fatty acids and a phosphate group connected to an amino group such as choline, serine, or ethanolamine.

glycosphingolipids The phospholipid that combines sphingosine with a fatty acid bonded to the nitrogen group and one or more monosaccharides bonded by a glycosidic link, which replaces the —OH group of sphingosine.

hydrogenation The addition of hydrogen to unsaturated fats.

lecithins Glycerophospholipids containing choline as the amino alcohol.

lipid bilayer A model of a cell membrane in which phospholipids are arranged in two rows interspersed with proteins arranged at different depths.

lipidoses Genetic diseases in which a deficiency of an enzyme for the hydrolysis of a lipid causes the accumulation of that lipid to toxic levels.

lipids A family of compounds that is nonpolar in nature and not soluble in water; includes fats, waxes, phospholipids, and steroids.

lipoprotein A combination of nonpolar lipids with glycerophospholipids and proteins to form a polar complex that can be transported through body fluids.

monounsaturated fatty acids Fatty acids with one double bond.

oil Another term for liquid triacylglycerols.

polyunsaturated fatty acids Fatty acids that contain two or more double bonds.

prostaglandins A number of compounds derived from arachidonic acid that regulate several physiological processes.

saturated fatty acids Fatty acids that have no double bonds; they have higher melting points than unsaturated lipids and are usually solid at room temperatures.

sphingolipids Phospholipids in which sphingosine has replaced glycerol.

steroids Types of lipid composed of a multicyclic ring system.

triacylglycerols A family of lipids composed of three fatty acids bonded through ester bonds to glycerol, a trihydroxy alcohol.

wax The ester of a long-chain alcohol and a long-chain saturated fatty acid.

Additional Problems

18.73 Among the ingredients of a lipstick are beeswax, carnauba wax, hydrogenated vegetable oils, and capric triglyceride. What types of lipids have been used? Draw the structure of capric triglyceride (capric acid is the saturated 10-carbon fatty acid).

18.74 Because peanut oil floats on the top of peanut butter, many brands of peanut butter are hydrogenated. A solid product then forms that is mixed into the peanut butter and does not separate. If a triacylglycerol in peanut oil that contains one palmitic acid, one oleic acid, and one linoleic acid is completely hydrogenated, what is the product?

18.75 Trans fats are produced during the hydrogenation of polyunsaturated oils.
   a. What is the typical configuration of a monounsaturated fatty acid?
   b. How does a trans fat compare with a cis fat?
   c. Draw the structure of *trans*-oleic acid.

18.76 One mole of triolein is completely hydrogenated. What is the product? How many moles of hydrogen are required? How many grams of hydrogen? How many liters of hydrogen are needed if the reaction is run at STP?

18.77 On the list of ingredients of a cosmetic product are glyceryl stearate and lecithin. What are the structures of these compounds?

18.78 Some typical meals at fast-food restaurants are listed here. Calculate the number of kilocalories from fat and the percentage of total kilocalories due to fat (1 gram of fat = 9 kcal). Would you expect the fats to be mostly saturated or unsaturated? Why?
   a. a chicken dinner, 830 kcal, 46 g of fat
   b. a quarter-pound cheeseburger, 518 kcal, 29 g of fat
   c. pepperoni pizza (three slices), 560 kcal, 18 g of fat
   d. beef burrito, 470 kcal, 21 g of fat
   e. deep-fried fish (three pieces), 480 kcal, 28 g of fat

18.79 Identify the following as fatty acids, soaps, triacyl-glycerols, wax, glycerophospholipid, sphingolipid, or steroid:

a. beeswax   **b.** cholesterol
c. lecithin
d. glyceryl tripalmitate (tripalmitin)
e. sodium stearate   **f.** safflower oil
g. sphingomyelin   **h.** whale blubber
i. adipose tissue   **j.** progesterone
k. cortisone   **l.** stearic acid

18.80 Why would an animal that lives in a cold climate have more unsaturated triacylglycerols in its body fat than an animal that lives in a warm climate?

18.81 Identify the components in each of the following lipids as:

1. glycerol   2. fatty acid   3. phosphate
4. amino alcohol   5. steroid nucleus
6. sphingosine

a. estrogen   **b.** cephalin
c. wax   **d.** triacylglycerol
e. glycerophospholipid
f. sphingomyelin

18.82 Which of the following are found in cell membranes?

a. cholesterol   **b.** triacylglycerols
c. carbohydrates   **d.** proteins
e. waxes   **f.** glycerophospholipids
g. sphingolipids   **h.** prostaglandins

18.83 Identify the lipoprotein in each description as:

1. chylomicrons   2. VLDL   3. LDL   4. HDL

a. "good" cholesterol
b. transports most of the cholesterol to the cells
c. carries triacylglycerols from the intestine to the fat cells
d. transports cholesterol to the liver
e. has the greatest abundance of protein
f. "bad" cholesterol
g. carries triacylglycerols synthesized in the liver to the muscles
h. has the lowest density

19 Amines and Amides

"The pharmacy is one of the many factors in the final integration of chemistry and medicine in patient care," says Dorothea Lorimer, pharmacist, Kaiser Hospital. "If someone is allergic to a medication, I have to find out if a new medication has similar structural features. For instance, some people are allergic to sulfur. If there is sulfur in the new medication, there is a chance it will cause a reaction."

A prescription indicates a specific amount of a medication. At the pharmacy, the chemical name, formula, and quantity in milligrams or micrograms are checked. Then the prescribed number of capsules is prepared and placed in a container. If it is a liquid medication, a specific volume is measured and poured into a bottle for liquid prescriptions.

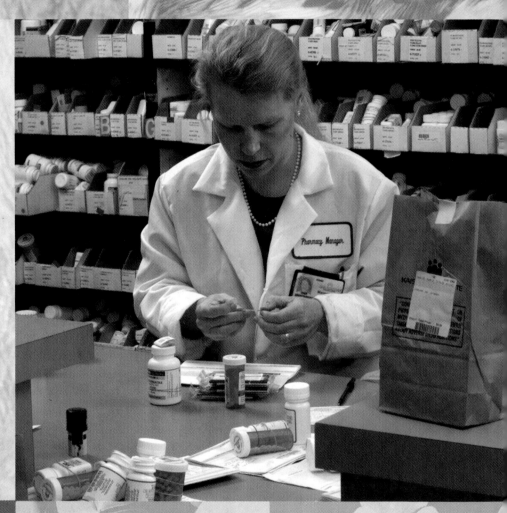

LOOKING AHEAD

the Chemistry place

www.chemplace.com/college

Visit the URL above or use the CD-ROM in the book for extra quizzing, interactive tutorials, and career resources.

In earlier chapters, we looked at organic compounds that contain carbon, hydrogen, and oxygen. Now we are ready to discuss amines and amides, which are organic compounds that contain nitrogen. Many nitrogen-containing compounds are important to life as components of amino acids, proteins, and nucleic acids: DNA and RNA. Many amines that exhibit strong physiological activity are used in medicine as decongestants, anesthetics, and sedatives. Examples include dopamine, histamine, epinephrine, and amphetamine.

Alkaloids such as caffeine, nicotine, cocaine, and digitalis, which demonstrate powerful physiological activity, are naturally occurring amines obtained from plants. Amides, which are derived from carboxylic acids, are important in biology as the types of bonds that link amino acids in proteins. Some amides used medically include acetaminophen (Tylenol) used to reduce fever; phenobarbital, a sedative and anticonvulsant medication; and penicillin, an antibiotic.

19.1 Amines

Amines are considered as derivatives of ammonia (NH_3), in which one or more hydrogen atoms is replaced with alkyl or aromatic groups. For example, in methylamine, a methyl group replaces one hydrogen atom in ammonia. The bonding of two methyl groups gives dimethylamine, and the three methyl groups in trimethylamine replace all the hydrogen atoms in ammonia. (See Figure 19.1.)

> **LEARNING GOAL**
>
> **Classify amines as primary, secondary, or tertiary.**

Classification of Amines

When we classified alcohols in Chapter 14, we looked at the number of carbon atoms bonded to the alcohol carbon. Amines are classified in a similar way by counting the number of carbon atoms directly bonded to a nitrogen atom. In a *primary (1°) amine*, one carbon is bonded to a nitrogen atom. In a *secondary (2°) amine*, two carbons are bonded to the nitrogen atom, and a *tertiary (3°) amine* has three carbon atoms bonded to the nitrogen.

Ammonia	Primary (1°) amine	Secondary (2°) amine	Tertiary (3°) amine

$$H - \overset{\cdot\cdot}{\underset{|}{N}} - H \qquad R - \overset{\cdot\cdot}{\underset{|}{N}} - H \qquad R - \overset{\cdot\cdot}{\underset{|}{N}} - R \qquad R - \overset{\cdot\cdot}{\underset{|}{N}} - R$$
$$\quad\; H \qquad\qquad\quad H \qquad\qquad\quad H \qquad\qquad\quad R$$

Alkyl or Aromatic group

Examples:

$$CH_3 - \overset{\cdot\cdot}{\underset{|}{N}} - H \qquad CH_3 - \overset{\cdot\cdot}{\underset{|}{N}} - CH_3 \qquad CH_3 - \overset{\cdot\cdot}{\underset{|}{N}} - CH_3$$
$$\qquad\quad H \qquad\qquad\qquad H \qquad\qquad\qquad CH_3$$

Line-Bond Formulas for Amines

We can draw line-bond formulas for amines just as we did for other organic compounds. For example, we can write the following line-bond formulas and classify each of the amines.

Ammonia

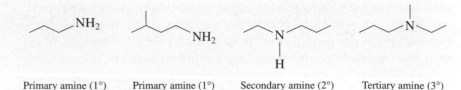

Primary amine (1°)   Primary amine (1°)   Secondary amine (2°)   Tertiary amine (3°)

Methylamine (primary amine 1°)

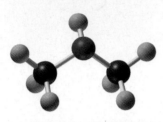

Dimethylamine (secondary amine 2°)

Trimethylamine (tertiary amine 3°)

Figure 19.1 Amines are classified according to the number of carbon atoms bonded to the N atom.
Q *How many carbon atoms are bonded to the nitrogen atom in dimethylamine?*

SAMPLE PROBLEM 19.1

Classifying Amines

Classify the following amines as primary, secondary, or tertiary:

a. NH₂

b.
$$CH_3 \!-\! \overset{\overset{\displaystyle H}{|}}{N} \!-\! CH_2CH_3$$

c. ⬡—N—CH₃
             |
             H

d.

Solution

a. This is a primary amine because there is one alkyl group (cyclohexyl) attached to a nitrogen atom.

b. This is a secondary amine. There are two alkyl groups (methyl and ethyl) attached to the nitrogen atom.

c. This is a secondary amine with two carbon groups, methyl and phenyl, bonded to the nitrogen atom.

d. The nitrogen atom in this line-bond formula is bonded to two carbon atoms, which makes it a secondary amine.

Study Check

Classify the following amine as primary, secondary, or tertiary:

$$CH_3CH_2 \!-\! \underset{\underset{\displaystyle CH_3}{|}}{N} \!-\! CH_2CH_3$$

QUESTIONS AND PROBLEMS

Amines

19.1 What is a primary amine?

19.2 What is a tertiary amine?

19.3 Classify each of the following amines as primary, secondary, or tertiary:

a. $CH_3\!-\!CH_2\!-\!CH_2\!-\!NH_2$

b. $CH_3 \!-\! \overset{\overset{\displaystyle H}{|}}{N} \!-\! CH_2\!-\!CH_3$

c. ╱╲╱NH₂

d. ⬡—N—CH₃
           |
           CH₃

e. $CH_3 \!-\! \underset{\underset{\displaystyle CH_3}{|}}{CH} \!-\! \overset{\overset{\displaystyle CH_3}{|}}{N} \!-\! CH_2\!-\!CH_3$

Amines in Health and Medicine

In response to allergic reactions or injury to cells, the body increases the production of histamine, which causes blood vessels to dilate and increases the permeability of the cells. Redness and swelling occur in the area. Administering an antihistamine such as diphenylhydramine helps block the effects of histamine.

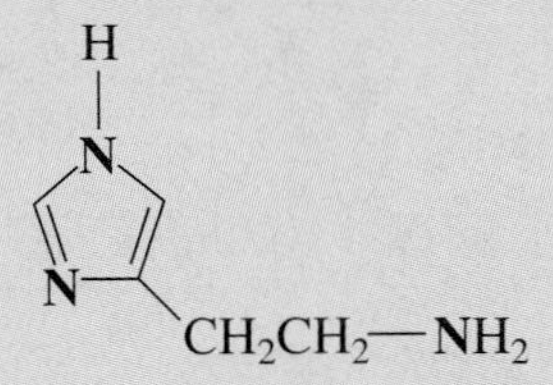

Histamine

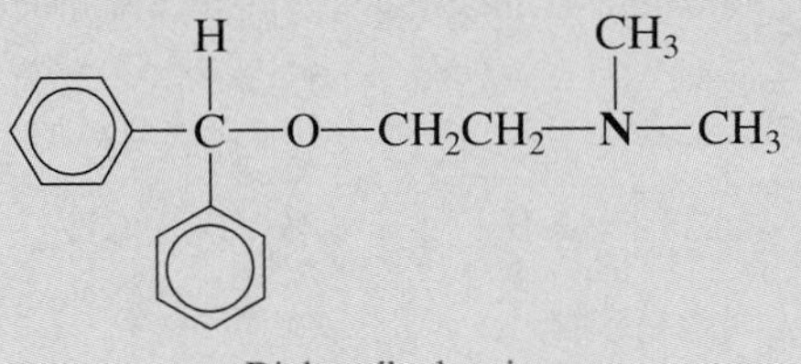

Diphenylhydramine

In the body, hormones called biogenic amines carry messages between the central nervous system and nerve cells. Epinephrine (adrenaline) and norepinephrine (noradrenaline) are released by the adrenal medulla in "fight or flight" situations to raise the blood glucose level and move the blood to the muscles. Used in remedies for colds, hay fever, and asthma, the norepinephrine contracts the capillaries in the mucous membranes of the respiratory passages. The prefix *nor–* in a drug name means there is one less CH_3— group on the nitrogen atom. Parkinson's disease is a result of a deficiency in another biogenic amine called dopamine.

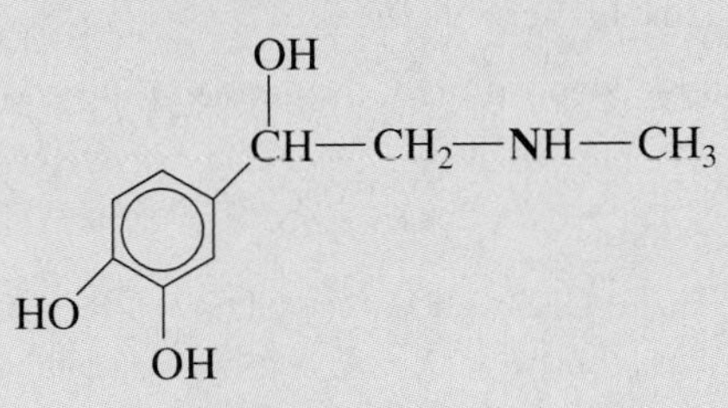

Epinephrine (adrenaline)

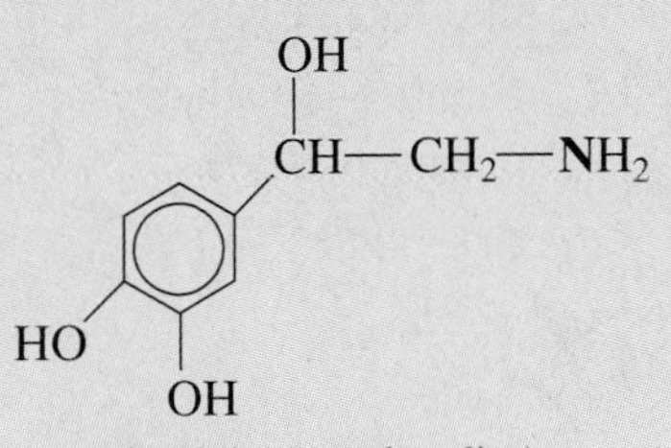

Norepinephrine (noradrenaline)

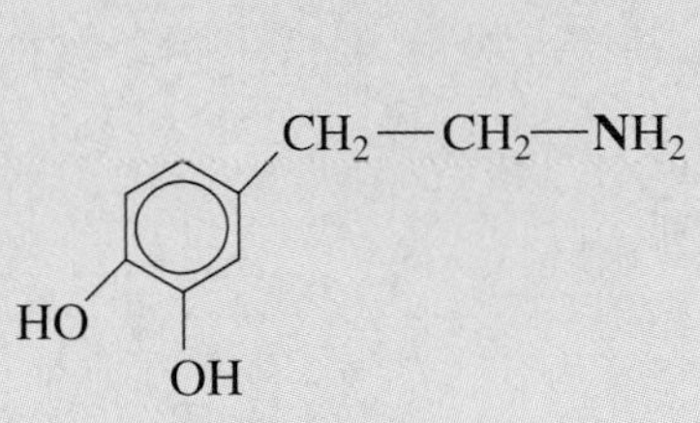

Dopamine

Produced synthetically, amphetamines (known as "uppers") are stimulants of the central nervous system much like epinephrine, but they also increase cardiovascular activity and depress the appetite. They are sometimes used to bring about weight loss, but they can cause chemical dependency. Benzedrine and Neo-Synephrine (phenylephrine) are used in medications to reduce respiratory congestion from colds, hay fever, and asthma. Sometimes, Benzedrine is taken internally to combat the desire to sleep, but it has side effects. Methedrine is used to treat depression and in the illegal form is known as "speed" or "crank." The prefix *meth–* means that there is one more methyl group on the nitrogen atom.

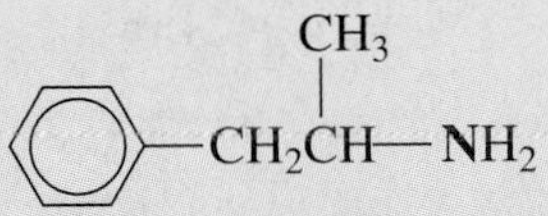

Benzedrine (amphetamine)

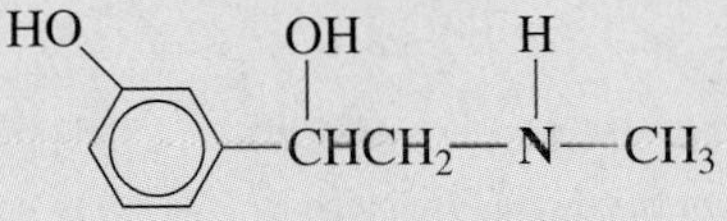

Neo-Synephrine (phenylephrine)

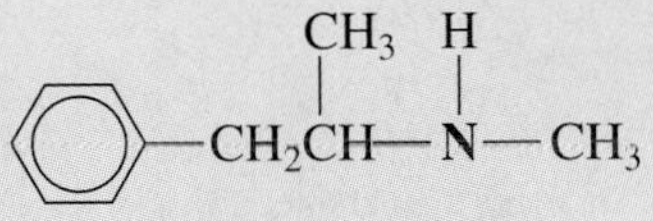

Methamphetamine (methedrine)

19.4 Classify each of the following amines as primary, secondary, or tertiary:

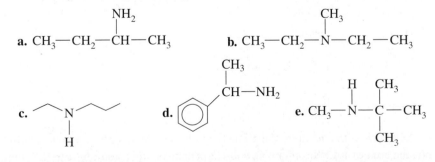

19.2 Naming Amines

There are several systems in use for naming amines. For simple amines, the common names are often used. In the common name, the alkyl groups bonded to the nitrogen atom are listed in alphabetical order. The prefixes *di-* and *tri-* are used to indicate two and three identical substituents.

CH_3—NH_2 CH_3—NH—CH_3 CH_3—CH_2—CH_2—N—CH_2—CH_3 (with CH_3 above N)

Methylamine   Dimethylamine   Ethylmethylpropylamine

The IUPAC names for amines are similar to the names we used for alcohols, except that the *–e* in the parent alkane name is replaced with *amine*.

CH_4 CH_3—OH CH_3—NH_2

Methane   Methan**ol**   Methan**amine**

Step 1   Identify the longest carbon chain bonded to the nitrogen atom. Replace the *–e* in the corresponding alkane name with *amine*.

Step 2   Number the carbon chain to show the position of the amine group and any other substituents.

CH_3—CH_2—NH_2 CH_3—CH_2—CH_2—NH_2 CH_3—CH—CH_3 (with NH_2 above CH)

Ethan**amine**   1-Propan**amine**   2-Propan**amine**

CH_3—CH—CH_2—CH_3 (with NH_2 above CH) CH_3—CH—CH_2—CH_2—NH_2 (with CH_3 above CH)

2-Butan**amine**   3-Methyl-1-butan**amine**

Step 3   In secondary and tertiary amines, the largest alkyl group attached to the nitrogen is named as the parent amine. The smaller alkyl groups are named with the prefix *N-* followed by the alkyl name, and listed alphabetically.

CH_3—CH_2—N—CH_3 (with CH_3 above N) CH_3—CH_2—CH_2—N—CH_3 (with CH_3 above N) CH_3—CH_2—CH_2—N—CH_2—CH_3 (with CH_3 above N)

N, N-Dimethylethanamine   *N, N*-Dimethyl-1-propanamine   *N*-Ethyl-*N*-methyl-1-propanamine

Heterocycles are cyclic amines that contain a nitrogen atom in the ring. For example, some of the pungent aroma and taste we associate with black pepper is due to a compound called piperidine, a six-atom ring containing a nitrogen atom. The fruit from the black pepper plant is dried and ground to give the black pepper we use to season our foods.

Piperidine

An amine with two amine functional groups is named as a *diamine*. For example, the amines 1,4-butanediamine and 1,5-pentanediamine contribute to the odors of decaying flesh.

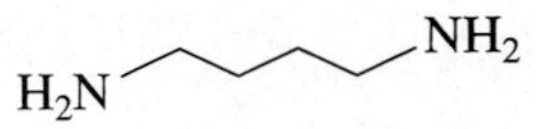

1,4-Butanediamine
(putrescine)

1,5-Pentanediamine
(cadaverine)

In amines where another functional group takes priority, the —NH_2 group is named as a substituent *amino* group and numbered to show its location. For the major functional groups we have studied, the increasing priority follows the increase in oxidation:

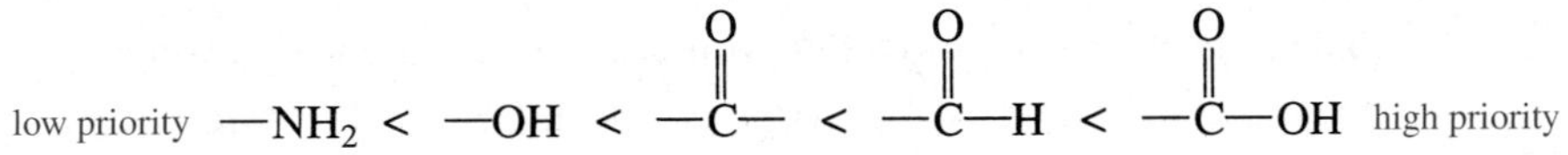

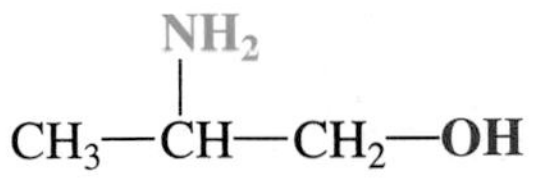

low priority —NH_2 < —OH < $-\overset{\overset{O}{\|}}{C}-$ < $-\overset{\overset{O}{\|}}{C}-H$ < $-\overset{\overset{O}{\|}}{C}-OH$ high priority

2-Amino-1-propanol 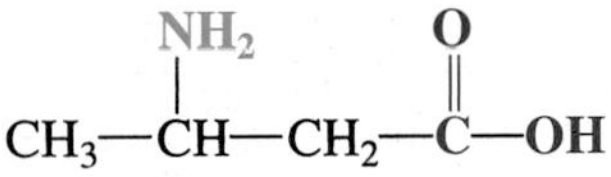   3-Aminobutanoic acid 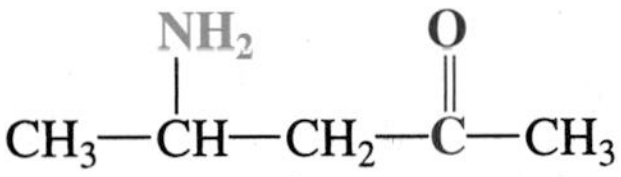   4-Amino-2-pentanone

Aromatic Amines

The aromatic amines use the name *aniline*, which is approved by IUPAC.

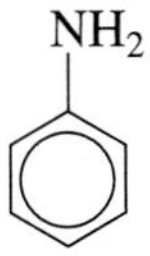

Aniline 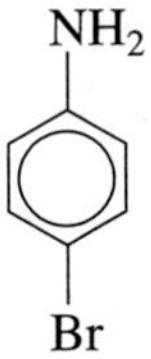   4-Bromoaniline   *N*-Methylaniline 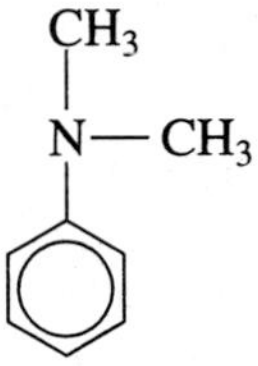   *N,N*-Dimethylaniline
               (*p*-bromoaniline)

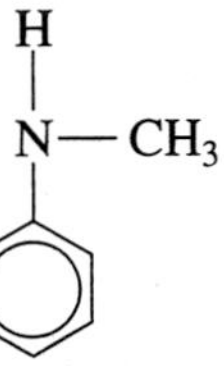

SAMPLE PROBLEM 19.2

Naming Amines

Give the common name for each of the following amines:

a. CH_3CH_2—NH_2 **b.** 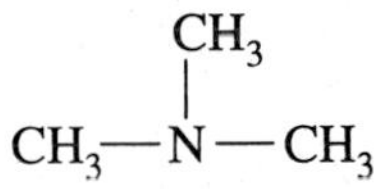 CH_3—$\overset{\overset{CH_3}{|}}{N}$—$CH_3$ **c.** 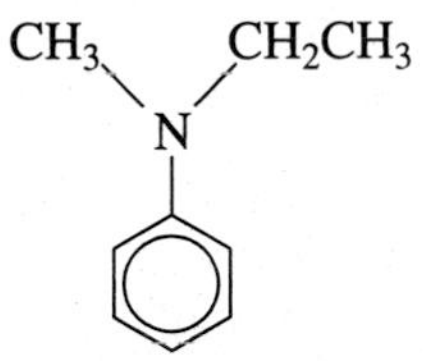 [structure]

Solution

a. This amine has one ethyl group attached to the nitrogen atom; its name is *ethylamine*.
b. This amine has three methyl groups attached to the nitrogen atom; its name is *trimethylamine*.
c. This amine is a derivative of aniline with a methyl and an ethyl group attached to the nitrogen atom; its name is *N-ethyl-N-methylaniline*.

Study Check

Draw the structure of ethylpropylamine.

QUESTIONS AND PROBLEMS

Naming Amines

19.5 Write the common and IUPAC names for each of the following:

    a. $CH_3-CH_2-NH_2$ **b.** $CH_3\ NH\ CH_2\ CH_2\ CH_3$

                      CH_3 NH_2

                     | |

    c. 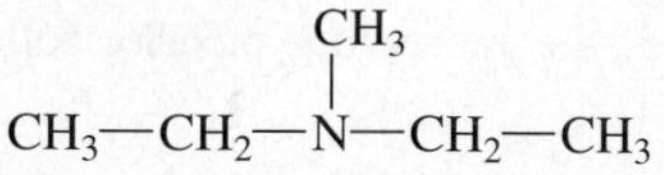 $CH_3-CH_2-N-CH_2-CH_3$ **d.** 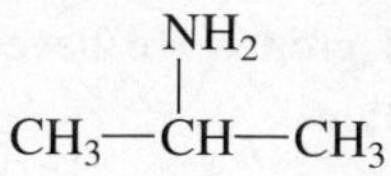 $CH_3-CH-CH_3$

19.6 Write the common and IUPAC names for each of the following:

    a. $CH_3-CH_2-CH_2-NH_2$ **b.** $CH_3-NH-CH_2-CH_3$

                                               CH_2-CH_3

                                                |

    c. $CH_3-CH_2-CH_2-CH_2-NH_2$ **d.** 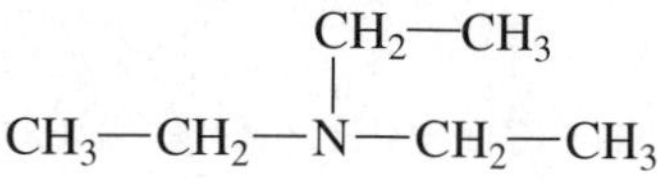 $CH_3-CH_2-N-CH_2-CH_3$

19.7 Write the IUPAC names for each of the following:

              NH_2 NH_2

             |

    a. 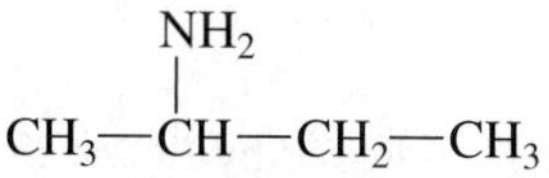 $CH_3-CH-CH_2-CH_3$ **b.** 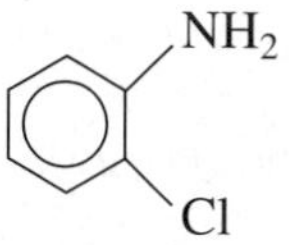 (benzene ring with NH_2 and Cl)

                          O $NH-CH_2-CH_3$

                          ||

    c. 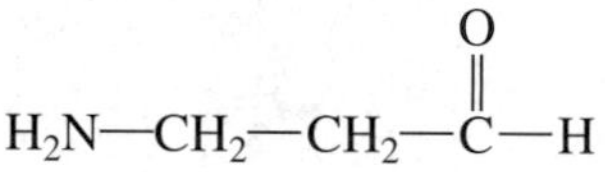 $H_2N-CH_2-CH_2-C-H$ **d.** 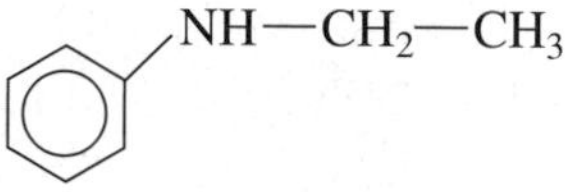 (benzene ring with $NH-CH_2-CH_3$)

19.8 Write the IUPAC names for each of the following:

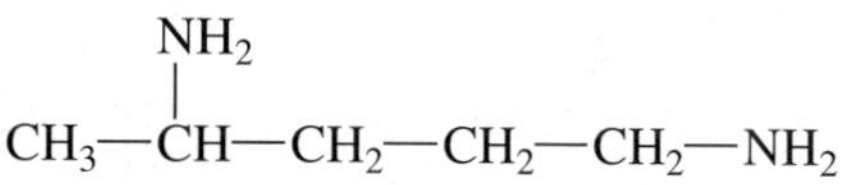

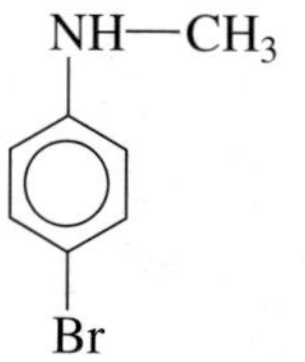

             O NH_2

             || |

    a. 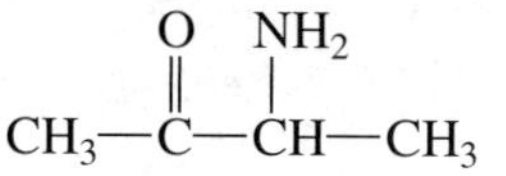 $CH_3-C-CH-CH_3$

                NH_2

                |

    b. $CH_3-CH-CH_2-CH_2-CH_2-NH_2$

        $NH-CH_3$ CH_3

    c. (benzene ring with $NH-CH_3$ and Br)   **d.** 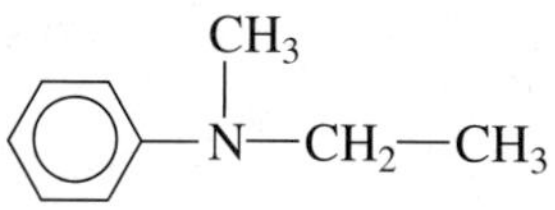 (benzene ring with $N-CH_2-CH_3$ and CH_3)

19.9 Draw the condensed structural formulas for each of the following amines:

    a. ethylamine
    b. *N*-methylaniline
    c. butylpropylamine
    d. 2-pentanamine

19.10 Draw the condensed structural formulas for each of the following amines:

    a. dimethylamine
    b. *p*-chloroaniline
    c. *N,N*-diethylaniline
    d. 1-amino-3-pentanone

19.3 Physical Properties of Amines

LEARNING GOAL

Describe the boiling points and solubility of amines.

Amines have higher boiling points than hydrocarbons of similar mass, but lower than the alcohols, as seen in Table 19.1.

Table 19.1 Comparison of Boiling Points (°C) of Amines, Alcohols, and Alkanes

NH_3	-33		
1 Carbon Atom		**3 Carbon Atoms**	
CH_4	-162	$CH_3-CH_2-CH_3$	-42
CH_3-NH_2	-7	$CH_3-N(CH_3)-CH_3$	3
CH_3-OH	65	$CH_3-CH_2-NH-CH_3$	36
2 Carbon Atoms		$CH_3-CH_2-CH_2-NH_2$	48
CH_3-CH_3	-89	$CH_3-CH_2-CH_2-OH$	97
$CH_3-NH-CH_3$	7		
$CH_3-CH_2-NH_2$	17		
CH_3-CH_2-OH	79		

Because amines contain a polar N—H bond, they form hydrogen bonds. (See Figure 19.2.) However, the nitrogen atom in amines is not as electronegative as the oxygen in alcohols, which makes the hydrogen bonds weaker in amines. The —NH_2 in primary amines can form more hydrogen bonds, which gives them higher boiling points than the secondary amines of the same mass. It is not possible for tertiary amines to hydrogen bond with each other (no N—H bonds), which makes their boiling points much lower and similar to those of alkanes and ethers.

$$CH_3-CH_2-CH_2-NH_2$$
Propylamine (1°)
bp 48°C

$$CH_3-CH_2-NH-CH_3$$
Ethylmethylamine (2°)
bp 36°C

$$CH_3-N(CH_3)-CH_3$$
Trimethylamine (3°)
bp 3°C

Solubility in Water

Like alcohols, the smaller amines, including tertiary ones, are soluble in water because they form hydrogen bonds with water. (See Figure 19.3.) However, in amines with more than six carbon atoms, the effect of hydrogen bonding is diminished, and their solubility in water decreases. Large amines are not soluble in water.

SAMPLE PROBLEM 19.3

Physical Properties of Amines

If the compounds trimethylamine and ethylmethylamine have the same molar mass, why is the boiling point of trimethylamine (3°C) lower than that of ethylmethylamine (37°C)?

Solution

With a polar N—H bond, hydrogen bonds form between ethylmethylamine molecules. Thus, a higher temperature is required to break the hydrogen bonds and form a gas. However, trimethylamine, which is a tertiary amine, has no N—H bond and cannot form hydrogen bonds between the amine molecules.

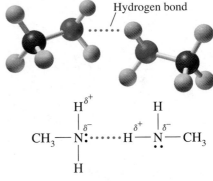

Figure 19.2 Hydrogen bonding occurs between 1° or 2° amines.
Q *Why don't tertiary (3°) amines form hydrogen bonds?*

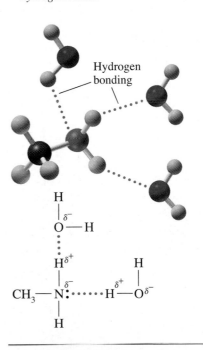

Figure 19.3 Hydrogen bonding occurs between amines and water molecules.
Q *Why are tertiary (3°) amines soluble in water?*

Study Check

Why is $CH_3CH_2NH_2$ more soluble in water than $CH_3CH_2CH_2CH_2NHCH_2CH_2CH_3$?

QUESTIONS AND PROBLEMS

Physical Properties of Amines

19.11 Indicate the compound in each pair that would have the higher boiling point:

a. $CH_3—CH_2—NH_2$ or $CH_3—CH_2—OH$ **b.** $CH_3—NH_2$ or $CH_3—CH_2—CH_2—NH_2$

c. $CH_3—\overset{\overset{\displaystyle CH_3}{|}}{N}—CH_3$ or $CH_3—CH_2—CH_2—NH_2$

19.12 Indicate the compound in each pair that would have the higher boiling point:

a. $CH_3—CH_2—CH_2—CH_3$ or $CH_3—CH_2—CH_2—NH_2$ **b.** $CH_3—NH_2$ or $CH_3—CH_2—NH_2$

c. $CH_3—CH_2—CH_2—OH$ or $CH_3—\overset{\overset{\displaystyle NH_2}{|}}{C}H—CH_3$

19.13 Propylamine (molar mass 59) has a boiling point of 48°C and ethylmethylamine (molar mass 59) has a boiling point of 37°C. Butane (molar mass 58) has a much lower boiling point −1°C. Explain.

19.14 Assign the boiling point of 3°C, 48°C, or 97°C to the appropriate compound: 1-propanol, propylamine, and trimethylamine.

19.15 Indicate if each of the following is soluble in water:
  a. $CH_3—CH_2—NH_2$

  b. $CH_3—NH—CH_3$

  c. $CH_3—CH_2—CH_2—\overset{\overset{\displaystyle CH_2—CH_2—CH_3}{|}}{N}—CH_2—CH_2—CH_3$

  d. $CH_3—\overset{\overset{\displaystyle NH_2}{|}}{C}H—CH_2—CH_3$

19.16 Indicate if each of the following is soluble in water:

  a. $CH_3—CH_2—CH_2—NH_2$ **b.** $CH_3—CH_2—CH_2—NH—CH_2—CH_3$

  c. $CH_3—\overset{\overset{\displaystyle CH_3}{|}}{N}—CH_3$ **d.**

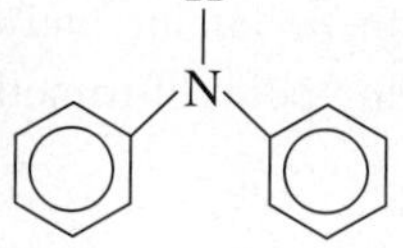

19.4 Amines React as Bases

LEARNING GOAL

Write equations for the ionization and neutralization of amines.

In Chapter 10, we saw that ammonia (NH_3) acts as a Brønsted–Lowry base because it accepts a proton (H^+) from water to produce an ammonium ion (NH_4^+) and a hydroxide ion (OH^-). Let's review that equation:

$$\ddot{N}H_3 + H_2O \rightleftharpoons NH_4^+ + OH^-$$

Ammonia — Ammonium ion — Hydroxide ion

Ionization of an Amine in Water

In water, amines also act as Brønsted–Lowry bases because the lone electron pair on the nitrogen atom accepts a proton. The products are an alkyl ammonium ion and hydroxide ion. The name of the alkyl ammonium ion is similar to the common amine name, but *amine* is replaced by *ammonium ion.*

$$CH_3{-}\ddot{N}H_2 + H_2O \rightleftharpoons CH_3{-}\overset{+}{N}H_3 + OH^-$$

Methylamine — Methylammonium ion — Hydroxide ion

Secondary amines also accept a proton to form dialkyl ammonium ions.

$$CH_3{-}\underset{\displaystyle CH_3}{\underset{|}{\ddot{N}H}} + H_2O \rightleftharpoons CH_3{-}\underset{\displaystyle CH_3}{\underset{|}{\overset{+}{N}H_2}} + OH^-$$

Dimethylamine — Dimethylammonium ion — Hydroxide ion

Basicity of Amines

Because amines act as weak bases by accepting protons from water and producing hydroxide ions, their aqueous solutions are basic. We can write the base dissociation constant K_b for methylamine as follows:

$$K_b = \frac{[CH_3{-}NH_3^+][OH^-]}{[CH_3{-}NH_2]} = 4.4 \times 10^{-4}$$

Most of the K_b values for amines are less than 10^{-3}, which means that the equilibrium favors the undissociated amine molecules. Aqueous solutions of amines have basic pH values and turn red litmus paper blue. We can compare the strengths of some amines by looking at their K_b values as follows:

	Ammonia	*1° amine*		*2° amine*	*3° amine*
	NH_3	$CH_3{-}NH_2$	$CH_3{-}CH_2{-}NH_2$	$CH_3{-}NH{-}CH_3$	$CH_3{-}\overset{\displaystyle CH_3}{\overset{\vert}{N}}{-}CH_3$
K_b	1.8×10^{-5}	4.4×10^{-4}	5.6×10^{-4}	5.1×10^{-4}	5.3×10^{-5}

Amine Salts

When you squeeze lemon juice on fish, the "fishy odor" of the amines is removed by converting them to amine salts. In a neutralization reaction, an amine acts as a base and reacts with an acid to form an **amine salt**. The lone pair of electrons on the nitrogen atom accepts a proton H^+ from an acid to give an amine salt; no water

is formed. An amine salt is named by replacing the *–amine* part of the amine name with *–ammonium* followed by the name of the negative ion.

Neutralization of an Amine

$$CH_3-\ddot{N}H_2 + HCl \rightleftharpoons CH_3-\overset{+}{N}H_3\,Cl^-$$

Amine　　　Acid　　　　　　　Amine salt

Methyl**amine**　　　　　Methyl**ammonium chloride**

$$CH_3-\underset{\underset{CH_3}{|}}{\ddot{N}H} + HCl \rightleftharpoons CH_3-\underset{\underset{CH_3}{|}}{\overset{+}{N}H_2}\,Cl^-$$

Dimethyl**amine**　　　　　Dimethyl**ammonium chloride**

The ammonium ions are classified as primary, secondary, and tertiary.

$$CH_3-\underset{\underset{H}{|}}{\overset{\overset{H}{|}}{\overset{+}{N}}}-H \qquad CH_3-\underset{\underset{H}{|}}{\overset{\overset{CH_3}{|}}{\overset{+}{N}}}-H \qquad CH_3-\underset{\underset{CH_3}{|}}{\overset{\overset{CH_3}{|}}{\overset{+}{N}}}-H$$

Primary　　　　　　Secondary　　　　　Tertiary

In a **quaternary ammonium ion**, a nitrogen atom bonds to four carbon groups. In the quaternary ion, the nitrogen atom has a positive charge just as it does in other amine salts. Choline, an amino alcohol present in glycerophospholipids is a quaternary ammonium ion.

$$CH_3-\underset{\underset{CH_3}{|}}{\overset{\overset{CH_3}{|}}{\overset{+}{N}}}-CH_3Cl^- \qquad HO-CH_2-CH_2-\underset{\underset{CH_3}{|}}{\overset{\overset{CH_3}{|}}{\overset{+}{N}}}-CH_3$$

Tetramethylammonium chloride　　　　　　　Choline

The quaternary salts differ from other amine salts because the nitrogen atom is not bonded to an H atom. Thus, quaternary salts do not react with bases.

Properties of Amine Salts

As ionic compounds, amine salts are solids at room temperature, odorless, and soluble in water and body fluids. For this reason, amines used as drugs are converted to their amine salts. The amine salt of ephedrine is used as a bronchodilator and in decongestant products such as Sudafed. The amine salt of diphenhydramine is used in products such as Benadryl for relief of itching and pain from skin irritations and rashes. (See Figure 19.4.) In pharmaceuticals, the naming of the amine salt follows an older method of giving the amine name followed by the name of the acid.

When an amine salt reacts with a strong base such as NaOH, it is converted back to the amine, which is also called the free amine or free base.

$$CH_3-NH_3^+\,Cl^- + NaOH \longrightarrow CH_3-NH_2 + NaCl + H_2O$$

The narcotic cocaine is typically extracted from coca leaves using an acidic HCl solution to give a white, solid amine salt, which is cocaine hydrochloride. This is the form in which cocaine is smuggled and sold illegally on the street to be snorted or injected. "Crack cocaine" is the free amine or free base of the amine obtained by treating the cocaine hydrochloride with NaOH and ether, a process known as "free-basing." The solid product is known as "crack cocaine" because it makes a crackling noise when heated. The free amine is rapidly absorbed when smoked and gives

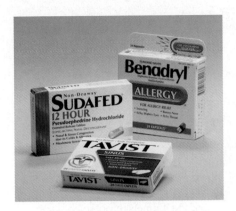

Figure 19.4 Decongestants and products that relieve itch and skin irritations can contain ammonium salts.
Q *Why are the ammoniuim salts used rather than the biologically active amines?*

Ephedrine hydrochloride
Ephedrine HCl
Sudafed

Diphenhydramine hydrochloride
Diphenhydramine HCl
Benadryl

stronger highs than the cocaine hydrochloride. Unfortunately, these effects of crack cocaine have caused a rise in addiction to cocaine.

+ NaOH ⟶

Cocaine hydrochloride

Cocaine ("free base")

SAMPLE PROBLEM 19.4

Reactions of Amines

Write an equation that shows ethylamine
a. ionizing as a weak base in water   **b.** neutralized by HCl

Solution

a. In water, ethylamine acts as a weak base by accepting a proton from water to produce ethylammonium hydroxide.

$$CH_3-CH_2-NH_2 + H-OH \rightleftharpoons CH_3-CH_2-NH_3^+ + OH^-$$

b. $CH_3-CH_2-NH_2 + HCl \longrightarrow CH_3-CH_2-NH_3^+Cl^-$

Study Check

What is the condensed structural formula of the salt formed by the reaction of trimethylamine and HCl?

QUESTIONS AND PROBLEMS

Amines React as Bases

19.17 Write an equation for the ionization of each of the following amines in water:

    a. methylamine   **b.** dimethylamine   **c.** aniline

19.18 Write an equation for the ionization of each of the following amines in water:

    a. ethylamine   **b.** propylamine   **c.** *N*-methylaniline

19.19 Write the condensed structural formula of the amine salt obtained when each of the amines in problem 19.17 reacts with HCl.

19.20 Write the condensed structural formula of the amine salt obtained when each of the amines in problem 19.18 reacts with HCl.

19.21 Novocain, a local anesthetic, is the hydrochloride salt of procaine.

$$H_2N-\bigcirc-\overset{\overset{\textstyle O}{\|}}{C}-O-CH_2-CH_2-\overset{\overset{\textstyle CH_2CH_3}{|}}{\underset{\underset{\textstyle CH_2CH_3}{|}}{N}}$$

Procaine

    a. What is the formula of the amine salt formed when procaine reacts with HCl?

    b. Why is Novocain, the amine salt, used rather than the amine procaine?

19.22 Lidocaine (xylocaine) is used as a local anesthetic and cardiac depressant.

$$\bigcirc\overset{\overset{\textstyle CH_3}{}}{\underset{\underset{\textstyle CH_3}{}}{}}-NH-\overset{\overset{\textstyle O}{\|}}{C}-CH_2-\overset{\overset{\textstyle CH_2-CH_3}{|}}{\underset{\underset{\textstyle CH_2-CH_3}{|}}{N}}$$

Lidocaine (xylocaine)

    a. What is the formula of the amine salt formed when lidocaine reacts with HCl?

    b. Why is the amine salt of lidocaine used rather than the amine?

19.5 Heterocyclic Amines and Alkaloids

A **heterocyclic amine** is a cyclic organic compound that contains one or more nitrogen atoms in the ring. The heterocyclic amine rings typically consist of 5 or 6 atoms and one or more nitrogen atoms. Of the five-atom rings, the simplest one is pyrrolidine, which is a ring of four carbon atoms and a nitrogen atom, all with single bonds. Pyrrole is a five-atom ring with one nitrogen atom and two double bonds. Imidazole is a five-atom ring that contains two nitrogen atoms.

Pyrrolidine   Pyrrole   Imidazole
(nicotine, etc.)   (hemoglobin)   (histidine)

Many of the six-atom heterocyclic amines are aromatic. Pyridine is similar to benzene except that it has a nitrogen atom in place of a carbon atom. Pyrimidine, which is found in nucleic acids, is also similar to benzene except that it has two nitrogen atoms. In purine, another component of nucleic acids, a pyrimidine ring is fused with imidazole.

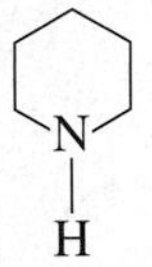

Piperidine
(quinine, drugs)

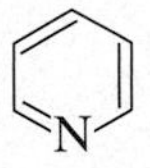

Pyridine
(nicotine, vitamins)

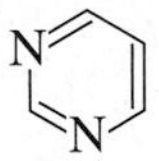

Pyrimidine
(nucleic acid: DNA, RNA)

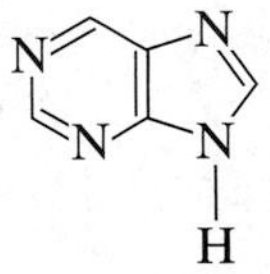

Purine
(nucleic acid: DNA, RNA)

SAMPLE PROBLEM 19.5

Identify each of the following heterocyclic amines:

a.   **b.**

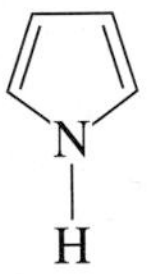

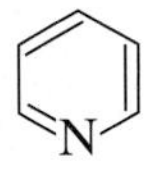

Solution

a. This five-atom ring with one nitrogen atom is pyrrole.
b. This aromatic ring with one nitrogen atom is pyridine.

Study Check

Identify the following heterocyclic amine.

Alkaloids: Amines in Plants

Alkaloids are physiologically active nitrogen-containing compounds produced by plants. The term *alkaloid* refers to the "alkali-like" or basic characteristics we have seen for amines. Certain alkaloids are used in anesthetics, in antidepressants, and as stimulants, although many are habit forming.

As a stimulant, nicotine increases the level of adrenaline in the blood, which increases the heart rate and blood pressure. It is well known that smoking cigarettes can damage the lungs and that exposure to tars and other carcinogens in cigarette smoke can lead to lung cancer. However, nicotine is responsible for the addiction of smoking. Nicotine has a simple alkaloid structure that includes a pyrrolidine ring. Coniine, which is obtained from hemlock, is an extremely toxic alkaloid that contains a piperidine ring.

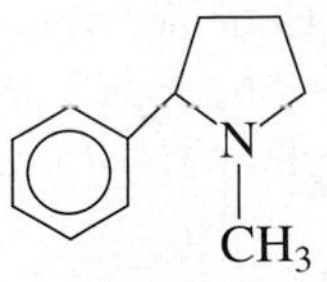

Nicotine

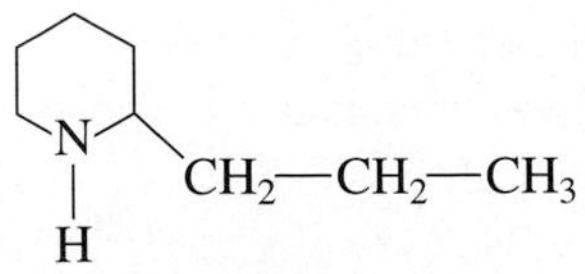

Coniine

Figure 19.5 Coffee beans contain caffeine, which is an alkaloid that is a stimulant of the central nervous system.
Q *Why is caffeine considered an alkaloid?*

Caffeine

Caffeine contains an imidazole ring and is a stimulant of the central nervous system (CNS). Present in coffee, tea, soft drinks, chocolate, and cocoa, caffeine increases alertness, but may cause nervousness and insomnia. Caffeine is also used in certain pain relievers to counteract the drowsiness caused by an antihistamine. (See Figure 19.5.)

Several alkaloids are used in medicine. Quinine obtained from the bark of the cinchona tree has been used in the treatment of malaria since the 1600s. Atropine from belladonna is used in low concentrations to accelerate slow heart rates and as an anesthetic for eye examinations.

Quinine

Atropine

For many centuries morphine and codeine, alkaloids found in the oriental poppy plant, have been used as effective painkillers. (See Figure 19.6.) Codeine, which is structurally similar to morphine, is used in some prescription painkillers and cough syrups. Heroin, obtained by a chemical modification of morphine, is strongly addicting and is not used medically.

Heroin

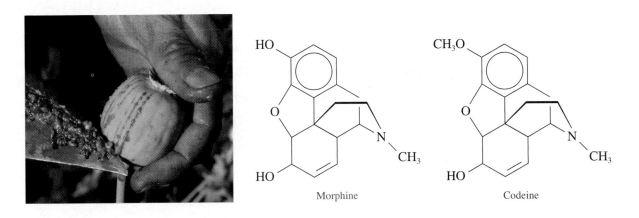

Morphine

Codeine

Figure 19.6 The green, unripe poppy seed capsule contains a milky sap (opium) that is the source of the alkaloids morphine and codeine.

Q *Where is the piperidine ring in the structures of morphine and codeine?*

HEALTH NOTE

Synthesizing Drugs

One area of research in pharmacology is the synthesis of compounds that retain the anesthetic characteristic of naturally occurring alkaloids such as cocaine and morphine without the addictive side effects. For example, cocaine is an effective anesthetic, but addictive. Research chemists modified the structure of cocaine, but kept the benzene group and nitrogen atom. The synthetic products procaine and lidocaine retain the anesthetic qualities of the natural alkaloid without the addictive side effects.

The structure of morphine was also modified to make a synthetic alkaloid, meperidine, or Demerol, which acts as an effective painkiller.

Meperidine (Demerol)

Cocaine

Procaine (Novocaine)

Lidocaine (Xylocaine)

SAMPLE PROBLEM 19.6

Heterocyclic Amines

Identify the heterocyclic amines in the alkaloids nicotine and caffeine.

Solution

In nicotine, the heterocyclic amine is the five-atom ring of pyrrolidine. Caffeine contains a purine, which is pyrimidine and imidazole fused together.

Study Check

What is the heterocyclic amine in meperidine (Demerol)?

QUESTIONS AND PROBLEMS

Heterocyclic Amines

19.23 Identify the following as amines or heterocyclic amines:

a. 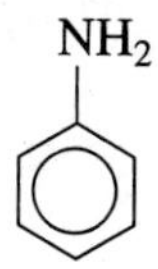 NH_2 b. $CH_3—CH_2—\overset{\overset{\displaystyle CH_3}{|}}{N}—CH_3$ c.  [pyrimidine ring structure]   d. 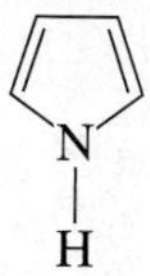 [pyrrole ring structure with N—H]

19.24 Identify the following as amines or heterocyclic amines:

a. 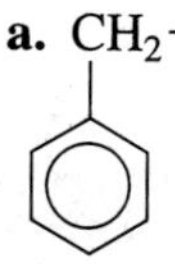 $CH_2—NH_2$ b. 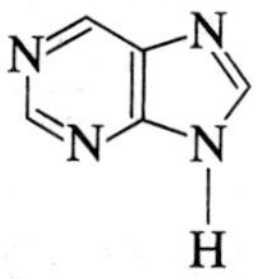 [purine ring structure with N—H]   c. 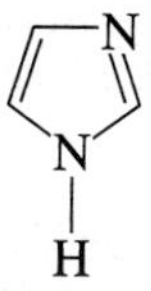 [imidazole ring structure with N—H]   d. 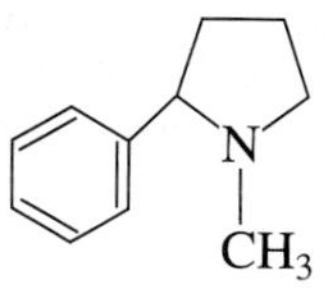 [2-phenyl pyrrolidine with N—CH_3]

19.25 Identify the type of heterocyclic amines in problem 19.23.
19.26 Identify the type of heterocyclic amines in problem 19.24.
19.27 Low levels of serotonin in the brain appear to be associated with depressed states. What type of heterocyclic amine is serotonin?

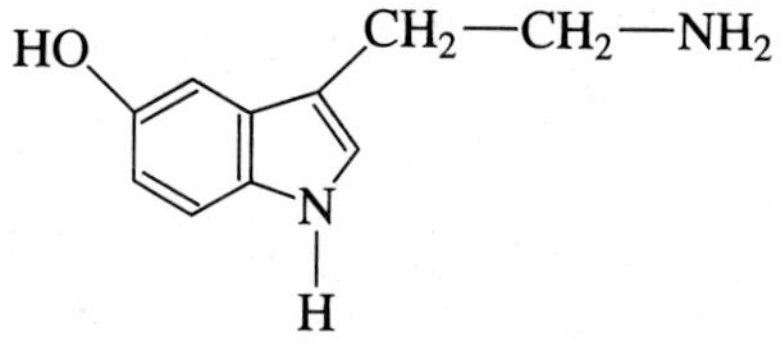

[structure of serotonin with HO, CH_2—CH_2—NH_2, and N—H]

Serotonin

19.28 LSD is made from lysergic acid, which is produced by a fungus that grows on rye. What types of heterocyclic amines are in lysergic acid?

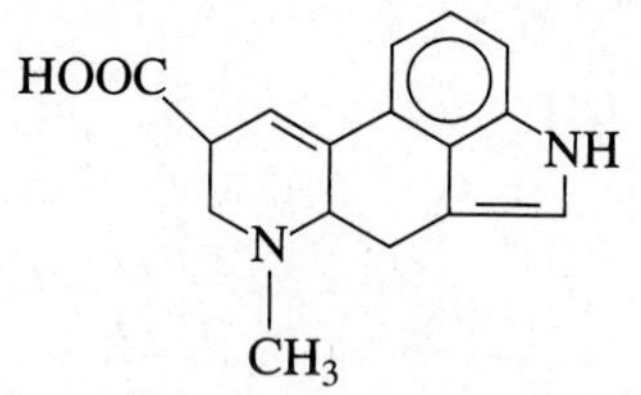

[structure of lysergic acid with HOOC, NH, N—CH_3]

Lysergic acid

19.6 Structures and Names of Amides

Write the amide products of amidation and give their common and IUPAC names.

The **amides** are derivatives of carboxylic acids in which an amino group replaces the hydroxyl group. (See Figure 19.7.) The amides are classified by the number of carbon atoms bonded to the nitrogen atom.

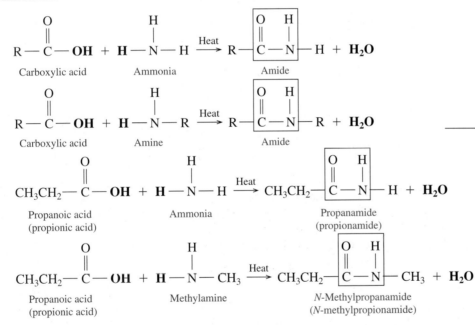

Preparation of Amides

An amide is produced in a reaction called **amidation,** in which a carboxylic acid reacts with ammonia or a primary or secondary amine. A molecule of water is eliminated, and the fragments of the carboxylic acid and amine molecules join to form the amide, much like the formation of ester. Because a hydrogen atom must be lost from the amines, only primary and secondary amines undergo amidation.

Amidation

Carboxylic acid

Ethanoic acid
(Acetic acid)

Amide

Ethanamide
(Acetamide)

Figure 19.7 Amides are derivatives of carboxylic acids in which an amino group ($-NH_2$, $-NHR$, or $-NR_2$) replaces the hydroxyl group ($-OH$).

Q *Why are amides with one to five carbon atoms soluble in water?*

SAMPLE PROBLEM 19.7

Amidation Products

Give the structural formula of the amide product in each of the following reactions:

a.

[benzoic acid structure] $-OH + NH_3 \xrightarrow{\text{Heat}}$

b.

$CH_3\overset{O}{\overset{\|}{C}}OH + NH_2CH_2CH_3 \xrightarrow{\text{Heat}}$

Solution

a. The structural formula of the amide product can be written by attaching the carbonyl group from the acid to the nitrogen atom of the amine. —OH is removed from the acid and —H from the amine to form water.

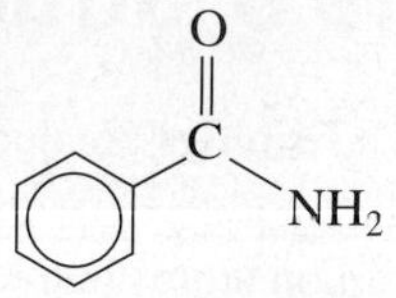

b.

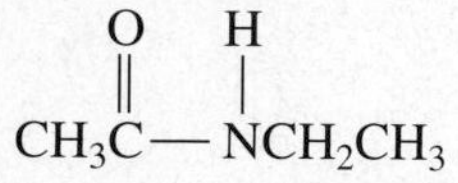

$$CH_3\overset{\overset{\displaystyle O}{\|}}{C}-\overset{\overset{\displaystyle H}{|}}{N}CH_2CH_3$$

Study Check

What are the condensed structural formulas of the carboxylic acid and amine needed to prepare the following amide?

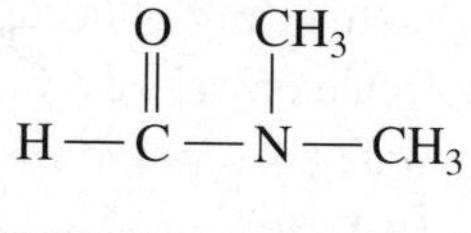

$$H-\overset{\overset{\displaystyle O}{\|}}{C}-\overset{\overset{\displaystyle CH_3}{|}}{N}-CH_3$$

Naming Amides

In both the common and IUPAC names, amides are named by dropping the *ic acid* or *oic acid* from the carboxylic acid names (IUPAC or common) and adding the suffix *amide*.

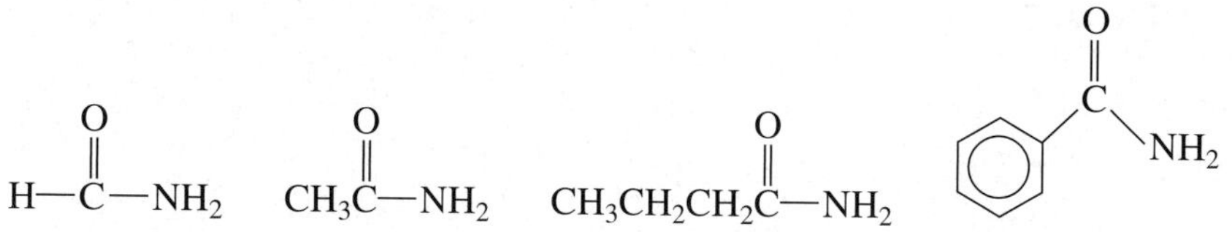

Methanamide (formamide)   Ethanamide (acetamide)   Butanamide (butyramide)   Benzamide

When alkyl groups are attached to the nitrogen atom, the name of an amide is preceded by *N-* or *N,N-* depending on whether there are one or two groups.

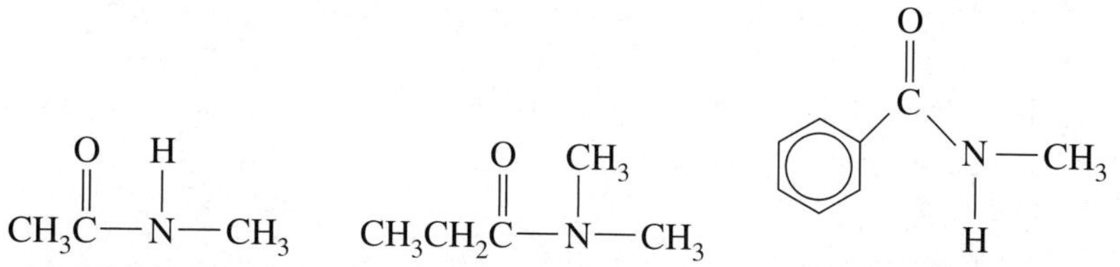

N-Methylethanamide (*N*-methylacetamide)   *N,N*-Dimethylpropanamide (*N,N*-dimethylpropionamide)   *N*-Methylbenzamide

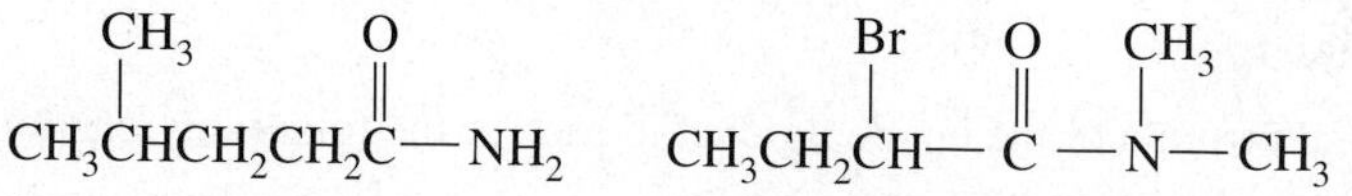

4-Methylpentanamide   *N,N*-Dimethyl-2-bromobutanamide (*N,N*-dimethyl-α-bromobutyramide)

SAMPLE PROBLEM 19.8

Naming Amides

Give the common and IUPAC names for each of the following amides:

a. 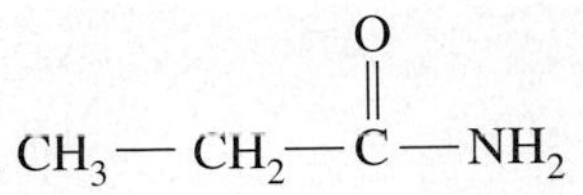 $CH_3—CH_2—\overset{\overset{\displaystyle O}{\|}}{C}—NH_2$ **b.** 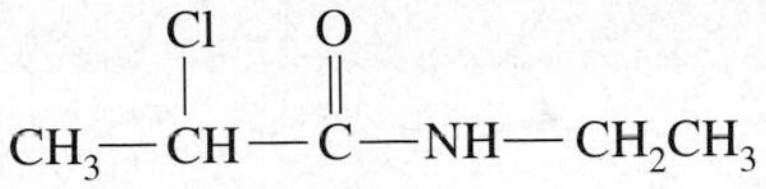 $CH_3—\overset{\overset{\displaystyle Cl}{|}}{CH}—\overset{\overset{\displaystyle O}{\|}}{C}—NH—CH_2CH_3$

Solution

a. The IUPAC name of the carboxylic acid is propanoic acid; the common name is propionic acid. Replacing the *oic acid* or *ic acid* ending with *amide* gives the IUPAC name of *propanamide* and common name of *propionamide*.

b. Using the same carboxylic acid as in part a, the ethyl group attached to the nitrogen atom is named *N-ethyl*. The amide is named *N-ethyl-2-chloro-propanamide* (IUPAC), and *N-ethyl-α-chloropropionamide* (common).

Study Check

Draw the condensed structural formula of *N,N*-dimethyl-*p*-chlorobenzamide.

Physical Properties of Amides

The amides do not have the properties of bases that we saw for the amines. Only foramide is a liquid at room temperature, while the other amides are solids. For primary amides, the —NH_2 group can form several hydrogen bonds, which gives primary amides high melting points. The melting points of the secondary amides are lower because the number of hydrogen bonds decreases. Tertiary amides have even lower melting points because they cannot form hydrogen bonds with other tertiary amides.

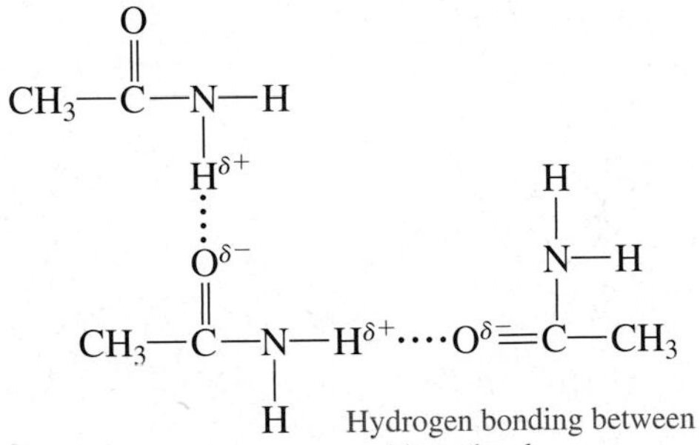

Hydrogen bonding between amide molecules

The amides with one to five carbon atoms are soluble in water because they can hydrogen bond with water molecules.

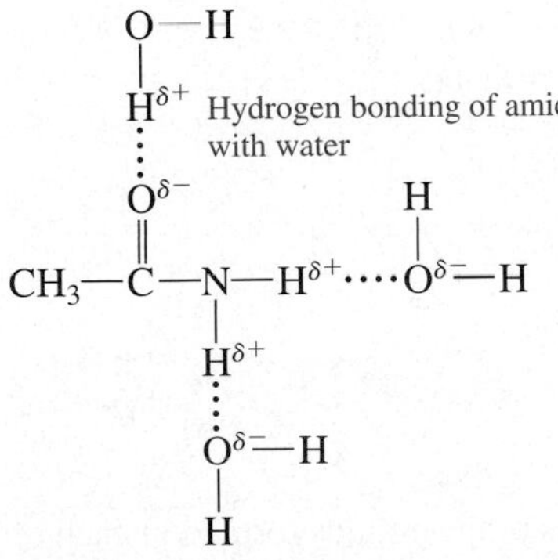

Hydrogen bonding of amides with water

Amides in Health and Medicine

The simplest natural amide is urea, an end product of protein metabolism in the body. The kidneys remove urea from the blood and provide for its excretion in urine. If the kidneys malfunction, urea is not removed and builds to a toxic level, a condition called uremia. Urea is also used as a component of fertilizer, to increase nitrogen in the soil.

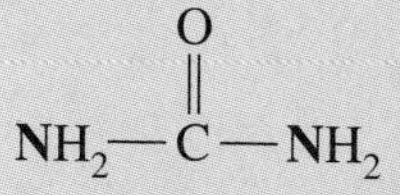

$$NH_2-\overset{\overset{\displaystyle O}{\|}}{C}-NH_2 \quad \text{Urea}$$

Synthetic amides are used as substitutes for sugar and aspirin. Saccharin is a very powerful sweetener and is used as a sugar substitute. The sweetener Aspartame is made from two amino acids, aspartic acid and phenylalanine.

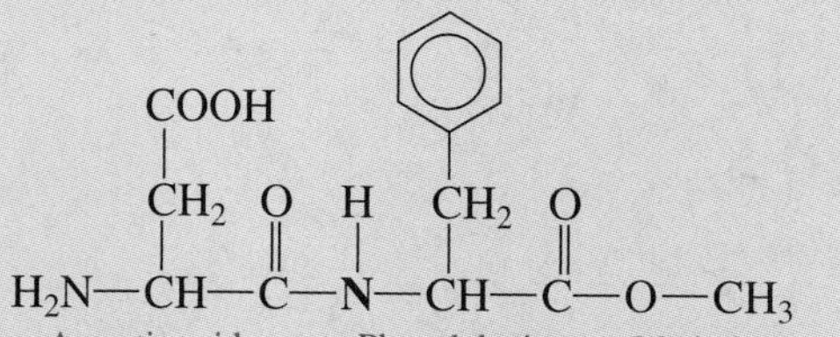

Aspirin substitutes contain phenacetin or acetaminophen, which is used in Tylenol. Like aspirin, acetaminophen reduces fever and pain, but it has little anti-inflammatory effect.

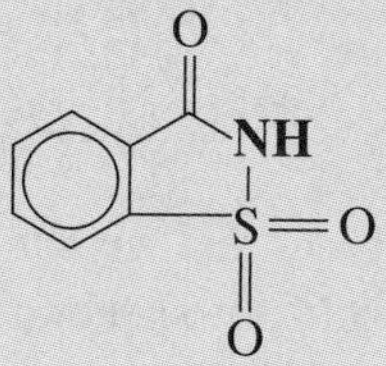

Aspartic acid   Phenylalanine   Methyl ester

Aspartame

Saccharin 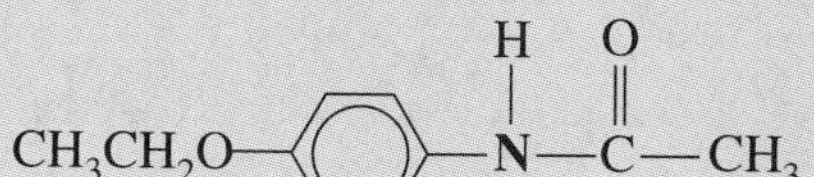   Phenacetin

Many barbiturates are cyclic amides of barbituric acid that act as sedatives in small dosages or sleep inducers in larger dosages. They are often habit forming. Barbiturate drugs include phenobarbital (Luminal), pentobarbital (Nembutal), and secobarbital (Seconal).

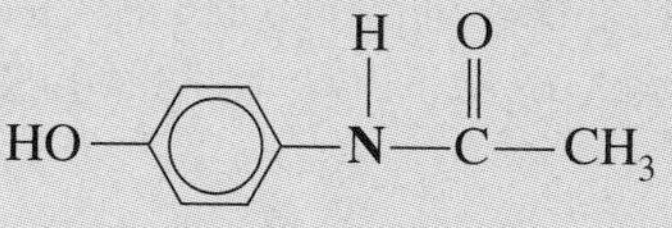

Acetaminophen

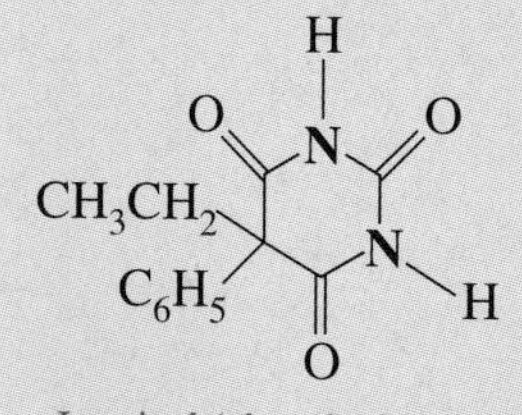

Luminal (phenobarbital) 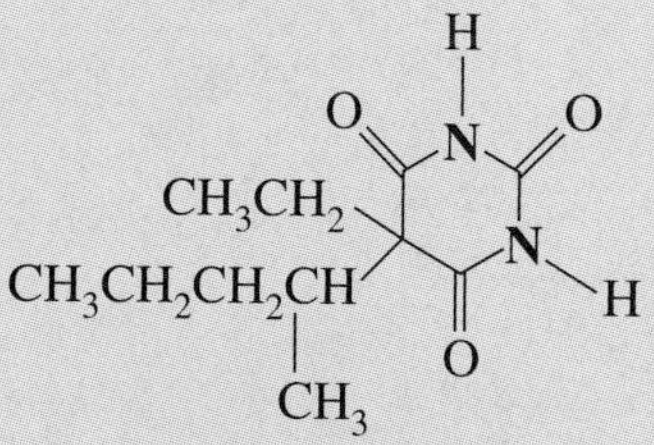   Nembutal (pentobarbital) 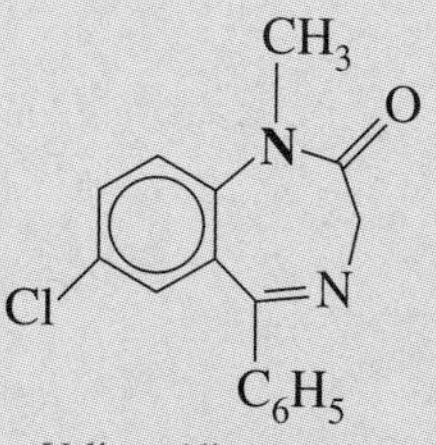   Valium (diazepam)

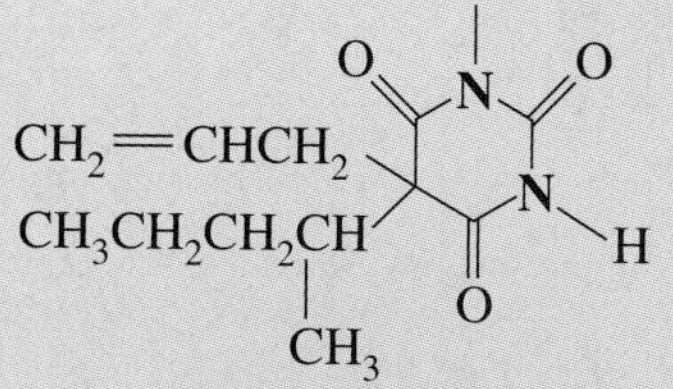

Seconal (secobarbital) 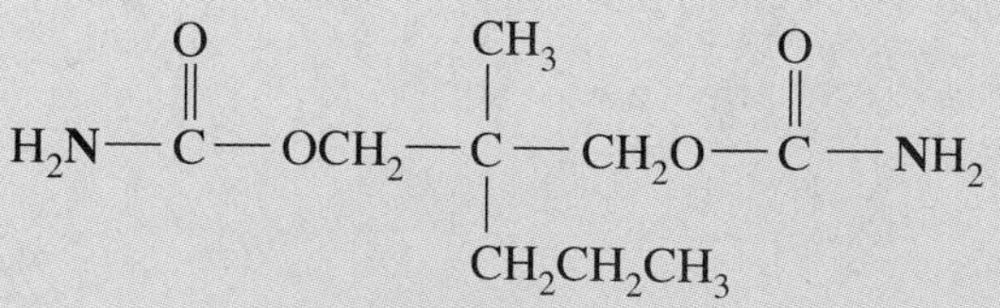   Equanil (meprobamate)

QUESTIONS AND PROBLEMS

Structure and Names of Amides

19.29 Draw the condensed structural formulas of the amides formed in each of the following reactions:

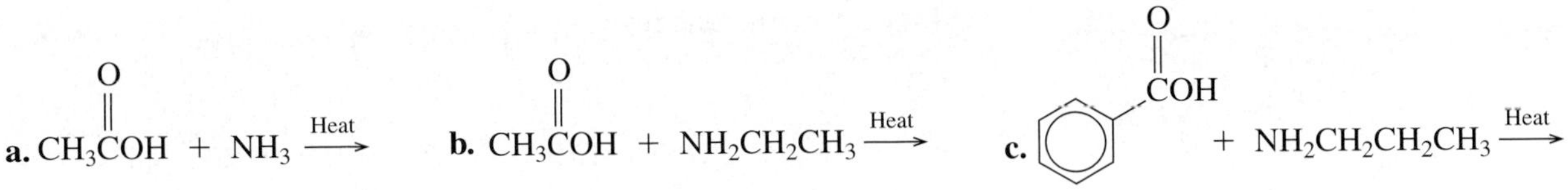

a. $CH_3\overset{\overset{\displaystyle O}{\|}}{C}OH$ + NH_3 $\xrightarrow{\text{Heat}}$ b. $CH_3\overset{\overset{\displaystyle O}{\|}}{C}OH$ + $NH_2CH_2CH_3$ $\xrightarrow{\text{Heat}}$ c. [benzene ring]$-\overset{\overset{\displaystyle O}{\|}}{C}OH$ + $NH_2CH_2CH_2CH_3$ $\xrightarrow{\text{Heat}}$

19.30 Draw the condensed structural formulas of the amides formed in each of the following reactions:

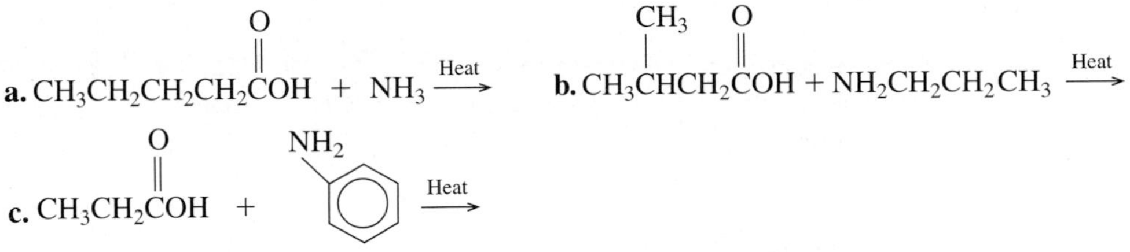

a. $CH_3CH_2CH_2CH_2\overset{\overset{\displaystyle O}{\|}}{C}OH$ + NH_3 $\xrightarrow{\text{Heat}}$ b. $CH_3\overset{\overset{\displaystyle CH_3}{|}}{C}HCH_2\overset{\overset{\displaystyle O}{\|}}{C}OH$ + $NH_2CH_2CH_2CH_3$ $\xrightarrow{\text{Heat}}$

c. $CH_3CH_2\overset{\overset{\displaystyle O}{\|}}{C}OH$ + [benzene ring with NH_2] $\xrightarrow{\text{Heat}}$

19.31 Give the IUPAC and common names (if any) for each of the following amides:

a. $CH_3-\overset{\overset{\displaystyle O}{\|}}{C}-NH-CH_3$ b. $CH_3CH_2CH_2-\overset{\overset{\displaystyle O}{\|}}{C}-NH_2$ c. $H-\overset{\overset{\displaystyle O}{\|}}{C}-NH_2$ d. [benzene ring]$-\overset{\overset{\displaystyle O}{\|}}{C}-\overset{\overset{\displaystyle H}{|}}{N}-CH_3$

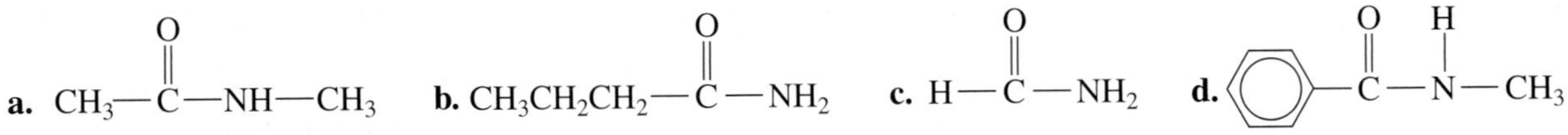

19.32 Give the IUPAC and common names (if any) for each of the following amides:

a. $CH_3CH_2\overset{\overset{\displaystyle O}{\|}}{C}-\overset{\overset{\displaystyle H}{|}}{N}-CH_2CH_3$ b. $CH_3CH_2CH_2CH_2CH_2\overset{\overset{\displaystyle O}{\|}}{C}-NH_2$

c. $CH_3\overset{\overset{\displaystyle O}{\|}}{C}-\overset{\overset{\displaystyle CH_3}{|}}{N}-CH_2CH_2CH_3$ d. [benzene ring]$-\overset{\overset{\displaystyle O}{\|}}{C}-\overset{\overset{\displaystyle CH_2CH_3}{|}}{N}-CH_2CH_3$

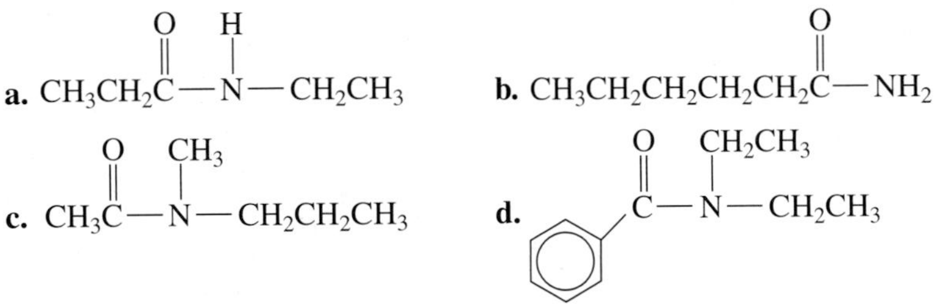

19.33 Draw the condensed structural formulas for each of the following amides:
   a. propionamide   **b.** 2-methylpentanamide   **c.** methanamide
   d. N-ethylbenzamide   **e.** N-ethylbutyramide

19.34 Draw the condensed structural formulas for each of the following amides:
   a. formamide   **b.** N,N-dimethylbenzamide
   c. 3-methylbutyramide   **d.** 2,2-dichlorohexanamide
   e. N-propyl-3-chloropentanamide

19.35 In each pair, identify the compound that has the higher melting point:
   a. acetamide or N-methylacetamide
   b. butane or propionamide
   c. N,N-dimethylpropanamide or N-methylpropanamide

19.36 In each pair, identify the compound that has the higher melting point:
   a. propane or acetamide
   b. N-methylacetamide or propanamide
   c. N,N-dimethylpropanamide or N-methylpropanamide

19.7 Hydrolysis of Amides

LEARNING GOAL

Write equations for the hydrolysis of amides.

As we have seen, amide bonds are formed by the elimination of water. The reverse reaction called **hydrolysis** occurs when water is added back to the amide bond to split the molecule. When an acid is used, the hydrolysis products of an amide are the carboxylic acid and the ammonium salt. In base hydrolysis, the amide produces the salt of the carboxylic acid and ammonia or amine.

Acid Hydrolysis of Amides

$$
\underset{\text{Amide}}{RC\!-\!NH_2} \; + \; HOH \; + \; HCl \;\longrightarrow\; \underset{\substack{\text{Carboxylic}\\\text{acid}}}{RC\!-\!OH} \; + \; \underset{\substack{\text{Ammonium chloride,}\\\text{a salt}}}{NH_4^+Cl^-}
$$

Example:

$$
\underset{\substack{\text{Ethanamide}\\\text{(acetamide)}}}{CH_3C\!-\!NH_2} \; + \; HOH \; + \; HCl \;\longrightarrow\; \underset{\substack{\text{Ethanoic acid}\\\text{(acetic acid)}}}{CH_3C\!-\!OH} \; + \; \underset{\substack{\text{Ammonium}\\\text{chloride}}}{NH_4^+Cl^-}
$$

Base Hydrolysis of Amides

$$
\underset{\text{Amide}}{R\!-\!C\!-\!NH\!-\!R} \; + \; NaOH \;\longrightarrow\; \underset{\substack{\text{Sodium carboxylate,}\\\text{a salt}}}{R\!-\!C\!-\!O^-Na^+} \; + \; \underset{\text{Alkyl amine}}{R\!-\!NH_2}
$$

Example:

$$
\underset{\substack{N\text{-Methylpropanamide}\\(N\text{-methylpropionamide})}}{CH_3\!-\!CH_2\!-\!C\!-\!NH\!-\!CH_3} \; + \; NaOH \;\longrightarrow\; \underset{\substack{\text{Sodium propanoate, a salt}\\\text{(sodium propionate)}}}{CH_3\!-\!CH_2\!-\!C\!-\!O^-Na^+} \; + \; \underset{\substack{\text{Methanamine}\\\text{(methylamine)}}}{NH_2CH_3}
$$

SAMPLE PROBLEM 19.9

Hydrolysis of Amides

Write the structural formulas for the products for the hydrolysis of *N*-methylpentanamide with NaOH.

Solution

Hydrolysis of the amide with a base produces a carboxylate salt (sodium pentanoate) and the corresponding amine (methylamine).

$$
\underset{}{CH_3CH_2CH_2CH_2CO^-Na^+} \; + \; NH_2CH_3
$$

Study Check

What are the structures of the products from the hydrolysis of *N*-methylbutyramide with HBr?

QUESTIONS AND PROBLEMS

Hydrolysis of Amides

19.37 Write the condensed structural formulas for the products of the acid hydrolysis of each of the following amides with HCl:

a. $CH_3\overset{\displaystyle O}{\overset{\displaystyle \|}{C}}-NH_2$

b. $CH_3CH_2\overset{\displaystyle O}{\overset{\displaystyle \|}{C}}-NH_2$

c. $CH_3CH_2CH_2\overset{\displaystyle O}{\overset{\displaystyle \|}{C}}-NH-CH_3$

d. 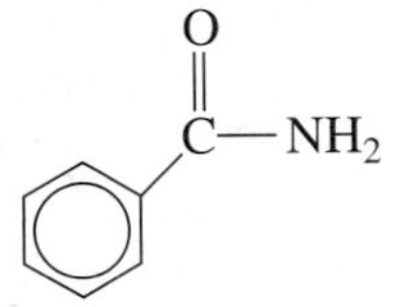 ⬡$-\overset{\displaystyle O}{\overset{\displaystyle \|}{C}}-NH_2$

e. *N*-ethylpentanamide

19.38 Write the condensed structural formulas for the products of the base hydrolysis of each of the following amides with NaOH:

a. $CH_3CH_2\overset{\displaystyle CH_3}{\overset{\displaystyle |}{C}}H-\overset{\displaystyle O}{\overset{\displaystyle \|}{C}}-NH_2$

b. $CH_3CH_2CH_2\overset{\displaystyle O}{\overset{\displaystyle \|}{C}}-\overset{\displaystyle CH_2CH_3}{\overset{\displaystyle |}{N}}-CH_2CH_3$

c. 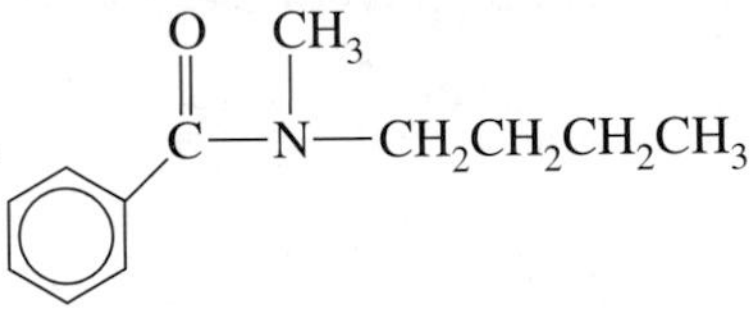 ⬡$-\overset{\displaystyle O}{\overset{\displaystyle \|}{C}}-\overset{\displaystyle CH_3}{\overset{\displaystyle |}{N}}-CH_2CH_2CH_2CH_3$

d. $CH_3\overset{\displaystyle Cl}{\overset{\displaystyle |}{C}}H-\overset{\displaystyle O}{\overset{\displaystyle \|}{C}}-\overset{\displaystyle CH_3}{\overset{\displaystyle |}{N}}-CH_2CH_3$

e. *N*-propyl benzamide

Chapter Review

19.1 Amines
A nitrogen atom attached to one, two, or three alkyl or aromatic groups forms a primary, secondary, or tertiary amine. Many amines, synthetic or naturally occurring, have physiological activity.

19.2 Naming Amines
In the common names of simple amines, the alkyl groups are listed alphabetically followed by the suffix –*amine*. In the IUPAC system, the –*amine* suffix is added to the alkane name of the longer carbon chain. Groups attached to the nitrogen atom use a *N*- prefix. When other functional groups are present, the —NH_2 is named as an amino group.

19.3 Physical Properties of Amines
Primary and secondary amines form hydrogen bonds, which make their boiling points higher than alkanes of similar mass, but lower than alcohols. Amines with up to six carbon atoms are soluble in water.

19.4 Amines React as Bases
In water, amines act as weak bases because the nitrogen atom accepts protons from water to produce ammonium and hydroxide ions. When amines react with acids, they form amine salts, which are named as ammonium salts. As ionic compounds, amine salts are solids, soluble in water, and odorless compared to the amines. Quaternary ammonium salts contain four carbon groups bonded to the nitrogen atom.

19.5 Heterocyclic Amines and Alkaloids
Heterocyclic amines are cyclic organic compounds that contain one or more nitrogen atoms in the ring. The amine rings typically consist of 5 or 6 atoms and one or more nitrogen atoms. Alkaloids such as caffeine and nicotine are naturally occurring amines derived from plants. Many are known for their physiological activity.

19.6 Structure and Names of Amides
Amides are derivatives of carboxylic acids in which the hydroxyl group is replaced by —NH_2, a primary, or secondary amine group. Amides are formed when carboxylic acids react with ammonia or primary or secondary amines in the presence of heat. Amides are named by replacing the –*ic acid* or –*oic acid* with –*amide*. Any carbon group attached to the nitrogen atom is named using the *N*- prefix.

19.7 Hydrolysis of Amides
Hydrolysis of an amide by an acid produces an amine salt. Hydrolysis by a base produces the salt of the carboxylic acid.

Summary of Naming

Family	Condensed Structural Formula	IUPAC name	Common Name
Amine	$CH_3CH_2-NH_2$	Ethanamine	Ethylamine
Amine salt	$CH_3CH_2-NH_3^+Cl^-$	Ethylammonium chloride	Ethylammonium chloride
Amide	$CH_3\overset{\overset{O}{\|\|}}{C}-NH_2$	Ethanamide	Acetamide

Summary of Reactions

Ionization of Amines in Water

$$R-NH_2 + HOH \rightleftharpoons R-NH_3^+ + OH^-$$

Amine — Alkyl ammonium hydroxide

$$CH_3-\overset{\overset{H}{\|}}{\underset{\underset{H}{\|}}{N}} + HOH \rightleftharpoons CH_3-\overset{\overset{H}{\|}}{\underset{\underset{H}{\|}}{\overset{+}{N}}}-H + OH^-$$

Methylamine — Methylammonium hydroxide

Formation of Amine Salts

$$R-NH_2 + HCl \longrightarrow R-NH_3^+ Cl^-$$

Amine — Ammonium salt

$$CH_3-\overset{\overset{H}{\|}}{\underset{\underset{H}{\|}}{N}} + HCl \longrightarrow CH_3-\overset{\overset{H}{\|}}{\underset{\underset{H}{\|}}{\overset{+}{N}}}-H \; Cl^-$$

Methylamine — Methylammonium chloride

Formation of Amides

$$R-\overset{\overset{O}{\|\|}}{C}-OH + H-\overset{\overset{H}{\|}}{N}-H \xrightarrow{\text{Heat}} R-\overset{\overset{O}{\|\|}}{C}-\overset{\overset{H}{\|}}{N}-H + H_2O$$

Carboxylic acid — Ammonia — Amide

$$R-\overset{\overset{O}{\|\|}}{C}-OH + H-\overset{\overset{H}{\|}}{N}-R \xrightarrow{\text{Heat}} R-\overset{\overset{O}{\|\|}}{C}-\overset{\overset{H}{\|}}{N}-R + H_2O$$

Carboxylic acid — Amine — Amide

$$CH_3CH_2-\overset{\overset{O}{\|\|}}{C}-OH + H-\overset{\overset{H}{\|}}{N}-H \xrightarrow{\text{Heat}} CH_3CH_2-\overset{\overset{O}{\|\|}}{C}-\overset{\overset{H}{\|}}{N}-H + H_2O$$

Propanoic acid (propionic acid) — Ammonia — Propanamide (propionamide)

$$CH_3CH_2-\overset{\overset{O}{\|\|}}{C}-OH + H-\overset{\overset{H}{\|}}{N}-CH_3 \xrightarrow{\text{Heat}} CH_3CH_2-\overset{\overset{O}{\|\|}}{C}-\overset{\overset{H}{\|}}{N}-CH_3 + H_2O$$

Propanoic acid (propionic acid) — Methanamine (methylamine) — N-Methylpropanamide (N-methylpropionamide)

Hydrolysis of Amides

$$\underset{\text{Amide}}{RC\overset{\displaystyle O}{\overset{\|}{-}}NH_2} + H_2O + HCl \longrightarrow \underset{\substack{\text{Carboxylic} \\ \text{acid}}}{RC\overset{\displaystyle O}{\overset{\|}{-}}OH} + \underset{\substack{\text{Ammonium} \\ \text{chloride}}}{NH_4{}^+Cl^-}$$

$$\underset{\substack{\text{Ethanamide} \\ \text{(acetamide)}}}{CH_3-\overset{\displaystyle O}{\overset{\|}{C}}-NH_2} + HOH + HCl \longrightarrow \underset{\substack{\text{Ethanoic acid} \\ \text{(acetic acid)}}}{CH_3-\overset{\displaystyle O}{\overset{\|}{C}}-OH} + \underset{\substack{\text{Ammonium} \\ \text{chloride}}}{NH_4{}^+Cl^-}$$

$$\underset{\substack{\textit{N}\text{-Methylpropanamide} \\ (\textit{N}\text{-methylpropionamide})}}{CH_3-CH_2-\overset{\displaystyle O}{\overset{\|}{C}}-NH-CH_3} + NaOH \longrightarrow \underset{\substack{\text{Sodium propanoate} \\ \text{(sodium propionate)}}}{CH_3-CH_2-\overset{\displaystyle O}{\overset{\|}{C}}-O^-Na^+} + \underset{\substack{\text{Methanamine} \\ \text{(methylamine)}}}{NH_2CH_3}$$

Key Terms

alkaloids Amines having physiological activity that are produced in plants.

amidation The formation of an amide from a carboxylic acid and ammonia or an amine.

amides Organic compounds containing the carbonyl group attached to an amino group or a substituted nitrogen atom:

$$\underset{}{R-\overset{\displaystyle O}{\overset{\|}{C}}-NH_2} \qquad R-\overset{\displaystyle O}{\overset{\|}{C}}-\overset{\displaystyle H}{\overset{|}{N}}-R$$

amines Organic compounds containing a nitrogen atom attached to one, two, or three hydrocarbon groups.

amine salt An ionic compound produced from an amine and an acid.

heterocyclic amine A cyclic organic compound that contains one or more nitrogen atoms in the ring.

hydrolysis The splitting of a molecule by the addition of water. Amides yield the corresponding carboxylic acid and amine or their salts.

quaternary ammonium ion An amine ion in which the nitrogen atom is bonded to four carbon groups.

Additional Problems

19.39 There are four amine isomers with the molecular formula C_3H_9N. Draw their condensed structural formulas. Name and classify each as a primary, secondary, or tertiary amine.

19.40 Name and classify each of the following compounds:

a. $CH_3-\overset{\displaystyle CH_2-CH_3}{\overset{|}{N}}-CH_2-CH_3$

b. $CH_3-CH_2-CH_2-CH_2-NH_2$

c. $CH_3-CH_2-CH_2-NH-CH_2-CH_3$

d.

e. $NHCH_3$

f. $CH_3-\overset{\displaystyle CH_3}{\overset{|}{CH}}-CH_2-\overset{\displaystyle CH_3}{\overset{|}{N}}-CH_2-CH_3$

g. $CH_3-\overset{\displaystyle CH_2-CH_3}{\overset{|+}{\underset{|}{\underset{\displaystyle CH_3}{N}}}}-CH_2-CH_3 \quad Cl^-$

19.41 Draw the structure of each of the following compounds:

a. 3-pentanamine   **b.** cyclohexylamine

c. dimethylammonium chloride

d. triethylamine

e. 3-amino-2-hexanol

f. tetramethylammonium bromide

g. *N,N*-dimethylaniline

19.42 In each pair, indicate the compound that has the higher boiling point:

a. 1-butanol or butanamine

b. trimethylamine or propylamine

c. butylamine or diethylamine

d. butane or propylamine

19.43 In each pair, indicate the compound that is more soluble in water:

a. ethylamine or butylamine

b. trimethylamine or *N*-ethylcyclohexylamine

c. butylamine or pentane

d. NH_2—CH_2—CH_2—CH_2—CH_2—CH_2—NH_2

or

CH_3—CH_2—CH_2—CH_2—CH_2—NH_2

19.44 Give the IUPAC name for each of the following amides:

a.
$$H-\overset{\overset{\displaystyle O}{\|}}{C}-NH_2$$

b.
$$CH_3-CH_2-\overset{\overset{\displaystyle O}{\|}}{C}-NH_2$$

c.
$$CH_3-\overset{\overset{\displaystyle O}{\|}}{C}-\overset{\overset{\displaystyle H}{|}}{N}-CH_3$$

d.
$$CH_3-CH_2-CH_2-\overset{\overset{\displaystyle O}{\|}}{C}-NH-CH_2-CH_3$$

e.
$$CH_3-\overset{\overset{\displaystyle O}{\|}}{C}-\overset{\overset{\displaystyle CH_3}{|}}{N}-CH_2-CH_2-CH_2-CH_3$$

19.45 Indicate the name of the alkaloid in each of the following:

a. malaria treatment   **b.** tobacco

c. coffee and tea

d. a painkiller in oriental poppy plant

19.46 Identify the heterocyclic amines in each of the following:

a. caffeine   **b.** Demerol

c. nicotine   **d.** quinine

19.47 Write the structure of each product of the following reactions:

a. CH_3—CH_2—NH_2 + H_2O $\longrightarrow$

b. CH_3—CH_2—NH_2 + HCl $\longrightarrow$

c. CH_3—CH_2—NH—CH_3 + H_2O $\longrightarrow$

d. CH_3—CH_2—NH—CH_3 + HCl $\longrightarrow$

e. CH_3—CH_2—CH_2—NH_3^+ Cl^- + NaOH $\longrightarrow$

f. CH_3—CH_2—$\overset{\overset{\displaystyle CH_3}{|}}{N}H_2^+Cl^-$ + NaOH $\longrightarrow$

19.48 Toradol is used in dentistry to relieve pain. Name the functional groups in this molecule.

19.49 Voltaren is indicated for acute and chronic treatment of the symptoms of rheumatoid arthritis. Name the functional groups in this molecule.

19.50 Many amine-containing drugs are given to patients in their salt form, such as hydrochloride or sulfate. What might be the reason?

19.51 Using a reference book such as the *Merck Index* or *Physicians' Desk Reference*, look up the structural formula of the following medicinal drugs and list the functional groups in the compounds. You may need to refer to the cross-index of names in the back of the reference book.

a. Keflex, an antibiotic

b. Inderal, a β-channel blocker used to treat heart irregularities

c. ibuprofen, an anti-inflammatory agent

d. Aldomet (methyldopa)

e. Percodan, a narcotic pain reliever

f. triamterene, a diuretic

20 Amino Acids and Proteins

"This lamb is fed with Lamb Lac, which is a chemically formulated replacement for ewe's milk," says part-time farmer Dennis Samuelson. "Its mother had triplets and didn't have enough milk to feed them all, so they weren't thriving the way the other lambs were. The Lamb Lac includes dried skim milk, dried whey, milk proteins, egg albumin, the amino acids methionine and lysine, vitamins, and minerals."

A veterinary technician diagnoses and treats diseases of animals, takes blood and tissue samples, and administers drugs and vaccines. Agricultural technologists assist in the study of farm crops to increase productivity and ensure a safe food supply. They look for ways to improve crop yields, develop safer methods of weed and pest control, and design methods to conserve soil and water.

LOOKING AHEAD

the Chemistry place

www.chemplace.com/college

Visit the URL above or use the CD-ROM in the book for extra quizzing, interactive tutorials, and career resources.

The word "protein" is derived from the Greek word *proteios*, meaning "first." Made of amino acids, proteins provide structure in membranes, build cartilage and connective tissue, transport oxygen in blood and muscle, direct biological reactions as enzymes, defend the body against infection, and control metabolic processes as hormones. They can even be a source of energy.

Compared with many of the compounds we have studied, protein molecules can be gigantic. Insulin has a molar mass of 5700, and hemoglobin has a molar mass of about 64,000. Some virus proteins are still larger, having molar masses of more than 40 million. Yet all proteins in humans are polymers made up of 20 different amino acids. Each kind of protein is composed of amino acids arranged in a specific order that determines the characteristics of the protein and its biological action.

Proteins perform many functions in the body: making up skin and hair, moving muscles, carrying oxygen, and regulating metabolism. All of these different functions of proteins depend on the structures and chemical behavior of amino acids, the building blocks of proteins. We will see how peptide bonds link amino acids and how the order of the amino acids in these protein polymers directs the formation of unique three-dimensional structures.

LEARNING GOAL

Classify proteins by their functions in the cells.

20.1 Functions of Proteins

The many kinds of proteins perform many different functions in the body. There are proteins that form structural components such as cartilage, muscles, hair, and nails. Wool, silk, feathers, and horns are some other proteins made by animals. Proteins called enzymes regulate biological reactions such as digestion and cellular metabolism. Still other proteins, hemoglobin and myoglobin, carry oxygen in the blood and muscle. (See Figure 20.1). Table 20.1 gives examples of proteins that are classified by their functions in biological systems.

SAMPLE PROBLEM 20.1

Classifying Proteins by Function

Give the class of protein that would perform each of the following functions.
a. catalyzes metabolic reactions of lipids
b. carries oxygen in the bloodstream
c. stores amino acids in milk

Solution

a. Enzymes catalyze metabolic reactions.
b. Transport proteins carry substances such as oxygen through the bloodstream.
c. A storage protein stores nutrients such as amino acids in milk.

Study Check

What proteins help regulate metabolism?

Table 20.1   Classification of Some Proteins and their Functions

Class of Protein	Function in the Body	Examples
Structural	Provide structural components	*Collagen* is in tendons and cartilage. *Keratin* is in hair, skin, wool, and nails.
Contractile	Movement of muscles	*Myosin* and *actin* contract muscle fibers.
Transport	Carry essential substances throughout the body	*Hemoglobin* transports oxygen. *Lipoproteins* transport lipids.
Storage	Store nutrients	*Casein* stores protein in milk. *Ferritin* stores iron in the spleen and liver.
Hormone	Regulate body metabolism and nervous system	*Insulin* regulates blood glucose level. *Growth hormone* regulates body growth.
Enzyme	Catalyze biochemical reactions in the cells	*Sucrase* catalyzes the hydrolysis of sucrose. *Trypsin* catalyzes the hydrolysis of proteins.
Protection	Recognize and destroy foreign substances	*Immunoglobulins* stimulate immune responses.

Figure 20.1 The feathers, horns, and wool of animals are made of proteins.
Q *What class of protein are wool, feathers, and horns?*

QUESTIONS AND PROBLEMS

Functions of Proteins

20.1 Classify each of the following proteins according to its function:
  a. hemoglobin, oxygen carrier in the blood
  b. collagen, a major component of tendons and cartilage

c. keratin, a protein found in hair
d. amylase, an enzyme that hydrolyzes starch
20.2 Classify each of the following proteins according to its function:
  a. insulin, a hormone needed for glucose utilization
  b. antibodies, proteins that disable foreign proteins
  c. casein, milk protein
  d. lipases that hydrolyze lipids

LEARNING GOAL

Draw the structure for an amino acid.

20.2 Amino Acids

Proteins are composed of molecular building blocks called amino acids. An **amino acid** contains two functional groups, an amino group ($-NH_2$) and a carboxylic acid group ($-COOH$). In all of the 20 amino acids found in proteins, the amino group, the carboxylic group, and a hydrogen atom are bonded to a central carbon atom. Amino acids with this structure are called 2-amino acids, or α (alpha) amino acids. Although there are many amino acids, only 20 different amino acids are present in the proteins in humans. The unique characteristics of the 20 amino acids are due to a side chain (R), which can be an alkyl, hydroxy, thiol, amino, sulfide, aromatic, or heterocyclic group.

General Structure of an α-Amino Acid

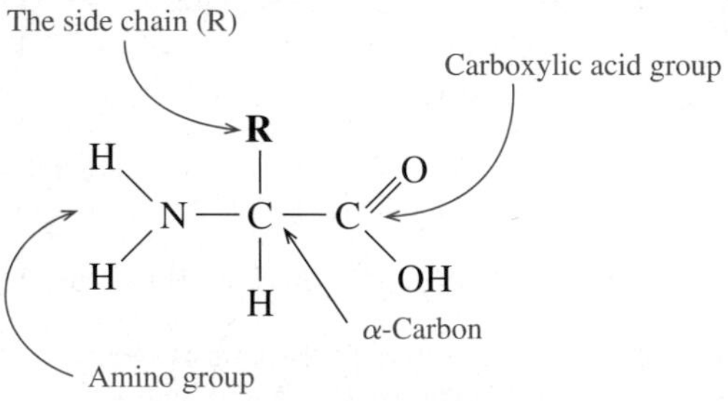

Classification of Amino Acids

A **nonpolar amino acid** contains an alkyl or aromatic side chain. Hydrocarbons are nonpolar and not soluble in water, which makes the nonpolar amino acids **hydrophobic** ("water-fearing"). **Polar amino acids** have side chains that contain polar groups such as hydroxyl ($-OH$), thiol ($-SH$), amide ($-CONH_2$), and heterocyclic amines. The electronegative atoms in the side chains make these amino acids **hydrophilic** ("water-attracting") by forming hydrogen bonds with water. The **acidic amino acids** have side chains that contain a carboxylic acid group ($-COOH$) and can ionize as a weak acid. The side chains of the **basic amino acids** contain an amino group that can ionize as a weak base. With the exception of proline, all the amino acids are primary amines. The structures of the side chains (R), common names, three-letter abbreviations, and isoelectric points, pI (see Section 20.3) of the 20 amino acids in proteins are listed in Table 20.2.

Table 20.2 The 20 Amino Acids in Proteins

Nonpolar Amino Acids

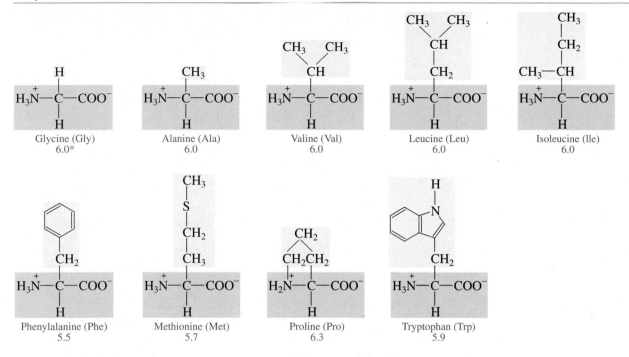

Glycine (Gly)
6.0*

Alanine (Ala)
6.0

Valine (Val)
6.0

Leucine (Leu)
6.0

Isoleucine (Ile)
6.0

Phenylalanine (Phe)
5.5

Methionine (Met)
5.7

Proline (Pro)
6.3

Tryptophan (Trp)
5.9

Polar Amino Acids (Neutral)

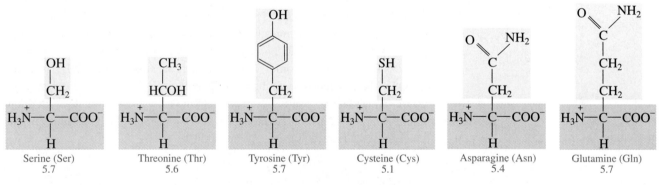

Serine (Ser)
5.7

Threonine (Thr)
5.6

Tyrosine (Tyr)
5.7

Cysteine (Cys)
5.1

Asparagine (Asn)
5.4

Glutamine (Gln)
5.7

Acidic Amino Acids   **Basic Amino Acids**

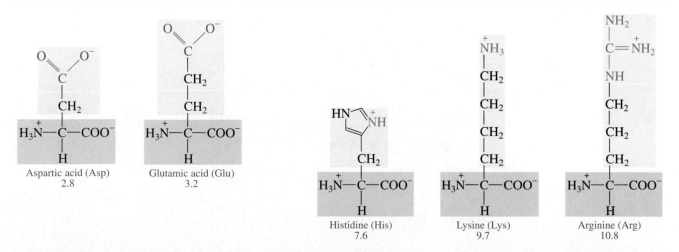

Aspartic acid (Asp)
2.8

Glutamic acid (Glu)
3.2

Histidine (His)
7.6

Lysine (Lys)
9.7

Arginine (Arg)
10.8

*Isoelectric points (pI)

Structural Formulas of Amino Acids

Write the structural formulas and abbreviations for the following amino acids:

a. alanine (R = — CH_3)   **b.** serine (R = — CH_2OH)

Solution

a. The structure of the amino acids is written by attaching the side group (R) to the central carbon atom of the general structure of an amino acid.

alanine (Ala)

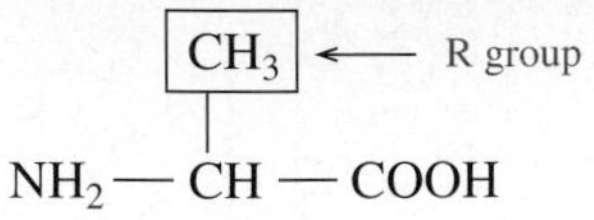

$$\boxed{CH_3} \longleftarrow \text{R group}$$
$$|$$
$$NH_2 - CH - COOH$$

b. serine (Ser)

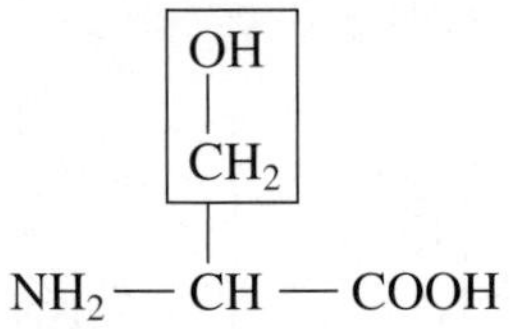

$$\boxed{\begin{array}{c} OH \\ | \\ CH_2 \end{array}}$$
$$|$$
$$NH_2 - CH - COOH$$

Study Check

Classify the amino acids in the sample problem as polar or nonpolar.

Amino Acid Stereoisomers

All of the α-amino acids except for glycine are chiral because the α carbon is attached to four different atoms. Thus amino acids can exist as D and L enantiomers. We can write Fischer projections for amino acids as we did in Chapter 15 for aldehydes with the carboxylic acid group (the most highly oxidized carbon) at the top and the R group at the bottom. For the L isomer, the amino group, NH_2, is on the left, and in the D isomer, it is on the right. In biological systems, only L amino acids are incorporated into proteins. There are D amino acids found in nature, but not in proteins. Let's take a look at the enantiomers for glyceraldehyde and how the stereoisomers of alanine and cysteine are similar.

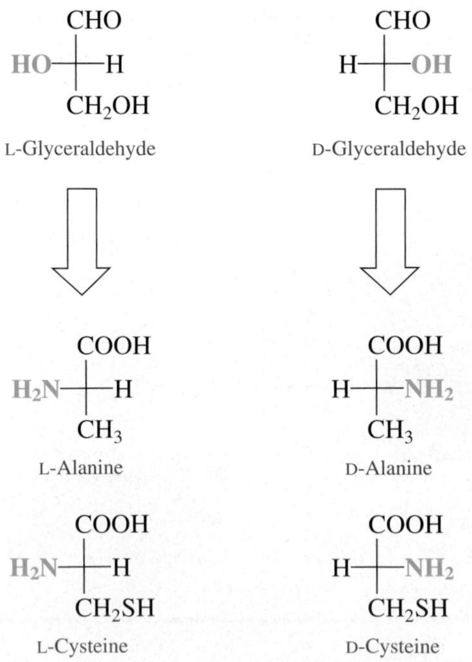

L-Glyceraldehyde   D-Glyceraldehyde

L-Alanine   D-Alanine

L-Cysteine   D-Cysteine

SAMPLE PROBLEM 20.3

Chiral Amino Acids

Write the Fischer projection for L-serine.

Solution

In L-serine, the —COOH is at the top and the R group —CH₂OH is at the bottom. The L-isomer has the —NH₂ on the left.

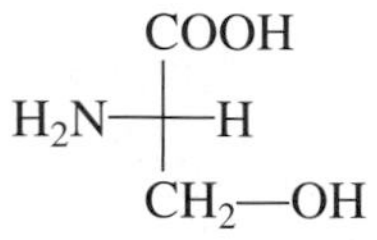

$$\begin{array}{c} COOH \\ | \\ H_2N\!-\!\!\!\!+\!\!\!\!-\!H \\ | \\ CH_2\!-\!OH \end{array}$$

L-Serine

Study Check

How does the Fischer projection for D-serine differ from L-serine?

QUESTIONS AND PROBLEMS

Amino Acids

20.3 Describe the functional groups found in all α-amino acids.

20.4 How does the polarity of the side chain in leucine compare to the side chain in serine?

20.5 Draw the structural formula for each of the following amino acids:
   a. alanine
   b. threonine
   c. glutamic acid
   d. phenylalanine

20.6 Draw the structural formula for each of the following amino acids:
   a. lysine
   b. aspartic acid
   c. leucine
   d. tyrosine

20.7 Classify the amino acids in problem 20.5 as hydrophobic (nonpolar), hydrophilic (polar, neutral), acidic, or basic.

20.8 Classify the amino acids in problem 20.6 as hydrophobic (nonpolar), hydrophilic (polar, neutral), acidic, or basic.

20.9 Give the name of the amino acid represented by each of the following three-letter abbreviations:
   a. Ala   **b.** Val   **c.** Lys   **d.** Cys

20.10 Give the name of the amino acid represented by each of the following three-letter abbreviations:
   a. Trp   **b.** Met   **c.** Pro   **d.** Gly

20.11 Draw the Fischer projections for the following amino acids.
   a. L-valine
   b. D-cysteine

20.12 Draw the Fischer projections for the following amino acids.
   a. L-threonine
   b. D-valine

LEARNING GOAL

Write the ionic form of an amino acid at pH values above, below, and equal to its isoelectric point.

20.3 Amino Acids as Acids and Bases

Although it is convenient to write amino acids with carboxyl (—COOH) group and amine (—NH_2) group, they are usually ionized. Depending on the pH, the carboxyl group loses a H^+, giving —COO^-, and the amino group accepts a H^+ to give an ammonium ion, —NH_3^+. The *dipolar* form of an amino acid called a **zwitterion** has a net charge of zero.

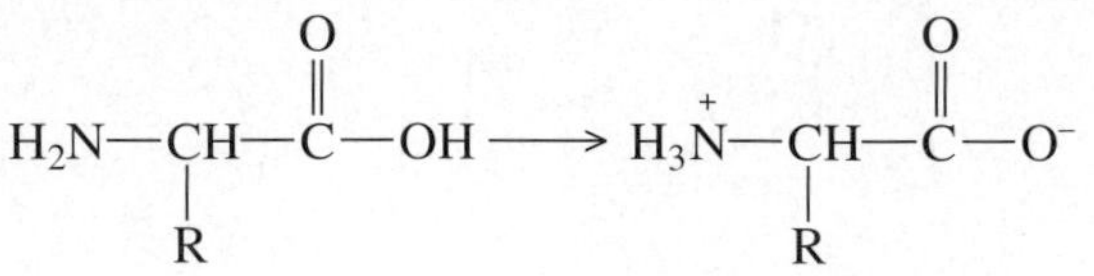

Amino Acid   Zwitterion (dipolar ion)

Solid amino acids have very high melting points because the zwitterion has the properties of a salt. For example, glycine melts at 260°C. However, most amino acids decompose when they are heated rather than melt. The ionic charges of the amino acids make them more soluble in water, but not in organic solvents.

At a certain pH known as the **isoelectric point (pI)**, the positive and negative charges are equal, which gives an overall charge of zero. However, when the pH is different from the pI, the zwitterion accepts or donates H^+. In a solution that is more acidic than the pI, the —COO^- group acts as a base and accepts an H^+, which gives an overall positive charge to the amino acid.

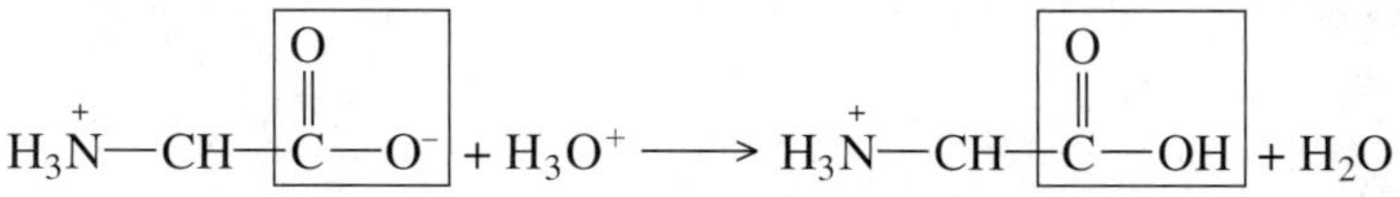

Zwitterion ion accepts H^+ Positively charged ion

In a solution more basic than the pI, the —NH_3^+ group acts as an acid and loses an H^+, which gives the amino acid an overall negative charge.

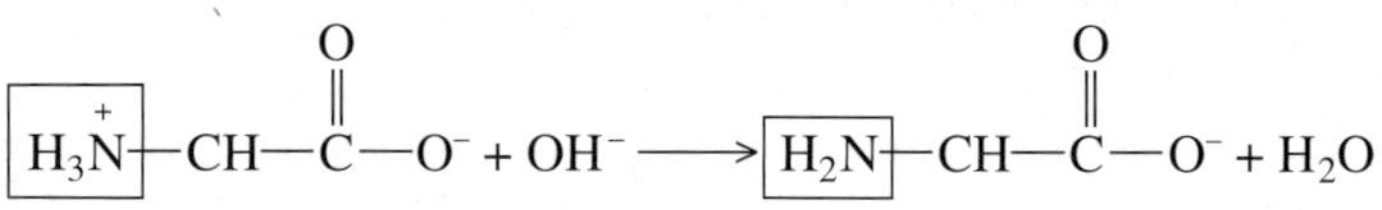

Zwitterion donates H^+ Negatively charged ion

Let's take a look at the changes in all the ionic forms of alanine from its zwitterion (pI = 6.0), to the positive ion in a more acidic solution, and the negative ion in basic solution.

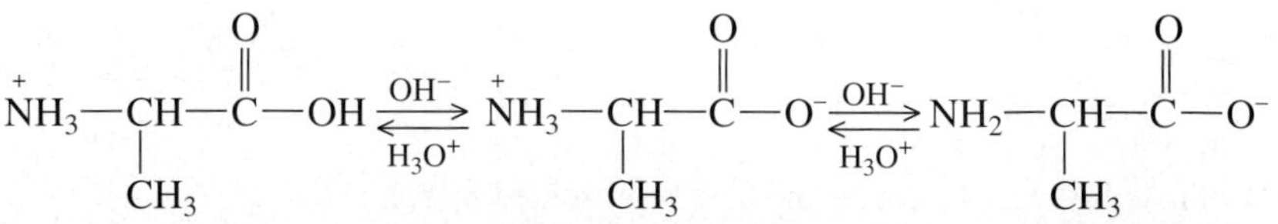

Alanine ion at acidic pH < 6 (charge = 1+)   Zwitterion of alanine pH = 6.0 (charge = 0)   Alanine ion at a pH > 6 (charge = 1–)

The zwitterions for polar and nonpolar amino acids typically exist at pH values of 5.0 to 6.0. However, the zwitterions of the acidic amino acids form at pH values of about 3 because the carboxyl group in their side chain must pick up H^+. The basic amino acids with amino groups in their side chains form zwitterions at high pI values from 7.6 to 10.8. The pI values are included in the list of the amino acids in Table 20.2.

$$\overset{+}{H_3N}-CH-\overset{\overset{O}{\parallel}}{C}-OH \underset{H_3O^+}{\overset{OH^-}{\rightleftharpoons}} \overset{+}{H_3N}-CH-\overset{\overset{O}{\parallel}}{C}-O^- \underset{H_3O^+}{\overset{OH^-}{\rightleftharpoons}} \overset{+}{H_3N}-CH-\overset{\overset{O}{\parallel}}{C}-O^- \underset{H_3O^+}{\overset{OH^-}{\rightleftharpoons}} H_2N-CH-\overset{\overset{O}{\parallel}}{C}-O^-$$

Aspartic acid
pH < 2
(charge = 1+)

Zwitterion
pH = 2.8
(charge = 0)

Aspartic acid
at pH 7
(charge = 1–)

Aspartic acid
basic pH
(charge = 2–)

Electrophoresis

It is possible to separate a mixture of amino acids by their isoelectric points using a laboratory method called **electrophoresis.** A buffered amino acid mixture is applied to a gel on a thin plate or piece of filter paper, which is connected to two electrodes. A voltage applied to the electrodes causes the positively charged amino acids to move toward the negative electrode and the negatively charged amino acids to move toward the positive electrode. Any amino acid at its isoelectric point with zero net charge would not move. After several hours, the sample is removed. It can be sprayed with a dye such as ninhydrin to make the amino acids visible. They are identified by their direction and rate of migration toward the electrodes. They are recovered separately by cutting up the filter paper or removing the amino acids from the gel.

Suppose we have a mixture of valine (pI 6.0), aspartic acid (pI 2.8), and lysine (pI 9.7) in a buffer of pH 6.0. When the mixture is placed between two electrodes at a high voltage, the aspartic acid, which would have a negative charge at pH 6.0, would move to the positive electrode (anode). (See Figure 20.2.) The lysine, which would be positively charged at a pH of 6.0, would move toward the negative electrode (cathode). Valine, which is a zwitterion at pH 6.0, would be neutral and not move in the presence of an electric field. Electrophoresis is a method used in medicine to screen for the sickle cell trait in newborn infants.

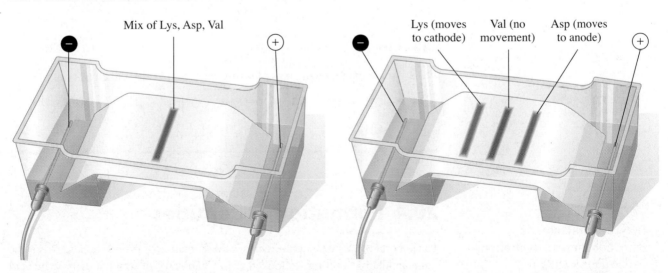

Figure 20.2 A positively charged amino acid (pH < pI) moves toward the negative electrode, a negatively charged amino acid (pH > pI) moves toward the positive electrode: An amino acid with no net charge (pH = pI) does not migrate.
Q *How would the three amino acids migrate if the mixture were buffered to pH 9.7, the pI of lysine?*

Amino Acids in Acid or Base

Serine exists in its zwitterion form at a pH of 5.7. Draw the structural formula for the zwitterion of serine.

Solution

As a zwitterion, both the carboxylic acid group and the amino group are ionized.

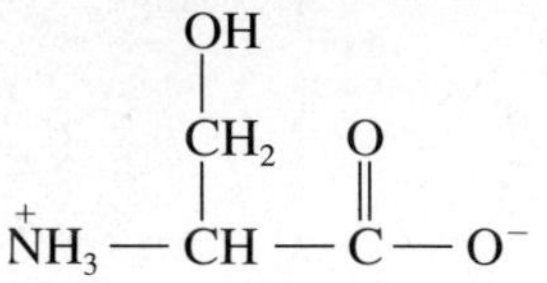

$$\underset{\overset{+}{N}H_3}{} - CH - \underset{\underset{O^-}{\overset{O}{\|}}}{C}$$

with CH$_2$ and OH above the CH

Zwitterion of serine

Study Check

Draw the structure of serine at a pH of 2.0.

QUESTIONS AND PROBLEMS

Amino Acids as Acids and Bases

20.13 Write the zwitterion of each of the following amino acids:
   a. glycine   **b.** cysteine   **c.** serine   **d.** alanine

20.14 Write the zwitterion of each of the following amino acids:
   a. phenylalanine   **b.** methionine   **c.** leucine   **d.** valine

20.15 Write the positive ion (acidic ion) of each of the amino acids in problem 20.13 at a pH below 1.0.

20.16 Write the negative ion (basic ion) of each of the amino acids in problem 20.13 at a pH above 12.0.

20.17 Would the following ions of valine exist at a pH above, below, or at pI?

a. $H_2N-CH-COO^-$ with CH and CH$_3$ CH$_3$ below

b. $H_3\overset{+}{N}-CH-COOH$ with CH and CH$_3$ CH$_3$ below

c. $H_3\overset{+}{N}-CH-COO^-$ with CH and CH$_3$ CH$_3$ below

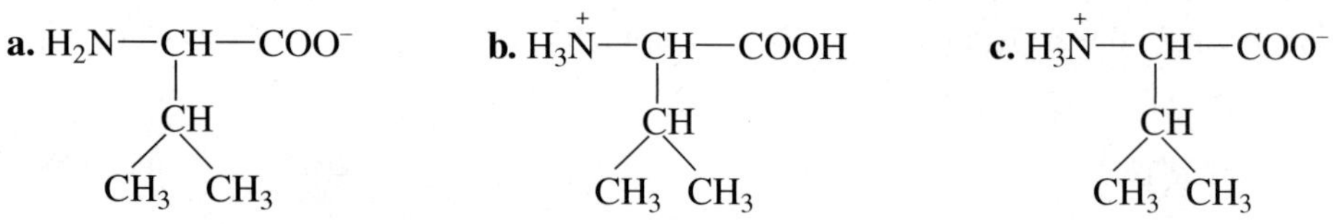

20.18 Would the following ions of serine exist at a pH above, below, or at pI?

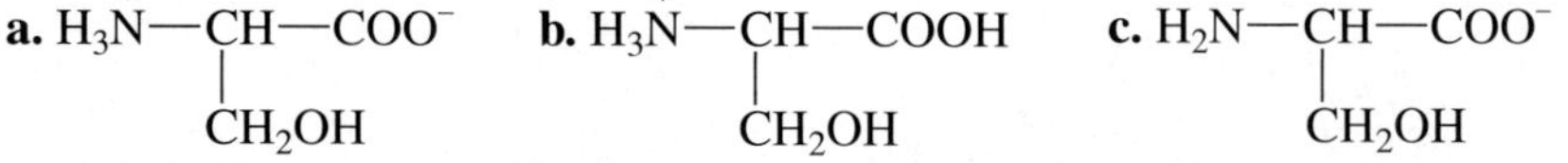

a. $H_3\overset{+}{N}-CH-COO^-$ with CH$_2$OH below

b. $H_3\overset{+}{N}-CH-COOH$ with CH$_2$OH below

c. $H_2N-CH-COO^-$ with CH$_2$OH below

Draw the structure of a dipeptide from the zwitterions of two amino acids.

20.4 Formation of Peptides

The important reaction of amino acids is the formation of peptide bonds. This is the same amidation reaction we looked at in Chapter 19 in which a carboxylic acid reacts with an amine to form an amide. Let's review this reaction.

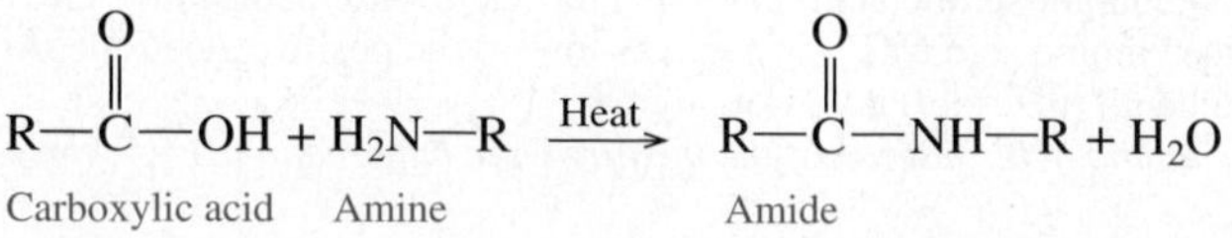

$$\underset{\text{Carboxylic acid}}{R-\overset{\overset{O}{\|}}{C}-OH} + \underset{\text{Amine}}{H_2N-R} \xrightarrow{\text{Heat}} \underset{\text{Amide}}{R-\overset{\overset{O}{\|}}{C}-NH-R} + H_2O$$

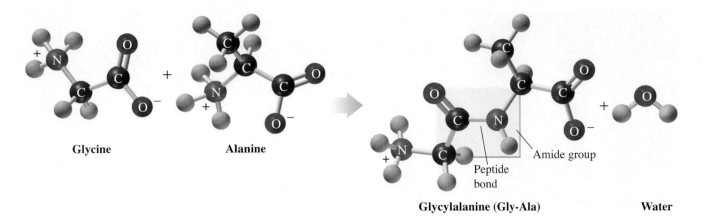

Figure 20.3 A peptide bond between glycine and alanine as zwitterions form the dipeptide glycylalanine.
Q *What functional groups in glycine and alanine form the peptide bond?*

The linking of two or more amino acids forms a **peptide.** A **peptide bond** is an amide bond that forms when the $-COO^-$ group of one amino acid reacts with the $-NH_3^+$ group of the next amino acid. We can write the amidation reaction for the zwitterion forms of two amino acids.

$$H_3\overset{+}{N}-CH-\boxed{\underset{}{\overset{O}{\overset{\|}{C}}}-O^-}+\boxed{H_3\overset{+}{N}}-CH-\overset{O}{\overset{\|}{C}}-O^- \longrightarrow H_3\overset{+}{N}-CH-\boxed{\overset{O}{\overset{\|}{C}}-\underset{}{\overset{H}{\overset{|}{N}}}}-CH-\overset{O}{\overset{\|}{C}}-O^-+H_2O$$

R	R	R	R
Amino acid 1	Amino acid 2	Dipeptide	

Two amino acids linked together by a peptide bond form a *dipeptide*. We can write the formation of the dipeptide glycylalanine between glycine and alanine as follows. (See Figure 20.3.) In a peptide, the amino acid written on the left with the unreacted or free amino group ($-NH_3^+$) is called the **N terminal** amino acid. The **C terminal** amino acid is the last amino acid in the chain with the unreacted or free carboxyl group ($-COO^-$).

N terminal   C terminal

$$H_3\overset{+}{N}-CH_2-\overset{O}{\overset{\|}{C}}-O^- + H_3\overset{+}{N}-CH-\overset{O}{\overset{\|}{C}}-O^- \longrightarrow H_3\overset{+}{N}-CH_2-\boxed{\overset{O}{\overset{\|}{C}}-\overset{H}{\overset{|}{N}}}-CH-\overset{O}{\overset{\|}{C}}-O^-+H_2O$$

Glycine CH_3 CH_3
Glycine   Alanine   Glycylalanine (Gly-Ala)

Naming Peptides

In naming a peptide, each amino acid beginning from the N terminal is named with a–*yl* ending followed by the full name of the amino acid at the C terminal. Tryptophan becomes tryptophyl, aspartic acid aspartyl, glutamic acid glutamyl, and glutamine is glutaminyl. For convenience, the order of amino acids in the peptide is often written as the sequence of three-letter abbreviations. For example, a tripeptide consisting of alanine, glycine, and serine is named as ala**nyl**glycy**l**serine.

Rehabilitation Specialist

"I am interested in the biomechanical part of rehabilitation, which involves strengthening activities that help people return to the activities of daily living," says Minna Robles, rehabilitation specialist. "Here I am fitting a patient with a wrist extension splint that allows her to lift her hand. This exercise will also help the muscles and soft tissues in that wrist area to heal. An understanding of the chemicals of the body, how they interact, and how we can affect the body on a chemical level is important in understanding our work. One technique we use is called myofacial release. We apply pressure to a part of the body, which helps to increase circulation. By increasing circulation, we can move the soft tissues better, which improves movement and range of motion."

$$
\underset{\substack{\text{From alanine}\\ \text{alanyl}}}{\boxed{\overset{+}{H_3N}-CH-\overset{\displaystyle O}{\overset{\|}{C}}}}\,\underset{\substack{\text{From glycine}\\ \text{glycyl}}}{\boxed{NH-CH_2-\overset{\displaystyle O}{\overset{\|}{C}}}}\,\underset{\substack{\text{From serine}\\ \text{serine}}}{\boxed{NH-CH-\overset{\displaystyle O}{\overset{\|}{C}}-O^-}}
$$

with CH_3 (from alanine) and CH_2OH (from serine)

Alanylglycylserine
(Ala-Gly-Ser)

SAMPLE PROBLEM 20.5

Writing Dipeptide Structures

Write a structural formula and three-letter abbreviation for the dipeptide valylserine.

Solution

Valine is joined to serine by a peptide bond; valine is the N terminal and serine is the C terminal.

$$
\overset{+}{H_3N}-CH-\overset{\displaystyle O}{\overset{\|}{C}}-NH-CH-\overset{\displaystyle O}{\overset{\|}{C}}-O^-
$$

CH—CH₃   CH₂OH

CH₃

From valine   From serine

Valylserine
Val-Ser

Study Check

Aspartame, an artificial sweetener 200 times sweeter than sucrose, contains the dipeptide Asp-Phe. Give the structure of the dipeptide in aspartame.

SAMPLE PROBLEM 20.6

Identifying a Tripeptide

Consider the following tripeptide.

$$
\overset{+}{H_3N}-CH-\overset{\displaystyle O}{\overset{\|}{C}}-NH-CH-\overset{\displaystyle O}{\overset{\|}{C}}-NH-CH-\overset{\displaystyle O}{\overset{\|}{C}}-O^-
$$

CHOH   CH₂   CH₂

CH₃   CHCH₃

CH₃

a. What amino acid is the N terminal? What amino acid is the C terminal?
b. What is the three-letter abbreviation name for the order of amino acids in the tripeptide?

Solution

a. Threonine is the N terminal; phenylalanine is the C terminal.
b. Thr-Leu-Phe

Study Check

What is the full name of the tripeptide in Sample Problem 20.6?

QUESTIONS AND PROBLEMS

Formation of Peptides

20.19 Draw the structural formula of each of the following peptides and give the abbreviation for their names:
  a. alanylcysteine   **b.** serylphenylalanine
  c. glycylalanylvaline   **d.** valylisoleucyltryptophan

20.20 Draw the structural formula of each of the following peptides and give the abbreviation for their names:
  a. methionylaspartic acid   **b.** alanyltryptophan
  c. methionylglutaminyllysine   **d.** histidylglycylglutamylalanine

---

20.5 Protein Structure: Primary and Secondary Levels

LEARNING GOAL
Identify the primary and secondary structures of a protein.

The particular sequence of amino acids in a peptide or protein is referred to as the **primary structure.** For example, a hormone that stimulates the thyroid to release thyroxine consists of a tripeptide Glu-His-Pro.

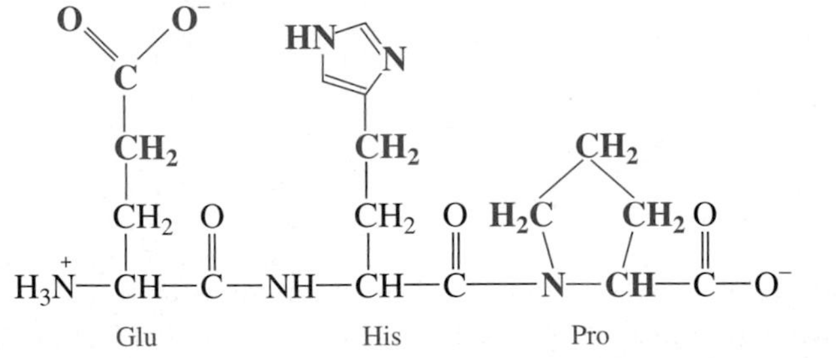

Although other sequences are possible for these three amino acids, only the tripeptide with the Glu-His-Pro sequence of amino acids has hormonal activity. Sequences such as His-Pro-Glu or Pro-His-Glu do not produce hormonal activity. Thus the biological function of peptides as well as proteins depends on the order of the amino acids.

When cells are damaged, a polypeptide called bradykinin is released at the site, which stimulates the release of prostaglandins. The presence of bradykinin, which contains nine amino acids, regulates blood pressure.

Arg — Pro — Pro — Gly — Phe — Ser — Pro — Phe — Arg
Bradykinin

Two hormones produced by the pituitary gland are the nonapeptides (nine-amino-acid peptide) oxytocin and vasopressin. Oxytocin stimulates uterine contractions in labor and vasopressin is an antidiuretic hormone that regulates blood pressure by adjusting the amount of water reabsorbed by the kidneys. The struc-

HEALTH NOTE

Natural Opiates in the Body

Painkillers known as enkephalins and endorphins are produced naturally in the body. They are polypeptides that bind to receptors in the brain to give relief from pain. This effect appears to be responsible for the runner's high, for the temporary loss of pain when severe injury occurs, and for the analgesic effects of acupuncture.

The *enkephalins*, found in the thalamus and the spinal cord, are pentapeptides, the smallest molecules with opiate activity.

The amino acid sequence of an enkephalin is found in the longer amino acid sequence of the endorphins.

Four groups of *endorphins* have been identified: α-endorphin contains 16 amino acids, β-endorphin contains 31 amino acids, γ-endorphin has 17 amino acids, and δ-endorphin has 27 amino acids. Endorphins may produce their sedating effects by preventing the release of substance P, a polypeptide with 11 amino acids, which has been found to transmit pain impulses to the brain.

←———————————————————— α-Endorphin ————————————————————→

Tyr — Gly — Gly — Phe — Met — Thr — Ser — Glu — Lys — Ser — Glu — Thr — Pro — Leu — Val — Thr

└——— Enkephalin ———┘

Leu

Glu — Gly — Lys — Lys — Tyr — Ala — Asn — Lys — Ile — Ile — Ala — Asn — Lys — Phe

β-Endorphin

Figure 20.4 The structures of the nonapeptides oxytocin and vasopressin are very similar except for two different amino acids at positions 3 and 8, shown in red for oxytocin and blue for vasopressin. In both structures an amide group replaces the C terminal oxygen.

Q *What amino acid has a side chain that can form a disulfide bond in oxytocin and vasopressin?*

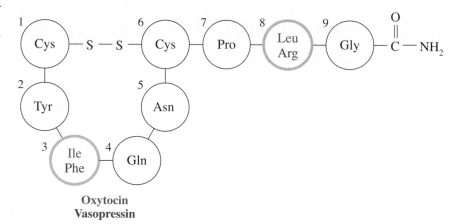

Oxytocin
Vasopressin

tures of these nonapeptides are very similar with nine amino acids in the same order and a cyclic structure with a disulfide bond. (See Figure 20.4.) Only the amino acids in positions 3 and 8 are different. However, the difference of two amino acids greatly affects how the two hormones function in the body.

Primary Structure of Proteins

When there are more than 50 amino acids in a chain, the polypeptide is usually called a **protein.** Each protein in our cells has a unique sequence of amino acids that determines its biological function. The primary structure of a protein is similar to the peptides we just looked at except that there are many more amino acids in the chain. In a protein, the primary structure is the order of the many amino acids held together by peptide bonds.

$$-NH-CH-\overset{O}{\overset{\|}{C}}-NH-CH-\overset{O}{\overset{\|}{C}}-NH-CH-\overset{O}{\overset{\|}{C}}-NH-CH-\overset{O}{\overset{\|}{C}}-$$
$$\qquad\ \ | \qquad\qquad\quad | \qquad\qquad\quad | \qquad\qquad\quad |$$
$$\qquad\ \ R \qquad\qquad\quad R \qquad\qquad\quad R \qquad\qquad\quad R$$

The first protein to have its primary structure determined was insulin, which is a hormone that regulates the glucose level in the blood. In the primary structure of human insulin, there are two polypeptide chains. In chain A, there are 21 amino acids, and chain B has 30 amino acids. The polypeptide chains are held together by disulfide bonds formed by the side chains of the cysteine amino acids in each of the chains. (See Figure 20.5.) This primary structure of insulin in humans is very similar to the primary structure of insulin in pigs and cows (bovine). Only the three amino acids at positions 8, 9, and 10 in chain A and position 30 in chain B vary from one species to another. For many years, bovine insulin obtained from the pancreas of cows was used to treat diabetics who lacked insulin. Today human insulin produced through genetic engineering is used in the treatment of diabetes.

Secondary Structure

The **secondary structure** of a protein describes the way the amino acids next to or near to each other along the polypeptide are arranged in space. The three most common types of secondary structure are the *alpha helix*, the *beta-pleated sheet*, and the *triple helix* found in collagen. In each type of secondary structure, we will look at the hydrogen bonding between the hydrogen atom of an amino group in the polypeptide chain and the oxygen atom of the carboxyl group in another part of the chain.

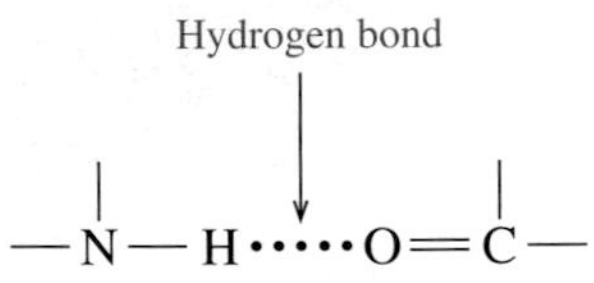

Hydrogen bond

$$-\overset{|}{N}-H\cdots\cdots O=\overset{|}{C}-$$

The Alpha Helix

The corkscrew shape of an **alpha helix** (α helix) is held in place by hydrogen bonds between each N—H group and the oxygen of a C=O group in the next turn of the helix, four amino acids down the chain. (See Figure 20.6.) Because many hydrogen bonds form along the peptide backbone, this portion of the protein takes the shape of a strong, tight coil that looks like a telephone cord or a Slinky toy. All the side chains (R groups) of the amino acids are located on the outside of the helix.

Beta-Pleated Sheet

Another type of secondary structure is known as the **beta-pleated sheet (β-pleated sheet)**. In a β-pleated sheet, polypeptide chains are held together side by side by hydrogen bonds between the peptide chains. In a β-pleated sheet of silk fibroin, the small R groups of the prevalent amino acids, glycine, alanine, and serine, extend above and below the sheet. This results in a series of β-pleated sheets that can be stacked close together. The hydrogen bonds holding the β-pleated sheets tightly in place account for the strength and durability of fibrous proteins such as silk. (See Figure 20.7.)

In some proteins, the polypeptide chain consists of mostly the α-helix secondary structure, whereas other proteins consist of mostly the β-pleated sheet structure. Another group of proteins have a mixture with some sections of the polypeptide

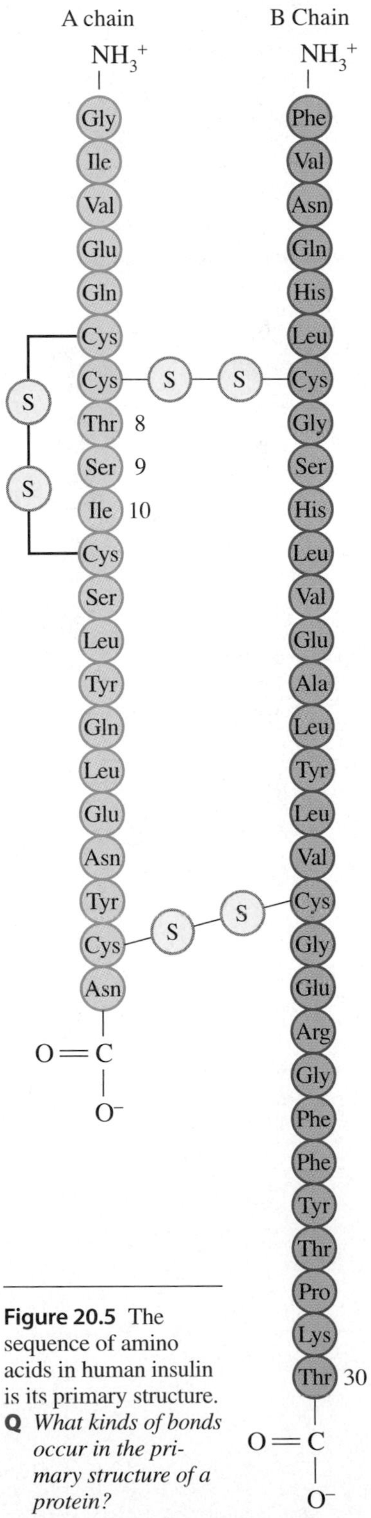

Change in Amino Acids of Insulin in Pigs and Cows		
	Pig	**Cow**
Chain A	8 Thr	Ala
	9 Ser	Ser
	10 Ile	Val
Chain B	30 Ala	Ala

Figure 20.5 The sequence of amino acids in human insulin is its primary structure.
Q *What kinds of bonds occur in the primary structure of a protein?*

Figure 20.6 The α (alpha) helix acquires a coiled shape from hydrogen bonds between the N—H of the peptide bond in one loop and the C=O of the peptide bond in the next loop.

Q *What are partial charges of the H in N—H and the O in C=O that permits hydrogen bonds to form?*

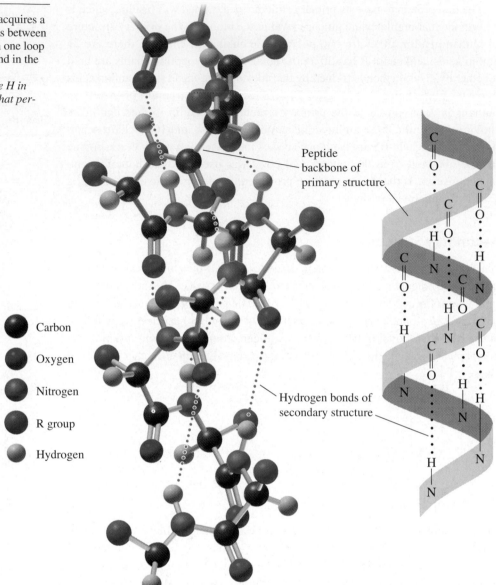

- Carbon
- Oxygen
- Nitrogen
- R group
- Hydrogen

Peptide backbone of primary structure

Hydrogen bonds of secondary structure

chain in α-helixes and other sections in the β-pleated sheet structure. The tendency to form a certain type of secondary structure depends on the amino acids in a particular segment of the polypeptide chain. The amino acids that tend to form an α-helix or β-pleated sheet follow.

Amino acids found in α-helix regions		Amino acids found in β-pleated sheets	
Alanine	Glutamic acid	Valine	Proline
Cysteine	Glutamine	Serine	Arginine
Leucine	Histidine	Aspartic acid	
Methionine	Lysine	Asparagine	

Triple Helix

Collagen is the most abundant protein; it makes up as much as one-third of all the protein in vertebrates. It is found in connective tissue, blood vessels, skin, tendons,

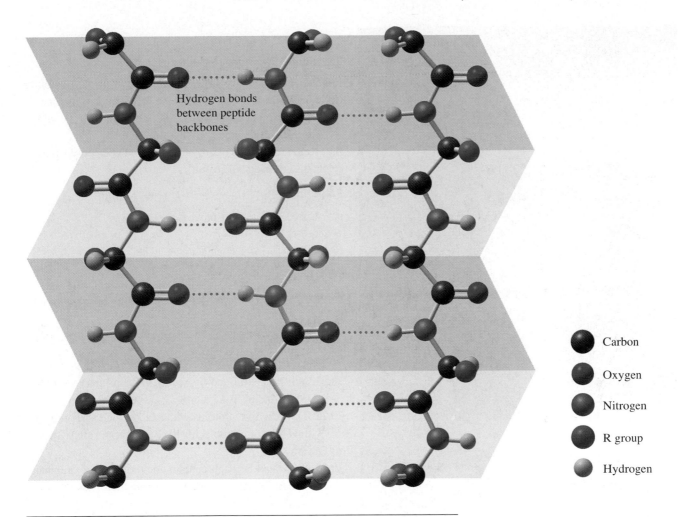

Hydrogen bonds
between peptide
backbones

Carbon

Oxygen

Nitrogen

R group

Hydrogen

Figure 20.7 In a β (beta)-pleated sheet secondary structure, hydrogen bonds form between the peptide chains.
Q *How do the hydrogen bonds differ in a β (beta)-pleated sheet from the alpha helix?*

ligaments, the cornea of the eye, and cartilage. The strong structure of collagen is a result of three polypeptides woven together like a braid to form a **triple helix**, as seen in Figure 20.8. Collagen has a high content of glycine (33%), proline (22%), alanine (12%), and smaller amounts of hydroxyproline, and hydroxylysine. The hydroxy forms of proline and lysine contain —OH groups that form hydrogen bonds across the peptide chains and give strength to the collagen triple helix. When several triple helixes wrap together, they form the fibrils that make up connective tissues and tendons.

When a diet is deficient in vitamin C, collagen fibrils are weakened because the enzymes needed to form hydroxyproline and hydroxylysine require vitamin C. Without the —OH groups of hydroxyproline and hydroxylysine, there is less hydrogen bonding between collagen fibrils. Sores appear on the skin and gums, and blood vessels may be weakened, which can lead to aneurisms and rupture. In a young person, collagen is elastic. As a person ages, additional cross links form between the fibrils, which make collagen less elastic. Bones, cartilage, and tendons become more brittle, and wrinkles are seen as the skin loses elasticity. In connective tissue disorders such as lupus and rheumatoid arthritis, an over-active immune system produces increased amounts of collagen. Organs in the body containing large amounts of connective tissues such as joints, skin, kidneys, lungs, and heart are affected.

$$
\begin{array}{c}
OH \\
| \\
CH \\
\diagup \quad \diagdown \\
CH_2 \quad CH_2 \\
\diagdown \quad \diagup \\
H_2N^+ - CH - COO^-
\end{array}
$$
Hydroxyproline

$$
\begin{array}{c}
\overset{+}{N}H_3 \\
| \\
CH_2 \\
| \\
H - C - OH \\
| \\
CH_2 \\
| \\
CH_2 \\
| \\
H_3N^+ - CH - COO^-
\end{array}
$$
Hydroxylysine

Triple helix   3 α-Helix peptide chains

Figure 20.8 Hydrogen bonds between polar R groups in three polypeptide chains form the triple helixes that combine to make fibers of collagen.

Q *What are some of the amino acids in collagen that form hydrogen bonds between the polypeptide chains?*

Identifying Secondary Structures

Indicate the secondary structure (α helix, β-pleated sheet, or triple helix) described in each of the following statements:

a. a coiled peptide chain held in place by hydrogen bonding between peptide bonds in the same chain

b. a structure that has hydrogen bonds between polypeptide chains arranged side by side.

Solution

a. α helix   **b.** β-pleated sheet

Study Check

What is the secondary structure in collagen?

QUESTIONS AND PROBLEMS

Protein Structure: Primary and Secondary Levels

20.21 What type of bonding occurs in the primary structure of a protein?

20.22 How can two proteins with exactly the same number and type of amino acids have different primary structures?

20.23 Two peptides each contain one molecule of valine and two molecules of serine. What are their possible primary structures?

20.24 What are three different types of secondary protein structure?

20.25 What happens to the primary structure of a protein when a protein forms a secondary structure?

20.26 In an α helix, how does bonding occur between the amino acids in the polypeptide chain?

20.27 What is the difference in bonding between an α helix and a β-pleated sheet?

20.28 How is the secondary structure of a β-pleated sheet different from that of a triple helix?

20.6 Protein Structure: Tertiary and Quaternary Levels

The **tertiary structure** of a protein involves attractions and repulsions between the side chain groups of the amino acids in the polypeptide chain. As interactions occur between different parts of the peptide chain, segments of the chain twist and bend until the protein acquires a specific three-dimensional shape.

Essential Amino Acids

Of the 20 amino acids used to build the proteins in the body, only 10 can be synthesized in the body. The other 10 amino acids, listed in Table 20.3, are **essential amino acids** that cannot be synthesized and must be obtained from the proteins in the diet.

Table 20.3   Essential Amino Acids

Arginine (Arg)	Methionine (Met)
Histidine (His)	Phenylalanine (Phe)
Isoleucine (Ile)	Threonine (Thr)
Leucine (Leu)	Tryptophan (Trp)
Lysine (Lys)	Valine (Val)

Complete proteins, which contain all of the essential amino acids are found in most animal products such as eggs, milk, meat, fish, and poultry. However, gelatin and plant proteins such as grains, beans, and nuts are *incomplete proteins* because they are deficient in one or more of the essential amino acids. Diets that rely on plant foods for protein must contain a variety of protein sources to obtain all the essential amino acids. For example, a diet of rice and beans contains all the essential amino acids because they are complementary proteins. Rice contains the methionine and tryptophan deficient in beans, while beans contain the lysine that is lacking in rice. (See Table 20.4.)

Table 20.4   Amino Acid Deficiency in Selected Vegetables and Grains

Food Source	Amino Acids Missing
Eggs, milk, meat, fish, poultry	None
Wheat, rice, oats	Lysine
Corn	Lysine, tryptophan
Beans	Methionine, tryptophan
Peas	Methionine
Almonds, walnuts	Lysine, tryptophan
Soy	Low in methionine

Cross-Links in Tertiary Structures

The tertiary structure of a protein is stabilized by interactions between the R groups of the amino acids in one region of the polypeptide chain with R groups of amino acids in other regions of the protein. (See Figure 20.9.) Table 20.5 lists the stabilizing interactions of tertiary structures.

1. **Hydrophobic interactions** are interactions between two nonpolar R groups. For example, hydrophobic interactions would occur between the aromatic group in phenylalanine and the alky group of valine or leucine. Within the compact shape of a globular protein, the amino acids with nonpolar side chains push as far away from the aqueous environment as possible, which forms a hydrophobic center at the interior of the protein molecule.

2. **Hydrophilic interactions** are attractions between the external aqueous environment and amino acids that have polar or ionized side chains. The polar side chains pull toward the outer surface of globular proteins to hydrogen bond with water. The presence of the hydrophilic side chains on the exterior surface makes globular proteins soluble in water.

3. **Salt bridges** are ionic bonds between side groups of basic and acidic amino acids, which have positive and negative charges. For example, at a pH of 7.4, the side chain of lysine has a positive charge, and the side chain of glutamic acid has a negative charge. The attraction of the oppositely charged side chains forms a strong bond called a salt bridge. If the pH changes, the basic and acidic side chains lose their ionic charges and cannot form salt bridges, which causes a change in the shape of the protein.

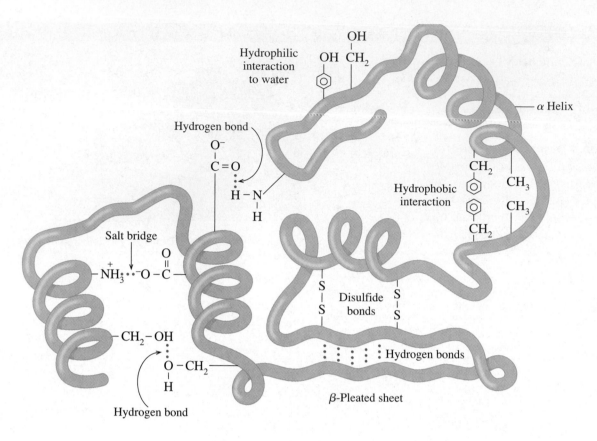

Figure 20.9 Interactions between amino acid R groups fold a protein into a specific three-dimensional shape called its tertiary structure.

Q *Why would one section of the protein chain move to the center, while another section remains on the surface of the tertiary structure?*

4. **Hydrogen bonds** form between polar amino acids. For example, a hydrogen bond can occur between the —OH of serine and the —NH_2 of glutamine.
5. **Disulfide bonds** (—S—S—) are covalent bonds that form between the —SH groups of cysteines in the polypeptide chain. In some proteins, there are several disulfide bonds between the R groups of cysteine in the polypeptide chain.

SAMPLE PROBLEM 20.8

Cross-Links in Tertiary Structures

What type of interaction would you expect between the R groups of the following amino acids?

a. cysteine and cysteine   **b.** glutamic acid and lysine

Solution

a. Because cysteine has an R group containing —SH, a disulfide bond will form.

b. An ionic bond (salt bridge) can form by the interaction of the —COO^- of glutamic acid and the —NH_3^+ of lysine.

Study Check

Would you expect to find valine and leucine in a globular protein on the outside or the inside of the tertiary structure? Why?

Table 20.5 Some Cross-Links in Tertiary Structures

	Nature of Bonding	Example
Hydrophobic interactions	Attractions between nonpolar alkyl and aromatic groups form a nonpolar center that is repelled by water	$-CH_3$ CH_3-
Hydrophilic Interactions	Attractions between polar or ionized R groups and water on the surface of the tertiary structure	$-CH_2OH \cdots\cdots O-H$ $\mid$ H
Salt bridges	Ionic interactions between ionized R groups of acidic and basic amino acids	$\underset{\parallel}{O}$ $\quad$ H $-CO^- \cdots\cdots H-\overset{+}{N}-$ $\mid$
Hydrogen bonds	Occur between polar side groups of amino acids	$\diagdown C=O \cdots\cdots HO-$ $\quad\quad\quad\quad H$ $\quad\quad\quad\quad\mid$ $\diagdown C=O \cdots\cdots H-N-$
Disulfide bonds	Strong covalent links between sulfur atoms of two cysteine amino acids	$-SH + HS- \longrightarrow -S-S-$

Protein Structure and Mad Cow Disease

Up until recently, researchers thought that only virus or bacteria were responsible for transmitting diseases. Now a group of diseases have been found in which the infectious agents are proteins called *prions*. Bovine spongiform encephalopathy (BSE), or "mad cow disease," is a fatal brain disease of cattle in which the brain fills with cavities resembling a sponge. In the noninfectious form of the prion PrP^c, the N-terminal portion is a random coil. Although the noninfectious form may be ingested from meat products, its structure can change to what is known as PrP^{sc} or *prion-related protein scrapie*. In this infectious form, the end of the peptide chain folds into a *β*-pleated sheet, which has disastrous effects on the brain and spinal cord. The conditions that cause this structural change are not yet known.

The human variant is called Creutzfeldt–Jakob (CJD) disease. Around 1955, Dr. Carleton Gajdusek was studying a disease known as Kuru, a neurological disease that was killing members of a tribe in Papua New Guinea. Because their diets were low in protein, it was a ritual to eat members of the tribe who died. As a result, the infectious agent Kuru was transmitted from one member to another. After Gajdusek identified the infectious agent in Kuru as similar to the prions that cause BSE, he received the Nobel prize.

BSE was diagnosed in Great Britain in 1986. The protein is present in nerve tissue, but is not found in meat. Control measures that exclude brain and spinal cord from animal feed are now in place to reduce the incidence of BSE. No cases of BSE have been found in the United States.

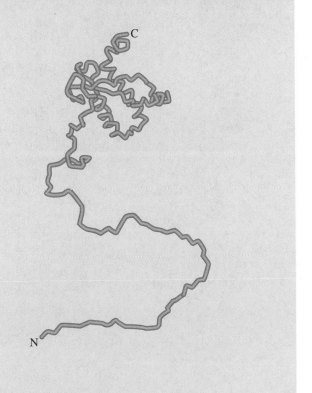

Globular and Fibrous Proteins

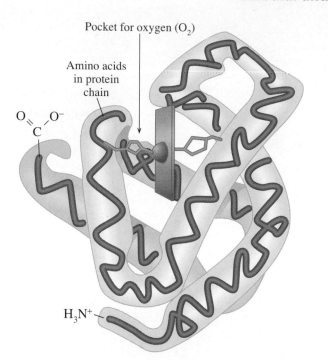

Pocket for oxygen (O₂)

Amino acids
in protein
chain

$O = C - O^-$

H_3N^+

Figure 20.10 Myoglobin is a globular protein with a heme pocket in its tertiary structure that binds oxygen to be carried to the tissues.

Q *Would hydrophilic amino acids be found on the outside or inside of the myoglobin structure?*

A group of proteins known as **globular proteins** have compact, spherical shapes because their secondary structures of the polypeptide chain fold over on top of each other. It is the globular proteins that carry out the work of the cells: functions such as synthesis, transport, and metabolism.

Myoglobin is a globular protein that stores oxygen in skeletal muscle. High concentrations of myoglobin have been found in the muscles of sea mammals, such as seals and whales, that stay under the water for long periods. Myoglobin contains 153 amino acids in a single polypeptide chain with about three-fourths of the chain in the α-helix secondary structure. The polypeptide chain, including its helical regions, forms a compact tertiary structure by folding upon itself. (See Figure 20.10.) Within the tertiary structure, a pocket of amino acids and a heme group binds and stores oxygen (O_2).

The **fibrous proteins** are proteins that consist of long, thin, fiber-like shapes. They are typically involved in the structure of cells and tissues. Two types of fibrous protein are the α- and β-keratins. The **α-keratins** are the proteins that make up hair, wool, skin, and nails. In hair, three α-helixes coil together like a braid to form a fibril. Within the fibril, the α-helices are held together by disulfide ($—S—S—$) linkages between the R groups of the many cysteine amino acids in hair. Several fibrils bind together to form a strand of hair. (Figure 20.11.) The β-keratins are the type of proteins found in the feathers of birds and scales of reptiles. In β-keratins, the proteins consist of large amounts of β-pleated sheet structure.

Quaternary Structure: Hemoglobin

When a biologically active protein consists of two or more polypeptide subunits, the structural level is referred to as a **quaternary structure.** Hemoglobin, a globular protein that transports oxygen in blood, consists of four polypeptide chains or

α helix

Alpha keratin

Figure 20.11 The fibrous proteins of α-keratin wrap together to form fibrils that make up hair and wool. The proteins called β-keratins are found in the feathers of birds and scales of reptiles.

Q *Why does hair have a large amount of cysteine amino acids?*

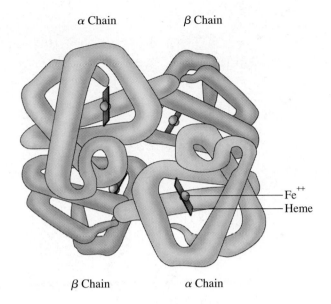

α Chain   β Chain

$$-Fe^{++}$$
$$-Heme$$

β Chain   α Chain

Figure 20.12 The quaternary structure of hemoglobin consists of four polypeptide subunits, each containing a heme group that binds an oxygen molecule.
Q *What is the difference between a tertiary structure and a quaternary structure?*

subunits, two α chains and two β chains. The subunits are held together in the quaternary structure by the same interactions that stabilize the tertiary structure, such as hydrogen bonds and salt bridges between side groups, disulfide links, and hydrophobic attractions. (See Figure 20.12.) Each subunit of the hemoglobin contains a heme group that binds oxygen. In the adult hemoglobin molecule, all four subunits $(\alpha_2\beta_2)$ must be combined for the hemoglobin to properly function as an oxygen carrier. Therefore, the complete quaternary structure of hemoglobin can bind and transport four molecules of oxygen.

Hemoglobin and myoglobin have similar biological functions. Hemoglobin carries oxygen in the blood, whereas myoglobin carries oxygen in muscle. Myoglobin, a single polypeptide chain with a molar mass of 17,000, has about one-fourth the molar mass of hemoglobin (64,000). The tertiary structure of the single polypeptide myoglobin is almost identical to the tertiary structure of each of the subunits of hemoglobin. Myoglobin stores just one molecule of oxygen, just as each subunit of hemoglobin carries one oxygen molecule. The similarity in tertiary structures allows each protein to bind and release oxygen in a similar manner. Table 20.6 and Figure 20.13 summarize the structural levels of proteins.

SAMPLE PROBLEM 20.9

Identifying Protein Structure

Indicate whether the following conditions are responsible for primary, secondary, tertiary, or quaternary protein structures:
a. Disulfide bonds form between portions of a protein chain.
b. Peptide bonds form a chain of amino acids.

Solution

a. Disulfide bonds help to stabilize the tertiary structure of a protein.
b. The sequence of amino acids in a polypeptide is a primary structure.

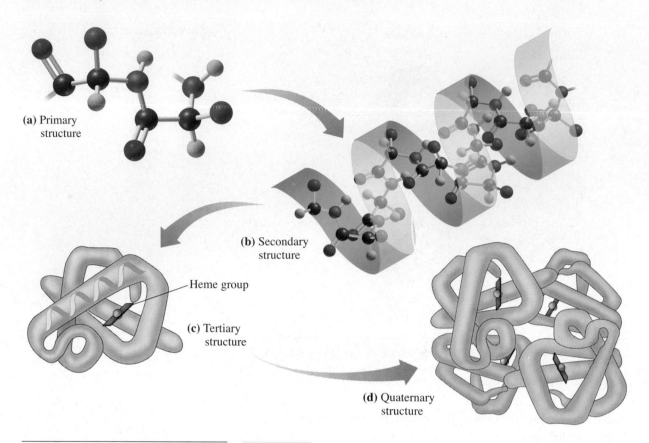

(a) Primary structure

(b) Secondary structure

Heme group

(c) Tertiary structure

(d) Quaternary structure

Figure 20.13 The quaternary structure of hemoglobin consists of four polypeptide subunits, each containing a heme group that binds an oxygen molecule.

Q *What is the difference between a tertiary structure and a quaternary structure?*

Study Check

What structural level is represented by the grouping of two subunits in insulin?

Table 20.6 Summary of Structural Levels in Proteins

Structural Level	Characteristics
Primary	The sequence of amino acids
Secondary	The coiled α-helix, β-pleated sheet, or a triple helix formed by hydrogen bonding between peptide bonds along the chain
Tertiary	A folding of the protein into a compact, three-dimensional shape stabilized by interactions between side R groups of amino acids
Quaternary	A combination of two or more protein subunits to form a larger, biologically active protein

QUESTIONS AND PROBLEMS

Protein Structure: Tertiary and Quaternary Levels

20.29 What type of interaction would you expect from the following R groups in a tertiary structure?
  a. two cysteine residues   **b.** glutamic acid and lysine
  c. serine and aspartic acid   **d.** two leucine residues

20.30 In myoglobin, about one-half of the 153 amino acids have nonpolar side chains.
  a. Where would you expect those amino acids to be located in the tertiary structure?

Sickle-Cell Anemia

Sickle-cell anemia is a disease caused by an abnormality in the shape of one of the subunits of the hemoglobin protein. In the β chain, the sixth amino acid, glutamic acid, which is polar, is replaced by valine, a nonpolar amino acid.

Because valine has a hydrophobic side chain, it draws the hydrophobic pocket that binds to oxygen to the surface of the hemoglobin. The affected red blood cells change from a rounded shape to a crescent shape, like a sickle, which interferes with their ability to transport adequate quantities of oxygen. Hydrophobic attractions cause several sickle-cell hemoglobin molecules to stick together, which forms long fibers of sickle-cell hemoglobin. The clumps of insoluble fibers clog capillaries, where they cause inflammation, pain, and organ damage. Critically low oxygen levels may occur in the affected tissues.

In sickle-cell anemia, both genes for the altered hemoglobin must be inherited. However, a few sickled cells are found in persons who carry one gene for sickle-cell hemoglobin, a condition that is also known to provide protection from malaria.

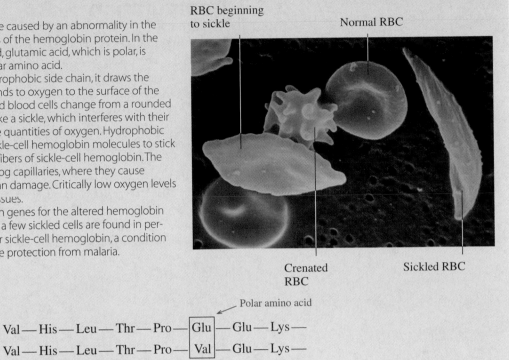

Normal β chain: Val — His — Leu — Thr — Pro — | Glu | — Glu — Lys —

Sickled β chain: Val — His — Leu — Thr — Pro — | Val | — Glu — Lys —

Polar amino acid

Nonpolar amino acid

b. Where would you expect the polar side chains to be?

c. Why is myoglobin more soluble in water than silk or wool?

20.31 A portion of a polypeptide chain contains the following sequence of amino acid residues:

  -Leu-Val-Cys-Asp-

a. Which R groups can form a disulfide cross-link?

b. Which amino acid residues are likely to be found on the inside of the protein structure? Why?

c. Which amino acid residues would be found on the outside of the protein? Why?

d. How does the primary structure of a protein affect its tertiary structure?

20.32 State whether the following statements apply to primary, secondary, tertiary, or quaternary protein structure:

a. Side groups interact to form disulfide bonds or ionic bonds.

b. Peptide bonds join amino acids in a polypeptide chain.

c. Several polypeptides are held together by hydrogen bonds between adjacent chains.

d. Hydrogen bonding between carbonyl oxygen atoms and nitrogen atoms of amide groups causes a polypeptide to coil.

e. Hydrophobic side chains seeking a nonpolar environment move toward the inside of the folded protein.

f. Protein chains of collagen form a triple helix.

g. An active protein contains four tertiary subunits.

LEARNING GOAL

Describe the hydrolysis and denaturation of proteins.

20.7 Protein Hydrolysis and Denaturation

We saw in Chapter 19 that amides can be hydrolyzed or split apart in the presence of an acid or base. Peptide bonds can be hydrolyzed too to give the individual amino acids. This is the process that occurs in the stomach when enzymes such as pepsin or trypsin catalyze the hydrolysis of proteins to give amino acids. This disrupts the primary structure by breaking the covalent amide bonds that link the amino acids. In the digestion of proteins, the amino acids are absorbed through the intestinal walls and carried to the cells where they can be used to synthesize proteins.

$$H_3\overset{+}{N}-CH-\overset{O}{\overset{\|}{C}}-NH-CH_2-\overset{O}{\overset{\|}{C}}-NH-CH-\overset{O}{\overset{\|}{C}}-O^-$$
$$\quad\quad\;\; CH_3 \quad\quad\quad\quad\quad\quad\quad\quad\quad\quad CH_2OH$$

Alanylglycylserine (Ala-Gly-Ser)

$$H_2O \;\big|\; \text{Enzyme}$$

$$H_3\overset{+}{N}-CH-\overset{O}{\overset{\|}{C}}-O^- + H_3\overset{+}{N}-CH_2-\overset{O}{\overset{\|}{C}}-O^- + H_3\overset{+}{N}-CH-\overset{O}{\overset{\|}{C}}-O^-$$
$$\quad\quad\;\; CH_3 \quad\quad\quad\quad\quad\quad\quad\quad\quad\quad\quad\quad\quad CH_2OH$$

Alanine (Ala)   Glycine (Gly)   Serine (Ser)

Denaturation of Proteins

Denaturation of a protein occurs when there is a disruption of any of the bonds that stabilize the secondary, tertiary, or quaternary structure. However, the covalent amide bonds of the primary structure are not affected.

When the interactions between the R groups are undone or altered, a globular protein unfolds like a loose piece of spaghetti. With the loss of its overall shape, the protein is no longer biologically active. (See Figure 20.14.)

Denaturing agents include heat, acids and bases, organic compounds, heavy metal ions, and mechanical agitation.

Heat

Heat denatures proteins by breaking apart hydrogen bonds and the hydrophobic attraction between nonpolar side groups. Few proteins can remain biologically active above 50°C. Whenever you cook food, you are using heat to denature protein. The nutritional value of the proteins in food is not changed, but they are made more digestible. High temperatures are also used to disinfect surgical instruments and gowns by denaturing the proteins of any bacteria present.

Acids and Bases

Placing a protein in an acid or base affects the hydrogen bonding between polar R groups and disrupts the ionic bonds (salt bridges). In the preparation of yogurt and cheese, a bacteria that produces lactic acid is added to denature the milk protein and produce solid casein. Tannic acid, a weak acid used in burn ointments, coagulates proteins at the site of the burn, forming a protective cover and preventing further loss of fluid from the burn.

Figure 20.14 Denaturation of a protein occurs when the bonds of the tertiary structure are disrupted, which destroys the shape and renders the protein biologically inactive.
Q *What are some ways in which proteins are denatured?*

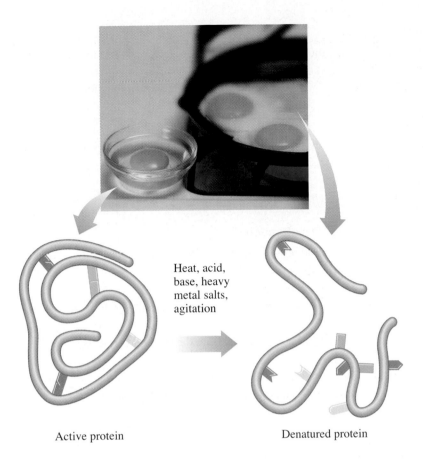

Heat, acid, base, heavy metal salts, agitation

Active protein   Denatured protein

EXPLORE YOUR WORLD

Denaturation of Milk Protein

Place some milk in five glasses. Add the following to the milk samples in glasses 1–4. The fifth glass of milk is a reference sample.

1. Vinegar, drop by drop. Stir.
2. One-half teaspoon of meat tenderizer. Stir.
3. One teaspoon of fresh pineapple juice. (Canned juice has been heated.)
4. One teaspoon of fresh pineapple juice that you have heated to boiling.

Questions

1. How did the appearance of the milk change in each of the samples?
2. What enzyme is listed on the package label of the tenderizer?
3. How does the effect of the heated pineapple juice compare with that of the fresh juice? Explain.
4. Why is cooked pineapple used when making gelatin (a protein) desserts?

Organic Compounds

Ethanol and isopropyl alcohol act as disinfectants by forming their own hydrogen bonds with a protein and disrupting the hydrophobic interactions. An alcohol swab is used to clean wounds or to prepare the skin for an injection because the alcohol passes through the cell walls and coagulates the proteins inside the bacteria.

Heavy Metal Ions

When heavy metal ions like Ag^+, Pb^{2+}, and Hg^{2+} form ionic bonds or react with the disulfide ($-S-S-$) bonds, the denatured protein solidifies. In hospitals, dilute (1%) solutions of $AgNO_3$ are placed in the eyes of newborn babies to destroy the bacteria that causes gonorrhea. These heavy metal ions can be toxic. If ingested, they act as poisons by severely disrupting body proteins, especially enzymes. An antidote is a high-protein food such as milk, eggs, or cheese that will tie up the heavy metal ions until the stomach can be pumped.

Agitation

The whipping of cream and the beating of egg whites are examples of using mechanical agitation to denature protein. The whipping action stretches the polypeptide chains until the stabilizing interactions are disrupted.

SAMPLE PROBLEM 20.10

Effects of Denaturation

What happens to the tertiary structure of a globular protein when it is placed in an acidic solution?

Solution

An acid causes denaturation by disrupting the hydrogen bonds and the ionic bonds between the R groups. A loss in interactions causes the tertiary structure to lose stability. As the protein unfolds, both the shape and biological function are lost.

Study Check

Why is a dilute solution of $AgNO_3$ used to disinfect the eyes of newborn infants?

QUESTIONS AND PROBLEMS

Protein Hydrolysis and Denaturation

20.33 What products would result from the complete hydrolysis of Gly-Ala-Ser?

20.34 Would the hydrolysis products of the tripeptide Ala-Ser-Gly be the same or different from the products in problem 20.33? Explain.

20.35 What dipeptides would be produced from the partial hydrolysis of His-Met-Gly-Val?

20.36 What tripeptides would be produced from the partial hydrolysis of Ser-Leu-Gly-Gly-Ala?

20.37 What structural level of a protein is affected by hydrolysis?

20.38 What structural level of a protein is affected by denaturation?

20.39 Indicate the changes in protein structure for each of the following:
   a. An egg placed in water at 100°C is soft boiled in about 3 min.
   b. Prior to giving an injection, the skin is wiped with an alcohol swab.
   c. Surgical instruments are placed in a 120°C autoclave.
   d. During surgery, a wound is closed by cauterization (heat).

20.40 Indicate the changes in protein structure for each of the following:
   a. Tannic acid is placed on a burn.
   b. Milk is heated to 60°C to make yogurt.
   c. To avoid spoilage, seeds are treated with a solution of $HgCl_2$.
   d. Hamburger is cooked at high temperatures to destroy *E. coli* bacteria that may cause intestinal illness.

Chapter Review

20.1 Functions of Proteins

Some proteins are enzymes or hormones, whereas others are important in structure, transport, protection, storage, and muscle contraction.

20.2 Amino Acids

A group of 20 amino acids provides the molecular building blocks of proteins. Attached to the central (alpha) carbon of each amino acid is an amino group, a carboxyl group, and a unique side group (R). The R group gives an amino acid the property of being nonpolar, polar, acidic, or basic.

20.3 Amino Acids as Acids and Bases

Amino acids exist as dipolar ions called zwitterions, as positive ions at low pH, and as negative ions at high pH levels. At the isoelectric point, zwitterions are neutral.

20.4 Formation of Peptides

Peptide bonds join amino acids to form peptides by forming an amide bond between the carboxyl group of one amino acid and the amino group of the second. Long chains of amino acids are called proteins.

20.5 Protein Structure: Primary and Secondary Levels

The primary structure of a protein is its sequence of amino acids. In the secondary structure, hydrogen bonds between peptide groups produce a characteristic shape such as an α-helix, β-pleated sheet, or a triple helix.

20.6 Protein Structure: Tertiary and Quaternary Levels

In globular proteins, the polypeptide chain including its α-helical and β-pleated sheet regions folds upon itself to form a tertiary structure. A tertiary structure is stabilized by interactions that move hydrophobic R groups to the inside and hydrophilic R groups to the outside surface, and by attrac-

tions between R groups that form hydrogen, disulfide, and salt bridges. In a quaternary structure, two or more tertiary subunits must combine for biological activity. They are held together by the same interactions found in tertiary structures.

20.7 Protein Hydrolysis and Denaturation

Denaturation of a protein occurs when heat or other denaturing agents destroy the structure of the protein (but not the primary structure) until biological activity is lost.

Key Terms

acidic amino acid An amino acid that has a carboxylic acid side chain (—COOH), which ionizes as a weak acid.

α (alpha) helix A secondary level of protein structure, in which hydrogen bonds connect the NH of one peptide bond with the C=O of a peptide bond later in the chain to form a coiled or corkscrew structure.

α-keratins Fibrous proteins that contain mostly α-helixes found in hair, nails, and skin.

amino acid The building block of proteins, consisting of an amino group, a carboxylic acid group, and a unique side group attached to the alpha carbon.

basic amino acid An amino acid that contains an amino (—NH$_2$) group that can ionize as a weak base.

β (beta)-pleated sheet A secondary level of protein structure that consists of hydrogen bonds between peptide links in parallel polypeptide chains.

C terminal The amino acid that is the last amino acid in a peptide chain with an unreacted or free carboxyl group (—COO$^-$).

collagen The most abundant form of protein in the body, which is composed of fibrils of triple helixes with hydrogen bonding between —OH groups of hydroxyproline and hydroxylysine.

denaturation The loss of secondary and tertiary protein structure caused by heat, acids, bases, organic compounds, heavy metals, and/or agitation.

disulfide bonds Covalent —S—S— bonds that form between the —SH group of cysteines in a protein to stabilize the tertiary structure.

electrophoresis The use of electrical current to separate proteins or other charged molecules with different isoelectric points.

essential amino acids Amino acids that must be supplied by the diet because they are not synthesized by the body.

fibrous protein A protein that is insoluble in water; consists of polypeptide chains with α-helixes or β-pleated sheets, that make up the fibers of hair, wool, skin, nails, and silk.

globular proteins Proteins that acquire a compact shape from attractions between the R group of the amino acid residues in the protein.

hydrogen bonds Attractions between the polar R groups such as —OH, —NH$_2$, and —COOH of amino acids in a polypeptide chain.

hydrophilic amino acid An amino acid having polar, acidic, or basic R groups that are attracted to water; "water-loving."

hydrophilic interactions the attraction between polar R groups on the protein surface and water

hydrophobic amino acid A nonpolar amino acid with hydrocarbon R groups; "water-fearing."

hydrophobic interactions The attraction between nonpolar R groups in a tertiary structure of a globular protein.

isoelectric point (pI) The pH at which an amino acid is in the neutral zwitterion form.

N terminal The amino acid in a peptide written on the left with the unreacted or free amino group (—NH$_3^+$).

nonpolar amino acids Amino acids that are not soluble in water because they contain a nonpolar R group.

peptide The combination of two or more amino acids joined by peptide bonds; dipeptide, tripeptide, and so on.

peptide bond The amide bond that joins amino acids in polypeptides and proteins.

polar amino acids Amino acids that are soluble in water because their R group is polar: hydroxyl (OH), thiol (SH), carbonyl (C=O), amino (NH$_2$), or carboxyl (COOH).

primary structure The sequence of the amino acids in a protein.

protein A term used for biologically active polypeptides that have many amino acids linked together by peptide bonds.

quaternary structure A protein structure in which two or more protein subunits form an active protein.

salt bridge The ionic bond formed between side groups of basic and acidic amino acids in a protein.

secondary structure The formation of an α-helix, β-pleated sheet, or triple helix.

tertiary structure The folding of the secondary structure of a protein into a compact structure that is stabilized

by the interactions of R groups such as ionic and disulfide bonds.

triple helix The protein structure found in collagen consisting of three polypeptide chains woven together like a braid.

zwitterion The dipolar form of an amino acid consisting of two oppositely charged ionic regions, $-NH_3^+$ and $-COO^-$.

Additional Problems

20.41 Seeds and vegetables are often deficient in one or more essential amino acids. Using the following table, state whether the following combinations would provide all the essential amino acids:

Source	Lysine	Tryptophan	Methionine
Oatmeal	No	Yes	Yes
Rice	No	Yes	Yes
Garbanzo beans	Yes	No	Yes
Lima beans	Yes	No	No
Cornmeal	No	Yes	Yes

    a. rice and garbanzo beans
    b. lima beans and cornmeal
    c. a salad of garbanzo beans and lima beans
    d. rice and lima beans
    e. rice and oatmeal
    f. oatmeal and lima beans

20.42 How does denaturation of a protein differ from its hydrolysis?

20.43 What are some differences between the following pairs?
    a. secondary and tertiary protein structures
    b. essential and nonessential amino acids
    c. polar and nonpolar amino acids
    d. di- and tripeptides
    e. an ionic bond (salt bridge) and a disulfide bond
    f. fibrous and globular proteins
    g. α-helix and β-pleated sheet
    h. tertiary and quaternary structures of proteins

20.44 The proteins placed on a gel for electrophoresis have the following isoelectric points: albumin, 4.9, hemoglobin, 6.8, and lysozyme, 11.0. A buffer of pH 6.8 is placed on the gel.
    a. Which protein will migrate toward the positive electrode?
    b. Which protein will migrate toward the negative electrode?

    c. Which protein will remain at the same place it was originally placed?

20.45 **a.** What are some functions of α-keratins?
    b. What amino acids give strength to the α-keratins?

20.46 **a.** Where is collagen found?
    b. What type of secondary structure is used to form collagen?

20.47 **a.** Draw the structure of Ser-Lys-Asp.
    b. Would you expect to find this segment at the center or at the surface of a globular protein? Why?

20.48 **a.** Draw the structure of Val-Ala-Leu.
    b. Would you expect to find this segment at the center or at the surface of a globular protein? Why?

20.49 Would you expect the following segments in a polypeptide to have an α-helix or β-pleated sheet secondary structure?
    a. a segment with a high content of Val, Pro, and Ser
    b. a segment with a high content of His, Met and Leu.

20.50 What type of interaction would you expect from the following R groups in a tertiary structure?
    a. threonine and asparagine
    b. valine and alanine
    c. arginine and aspartic acid

20.51 If serine were replaced by valine in a protein, how would the tertiary structure be affected?

20.52 If you eat rice, what other vegetable protein source(s) could you eat to ingest all essential amino acids?

20.53 Draw the structure of each of the following amino acids at pH 4.
    a. serine   **b.** alanine   **c.** lysine

20.54 Draw the structure of each of the following amino acids at pH 11.
    a. cysteine   **b.** aspartic acid   **c.** valine

21 Enzymes and Vitamins

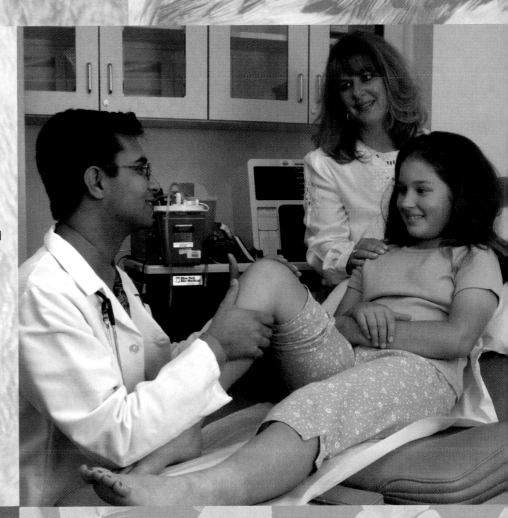

"At a time when we have a shortage of health care professionals, I think of myself as a physician extender," says Pushpinder Beasley, orthopedic physician assistant, Kaiser Hospital. "We can put a significant amount of time into our patient care. Just today, I examined a child's knee. One of the most common injuries to children is disruption of either knee ligaments or the soft tissue around the knees. In this child's case, we were checking her anterior ligaments, also known as ACL. I think an important role of the health care professional is to earn the trust of young people."

As part of a health care team, physician assistants examine patients, order laboratory tests, make diagnoses, report patient progress, order therapeutic procedures and, in most states, prescribe medications.

LOOKING AHEAD

the Chemistry place

www.chemplace.com/college

Visit the URL above or use the CD-ROM in the book for extra quizzing, interactive tutorials, and career resources.

Every second, thousands of chemical reactions occur in the cells of the human body. For example, many reactions occur to digest the food we eat, convert the products to chemical energy, and synthesize proteins and other macromolecules in our cells. In the laboratory, we can carry out reactions that hydrolyze polysaccharides, fats, or proteins, but we must use a strong acid or base, high temperatures, and long reaction times. In the cells of our body, these reactions must take place at rates that meet our physiological and metabolic needs. To make this happen, enzymes catalyze the chemical reactions in our cells, with a different enzyme for every reaction. Digestive enzymes in the mouth, stomach, and small intestine catalyze the hydrolysis of carbohydrate, fats, and proteins. Enzymes in the mitochondria extract energy from biomolecules to give us energy.

Every enzyme responds to what comes into the cells and to what the cells need. Enzymes keep reactions going when our cells need certain products, and turn off reactions when they don't need those products.

Many enzymes require cofactors to function properly. A cofactor can be an inorganic metal ion or an organic compound such as a vitamin. We obtain minerals such as zinc (Zn^{2+}) and iron (Fe^{3+}) and vitamins from our diets. A lack of minerals and vitamins can lead to certain nutritional diseases. For example, rickets is a deficiency of vitamin D and scurvy occurs when a diet is low in vitamin C.

21.1 Biological Catalysts

As a **catalyst,** an enzyme increases the rate of a reaction by changing the way a reaction takes place, but is itself not changed at the end of the reaction. An uncatalyzed reaction in a cell may take place eventually, but not at a rate fast enough for survival. For example, the hydrolysis of proteins in our diet would eventually occur without a catalyst, but not fast enough to meet the body's requirements for amino acids. The chemical reactions in our cells must occur at incredibly fast rates under mild conditions of pH 7.4 and a body temperature of 37°C. To do this, biological catalysts known as **enzymes** catalyze nearly all the chemical reactions that take place in the body. Because reactions in the cells are catalyzed, they produce only those products that are useful in the cell. Enzymes permit cells to use energy and materials efficiently while responding to cellular needs.

As catalysts, enzymes lower the activation energy for the reaction. (See Figure 21.1.) As a result, less energy is required to convert reactant molecules to products, which allows more reacting molecules to form product. However, as a catalyst an enzyme does not affect the equilibrium position, which means that there is an increase in the rates of both the forward and reverse directions. The rates of enzyme-catalyzed reactions are much faster than the rates of the uncatalyzed reactions. Some enzymes can increase the rate of a biological reaction by a factor of a billion or trillion or even a hundred million trillion compared to the rate of the uncatalyzed reaction. For example, an enzyme in the blood called carbonic anhydrase converts carbon dioxide (CO_2) and water (H_2O) to carbonic acid (H_2CO_3). In one minute, one molecule of anhydrase catalyzes the reaction of about one million (10^6) molecules.

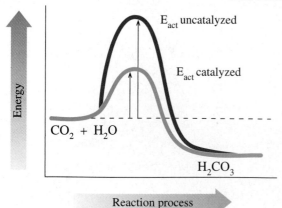

Figure 21.1 The activation energy needed for the reaction of CO_2 and H_2O is lowered by the enzyme carbonic anhydrase.

Q *Why does an enzyme accelerate a reaction?*

$$\text{CO}_2 + \text{H}_2\text{O} \underset{}{\overset{\text{Carbonic anhydrase}}{\rightleftharpoons}} \text{H}_2\text{CO}_3$$

SAMPLE PROBLEM 21.1

Biological Catalysts

How do enzymes differ from catalysts used in the laboratory?

Solution

Enzymes function under mild physiological conditions of about pH 7 and 37°C. Catalysts in the laboratory are used at high temperatures and low or high pH.

Study Check

Why are enzymes called biological catalysts?

QUESTIONS AND PROBLEMS

Biological Catalysts

21.1 Why do cellular reactions require enzymes?

21.2 How do enzymes make cellular reactions proceed at faster rates?

21.2 Names and Classification of Enzymes

LEARNING GOAL

Classify enzymes according to the reaction they catalyze.

The names of enzymes describe the compound or the reaction that is catalyzed. The actual names of enzymes are derived by replacing the end of the name of the reaction or reacting compound with the suffix *ase*. For example, an *oxidase* catalyzes an oxidation reaction, and a *dehydrogenase* removes hydrogen atoms. The compound sucrose is hydrolyzed by the enzyme *sucrase,* and a lipid is hydrolyzed by a *lipase*. Some early known enzymes use names that end in the suffix *in,* such as *papain* found in papaya, *rennin* found in milk, and *pepsin* and *trypsin,* enzymes that catalyze the hydrolysis of proteins.

The International Commission on Enzymes has classified enzymes according to the six general types of reactions they catalyze. (See Table 21.1.)

SAMPLE PROBLEM 21.2

Naming Enzymes

What chemical reaction do the following enzymes catalyze?
a. amino transferase   **b.** lactate dehydrogenase

Solution

a. catalyzes the transfer of an amino group
b. catalyzes the removal of hydrogen from lactate

Study Check

What is the class of the enzyme lipase that catalyzes the hydrolysis of ester bonds in triglycerides?

QUESTIONS AND PROBLEMS

Names and Classification of Enzymes

21.3 What types of reaction are catalyzed by each of the following classes of enzymes?
  a. oxidoreductases   **b.** transferases
  c. hydrolases

21.4 What types of reaction are catalyzed by each of the following classes of enzymes?
  a. lyases   **b.** isomerases
  c. ligases

21.5 What class of enzyme catalyzes each of the following reactions?
  a. hydrolysis of sucrose   **b.** addition of oxygen
  c. converting glucose ($C_6H_{12}O_6$) to fructose ($C_6H_{12}O_6$)
  d. moving an amino group from one molecule to another

21.6 What class of enzyme catalyzes each of the following reactions?
  a. addition of water to a double bond
  b. removing hydrogen atoms
  c. splitting peptide bonds in proteins
  d. removing CO_2 from pyruvate

21.7 Identify the class of enzyme that catalyzes each of the following reactions:

a. $CH_3-\overset{\overset{\displaystyle O}{\|}}{C}-COO^- + H^+ \longrightarrow CH_3-\overset{\overset{\displaystyle O}{\|}}{C}-H + CO_2$

b. $CH_3-\overset{\overset{\displaystyle NH_3^+}{|}}{CH}-COO^- + {}^-OOC-\overset{\overset{\displaystyle O}{\|}}{C}-CH_2-CH_3 \longrightarrow$

$CH_3-\overset{\overset{\displaystyle O}{\|}}{C}-COO^- + {}^-OOC-\overset{\overset{\displaystyle NH_3^+}{|}}{CH}-CH_2-CH_3$

21.8 Identify the class of enzyme that would catalyze each of the following reactions:

a. $CH_3-\overset{\overset{\displaystyle O}{\|}}{C}-COO^- + CO_2 + ATP \longrightarrow {}^-OOC-CH_2-\overset{\overset{\displaystyle O}{\|}}{C}-COO^- + ADP + P_i$

b. $CH_3-CH_2-OH + NAD^+ \longrightarrow CH_3-\overset{\overset{\displaystyle O}{\|}}{C}-H + NADH + H^+$

21.9 Assign a name to an enzyme that catalyzes each of the following reactions:
  a. oxidizes succinate   **b.** adds water to fumarate
  c. removes 2H from alcohol

21.10 Assign a name to an enzyme that catalyzes each of the following reactions:
  a. hydrolyzes sucrose
  b. transfers an amino group from aspartate
  c. removes a carboxylate group from pyruvate

Table 21.1 Classification of Enzymes

Class	General Reactions Catalyzed	Typical Subclasses	Function
1. Oxidoreductases	Oxidation–reduction reactions	Oxidases Reductases Dehydrogenases	Oxidation Reduction Remove 2H to form double bonds

$$CH_3-CH_2-OH \;+\; NAD^+ \xrightarrow{\text{Alcohol dehydrogenase}} CH_3-\overset{\displaystyle O}{\overset{\|}{C}}-H \;+\; NADH \;+\; H^+$$

Ethanol Coenzyme Acetaldehyde Coenzyme

Class	General Reactions Catalyzed	Typical Subclasses	Function
2. Transferases	Transfer of functional groups	Transaminases Kinases	Transfer amino groups Transfer phosphate groups

$$CH_3-\overset{\displaystyle NH_3^+}{\overset{\|}{CH}}-COO^- \;+\; {}^-OOC-\overset{\displaystyle O}{\overset{\|}{C}}-CH_2CH_2-COO^- \underset{}{\overset{\text{Alanine transaminase}}{\rightleftharpoons}} CH_3-\overset{\displaystyle O}{\overset{\|}{C}}-COO^- \;+\; {}^-OOC-\overset{\displaystyle NH_3^+}{\overset{\|}{CH}}-CH_2CH_2-COO^-$$

Alanine α-Ketoglutarate Pyruvate Glutamate

Class	General Reactions Catalyzed	Typical Subclasses	Function
3. Hydrolases	Hydrolysis reactions	Peptidases Lipases Amylases	Hydrolyze peptide bonds Hydrolyze ester bonds in lipids Hydrolyze 1,4-glycosidic bonds in amylose

$$-\overset{}{\underset{H}{N}}-\overset{R}{\underset{}{CH}}-\overset{O}{\overset{\|}{C}}-\overset{}{\underset{H}{N}}-\overset{R}{\underset{}{CH}}-COO^- \;+\; H_2O \xrightarrow{\text{Peptidase}} -\overset{}{\underset{H}{N}}-\overset{R}{\underset{}{CH}}-\overset{O}{\overset{\|}{C}}-O^- \;+\; H_3\overset{+}{N}-\overset{R}{\underset{}{CH}}-COO^-$$

Polypeptide C terminal Shorter polypeptide Amino acid from C terminal

Class	General Reactions Catalyzed	Typical Subclasses	Function
4. Lyases	Addition of a group to a double bond or removal of a group from a double bond without hydrolysis or oxidation	Decarboxylases Dehydrases Deaminases	Remove CO_2 Remove H_2O Remove NH_3

$$CH_3-\overset{\displaystyle O}{\overset{\|}{C}}-COO^- \;+\; H^+ \xrightarrow{\text{Pyruvate decarboxylase}} CH_3-\overset{\displaystyle O}{\overset{\|}{C}}-H \;+\; CO_2$$

Pyruvate Acetaldehyde Carbon dioxide

Class	General Reactions Catalyzed	Typical Subclasses	Function
5. Isomerases	Rearrangement of atoms to form isomers	Isomerases Epimerases	Convert cis and trans Convert D and L isomers

$$\overset{{}^-OOC}{\underset{H}{\diagdown}}C=C\overset{COO^-}{\underset{H}{\diagup}} \underset{}{\overset{\text{Maleate isomerase}}{\rightleftharpoons}} \overset{{}^-OOC}{\underset{H}{\diagdown}}C=C\overset{H}{\underset{COO^-}{\diagup}}$$

Maleate Fumarate

Class	General Reactions Catalyzed	Typical Subclasses	Function
6. Ligases	Bonding of molecules using ATP energy	Synthetases Carboxylases	Combine molecules Add CO_2

$${}^-OOC-\overset{\displaystyle O}{\overset{\|}{C}}-CH_3 \;+\; CO_2 \;+\; ATP \xrightarrow{\text{Pyruvate carboxylase}} {}^-OOC-\overset{\displaystyle O}{\overset{\|}{C}}-CH_2-COO^- \;+\; ADP \;+\; P_i \;+\; H^+$$

Pyruvate Oxaloacetate

Describe the role of an enzyme in an enzyme-catalyzed reaction.

21.3 Enzymes as Catalysts

Nearly all enzymes are globular proteins. Each has a unique three-dimensional shape that recognizes and binds a small group of reacting molecules, which are called **substrates.** The tertiary structure of an enzyme plays an important role in how that enzyme catalyzes reactions.

Active Site

In a catalyzed reaction, an enzyme must first bind to a substrate in a way that favors catalysis. A typical enzyme is much larger than its substrate. However, within its large tertiary structure, there is a region called the **active site** where the enzyme binds a substrate or substrates and catalyzes the reaction. This active site is often a small pocket that closely fits the structure of the substrate. (See Figure 21.2) Within the active site, the side chains of amino acids bind the substrate with hydrogen bonds, salt bridges, or hydrophobic attractions. The active site of a particular enzyme fits the shape of only a few types of substrates, which makes enzymes very specific about the type of substrate they bind.

Some enzymes show absolute specificity by catalyzing one reaction of one specific substrate. Other enzymes catalyze a reaction for a group of substrates. Still other enzymes catalyze a reaction for a specific type of bond in a substrate. Types of enzyme specificity are listed in Table 21.2.

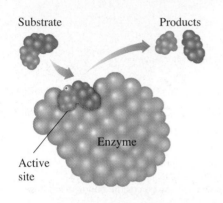

Substrate   Products

Enzyme

Active
site

Figure 21.2 On the surface of an enzyme, a small region called an active site binds a substrate and catalyzes a reaction of that substrate.

Q *Why does an enzyme catalyze a reaction of only certain substrates?*

Table 21.2	Types of Enzyme Specificity	
Type	**Reaction Type**	**Example**
Absolute	Catalyze one type of reaction for a single substrate	Urease catalyzes only the hydrolysis of urea
Group	Catalyze one type of reaction for similar substrates	Hexokinase adds a phosphate group to hexoses
Linkage	Catalyze one type of reaction for a specific type of bond	Chymotrypsin catalyzes the hydrolysis of peptide bonds

Lock-and-Key and Induced Fit Models

In an early theory of enzyme action called the **lock-and-key model,** the active site is described as having a rigid, nonflexible shape. Thus only those substrates with shapes that fit exactly into the active site are able to bind with that enzyme. The shape of the active site is analogous to a lock, and the proper substrate is the key that fits into the lock. (See Figure 21.3a.)

While the lock-and-key model explains the binding of substrates for many enzymes, certain enzymes have a broader range of specificity than the lock and key model allows. In the **induced-fit model,** there is an interaction between both the enzyme and substrate. (See Figure 21.3b.) The active site adjusts to fit the shape of the substrate more closely. At the same time the substrate adjusts its shape to better adapt to the geometry of the active site. As a result, the reacting section of the substrate becomes aligned exactly with the groups in the active site that catalyze the reaction.

In the induced-fit model, substrate and enzyme work together to acquire a geometrical arrangement that lowers the activation energy. A different substrate could not induce these structural changes and no catalysis would occur. (See Figure 21.3c.)

What is the function of the active site in an enzyme?

Solution

The active site in an enzyme binds the substrate and contains the amino acid side chains that bind the substrate and catalyze the reaction.

Study Check

How do the lock-and-key and the induced-fit models differ in their description of the active site in an enzyme?

Enzyme Catalyzed Reaction

The proper alignment of a substrate within the active site forms an **enzyme–substrate (ES) complex.** This combination of enzyme and substrate provides an alternative pathway for the reaction that has a lower activation energy. Within the active site, the amino acid side chains take part in catalyzing the chemical reaction. For example, acidic and basic side chains remove protons from or provide protons for the substrate. As soon as the catalyzed reaction is complete, the products are quickly released from the enzyme so it can bind to a new substrate molecule. We can write the catalyzed reaction of an enzyme (E) with a substrate (S) to form product (P) as follows:

Step 1   E + S $\rightleftharpoons$ ES

Step 2   ES $\longrightarrow$ E + P
_____
E + S $\rightleftharpoons$ ES $\longrightarrow$ E + P
Enzyme + substrate   ES complex   Enzyme + product

Let's consider the hydrolysis of sucrose by sucrase. When sucrose binds to the active site of sucrase, the glycosidic bond of sucrose is placed into a geometry favorable for reaction. The amino acid side chains catalyze the cleavage of the sucrose to give the products glucose and fructose.

Sucrase + sucrose $\rightleftharpoons$ sucrase-sucrose complex $\longrightarrow$ sucrase + glucose + fructose

E   +   S   ES complex   E   +   P$_1$ +   P$_2$

Because the structures of the products are no longer attracted to the active site, they are released and the sucrase binds another sucrose substrate. (See Figure 21.4).

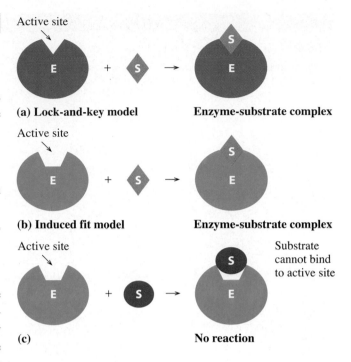

(a) Lock-and-key model   Enzyme-substrate complex

(b) Induced fit model   Enzyme-substrate complex

(c)   No reaction

Figure 21.3 (a) In the lock-and-key model, a substrate fits the shape of the active site and forms an enzyme–substrate complex. **(b)** In the induced-fit model, a flexible active site and substrate adapt to provide a close fit to a substrate and proper orientation for reaction. **(c)** A substrate that does not fit or induce a fit in the active site cannot undergo catalysis by the enzyme.
Q *How does the induced-fit model differ from the lock-and-key model?*

Enzymes as Catalysts

21.11 Match the following three terms, (1) enzyme–substrate complex, (2) enzyme, and (3) substrate, with these phrases:
  a. has a tertiary structure that recognizes the substrate
  b. the combination of an enzyme with the substrate
  c. has a structure that fits the active site of an enzyme

21.12 Match the following three terms, (1) active site, (2) lock-and-key model, and (3) induced-fit model, with these phrases:
  a. the portion of an enzyme where catalytic activity occurs

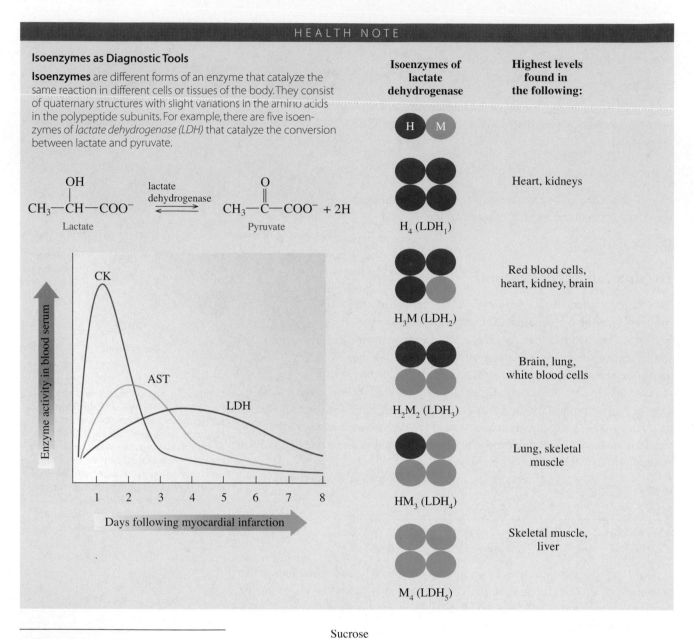

HEALTH NOTE

Isoenzymes as Diagnostic Tools

Isoenzymes are different forms of an enzyme that catalyze the same reaction in different cells or tissues of the body. They consist of quaternary structures with slight variations in the amino acids in the polypeptide subunits. For example, there are five isoenzymes of *lactate dehydrogenase (LDH)* that catalyze the conversion between lactate and pyruvate.

$$CH_3-\underset{\underset{Lactate}{|}}{\overset{\overset{OH}{|}}{CH}}-COO^- \underset{\underset{dehydrogenase}{}}{\overset{lactate}{\rightleftharpoons}} CH_3-\underset{\underset{Pyruvate}{}}{\overset{\overset{O}{\parallel}}{C}}-COO^- + 2H$$

Enzyme activity in blood serum (vertical axis)

CK, AST, LDH

Days following myocardial infarction (1–8)

Isoenzymes of lactate dehydrogenase	Highest levels found in the following:
H, M	
H_4 (LDH$_1$)	Heart, kidneys
H_3M (LDH$_2$)	Red blood cells, heart, kidney, brain
H_2M_2 (LDH$_3$)	Brain, lung, white blood cells
HM_3 (LDH$_4$)	Lung, skeletal muscle
M_4 (LDH$_5$)	Skeletal muscle, liver

Figure 21.4 After sucrose binds to sucrase at the active site, it is properly aligned for the hydrolysis reaction by the active site. The monosaccharide products dissociate from the active site and the enzyme is ready to bind to another sucrose molecule.

Q *Why does the enzyme-catalyzed hydrolysis of sucrose go faster than the hydrolysis of sucrose in the chemistry laboratory?*

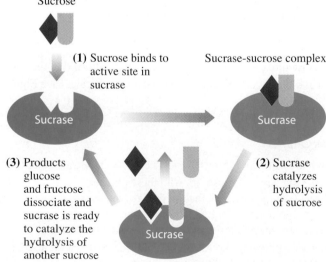

Sucrose

(1) Sucrose binds to active site in sucrase

Sucrase-sucrose complex

(2) Sucrase catalyzes hydrolysis of sucrose

(3) Products glucose and fructose dissociate and sucrase is ready to catalyze the hydrolysis of another sucrose

Each LDH isoenzyme contains a mix of polypeptide subunits, M and H. In the liver and muscle, lactate is converted to pyruvate by a LDH$_5$ isoenzyme with M subunits designated M$_4$. In the heart, the same reaction is catalyzed by a LDH$_1$ isoenzyme (H$_4$) containing four H subunits. Different combinations of the M and H subunits are found in the LDH isoenzymes of the brain, red blood cells, kidney, and white blood cells.

The different forms of an enzyme allow a medical diagnosis of damage or disease to a particular organ or tissue. In healthy tissues, isoenzymes function within the cells. However, when a disease damages a particular organ, cells die, which releases cell contents including the isoenzymes into the blood. Measurements of the elevated levels of specific isoenzymes in the blood serum help to identify the disease and its location in the body. For example, an

elevation in the serum LDH$_5$, which is the M$_4$ isoenzyme of lactate dehydrogenase, indicates liver damage or disease. When a myocardial infarction (MI), or heart attack, damages the cells in heart muscle, an increase in the level of LDH$_1$ (H$_4$) isoenzyme is detected in the blood serum. (See Table 21.3.)

Another isoenzyme used diagnostically is creatine kinase (CK), which consists of two types of polypeptide subunits. One subunit (B) is prevalent in the brain and the other predominates in skeletal muscle (M). Normally only the CK$_3$ is present in low amounts in the blood serum. However, in a patient who has suffered an MI, the levels of CK$_2$ will be elevated soon after the heart attack. Table 21.4 lists some enzymes used to diagnose tissue damage and diseases of certain organs.

Table 21.3   Isoenzymes of Lactate Dehydrogenase and Creatinine Kinase

Isoenzyme	Abundant in	Subunits
Lactate Dehydrogenase (LDH)		
LDH$_1$	Heart, kidneys	H$_4$
LDH$_2$	Red blood cells, heart, kidney, brain	MH$_3$
LDH$_3$	Brain, lung, white blood cells	M$_2$H$_2$
LDH$_4$	Lung, skeletal muscle	M$_3$H
LDH$_5$	Skeletal muscle, liver	M$_4$
Creatinine Kinanse (CK)		
CK$_1$	Brain, lung	BB
CK$_2$	Heart	MB
CK$_3$	Skeletal muscle, red blood cells	MM

Table 21.4   Serum Enzymes Used in Diagnosis of Tissue Damage

Condition	Diagnostic Enzymes Elevated
Heart attack, or liver disease (cirrhosis, hepatitis)	Lactate dehydrogenase (LDH) Aspartate transaminase (AST)
Heart attack	Creatine kinase (CK)
Hepatitis	Alanine transaminase (ALT)
Liver (carcinoma) or bone disease (rickets)	Alkaline phosphatase (ALP)
Pancreatic disease lipase (LPS)	Amylase, cholinesterase,
Prostate carcinoma	Acid phosphatase (ACP)

    b. an active site that adapts to the shape of a substrate
    c. an active site that has a rigid shape

21.13 **a.** Write an equation that represents an enzyme-catalyzed reaction.
    b. How is the active site different from the whole enzyme structure?

21.14 **a.** Why does an enzyme speed up the reaction of a substrate?
    b. After the products have formed, what happens to the enzyme?

21.15 What are isoenzymes?

21.16 How is the LDH isoenzyme in the heart different from LDH isoenzyme in the liver?

21.17 A patient arrives in emergency complaining of chest pains. What enzymes would you test for in the blood serum?

21.18 A patient who is an alcoholic has elevated levels of LDH and AST. What condition might be indicated?

21.4 Factors Affecting Enzyme Activity

LEARNING GOAL

Describe the effect of temperature, pH, concentration of enzyme, and concentration of substrate on enzyme activity.

The **activity** of an enzyme describes how fast an enzyme catalyzes the reaction that converts a substrate to product. This activity is strongly affected by reaction conditions, which include the temperature, pH, concentration of the substrate, and concentration of the enzyme.

Enzyme Activity

The enzymes on the surface of a freshly cut apple, avocado, or banana react with oxygen in the air to turn the surface brown. An antioxidant, such as vitamin C in lemon juice, prevents the oxidation reaction. Cut an apple, an avocado, or a banana into several slices. Place one slice in a plastic zipper lock bag, squeeze out all the air, and close the zipper lock. Dip another slice in lemon juice and place it on a plate. Sprinkle another slice with a crushed vitamin C tablet. Leave another slice alone as a control. Observe the surface of each of your samples. Record your observations immediately, then every hour for 6 hours or longer.

Questions

1. Which slice(s) shows the most oxidation (turns brown)?
2. Which slice(s) shows little or no oxidation?
3. How was the oxidation reaction on each slice affected by treatment with an antioxidant?

Temperature

Enzymes are very sensitive to temperature. At low temperatures, most enzymes show little activity because there is not a sufficient amount of energy for the catalyzed reaction to take place. At higher temperatures, enzyme activity increases as reacting molecules move faster to cause more collisions with enzymes. Enzymes are most active at **optimum temperature,** which is 37°C or body temperature for most enzymes. (See Figure 21.5.) At temperatures above 50°C, the tertiary structure and thus the shape of most proteins is destroyed causing a loss in enzyme activity. For this reason, equipment in hospitals and laboratories is sterilized in autoclaves where the high temperatures denature the enzymes in harmful bacteria.

pH

Enzymes are most active at their **optimum pH,** the pH that maintains the proper tertiary structure of the protein. (See Figure 21.6.) A pH value above or below the optimum pH causes a change in the three-dimensional structure of the enzyme that disrupts the active site. As a result the enzyme cannot bind substrate properly and no reaction occurs.

Enzymes in most cells have optimum pH values at physiological pH values around 7.4. However, enzymes in the stomach have a low optimum pH because they hydrolyze proteins at the acidic pH in the stomach. For example, pepsin, a digestive enzyme in the stomach has an optimum pH of 2. Between meals, the pH in the stomach is 4 or 5 and pepsin shows little or no digestive activity. When food enters the stomach, the secretion of HCl lowers the pH to about 2, which activates pepsin.

If small changes in pH are corrected, an enzyme can regain its structure and activity. However, large variations from optimum pH permanently destroy the structure of the enzyme. Table 21.5 lists the optimum pH values for selected enzymes.

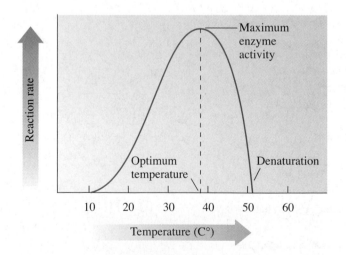

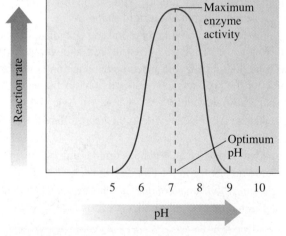

Figure 21.5 An enzyme attains maximum activity at its optimum temperature, usually 37°C. Lower temperatures slow the rate of reaction and temperatures above 50°C cause denaturation of the enzyme protein and loss of catalytic activity.
Q *Why is 37°C the optimum temperature for many enzymes?*

Figure 21.6 Enzymes are most active at their optimum pH. At a higher or lower pH, denaturation of the enzyme causes a loss of catalytic activity.
Q *Why does the digestive enzyme pepsin have an optimum pH of 2?*

Table 21.5   Optimum pH for Selected Enzymes

Enzyme	Location	Substrate	Optimum pH
Pepsin	Stomach	Peptide bonds	2
Urease	Liver	Urea	5
Sucrase	Small intestine	Sucrose	6.2
Pancreatic amylase	Pancreas	Amylose	7
Trypsin	Small intestine	Peptide bonds	8
Arginase	Liver	Arginine	9.7

Enzyme and Substrate Concentration

In any catalyzed reaction, the substrate must first bind with the enzyme to form the substrate–enzyme complex. Increasing the enzyme concentration when the substrate concentration remains constant increases the rate of the catalyzed reaction and thus enzyme activity. At higher concentrations more enzyme molecules are available to bind and catalyze the reaction of substrate molecules. When the enzyme concentration is increased to twice the initial concentration, the rate of the catalyzed reaction is twice as fast. If the enzyme concentration is increased to three times the initial enzyme concentration, the rate of reaction will increase also to three times as fast. There is a direct relationship between the enzyme concentration and enzyme activity. (See Figure 21.7a).

When enzyme concentration is kept constant, increasing the substrate concentration increases the rate of the catalyzed reaction as long as there are more enzyme molecules present than substrate molecules. At some point an increase in substrate concentration saturates the enzyme. With all the available enzyme molecules bonded to substrate, the rate of the catalyzed reaction reaches its maximum. Adding more substrate molecules cannot increase the rate further. (See Figure 21.7b.)

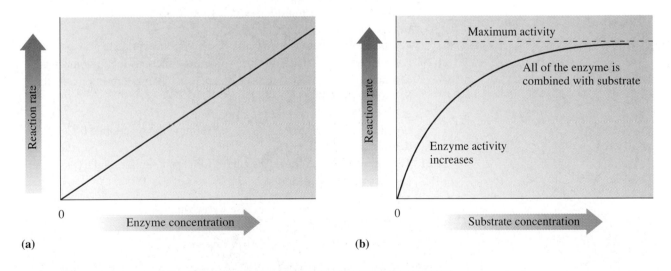

Figure 21.7 (a) The rate of reaction increases when the enzyme concentration increases at a fixed substrate concentration. **(b)** Increasing the substrate concentration increases the rate of reaction until the enzyme molecules are saturated with substrate.
Q *What happens to the rate of reaction when substrate saturates enzyme?*

SAMPLE PROBLEM 21.4

Factors Affecting Enzymatic Activity

Describe what effect the changes in parts a and b would have on the rate of the reaction catalyzed by urease.

$$\underset{\text{Urea}}{H_2N-\overset{\overset{\displaystyle O}{\|}}{C}-NH_2} + H_2O \xrightarrow{\text{Urease}} 2NH_3 + CO_2$$

a. increasing the urea concentration
b. lowering the temperature to 10°C

Solution

a. An increase in urea concentration will increase the rate of reaction until all of the enzyme molecules are bound to the urea substrate. No further increase in rate occurs.

b. Because 10°C is lower than the optimum temperature of 37°C, the lower temperature will decrease the rate of the reaction.

Study Check

If urease has an optimum pH of 5, what is the effect of lowering the pH to 3?

QUESTIONS AND PROBLEMS

Factors Affecting Enzyme Action

21.19 Trypsin, a peptidase that hydrolyzes polypeptides, functions in the small intestine at an optimum pH of 8. How is the rate of a trypsin-catalyzed reaction affected by each of the following conditions?
a. lowering the concentration of polypeptides
b. changing the pH to 3
c. running the reaction at 75°C **d.** adding more trypsin

21.20 Pepsin, a peptidase that hydrolyzes proteins, functions in the stomach at an optimum pH of 2. How is the rate of a pepsin-catalyzed reaction affected by each of the following conditions?
a. increasing the concentration of proteins
b. changing the pH to 5 **c.** running the reaction at 0°C
d. using less pepsin

21.21 The following graph shows the curves for pepsin, urease, and trypsin. Estimate the optimum pH for each.

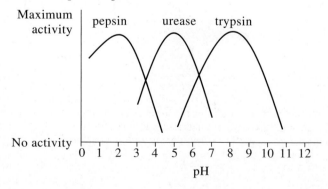

21.22 Refer to the graph in problem 21.21 to determine if the reaction rate in each condition will be at the optimum rate or not.
a. trypsin, pH 5   **b.** urease, pH 5   **c.** pepsin, pH 4
d. trypsin, pH 8   **e.** pepsin, pH 2

21.5 Enzyme Inhibition

LEARNING GOAL

Describe reversible and irreversible inhibition.

Many kinds of molecules called **inhibitors** cause enzymes to lose catalytic activity. Although inhibitors act differently, they all prevent the active site from binding with a substrate. Some inhibitors cause a **reversible inhibition,** which means that the enzyme regains activity when the inhibitor dissociates from the enzyme. In an **irreversible inhibition,** an inhibitor bonds covalently with an enzyme and cannot be removed, which makes the loss of enzyme activity irreversible.

Reversible Competitive Inhibition

Reversible inhibition can be competitive or noncompetitive. In competitive inhibition, an inhibitor competes for the active site, whereas in noncompetitive inhibition, the inhibitor acts on a site that is not the active site.

A **competitive inhibitor** has a structure that is so similar to the substrate it can bond to the enzyme just like the substrate. Thus the competitive inhibitor competes with the substrate for the active site on the enzyme. As long as the inhibitor occupies the active site, the substrate cannot bind to the enzyme and no reaction takes place. (See Figure 21.8.)

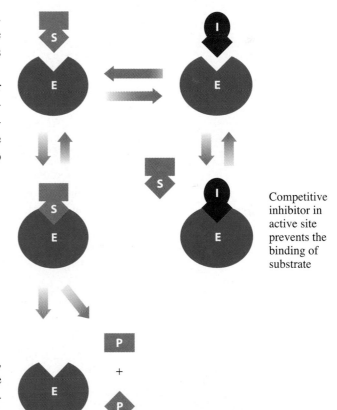

Competitive inhibitor in active site prevents the binding of substrate

$$S \rightleftharpoons ES \longrightarrow E + P$$

Favored by increasing [S]

Enzyme–substate complex

E

+

$$I \rightleftharpoons EI$$ Favored when [S] is low

Enzyme–inhibitor complex

As long as the concentration of the inhibitor is substantial, there is a loss of enzyme activity. However, increasing the substrate concentration displaces more of the inhibitor molecules. As more enzyme molecules bind to substrate (ES), enzyme activity is regained.

Malonate is a competitive inhibitor of the enzyme succinate dehydrogenase. Because malonate has a structure similar to the substrate, the two substances compete for the active site on the dehydrogenase. As long as the inhibitor occupies the active site on the enzyme, no reaction occurs. When more succinate is added, more active sites will fill with substrate, and there will be less inhibition.

Figure 21.8 With a structure similar to the substrate for an enzyme, a competitive inhibitor also fits the active site and competes with the substrate when both are present.
Q *Why does increasing the substrate concentration reverse the inhibition by a competitive inhibitor?*

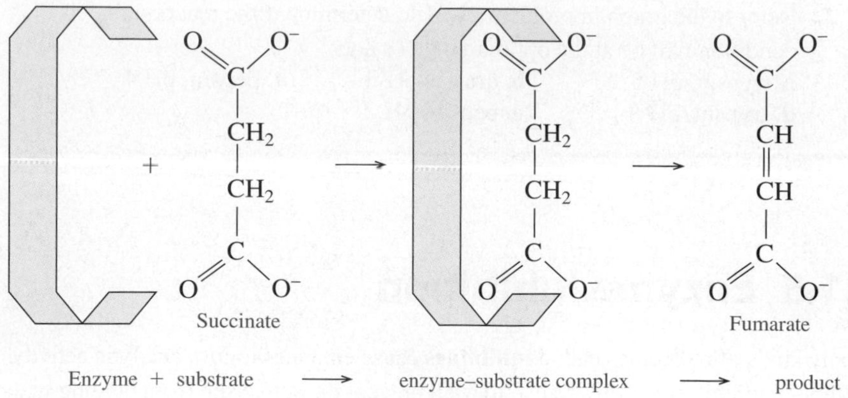

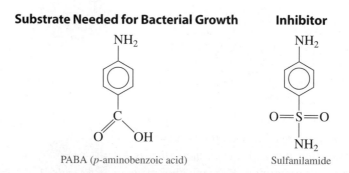

Enzyme + substrate ⟶ enzyme–substrate complex ⟶ product

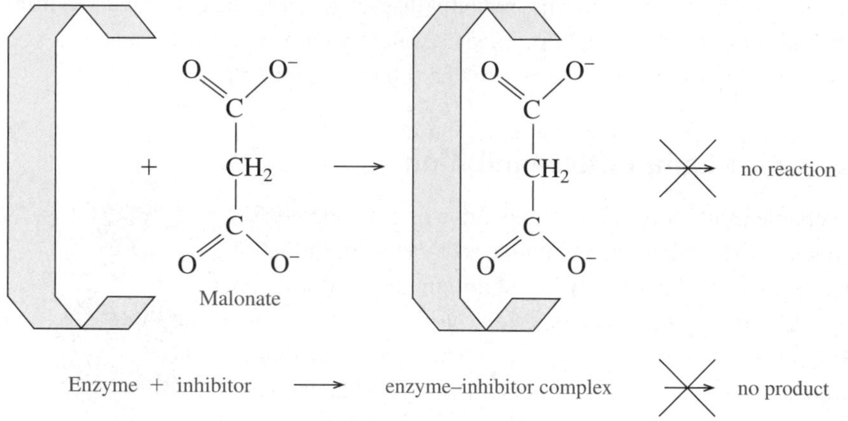

Enzyme + inhibitor ⟶ enzyme–inhibitor complex ⟶ no product

Some bacterial infections are treated with competitive inhibitors called antimetabolites. Sulfanilamide, one of the first sulfa drugs, competes with PABA (*p*-aminobenzoic acid), which is an essential substance (metabolite) in the growth cycle of bacteria.

Substrate Needed for Bacterial Growth

PABA (*p*-aminobenzoic acid)

Inhibitor

Sulfanilamide

Reversible Noncompetitive Inhibition

The structure of a **noncompetitive inhibitor** does not resemble the substrate and does not compete for the active site. Instead a noncompetitive inhibitor binds to a site on the enzyme that is not the active site. When the noncompetitive inhibitor is bonded to the enzyme, the shape of the enzyme is distorted. Inhibition occurs because the substrate cannot fit in the active site, or it does not fit properly. Without the proper alignment of substrate with the amino acid side groups, no catalysis can take place. (See Figure 21.9.)

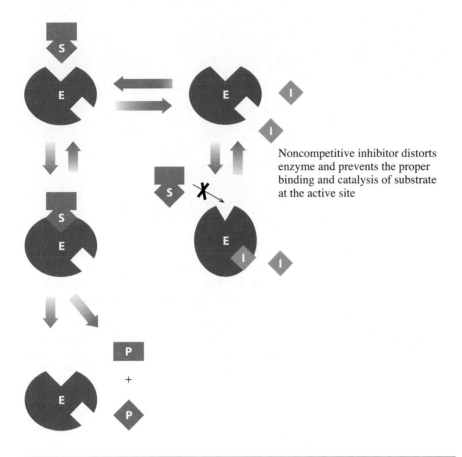

Noncompetitive inhibitor distorts enzyme and prevents the proper binding and catalysis of substrate at the active site

Figure 21.9 A noncompetitive inhibitor binds to an enzyme at a site other than the active site, which distorts the enzyme and prevents the proper binding and catalysis of the substrate at the active site.

Q *Why won't an increase in substrate concentration reverse the inhibition by a noncompetitive inhibitor?*

Because a noncompetitive inhibitor is not competing for the active site, the addition of more substrate does not reverse this type of inhibition. However, enzyme activity can be regained by lowering the concentration of the noncompetitive inhibitor making more enzyme molecules available. Examples of noncompetitive inhibitors are the heavy metal ions Pb^{2+}, Ag^+, and Hg^{2+} that bond with amino acid side groups such as $-COO^-$, or $-OH$. Catalytic activity is restored when chemical reagents remove the inhibitors.

Irreversible Inhibition

In irreversible inhibition, a molecule causes an enzyme to lose all enzymatic activity. Most irreversible inhibitors are toxic substances that destroy enzymes. Usually an irreversible inhibitor forms a covalent bond with an amino acid side group within the active site, which prevents the substrate from entering the active site or prevents catalytic activity.

Insecticides and nerve gases act as irreversible inhibitors of acetylcholinesterase, an enzyme needed for nerve conduction. The compound DFP (diisopropyl fluorophosphate) forms a covalent bond with the side chain $-CH_2OH$ of serine in the active site. When acetylcholinesterase is inhibited, the transmission of nerve impulses is blocked, and paralysis occurs.

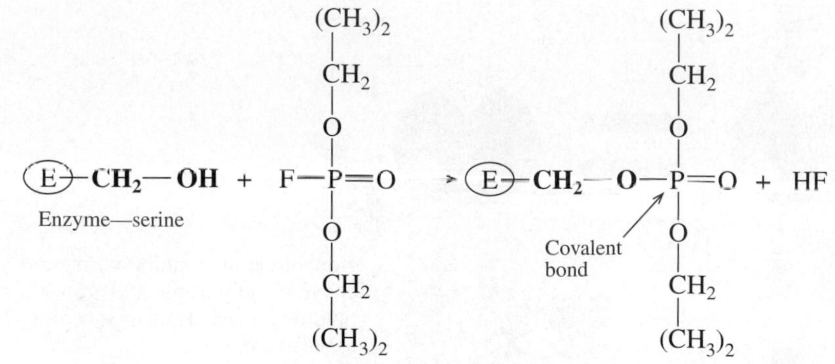

$$\boxed{E}\text{—CH}_2\text{—OH} + \text{F—P=O} \longrightarrow \boxed{E}\text{—CH}_2\text{—O—P=O} + \text{HF}$$

Enzyme—serine

DFP (Diisopropyl fluorophosphate)   Serine covalently bonded to DFP

Antibiotics produced by bacteria, mold, or yeast are irreversible inhibitors used to inhibit bacterial growth. For example, penicillin inhibits an enzyme needed for the formation of cell walls in bacteria, but not human cell membranes. With an incomplete cell wall, bacteria cannot survive, and the infection is stopped. However, some bacteria are resistant to penicillin because they produce penicillinase, an enzyme that breaks down penicillin. Over the years, derivatives of penicillin to which bacteria have not yet become resistant have been produced. Some irreversible enzyme inhibitors are listed in Table 21.6.

Table 21.6   Selected Irreversible Enzyme Inhibitors

Name	Structure	Natural/Synthetic Source	Inhibitory Action
Cyanide	CN^-	Bitter almonds	Bonds to metal ions in enzymes in the electron transport chain
Sarin	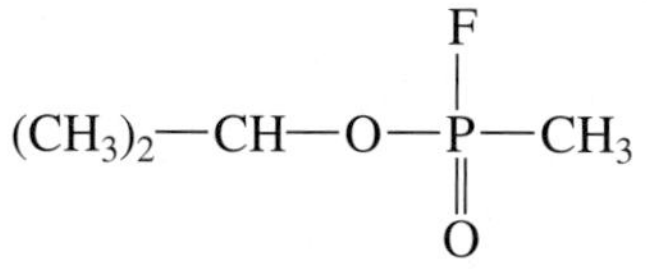 $(CH_3)_2\text{—CH—O—P(F)(O)—CH}_3$	Nerve gas	Similar to DFP
Parathion	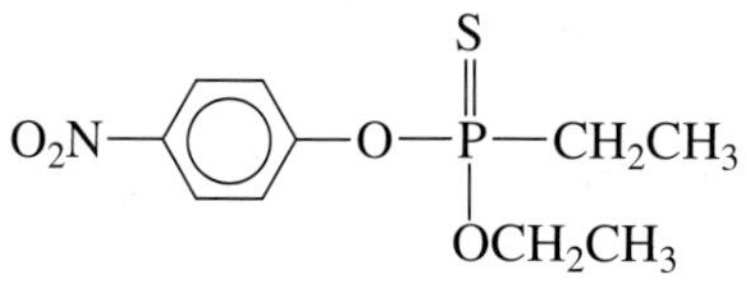 $O_2N\text{—}\langle\rangle\text{—O—P(S)(OCH}_2CH_3)\text{—CH}_2CH_3$	Insecticide	Similar to DFP
Penicillin	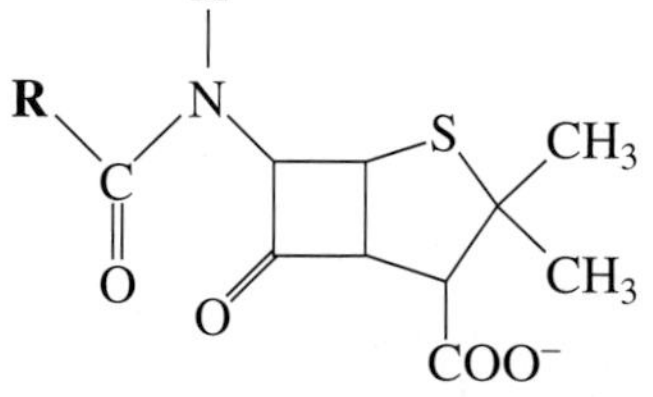 *(structure shown)*	*Penicillium* fungus	Inhibits enzymes that build cell walls in bacteria

R Groups for Penicillin Derivatives

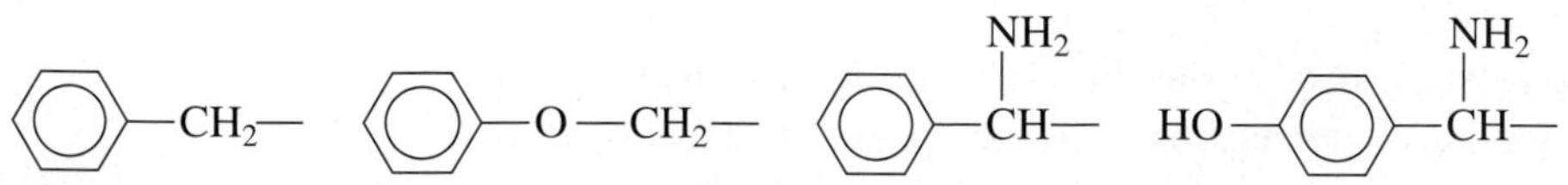

SAMPLE PROBLEM 21.5

Enzyme Inhibition

State the type of reversible inhibition in the following:

a. The inhibitor has a structure that is similar to the substrate.

b. This inhibitor binds to the surface of the enzyme, changing its shape in such a way that it cannot bind to substrate.

Solution

a. competitive inhibition

b. noncompetitive inhibition

Study Check

Hydrogen cyanide (HCN) forms covalent bonds with catalase, an enzyme that contains iron (Fe^{3+}). What type of inhibitor is HCN?

QUESTIONS AND PROBLEMS

Enzyme Inhibition

21.23 Indicate whether the following describe a reversible competitive or a reversible noncompetitive enzyme inhibitor:

    a. The inhibitor has a structure similar to the substrate.

    b. The effect of the inhibitor cannot be reversed by adding more substrate.

    c. The inhibitor competes with the substrate for the active site.

    d. The structure of the inhibitor is not similar to the substrate.

    e. The addition of more substrate reverses the inhibition.

21.24 Oxaloacetate is an inhibitor of succinate dehydrogenase:

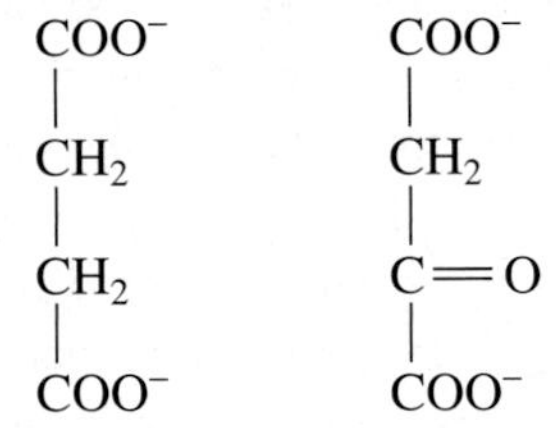

$$
\begin{array}{ll}
COO^- & COO^- \\
| & | \\
CH_2 & CH_2 \\
| & | \\
CH_2 & C{=}O \\
| & | \\
COO^- & COO^- \\
\text{Succinate} & \text{Oxaloacetate}
\end{array}
$$

    a. Would you expect oxaloacetate to be a reversible competitive or a noncompetitive inhibitor? Why?

    b. Would oxaloacetate bind to the active site or elsewhere on the enzyme?

    c. How would you reverse the effect of the inhibitor?

21.25 Methanol and ethanol are oxidized by alcohol dehydrogenase. In methanol poisoning, a high concentration of ethanol is given intravenously as an antidote.

    a. Compare the structure of methanol and ethanol.

    b. Would ethanol compete for the active site or bind to a different site?

    c. Would ethanol be a competitive or noncompetitive inhibitor of methanol oxidation?

21.26 In humans, the antibiotic amoxycillin (a type of penicillin) is used to treat certain bacterial infections.

    a. Does the antibiotic inhibit enzymes in humans?

    b. Why does the antibiotic kill bacteria, but not humans?

    c. Are antibiotics reversible or irreversible inhibitors?

21.6 Regulation of Enzyme Activity

In any enzyme-catalyzed reaction, compounds are produced only in the amounts and at the times they are needed in the cell. This means that the rate of a catalyzed reaction must be controlled so it can speed up when more molecules of a compound are needed and slow down when an accumulation of that compound occurs. There are several factors in the cell that regulate enzyme activity.

Zymogens

Although most enzymes are active as soon as they are synthesized and acquire their tertiary structure, some are produced as inactive precursors called **zymogens** or *proenzymes*. Zymogens, which contain longer protein chains, are activated by the removal of peptide sections from the protein. Zymogens are often produced in an organ where they are stored and then transported to where they are needed, at which time they are activated.

Insulin

The protein hormone insulin is synthesized as inactive proinsulin. Recall that insulin contains two polypeptide chains (Chapter 20) linked by disulfide bonds. In proinsulin, the two chains are connected by a polypeptide of 33 amino acids. This peptide section is removed to form the active insulin hormone. (See Figure 21.10.)

Digestive Enzymes

Several of the digestive enzymes including trypsinogen, chymotrypsinogen, and procarboxypeptidase are produced as inactive forms and stored in the pancreas. After food is ingested, it reaches the small intestine. Hormones trigger the release of the zymogens from the pancreas. In the small intestine, the zymogens are converted into active forms by proteases that remove peptide sections from their protein chains. The change in tertiary structure activates the enzyme. For example, an enzyme called enteropeptidase removes a hexapeptide from trypsinogen to give active trypsin. Trypsin in turn cleaves peptide sections from chymotrypsinogen to

Figure 21.10 The removal of a peptide with 33 amino acids converts the zymogen proinsulin to active insulin.
Q *Why is proinsulin called a zymogen?*

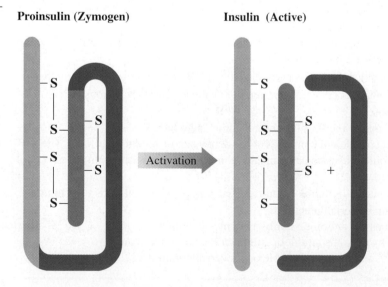

Proinsulin (Zymogen)   Insulin (Active)

form active chymotrypsin and from procarboxypeptidase to yield active carboxypeptidase.

$$\text{Chymotrypsinogen} \xrightarrow{\text{trypsin}} \text{Chymotrypsin} + 2 \text{ dipeptides}$$

If the digestive zymogens were active in the pancreas, their catalytic action as proteases would digest the proteins of the pancreas. This can lead to conditions such as pancreatitis, which is an inflammation of the pancreas. Another zymogen, pepsinogen, is produced in the gastric mucosal cells that line the stomach. As food enters the stomach, HCl is secreted. Low pH levels cleave a peptide containing 42 amino acids from the pepsinogen protein to form pepsin, which digests proteins.

$$\text{Pepsinogen} \xrightarrow{\text{H}^+} \text{pepsin} + \text{peptides and amino acids}$$

Most protein hormones such as insulin, enzymes involved in digestion, and enzymes that catalyze blood clotting are initially synthesized as zymogens. (See Table 21.7.)

Table 21.7 Example of Zymogens and Their Active Forms

Zymogen	Produced in	Activated in	Active Form
Proinsulin	Pancreas	Blood	Insulin
Chymotrypsinogen	Pancreas	Small intestine	Chymotrypsin
Pepsinogen	Gastric mucosa	Stomach	Pepsin
Trypsinogen	Pancreas	Small intestine	Trypsin
Fibrinogen	Blood	Damaged tissues	Fibrin
Prothrombin	Blood	Damaged tissues	Thrombin

Allosteric Enzymes

Certain enzymes known as **allosteric enzymes** are capable of binding a regulator molecule that is different from the substrate. The binding of the regulator causes a change in the shape of the enzyme and therefore in the active site. There are both positive and negative regulators. A **positive regulator** speeds up a reaction by causing a change in the shape of the active site that permits the substrate to bind more effectively. A **negative regulator** slows down the rate of catalysis by preventing the proper binding of the substrate. In **feedback control,** the end product acts as a negative regulator. When the end product is produced in sufficient amounts for the cell, some binds to the first enzyme (E_1), which is an allosteric enzyme. (See Figure 21.11.) By inhibiting the reaction of the initial substrate, no intermediate compounds are produced for the other enzymes in the reaction pathway. The entire enzyme-catalyzed reaction sequence shuts down.

Eventually the concentration of end product becomes too low, which causes the end product inhibitor to dissociate from the allosteric enzyme (E_1). As its shape returns to its active form, the catalysis of initial substrate begins once again. Through feedback control, the catalysis of the initial substrate undergoes reaction only when the end product is needed somewhere in the cell. There is no accumulation of the end product, which conserves the materials needed in other reactions.

Let's look at the feedback control in a reaction pathway with five enzymes that converts the amino acid threonine to isoleucine, another amino acid.

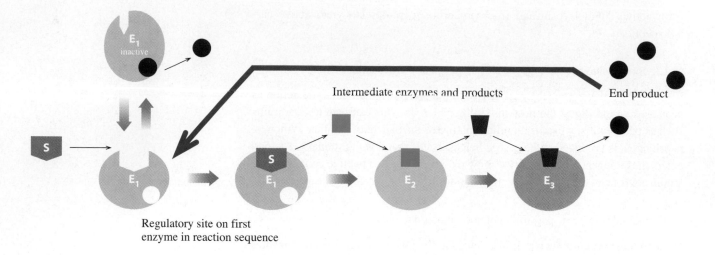

Figure 21.11 In feedback control, end product binds to a regulatory site on the first enzyme in the reaction pathway, which prevents the formation of all intermediate compounds needed in the synthesis of the end product.

Q *Why don't the intermediate enzymes in a reaction sequence have regulatory sites?*

When isoleucine begins to accumulate in the cell, it binds with the first enzyme (E_1) threonine deaminase in the pathway. The binding of isoleucine changes the shape of the deaminase, which prevents the substrate threonine from binding with the active site. The entire reaction pathway is turned off. None of the intermediate products from the other enzymes in the pathway can inhibit the first enzyme. As isoleucine is utilized in the cell, its concentration decreases, which causes the threonine deaminase to release the end product inhibitor. The tertiary shape of the deaminase returns to its active form and the reaction sequence once again converts threonine to isoleucine.

SAMPLE PROBLEM 21.6

Enzyme Regulation

How is the rate of a reaction sequence regulated in feedback control?

Solution

When the end product of a reaction sequence is produced at sufficient levels for the cell, some product molecules bind to the first enzyme in the sequence, which shuts down all the reactions that follow and stops the production of end product.

Study Check

Why is pepsin, a digestive enzyme, produced as a zymogen?

QUESTIONS AND PROBLEMS

Regulation of Enzyme Activity

21.27 Why are many of the enzymes that act on proteins synthesized as zymogens?

21.28 The zymogen trypsinogen produced in the pancreas is activated in the small intestine where it catalyzes the digestion and hydrolysis of proteins. Explain how the activation of the zymogen while still in the pancreas can lead to an inflammation of the pancreas called pancreatitis.

21.29 In feedback inhibition, how does the end product of a reaction sequence regulate enzyme activity?

21.30 Why don't the second or third enzymes in a reaction sequence function as regulatory enzymes?

21.31 How does an allosteric enzyme function as a regulatory enzyme?

21.32 What is the difference between a negative regulator and a positive regulator?

21.33 Indicate if the following statements describe (1) a zymogen, (2) a positive regulator, (3) a negative regulator, or (4) allosteric enzyme.
   a. It slows down a reaction, but its shape is different from that of the substrate.
   b. It is the first enzyme in a sequence of reactions leading to an end product.
   c. It is produced as an inactive enzyme.

21.34 Indicate if the following statements describe (1) a zymogen, (2) a positive regulator, (3) a negative regulator, or (4) allosteric enzyme.
   a. It is activated when a peptide section is removed from its protein chain.
   b. It speeds up a reaction, but it is not the substrate.
   c. When it binds to end product, it stops the formation of more end product.

21.7 Enzyme Cofactors

LEARNING GOAL
Describe the types of cofactors found in enzymes.

Enzymes are known as **simple enzymes** when their functional forms consist only of proteins with tertiary structures. However, many enzymes require small molecules or metal ions called **cofactors** to catalyze reactions properly. When the cofactor is a small organic molecule, it is known as a **coenzyme.** If an enzyme requires a cofactor, neither the protein structure nor the cofactor alone has catalytic activity.

Forms of Active Enzymes

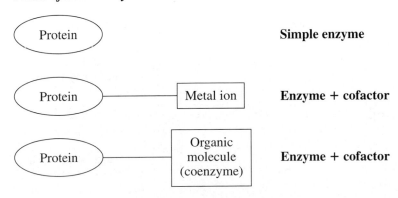

Metal Ions

Many enzymes must contain a metal ion to carry out their catalytic activity. The metal ions are bonded to one or more of the amino acid side chains. The metal ions from the minerals that we obtain from foods in our diet have various functions in catalysis. Ions such as Fe^{2+} and Cu^{2+} are used by oxidases where they lose or gain electrons in oxidation and reduction reactions. Other metals ions such as Zn^{2+} stabilize the amino acid side chains during hydrolysis reactions. Some metal cofactors required by enzymes are listed in Table 21.8.

Let's look at an example of a metal ion in an enzyme-catalyzed reaction. The enzyme carboxypeptidase A cleaves the C terminal amino acid of a protein when that amino acid has a bulky hydrophobic or aromatic side chain. (See Figure 21.12.) With the substrate in the active site, the Zn^{2+} helps to stabilize the negative charge on the oxygen atom of the carbonyl group and promotes the hydrolysis of the peptide bond.

Table 21.8   Enzymes and the Metal Ions Required as Cofactors

Metal Ion Cofactor	Function	Enzyme
Cu^{2+}	Oxidation-reduction	Cytochrome oxidase
Fe^{2+}/Fe^{3+}	Oxidation-reduction	Catalase
	Oxidation-reduction	Cytochrome oxidase
Zn^{2+}	Used with NAD^+	Alcohol dehydrogenase
		Carbonic anhydrase
		Carboxypeptidase A
Mg^{2+}	Hydrolyzes phosphate esters	Glucose-6-phosphatase
Mn^{2+}	Removes electrons	Arginase
Ni^{2+}	Hydrolyzes amides	Urease

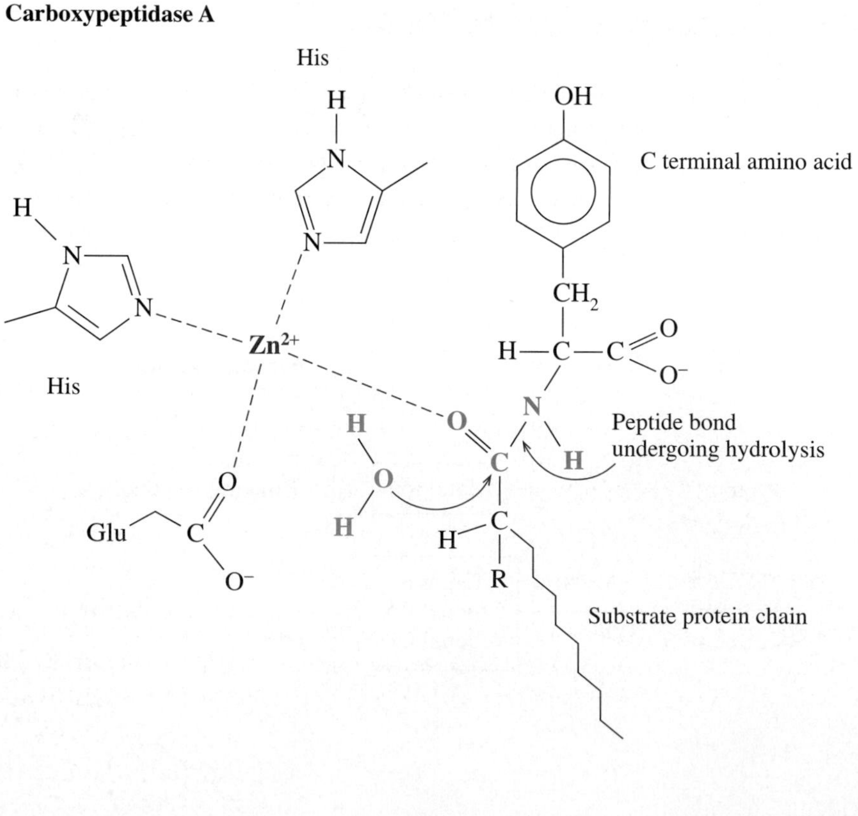

Figure 21.12 A Zn^{2+} cofactor aids in the hydrolysis of the peptide bond to a bulky C terminal amino acid by helping to stabilize the carbonyl oxygen.

Q *When would the Zn^{2+} be utilized as a cofactor by other enzymes?*

Carboxypeptidase A

SAMPLE PROBLEM 21.7

Cofactors

Indicate whether each of the following enzymes are active as a simple enzyme or require a cofactor.

a. a polypeptide that needs Mg^{2+} for catalytic activity
b. an active enzyme composed only of a polypeptide chain
c. an enzyme that consists of a quaternary structure attached to vitamin B_6

Solution

a. The enzyme requires a cofactor.
b. An active enzyme that consists of only a polypeptide chain is a simple enzyme.
c. The enzyme requires a cofactor.

Study Check

Which of the nonprotein portions of the enzymes in Sample Problem 21.7 is a coenzyme?

QUESTIONS AND PROBLEMS

Enzyme Cofactors

21.35 Is the enzyme described in each of the following statements a simple enzyme or one that requires a cofactor?
a. requires vitamin B_1 (thiamine)
b. needs Zn^{2+} for catalytic activity
c. its active form consists of two polypeptide chains

21.36 Is the enzyme described in each of the following statements a simple enzyme or one that requires a cofactor?
a. requires vitamin B_2 (riboflavin)
b. its active form is composed of 155 amino acids
c. uses Cu^{2+} during catalysis

21.8 Vitamins and Coenzymes

LEARNING GOAL

Describe the functions of the water-soluble and fat-soluble vitamins.

Vitamins are organic molecules that are essential for normal health and growth. They are required in trace amounts and must be obtained from the diet because they are not synthesized in the body. Before vitamins were discovered, it was known that lime juice prevented the disease scurvy in sailors and that cod liver oil could prevent rickets. In 1912, scientists found that, in addition to carbohydrates, fats, and proteins, certain other factors called vitamins must be obtained from the diet.

Classification of Vitamins

Vitamins are classified into two groups by solubility: water-soluble and fat-soluble. **Water-soluble vitamins** have polar groups such as —OH and —COOH, which make them soluble in the aqueous environment of the cells. The **fat-soluble vita-**

mins are nonpolar compounds, which are soluble in the fat (lipid) components of the body such as fat deposits and cell membranes.

Most water-soluble enzymes are not stored in the body and excess amounts are eliminated in the urine each day. Therefore, the water-soluble vitamins must be in the foods of our daily diets. Because many water-soluble vitamins are easily destroyed by heat, oxygen, and ultraviolet light, care must be taken in food preparation, processing, and storage. In the 1940s, a Committee on Food and Nutrition of the National Research Council began to recommend dietary enrichment of cereal grains. It was known that refining grains such as wheat caused a loss of vitamins. Thiamine (B_1), riboflavin (B_2), niacin, and iron were in the first group of added nutrients recommended . We now see the Recommended Daily Allowance (RDA) for many vitamins and minerals on food product labels such as cereals and bread.

The water-soluble vitamins are required by many enzymes as cofactors to carry out certain aspects of catalytic action. (See Table 21.9.) The coenzymes do not remain bonded to a particular enzyme, but are used over and over again by differ-

Table 21.9   Vitamins and Function

Water-Soluble Vitamins	Coenzyme	Function
Thiamine (vitamin B_1)	Thiamine pyrophosphate	Decarboxylation
Riboflavin (vitamin B_2)	Flavin adenine dinucleotide (FAD); Flavin mononucleotide (FMN)	Electron transfer
Niacin (vitamin B_3)	Nicotinamide adenine dinucleotide (NAD^+); Nicotinamide adenine dinucleotide phosphate ($NADP^+$);	Oxidation–reduction
Pantothenic acid (vitamin B_5)	Coenzyme A	Acetyl group transfer
Pyridoxine (vitamin B_6)	Pyridoxal phosphate	Transamination
Cobalamin (vitamin B_{12})	Methylcobalamin	Methyl group transfer
Ascorbic acid (vitamin C)	Vitamin C	Collagen synthesis, healing of wounds
Biotin	Biocytin	Carboxylation
Folic acid	Tetrahydrofolate	Methyl group transfer
Fat-Soluble Vitamins		
Vitamin A	Formation of visual pigments; development of epithelial cells	
Vitamin D	Absorption of calcium and phosphate; deposition of calcium and phosphate in bone	
Vitamin E	Antioxidant; prevents oxidation of vitamin A and unsaturated fatty acids	
Vitamin K	Synthesis of prothrombin for blood clotting	

Figure 21.13 The active forms of many enzymes require the combination of the protein with a coenzyme.

Q *What is the function of water-soluble vitamins in enzymes?*

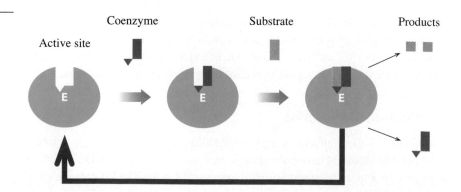

ent enzymes to facilitate an enzyme-catalyzed reaction. (See Figure 21.13.) Thus, only small amounts of coenzymes are required in the cells.

The fat-soluble vitamins A, D, E, and K are not involved as coenzymes, but they are important in processes such as vision, formation of bone, protection from oxidation, and proper blood clotting. Because the fat-soluble vitamins are stored in the body and not eliminated, it is possible to take too much, which could be toxic.

Water Soluble Vitamins

Thiamin (Vitamin B$_1$)

Thiamin (vitamin B$_1$) was the first B vitamin to be identified, thus the abbreviation B$_1$. The coenzyme thiamine pyrophosphate (TPP) is obtained when a synthetase adds two phosphate groups to the alcohol group of thiamine.

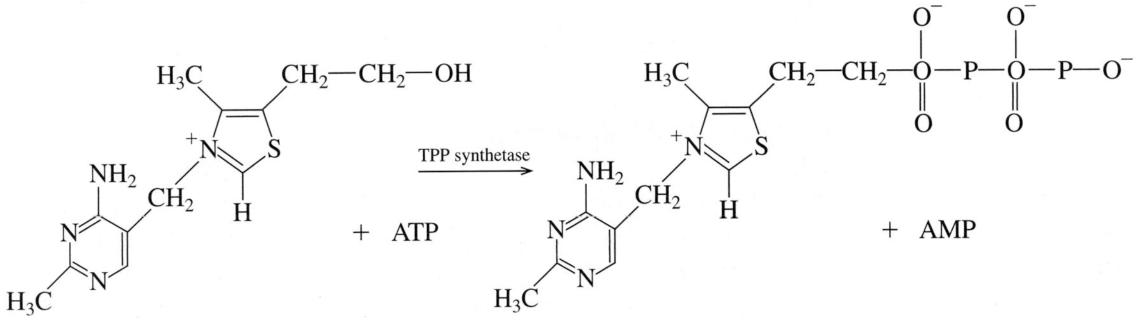

Thiamine (vitamin B$_1$)   Thiamine pyrophosphate (TPP)

The TPP coenzyme is involved in the decarboxylation reactions of α-keto carboxylic acids and reactions that cleave bonds to carbonyl carbons of α-hydroxyketones. A deficiency of thiamin may result in beriberi, which is characterized by fatigue, poor appetite, weight loss, nerve degeneration, and heart failure. The RDA (recommended daily allowance) for adults is about 2 mg. Dietary sources of thiamine include liver, yeast, whole grain bread, cereals, and milk.

Riboflavin (Vitamin B$_2$)

Riboflavin or vitamin B$_2$ is used to make the coenzymes flavin adenine dinucleotide (FAD) and flavin mononucleotide (FMN). The *ribo* part of the name comes from the sugar alcohol ribitol in the *riboflavin* molecule. The coenzymes FAD and FMN are used by enzymes called *flavoenzymes* to catalyze oxidation–reduction reactions of carbohydrates, fats, and proteins. Riboflavin is needed for good vision and healthy skin and hair. The RDA for adults is 1.7 mg. Deficiency symptoms of riboflavin include dermatitis, dry skin, tongue inflammation, and cataracts. Riboflavin is provided in the diet by beef liver, chicken, eggs, green leafy vegetables, dairy foods, peanuts, and whole grains.

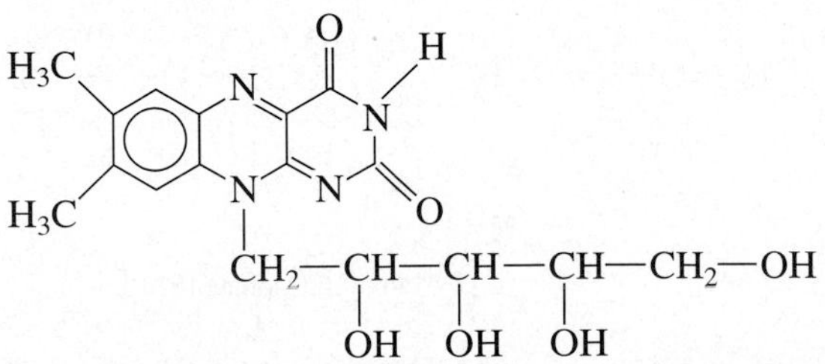

Riboflavin (vitamin B$_2$)   D-Ribitol

Niacin (Vitamin B₃)

Niacin or vitamin B_3 is a component of coenzymes nicotinamide adenine dinucleotide (NAD^+) and $NADP^+$, the phosphate form of NAD^+. The name niacin was assigned to the vitamin because its actual name, *nicotinic acid,* might be confused with nicotine. These coenzymes participate in oxidation–reduction, energy-production reactions in carbohydrate, fat, and protein metabolism. The RDA for niacin for adults is from 13–18 mg. A deficiency of niacin can result in *pellagra* characterized by dermatitis, muscle fatigue, loss of appetite, diarrhea, mouth sores, and mental disorders. In the diet, niacin is found in brewer's yeast, chicken, beef, fish, liver, brown rice, and whole grains.

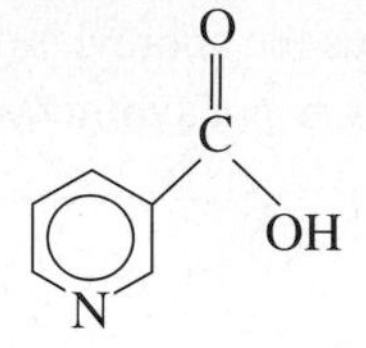

Niacin (vitamin B_3)

Pantothenic Acid (Vitamin B₅)

Pantothenic acid or vitamin B_5 is part of a complex coenzyme known as coenzyme A. Coenzyme A transfers a two-carbon acetyl group from pyruvate to the citric acid cycle for the production of energy. The A in the name refers to the acetylation reaction. Coenzyme A is also involved in the conversion of amino acids and lipids to glucose as well as the synthesis of cholesterol and steroid hormones. The RDA of pantothenic acid for adults is 10 mg per day. A deficiency of pantothenic acid is characterized by fatigue, retarded growth, muscle cramps, and anemia. Dietary sources of pantothenic acid include salmon, beef, liver, eggs, brewer's yeast, whole grains, and fresh vegetables.

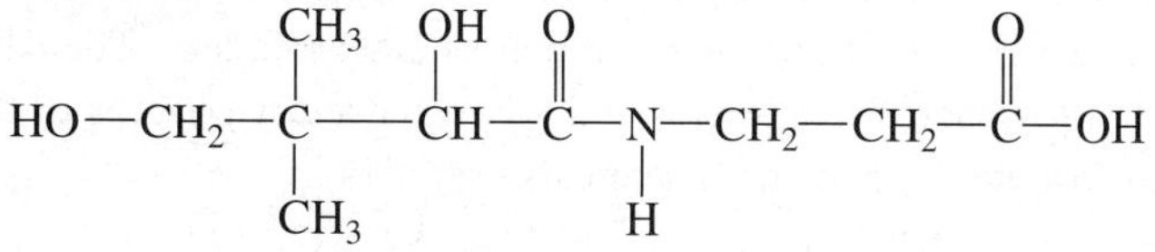

Pantothenic acid (vitamin B_5)

Pyridoxine (Vitamin B₆)

Pryidoxine and pyridoxal (an aldehyde) are forms of vitamin B_6 that are converted to the coenzymes pyridoxal phosphate (PLP). The PLP coenzyme participates in many enzyme-catalyzed reactions such as transaminations of amino acids and decarboxylations. The RDA for adults is 2 mg per day. Dietary sources include meat, liver, fish, nuts, whole grains, and spinach. Deficiency of pyridoxine may lead to dermatitis, fatigue, anemia, and retarded growth.

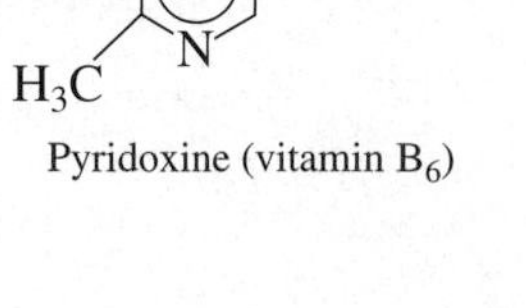

Pyridoxine (vitamin B_6) 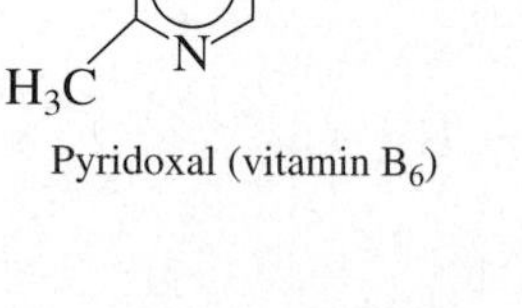   Pyridoxal (vitamin B_6) 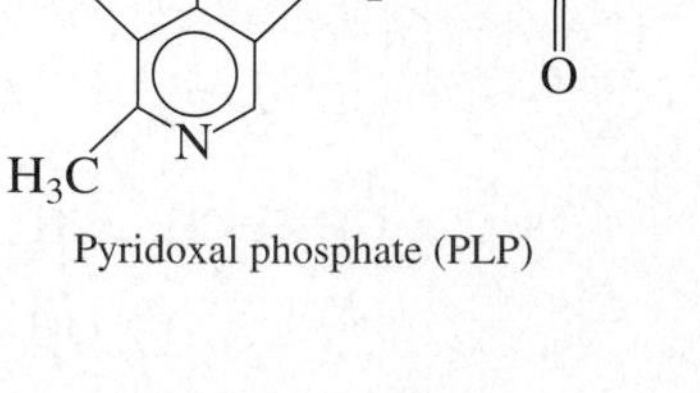   Pyridoxal phosphate (PLP)

Vitamin B$_{12}$ (cobalamin)

Cobalamin (Vitamin B$_{12}$)

Cobalamin or vitamin B$_{12}$ is a coenzyme consisting of four pyrrole rings with a cobalt ion (Co^{2+}) in the center. In its coenzyme form cobalamin participates in the transfer of methyl groups, molecular rearrangements, the formation of red blood cells, and the synthesis of acetylcholine for nerve cells. The RDA is about 3 μg per day for adults. A deficiency of cobalamin is found in pernicious anemia and results in malformed red blood cells, nerve damage, and some mental disorders. Dietary sources include liver, beef, kidney, chicken, fish such as salmon, halibut and tuna, yogurt, and milk. Because vitamin B$_{12}$ is not present in plants, strict vegetarians can experience symptoms of pernicious anemia.

Ascorbic Acid (Vitamin C)

Ascorbic acid or vitamin C has a simple chemical structure compared to most of the other vitamins. Its major function in the cells is its role in the synthesis of hydroxyproline and hydroxylysine, which are needed to form collagen. Collagen is the protein found in tendons, connective tissue, bone structure, and skin. The RDA for adults is 60 mg daily. A deficiency of ascorbic acid can lead to scurvy characterized by bleeding gums, weakened connective tissues, slow-healing wounds, and anemia. Dietary sources of ascorbic acid include blueberries, oranges, strawberries, cantaloupe, tomatoes, and vegetables especially red and green peppers, broccoli, cabbage, and spinach. (See Figure 21.14.)

Ascorbic acid (vitamin C)

Figure 21.14 Oranges, lemons, peppers, and tomatoes contain vitamin C or ascorbic acid.

Q *What happens to excess vitamin C that may be consumed in a day?*

Biotin

Biotin is a coenzyme for enzymes that transfer a carboxyl group in the reaction of pyruvate to oxaloacetate or acetyl-CoA to malonyl-CoA, which occurs in the synthesis of fatty acids. The RDA for adults is 0.3 mg per day. A deficiency of biotin can lead to dermatitis, loss of hair, fatigue, anemia, nausea, and depression. Dietary resources of biotin include liver, yeast, nuts, and eggs.

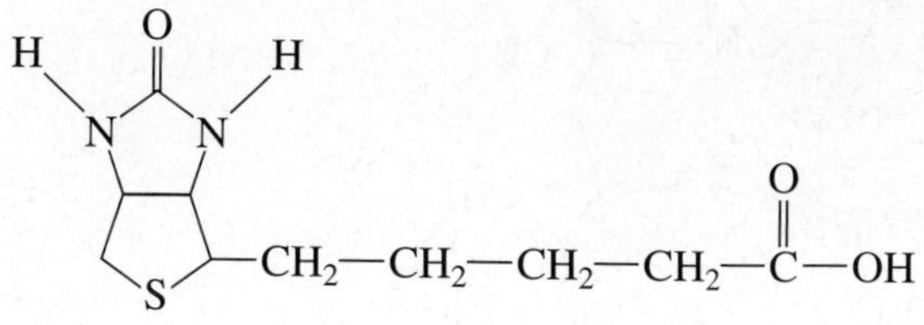

Biotin

Folic Acid (Folate)

Folic acid or folate is composed of a pyrimidine ring, *p*-aminobenzoic acid (PABA), and glutamate. The vitamin was discovered in the 1930s when people with a form of anemia were cured with extracts from liver or yeast. Folic acid is also found in spinach leaves, hence the name *folium,* Latin for leaf. In the cells, an enzyme called dihydrofolate reductase adds hydrogen atoms to the atoms in the heterocyclic ring of folate to yield the coenzyme tetrahydrofolate (THF). This coenzyme is used in reactions that transfer single-carbon groups and synthesize purines and pyrimidine to make DNA and RNA. It also plays a role with cobalamin in the production of red blood cells.

The RDA for folic acid is 0.4 mg for adults. A deficiency of folic acid can lead to abnormal red blood cells, anemia, intestinal-tract disturbances, loss of hair, growth impairment, depression, and spina bifida when there is a deficiency of folic acid during pregnancy. Dietary resources include green leafy vegetables, beans, meat, seafood, yeast, asparagus, and whole grain products, which are now enriched with folic acid.

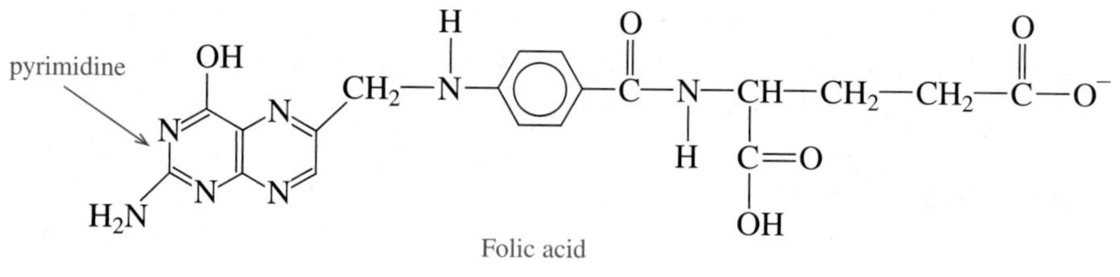

Folic acid

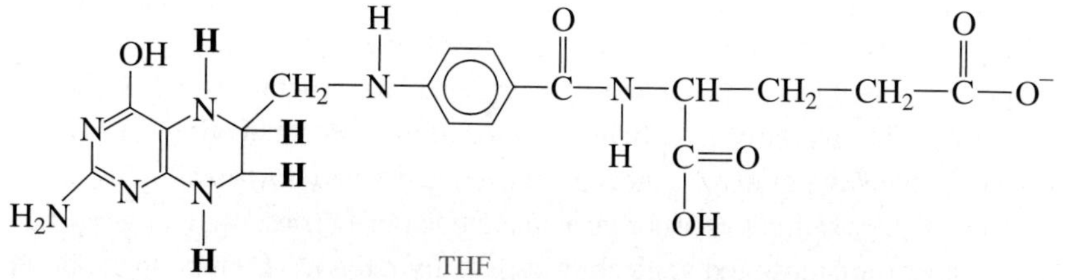

THF

Some compounds related to folate have been found that bring about remissions in people with leukemia. For example, 4-aminofolate referred to medically as *methotrexate* acts as a competitive inhibitor of the dihydrofolate reductase that forms THF. The growth of cells and tumor cells depends on THF to build purines and thymine. By inhibiting the reductase enzyme with methotrexate, THF cannot be produced, and the growth of tumor cells is blocked.

4- Aminofolate (methotrexate)

Fat-Soluble Vitamins

The fat-soluble vitamins are not used as coenzymes, but have important roles in vision, bone growth, and blood clotting.

Vitamin A

Vitamin A consists of three different forms depending on the oxidation of the functional group: retinol (alcohol), retinal (aldehyde), and retinoic acid (carboxylic acid). Vitamin A is obtained from animal sources in the diet or the β-carotenes of plants, which are converted to vitamin A in the liver. The retinol in the retinas of the eyes accumulates in the rod and cone cells where it plays a role in vision. Vitamin A is also involved in the synthesis of RNA and glycoproteins. The RDA for adults is 3 mg. A deficiency of vitamin A can cause night blindness, depress the immune response, and inhibit growth. Dietary resources of vitamin A include yellow and green fruits and vegetables such as apricots, broccoli, carrots, papaya, peaches, and spinach. (See Figure 21.15.)

Figure 21.15 Yellow and green fruits and vegetables contain vitamin A.
Q *Why is vitamin A called a fat-soluble vitamin?*

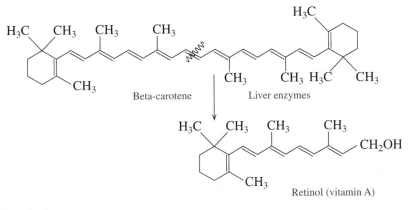

Beta-carotene   Liver enzymes

Retinol (vitamin A)

Vitamin D

The most prevalent form of vitamin D is vitamin D_3 or cholecalciferol. Technically, this is not a vitamin because it is not required in the diet. In skin, vitamin D_3 is synthesized from 7-dehydrocholesterol by the ultraviolet rays from sunlight. In regions of limited sunlight, vitamin D_3 is added to milk products to avoid vitamin D_3 deficiency. Its function in the body is to regulate the absorption of phosphorus and calcium during bone growth. The RDA for adults is 10 μg. A deficiency in vitamin D can result in weak bone structure, rickets in children, and osteomalacia in adults. The main source of vitamin D is sunlight with dietary sources including cod liver oil, egg yolk, and milk products enriched with vitamin D.

7-Dehydrocholesterol

UV (sunlight)

Vitamin D_3 (cholecalciferol)

Vitamin E

Vitamin E or tocopherol has a major role in cells as an antioxidant but not much is known about the mechanism of its activity. It appears to protect the cells in the body by removing damaging chemicals and by preventing the oxidation of unsaturated fatty acids. It has been used to reduce the damage to the retinas that can be caused by the high oxygen levels needed for respiration by premature infants. The RDA for adults is about 10 mg. A deficiency of vitamin E can cause hemolysis of red blood cells and anemia. Dietary resources of vitamin E include meat, whole grains, vegetables, and vegetable oils.

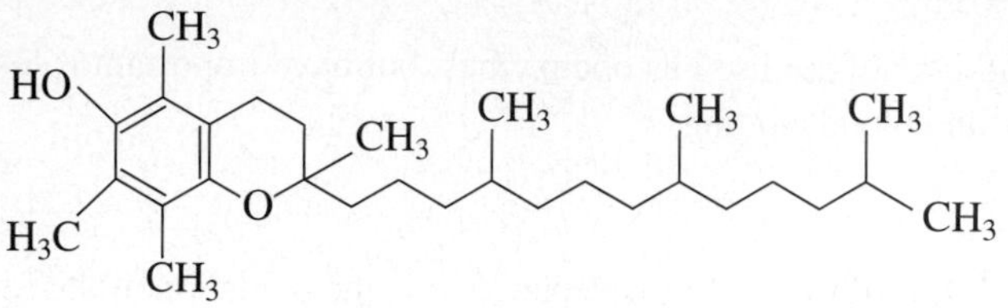

Vitamin E (tocopherol)

Vitamin K

Vitamin K_1, or phylloquinone, a substance found in plants, has a large saturated side chain. Vitamin K_2 or menaquinone, found in animals, has a very long unsaturated side chain. Vitamin K_2 takes part in the synthesis of zymogens needed for blood clotting. The RDA for vitamin K in adults is 80 μg. A deficiency of vitamin K can lead to bleeding from minor cuts, delayed clotting times, and bruising. Dietary resources of vitamin K include many foods such as liver, spinach, and cauliflower.

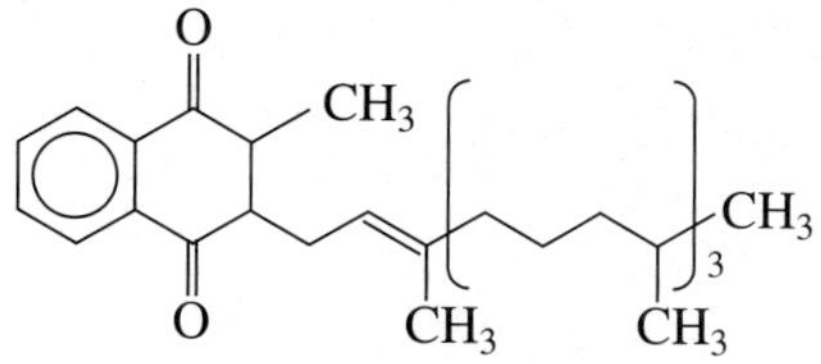

Vitamin K_1 (phylloquinone)

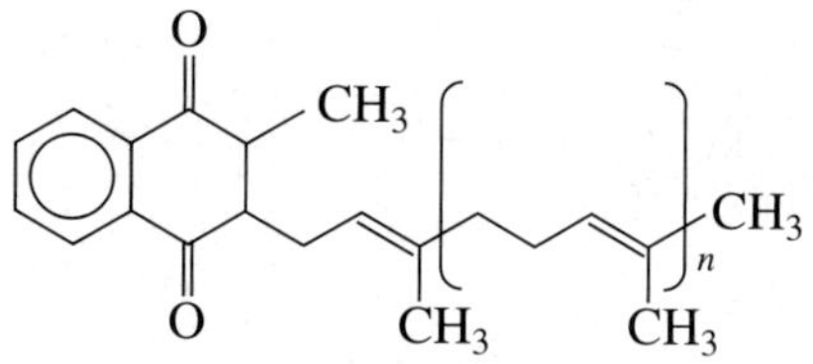

Vitamin K_2 (menaquinone)

SAMPLE PROBLEM 21.8

Vitamins

Why do you need a certain amount of thiamine and riboflavin in your diet every day, but not vitamin A or D?

Solution

Water-soluble vitamins like thiamine and riboflavin are not stored in the body whereas fat-soluble vitamins such as A and D are stored in the liver. Any excess of thiamine or riboflavin are eliminated in the urine and must be replenished each day from the diet.

Study Check

Why are fresh fruits rather than cooked fruits recommended as a source of vitamin C?

QUESTIONS AND PROBLEMS

Vitamins and Coenzymes

21.37 Give the abbreviation for each of the following coenzymes:
a. tetrahydrofolate
b. nicotinamide adenine dinucleotide

21.38 Give the abbreviation for each of the following coenzymes:
a. flavin adenine dinucleotide   b. thiamine pyrophosphate

21.39 Identify a vitamin that is a component of each of the following coenzymes:
a. coenzyme A   b. tetrahydrofolate (THF)
c. NAD$^+$

21.40 Identify a vitamin that is a component of each of the following coenzymes:
a. thiamine pyrophosphate   b. FAD
c. pyridoxal phosphate

21.41 What vitamin may be deficient in the following conditions?
a. rickets   b. scurvy   c. pellagra

21.42 What vitamin may be deficient in the following conditions?
a. poor night vision   b. pernicious anemia   c. beriberi

21.43 The RDA for pyridoxine (vitamin B$_6$) is 2 mg daily. Why will it not improve your nutrition to take 100 mg of pyridoxine daily?

21.44 The RDA for vitamin A is 3 mg daily. Why would an excess of vitamin A cause hypervitaminosis?

21.45 What is the change in the structure of pryidoxine (B$_6$) that yields the coenzyme PLP?

21.46 What is the change in the structure of folate that yields the coenzyme THF?

Chapter Review

21.1 Biological Catalysts

Enzymes are globular proteins that act as biological catalysts by lowering activation energy and accelerating the rate of cellular reactions.

21.2 Names and Classification of Enzymes

The names of most enzymes ending in *ase* describe the compound or reaction catalyzed by the enzyme. Enzymes are classified by the main type of reaction they catalyze, such as oxidoreductase, transferase, or isomerase.

21.3 Enzymes as Catalysts

Within the tertiary structure of an enzyme, a small pocket called the active site binds the substrates. In the lock-and-key model, a substrate precisely fits the shape of the active site. In the induced-fit model, substrates induce the active site to change structure to give an optimal fit by the substrate. In the enzyme–substrate complex, catalysis takes place when amino acid side chains react with a substrate. The products are released and the enzyme is available to bind another substrate molecule.

21.4 Factors Affecting Enzyme Action

Enzymes are most effective at optimum temperature and pH, usually 37°C and 7.4. The rate of an enzyme-catalyzed reaction decreases considerably at temperature and pH values above or below the optimum. An increase in substrate concentration increases the reaction rate of an enzyme-catalyzed reaction. If an enzyme is saturated, adding more substrate will not increase the rate further.

21.5 Enzyme Inhibition

An inhibitor reduces the activity of an enzyme or makes it inactive. A competitive inhibitor has a structure similar to the substrate and competes for the active site. When the active site is occupied, the enzyme cannot catalyze the reaction of the substrate. A noncompetitive inhibitor attaches elsewhere on the enzyme, changing the shape of both the enzyme and its active site. As long as the noncompetitive inhibitor is attached, the altered active site cannot bind with substrate.

21.6 Regulation of Enzyme Activity

Insulin and digestive enzymes are produced as inactive forms called zymogens. They are converted to active forms by removing a peptide portion from their protein chains. The rate of an enzyme-catalyzed reaction can be increased or decreased by regulator molecules that bind to a regulator site on an allosteric enzyme. The regulator molecule changes the shape of the enzyme and therefore the shape of the active site. A positive regulator increases the rate whereas a negative regulator decreases the rate. In feedback inhibition, the end product of a reaction sequence binds to a regulator site on the first enzyme, which is an allosteric enzyme, to decrease product formation.

21.7 Enzyme Cofactors

Simple enzymes are biologically active as a protein only, whereas other enzymes require small organic molecules or metals ions called cofactors. A cofactor may be a metal ion such as Cu^{2+} or Fe^{2+}, or an organic molecule called a coenzyme.

21.8 Vitamins and Coenzymes

A vitamin is a small organic molecule needed for health and normal growth. Vitamins are obtained in small amounts through the foods in the diet. The water-soluble vitamins B and C function as coenzymes. The fat-soluble vitamins are A, D, E, and K. Vitamin A is important in vision, vitamin D for proper bone growth, vitamin E is an antioxidant, and vitamin K is required for proper blood clotting.

Key Terms

active site A pocket in a part of the tertiary enzyme structure that binds substrate and catalyzes a reaction.

activity The rate at which an enzyme catalyzes the reaction that converts substrate to product.

allosteric enzyme An enzyme that regulates the rate of a reaction when a regulator molecule attaches to a site other than the active site.

antibiotic An irreversible inhibitor produced by bacteria, mold, or yeast that is toxic to bacteria.

catalyst A substance that takes part in a reaction to lower the activation energy, but is not changed.

coenzyme An organic molecule, usually a vitamin, required as a cofactor in enzyme action.

cofactor A nonprotein metal ion or an organic molecule that is necessary for a biologically functional enzyme.

competitive inhibitor A molecule with a structure similar to the substrate that inhibits enzyme action by competing for the active site.

enzymes Globular proteins that catalyze biological reactions.

enzyme–substrate (ES) complex An intermediate consisting of an enzyme that binds to a substrate in an enzyme-catalyzed reaction.

fat-soluble vitamins Vitamins that are not soluble in water and can be stored in the liver and body fat.

feedback control A type of inhibition in which an end product inhibits the first enzyme in a sequence of enzyme-catalyzed reactions.

induced-fit model A model of enzyme action in which a substrate induces an enzyme to modify its shape to give an optimal fit with the substrate structure.

inhibitors Substances that make an enzyme inactive by interfering with its ability to react with a substrate.

irreversible inhibition An inhibition caused by the covalent binding of an inhibitor to a part of the active site that cannot be reversed by adding more substrate.

isoenzymes Isoenzymes catalyze the same reaction, but have different combinations of polypeptide subunits.

lock-and-key model A model of an enzyme in which the substrate, like a key, exactly fits the shape of the lock, which is the specific shape of the active site.

negative regulator A molecule or end product that slows down or stops a catalytic reaction by binding to an allosteric enzyme.

noncompetitive inhibitor A substance that changes the shape of the enzyme, which prevents the active site from binding substrate properly.

optimum pH The pH at which an enzyme is most active.

optimum temperature The temperature at which an enzyme is most active.

positive regulator A molecule that increases the rate of an enzyme-catalyzed reaction by making the catalysis more favorable.

reversible inhibition Inhibition of an enzyme that is reversed by increasing the substrate concentration.

simple enzyme An enzyme that is active as a polypeptide only.

substrate The molecule that reacts in the active site in an enzyme-catalyzed reaction.

vitamins Organic molecules, which are essential for normal health and growth, obtained in small amounts from the diet.

water-soluble vitamins Vitamins that are soluble in water, cannot be stored in the body, are easily destroyed by heat, ultraviolet light, and oxygen, and function as coenzymes.

zymogen An inactive form of an enzyme that is activated by removing a peptide portion from one end of the protein.

Additional Problems

21.47 Why do the cells in the body have so many enzymes?

21.48 Are all the possible enzymes present at the same time in a cell?

21.49 How are enzymes different from the catalysts used in chemistry laboratories?

21.50 Why do enzymes function only under mild conditions?

21.51 Lactase is an enzyme that hydrolyzes lactose to glucose and galactose.
  a. What are the reactants and products of the reaction?
  b. Draw an energy diagram for the reaction with and without lactase.
  c. How does lactase make the reaction go faster?

21.52 Maltase is an enzyme that hydrolyzes maltose to two glucose molecules.
  a. What are the reactants and products of the reaction?
  b. Draw an energy diagram for the reaction with and without maltase.
  c. How does maltase make the reaction go faster?

21.53 Indicate whether each of the following would be a substrate (S) or an enzyme (E):
  a. lactose   **b.** lactase
  c. urease   **d.** trypsin
  e. pyruvate   **f.** transaminase

21.54 Indicate whether each of the following would be a substrate (S) or an enzyme (E):
  a. glucose   **b.** hydrolase
  c. maleate isomerase   **d.** alanine
  e. amylose   **f.** amylase

21.55 Give the substrate of each of the following enzymes:
  a. urease   **b.** lactase

  c. aspartate transaminase
  d. tyrosine synthetase

21.56 Give the substrate of each of the following enzymes:
  a. maltase
  b. fructose oxidase
  c. phenolase
  d. sucrase

21.57 Predict the major class of each of the following enzymes:
  a. acyltransferase
  b. oxidase
  c. lipase
  d. decarboxylase

21.58 Predict the major class for each of the following enzymes:
  a. cis–trans isomerase
  b. reductase
  c. carboxylase
  d. peptidase

21.59 What is the class of the enzyme that would catalyze each of the following reactions?

$$\textbf{a.} \quad CH_3\overset{\overset{\displaystyle O}{\|}}{C}H \longrightarrow CH_3\overset{\overset{\displaystyle O}{\|}}{C}OH$$

$$\textbf{b.} \quad NH_2-CH_2-\overset{\overset{\displaystyle O}{\|}}{C}-NH-\overset{\overset{\displaystyle CH_3}{|}}{C}H-\overset{\overset{\displaystyle O}{\|}}{C}OH + H_2O \longrightarrow$$
$$NH_2-CH_2-\overset{\overset{\displaystyle O}{\|}}{C}OH + NH_2-\overset{\overset{\displaystyle CH_3}{|}}{C}H-\overset{\overset{\displaystyle O}{\|}}{C}OH$$

$$\textbf{c.} \quad CH_3-CH{=}CH-CH_3 + H_2O \longrightarrow$$
$$CH_3-CH_2-\overset{\overset{\displaystyle OH}{|}}{C}H-CH_3$$

21.60 What is the class of the enzyme that would catalyze each of the following reactions?

a. $CH_3 - \overset{\overset{O}{\|}}{C} - \overset{\overset{O}{\|}}{C}OH \longrightarrow$

$CH_3 - \overset{\overset{O}{\|}}{C} - OH + CO_2$

b. $CH_3 - \overset{\overset{O}{\|}}{C} - \overset{\overset{O}{\|}}{C}OH + CO_2 + ATP \longrightarrow$

$HO - \overset{\overset{O}{\|}}{C} - CH_2 - \overset{\overset{O}{\|}}{C} - \overset{\overset{O}{\|}}{C} - OH + ADP + P_i$

c. glucose-6-phosphate $\longrightarrow$ fructose-6-phosphate

21.61 How would the lock-and-key theory explain that sucrase hydrolyses sucrose, but not lactose?

21.62 How does the induced-fit model of enzyme action allow an enzyme to catalyze a reaction of a group of substrates?

21.63 If a blood test indicates a high level of LDH and CK, what could be the cause?

21.64 If a blood test indicates a high level of ALT, what could be the cause?

21.65 Indicate whether an enzyme is saturated or unsaturated in each of the following conditions:
  a. adding more substrate does not increase the rate of reaction
  b. doubling the substrate concentration doubles the rate of reaction

21.66 Indicate whether each of the following enzymes would be functional:
  a. pepsin, a digestive enzyme, at pH 2
  b. an enzyme at 37°C, if the enzyme is from a type of thermophilic bacteria that have an optimum temperature of 100°C

21.67 How does a reversible inhibition differ from an irreversible inhibition?

21.68 How does a competitive reversible inhibition differ from a noncompetitive reversible inhibition?

21.69 Ethylene glycol ($HO - CH_2 - CH_2 - OH$) is a major component of antifreeze. In the body, it is first converted to $HOOC - CHO$ (oxoethanoic acid) and then to $HOOC - COOH$ (oxalic acid), which is toxic.

  a. What class of enzyme catalyzes both of the reactions of ethylene glycol?
  b. The treatment for the ingestion of ethylene glycol is an intravenous solution of ethanol. How might this help prevent toxic levels of oxalic acid in the body?

21.70 Adults who are lactose intolerant cannot break down the disaccharide in milk products. To help digest dairy food, a product known as Lactaid can be added to milk and refrigerated for 24 hours.
  a. What enzyme is present in Lactaid and what is the major class?
  b. What might happen to the enzyme if the digestion product were stored in a warm area?

21.71 a. What type of an inhibitor is the antibiotic amoxicillin?
  b. Why can antibiotics be used to treat bacterial infections?

21.72 a. A gardener using Parathion develops a headache, dizziness, nausea, blurred vision, excessive salivation, and muscle twitching. What might be happening to the gardener?
  b. Why must humans be careful when using insecticides?

21.73 The enzyme pepsin is produced in the pancreas as the zymogen pepsinogen.
  a. How and where does pepsinogen become the active form pepsin?
  b. Why are proteases such as pepsin produced in inactive forms?

21.74 Thrombin is an enzyme that helps produce blood clotting when an injury and bleeding occurs.
  a. What would be the name of the zymogen of thrombin?
  b. Why would the active form of thrombin be produced only when an injury occurs to tissue?

21.75 What is an allosteric enzyme?

21.76 Why can some regulator molecules speed up a reaction, while others slow it down?

21.77 In feedback control, what type of regulator modifies the catalytic activity of the reaction pathway?

21.78 Why aren't the intermediate products in a reaction sequence used in feedback control?

21.79 Are each of the following statements describing a simple enzyme or one that requires a cofactor?

a. contains Mg^{2+} in the active site

b. has catalytic activity as a tertiary protein structure

c. requires folic acid for catalytic activity

21.80 Are each of the following statements describing a simple enzyme or one that contains a cofactor?

a. contains riboflavin or vitamin B_2

b. has four subunits of polypeptide chains

c. requires Fe^{3+} in the active site for catalytic activity

21.81 Match the following vitamins with their coenzymes:

a. pantothenic acid (B_5) NAD^+

b. niacin (B_3)   biocytin

c. biotin   coenzyme A

21.82 Match the following vitamins with their coenzymes:

a. folate   pyridoxal phosphate

b. riboflavin (B_2)   THF

c. pyridoxine   FAD

21.83 Why are only small amounts of vitamins needed in the cells when there are several enzymes that require coenzymes?

21.84 Why is there a daily requirement for vitamins?

21.85 Match each of the following vitamins with their deficiency symptoms or conditions:

a. niacin   night blindness

b. vitamin A   weak bone structure

c. vitamin D   pellagra

21.86 Match each of the following vitamins with their deficiency symptoms or conditions:

a. cobalamin   bleeding

b. vitamin C   anemia

c. vitamin K   scurvy

22 Nucleic Acids and Protein Synthesis

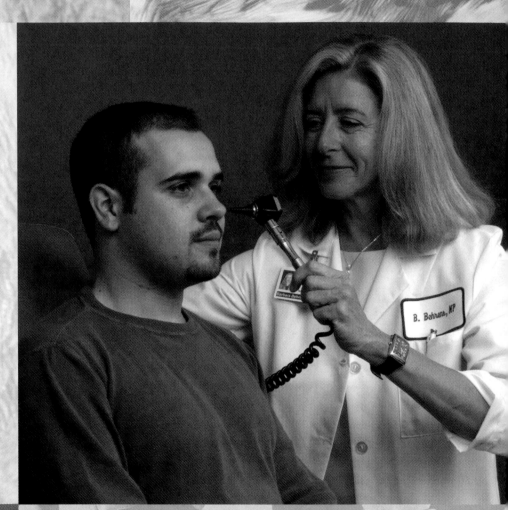

"I run the Hepatitis C Clinic, where patients are often anxious when diagnosed," says Barbara Behrens, nurse practitioner, Hepatitis C Clinic, Kaiser Hospital. "The treatment for Hepatitis C can produce significant reactions such as a radical drop in blood count. When this happens, I get help within 24 hours. I monitor our patients very closely and many call me whenever they need to."

Hepatitis C is an RNA virus or retrovirus that causes liver inflammation, often resulting in chronic liver disease. Unlike many viruses to which we eventually develop immunity, the Hepatitis C virus undergoes mutations so rapidly that scientists have not been able to produce vaccines. People who carry the virus are contagious throughout their lives and are able to pass the virus to other people.

LOOKING AHEAD

the Chemistry place

www.chemplace.com/college

Visit the URL above or use the CD-ROM in the book for extra quizzing, interactive tutorials, and career resources.

Nucleic acids are the molecules in our cells that store and direct information for cellular growth and reproduction. Deoxyribonucleic acid (DNA), the genetic material in the nucleus of a cell, contains all the information needed for the development of a complete living system. The way you grow, your hair, your eyes, your physical appearance, the activities of the cells in your body are all determined by a set of directions contained in the DNA of your cells.

The nucleic acids are large molecules found in the nuclei of cells that contain all the information needed to direct the activities of a cell and its reproduction. All of the genetic information in the cell is called the *genome*. Every time a cell divides, the information in the genome is copied and passed on to the new cells. This replication process must duplicate the genetic instructions exactly. Some sections of DNA called *genes* contain the information to make a particular protein.

As a cell requires protein, another type of nucleic acid, RNA, translates the genetic information in DNA, and carries that information to the ribosomes where the synthesis of protein takes place. However, mistakes sometimes occur that lead to mutations that affect the synthesis of a certain protein.

22.1 Components of Nucleic Acids

There are two closely related types of nucleic acids: *deoxyribonucleic acid* (**DNA**), and *ribonucleic acid* (**RNA**). Both are linear polymers of repeating monomer units known as *nucleotides*. A DNA molecule may contain several million nucleotides; smaller RNA molecules may contain up to several thousand. Each nucleotide has three components: a nitrogenous base, a five-carbon sugar, and a phosphate group. The sugar is bonded to a nitrogen base at carbon 1' (C1'), and to a phosphate group at carbon 5' (C5'). (See Figure 22.1.)

Nitrogen Bases

The **nitrogen-containing bases** in nucleic acids are derivatives of *pyrimidine* or *purine*.

Pyrimidine   Purine

In DNA, there are two purines, adenine (A) and guanine (G), and two pyrimidines, cytosine (C) and thymine (T). RNA contains the same bases, except thymine (5-methyluracil) is replaced by uracil (U). (See Figure 22.2.)

Ribose and Deoxyribose Sugars

The nucleotides of RNA and DNA contain five-carbon pentose sugars. In RNA, the five-carbon sugar is *ribose,* which gives the letter R in the abbreviation RNA. In DNA, the five-carbon sugar is *deoxyribose,* which is similar to ribose except that

Describe the nitrogen bases and ribose sugars that make up the nucleic acids DNA and RNA.

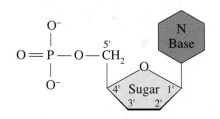

Figure 22.1 A diagram of the general structure of a nucleotide found in nucleic acids.

Q *In a nucleotide, what types of groups are bonded to a five-carbon sugar?*

Pyrimidines

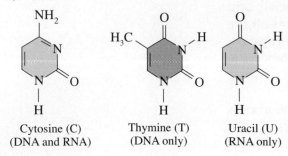

Cytosine (C)
(DNA and RNA)

Thymine (T)
(DNA only)

Uracil (U)
(RNA only)

Purines

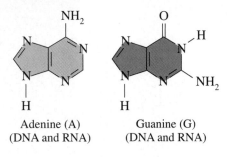

Adenine (A)
(DNA and RNA)

Guanine (G)
(DNA and RNA)

Figure 22.2 DNA contains the nitrogen bases A, G, C, and T; RNA contains A, G, C, and U.
Q *Which nitrogen bases are found in DNA?*

Pentose sugars in RNA and DNA

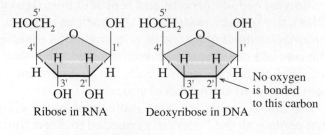

Ribose in RNA

Deoxyribose in DNA

No oxygen is bonded to this carbon

Figure 22.3 The five-carbon pentose sugar found in RNA is ribose and deoxyribose in DNA.
Q *What is the difference between ribose and deoxyribose?*

there is no hydroxyl group (—OH) on C2' of ribose. The *deoxy–* prefix means "without oxygen" and provides the D in DNA.

The atoms in the pentose sugars are numbered with primes (1', 2', 3', 4', and 5') to differentiate them from the atoms in the nitrogen bases. (See Figure 22.3.) In the following sections, we will see that the presence or absence of a hydroxyl group (—OH) on the sugar strongly influences the structures and properties of RNA and DNA.

SAMPLE PROBLEM 22.1

Components of Nucleic Acids

Identify each of the following bases as a purine or pyrimidine.

a.

b.

Solution

a. Guanine is a purine.
b. Uracil is a pyrimidine.

Study Check

Indicate if the bases in problem 22.1 are found in RNA, DNA, or both.

QUESTIONS AND PROBLEMS

Components of Nucleic Acids

22.1 Identify each of the following bases as a purine or pyrimidine:
   a. thymine
   b. NH$_2$

22.2 Identify each of the following bases as a purine or pyrimidine:

a. guanine

b.

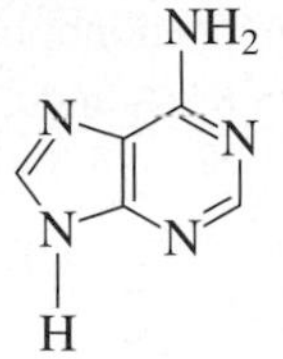

22.3 Identify the bases in problem 22.1 as present in RNA, DNA, or both.

22.4 Identify the bases in problem 22.2 as present in RNA, DNA, or both.

22.2 Nucleosides and Nucleotides

LEARNING GOAL

Identify the nucleosides and nucleotides in RNA and DNA.

A **nucleoside** is produced when a nitrogen base of pyrimidine or a purine forms a glycosidic bond to C1' of a sugar, either ribose or deoxyribose. The name of a nucleoside is obtained from the name of the base and the ending *–osine* for purines and *–idine* for pyrimidines. For example, adenine, a purine, and ribose form a nucleoside called aden**osine**. When ribose combines with the base cytosine, a pyrimidine, the nucleoside is named cyt**idine.**

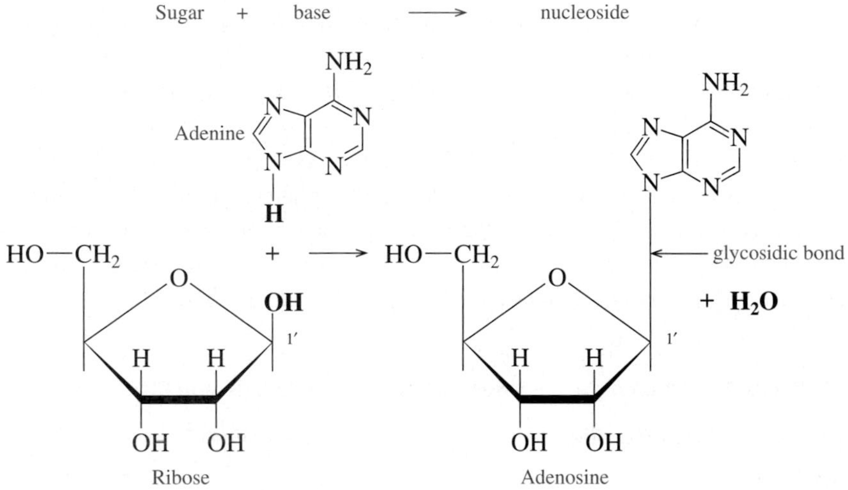

Formation of Nucleotides

In Chapter 18, we saw that phosphate esters form between a hydroxyl group and phosphoric acid to give ester products.

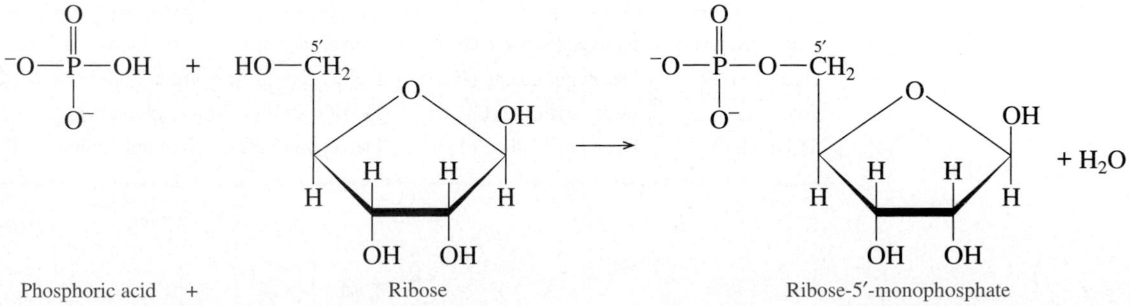

Nucleotides are formed when the C5' —OH group of ribose or deoxyribose in a nucleoside forms phosphate esters with phosphoric acid. Other hydroxyl groups on ribose can form phosphate esters too, but only the 5'-monophosphate nucleotides are found in RNA and DNA. All the nucleotides in RNA and DNA are shown in Figure 22.4.

Phosphoric acid   +   nucleoside   ⟶   nucleotide

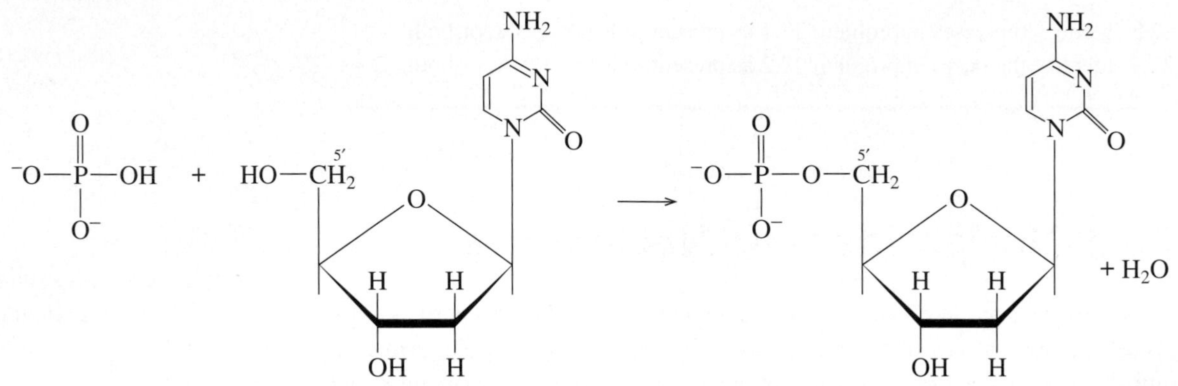

phosphoric acid   deoxycytidine   deoxycytidine 5'-monophosphate (dCMP)

Naming Nucleotides

The name of a nucleotide is obtained from the name of the nucleoside followed by 5'-monophosphate. A nucleotide in RNA is named as a *nucleoside 5'-monophosphate* or NMP. Nucleotides of DNA have the prefix *deoxy* added to the beginning of the nucleoside name, *deoxynucleoside 5'-monophosphate* or dNMP. For example, the nucleotide of adenosine in RNA is adenosine 5'-monophosphate or AMP. In DNA, it is named deoxyadenosine 5'-monophosphate or dAMP. Although the letters A, G, C, U, and T represent the bases, they are often used in the abbreviations of the respective nucleotides as well. The names and abbreviations of the bases, nucleosides, and nucleotides in DNA and RNA are listed in Table 22.1.

Table 22.1   Names of Nucleosides and Nucleotides in DNA and RNA

Base	Nucleosides	Nucleotides
RNA		
Adenine (A)	Adenosine (A)	Adenosine 5'-monophosphate (AMP)
Guanine (G)	Guanosine (G)	Guanosine 5'-monophosphate (GMP)
Cytosine (C)	Cytidine (C)	Cytidine 5'-monophosphate (CMP)
Uracil (U)	Uridine (U)	Uridine 5'-monophosphate (UMP)
DNA		
Adenine (A)	Deoxyadenosine (A)	Deoxyadenosine 5'-monophosphate (dAMP)
Guanine (G)	Deoxyguanosine (G)	Deoxyguanosine 5'-monophosphate (dGMP)
Cytosine (C)	Deoxycytidine (C)	Deoxycytidine 5'-monophosphate (dCMP)
Thymine (T)	Deoxythymidine (T)	Deoxythymidine 5'-monophosphate (dTMP)

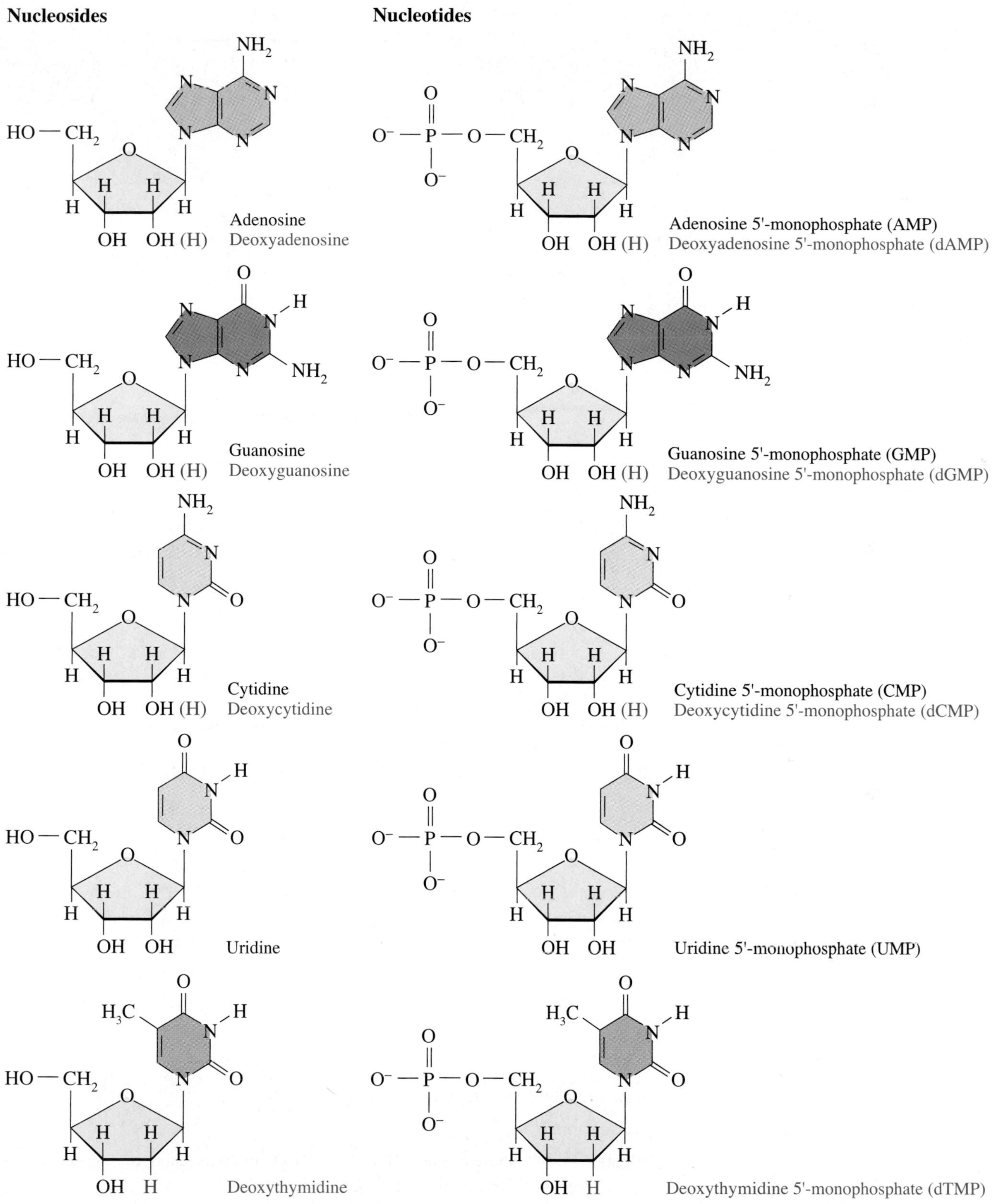

Nucleosides   **Nucleotides**

Adenosine
Deoxyadenosine

Adenosine 5'-monophosphate (AMP)
Deoxyadenosine 5'-monophosphate (dAMP)

Guanosine
Deoxyguanosine

Guanosine 5'-monophosphate (GMP)
Deoxyguanosine 5'-monophosphate (dGMP)

Cytidine
Deoxycytidine

Cytidine 5'-monophosphate (CMP)
Deoxycytidine 5'-monophosphate (dCMP)

Uridine

Uridine 5'-monophosphate (UMP)

Deoxythymidine

Deoxythymidine 5'-monophosphate (dTMP)

Figure 22.4 The nucleosides and nucleotides of RNA are identical to those of DNA except in DNA the sugar is deoxyribose and deoxythymidine replaces uridine.
Q *What are two differences in the nucleotides of RNA and DNA?*

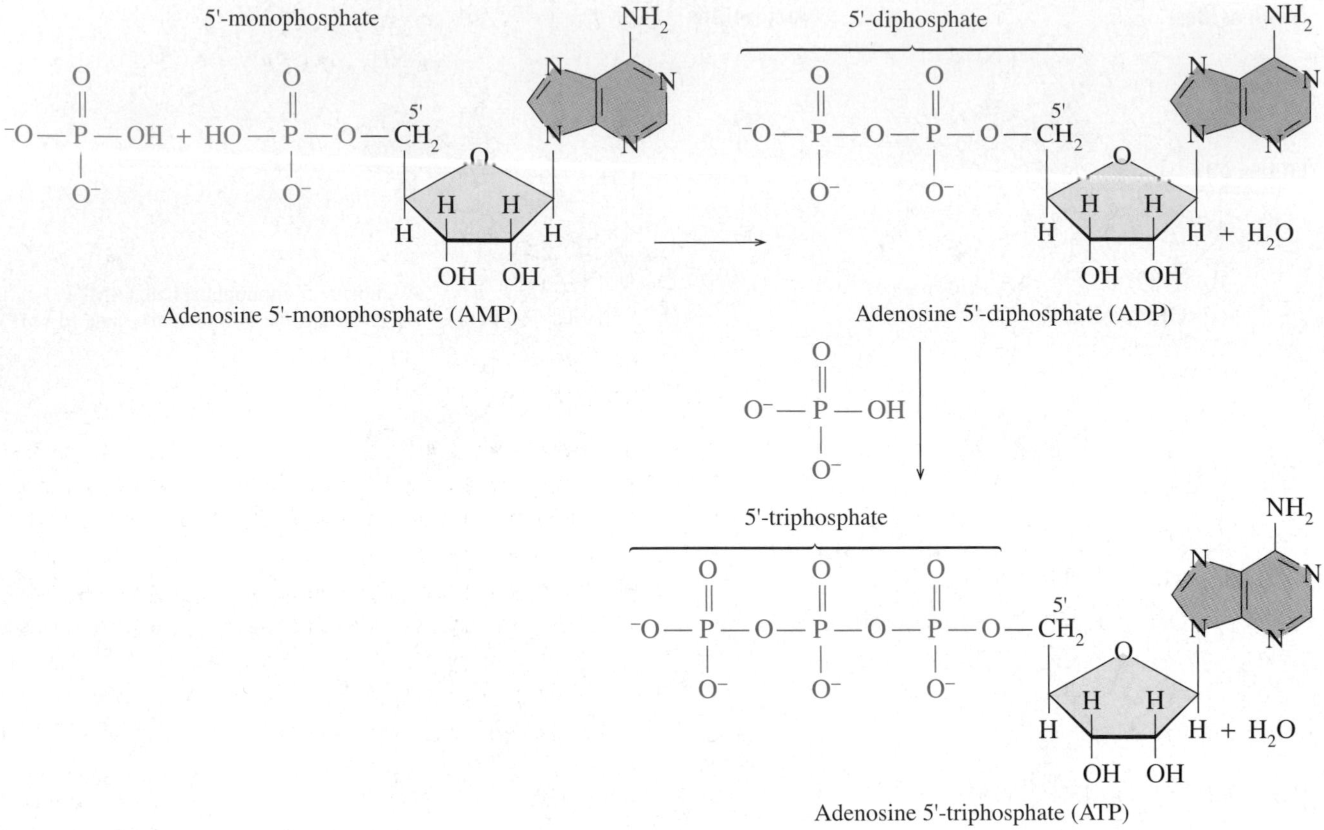

Figure 22.5 The addition of more phosphate groups forms adenosine 5'-diphosphate (ADP) and adenosine 5'-triphosphate (ATP).
Q *How does the structure of deoxyguanidine triphosphate (dGTP) differ from ATP?*

Formation of Nucleoside Diphosphates and Triphosphates

Any of the nucleoside 5'-monophosphates such as AMP can bond to additional phosphate groups. For example, adding another phosphate to AMP gives *ADP* (*adenosine 5'-diphosphate*) and ATP (*adenosine 5'-triphosphate*) when there are a total of three phosphates. (See Figure 22.5.) The other nucleotides also form di- and triphosphates. For example, GMP adds phosphate to yield GDP and GTP, dCMP forms dCDP and dCTP, and so on. Of the triphosphates, ATP is of particular interest because it is the major source of energy for most energy-requiring activities in the cell. GTP is an energy source for protein synthesis, and CTP is an intermediate in phospholipid synthesis.

SAMPLE PROBLEM 22.2

Nucleotides

Identify which nucleic acid (DNA or RNA) contains each of the following nucleotides; state the components of each nucleotide:

a. deoxyguanosine 5'-monophosphate (dGMP)

b. adenosine 5'-monophosphate (AMP)

Solution

a. This DNA nucleotide consists of deoxyribose, guanine, and phosphate.

b. This RNA nucleotide contains ribose, adenine, and phosphate.

Study Check

What is the name and abbreviation of the DNA nucleotide of cytosine?

QUESTIONS AND PROBLEMS

Nucleosides and Nucleotides

22.5 What are the names and abbreviations of the four nucleotides in DNA?

22.6 What are the names and abbreviations of the four nucleotides in RNA?

22.7 Identify each of the following as a nucleoside or nucleotide:
  a. adenosine   **b.** deoxycytidine
  c. uridine   **d.** cytidine 5'-monophosphate

22.8 Identify each of the following as a nucleoside or nucleotide in DNA or RNA:
  a. deoxythymidine   **b.** guanosine
  c. adenosine   **d.** uridine 5'-monophosphate

22.9 Draw the structure of deoxyadenosine 5'-monophosphate (dAMP).

20.10 Draw the structure of uridine 5'-monophosphate (UMP).

22.3 Primary Structure of Nucleic Acids

LEARNING GOAL

Describe the primary structures of RNA and DNA.

The **nucleic acids** consist of polymers of many nucleotides. In a polymer of nucleotides, the 3'—OH group of the sugar in one nucleotide bonds to the phosphate group on the 5'—carbon atom in the sugar of the next nucleotide. This phosphate link between the sugars in adjacent nucleotides is referred to as a **phosphodiester bond.** As more nucleotides are added through phosphodiester bonds, a backbone forms that consists of alternating sugar and phosphate groups.

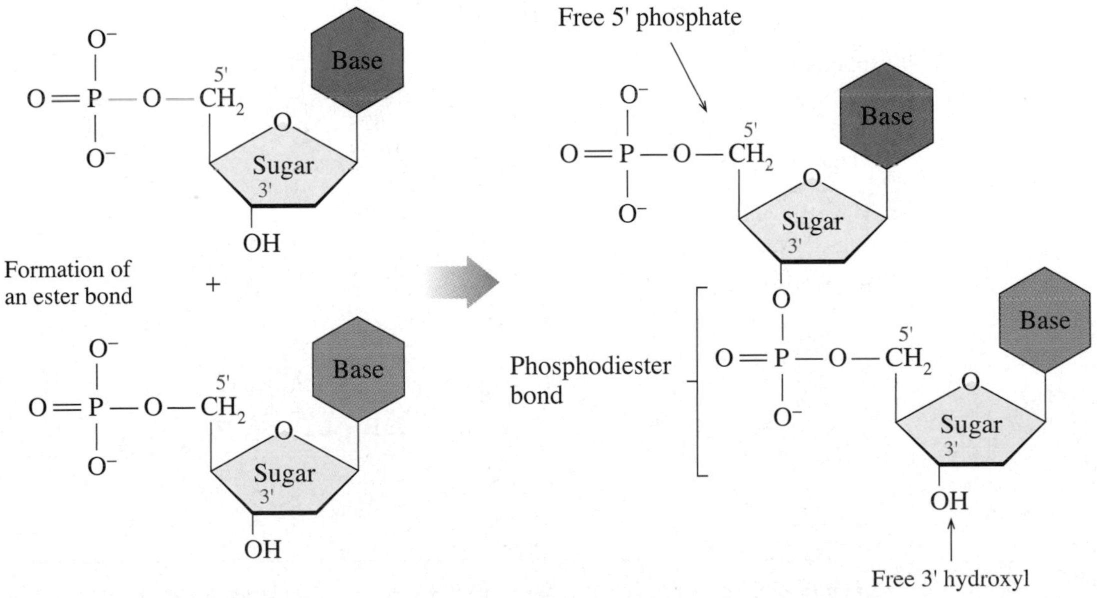

Free 5' end

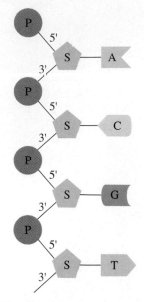

Free 3' end

Each nucleic acid has its own unique sequence of bases, which is known as its **primary structure.** It is this sequence of bases that carries the genetic information from one cell to the next. Along a polynucleotide of a DNA or RNA chain, the bases attached to each of the sugars extend out from the nucleic acid backbone. In RNA, ribose is the sugar used in the polynucleotide, whereas in DNA, the sugar is deoxyribose. In any nucleic acid, the sugar at the one end has an unreacted or free 5'-phosphate terminal end, and the 3'— terminal sugar at the other end has a free 3'-hydroxyl group.

A nucleic acid sequence is read from the sugar with free 5'-phosphate to the sugar with the free 3'-hydroxyl group. The order of bases is often written using only the letters of the bases. Thus, the nucleotides in a section of RNA shown in Figure 22.6 is read as 5'-ACGU-3'. This abbreviation starts with the base attached to the ribose with a free 5' phosphate, and ends with uracil, the base attached to the sugar with a free 3' hydroxyl group. In the nucleotide sequence in DNA, 5'-ACGT-3' (see Figure 22.7), adenine is the base attached to the deoxyribose with the free 5' phosphate and thymine is the base attached to the sugar with the free 3' hydroxyl group.

RNA (ribonucleic acid)

Adenine (A)

Cytosine (C)

3'–5' Phosphodiester bond

Guanine (G)

Uracil (U)

Free 3'

Figure 22.6 In the primary structure of an RNA, A, C, G, and U are linked by 3'–5' phosphodiester bonds.

Q *Where are the free 5' phosphate and 3' hydroxyl groups of ribose?*

DNA (deoxyribonucleic acid)

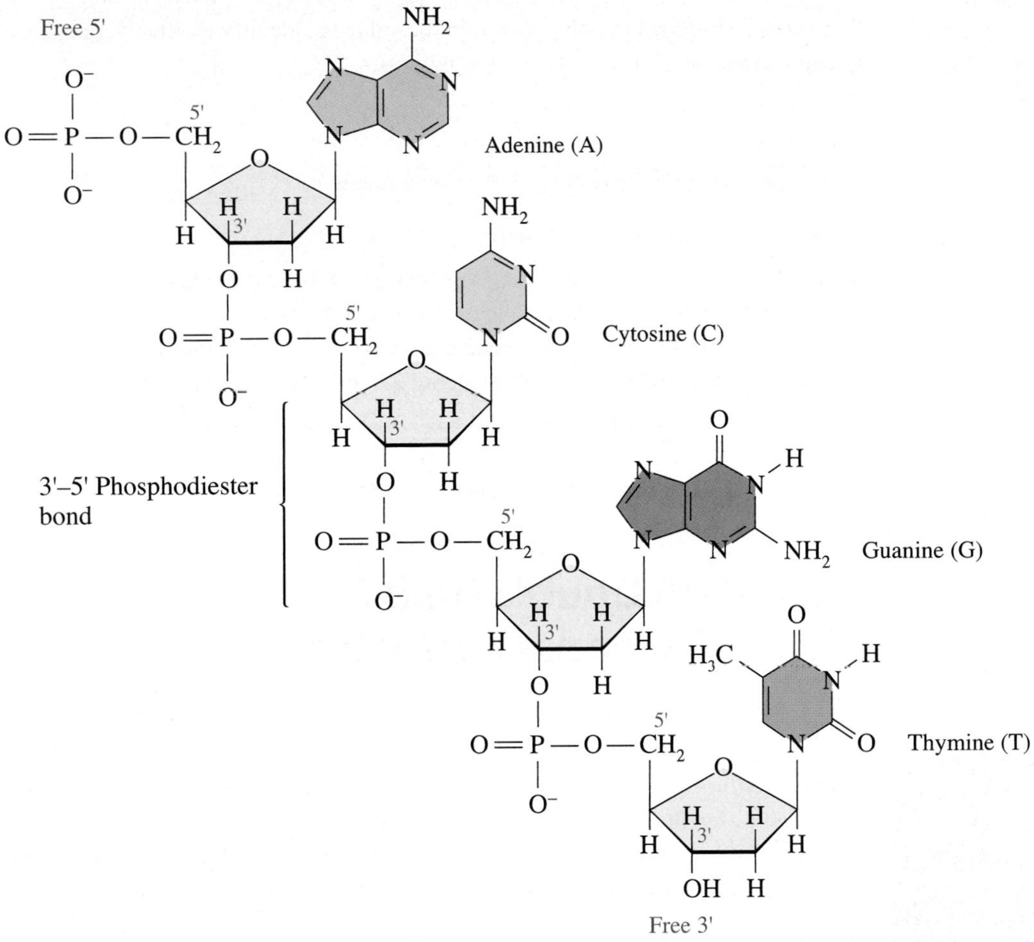

Free 5'

Adenine (A)

Cytosine (C)

3'–5' Phosphodiester bond

Guanine (G)

Thymine (T)

Free 3'

Figure 22.7 In the primary structure of a single DNA strand, A, C, G, and T are linked by 3'–5' phosphodiester bonds.
Q *What hydroxyl groups does deoxyribose use to form a polymer of DNA?*

SAMPLE PROBLEM 22.3

Bonding of Nucleotides

Draw the structure of an RNA dinucleotide formed by two cytidine monophosphates.

Solution

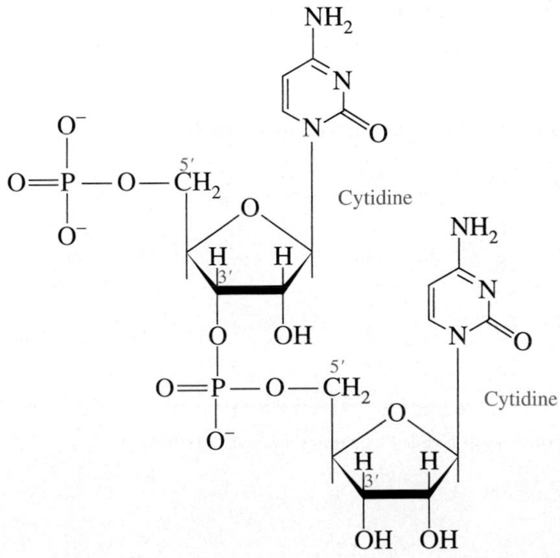

Cytidine

Cytidine

Study Check

In the dinucleotide of cytidine shown in the solution, identify the free 5' phosphate group and the free 3' hydroxyl (—OH) group.

QUESTIONS AND PROBLEMS

Primary Structure of Nucleic Acids

22.11 How are the nucleotides held together in a nucleic acid chain?

22.12 How do the ends of a nucleic acid polymer differ?

22.13 Write the structure of the dinucleotide 5'-GC-3' that would be in RNA.

22.14 Write the structure of the dinucleotide 5'-AT-3' that would be in DNA.

LEARNING GOAL

Describe the double helix of DNA.

Figure 22.8 This space-filling model shows the double helix that is the characteristic shape of DNA molecules.

Q *What is meant by the term double-helix?*

22.4 DNA Double Helix: A Secondary Structure

In the 1940s, biologists determined that DNA in a variety of organisms had a specific relationship between bases: the percent of adenine (A) was equal to the percent of thymine (T), and the percent of guanine (G) was equal to cytosine (C). Although different organisms vary in the amounts of bases, adenine is paired only with thymine, and guanine is paired only with cytosine in 1:1 ratios. (See Table 22.2.) This relationship known as *Chargaff's rules* can be summarized as follows:

Amount of purines = Amount of pyrimidines

$$A = T$$

$$G = C$$

In 1953, James Watson and Francis Crick proposed that DNA was a **double helix** that consists of two polynucleotide strands winding about each other like a spiral staircase. (See Figure 22.8.) The hydrophilic sugar-phosphate backbones are analogous to the outside railings with the hydrophobic bases arranged like steps along the inside. The two strands run in opposite directions. One strand goes in the 5'-3' direction, and the opposite strand goes in the 3'-5' direction.

Table 22.2	Percent Bases in the DNAs of Selected Organisms			
Organism	%A	%T	%G	%C
Humans	30	30	20	20
Chicken	28	28	22	22
Salmon	28	28	22	22
Corn (maize)	27	27	23	23
Neurospora	23	23	27	27

Source: Handbook of Biochemistry, 2nd ed. Sober, H.E., Ed.; CRC Press: Cleveland, OH, 1970.

Complementary Base Pairs

Each of the bases along one polynucleotide strand forms hydrogen bonds to a specific base on the opposite DNA strand. Adenine only bonds to thymine, and guanine only bonds to cytosine. (See Figure 22.9.) The pairs A—T and G—C are called **complementary base pairs.** The specificity of the base pairing is due to the fact that adenine and thymine form two hydrogen bonds, while cytosine and guanine form three hydrogen bonds. This explains why DNA has equal amounts of A and T bases and equal amounts of G and C.

Since each base pair contains a purine and a pyrimidine, the total width of the two pairs of bases A—T and G—C is the same. Thus the two polynucleotide strands of a DNA are the same distance apart all along the DNA polymer. There are no C—T, C—A, G—T, or G—A pairs in DNA because the other base-pair combinations cannot form as many hydrogen bonds or maintain a constant width

Figure 22.9 Hydrogen bonds between complementary base pairs hold the polynucleotide strands together in the double helix of DNA.
Q *Why are G—C base pairs more stable than A—T base pairs?*

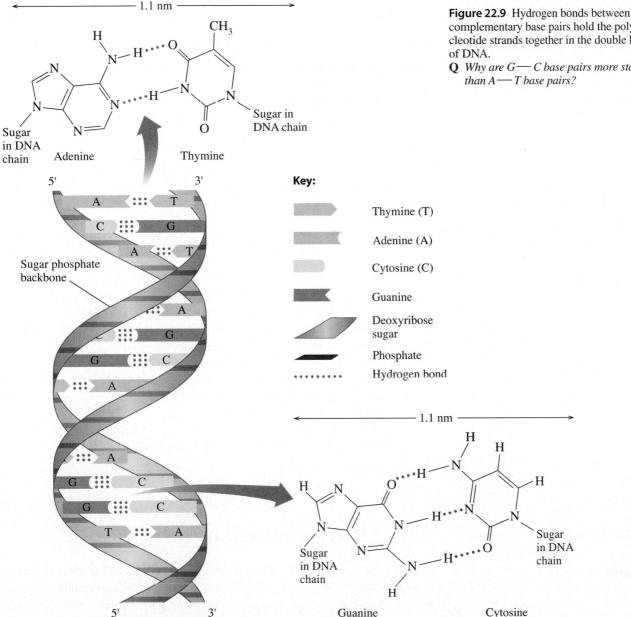

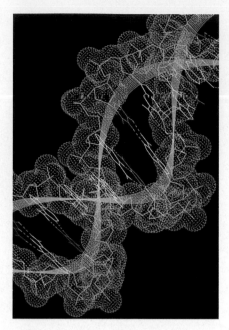

Figure 22.10 A computer-generated model of a DNA molecule.

Q *What is the complementary strand of a DNA section of 5'—GGCCTT—3'.*

between the two DNA backbones. As we shall see, complementary base pairing plays a crucial role in cell replication and the transfer of hereditary information.

X-ray diffraction patterns of DNA indicate that DNA is a right-handed or alpha (α) helix. In one complete turn, there are about 10 pairs of nucleotides. (See Figure 22.10.) In mitochondria, bacteria, and viruses, DNA molecules are compact, highly coiled molecules. In the chromosomes, DNA strands are wrapped around proteins called *histones*, a structure that provides the most stable and orderly arrangement for the long DNA molecules.

SAMPLE PROBLEM 22.4

Complementary Base Pairs

Write the base sequence of the complementary segment for the following segment of a strand of DNA.

5'—A—C—G—A—T—C—T—3'

Solution

In the complementary strand of DNA, the base A pairs with T, and G pairs with C.

Given segment of DNA: 5'—A—C—G—A—T—C—T—3'
                                                   :   :   :   :   :   :   :
Complementary segment: 3'—T—G—C—T—A—G—A—5'

Study Check

What is the sequence of bases that is complementary to a portion of DNA with a base sequence of 5'—G—G—T—T—A—A—C—C—3'?

QUESTIONS AND PROBLEMS

DNA Double Helix: A Secondary Structure

22.15 How are the two strands of nucleic acid in DNA held together?

22.16 What is meant by complementary base pairing?

22.17 Complete the base sequence in a second DNA strand if a portion of one strand has the following base sequence:
   a. 5'—AAAAAA—3'   **b.** 5'—GGGGGG—3'
   c. 5'—AGTCCAGGT—3'   **d.** 5'—CTGTATACGTTA—3'

22.18 Complete the base sequence in a second DNA strand if a portion of one strand has the following base sequence:
   a. 5'—TTTTTT—3'   **b.** 5'—CCCCCCCCC—3'
   c. 5'—ATGGCA—3'   **d.** 5'—ATATGCGCTAAA—3'

22.5 DNA Replication

DNA is found in *eukaryotic cells* of animals and plants including algae as well as in *prokaryotic cells* of bacteria. DNA from both types of cells is chemically similar and has the same function, which is to preserve genetic information. However, there are several differences in the cell types. Eukaryotic cells contain DNA in a nucleus with a nuclear membrane and combined with proteins called histones, have several organelles within membranes such as mitochondria and lysosomes, and divide by mitosis. Prokaryotic cells contain DNA as a circular chromosome with no membrane and not associated with histones, contain organelles without membranes, and divide by a simpler process called binary fission. As cells divide, copies of DNA must be produced in order to transfer the genetic information to the new cells. This is the process of DNA replication.

Replication and Energy

In DNA **replication**, the strands in the parent DNA separate, which allows each of the original strands to makes copies by synthesizing complementary strands. The

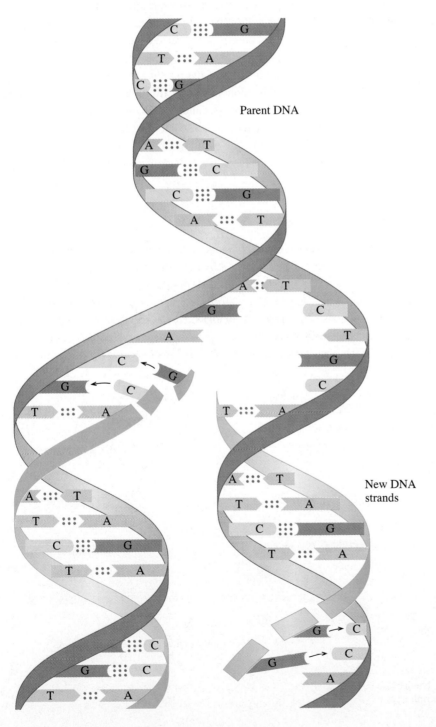

Figure 22.11 In DNA replication, the separate strands of the parent DNA are the templates for the synthesis of complementary strands, which produces two exact copies of DNA.

Q *How many strands of the parent DNA are in each of the new double-stranded copies of DNA?*

replication process begins when an enzyme called *helicase* catalyzes the unwinding of a portion of the double helix by breaking the hydrogen bonds between the complementary bases. These single strands now act as templates for the synthesis of new complementary strands. (See Figure 22.11.) Within the nucleus, nucleoside triphosphates for each base are available so that each exposed base on the template strand can form hydrogen bonds with its complementary base in the nucleoside triphosphate.

When a nucleoside triphosphate bonds to a sugar at the end of a growing strand, two phosphate groups are cleaved, which provides the energy for the reaction. For example, a T in the template strand forms a hydrogen bond with an A in ATP, and a G on the template strand forms hydrogen bonds with a CTP. After the base pairs are formed, *DNA polymerase* catalyzes the formation of phosphodiester bonds between the nucleotides. The hydrolysis of two phosphate groups releases energy for the new bonds. In this way, energy is provided to join each new nucleotide to the backbone of a growing DNA daughter strand. (See Figure 22.12.)

Eventually the entire double helix of the parent DNA is copied. In each new DNA molecule, one strand of the double helix is from the original DNA and one is a newly synthesized strand. This process, called *semi-conservative replication,* produces two new DNAs called *daughter DNA* that are identical to each other and exact copies of the original parent DNA. In the process of DNA replication, complementary base pairing ensures the correct placements of bases in the new DNA strands.

Direction of Replication

Now that we have seen the overall process, we can take a look at some of the details that are important in understanding DNA replication. The unwinding of DNA by *helicase* occurs simultaneously in several sections along the parent DNA molecule. As a result *DNA polymerase* can catalyze the replication process at each of these open DNA sections called **replication forks.** However, DNA polymerase catalyzes only phosphodiester bonds between the 5' phosphate of one nucleotide and the 3' hydroxyl of the next. That means that DNA polymerases have to move in opposite

Figure 22.12 As thymidine triphosphate bonds to the 3' — OH of the adjacent sugar, two phosphates are removed, which provides energy for the synthesis reaction.

Q *Why are nucleoside triphosphates used to provide the complementary bases instead of nucleoside monophosphates?*

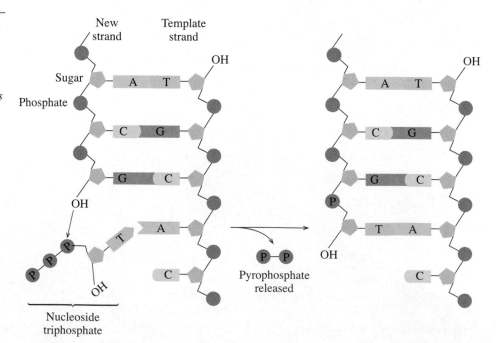

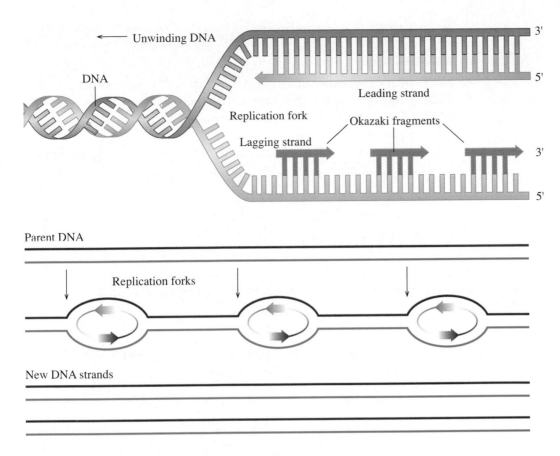

Figure 22.13 At each replication fork, DNA polymerase synthesizes a continuous DNA strand in the 5' to 3' direction. In the new 3' to 5' DNA strand, small Okazaki fragments are produced that are joined by DNA ligase.
Q *Why is only one of the new DNA strands synthesized in a continuous direction?*

directions along the separated strands of DNA. The new DNA strand that grows in the 5' to 3' direction, the *leading strand,* is synthesized continuously. The other new DNA, the *lagging strand,* is synthesized in the opposite direction, which is in the reverse 3' to 5' direction. In this lagging strand, short sections called **Okazaki fragments** are synthesized at the same time by several DNA polymerases and connected by DNA *ligase* to give a single 3' to 5' DNA strand. (See Figure 22.13.)

SAMPLE PROBLEM 22.5

DNA Replication

In an original DNA strand, there is a portion with the bases 5'—A—G—T—3'. What nucleotides are placed in a growing DNA daughter strand for this sequence?

Solution

Complementary base pairing allows only one possible nucleotide to pair with each base on the original strand. Thymine will pair only with adenine, cytosine with guanine, and adenine with thymine to give the sequence 3'—T—C—A—5'.

Study Check

Why would the new DNA strand you wrote for problem 22.5 be synthesized as Okazaki fragments that require a DNA ligase?

QUESTIONS AND PROBLEMS

DNA Replication

22.19 What is the function of the enzyme helicase in DNA replication?

22.20 What is the function of the enzyme DNA polymerase in DNA replication?

22.21 What process ensures that the replication of DNA produces identical copies?

22.22 Why are Okazaki fragments needed in the synthesis of the lagging strand?

LEARNING GOAL

Describe the structures and functions of the different RNAs.

22.6 Types of RNA

Ribonucleic acid, RNA, which makes up most of the nucleic acid found in the cell, is involved with transmitting the genetic information needed to operate the cell. Similar to DNA, RNA molecules are linear polymers of nucleotides. However, there are several important differences.

1. The sugar in RNA is ribose rather than the deoxyribose found in DNA.
2. The nitrogen base uracil replaces thymine.
3. RNA molecules are single, not double stranded.
4. RNA molecules are much smaller than DNA molecules.

There are three major types of RNA in the cells: *messenger RNA, ribosomal RNA,* and *transfer RNA.* Messenger RNA (**mRNA**) is a group of RNA molecules that carry instructions from the genes in the DNA to the ribosomes in the cell, where that information is used to synthesize proteins. Ribosomal RNA (**rRNA**), which makes up the largest percentage (about 75%) of RNA in the cell, is the RNA contained in the ribosomes. Transfer RNA (**tRNA**), which makes up about 15% of the RNA in a cell, is a group of small RNA molecules that bond to and carry specific amino acids to the ribosomes to be placed in peptide chains. The different types of RNA molecules found in a cell are classified according to their location and function in Table 22.3.

Table 22.3   Types of RNA Molecules

Type	Abbreviation	Percentage of Total RNA	Function in the Cell
Ribosomal RNA	rRNA	75	Major component of the ribosomes
Messenger RNA	mRNA	5–10	Carries information for protein synthesis from the DNA in the nucleus to the ribosomes
Transfer RNA	tRNA	10–15	Brings amino acids to the ribosomes for protein synthesis

Ribosomal RNA

Ribosomal RNA (rRNA), the most abundant type of RNA, makes up 65% of the structural material of the ribosomes; the other 35% is protein. Ribosomes, which are the sites for protein synthesis, consist of two subunits, a large subunit and a small subunit, which differ in their sedimentation rates (S values) when cen-

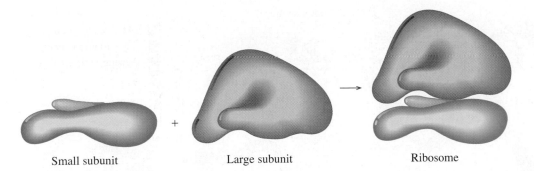

Small subunit   +   Large subunit   →   Ribosome

trifuged. In *E. coli* the small subunit is 30S and the large subunit is 50S. (See Figure 22.14.) Eukaryotic cells have larger ribosomes with 40S and 60S subunits. Cells that synthesize large numbers of proteins have thousands of ribosomes in their cells.

Figure 22.14 A typical prokaryotic ribosome consists of a small subunit and a large subunit.
Q *Why would there be many thousands of ribosomes in a cell?*

Messenger RNA

Messenger RNA (mRNA) carries genetic information from the DNA in the nucleus to the ribosomes in the cytoplasm for protein synthesis. Each gene, a segment of DNA, produces a separate mRNA molecule when a certain protein is needed in the cell, but then the mRNA is broken down quickly. The size of an mRNA depends on the number of nucleotides in that particular gene.

Transfer RNA

Transfer RNA (tRNA), the smallest of the RNA molecules, interprets the genetic information in DNA and brings specific amino acids to the ribosome for protein synthesis. Only the tRNAs can translate the genetic information into amino acids for proteins. There are one or more different tRNAs for each of the 20 amino acids. The structures of the transfer RNAs are similar, consisting of 70–90 nucleotides. Hydrogen bonds between some of the complementary bases in the chain produce loops that give some double-stranded regions.

The actual structure of a tRNA is a three-dimensional L shape, but we often draw tRNA as a cloverleaf to illustrate its features. All tRNA molecules have a 3' end with the nucleotide sequence 3'—ACC, which is known as the *acceptor stem*. An enzyme attaches an amino acid by forming an ester bond with the free —OH at the end of the acceptor stem. Each tRNA contains an **anticodon**, which is a series of three bases that complements three bases on a mRNA. (See Figure 22.15.)

SAMPLE PROBLEM 22.6

Types of RNA

What is the function of mRNA in a cell?

Solution

mRNA carries the instructions for the synthesis of a protein from the DNA to the ribosomes.

Study Check

What is the function of tRNA in a cell?

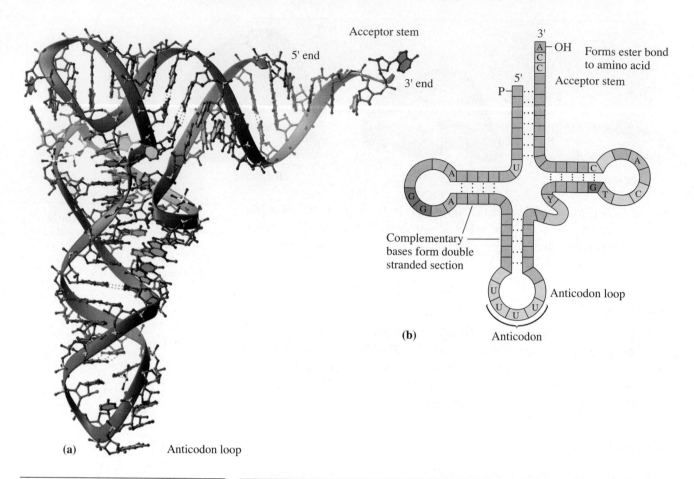

(a)

(b)

Figure 22.15 (a) In the L shape of a transfer RNA, some sections of the ribose–phosphate backbone form regions of complementary base bonding. (b) A typical tRNA molecule has an acceptor stem at the 3' end of the nucleic acid where an amino acid attaches, and an anticodon loop that complements three bases on mRNA.

Q *Why will different tRNAs have different bases in the anticodon loop?*

QUESTIONS AND PROBLEMS

Types of RNA

22.23 What are the three different types of RNA?

22.24 What are the functions of each type of RNA?

22.25 What is the composition of a ribosome?

22.26 What is the smallest RNA?

22.7 Transcription: Synthesis of mRNA

We now look at the processes involved in transferring genetic information encoded in the DNA to the production of proteins. First, the genetic information is copied from a gene in DNA to make a messenger RNA (mRNA), a process called **transcription**. As a carrier molecule, the mRNA moves out of the nucleus and goes to the ribosomes. During **translation,** tRNA molecules convert the information in the mRNA into amino acids, which are placed in the proper sequence to synthesize a protein. (See Figure 22.16.) The language that relates the series of nucleotides in mRNA with the amino acids specified is the *genetic code*.

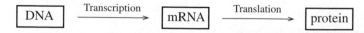

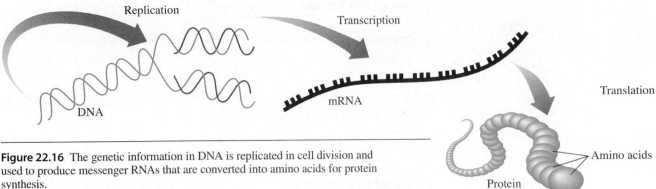

Figure 22.16 The genetic information in DNA is replicated in cell division and used to produce messenger RNAs that are converted into amino acids for protein synthesis.

Q *What is the difference between transcription and translation?*

Transcription begins when the section of a DNA that contains the gene to be copied unwinds. Within this unwound DNA, a polymerase enzyme identifies a nucleotide sequence of TATAAA, which is the starting or initiation point that is the signal to begin mRNA synthesis. Just as in DNA synthesis, C is paired with G, T pairs with A, and A pairs with U (not T). The polymerase enzyme moves along the unwound DNA in a 3' to 5' direction, forming bonds between the complementary bases in nucleoside triphosphates. Eventually the RNA polymerase reaches the termination point that signals the end of transcription, the new mRNA is released, and the DNA returns to its double helix structure. (See Figure 22.17.)

RNA polymerase ⟹

Section of bases on DNA template: 3'—G—A—A—C—T—5'
                                                                     ↓   ↓   ↓   ↓   ↓
Complementary base sequence in mRNA: 5'—C—U—U—G—A—3'

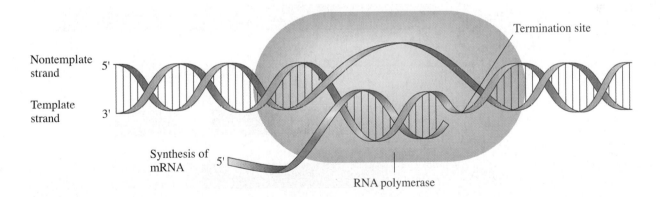

Figure 22.17 DNA undergoes transcription when RNA polymerase makes a complementary copy of a gene using the 3' to 5' strand as the template.

Q *Why is the mRNA connected in a 5' to 3' direction?*

RNA Synthesis

The sequence of bases in a part of the DNA template for mRNA is 3'—CGATCA—5'. What is the corresponding mRNA produced?

Solution

The nucleotides in DNA pair up with the ribonucleotides as follows: G ⟶ C, C ⟶ G, T ⟶ A, and A ⟶ U.

Portion of DNA template: 3′—C—G—A—T—C—A—5′

                                                 ↓   ↓   ↓   ↓   ↓   ↓

Complementary bases in mRNA: 5′—G—C—U—A—G—U—3′

Study Check

What is the DNA template that codes for the mRNA having the ribonucleotide sequence 5′—GGGUUUAAA—3′?

Processing of mRNA

In eukaryotes, the genes contain sections known as **exons** that code for proteins that are mixed in with sections called **introns** that do not code for protein. A newly formed mRNA is called a pre-mRNA because it is a copy of the entire DNA template including the noncoding introns. Before the newly synthesized pre-mRNA leaves the nucleus, it undergoes processing to remove the intron sections. The splicing of the pre-RNA produces a mature, functional mRNA that leaves the nucleus to deliver the genetic information to the ribosomes for the synthesis of protein. (See Figure 22.18.)

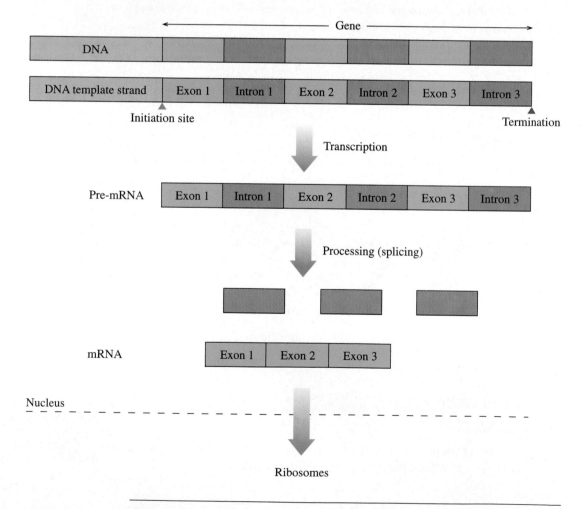

Figure 22.18 A pre-RNA, containing the exons' and introns' copies from the gene, is processed to remove the introns and form a mature mRNA that codes for a protein.
Q *What is the difference between exons and introns?*

Regulation of Transcription

The synthesis of mRNA does not occur randomly; rather, mRNA is synthesized in response to cellular needs for a particular protein. This regulation takes place at the transcription level where the absence or presence of end products speeds up or slows down the synthesis of mRNAs for specific enzymes. For example, *E. coli* that grow on lactose need β-galactosidase to hydrolyze lactose to glucose and galactose. However, when lactose is low or not present in the cell, there is little or no β-galactosidase present. The transcription of the mRNA for the enzyme is turned off. When lactose enters the cell, β-galactosidase levels rise because lactose induces the synthesis of its enzymes. In this process known as **enzyme induction:** lactose activates the transcription of the genes that produce the mRNAs that code for β-galactosidase.

In prokaryotes, each group of related proteins is regulated by an **operon,** which is a section of DNA containing a control site preceding the **structural genes** that code for proteins. The **control site** consists of a *promoter* and an *operator.* Preceding the operon is a regulatory gene. (See Figure 22.19.) A **regulatory gene** produces a mRNA for the synthesis of a **repressor** protein that binds to the operator and blocks the synthesis of β-galactosidase by RNA polymerase. When lactose enters the cell, lactose inactivates the repressor and it dissociates from the operator.

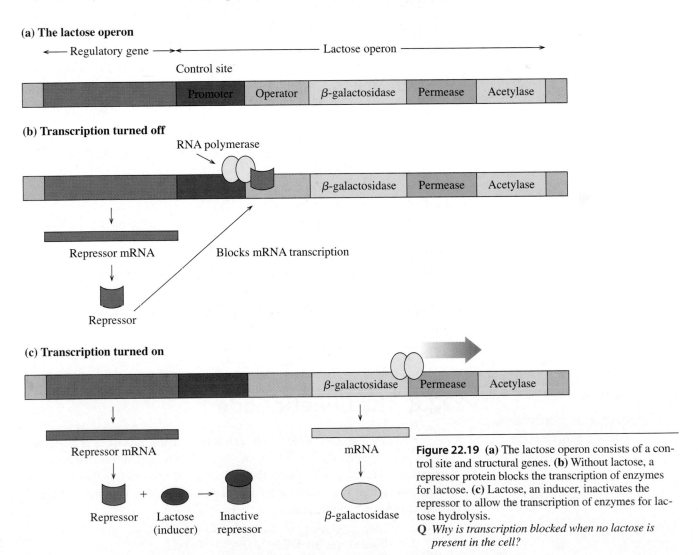

(a) The lactose operon

← Regulatory gene → ← Lactose operon →

Control site

| Promoter | Operator | β-galactosidase | Permease | Acetylase |

(b) Transcription turned off

RNA polymerase

| β-galactosidase | Permease | Acetylase |

Repressor mRNA   Blocks mRNA transcription

Repressor

(c) Transcription turned on

| β-galactosidase | Permease | Acetylase |

Repressor mRNA

Repressor + Lactose (inducer) → Inactive repressor

mRNA

β-galactosidase

Figure 22.19 (a) The lactose operon consists of a control site and structural genes. **(b)** Without lactose, a repressor protein blocks the transcription of enzymes for lactose. **(c)** Lactose, an inducer, inactivates the repressor to allow the transcription of enzymes for lactose hydrolysis.

Q *Why is transcription blocked when no lactose is present in the cell?*

Without a repressor, RNA polymerase transcribes the genes that codes for the lactose enzymes.

SAMPLE PROBLEM 22.8

Transcription

Indicate whether transcription takes place in each of the following conditions:

a. A repressor binds to the operator in the control site.
b. An inducer binds to the repressor protein.

Solution

a. No. A repressor blocks the synthesis of mRNA by RNA polymerase.
b. Yes. The lactose inactivates the repressor, which activates the transcription of the genes.

Study Check

How does a regulatory gene control the use of materials in a cell?

QUESTIONS AND PROBLEMS

Transcription: Synthesis of mRNA

22.27 What is meant by the term "transcription"?

22.28 What bases in mRNA are used to complement the bases A, T, G, and C in DNA?

22.29 Write the corresponding section of mRNA produced from the following section of DNA template:
3'—CCGAAGGTTCAC—5'

22.30 Write the corresponding section of mRNA produced from the following section of DNA template:
3'—TACGGCAAGCTA—5'

22.31 What are introns and exons?

22.32 What kind of processing do mRNA molecules undergo before they leave the nucleus?

22.33 What is an operon?

22.34 Why does the operon model control protein synthesis at the transcription level?

22.35 How is the operon turned off in *E. coli* that grow on lactose?

22.36 How is the operon activated in *E. coli* that grow on lactose?

LEARNING GOAL

Describe the function of the codons in the genetic code.

22.8 The Genetic Code

The overall function of the RNAs in the cell is to facilitate the task of synthesizing protein. After the genetic information encoded in DNA is transcribed into mRNA molecules, the mRNAs move out of the nucleus to the ribosomes in the cytoplasm. At the ribosomes, the genetic information in the mRNAs is converted into a sequence of amino acids in protein.

$$\text{mRNA} \xrightarrow[\text{}]{\text{Translation at the ribosomes in the cytoplasm}} \text{protein}$$

Codons

Genetic information from DNA is encoded in the mRNA as a sequence of nucleotides. In the **genetic code,** a sequence of three bases (triplet), called a **codon,** specifies each amino acid in the protein. Early work on protein synthesis showed that repeating triplets of uracil (UUU) produced a polypeptide that contained only phenylalanine. Therefore, a sequence of 5'—UUU—UUU—UUU—3' codes is for three phenylalanines.

Codons in mRNA	5'—UUU—UUU—UUU—3'
Translation	↓ ↓ ↓
Amino acid sequence	Phe — Phe — Phe

The codons have now been determined for all 20 amino acids. A total of 64 codons are possible from the triplet combinations of A, G, C, and U. Three of these, UGA, UAA, and UAG, are stop signals that code for the termination of protein synthesis. All the other three-base codons shown in Table 22.4 specify amino acids, which means that one amino acid can have several codons. For example, glycine is the amino acid coded by the codons GGU, GGC, GGA, and GGG. The triplet AUG has two roles in protein synthesis. At the beginning of an mRNA, the codon AUG signals the start of protein synthesis. In the middle of a series of codons, the AUG codon specifies the amino acid methionine.

Table 22.4   mRNA Codons: The Genetic Code for Amino Acids

First Letter	U (Second)	C	A	G	Third Letter
U	UUU UUC Phe; UUA UUG Leu	UCU UCC UCA UCG Ser	UAU UAC Tyr; UAA STOP; UAG STOP	UGU UGC Cys; UGA STOP; UGG Trp	U C A G
C	CUU CUC CUA CUG Leu	CCU CCC CCA CCG Pro	CAU CAC His; CAA CAG Gln	CGU CGC CGA CGG Arg	U C A G
A	AUU AUC AUA Ile; AUG Met/start	ACU ACC ACA ACG Thr	AAU AAC Asn; AAA AAG Lys	AGU AGC Ser; AGA AGG Arg	U C A G
G	GUU GUC GUA GUG Val	GCU GCC GCA GCG Ala	GAU GAC Asp; GAA GAG Glu	GGU GGC GGA GGG Gly	U C A G

[a]Codon that signals the start of a peptide chain.
STOP codons signal the end of a peptide chain.

SAMPLE PROBLEM 22.9

Codons

What is the sequence of amino acids coded by the following codons in mRNA?
5'—GUC—AGC—CCA—3'

Solution

According to Table 22.4, GUC codes for valine, AGC for serine, and CCA for proline. The sequence of amino acids is Val-Ser-Pro.

Study Check

The codon UGA does not code for an amino acid. What is its function in mRNA?

QUESTIONS AND PROBLEMS

The Genetic Code

22.37 What is a codon?

22.38 What is the genetic code?

22.39 What amino acid is coded for by each codon?
  a. CUU   **b.** UCA
  c. GGU   **d.** AGG

22.40 What amino acid is coded for by each codon?
  a. AAA   **b.** GUC
  c. CGG   **d.** GCA

22.41 When does the codon AUG signal the start of a protein, and when does it code for the amino acid methionine?

22.42 The codons UGA, UAA, and UAG do not code for amino acids. What is their role as codons in mRNA?

LEARNING GOAL
Describe the process of protein synthesis from mRNA.

22.9 Protein Synthesis: Translation

Once the mRNA is synthesized, it migrates out of the nucleus into the cytoplasm to the ribosomes. At the ribosomes, the **translation** process involves tRNA molecules, amino acids, and enzymes, all which convert the codons on mRNA into amino acids to make a protein.

Activation of tRNA

We have seen that the genetic code carried by the mRNA consists of triplets of nucleotides that correspond to one of 20 different amino acids. Now the tRNA are utilized to translate the codons into specific amino acids. This adaptor role of tRNA is possible because it reads both the triplet language of mRNA and picks up the amino acid that corresponds to that codon. The anticodon in the loop at the bottom of a tRNA contains a triplet of bases that complement a codon on mRNA. An amino acid is attached to the acceptor stem of the tRNA by enzymes called *aminoacyl-tRNA synthetases*. There is a different synthetase for each amino acid. The synthetase enzymes use energy from ATP to catalyze the formation of ester bonds between the carboxylic acid groups on the amino acids and the hydroxyl groups of the acceptor stem. (See Figure 22.20.) The matching of a tRNA to the correct amino acid is essential. If the wrong amino acid is attached, it would be placed into the protein and make an incorrect protein. However, each synthetase checks the attachment of an amino acid and tRNA and hydrolyzes any incorrect combinations.

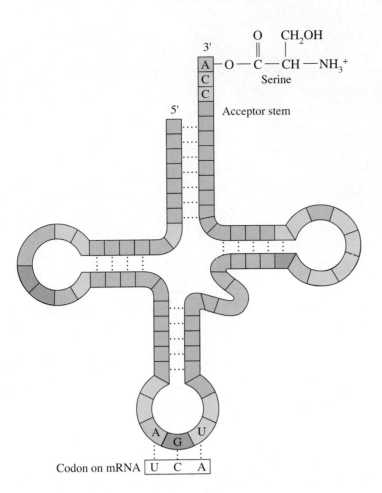

Figure 22.20 An activated tRNA with anti-codon AGU bonds to serine at the acceptor stem.
Q *What is the codon for serine for this tRNA?*

Initiation

Protein synthesis begins when a mRNA combines with the smaller subunit of a ribosome. Because the first codon in an mRNA is a *start* codon, AUG, a tRNA with anticodon UAC and carrying the amino acid methionine forms hydrogen bonds with the AUG codon. The larger ribosome completes the ribosomal unit and translation is ready to begin. (See Figure 22.21.) Protein synthesis takes place on the ribosome at two adjacent sites on the larger subunit. A tRNA that has the anticodon for the second codon with the second amino acid bonds to mRNA. A peptide bond forms between the amino acids and the first tRNA dissociates from the ribosome and returns to the cytoplasm. After the first tRNA detaches from the ribosome, the ribosome shifts to the next codon on the mRNA, a process called **translocation.** A new tRNA can now attach to the open binding site so that its amino acid forms a peptide bond with the previous amino acid. Again the earlier tRNA detaches and the ribosome shifts down the mRNA to read the next codon. Each time a peptide bond joins the new amino acid to the growing polypeptide chain. Sometimes several ribosomes, called a polysome, translate the same strand of mRNA at the same time to produce several copies of the peptide chain.

Termination

After all the amino acids for a particular protein have been linked together by peptide bonds, the ribosome encounters a stop codon. Because there are no tRNAs to complement the termination codon, protein synthesis ends. An enzyme releases the completed polypeptide chain from the ribosome. The initiating amino acid methio-

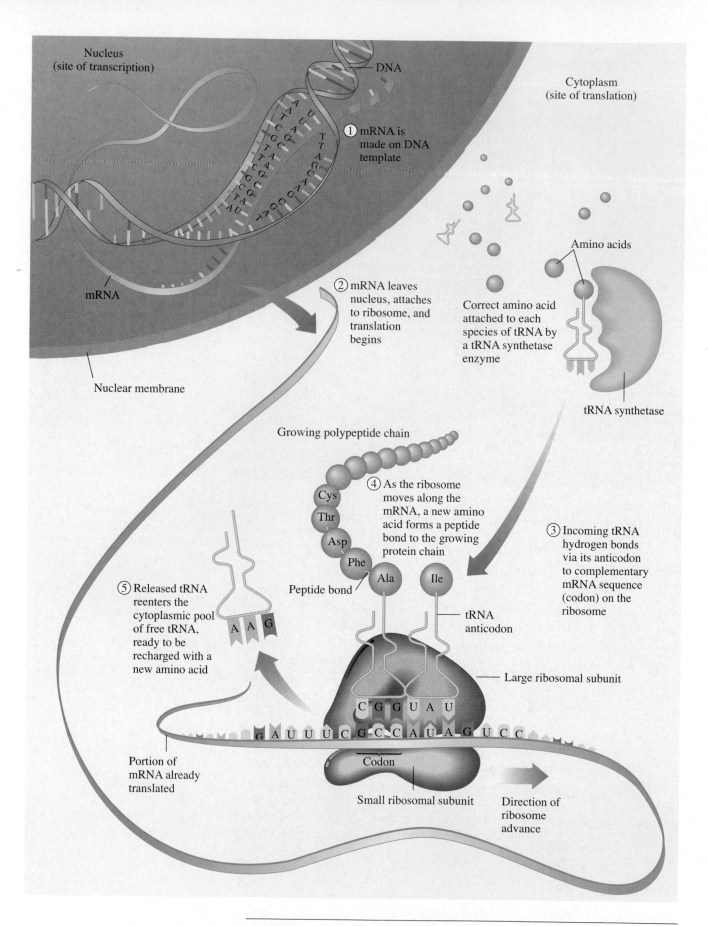

Figure 22.21 In the translation process, the mRNA synthesized by transcription attaches to a ribosome and tRNAs pick up their amino acids and place them in a growing peptide chain.
Q *How is the correct amino acid placed in the peptide chain?*

nine is often removed from the beginning of the peptide chain. Now the amino acids in the chain form the three-dimensional structure that makes the polypeptide into a biologically active protein.

SAMPLE PROBLEM 22.10

Protein Synthesis: Translation

What order of amino acids would you expect in a peptide for the mRNA sequence of 5'—UCA—AAA—GCC—CUU—3'?

Solution

Each of the codons specifies a particular amino acid. Using Table 22.4, we write a peptide with the following amino acid sequence:

RNA codons: 5'—UCA—AAA—GCC—CUU—3'
                          ↓   ↓   ↓   ↓
Amino acid sequence: Ser — Lys — Ala— Leu

Study Check

Where would protein synthesis stop in the following series of bases in an mRNA?
5'—GGG—AGC—AGU—UAG—GUU—3'

QUESTIONS AND PROBLEMS

Protein Synthesis: Translation

22.43 What is the difference between a *codon* and an *anticodon*?

22.44 Why are there at least 20 different tRNAs?

22.45 What are the three steps of translation?

22.46 Where does protein synthesis take place?

22.47 What amino acid sequence would you expect from each of the following mRNA segments?
  a. 5'—AAA—AAA—AAA—3'
  b. 5'—UUU—CCC—UUU—CCC—3'
  c. 5'—UAC—GGG—AGA—UGU—3'

22.48 What amino acid sequence would you expect from each of the following mRNA segments?
  a. 5'—AAA—CCC—UUG—GCC—3'
  b. 5'—CCU—CGA—AGC—CCA—UGA—3'
  c. 5'—AUG—CAC—AAA—GAA—GUA—CUU—3'

22.49 How is a peptide chain extended?

22.50 What is meant by "translocation"?

22.51 The following portion of DNA is in the template DNA strand:

  3'—GCT—TTT—CAA—AAA—5'

  a. What is the corresponding mRNA section?
  b. What are the anticodons of the tRNAs?
  c. What amino acids will be placed in the peptide chain?

22.52 The following portion of DNA is in the template DNA strand:

  3'—TGT—GGG—GTT—ATT—5'

  a. What is the corresponding mRNA section?
  b. What are the anticodons of the tRNAs?
  c. What amino acids will be placed in the peptide chain?

HEALTH NOTE

Many Antibiotics Inhibit Protein Synthesis

Several antibiotics stop bacterial infections by interfering with the synthesis of proteins needed by the bacteria. Some antibiotics act only on bacterial cells by binding to the ribosomes in bacteria but do not act on human cells. A description of some of these antibiotics is given in Table 22.5.

Table 22.5   Antibiotics That Inhibit Protein Synthesis in Bacterial Cells

Antibiotic	Effect on Ribosomes to Inhibit Protein Synthesis
Chloramphenicol	Inhibits peptide bond formation and prevents the binding of tRNA
Erythromycin	Inhibits peptide chain growth by preventing the translocation of the ribosome along the mRNA
Puromycin	Causes release of an incomplete protein by ending the growth of the polypeptide early
Streptomycin	Prevents the proper attachment of tRNAs
Tetracycline	Prevents the binding of tRNAs

LEARNING GOAL

Describe some ways in which DNA is altered to cause mutations.

22.10 Genetic Mutations

A **mutation** is a change in the DNA nucleotide sequence that alters the sequence of amino acids, which may alter the structure and function of a protein in a cell. Some mutations are known to result from X rays, overexposure to sun (ultraviolet or UV light), chemicals called mutagens, and possibly some viruses. If a change in DNA occurs in a somatic cell (a cell other than a reproductive cell) the altered DNA will be limited to that cell and its daughter cells. If there is uncontrolled growth, the mutation could lead to cancer. If the mutation occurs in germ cell DNA (egg or sperm), then all the DNA produced in a new individual will contain the same genetic change. If the genetic change greatly affects the catalysis of metabolic reactions or the formation of important structural proteins, the new cells may not survive or the person may exhibit a genetic disease.

Consider a triplet of bases CCG in the coding strand of DNA, which produces the codon GGC in mRNA. At the ribosome, tRNA would place the amino acid glycine in the peptide chain. (See Figure 22.22a.) Now, suppose that T replaces the first C in the DNA triplet, which gives TCG as the triplet. Then the codon produced in the mRNA is AGC, which brings the amino acid serine to the peptide chain. The replacement of one base in the coding strand of DNA with another is called a **substitution** mutation. The change in the codon can lead to the insertion of a different amino acid at that point in the polypeptide. Substitution is the most common way in which mutations occur. (See Figure 22.22b.)

In a **frame shift mutation,** a base is added to or deleted from the normal order of bases in the coding strand of DNA. Suppose now an A is deleted from the triplet AAA, which gives a new triplet of AAC. The next triplet becomes CGA rather than CCG and so on. All the triplets shift over by one base, which changes all the codons that follow and leads to a different sequence of amino acids from that point. Figure 22.22c illustrates a frame shift mutation by deletion.

EXPLORE YOUR WORLD

A Model for DNA Replication and Mutation

1. Cut out 16 rectangular pieces of paper. For strand 1 of a segment of DNA, write on two pieces this nucleotide symbol with hydrogen bonds: A ═══, T ═══, G ≡≡≡, and C ≡≡≡. For strand 2, write on two pieces each: ═══ A, ═══ T, ≡≡≡ G, and ≡≡≡ C. Mix up the pieces for strand 1 and randomly place in vertical order. From the second group, strand 2 nucleotides, select the complementary bases to complete this section of DNA.
2. Using the nucleotides from experiment 1, make a strand 1 of A—T—T—G—C—C. Arrange the corresponding bases to complete the DNA section. In the original column, change the G to an A. What does this do to the complementary strand? How could this change result in a mutation?

Effect of Mutations

When a mutation causes a change in the amino acid sequence, the structure of the resulting protein can be altered severely and it may lose biological activity. If the protein is an enzyme, it may no longer bind to its substrate or react with the substrate at the active site. When an altered enzyme cannot catalyze a reaction, certain substances may accumulate until they act as poisons in the cell, or substances vital to survival may not be synthesized. If a defective enzyme occurs in a major metabolic pathway or in the building of a cell membrane, the mutation can be lethal. When a protein deficiency is genetic, the condition is called a **genetic disease.**

$$
\text{DNA} \xrightarrow[\text{viruses}]{\substack{\text{X rays,} \\ \text{UV sunlight,} \\ \text{mutagens,}}} \begin{array}{c}\text{alteration of}\\\text{DNA}\end{array} \longrightarrow \begin{array}{c}\text{defective}\\\text{protein}\end{array} \longrightarrow \begin{array}{c}\text{genetic disease (germ cells)}\\\text{or cancer (somatic cells)}\end{array}
$$

SAMPLE PROBLEM 22.11

Mutations

An mRNA has the sequence of codons 5'—CCC—AGA—GGG—3'. If a base substitution in the DNA changes the mRNA codon of AGA to ACA, how is the amino acid sequence affected in the resulting protein?

(a) Normal DNA and protein synthesis

DNA (coding strand)

T A C T T C A A A C C G A T T

Transcription

A U G A A G U U U G G C U A A

mRNA

Translation

Amino acid sequence

Met — Lys — Phe — Gly — Stop

Figure 22.22 An alteration in the DNA coding strand (template) produces a change in the sequence of amino acids in the protein, which may lead to a mutation. (**a**) A normal DNA leads to the correct amino order in a protein. (**b**) The substitution of a base in DNA leads to a change in the mRNA codon and a change in the amino acid. (**c**) The deletion of a base causes a frame shift mutation, which changes the amino acid order.

Q *When would a substitution mutation cause protein synthesis to stop?*

(b) Substitution of one base

Substitution of C by T

DNA (coding strand)

T A C T T C A A A T C G A T T

A U G A A G U U U A G C U A A

mRNA

Amino acid sequence

Met — Lys — Phe — Ser — Stop

(c) Frameshift mutation caused by the deletion of a base

A

DNA (coding strand)

T A C T T C A A C C G A T T

A U G A A G U U G G C U A A .....

mRNA

Met — Lys — Leu — Ala .....

Solution

The initial mRNA sequence of CCC—AGA—GGG codes are for the amino acids proline, arginine, and glycine. When the mutation occurs, the new sequence of mRNA codons CCC—ACA—GGG are for proline, threonine, and glycine. The amino acid arginine is replaced by threonine.

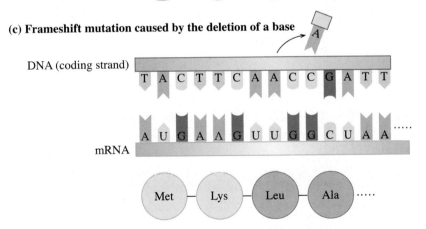

	Normal	*After Mutation*
mRNA codons	CCC — AGA — GGG	CCC — ACA — GGG
Amino acids	Pro — Arg — Gly	Pro — Thr — Gly

Study Check

How might the protein made from this mRNA be affected by this mutation?

Genetic Diseases

A genetic disease is the result of a defective enzyme caused by a mutation in its genetic code. For example, phenylketonuria, PKU, results when DNA cannot direct the synthesis of the enzyme phenylalanine hydroxylase, required for the conversion of phenylalanine to tyrosine. In an attempt to break down the phenylalanine, other enzymes in the cells convert it to phenylpyruvate. The accumulation of phenylalanine and phenylpyruvate in the blood can lead to severe brain damage and mental retardation. If PKU is detected in a newborn baby, a diet is prescribed that eliminates all the foods that contain phenylalanine. Preventing the buildup of the phenylpyruvate ensures normal growth and development.

The amino acid tyrosine is needed in the formation of melanin, the pigment that gives the color to our skin and hair. If the enzyme that converts tyrosine to melanin is defective, no melanin is produced, a genetic disease known as albinism. Persons and animals with no melanin have no skin or hair pigment. (See Figure 22.23.) Table 22.6 lists some other common genetic diseases and the type of metabolism or area affected.

Figure 22.23 This peacock with albinism does not produce the melanin needed to make bright colors of its feathers.
Q *Why are traits such as albinism related to the gene?*

(Reaction diagram: Phenylalanine → Phenylpyruvate → Phenylketonuria (PKU); Phenylalanine hydroxylase converts Phenylalanine to Tyrosine; Tyrosine → Melanin (pigments) → Albinism)

QUESTIONS AND PROBLEMS

Genetic Mutations

22.53 What is a substitution mutation?

22.54 How does a substitution mutation in the genetic code for an enzyme affect the order of amino acids in that protein?

22.55 What is the effect of a mutation on the amino acids in the polypeptide?

22.56 How can a mutation decrease the activity of a protein?

22.57 How is protein synthesis affected if the normal base sequence TTT in the DNA template is changed to TTC?

22.58 How is protein synthesis affected if the normal base sequence CCC is changed to ACC?

22.59 Consider the following portion of mRNA produced by the normal order of DNA nucleotides:

Table 22.6   Some Genetic Diseases

Genetic Disease	Result
Galactosemia	The transferase enzyme required for the metabolism of galactose-1-phosphate is absent. Accumulation of Gal-1-P leads to cataracts and mental retardation.
Cystic fibrosis	The most common inherited disease. Thick mucus secretions make breathing difficult and block pancreatic function.
Down syndrome	The leading cause of mental retardation, occurring in about 1 of every 800 live births. Mental and physical problems including heart and eye defects are the result of the formation of three chromosomes, usually chromosome 21, instead of a pair.
Familial hypercholesterolemia	A mutation of a gene on chromosome 19 results in high cholesterol levels that lead to early coronary heart disease in people 30–40 years old.
Muscular dystrophy (Duchenne)	One of 10 forms of MD. A mutation in the X chromosome results in the low or abnormal production of *dystrophin* by the X gene. This muscle-destroying disease appears at about age 5 with death by age 20 and occurs in about 1 of 10,000 males.
Huntington's disease (HD)	Appearing in middle age, HD affects the nervous system, leading to total physical impairment. It is the result of a mutation in a gene on chromosome 4, which can now be mapped to test people in families with HD.
Sickle-cell anemia	A defective hemoglobin from a mutation in a gene on chromosome 11 decreases the oxygen-carrying ability of red blood cells, which take on a sickled shape, causing anemia and plugged capillaries from red blood cell aggregation.
Hemophilia	One or more defective blood-clotting factors lead to poor coagulation, excessive bleeding, and internal hemorrhages.
Tay-Sachs disease	Hexosaminidase A is defective, causing an accumulation of gangliosides resulting in mental retardation, loss of motor control, and early death.

5' — ACA — UCA — CGG — GUA — 3'

a. What is the amino acid order produced for normal DNA?
b. What is the amino acid order if a mutation changes UCA to ACA?
c. What is the amino acid order if a mutation changes CGG to GGG?
d. What happens to protein synthesis if a mutation changes UCA to UAA?
e. What happens if a G is added to the beginning of a chain?
f. What happens if the A is removed from the beginning of a chain?

22.60 Consider the following portion of mRNA produced by the normal order of DNA nucleotides:

5' — CUU — AAA — CGA — GUU — 3'

a. What is the amino acid order produced for normal DNA?
b. What is the amino acid order if a mutation changes CUU to CCU?
c. What is the amino acid order if a mutation changes CGA to AGA?
d. What happens to protein synthesis if a mutation changes AAA to UAA?

22.61 a. A base substitution changes a codon for an enzyme from GCC to GCA. Why is there no change in the amino acid order in the protein?
b. In sickle-cell anemia, a base substitution in hemoglobin replaces glutamine (a polar amino acid) with valine. Why does the replacement of one amino acid cause such a drastic change in biological function?

22.62 a. A base substitution for an enzyme replaces leucine (a nonpolar amino acid) with alanine. Why does this change in amino acids have little effect on the biological activity of the enzyme?
b. A base substitution replaces cytosine in the codon UCA with adenine. How would this substitution affect the amino acids in the protein?

LEARNING GOAL
Describe the preparation and
uses of recombinant DNA.

22.11 Recombinant DNA

Over the past two decades, new techniques called genetic engineering have permitted scientists to cut, splice, and recombine DNA from different kinds of cells. The new synthetic forms of DNA, which contain a DNA fragment from another organism, are known as **recombinant DNA.** Recombinant DNA technology is now used to produce human insulin for diabetics, the antiviral substance interferon, blood-clotting factor VIII, and human growth hormone.

Restriction Enzymes

Most of the work in recombinant DNA is done with *Escherichia coli (E. coli)* bacteria. The DNA in prokaryotes is present in the bacterial cells as several small circular molecules called *plasmids*. These plasmids, which are easy to isolate, are also capable of replication. The ability to insert foreign DNA into cells developed after scientists discovered that enzymes called **restriction enzymes** can cut DNA between a specific pair of nucleotides along the chain. The resulting DNA fragments have "sticky" ends that can form phosphodiester bonds when they are mixed with plasmids that are also cut open using restriction enzymes. After the foreign DNA fragment joins with the plasmid DNA, the recombined DNA in the plasmids is able to synthesize the proteins coded for by the new DNA. In one day, one *E. coli* bacterium is capable of producing a million copies of itself including the foreign DNA, a process known as gene cloning.

Preparing Recombinant DNA

Initially, *E. coli* cells are soaked in a detergent solution to dissolve the plasma membrane. The contents of the cells including the plasmids are released and collected.

Figure 22.24 Recombinant DNA is formed by placing a gene from another organism in a plasmid DNA of the bacterium, which causes the bacterium to produce a non-bacterial protein such as insulin or growth hormone.
Q *How can recombinant DNA help a person with a genetic disease?*

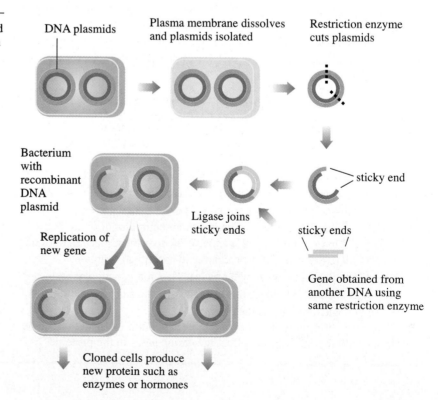

DNA plasmids

Plasma membrane dissolves and plasmids isolated

Restriction enzyme cuts plasmids

sticky end

sticky ends

Gene obtained from another DNA using same restriction enzyme

Ligase joins sticky ends

Bacterium with recombinant DNA plasmid

Replication of new gene

Cloned cells produce new protein such as enzymes or hormones

A *restriction enzyme,* which breaks phosphodiester bonds in DNA between specific nucleotides, is used to cut open the circular DNA strands in the plasmids. (See Figure 22.24.) The same enzymes are also used to cut out a piece of DNA from the chromosome of a different organism, such as the gene that produces insulin or growth hormone. The cut-out genes are then mixed with the plasmids that were cut open. The ends of the foreign DNA piece and the ends of the opened plasmids are joined by a DNA ligase. Then the altered plasmids containing the recombined DNA are placed in a fresh culture of *E. coli* bacteria where they can be reabsorbed into the bacterial cells. The new gene that was inserted in the plasmids is copied as the genetically engineered *E. coli* cells start to replicate. If the inserted DNA coded for the human insulin protein, the altered plasmids would begin to synthesize human insulin. Eventually, a large number of cells with the new DNA produce the insulin protein. Table 22.7 lists some of the products developed through recombinant DNA technology that are now used therapeutically.

Table 22.7   Therapeutic Products of Recombinant DNA

Product	Therapeutic Use
Human insulin	Treat diabetes
Erythropoietin (EPO)	Treat anemia; stimulate production of erythrocytes
Human growth hormone (HGH)	Stimulate growth
Interferon	Treat cancer and viral disease
Tumor necrosis factor (TNF)	Destroy tumor cells
Monoclonal antibodies	Transport drugs needed to treat cancer and transplant rejection
Epidermal growth factor (EGF)	Stimulate healing of wounds and burns
Interleukins	Stimulate immune system; treat cancer
Prourokinase	Destroy blood clots; treat myocardial infarctions

DNA Fingerprinting

In a process referred to as DNA fingerprinting or Southern transfer, restriction enzymes are used to cut fragments from DNA. The resulting DNA fragments called RFLPs (restriction fragment length polymorphisms) are then sorted by size using gel electrophoresis. The gel is treated with a radioactive isotope that adheres to specific base sequences in the RFLPs. A piece of x-ray film placed over the gel is exposed by the radioactivity emitted from RFLPs, which gives a pattern of dark and light bands known as a DNA fingerprint. (See Figure 22.25.) It has been estimated that the odds of two persons (who are not identical twins) producing the same DNA fingerprint is less than one in a billion.

One application of Southern transfer is genetic screening in which an individual can be screened for genes that are responsible for genetic diseases. It is being used to screen for genes responsible for sickle-cell disease, cystic fibrosis, breast cancer, colon cancer, Huntington's disease, and amyotropic lateral sclerosis also called Lou Gehrig's disease.

Human Genome Project

In the 1970s, scientists began to use restriction enzymes to map the location of genomes in viruses and plasmids. By 1987, the genome of *E. coli* was determined.

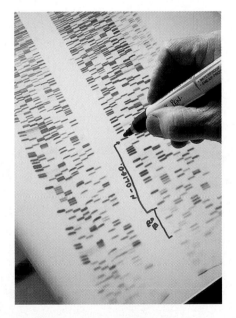

Figure 22.25 A scientist analyzes a nucleotide sequence in DNA from a human gene. Dark and light bands on the film represent the order of nucleotides. The marked sequences are involved in the growth of melanoma cancer cells.
Q *What causes DNA fragments to appear on x-ray film?*

Recently, these techniques combined with new computer programs have helped scientists compile the map of the human genome. In some recent reports, the total number of human genes is about 30,000. It now appears that most of the genome is not functional and perhaps carried over for millions of years. Large blocks of genes are copied from one human chromosome to another even though they no longer code for needed proteins. Thus, the coding portions of the genes seems to be even less than previously thought making up only about 1% of the total genome. The results of the genome project will help us identify defective genes that lead to genetic disease.

Polymerase Chain Reaction

The process of gene cloning using recombinant DNA requires living cells such as *E. coli*. In 1987 a process called **polymerase chain reaction (PCR)** made it possible to produce multiple copies of (amplify) the DNA in a short time. In the PCR technique, a sequence of a DNA is selected to copy, and the DNA is heated to separate the strands. Primers that are complementary to a small group of nucleotides on each side of the sequence to be copied are added to the ends of the templates. The DNA strands with their primers are mixed with DNA polymerase and a mixture of deoxyribonucleotides and complementary strands for the DNA section are produced. Then the process is repeated with the new batch of DNA. After several cycles of the PCR process, millions of copies of the initial DNA section are produced. (See Figure 22.26.)

SAMPLE PROBLEM 22.12

Recombinant DNA

What is the function of restriction enzymes in recombinant DNA?

Solution

Restriction enzymes are used to cut out a particular piece of DNA from a gene and to cut open the circular plasmids in a bacterium where the foreign DNA attaches.

Study Check

What is gene cloning?

QUESTIONS AND PROBLEMS

Recombinant DNA

22.63 Why are *E. coli* bacteria used in recombinant DNA procedures?

22.64 What is a plasmid?

22.65 How are plasmids obtained from *E. coli*?

22.66 Why are restriction enzymes mixed with the plasmids?

22.67 How is a gene for a particular protein inserted into a plasmid?

22.68 Why is DNA polymerase useful in criminal investigations?

22.69 What is a DNA fingerprint?

22.70 What beneficial proteins are produced from recombinant DNA technology?

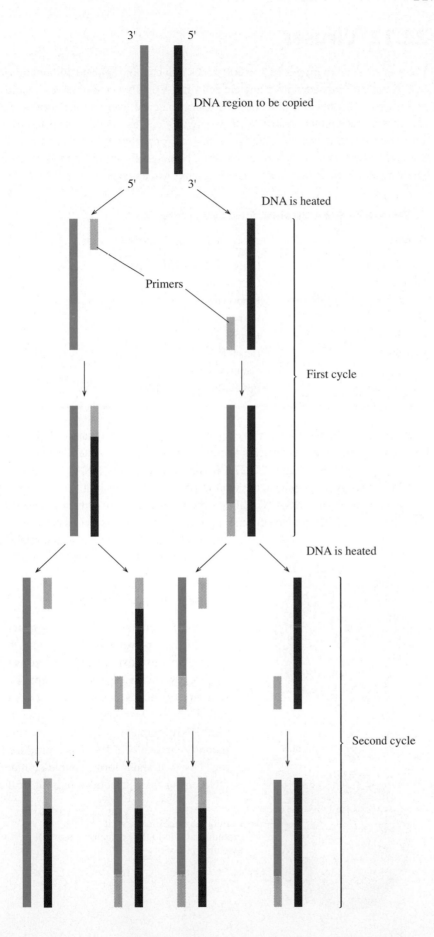

Figure 22.26 Each cycle of the polymerase chain reaction doubles the number of copies of the DNA section.

Q *Why are the DNA strands heated at the start of each cycle?*

3′   5′

DNA region to be copied

5′   3′

DNA is heated

Primers

First cycle

DNA is heated

Second cycle

22.12 Viruses

Viruses are small particles of 3 to 200 genes that cannot replicate without a host cell. A typical virus contains a nucleic acid, DNA or RNA, but not both, inside a protein coat. It does not have the necessary material such as nucleotides and enzymes to make proteins and grow. The only way a virus can replicate is to invade a host cell and take over the materials necessary for protein synthesis and growth. Some infections caused by viruses invading human cells are listed in Table 22.8. There are also viruses that attack bacteria, plants, and animals.

Table 22.8   Some Diseases Caused by Viral Infection

Disease	Virus
Common cold	Coronavirus (over 100 types)
Influenza	Orthomyxovirus
Warts	Papovavirus
Herpes	Herpesvirus
Leukemia, cancers, AIDS	Retrovirus
Hepatitis	Hepatitis A virus (HAV), hepatitis B virus (HBV)
Mumps	Paramyxovirus
Epstein-Barr	Epstein-Barr virus (EBV)

Figure 22.27 After a virus attaches to the host cell, it injects its viral DNA and uses the host cell's amino acids to synthesize viral protein and nucleic acids, enzymes, and ribosomes to make viral RNA. When the cell bursts, the new viruses are released to infect other cells.
Q *Why does a virus need a host cell for replication?*

A viral infection begins when an enzyme in the protein coat makes a hole in the host cell, allowing the nucleic acids to enter and mix with the materials in the host cell. (See Figure 22.27.) If the virus contains DNA, the host cell begins to replicate the viral DNA in the same way it would replicate its normal DNA. Viral DNA produces viral RNA, which proceeds to make the proteins for more protein coats. So many virus particles are synthesized that the cell bursts and releases new viruses to infect more cells.

Reverse Transcription

A virus that contains RNA as its genetic material is a **retrovirus.** Once inside the host cell, it must first make viral DNA using a process known as reverse transcriptase. A retrovirus contains a polymerase enzyme called *reverse transcriptase* that uses the viral RNA template to synthesize complementary strands of DNA. Once produced, the DNA strands form double-stranded DNA using the nucleotides present in the

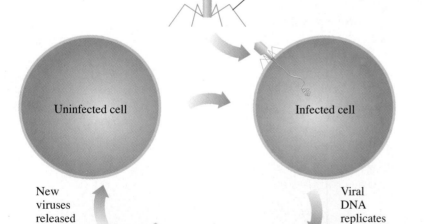

Head   DNA
Tail   Tail fiber

Uninfected cell   Infected cell

New viruses released

Viral DNA replicates

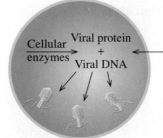

Cellular enzymes → Viral protein + Viral DNA

Cellular enzymes make viral proteins and viral DNA, which assemble into viruses.

Cell bursts

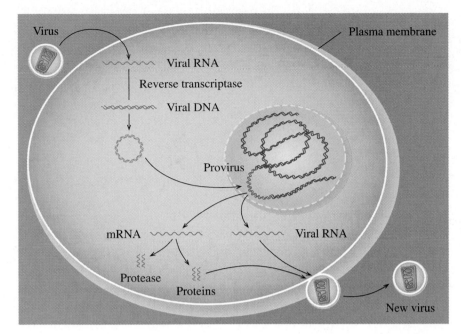

Figure 22.28 After a retrovirus injects its viral RNA into a cell, it forms a DNA strand by reverse transcription. The DNA forms a double-stranded DNA called a provirus, which joins the host cell DNA. When the cell replicates, the provirus produces the viral RNA needed to produce more virus particles.
Q *What is reverse transcription?*

host cell. This newly formed viral DNA called a provirus joins the DNA of the host cell. (See Figure 22.28.)

Vaccines are inactive forms of viruses that boost the immune response by causing the body to produce antibodies to the virus. Several childhood diseases such as polio, mumps, chicken pox, and measles can be prevented through the use of vaccines.

AIDS

In the early 1980s, a disease called AIDS (acquired immune deficiency syndrome) began to claim an alarming number of lives. An HIV-1 virus (human immunodeficiency virus type 1) is now known to be the AIDS-causing agent. (See Figure 22.29.) HIV is a retrovirus that infects and destroys T4 lymphocyte cells, which are involved in the immune response. After the HIV-1 virus binds to receptors on the surface of a T4 cell, the virus injects viral RNA into the host cell. As a retrovirus, the genes of the viral RNA direct the formation of viral DNA. The gradual depletion of T4 cells reduces the ability of the immune system to destroy harmful organisms. The AIDS syndrome is characterized by opportunistic infections such as *Pneumocystis carinii,* which causes pneumonia, and *Kaposi's sarcoma,* a skin cancer.

Treatment for AIDS is based on attacking the HIV-1 at different points in its life cycle including reverse transcription and protein synthesis. Nucleoside analogs mimic the structures of the nucleosides used for DNA synthesis. For example, the drug AZT (azidothymine) is similar to thymine, and ddI (dideoxyinosine) is similar to guanosine. Two other drugs include dideoxycytidine (ddC) and didehydro-3'-deoxythymidine (d4T). Such compounds are found in the "cocktails" that are providing extended remission of HIV infections. When a nucleoside analog is incorporated into viral DNA, the lack of a hydroxyl group on the 3' carbon in the sugar prevents the formation of the sugar–phosphate bonds, which prevents the replication of the virus.

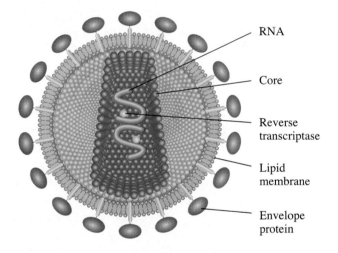

Figure 22.29 The HIV virus that causes AIDS syndrome destroys the immune system in the body.
Q *Is the HIV virus a DNA virus or an RNA retrovirus?*

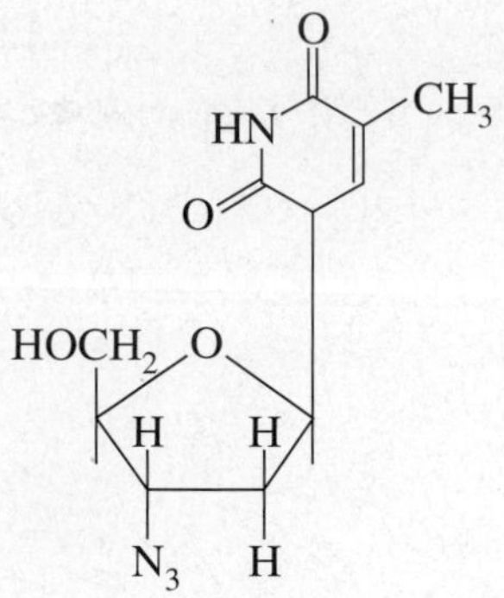

AZT
Azidothymine

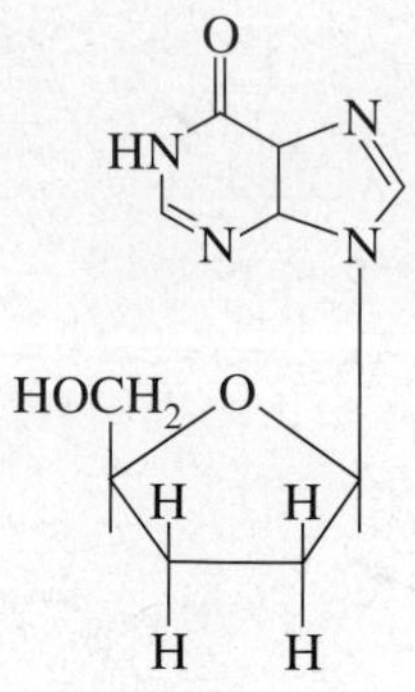

Dideoxyinosine (ddI)

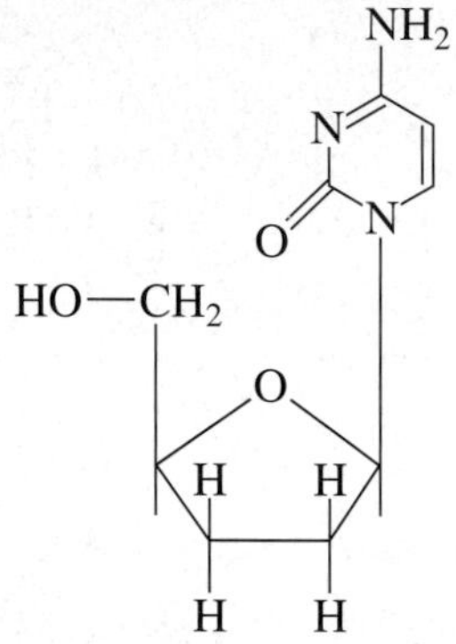

Dideoxycytidine (ddC)

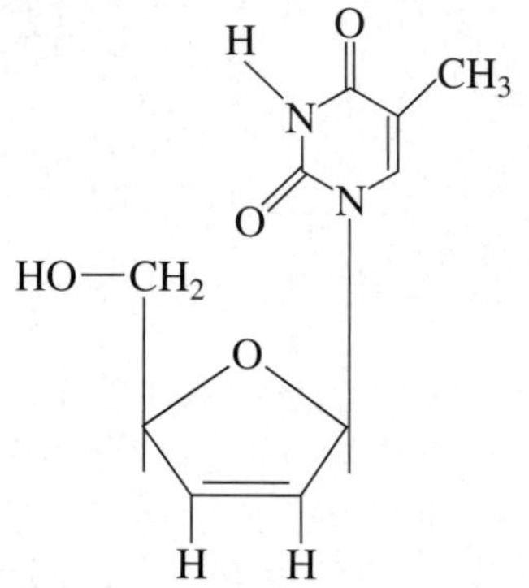

Didehydro-3'-deoxythymidine (d4T)

The newest and most powerful anti-HIV drugs are the protease inhibitors such as saquinavir (Invirase), indinavir, and ritonavir, which modify the three-dimensional structure of the protease active site. The inhibition of protease prevents the synthesis of proteins needed to make more copies of the virus. Researchers are not yet certain how long protease inhibitors will be beneficial for a person with AIDS, but they are encouraged by the current studies.

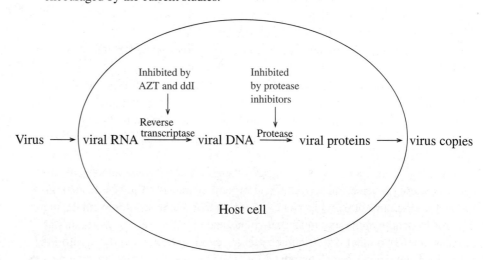

Cancer

In an adult body, most cells do not continue to reproduce. When cells in the body begin to grow and multiply without control, they invade neighboring cells and appear as a tumor or growth (neoplasm). When tumors interfere with normal functions of the body, they are cancerous. If they are limited, they are benign. Cancer can be caused by chemical and environmental substances, by radiation, or by oncogenic viruses.

Some reports estimate that 70–80% of all human cancers are initiated by chemical and environmental substances. A carcinogen is any substance that increases the chance of inducing a tumor. Known carcinogens include dyes, cigarette smoke, and asbestos. More than 90% of all persons with lung cancer are smokers. A carcinogen causes cancer by reacting with molecules in a cell, probably DNA, and altering the growth of that cell. Some known carcinogens are listed in Table 22.9.

Radiant energy from sunlight or medical radiation is another type of environmental factor. Skin cancer has become one of the most prevalent forms of cancer. It appears that DNA damage in the exposed areas of the skin causes mutations. The cells lose their ability to control protein synthesis, and uncontrolled cell division leads to cancer. The incidence of malignant melanoma, one of the most serious skin cancers, has been rapidly increasing. Some possible factors for this increase may be the popularity of suntanning as well as the reduction of the ozone layer, which absorbs much of the harmful radiation from sunlight.

Oncogenic viruses cause cancer when cells are infected. Several viruses associated with human cancers are listed in Table 22.10. Some cancers such as retinoblastoma and breast cancer appear to occur more frequently in families. There is some indication that a missing or defective gene may be responsible.

Table 22.9   Some Chemical and Environmental Carcinogens

Carcinogen	Tumor Site
Asbestos	Lung, respiratory tract
Arsenic	Skin, lung
Cadmium	Prostate, kidneys
Chromium	Lung
Nickel	Lung, sinuses
Aflatoxin	Liver
Nitrites	Stomach
Aniline dyes	Bladder
Vinyl chloride	Liver

Table 22.10   Human Cancers Caused by Oncogenic Viruses

Virus	Disease
RNA viruses	
Human T-cell lymphotropic virus-type I (HTLV-I)	Leukemia
DNA viruses	
Epstein-Barr virus (EBV)	Burkitt's lymphoma (cancer of white blood B cells)
	Nasopharyngeal carcinoma
	Hodgkin's disease
Hepatitis B virus (HBV)	Liver cancer
Herpes simplex virus (type 2)	Cervical and uterine cancer
Papilloma virus	Cervical and colon cancer, genital warts

SAMPLE PROBLEM 22.13

Viruses

Why are viruses unable to replicate on their own?

Solution

Viruses only contain packets of DNA or RNA, but not the necessary replication machinery that includes enzymes and nucleosides.

Study Check

How do protease inhibitors affect the life cycle of the HIV-1 virus?

QUESTIONS AND PROBLEMS

Viruses

22.71 What type of genetic information is found in a virus?

22.72 Why do viruses need to invade a host cell?

22.73 A virus contains viral RNA.
  a. Why would reverse transcription be used in the life cycle of this type of virus?
  b. What is the name of this type of virus?

22.74 What is the purpose of a vaccine?

22.75 How do nucleoside analogs disrupt the life cycle of the HIV-1 virus?

22.76 How do protease inhibitors disrupt the life cycle of the HIV-1 virus?

Chapter Review

22.1 Components of Nucleic Acids

Nucleic acids, deoxyribonucleic acid (DNA), and ribonucleic acid (RNA) are polymers of nucleotides. A nucleoside is a combination of a pentose sugar and a nitrogen base.

22.2 Nucleosides and Nucleotides

A nucleotide is composed of three parts: a nitrogenous base, a sugar, and a phosphate group. In DNA, the sugar is deoxyribose and the nitrogen-containing base can be adenine, thymine, guanine, or cytosine. In RNA, the sugar is ribose and uracil replaces thymine.

22.3 Primary Structure of Nucleic Acids

Each nucleic acid has its own unique sequence of bases known as its primary structure. In a nucleic acid polymer, the 3'OH of each ribose sugar in RNA or deoxyribose sugar in DNA forms a phosphodiester bond to the phosphate group of the 5'-carbon atom group of the sugar in the next nucleotide to give a backbone of alternating sugar and phosphate groups. There is a free 5'-phosphate at one end of the polymer, and a free 3' OH group at the other end.

22.4 DNA Double Helix: A Secondary Structure

A DNA molecule consists of two strands of nucleotides that are wound around each other like a spiral staircase. The two strands are held together by hydrogen bonds between complementary base pairs, A with T, and G with C.

22.5 DNA Replication

During DNA replication, DNA polymerase makes new DNA strands along each of the original DNA strands that serve as templates. Complementary base pairing ensures the correct pairing of bases to give identical copies of the original DNA.

22.6 Types of RNA

The three types of RNA differ by function in the cell: ribosomal RNA makes up most of the structure of the ribosomes, messenger RNA carries genetic information from the DNA to the ribosomes, and transfer RNA places the correct amino acids in the protein.

22.7 Transcription: Synthesis of mRNA

Transcription is the process by which RNA polymerase produces mRNA from one strand of DNA. The bases in the mRNA are complementary to the DNA, except U is paired with A in RNA. The production of mRNA occurs when certain proteins are needed in the cell. In enzyme induction, the appearance of a substrate in a cell removes a repressor, which allows RNA polymerase to produce mRNA at the structural genes.

22.8 The Genetic Code

The genetic code consists of a sequence of three bases (triplet) that specifies the order for the amino acids in a protein. There are 64 codons for the 20 amino acids, which means there are several codons for some amino acids. The codon AUG signals the start of transcription and codons UAG, UGA, and UAA signal it to stop.

22.9 Protein Synthesis: Translation

Proteins are synthesized at the ribosomes in a translation process that includes three steps: initiation, translocation, and termination. During translation, tRNAs bring the appropriate amino acids to the ribosome and peptide bonds form. When the polypeptide is released, it takes on its secondary and tertiary structures and becomes a functional protein in the cell.

22.10 Genetic Mutations

A genetic mutation is a change of one or more bases in the DNA sequence that alters the structure and ability of the resulting protein to function properly. In a substitution, one base may be altered, and a frame shift mutation inserts or deletes a base and changes all the codons after the base change.

22.11 Recombinant DNA

A recombinant DNA is prepared by inserting a DNA segment, a gene, into the DNA present in plasmids of *E. coli* bacteria. As the altered bacterial cells replicate, the protein expressed by the foreign DNA segment is produced. In criminal investigation, large quantities of DNA are obtained from smaller amounts by DNA polymerase chain reactions.

22.12 Viruses

Viruses containing DNA or RNA must invade host cells to use the machinery within the cell for the synthesis of more viruses. For a retrovirus containing RNA, a viral DNA is synthesized by reverse transcription using the nucleotides and enzymes in the host cell. In the treatment of AIDS, nucleoside analogs inhibit the reverse transcriptase of the HIV-1 virus, and protease inhibitors disrupt the catalytic activity of protease needed to produce proteins for the synthesis of more viruses.

Key Terms

anticodon The triplet of bases in the center loop of tRNA that is complementary to a codon on mRNA.

codon A sequence of three bases in mRNA that specifies a certain amino acid to be placed in a protein. A few codons signal the start or stop of transcription.

complementary base pairs In DNA, adenine is always paired with thymine (A—T or T—A), and guanine is always paired with cytosine (G—C or C—G). In forming RNA, adenine is paired with uracil (A—U).

control site A section of DNA composed of a promoter and operator that regulates protein synthesis.

DNA Deoxyribonucleic acid; the genetic material of all cells containing nucleotides with deoxyribose sugar, phosphate, and the four nitrogenous bases adenine, thymine, guanine, and cytosine.

double helix The helical shape of the double chain of DNA that is like a spiral staircase with a sugar–phosphate backbone on the outside and base pairs like stair steps on the inside.

enzyme induction A model of cellular regulation in which protein synthesis is induced by a substrate.

exons The sections in a DNA template that code for proteins.

frame shift mutation A mutation that inserts or deletes a base in a DNA sequence.

genetic code The sequence of codons in mRNA that specifies the amino acid order for the synthesis of protein.

genetic disease A physical malformation or metabolic dysfunction caused by a mutation in the base sequence of DNA.

introns The sections in DNA that do not code for protein.

mRNA Messenger RNA; produced in the nucleus by DNA to carry the genetic information to the ribosomes for the construction of a protein.

mutation A change in the DNA base sequence that alters the formation of a protein in the cell.

nitrogen-containing base Nitrogen-containing compounds found in DNA and RNA: adenine (A), thymine (T), cytosine (C), guanine (G), and uracil (U).

nucleic acids Large molecules composed of nucleotides, found as a double helix in DNA, and as the single strands of RNA.

nucleoside The combination of a pentose sugar and a nitrogen-containing base.

nucleotides Building blocks of a nucleic acid consisting of a nitrogen-containing base, a pentose sugar (ribose or deoxyribose), and a phosphate group.

Okazaki fragments The short segments formed by DNA polymerase in the daughter DNA strand that runs in the 3' to 5' direction.

operon A group of genes, including a control site and structural genes, whose transcription is controlled by the same regulatory gene.

phosphodiester bond The phosphate link that joins the 3' hydroxyl group in one nucleotide to the phosphate group on the 5'-carbon atom in the next nucleotide.

polymerase chain reaction (PCR) A strand of DNA is copied many times by mixing it with DNA polymerase and a mixture of deoxyribonucleotides.

primary structure The sequences of nucleotides in nucleic acids.

recombinant DNA DNA spliced from different organisms to form new, synthetic DNA.

regulatory gene A gene in front of the control site that produces a repressor.

replication The process of duplicating DNA by pairing the bases on each parent strand with their complementary base.

replication forks The open sections in unwound DNA strands where DNA polymerase begins the replication process.

repressor A protein that interacts with the operator gene in an operon to prevent the transcription of mRNA.

restriction enzyme An enzyme that cuts a large DNA strand into smaller fragments to isolate a gene or to remove a portion of the DNA in the plasmids of *E. coli*.

retrovirus A virus that contains RNA as its genetic material and that synthesizes a complementary DNA strand inside a cell.

RNA Ribonucleic acid, a type of nucleic acid that is a single strand of nucleotides containing adenine, cytosine, guanine, and uracil.

rRNA Ribosomal RNA; the most prevalent type of RNA; a major component of the ribosomes.

structural genes The sections of DNA that code for the synthesis of proteins.

substitution A mutation that replaces one base in a DNA with a different base.

transcription The transfer of genetic information from DNA by the formation of mRNA.

translation The interpretation of the codons in mRNA as amino acids in a peptide.

translocation The shift of a ribosome along mRNA from one codon (three bases) to the next codon during translation.

tRNA Transfer RNA; an RNA that places a specific amino acid into a peptide chain at the ribosome. There is one or more tRNA for each of the 20 different amino acids.

virus Small particles containing DNA or RNA in a protein coat that require a host cell for replication.

Additional Problems

22.77 Identify each of the following nitrogen bases as a pyrimidine or a purine:
   a. cytosine   **b.** adenine   **c.** uracil
   d. thymine   **e.** guanine

22.78 Indicate if each of the nitrogen bases in problem 22.77 are found in DNA only, RNA only, or both DNA and RNA.

22.79 Identify the nitrogen base and sugar in each of the following nucleosides:
   a. deoxythymidine   **b.** adenosine
   c. cytidine   **d.** deoxyguanosine

22.80 Identify the nitrogen base and sugar in each of the following nucleotides:
   a. CMP   **b.** dAMP
   c. dGMP   **d.** UMP

22.81 How do the bases of thymine and uracil differ?

22.82 How do the bases of cytosine and uracil differ?

22.83 Draw the structure of CMP.

22.84 Draw the structure of dGMP.

22.85 What is similar about the primary structure of RNA and DNA?

22.86 What is different about the primary structure of RNA and DNA?

22.87 If the DNA double helix in salmon contains 28% adenine, what is the percent of thymine, guanine, and cytosine?

22.88 If the DNA double helix in humans contains 20% cytosine, what is the percent of guanine, adenine, and thymine?

22.89 In DNA, how many hydrogen bonds form between each of the following?
   a. adenine and thymine
   b. guanine and cytosine

22.90 Why are there no base pairs in DNA between adenine and guanine, or thymine and cytosine?

22.91 Write the complementary base sequence for each of the following DNA segments:
   a. 5'—GACTTAGGC—3'
   b. 3'—TGCAAACTAGCT—5'
   c. 5'—ATCGATCGATCG—3'

22.92 Write the complementary base sequence for each of the following DNA segments:
   a. 5'—TTACGGACCGC—3'
   b. 5'—ATAGCCCTTACTGG—3'
   c. 3'—GGCCTACCTTAACGACG—5'

22.93 In DNA replication, what is the difference in the synthesis of the leading strand and the lagging strand?

22.94 How are the Okazaki fragments joined to the growing DNA strand?

22.95 Where are the DNA strands of the original DNA found in the double helix of each of the daughter DNA molecules?

22.96 How can replication occur at several places along a DNA double helix?

22.97 Match the following statements with rRNA, mRNA, or tRNA.
  a. is the smallest type of RNA
  b. makes up the highest percent of RNA in the cell
  c. carries genetic information from the nucleus to the ribosomes

22.98 Match the following statements with rRNA, mRNA, or tRNA.
  a. have two subunits
  b. brings amino acids to the ribosomes for protein synthesis
  c. acts as a template for protein synthesis

22.99 What are the possible codons for each of the following amino acids?
  a. threonine   **b.** serine   **c.** cysteine

22.100 What are the possible codons for each of the following amino acids?
  a. valine   **b.** proline   **c.** histidine

22.101 What is the amino acid for each of the following codons?
  a. AAG   **b.** AUU   **c.** CGG

22.102 What is the amino acid for each of the following codons?
  a. CAA   **b.** GGC   **c.** AAC

22.103 Endorphins are polypeptides that reduce pain. What is the amino acid order for the following mRNA that codes for a pentapeptide that is an endorphin called leucine enkephalin?

5′—AUG—UAC—GGU—GGA—UUU—CUA—UAA—3′

22.104 Endorphins are polypeptides that reduce pain. What is the amino acid order for the following mRNA that codes for a pentapeptide that is an endorphin called methionine enkephalin?

5′—AUG—UAC—GGU—GGA—UUU—AUG—UAA—3′

22.105 What is the anticodon on tRNA for each of the following codons in an mRNA?
  a. AGC   **b.** UAU   **c.** CCA

22.106 What is the anticodon on tRNA for each of the following codons in an mRNA?
  a. GUG   **b.** CCC   **c.** GAA

22.107 Oxytocin is a nonapeptide with nine amino acids. How many nucleotides would be found in the mature mRNA for this protein?

22.108 A protein contains 35 amino acids. How many nucleotides would be found in the mature mRNA for this protein?

22.109 What is the difference between a DNA virus and a retrovirus?

22.110 How does polymerase chain reaction (PCR) produce many copies of a DNA section?

23 Metabolic Pathways for Carbohydrates

"I was checking this dog's ears for foxtails and her eyes for signs of conjunctivitis," says Joyce Rhodes, veterinary assistant at the Sonoma Animal Hospital. "We always check a dog's teeth for tartar, because dental care is very important to the well-being of the animal. When I do need to give a medication to an animal, I use my chemistry to prepare the proper dose that the pet should take. Dosages may be in milligrams, kilograms, or milliliters."

As a member of the veterinary health care team, a veterinary technician (VT) assists a veterinarian in the care and handling of animals. A VT takes medical histories, collects specimens, performs laboratory procedures, prepares an animal for surgery, assists in surgical procedures, takes X rays, talks with animal owners, and cleans teeth.

LOOKING AHEAD

the Chemistry place

www.chemplace.com/college

Visit the URL above or use the CD-ROM in the book for extra quizzing, interactive tutorials, and career resources.

When we eat food such as a tuna fish sandwich, the polysaccharides, lipids, and proteins are digested to smaller molecules that are absorbed into the cells of our body. As these molecules of glucose, fatty acids, and amino acids are broken down further, energy is released. This energy is used in the cells to synthesize high-energy compounds such as adenosine triphosphate (ATP). Our cells utilize ATP energy when they do work such as contracting muscles, synthesizing large molecules, sending nerve impulses, and moving substances across cell membranes.

All the chemical reactions that take place in living cells to break down or build molecules are known as *metabolism.* In a metabolic pathway, reactions are linked together in a series, each catalyzed by a specific enzyme to produce an end product. In this and the following chapters, we will look at these pathways and the way they produce energy and cellular compounds.

23.1 Metabolism and Cell Structure

The term **metabolism** refers to all the chemical reactions that provide energy and the substances required for continued cell growth. There are two types of metabolic reactions: catabolic and anabolic. In **catabolic reactions,** complex molecules are broken down to simpler ones with an accompanying release of energy. **Anabolic reactions** utilize energy available in the cell to build large molecules from simple ones.

We can think of the catabolic processes in metabolism as consisting of three stages. (See Figure 23.1.) Let's use that tuna fish sandwich for our example. In stage 1 of metabolism, the processes of digestion break down the large macromolecules into small monomer units. For example, the polysaccharides in bread break down to monosaccharides, the lipids in the mayonnaise break down to glycerol and fatty acids, and the proteins from the tuna yield amino acids. These digestion products diffuse into the bloodstream for transport to cells. In stage 2, they are broken down to two- and three-carbon compounds such as pyruvate and acetyl CoA in the cells. Stage 3 begins with the oxidation of the two-carbon acetyl CoA in the citric acid cycle, which produces several reduced coenzymes. As long as the cells have oxygen, the hydrogen ions and electrons are transferred to the electron transport chain, where most of the energy in the cell is produced. This energy is used to synthesize adenosine triphosphate (ATP), which provides energy for the anabolic pathways in the cell.

Cell Structure for Metabolism

To understand the relationship of metabolic reactions, we need to look at where metabolic reactions take place in the cells. (See Figure 23.2.) In Chapter 22, we described two types of cells: prokaryotic and eukaryotic. *Prokaryotic* (before nucleus) cells have no nucleus and form single-celled organisms such as bacteria. The cells in plants and animals are *eukaryotic* cells, which have a nucleus.

In a eukaryotic cell, a *plasma membrane* is a lipid bilayer that separates the materials inside the cell from the aqueous environment surrounding the cell. In addition, the outer surface of the plasma membrane contains structures that allow

Stages of Metabolism

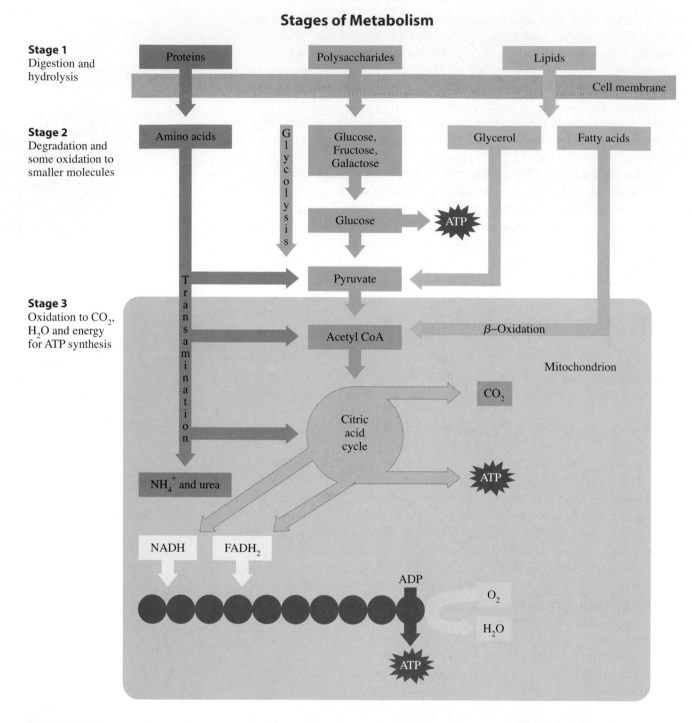

Stage 1
Digestion and
hydrolysis

Stage 2
Degradation and
some oxidation to
smaller molecules

Stage 3
Oxidation to CO_2,
H_2O and energy
for ATP synthesis

Figure 23.1 In the three stages of catabolic metabolism, foods are digested and degraded into smaller molecules, which are oxidized to produce energy.
Q *Where is most of the ATP energy produced in the cells?*

cells to communicate with each other. The *nucleus* contains the genes that control DNA replication and protein synthesis of the cell. The **cytoplasm** consists of all the materials between the nucleus and the plasma membrane. The **cytosol,** which is the fluid part of the cytoplasm, is an aqueous solution of electrolytes and enzymes that catalyze many of the cell's chemical reactions.

Within the cytoplasm are specialized structures called *organelles* that carry out specific functions in the cell. We have already seen (Chapter 22) that the *ribosomes* are the sites of protein synthesis using mRNA templates. The *endoplasmic reticulum* consists of two forms, a rough endoplasmic reticulum where proteins are

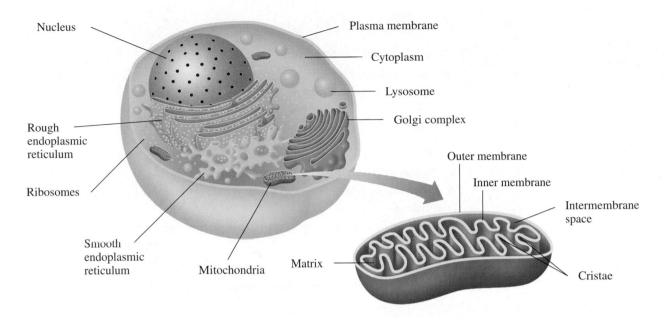

processed for secretion and phospholipids are synthesized, and a smooth endoplasmic reticulum where fats and steroids are synthesized. The *Golgi complex* modifies proteins it receives from the rough endoplasmic reticulum and secretes these modified proteins into the extracellular fluid as well as forming glycoproteins and cell membranes. *Lysosomes* contain many hydrolytic enzymes that break down recyclable cellular structures that are no longer needed by the cell. The **mitochondria** are the energy-producing factories of the cells. A mitochondrion consists of an outer membrane and an inner membrane with an intermembrane space between them. The fluid section surrounded by the inner membrane is called the *matrix.* Enzymes located in the matrix and along the inner membrane catalyze the oxidation of carbohydrates, fats, and amino acids. All of these oxidation pathways lead to CO_2, H_2O, and energy, which is used to form energy-rich compounds. Table 23.1 summarizes some of the functions of the cellular components in eukaryotic cells.

Figure 23.2 The diagram illustrates the major components of a typical eukaryotic cell.
Q *What is the cytoplasm in a cell?*

Table 23.1   Locations and Functions of Components in Eukaryotic Cells

Component	Description and Function
Plasma membrane	Separates the contents of a cell from the external environment and contains structures that communicate with other cells
Cytoplasm	Consists of all of the cellular contents between the plasma membrane and nucleus
Cytosol	Is the fluid part of the cytoplasm that contains enzymes for many of the cell's chemical reactions including glycolysis, glucose, and fatty acid synthesis
Endoplasmic reticulum	Rough type processes proteins for secretion and synthesizes phospholipids; smooth type synthesizes fats and steroids
Golgi complex	Modifies and secretes proteins from the endoplasmic reticulum and synthesizes glycoproteins and cell membranes
Lysosomes	Contain hydrolytic enzymes that digest and recycle old cell structures
Mitochondria	Contain the structures for the synthesis of ATP from energy-producing reactions
Nucleus	Contains genetic information for the replication of DNA and the synthesis of protein
Ribosomes	Are the sites of protein synthesis using mRNA templates

SAMPLE PROBLEM 23.1

Metabolism and Cell Structure

Identify the following as catabolic or anabolic reactions.

a. Digestion of polysaccharides
b. Synthesis of proteins
c. Oxidation of glucose to CO_2 and H_2O

Solution

a. The breakdown of large molecules is a catabolic reaction.
b. The synthesis of large molecules requires energy and involves anabolic reactions.
c. The breakdown of monomers such as glucose involves catabolic reactions.

Study Check

What is the difference between the cytoplasm and cytosol?

QUESTIONS AND PROBLEMS

Metabolism and Cell Structure

23.1 What stage of metabolism involves the digestion of polysaccharides?

23.2 What stage of metabolism involves the conversion of small molecules to CO_2, H_2O, and energy for the synthesis of ATP?

23.3 What is meant by a catabolic reaction in metabolism?

23.4 What is meant by an anabolic reaction in metabolism?

23.5 Match each of the following organelles with their function:
(1) lysosome (2) Golgi complex (3) smooth endoplasmic reticulum
a. synthesis of fats and steroids **b.** contains hydrolytic enzymes
c. modifies products from rough endoplasmic reticulum

23.6 Match each of the following organelles with their function:
(1) mitochondria (2) rough endoplasmic reticulum
(3) plasma membrane
a. separates cell contents from external surroundings
b. sites of energy production **c.** synthesizes proteins for secretion

LEARNING GOAL

Describe the role of ATP in catabolic and anabolic reactions.

23.2 ATP and Energy

In our cells, the energy released from the oxidation of the food we eat is used to form a compound called *adenosine triphosphate* **(ATP).** As we saw in Chapter 22, the ATP molecule is composed of the nitrogen base adenine, a ribose sugar, and three phosphate groups replacing the 5'—OH of ribose. (See Figure 23.3.)

Hydrolysis of ATP Yields Energy

In the cells, there are a variety of compounds known as "high-energy" compounds. The most important of these is ATP, which transfers a phosphate group to water when it undergoes hydrolysis. The cleavage of one phosphate group releases energy of 7.3 kcal per mole of ATP or 31 kJ per mole of ATP. The equilibrium strongly favors the products, which are adenosine diphosphate **(ADP)** and hydrogen phosphate ion (HPO_4^{2-}) abbreviated as P_i (inorganic phosphate).

Adenosine$-$O$-$P$-$O$-$P$-$O$-$P$-$O$^-$ + H_2O $\longrightarrow$

ATP (Adenosine triphosphate)

Adenosine$-$O$-$P$-$O$-$P$-$O$^-$ + HO$-$P$-$O$^-$ + H^+ + 7.3 kcal (31 kJ)/mole

ADP (Adenosine diphosphate)   Inorganic phosphate P_i

We can write this reaction in its simplified form as

ATP + H_2O $\longrightarrow$ ADP + P_i + 7.3 kcal (31 kJ)/mole

Every time we contract muscles, move substances across cellular membranes, send nerve signals, or synthesize an enzyme, we use energy from ATP hydrolysis. In a cell that is doing work (anabolic processes), 1–2 million ATP molecules may be hydrolyzed in one second. The amount of ATP hydrolyzed in one day can be as much as our body mass, even though only about 1 gram of ATP is present in all our cells at any given time.

When we take in food, the resulting catabolic reactions provide energy to regenerate ATP in our cells. Then 7.3 kcal (31 kJ)/mole is used to make ATP from ADP and P_i. (See Figure 23.4.)

ADP + P_i + 7.3 (31 kJ) kcal/mole $\longrightarrow$ ATP

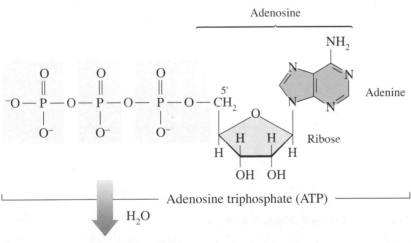

Figure 23.3 Adenosine triphosphate hydrolyzes to form ADP and AMP, along with a release of energy.

Q *How much energy is released when a phosphate group is cleaved from one mole of ATP?*

Adenosine triphosphate (ATP)

H_2O

Adenosine$-$ P $-$ P + P_i + 7.3 kcal (31 kJ)/mole

Adenosine diphosphate (ADP)

H_2O

Adenosine$-$ P + P_i + 7.3 kcal (31 kJ)/mole

Adenosine monophosphate (AMP)

P = Phosphate group

P_i = Inorganic phosphate

Figure 23.4 ATP, the energy-storage molecule, connects the energy-producing reactions with the energy-requiring reactions that do work in the cells.
Q *What type of reaction provides energy for ATP synthesis?*

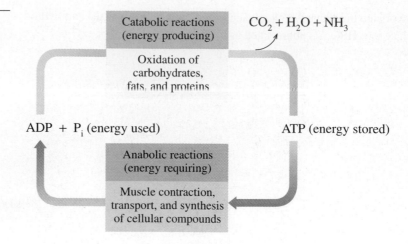

In some pathways, ATP is hydrolyzed to yield AMP and pyrophosphate (PP_i).

ATP (Adenosine triphosphate)

AMP
(Adenosine
monophosphate)   Pyrophosphate (PP_i)

Usually the PP_i, which is also an energy-rich compound, is immediately hydrolyzed to give $2 P_i$. The combination of these two reactions releases energy that is similar to the hydrolysis of two ATP to ADP and P_i.

PP_i $2 P_i$

ATP Is Coupled with Reactions

ATP and other energy-rich compounds are coupled with metabolic reactions and processes that require energy. ATP is very useful in the body because it can be used to drive reactions that require energy and do not occur on their own. For example, the glucose obtained from carbohydrates must add a phosphate group to start its breakdown in the cell. However, the cost of adding a phosphate to glucose is 3.3 kcal (14 kJ)/mole, which means that the reaction does not occur spontaneously in the cell. By coupling the reaction with the hydrolysis of an energy-rich compound such as ATP, the reaction takes place because the energy from hydrolysis pushes or "drives" the energy-requiring reaction.

Glucose + P_i + 3.3 kcal (14 kJ)/mole ⟶ glucose-6-phosphate   Requires energy

ATP   ⟶ ADP + P_i + 7.3 kcal (31 kJ)/mole   Provides energy

Glucose + ATP   ⟶ ADP + glucose-6-phosphate + 4.0 kcal (17 kJ)/mole

The coupling of a reaction that requires energy with a reaction that supplies energy is a very important concept in biochemical pathways. Many of the reactions essential to a cell for survival cannot proceed by themselves, but can be made to proceed by coupling them with a reaction that releases energy. Similar kinds of coupled reactions are also used to transmit nerve impulses, transport substances across membranes to higher concentrations, and to contract muscles.

SAMPLE PROBLEM 23.2

Hydrolysis of ATP

Write an equation for the hydrolysis of ATP.

Solution

The hydrolysis of ATP produces ADP, P_i, and energy.

$$ATP + H_2O \longrightarrow ADP + P_i + energy$$

HEALTH NOTE

ATP Energy and Ca^{2+} Needed to Contract Muscles

Our muscles consist of thousands of parallel fibers. Within these muscle fibers are fibrils composed of two kinds of proteins called filaments. Arranged in alternating rows, the thick filaments of myosin overlap the thin filaments containing actin. During a muscle contraction, the thin filaments slide inward over the thick filaments causing a shortening of the muscle fibers.

Calcium ion, Ca^{2+}, and ATP play an important role in muscle contraction. An increase in the Ca^{2+} concentration in the muscle fibers causes the filaments to slide, while a decrease stops the process. In a relaxed muscle, the Ca^{2+} concentration is low. When a nerve impulse reaches the muscle, calcium channels in the membrane open and Ca^{2+} flows into the fluid surrounding the filaments. The Ca^{2+} combines with troponin, a protein that covers the binding site on actin. When the troponin detaches from actin, myosin can bind to actin, and start to pull the actin filaments inward. The energy used for the contraction is provided by splitting ATP that is attached to myosin to ADP + P_i and energy.

Muscle contraction continues as long as both ATP and Ca^{2+} levels are high around the filaments. Relaxation of the muscle occurs as the nerve impulse ends and the Ca^{2+} channels close. The Ca^{2+} concentration is lowered as ATP energy is used to pump the remaining calcium ions out of the area. Troponin re-attaches to the actin, which causes the thin filaments to return to their relaxed state. In rigor mortis, Ca^{2+} concentration remains high within the muscle fibers causing a continued state of rigidity for approximately 24 hours. After that, cellular deterioration causes a decrease in calcium ions and muscles relax.

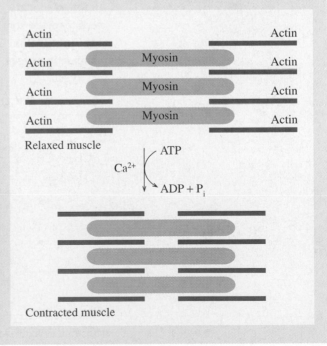

Study Check

What are the components of the ATP molecule?

QUESTIONS AND PROBLEMS

ATP and Energy

23.7 Why is ATP considered an energy-rich compound?

23.8 What is meant when we say that the hydrolysis of ATP is used to "drive" a reaction?

23.9 Phosphoenolpyruvate (PEP) is a high-energy phosphate compound that releases 14.8 kcal/mole of energy when it hydrolyzes to pyruvate and P_i. This reaction can be coupled with the synthesis of ATP from ADP and P_i.

   a. Write an equation for the energy-releasing reaction of PEP.

   b. Write an equation for the energy-requiring reaction that forms ATP.

   c. Write the overall equation for the coupled reaction including the net energy change.

23.10 The phosphorylation of glycerol to glycerol-3-phosphate requires 2.2 kcal/mole and is driven by the hydrolysis of ATP.

   a. Write an equation for the energy-releasing reaction of ATP.

   b. Write an equation for the energy-requiring reaction that forms glycerol-3-phosphate.

   c. Write the overall equation for the coupled reaction including the net energy change.

23.3 Important Coenzymes in Metabolic Pathways

Before we look at the metabolic reactions that extract energy from our food, we need to review some ideas about oxidation and reduction reactions. (Chapters 6 and 14). An **oxidation** involves the loss of hydrogen and electrons by a substance or an increase in oxygen. **Reduction** is the gain of hydrogen and electrons or a decrease in oxygen.

Oxidation: Loss of H, loss of e$^-$, or increase of O

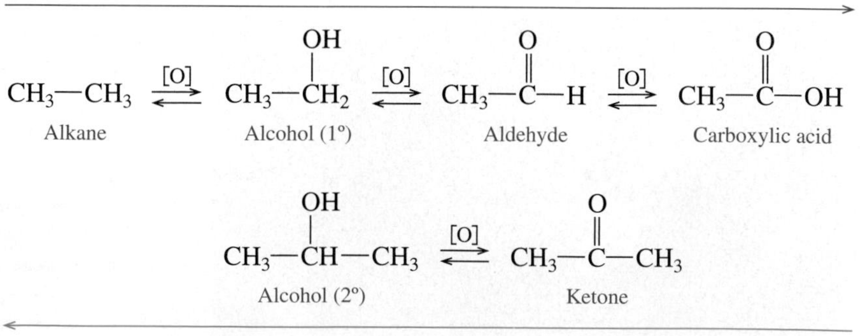

$$CH_3-CH_3 \underset{}{\overset{[O]}{\rightleftharpoons}} \underset{OH}{CH_3-CH_2} \overset{[O]}{\rightleftharpoons} \underset{O}{CH_3-\overset{\|}{C}-H} \overset{[O]}{\rightleftharpoons} \underset{O}{CH_3-\overset{\|}{C}-OH}$$

Alkane   Alcohol (1°)   Aldehyde   Carboxylic acid

$$\underset{OH}{CH_3-\overset{|}{C}H-CH_3} \overset{[O]}{\rightleftharpoons} \underset{O}{CH_3-\overset{\|}{C}-CH_3}$$

Alcohol (2°)   Ketone

Reduction: Gain of H, gain of e$^-$, or decrease of O

When an enzyme catalyzes an oxidation, hydrogen atoms are removed from a substrate as hydrogen ions, $2H^+$, and electrons, $2e^-$.

$$2 \text{ H atoms (removed in oxidation)} \longrightarrow 2H^+ + 2e^-$$

Those hydrogen ions and electrons are picked up by a coenzyme for the reaction, which is reduced. As we saw in Chapter 22, the structures of many coenzymes include the water-soluble B vitamins we obtain from the foods in our diets. Now we can look at the structure of several important coenzymes.

NAD⁺

NAD⁺ (nicotinamide adenine dinucleotide) is an important coenzyme in which the vitamin *niacin* provides the *nicotinamide* group, which is bonded to adenosine diphosphate (ADP). (See Figure 23.5.) In the oxidized form of NAD⁺, the nitrogen atom in nicotinamide has a positive charge because it has four bonds rather than three. The NAD⁺ coenzyme participates in reactions that produce a carbon–oxygen ($C=O$) double bond such as the oxidation of alcohols to aldehydes and ketones. The NAD⁺ is reduced when the carbon in the pyridine ring of nicotinamide accepts a hydrogen ion and two electrons leaving one H^+. Let's look at the reactions that take place when ethanol is oxidized in the liver to acetaldehyde using NAD⁺.

Oxidation

$$CH_3-CH_3-OH \underset{\text{Alcohol dehydrogenase}}{\rightleftharpoons} CH_3-\overset{\displaystyle O}{\overset{\|}{C}}-H + 2H^+ + 2e^-$$

Ethanol   Acetaldehyde

Reduction

$$NAD^+ + 2H^+ + 2e^- \rightleftharpoons NADH + H^+$$

Overall oxidation–reduction reaction

$$CH_3-CH_2-OH + NAD^+ \underset{\text{Alcohol dehydrogenase}}{\rightleftharpoons} CH_3-\overset{\displaystyle O}{\overset{\|}{C}}-H + NADH + H^+$$

Ethanol   Acetaldehyde

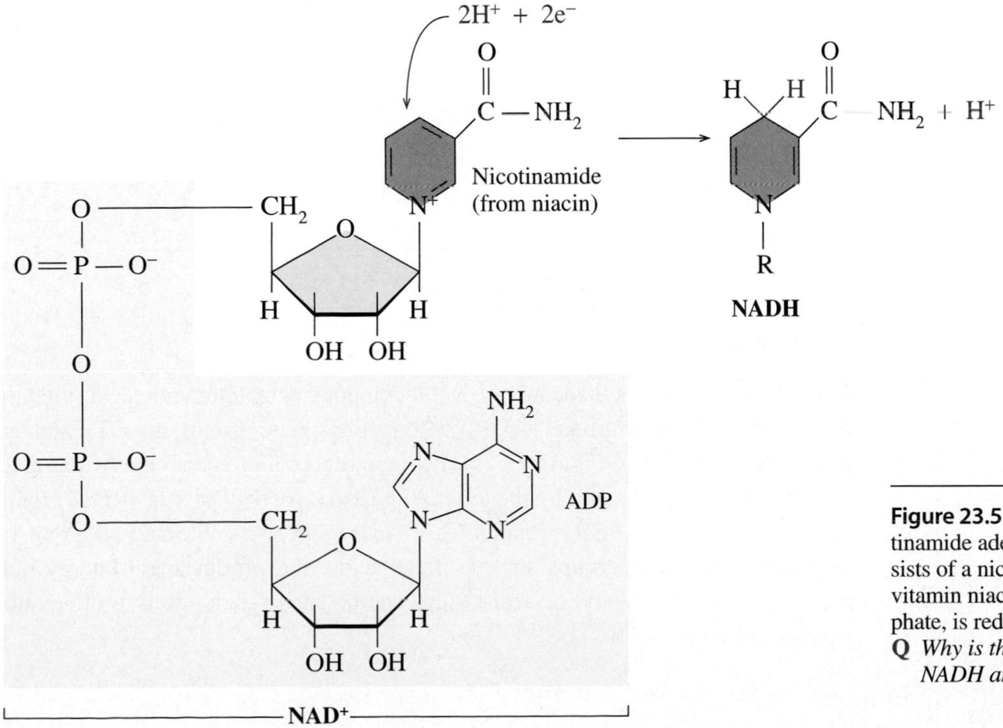

Figure 23.5 The coenzyme NAD⁺ (nicotinamide adenine dinucleotide), which consists of a nicotinamide portion from the vitamin niacin, ribose, and adenine diphosphate, is reduced to NADH + H⁺.
Q *Why is the conversion of NAD⁺ to NADH and H⁺ called a reduction?*

FAD

FAD (flavin adenine dinucleotide) is a coenzyme derived from adenosine diphosphate (ADP) and riboflavin. Riboflavin, also known as vitamin B_2, consists of ribitol, a sugar alcohol, and flavin, which is a ring structure found in flavoenzymes. (See Figure 23.6.) As a coenzyme, two nitrogen atoms in the flavin part of the FAD coenzyme accept the hydrogen, which reduces the FAD to $FADH_2$.

$$FAD + 2H^+ + 2e^- \rightleftharpoons FADH_2$$

FAD typically participates in oxidation reactions that produce a carbon–carbon (C=C) double bond.

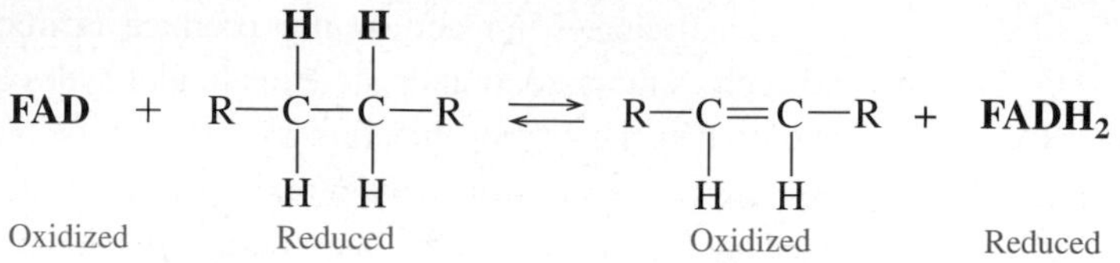

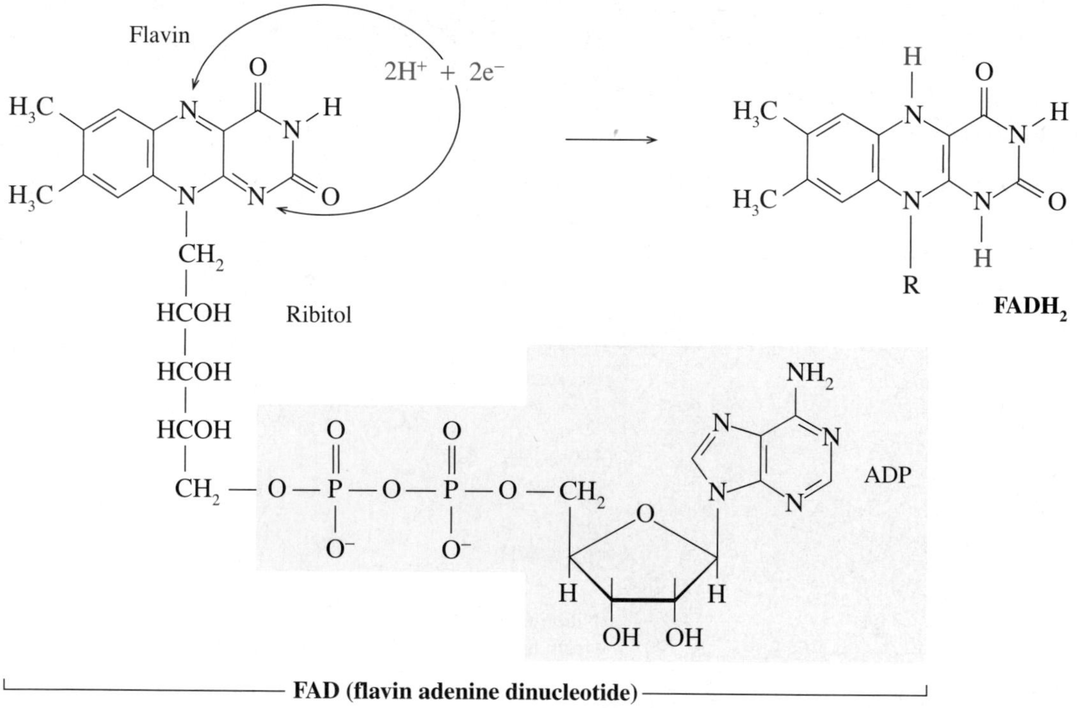

Figure 23.6 The coenzyme FAD (flavin adenine dinucleotide) made from riboflavin (vitamin B_2) and adenine diphosphate is reduced to $FADH_2$.
Q *What is the type of reaction in which FAD accepts hydrogen?*

Coenzyme A

Coenzyme A (CoA) is made up of several components: pantothenic acid (vitamin B_3), adenosine diphosphate (ADP) with a phosphate group on C3′, and an aminoethanethiol. (See Figure 23.7.) The main function of coenzyme A is to activate acyl groups (indicated by the letter A in CoA), particularly the acetyl group. When the free thiol (—SH) group of CoA bonds to the two-carbon acetyl group, or to longer chain acyl groups such as fatty acids, the products are energy-rich thioesters. Then the acetyl or acyl groups can be easily transferred to other substrates.

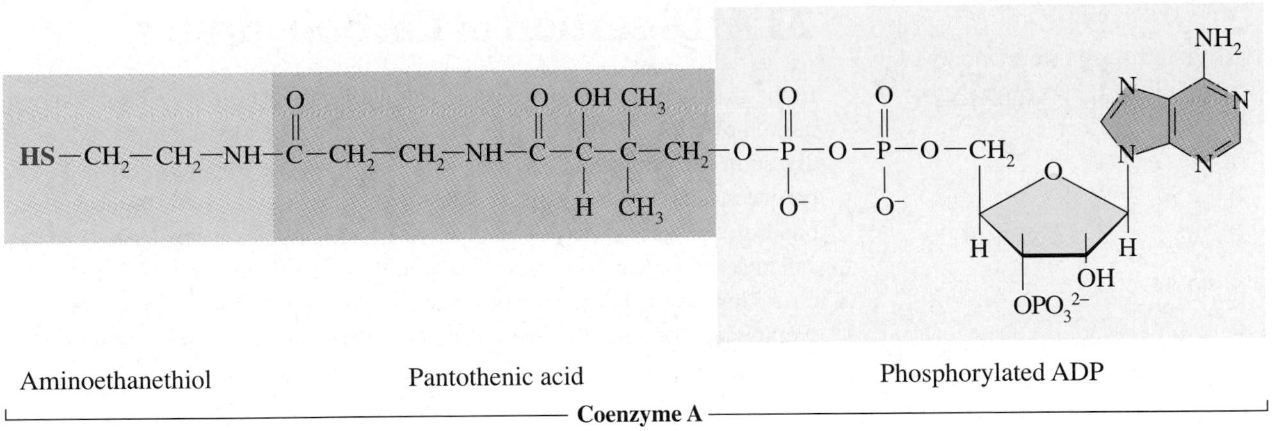

Aminoethanethiol   Pantothenic acid   Phosphorylated ADP

Coenzyme A

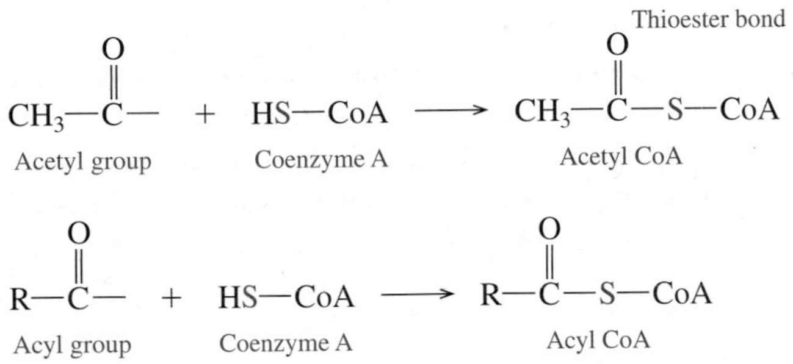

Figure 23.7 Coenzyme A is derived from a phosphorylated (3') adenine diphosphate (ADP) and pantothenic acid bonded by an amide linked to aminoethanethiol, which contains the —SH reactive part of the molecule.

Q *What part of coenzyme A reacts with a two-carbon acetyl group?*

SAMPLE PROBLEM 23.3

Coenzymes in Metabolic Pathways

What vitamin is part of the coenzyme FAD?

Solution

FAD, flavin adenine dinucleotide, is made from the vitamin riboflavin.

Study Check

What is the abbreviation of the reduced form of FAD?

QUESTIONS AND PROBLEMS

Important Coenzymes in Metabolic Pathways

23.11 Identify one or more coenzymes with each of the following components:
   a. pantothenic acid   **b.** niacin   **c.** ribitol

23.12 Identify one or more coenzymes with each of the following components:
   a. riboflavin   **b.** adenine   **c.** aminoethanethiol

23.13 Give the abbreviation for the following:
   a. the reduced form of NAD^+ **b.** the oxidized form of $FADH_2$

23.14 Give the abbreviation for the following:
   a. the reduced form of FAD   **b.** the oxidized form of NADH

23.15 What coenzyme picks up hydrogen when a carbon–carbon double bond is formed?

23.16 What coenzyme picks up hydrogen when a carbon–oxygen double bond is formed?

LEARNING GOAL

Give the sites and products of digestion for carbohydrates.

23.4 Digestion of Carbohydrates

In the first stage of catabolism, foods undergo **digestion,** a process that converts large molecules to smaller ones that can be absorbed by the body. We begin the digestion of carbohydrates as soon as we chew food. (See Figure 23.8.) An enzyme produced in the salivary glands called **amylase** hydrolyzes some of the α-glycosidic bonds in the starches amylose and amylopectin, producing smaller polysaccharides called dextrins, which contain three to eight glucose units, maltose, and some glucose. However, amylase does not cleave the branched chains of amylopectin. After swallowing, the partially digested starches enter the acid environment of the stomach, where the low pH soon stops further carbohydrate digestion.

Digestion of Disaccharides

In the small intestine, which has a pH of about 8, an α-amylase produced in the pancreas hydrolyzes the remaining polysaccharides to maltose and glucose. A branching enzyme hydrolyzes the glycosidic bonds to the branches in amylopectin. Then enzymes produced in the mucosal cells that line the small intestine hydrolyze maltose as well as lactose and sucrose from milk products and natural sugar. The hydrolysis reactions for the three common dietary disaccharides are written as follows.

$$\text{Lactose} + \text{H}_2\text{O} \xrightarrow{\text{Lactase}} \text{galactose} + \text{glucose}$$

$$\text{Sucrose} + \text{H}_2\text{O} \xrightarrow{\text{Sucrase}} \text{fructose} + \text{glucose}$$

$$\text{Maltose} + \text{H}_2\text{O} \xrightarrow{\text{Maltase}} \text{glucose} + \text{glucose}$$

The monosaccharides are absorbed through the intestinal wall into the bloodstream, which carries them to the liver where fructose and galactose are converted to glucose.

SAMPLE PROBLEM 23.4

Digestion of Carbohydrates

Indicate the carbohydrate that undergoes digestion in each of the following sites:
a. mouth   **b.** stomach   **c.** small intestine

Solution

a. Starches amylose and amylopectin (α-1,4-glycosidic bonds only).
b. Essentially no digestion occurs in the stomach.
c. Dextrins, maltose, sucrose, and lactose.

Study Check

Describe the digestion of amylose, a polymer of glucose molecules joined by α-glycosidic bonds.

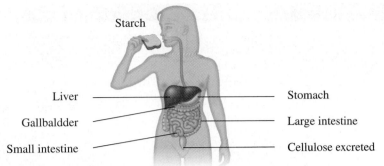

Figure 23.8 In stage 1 of metabolism, the digestion of carbohydrates begins in the mouth with the action of salivary amylase to yield dextrins (small polysaccharides), maltose, and some glucose and is completed in the small intestine by pancreatic amylase and enzymes for the disaccharides.
Q *Why is there little or no digestion of carbohydrates in the stomach?*

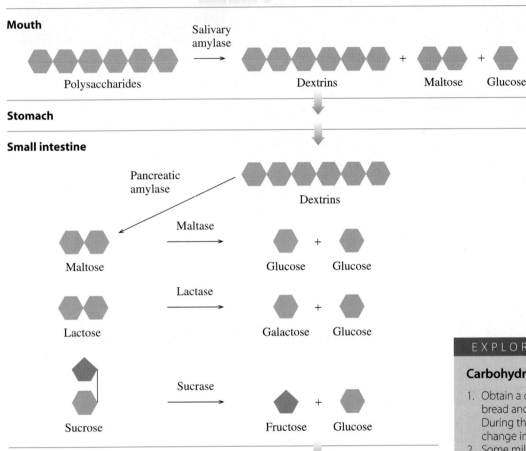

QUESTIONS AND PROBLEMS

Digestion of Foods

23.17 What is the general type of reaction that occurs during the digestion of carbohydrates?

23.18 Why is α-amylase produced in the salivary glands and in the pancreas?

23.19 Complete the following equation by filling in the missing words:
  a. _____ + H₂O ⟶ galactose + glucose
  b. Sucrose + H₂O ⟶ _____ + _____
  c. Maltose + H₂O ⟶ glucose + _____

23.20 Give the site and the enzyme for each of the reactions in problem 23.19.

EXPLORE YOUR WORLD

Carbohydrate Digestion

1. Obtain a cracker or small piece of bread and chew it for 4–5 minutes. During that time observe any change in the taste.
2. Some milk products contain Lactaid, which is the lactase that digests lactose. Look for the brands of milk and ice cream that contain Lactaid or lactase enzyme.

Questions

1a. How does the taste of the cracker or bread change after you have chewed it for 4–5 minutes? What could be an explanation?
1b. What part of carbohydrate digestion occurs in the mouth?
2a. Write an equation for the digestion of lactose.
2b. Where does lactose undergo digestion?

HEALTH NOTE

Lactose Intolerance

The disaccharide in milk is lactose, which is broken down by *lactase* in the intestinal tract to monosaccharides that are a source of energy. Infants and small children produce lactase to break down the lactose in milk. It is rare for an infant to lack the ability to produce lactase. As they mature, many people experience a decrease in the production of the lactase enzyme. By the time they are adults, many people have little or no lactase production, causing lactose intolerance. This condition may affect 25% of the people in the United States. A deficiency of lactase occurs in adults throughout the world, but in the United States it is more prevalent among the African–American, Hispanic, and Asian populations.

When lactose is not broken down into glucose and galactose, it cannot be absorbed through the intestinal wall and remains in the intestinal tract. In the intestines, the lactose undergoes fermentation to products that include lactic acid and gases such as methane (CH_4) and CO_2. Symptoms of lactose intolerance, which appear approximately $\frac{1}{2}$ to 1 hour after ingesting milk or milk products, include nausea, abdominal cramps, and diarrhea. The severity of the symptoms depends on how much lactose is present in the food and how much lactase a person produces.

Treatment of Lactose Intolerance

One way to reduce the reaction to lactose is to avoid products that contain lactose. However, it is important to consume foods that provide the body with calcium. Many people with lactose intolerance seem to tolerate yogurt, which is a good source of

calcium. Although there is lactose in yogurt, the bacteria in yogurt may produce some lactase, which helps to digest the lactose. A person who is lactose intolerant should also know that some foods that may not seem to be dairy products contain lactose. For example, baked goods, cereals, breakfast drinks, salad dressings, and even lunchmeat can contain lactose in their ingredients. Labels must be read carefully to see if the ingredients include "milk" or "lactose."

The enzyme lactase is now available in many forms such as tablets that are taken with meals, drops that are added to milk, or as additives in many dairy products such as milk. When lactase is added to milk that is left in the refrigerator for 24 hours, the lactose level is reduced by 70–90%. Lactase pills or chewable tablets are taken when a person begins to eat a meal that contains dairy foods. If taken too far ahead of the meal, too much of the lactase will be degraded by stomach acid. If taken following a meal, the lactose will have entered the lower intestine.

LEARNING GOAL

Describe the conversion of glucose to pyruvate in glycolysis.

23.5 Glycolysis: Oxidation of Glucose

Our major source of energy is the glucose produced when we digest the carbohydrates in our food, or from glycogen, a polysaccharide stored in the liver and skeletal muscle. Glucose in the bloodstream enters our cells for further degradation in a pathway called glycolysis. Early organisms used this pathway to produce energy from simple nutrients long before there was any oxygen in Earth's atmosphere. Glycolysis is an **anaerobic** process; no oxygen is required.

All the reactions in glycolysis take place in the cytoplasm of the cell where all the enzymes for glycolysis are located. We can now look at the details of the individual steps in glycolysis, which is one of the metabolic pathways in stage 2. In **glycolysis,** a six-carbon glucose molecule is broken down to yield two three-carbon pyruvate molecules. (See Figure 23.9.) In glycolysis glucose is converted to pyruvate in a sequence of 10 reactions. In the first five reactions (1–5), the energy of 2 ATPs is invested to add phosphate groups to form sugar phosphates. Then the six-carbon sugar phosphate is cleaved to yield two three-carbon sugar phosphate molecules. In the last five reactions (6–10), the phosphate groups in these energy-rich trioses are hydrolyzed, which generates energy in the form of 4 ATPs. The final products are 2 pyruvate molecules and 2 reduced NADH molecules. (See Figure 23.10.)

Stage 1

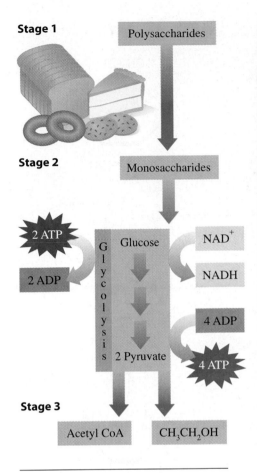

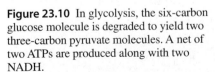

Figure 23.9 Glucose obtained from the digestion (stage 1) of polysaccharides is degraded in a metabolic pathway called glycolysis to give pyruvate.
Q *What is the end product of glycolysis?*

Figure 23.10 In glycolysis, the six-carbon glucose molecule is degraded to yield two three-carbon pyruvate molecules. A net of two ATPs are produced along with two NADH.
Q *Where in the glycolysis pathway is glucose cleaved to yield two three-carbon compounds?*

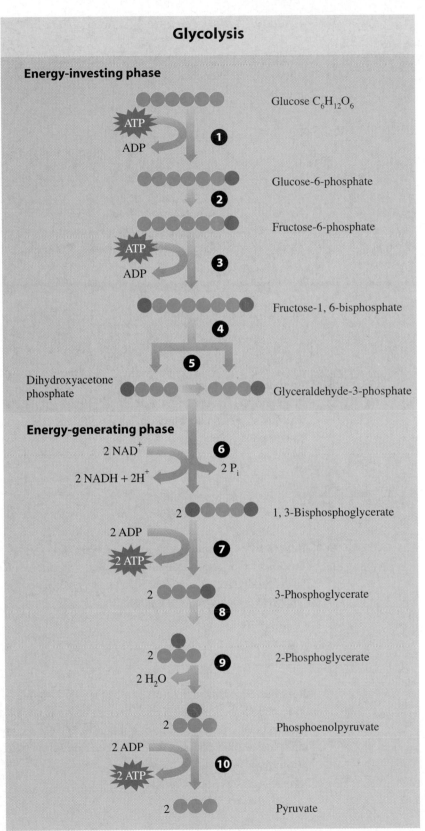

Energy-Investing Reactions: Steps 1–5

Reaction 1   Phosphorylation: first ATP invested

Glucose is converted to glucose-6-phosphate by a reaction with ATP catalyzed by *hexokinase*.

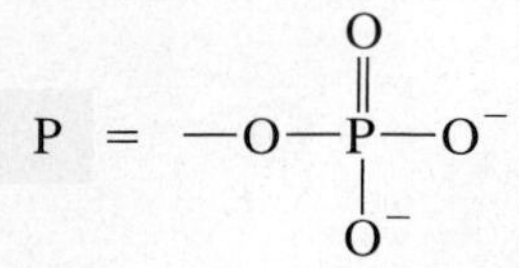

$$P = -O-\overset{\displaystyle O}{\underset{\displaystyle O^-}{\overset{\|}{P}}}-O^-$$

Reaction 2   Isomerization

The enzyme *phosphoglucoisomerase* converts glucose-6-phosphate, an aldose, to fructose-6-phosphate, a ketose.

Reaction 3   Phosphorylation: second ATP invested

A second ATP reacts with fructose-6-phosphate to give fructose-1,6-bisphosphate. The word *bisphosphate* is used to show that the phosphates are on different carbons in fructose and not connected to each other.

Reaction 4   Cleavage: two trioses form

Fructose-1,6-bisphosphate is "split" into two triose phosphates—dihydroxyacetone phosphate and glyceraldehyde-3-phosphate—catalyzed by *aldolase*.

Reaction 5   Isomerization of a triose

In reaction 5, *triose phosphate isomerase* converts one of the triose products, dihydroxyacetone phophate, to the other, glyceraldehyde-3-phosphate. Now all 6 carbon atoms from glucose are in two identical triose phosphates.

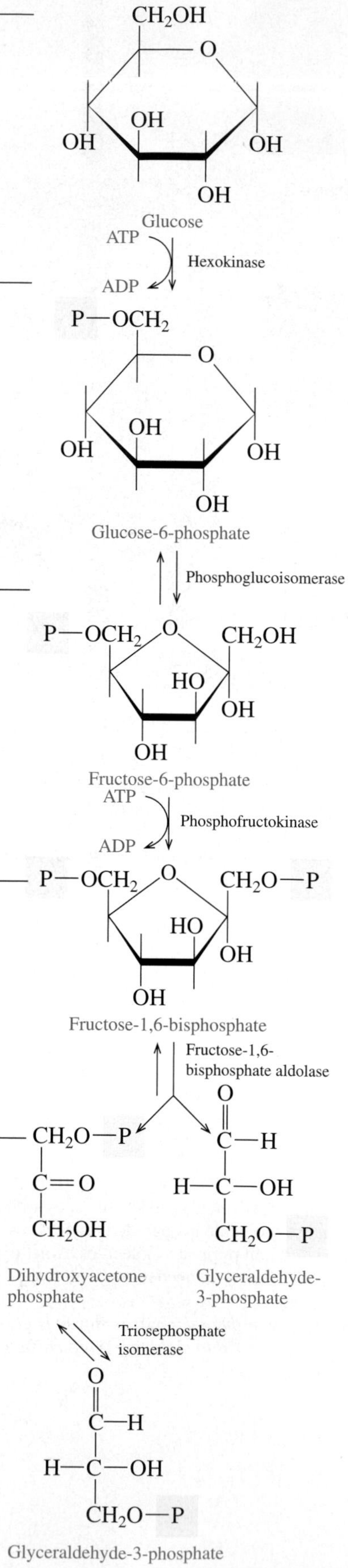

Glucose

ATP
Hexokinase
ADP

Glucose-6-phosphate

Phosphoglucoisomerase

Fructose-6-phosphate

ATP
Phosphofructokinase
ADP

Fructose-1,6-bisphosphate

Fructose-1,6-bisphosphate aldolase

Dihydroxyacetone phosphate

Glyceraldehyde-3-phosphate

Triosephosphate isomerase

Glyceraldehyde-3-phosphate

Energy-Generating Reactions: Steps 6–10

Reaction 6   First energy-rich compound

The aldehyde group of glyceraldehyde-3-phosphate is oxidized and phosphorylated by *glyceraldehyde-3-phosphate dehydrogenase*. The coenzyme NAD^+ is reduced to NADH and H^+. The 1,3-bisphosphoglycerate is an extremely high-energy compound that releases 12 kcal (49 kJ) per mole when the phosphate bond is broken.

Reaction 7   Formation of first ATP

The energy-rich 1,3-bisphosphoglycerate now drives the formation of ATP when *phosphoglycerate kinase* transfers a phosphate group to ADP. This process is called a **substrate-level phosphorylation.** At this point, glycolysis pays back the debt of two ATP invested in the early reactions.

Reaction 8   Formation of 2-phosphoglycerate

A *phosphoglycerate mutase* transfers the phosphate group from carbon 3 to carbon 2 to yield 2-phosphoglycerate.

Reaction 9   Second energy-rich compound

An *enolase* catalyzes the removal of water to yield phosphoenolpyruvate, a high-energy compound that releases 15 kcal or 62 kJ per mole when the phosphate bond is hydrolyzed.

Reaction 10   Formation of second ATP

This last reaction catalyzed by *pyruvate kinase* transfers the phosphate to ADP in a second direct substrate phosphorylation to yield pyruvate and ATP.

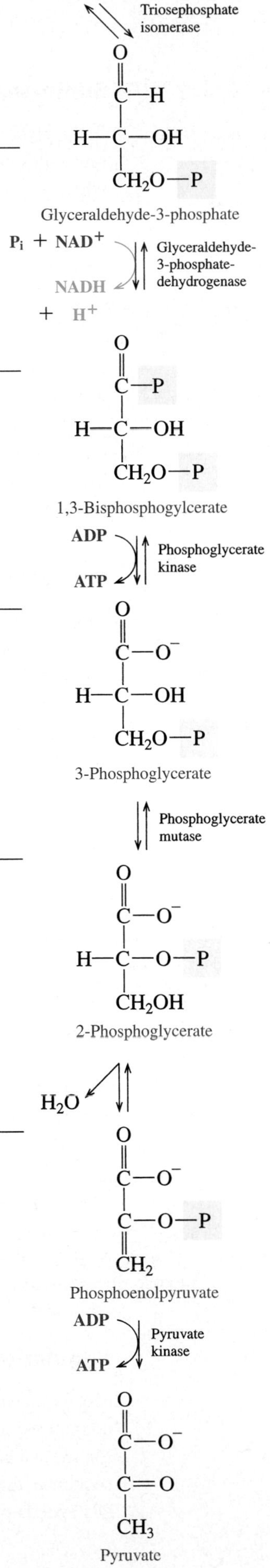

Triosephosphate isomerase

Glyceraldehyde-3-phosphate

P_i + NAD^+ Glyceraldehyde-3-phosphate-dehydrogenase

NADH + H^+

1,3-Bisphosphoglycerate

ADP   Phosphoglycerate kinase

ATP

3-Phosphoglycerate

Phosphoglycerate mutase

2-Phosphoglycerate

H_2O

Phosphoenolpyruvate

ADP   Pyruvate kinase

ATP

Pyruvate

Summary of Glycolysis

In the glycolysis pathway, glucose is converted to two pyruvates. Initially, two ATP molecules were required to form sugar phosphate. Later, four ATP were generated, which gives a net gain of two ATP. Overall, glycolysis yields two ATP and two NADH for each glucose that is converted to two pyruvates.

$$C_6H_{12}O_6 + 2\,NAD^+ \xrightarrow[\qquad\qquad]{2\,ADP\ +\ 2\,P_i \quad 2\,ATP} 2\,CH_3\!-\!\overset{\overset{\textstyle O}{\|}}{C}\!-\!COO^- + 2\,NADH + 4H^+$$

Glucose   Pyruvate

It appears right now that glycolysis does a lot of work to produce only 2 ATP, 2 NADH, and pyruvate. However, under aerobic conditions, stage 3 operates to reoxidize NADH to produce more ATP, and pyruvate enters the citric acid cycle where it generates considerably more energy. We will look at the oxidative pathways of stage 3 in Chapter 24.

Other Hexoses Enter Glycolysis

Other monosaccharides can enter glycolysis if they are first converted to intermediates of the pathway. Digestion of carbohydrates produces galactose from lactose in milk products and fructose from fruits and sucrose. Galactose reacts with ATP to yield galactose-1-phosphate, which is converted to glucose-6-phosphate, an intermediate of glycolysis in reaction 2. In the liver, fructose is reacted with ATP to yield fructose-1-phosphate, which is cleaved in a reaction similar to reaction 4 to give dihydroxyacetone phosphate and glyceraldehyde. Dihydroxyacetone phosphate isomerizes to glyceraldehyde-3- phosphate, and glyceraldehyde is phosphorylated to glyceraldehyde-3-phosphate, an intermediate in reaction 6. In muscle and kidney, fructose is phosphorylated to fructose-6-phosphate, which enters glycolysis in reaction 3.

$$\text{Galactose} \xrightarrow[\qquad]{ATP \quad ADP + P_i} \text{galactose-1-phosphate} \longrightarrow \text{glucose-6-phosphate}$$

Muscle and Kidney

Fructose ⟶ fructose-6-phosphate ———3———⟶ │ Gycolysis │

glyceraldehyde-3-phosphate   6   2   5

Liver ATP   ADP + P$_i$ ADP + P$_i$ ATP

Fructose ⟶ fructose-1-phosphate ⟶ glyceraldehyde + dihydroxyacetone phosphate

Regulation of Glycolysis

Metabolic pathways such as glycolysis do not run at the same rates all the time. The amount of glucose that is broken down is controlled by the requirements in the cells for pyruvate, ATP, and other intermediates of glycolysis. Within the glycolysis sequence, three enzymes respond to the levels of ATP and other products continually speed up or slow down the flow of glucose into the pathway.

Reaction 1   Hexokinase

The amount of glucose entering the glycolysis pathway decreases when high levels of glucose-6-phosphate are present in the cell. This phosphorylation product inhibits hexokinase, which prevents glucose from reacting with ATP. This is a feedback inhibition, which is a type of enzyme regulation we discussed in Chapter 21.

Reaction 3   Phosphofructokinase

The reaction catalyzed by phosphofructokinase is a very important control point for glycolysis. Once fructose-1,6-bisphosphate is formed, it must continue through the remaining reactions to pyruvate. As an allosteric enzyme, phosphofructokinase is inhibited by high levels of ATP and activated by high levels of ADP and AMP. High levels of ADP and AMP indicate that the cell has used up much of its ATP. As a regulator, phosphofructokinase increases the rate of pyruvate production for ATP synthesis when the cell needs to replenish ATP and slows or stops the reaction when ATP is plentiful.

Reaction 10   Pyruvate Kinase

In the last reaction of glycolysis, high levels of ATP as well as acetyl CoA inhibit pyruvate kinase, which is another allosteric enzyme.

Summary of Regulation

These are examples of how metabolic pathways shut off enzymes to stop the production of molecules that are not needed. Since pyruvate can be used to synthesize ATP, several enzymes in glycolysis respond to ATP levels in the cell. When ATP levels are high, enzymes in glycolysis slow or stop the synthesis of pyruvate. With phosphofructokinase and pyruvate kinase inhibited by ATP, glucose-6-phosphate accumulates and inhibits the first reaction and glucose does not enter the glycolysis pathway. The glycolysis pathway is shut down until ATP is once again needed in the cell. When ATP levels are low or AMP/ADP levels are high, these enzymes are activated and pyruvate production starts again.

SAMPLE PROBLEM 23.5

Glycolysis

What are the steps in glycolysis that generate ATP?

Solution

ATP is produced when phosphate groups are transferred directly to ADP from 1,3-bisphosphoglycerate (step 7) and from phosphoenolpyruvate (step 10).

Study Check

If four ATP molecules are produced in glycolysis, why is there a net yield of two ATP?

QUESTIONS AND PROBLEMS

Glycolysis: Oxidation of Glucose

23.21 What is the starting product of glycolysis?

23.22 What is the end product of glycolysis?

23.23 How is ATP used in the initial steps of glycolysis?

23.24 How many ATP molecules are used in the initial steps of glycolysis?

23.25 What trioses are obtained when fructose-1,6-diphosphate splits?

23.26 Why does one of the triose products undergo isomerization?

23.27 How does direct phosphorylation account for the production of ATP in glycolysis?

23.28 Why are there two ATP molecules formed for one molecule of glucose?

23.29 Indicate the enzyme that catalyzes the following reactions in glycolysis:
a. phosphorylation
b. direct transfer of a phosphate group

23.30 Indicate the enzyme that catalyzes the following reactions in glycolysis:
a. isomerization
b. formation of a ketotriose and an aldotriose

23.31 How many ATP or NADH are produced (or required) in each of the following steps in glycolysis?
a. glucose to glucose-6-phosphate
b. glyceraldehyde-3-phosphate to 1,3-bisphosphoglycerate
c. glucose to pyruvate

23.32 How many ATP or NADH are produced (or required) in each of the following steps in glycolysis?
a. 1,3-bisphosphoglycerate to 3-phosphoglycerate
b. fructose-6-phosphate to fructose-1,6-bisphosphate
c. phosphoenolpyruvate to pyruvate

23.33 Which step in glycolysis involves the following?
a. the first ATP molecule is hydrolyzed
b. direct substrate phosphorylation occurs
c. six-carbon sugar splits into two three-carbon molecules

23.34 Which step in glycolysis involves the following?
a. isomerization takes place   **b.** NAD^+ is reduced
c. a second ATP molecule is synthesized

23.35 How do galactose and fructose, obtained from the digestion of carbohydrates, enter glycolysis?

23.36 What are three enzymes that regulate glycolysis?

23.37 Indicate whether each of the following would activate or inhibit phosphofructokinase.
a. low levels of ATP   **b.** high levels of ATP

23.38 Indicate whether each of the following would activate or inhibit pyruvate kinase.
a. low levels of ATP
b. high levels of fructose-1,6-bisphosphate

LEARNING GOAL

Give the conditions for the conversion of pyruvate to lactate, ethanol, and acetyl coenzyme A.

23.6 Pathways for Pyruvate

The pyruvate produced from glucose can now enter one of three pathways that continue to extract energy. The available pathway depends on whether there is sufficient oxygen in the cell. During **aerobic** conditions, oxygen is available to convert pyruvate to acetyl coenzyme A (CoA). When oxygen levels are low, pyruvate is reduced to lactate. In yeast cells, which are anaerobic, pyruvate is converted to ethanol.

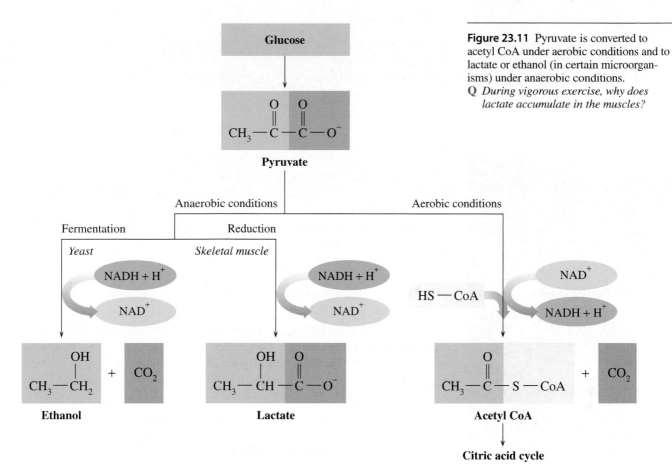

Figure 23.11 Pyruvate is converted to acetyl CoA under aerobic conditions and to lactate or ethanol (in certain microorganisms) under anaerobic conditions.
Q *During vigorous exercise, why does lactate accumulate in the muscles?*

Pyruvate to Acetyl CoA (Aerobic Conditions)

In glycolysis, two ATP molecules were generated when glucose was converted to pyruvate. However, much more energy is still available. The greatest amount of the energy is obtained from glucose when oxygen levels are high in the cells. Under aerobic conditions, pyruvate moves from the cytoplasm (where glycolysis took place) into the matrix of the mitochondria to be oxidized further. In a complex reaction, pyruvate is oxidized and a carbon atom is removed from pyruvate as CO_2. The coenzyme NAD^+ is required for the oxidation. The resulting two-carbon acetyl compound is attached to CoA, producing **acetyl CoA,** an important intermediate in many metabolic pathways. (See Figure 23.11.)

$$CH_3-\overset{O}{\overset{\|}{C}}-\overset{O}{\overset{\|}{C}}-O^- + HS-CoA + NAD^+ \xrightarrow{\text{Pyruvate dehydrogenase}} CH_3-\overset{O}{\overset{\|}{C}}-S-CoA + CO_2 + NADH + H^+$$

Pyruvate $\qquad\qquad\qquad\qquad\qquad\qquad\qquad\qquad\qquad\qquad$ Acetyl CoA

Pyruvate to Lactate (Anaerobic Conditions)

When we engage in strenuous exercise, the oxygen stored in our muscle cells is quickly depleted. Under anaerobic conditions, pyruvate remains in the cytoplasm where it is reduced to lactate. NAD^+ is produced and used to oxidize more glyceraldehyde-3-phosphate in the glycolysis pathway, which produces a small but needed amount of ATP.

$$CH_3-\overset{\overset{\displaystyle O}{\|}}{C}-\overset{\overset{\displaystyle O}{\|}}{C}-O^- \underset{\text{Lactate dehydrogenase}}{\overset{\text{NADH} + \text{H}^+ \quad \text{NAD}^+}{\rightleftharpoons}} CH_3-\overset{\overset{\displaystyle OH}{|}}{\underset{\underset{\displaystyle H}{|}}{C}}-\overset{\overset{\displaystyle O}{\|}}{C}-O^-$$

Pyruvate   Lactate
(oxidized)   (reduced)

The accumulation of lactate causes the muscles to tire rapidly and become sore. After exercise, a person continues to breathe rapidly to repay the oxygen debt incurred during exercise. Most of the lactate is transported to the liver where it is converted back into pyruvate. Under anaerobic conditions, the only ATP production in glycolysis occurs during the steps that phosphorylate ADP directly, giving a net gain of only two ATP molecules:

$$C_6H_{12}O_6 + 2\,ADP + 2\,P_i \longrightarrow 2\,CH_3-\overset{\overset{\displaystyle OH}{|}}{CH}-COO^- + 2\,ATP$$

Glucose   Lactate

Bacteria also convert pyruvate to lactate under anaerobic conditions. In the preparation of kimchee and sauerkraut, cabbage is covered with a salt brine. The glucose from the starches is converted to lactate. This acid environment acts as a preservative that prevents the growth of other bacteria. The pickling of olives and cucumbers gives similar products. When cultures of bacteria that produce lactate are added to milk, the acid denatures the milk proteins to give sour cream and yogurt.

Pyruvate to Ethanol (Anaerobic)

Some microorganisms, particularly yeast, convert sugars to ethanol under anaerobic conditions by a process called **fermentation.** After pyruvate is formed in glycolysis, a carbon atom is removed in the form of CO_2 (**decarboxylation**). The NAD^+ for continued glycolysis is regenerated when the acetaldehyde is reduced to ethanol.

$$CH_3-\overset{\overset{\displaystyle O}{\|}}{C}-\overset{\overset{\displaystyle O}{\|}}{C}-O^- \underset{\underset{\text{decarboxylase}}{\text{Pyruvate}}}{\overset{\text{CO}_2}{\longrightarrow}} CH_3-\overset{\overset{\displaystyle O}{\|}}{C}-H \underset{\text{Alcohol dehydrogenase}}{\overset{\text{NADH} + \text{H}^+ \quad \text{NAD}^+}{\longrightarrow}} CH_3-\overset{\overset{\displaystyle H}{|}}{\underset{\underset{\displaystyle H}{|}}{C}}-OH$$

Pyruvate   Acetaldehyde   Ethanol

The process of fermentation by yeast is one of the oldest known chemical reactions. Enzymes in the yeast convert the sugars in a variety of carbohydrate sources to glucose and then to ethanol. The evolution of CO_2 gas produces the bubbles in beer, sparkling wines, and champagne. The type of carbohydrate used determines the taste associated with a particular alcoholic beverage. Beer is made from the fermentation of barley malt, wine from the sugars in grapes, vodka from potatoes, sake from rice, and whiskeys from corn or rye. Fermentation produces solutions up to about 15% alcohol by volume. At this concentration, the alcohol kills the yeast, and fermentation stops. Higher concentrations are obtained by distilling the alcohol.

SAMPLE PROBLEM 23.6

Fates of Pyruvate

Is each of the following products from pyruvate produced under anaerobic or aerobic conditions?

a. acetyl CoA   **b.** lactate

Solution

a. aerobic conditions   **b.** anaerobic conditions

Study Check

After strenuous exercise, some lactate is oxidized back to pyruvate by lactate dehydrogenase using NAD^+. Write an equation to show this reaction.

QUESTIONS AND PROBLEMS

Pathways for Pyruvate

23.39 What condition is needed in the cell to convert pyruvate to acetyl CoA?

23.40 What coenzymes are needed for the oxidation of pyruvate to acetyl CoA?

23.41 Write the overall equation for the conversion of pyruvate to acetyl CoA.

23.42 What are the possible products of pyruvate under anaerobic conditions?

23.43 How does the formation of lactate permit glycolysis to continue under anaerobic conditions?

23.44 After running a marathon, a runner has muscle pain and cramping. What might have occurred in the muscle cells to cause this?

23.45 In fermentation, a carbon atom is removed from pyruvate. What is the compound formed by that carbon atom?

23.46 Some students decided to make some wine by placing yeast and grape juice in a container with a tight lid. A few weeks later, the container explodes. What reaction could account for the explosion?

23.7 Glycogen Metabolism

LEARNING GOAL

Describe the breakdown and synthesis of glycogen.

We have just eaten a large meal that has supplied us with all the glucose we need to produce pyruvate and ATP by glycolysis. Then we use excess glucose to replenish our energy reserves by synthesizing glycogen that is stored in limited amounts in our skeletal muscle and liver. When glycogen stores are full, any remaining glucose is converted to triacylglycerols and stored as body fat as we will see in Chapter 25. When our diet does not supply sufficient glucose, or we have utilized our blood glucose, we degrade the stored glycogen and release glucose.

Glycogenesis

Glycogen is a polymer of glucose with α-1,4 glycosidic bonds and multiple branches attached by α-1,6 glycosidic bonds, as seen in Chapter 16. **Glycogenesis** is the synthesis of glycogen from glucose molecules, which occurs when the digestion of polysaccharides produces high levels of glucose. The synthesis of glycogen starts with glucose-6-phosphate, which can be obtained from the first reaction in glycolysis. (See Figure 23.12.) It is converted to an isomer glucose-1-phosphate, which is activated using high-energy UTP (uridine triphosphate) to yield UDP-glucose. The reaction is driven by the energy released from the hydrolysis of pyrophosphate (PP_i).

Glucose-6-phosphate $\rightleftharpoons$ glucose-1-phosphate

Glucose-1-phosphate + UTP $\longrightarrow$ UDP-glucose + PP_i

$$PP_i \longrightarrow 2 P_i$$

UDP-Glucose (uridine diphosphate glucose)

The UDP-glucose attaches to the end glucose of a glycogen chain releasing UDP, which reacts with ATP to regenerate UTP.

$$\text{UDP-Glucose} + \text{glycogen} \xrightarrow[\text{synthase}]{\text{Glycogen}} \text{glucose—glycogen} + \text{UDP}$$

$$\text{UDP} + \text{ATP} \longrightarrow \text{UTP} + \text{ADP}$$

The overall reaction of glycogenesis starting with glucose is written simply as

$$\text{Glucose} \longrightarrow \text{glycogen}$$

Figure 23.12 In glycogenesis, glucose is used to synthesize glycogen.
Q *What is the function of UTP in glycogen synthesis?*

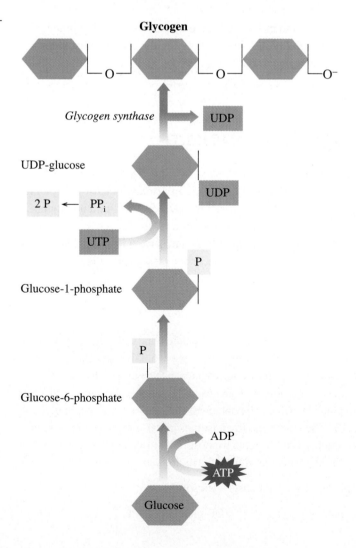

Glycogenolysis

Glucose is the primary energy source for muscle contraction and the brain. When blood glucose is depleted, glycogen breaks down to glucose in a process called **glycogenolysis.** Glucose molecules are removed one by one from the end of the glycogen chain and phosphorylated to yield glucose-1-phosphate.

$$\text{Glucose — glycogen} + P_i \xrightarrow{\substack{\text{Glycogen} \\ \text{phosphorylase}}} \text{glucose-1-phosphate} + \text{glycogen}$$

Glucose-1-phosphate may be converted to glucose-6-phosphate, which enters the glycolysis pathway to replenish ATP.

$$\text{Glucose-1-phosphate} \xrightleftharpoons{\text{Phosphoglucomutase}} \text{glucose-6-phosphate}$$

Free glucose is needed for energy by the brain and muscle. While glucose can diffuse across cell membranes, glucose phosphates cannot. Only cells in the liver and kidneys have a *gluco*se-6-*phosphatase* that hydrolyzes the phosphate to yield free glucose.

$$\text{Glucose-6-phosphate} \xrightarrow{\text{Glucose-6-phosphatase}} \text{Glucose} + P_i$$

The overall reaction of glycogenolysis, which converts glycogen to glucose, is written as

$$\text{Glycogen} \longrightarrow \text{glucose}$$

Let's summarize the events and compounds involved in the breakdown and synthesis of glycogen as follows.

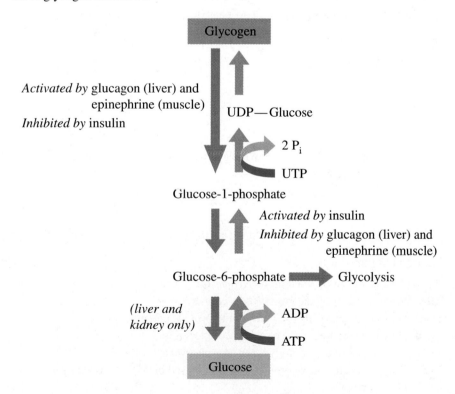

Regulation of Glycogen Metabolism

The brain, skeletal muscles, and red blood cells require large amounts of glucose every day to function properly. To protect the brain, hormones with opposing actions control blood glucose levels. When glucose is low, glucagon, a hormone produced in the pancreas, is secreted into the bloodstream. In the liver, glucagon accelerates the rate of glycogenolysis, which raises blood glucose levels. At the same time, glucagon inhibits the synthesis of glycogen.

Glycogen in skeletal muscle is broken down quickly when the body requires a "burst of energy" often referred to as "fight or flight." Epinephrine released from the adrenal glands converts *glycogen phosphorylase* from an inactive to an active form. The secretion of only a few molecules of epinephrine can break down a huge number of glycogen molecules.

Soon after we have eaten and digested a meal, our blood glucose levels rise. Rising glucose levels stimulate the pancreas to secrete the hormone insulin into our bloodstreams. Insulin promotes the use of glucose in the cells by accelerating glycogen synthesis as well as degradation reactions such as glycolysis. At the same time, insulin inhibits the synthesis of glucose, which we will discuss in the next section.

SAMPLE PROBLEM 23.7

Glycogen Metabolism

Identify each of the following as part of the reaction pathways of (1) glycolysis, (2) glycogenolysis, or (3) glycogenesis.

a. Glucose-1-phosphate is converted to glucose-6-phosphate.
b. Glucose-1-phosphate forms UDP-glucose.
c. An isomerase converts glucose-6-phosphate to fructose-6-phosphate.

Solution

a. (2) glycogenolysis
b. (3) glycogenesis
c. (1) glycolysis

Study Check

Why do cells in liver and kidney provide glucose to raise blood glucose levels, but cells in skeletal muscle do not?

QUESTIONS AND PROBLEMS

Glycogen Metabolism

23.47 What is meant by the term *glycogenesis*?
23.48 What is meant by the term *glycogenolysis*?
23.49 How do muscle cells use glycogen to provide energy?
23.50 How does the liver raise blood glucose levels?
23.51 What is the function of *glycogen phosphorylase*?
23.52 Why is the enzyme *phosphoglucomutase* used in both glycogenolysis and glycogenesis?

23.8 Gluconeogenesis: Glucose Synthesis

Glycogen stored in our liver and muscles can supply us with about one day's requirement of glucose. However, glycogen stores are quickly depleted if we fast for more than one day, run a marathon, or participate in other heavy exercise. Then glucose is synthesized from carbon atoms obtained from noncarbohydrate compounds in a process called **gluconeogenesis** ("new glucose"). Most glucose is synthesized in the cytosol of liver cells. (See Figure 23.13.)

Carbon atoms for glucose can be obtained from lactate and other food sources such as amino acids, and glycerol from fats. Each is converted to pyruvate or an intermediate for the synthesis of glucose. Most of the reactions in gluconeogenesis are the reverse of glycolysis and catalyzed by the same enzymes. However, three of the reactions are not reversible: the ones catalyzed by hexokinase, phosphofructokinase, and pyruvate kinase, glycolysis reactions 1, 3, and 10, respectively. Different enzymes are used to replace them, but all the other reactions simply reverse glycolysis and use the same enzymes. We will now look at these three reactions in gluconeogenesis that differ from the reverse of glycolysis.

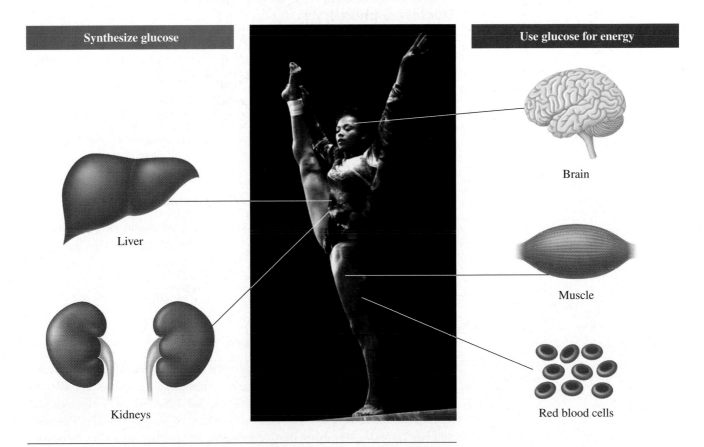

Figure 23.13 Glucose is synthesized in the tissues of the liver and kidneys. Tissues that use glucose as their main energy source are the brain, skeletal muscle, and red blood cells.
Q *Why does the body need a pathway for the synthesis of glucose from noncarbohydrate sources?*

Converting Pyruvate to Phosphoenolpyruvate

To start the synthesis of glucose, two steps are needed. The first step converts pyruvate to oxaloacetate, and the second step converts oxaloacetate to phosphoenolpyruvate. The hydrolysis of ATP and GTP are used to drive the reactions. (See Figure 23.14.)

$$CH_3-\overset{\overset{O}{\|}}{C}-COO^- \ + \ CO_2 \ + \ ATP \ + \ H_2O \ \xrightarrow{\text{Pyruvate carboxylase}} \ ^-OOC-CH_2-\overset{\overset{O}{\|}}{C}-COO^- \ + \ ADP \ + \ P_i$$

Pyruvate   Oxaloacetate

$$^-OOC-CH_2-\overset{\overset{O}{\|}}{C}-COO^- \ + \ GTP \ \xrightarrow{\text{Phosphoenolpyruvate carboxykinase}} \ CH_2=\overset{\overset{P}{|}}{C}-COO^- \ + \ CO_2 \ + \ GDP$$

Oxaloacetate   Phosphoenolpyruvate

Molecules of phosphoenolpyruvate now enter the next five reverse reactions in glycolysis using the same enzymes to form fructose-1,6-bisphosphate.

Converting Fructose-1,6-bisphosphate to Fructose-6-phosphate

The second irreversible reaction in glycolysis is bypassed using *fructose-1,6-bisphosphatase* to cleave a phosphate from fructose-1,6-bisphosphate by hydrolysis with water and releasing the energy that drives the reaction.

$$\text{Fructose-1,6-bisphosphate} \ + \ H_2O \ \xrightarrow{\text{Fructose-1,6-bisphosphatase}} \ \text{fructose-6-phosphate} \ + \ P_i$$

Then fructose-6-phosphate undergoes a reversible reaction to yield glucose-6-phosphate.

Converting Glucose 6-phosphate to Glucose

In the final reaction, glucose-6-phosphate is converted to glucose by a different enzyme than used in glycolysis. *Glucose-6-phosphatase* catalyzes the hydrolysis of glucose-6-phosphate with water.

$$\text{Glucose-6-phosphate} \ + \ H_2O \ \xrightarrow{\text{Glucose-6-phosphatase}} \ \text{glucose} \ + \ P_i$$

Energy Cost of Gluconeogensis

The pathway of gluconeogenesis consists of seven reversible reactions of glycolysis and four new reactions that replace the three irreversible reactions. Overall, this synthesis of glucose requires 4 ATPs, 2GTPs, and 2 NADH. If all the reactions were simply the reverse of glycolysis, the synthesis of glucose would not be energetically favorable. By using the energy resources and bypassing the three irreversible and energy-requiring reactions, gluconeogenesis becomes favorable in terms of energy. The overall equation for gluconeogenesis is written as follows:

$$2\,\text{Pyruvate} \ + \ 4\,ATP \ + \ 2\,GTP \ + \ 2\,NADH \ + \ 2H^+ \ + \ 6H_2O \ \longrightarrow$$
$$\text{glucose} \ + \ 4\,ADP \ + \ 2\,GDP \ + \ 6\,P_i \ + \ 2\,NAD^+$$

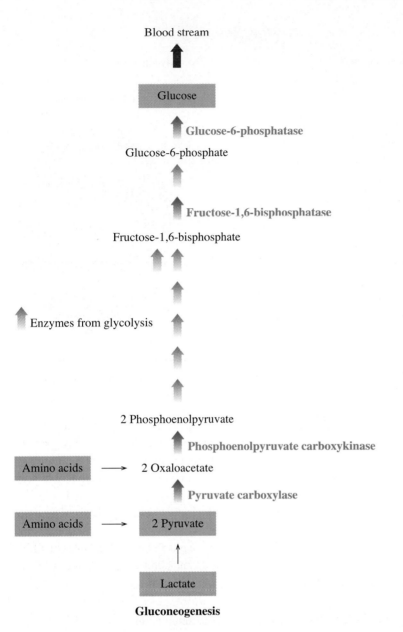

Figure 23.14 In gluconeogenesis, three irreversible reactions of glycolysis are bypassed using four different enzymes.
Q *Why are 11 enzymes required for gluco-neogenesis, and only 10 for glycolysis?*

Lactate and the Cori Cycle

When a person exercises vigorously, the anaerobic conditions cause the reduction of pyruvate to lactate, which accumulates in the muscle. This reaction is necessary to oxidize NADH to NAD^+, which allows glycolysis to continue to produce a small amount of ATP. Lactate is an important source of carbon for gluconeogenesis. Lactate is transported to the liver where it is oxidized to pyruvate, which is used to synthesize glucose. Glucose enters the bloodstream and returns to the muscle to rebuild glycogen stores. This flow of lactate and glucose between the muscle and liver, known as the **Cori cycle,** is very active when a person has just completed a period of vigorous exercise. (See Figure 23.15.)

Regulation of Gluconeogenesis

Gluconeogenesis is a pathway that protects the brain and nervous system from experiencing a loss of glucose, which causes impairment of function. It is also a

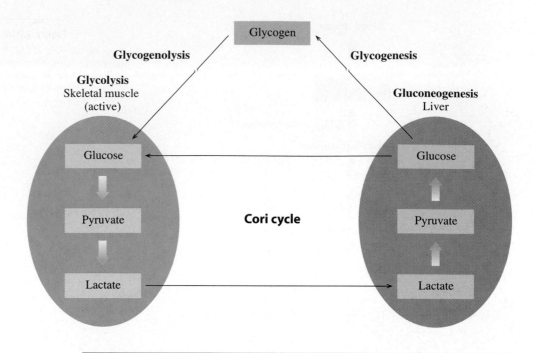

Figure 23.15 Different pathways connect the utilization and synthesis of glucose.
Q *Why is lactate formed in the muscle converted to glucose in the liver?*

pathway that is utilized when vigorous activity depletes blood glucose and glycogen stores. Thus, the level of carbohydrate available from the diet controls gluconeogenesis. When a diet is high in carbohydrate, the gluconeogenesis pathway is not utilized. However, when a diet is low in carbohydrate, the pathway is very active.

As long as conditions in a cell favor glycolysis, there is no synthesis of glucose, but when the cell requires synthesis of glucose, glycolysis is turned off. The same three reactions that control glycolysis also control gluconeogenesis, but with different enzymes. Let's look at how high levels of certain compounds activate or inhibit the two processes. (See Table 23.2.)

Table 23.2	Regulation of Glycolysis and Gluconeogenesis	
	Glycolysis	**Gluconeogenesis**
Enzyme	**Hexokinase**	**Glucose-6-phosphatase**
Activated by	High glucose levels Insulin, epinephrine	Low glucose levels Glucose-6-phosphate
Inhibited by	Glucose-6-phosphate	
Enzyme	**Phosphofructokinase**	**Fructose-1,6-bisphosphatase**
Activated by	AMP	Low glucose levels, glucagon
Inhibited by	ATP	AMP, insulin
Enzyme	**Pyruvate kinase**	**Pyruvate carboxylase**
Activated by	Fructose-1,6-bisphosphate	Low glucose level, glucagon
Inhibited by	ATP Acetyl CoA	Insulin

SAMPLE PROBLEM 23.8

Gluconeogenesis

The conversion of fructose-1,6-bisphosphate to fructose-6-phosphate is an irreversible reaction using the glycolytic enzyme. How does gluconeogenesis make this reaction happen?

Solution

This reverse reaction is catalyzed by a different enzyme *fructose-1,6-bisphosphatase,* which cleaves a phosphate using a hydrolysis reaction, a reaction that is energetically favorable.

Study Check

Why is hexokinase in glycolysis replaced by glucose-6-phosphatase in gluconeogenesis?

QUESTIONS AND PROBLEMS

Gluconeogenesis: Glucose Synthesis

23.53 What is the function of gluconeogenesis in the body?

23.54 What enzymes in glycolysis are not used in gluconeogenesis?

23.55 What enzymes in glycolysis are used in gluconeogenesis?

23.56 How is the lactate produced in skeletal muscle used for glucose synthesis?

23.57 Indicate whether each of the following activates or inhibits gluconeogenesis:
  a. low glucose levels   **b.** glucagon   **c.** insulin

23.58 Indicate whether each of the following activates or inhibits glycolysis?
  a. low glucose levels   **b.** insulin   **c.** glucagon

Chapter Review

23.1 Metabolism and Cell Structure

Metabolism includes all the catabolic and anabolic reactions that occur in the cells. Catabolic reactions degrade large molecules into smaller ones with an accompanying release of energy. Anabolic reactions require energy to synthesize larger molecules from smaller ones. The 3 stages of metabolism are digestion of food, degradation of monomers such as glucose to pyruvate, and the extraction of energy from the two- and three-carbon compounds from stage 2. Many of the metabolic enzymes are present in the cytosol of the cell where metabolic reactions take place.

23.2 ATP and Energy

Energy obtained from catabolic reactions is stored primarily in adenosine triphosphate (ATP), a high-energy compound that is hydrolyzed when energy is required by anabolic reactions.

23.3 Important Coenzymes in Metabolic Pathways

FAD and NAD^+ are the oxidized forms of coenzymes that participate in oxidation–reduction reactions. When they pick up hydrogen ions and electrons, they are reduced to $FADH_2$ and $NADH + H^+$. Coenzyme A contains a thiol group that bonds with a two-carbon acetyl group (acetyl CoA) or a longer chain acyl group (acyl CoA).

23.4 Digestion of Carbohydrates

The digestion of carbohydrates is a series of reactions that breaks down polysaccharides into hexose monomers such as glucose, galactose, and fructose. These monomers can be absorbed through the intestinal wall into the bloodstream to be carried to cells where they provide energy and carbon atoms for synthesis of new molecules.

23.5 Glycolysis: Oxidation of Glucose

Glycolysis, which occurs in the cytosol, consists of 10 reactions that degrade glucose (six carbons) to two pyruvate molecules (three carbons each). The overall series of reactions yields two molecules of the reduced coenzyme NADH and two ATP.

23.6 Pathways for Pyruvate

Under aerobic conditions, pyruvate is oxidized in the mitochondria to acetyl CoA. In the absence of oxygen, pyruvate is reduced to lactate and NAD^+ is regenerated for the continuation of glycolysis, while microorganisms such as yeast reduce pyruvate to ethanol, a process known as fermentation.

23.7 Glycogen Metabolism

Glycogenolysis breaks down glycogen to glucose when glucose and ATP levels are low. When blood glucose levels are high, glycogenesis converts glucose to glycogen, which is stored in the liver.

23.8 Gluconeogenesis: Glucose Synthesis

When blood glucose levels are low and glycogen stores in the liver are depleted, glucose is synthesized from compounds such as pyruvate and lactate.

Summary of Important Reactions

Hydrolysis of ATP

$$ATP + H_2O \longrightarrow ADP + P_i + 7.3 \text{ kcal } (31 \text{ kJ})/\text{mole}$$

$$ATP + H_2O \longrightarrow AMP + PP_i + 7.3 \text{ kcal } (31 \text{ kJ})/\text{mole}$$

Formation of ATP

$$ADP + P_i + 7.3 \text{ kcal}/ (31 \text{ kJ})/\text{mole} \longrightarrow ATP$$

Reduction of FAD and NAD^+

$$FAD + 2H^+ + 2e^- \longrightarrow FADH_2$$

$$NAD^+ + 2H^+ + 2e^- \longrightarrow NADH + H^+$$

Hydrolysis of Disaccharides

$$\text{Lactose} + H_2O \xrightarrow{\text{Lactase}} \text{galactose} + \text{glucose}$$

$$\text{Sucrose} + H_2O \xrightarrow{\text{Sucrase}} \text{fructose} + \text{glucose}$$

$$\text{Maltose} + H_2O \xrightarrow{\text{Maltase}} \text{glucose} + \text{glucose}$$

Glycolysis

$$C_6H_{12}O_6 + 2\,ADP + 2\,P_i + 2\,NAD^+ \rightarrow 2\,CH_3{-}\overset{\displaystyle O}{\overset{\|}{C}}{-}COO^- + 2\,ATP + 2\,NADH + 4H^+$$

Glucose   Pyruvate

Oxidation of Pyruvate to Acetyl-CoA

$$CH_3{-}\overset{\displaystyle O}{\overset{\|}{C}}{-}COO^- + NAD^+ + HS{-}CoA \longrightarrow CH_3{-}\overset{\displaystyle O}{\overset{\|}{C}}{-}S{-}CoA + NADH + H^+ + CO_2$$

Pyruvate   Acetyl CoA

Reduction of Pyruvate to Lactate

$$\text{Pyruvate} + \text{NADH} + \text{H}^+ \longrightarrow \text{Lactate} + \text{NAD}^+$$

$$\underset{\text{Pyruvate}}{CH_3-\overset{\displaystyle O}{\overset{\|}{C}}-COO^-} + NADH + H^+ \longrightarrow \underset{\text{Lactate}}{CH_3-\overset{\displaystyle OH}{\overset{|}{CH}}-COO^-} + NAD^+$$

Oxidation of Glucose to Lactate

$$\text{Glucose} + 2\,\text{ADP} + 2\,P_i \longrightarrow 2\,\text{Lactate} + 2\,\text{ATP}$$

Reduction of Pyruvate to Ethanol

$$\underset{\text{Pyruvate}}{CH_3-\overset{\displaystyle O}{\overset{\|}{C}}-COO^-} + NADH + H^+ \longrightarrow \underset{\text{Ethanol}}{CH_3-CH_2-OH} + NAD^+ + CO_2$$

Glycogenesis

$$\text{Glucose} \longrightarrow \text{glycogen}$$

Glycogenolysis

$$\text{Glycogen} \longrightarrow \text{glucose}$$

Gluconeogensis

$$\text{Pyruvate (or lactate)} \longrightarrow \text{glucose}$$

$$2\,\text{Pyruvate} + 4\,\text{ATP} + 2\,\text{GTP} + 2\,\text{NADH} + 2\text{H}^+ + 6\text{H}_2\text{O} \longrightarrow$$
$$\text{Glucose} + 4\,\text{ADP} + 2\,\text{GDP} + 6\,P_i + 2\,\text{NAD}^+$$

Key Terms

acetyl CoA A two-carbon acetyl unit from oxidation of pyruvate that bonds to coenzyme A.

ADP Adenosine diphosphate, a compound of adenine, a ribose sugar, and two phosphate groups, it is formed by the hydrolysis of ATP.

aerobic An oxygen-containing environment in the cells.

amylase An enzyme that hydrolyzes the glycosidic bonds in polysaccharides during digestion.

anabolic reaction A metabolic reaction that requires energy.

anaerobic A condition in cells when there is no oxygen.

ATP Adenosine triphosphate, a high-energy compound that stores energy in the cells, consists of adenine, a ribose sugar, and three phosphate groups.

catabolic reaction A metabolic reaction that produces energy for the cell by the degradation and oxidation of glucose and other molecules.

coenzyme A (CoA) A coenzyme that transports acyland acetyl groups.

Cori cycle A cyclic process in which lactate produced in muscle is transferred to the liver to be synthesized to glucose, which can be used again by muscle.

cytoplasm The material in eukaryotic cells between the nucleus and the plasma membrane.

cytosol The fluid of the cytoplasm, which is an aqueous solution of electrolytes and enzymes.

decarboxylation The loss of a carbon atom in the form of CO_2.

digestion The processes in the gastrointestinal tract that break down large food molecules to smaller ones that pass through the intestinal membrane into the blood stream.

FAD A coenzyme (flavin adenine dinucleotide) for dehydrogenase enzymes that form carbon–carbon double bonds.

fermentation The anaerobic conversion of glucose by enzymes in yeast to yield alcohol and CO_2.

gluconeogenesis The synthesis of glucose from noncarbohydrate compounds.

glycogenesis The synthesis of glycogen from glucose molecules.

glycogenolysis The breakdown of glycogen into glucose molecules.

glycolysis The ten oxidation reactions of glucose that yield two pyruvate molecules.

metabolism All the chemical reactions in living cells that carry out molecular and energy transformations.

mitochondria The organelles of the cells where energy-producing reactions take place.

NAD$^+$ The hydrogen acceptor used in oxidation reactions that form carbon–oxygen double bonds.

oxidation The loss of hydrogen as hydrogen ions and electrons or the gain of oxygen by a substrate that is degraded to smaller molecules or a coenzyme.

reduction The gain of hydrogen ions and electrons or the loss of oxygen by a substrate or a coenzyme.

Additional Problems

23.59 What is meant by the term metabolism?

23.60 How do catabolic reactions differ from anabolic reactions?

23.61 What stage of metabolism involves the digestion of large food polymers?

23.62 What state of metabolism degrades monomers such as glucose into smaller molecules?

23.63 What type of cell has a nucleus?

23.64 What are the organelles in a cell?

23.65 What is the full name of ATP?

23.66 What is the full name of ADP?

23.67 Write an equation for the hydrolysis of ATP to ADP.

23.68 At the gym, you expend 300 kcal riding the stationary bicycle for 1 hr. How many moles of ATP will this require?

23.69 What is the full name of FAD?

23.70 What type of reaction uses FAD as the coenzyme?

23.71 What is the full name of NAD$^+$?

23.72 What type of reaction uses NAD$^+$ as the coenzyme?

23.73 Write the letters for the reduced forms of
a. FAD   **b.** NAD$^+$

23.74 What is the name of the vitamin in the structure of each of the following?
a. FAD   **b.** NAD$^+$ **c.** coenzyme A

23.75 How and where does lactose undergo digestion in the body? What are the products?

23.76 How and where does sucrose undergo digestion in the body? What are the products?

23.77 How do galactose and fructose enter glycolysis?

23.78 What is the general type of reaction that takes place in the digestion of carbohydrates?

23.79 What are the reactant and product of glycolysis?

23.80 What is the coenzyme used in glycolysis?

23.81 In glycolysis, which reactions involve phosphorylation and which reactions involve a direct substrate phosphorylation to generate ATP?

23.82 How do ADP and ATP regulate the glycolysis pathway?

23.83 What reaction and enzyme in glycolysis convert a hexose bisphosphate into two triose phosphates?

23.84 How does the investment and generation of ATP give a net gain of ATP for glycolysis?

23.85 What compound is converted to fructose-6-phosphate by phosphoglucoisomerase?

23.86 What product forms when glyceraldehyde-3-phosphate adds a phosphate group?

23.87 When is pyruvate converted to lactate in the body?

23.88 When pyruvate is used to form acetyl CoA or ethanol in fermentation, the product has only two carbon atoms. What happened to the third carbon?

23.89 How does phosphofructokinase regulate the rate of glycolysis?

23.90 How does pyruvate kinase regulate the rate of glycolysis?

23.91 When does the rate of glycogenolysis increase in the cells?

23.92 If glucose-1-phosphate is the product from glycogen, how does it enter glycolysis?

23.93 What is the end product of glycogenolysis in the liver?

23.94 What is the end product of glycogenolysis in skeletal muscle?

23.95 Why is glucose for blood glucose provided by glycogenolysis in the liver, but not in skeletal muscle?

23.96 When does the rate of glycogenesis increase in the cells?

23.97 How do the hormones insulin and glucagon affect the rate of glycogenesis, glycogenolysis, and glycolysis?

23.98 What is the function of gluconeogenesis?

23.99 Where does the Cori cycle operate?

23.100 Identify each of the following as part of glycolysis, glycogenolysis, glycogenesis, or gluconeogenesis:
a. Glycogen is broken down to glucose in the liver.
b. Glucose is synthesized from noncarbohydrate sources.
c. Glucose is degraded to pyruvate.
d. Glycogen is synthesized from glucose.

23.101 Indicate whether each of the following conditions would increase or decrease the rate of glycogenolysis in the liver:
a. low blood glucose level
b. secretion of insulin
c. secretion of glucagon
d. high levels of ATP

23.102 Indicate whether each of the following conditions would increase or decrease the rate of glycogenesis in the liver:
a. low blood glucose level
b. secretion of insulin
c. secretion of glucagon
d. high levels of ATP

23.103 Indicate whether each of the following conditions would increase or decrease the rate of gluconeogenesis:
a. high blood-glucose level
b. secretion of insulin
c. secretion of glucagon
d. high levels of ATP

23.104 Indicate whether each of the following conditions would increase or decrease the rate of glycolysis:
a. high blood-glucose level
b. secretion of insulin
c. secretion of glucagon
d. high levels of ATP

24 Metabolism and Energy Production

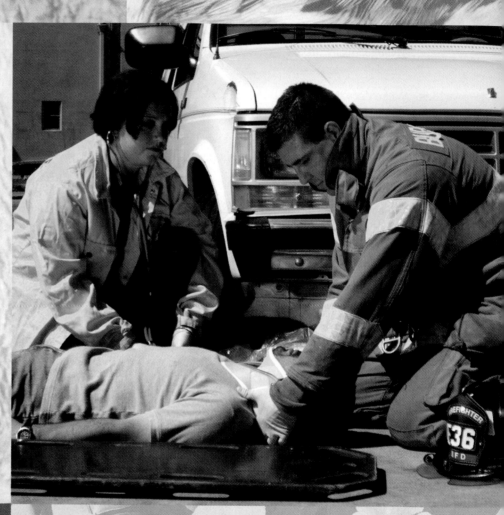

"I am trained in basic life support. I work with the ER staff to assist in patient care," says Mandy Dornell, emergency medical technician at Seaton Medical Center. "In the ER, I take vital signs, patient assessment, and do CPR. If someone has a motor vehicle accident, I may suspect a neck or back injury. Then I may use a backboard or a cervical collar, which prevents the patient from moving and causing further damage. When people have difficulty breathing, I insert an airway—nasal or oral—to assist ventilation. I also set up and monitor IVs, and I am trained in childbirth."

When someone is critically ill or injured, the quick reactions of emergency medical technicians (EMTs) and paramedics provide immediate medical care and transport to an ER or trauma center.

LOOKING AHEAD

the Chemistry place

www.chemplace.com/college

Visit the URL above or use the CD-ROM in the book for extra quizzing, interactive tutorials, and career resources.

In Chapter 23 we described the digestion of carbohydrates to glucose and the degradation of glucose to pyruvate. We saw that pyruvate is converted to lactate when no oxygen is available in the cell or to two-carbon acetyl CoA when oxygen is plentiful. Although glycolysis produces a small amount of ATP, most of the ATP in the cells is produced during the conversion of pyruvate, when oxygen is available in the cell. In a process known as *respiration,* oxygen is required to complete the oxidation of glucose to CO_2 and H_2O.

In the *citric acid cycle,* a series of metabolic reactions in the mitochondria oxidize acetyl CoA to carbon dioxide, which releases energy to produce NADH and $FADH_2$. These reduced coenzymes enter the *electron transport chain* or *respiratory chain* where they provide hydrogen ions and electrons that combine with oxygen (O_2) to form H_2O. The energy released during electron transport is used to synthesize ATP from ADP.

24.1 The Citric Acid Cycle

The **citric acid cycle** is a series of reactions that degrades acetyl CoA to yield CO_2 and energy, which is used to produce $NADH + H^+$ and $FADH_2$. (See Figure 24.1.) The citric acid cycle connects the products from stages 1 and 2 with the electron transport chain and the synthesis of ATP of stage 3. As a central pathway in metabolism, the citric acid cycle uses acetyl CoA from the degradation of carbohydrates as well as lipids and proteins.

The citric acid cycle is also called the tricarboxlyic acid cycle because citric acid is the product that forms in the first reaction. Although citric acid is present as citrate at cellular pH, its acid name is retained. The citric acid cycle is also known as the Krebs cycle for H. A. Krebs, who recognized it as the major pathway for the oxidation of carbohydrate.

Overview of the Citric Acid cycle

The fuel for the citric acid cycle is acetyl CoA, which is primarily produced from the oxidative decarboxylation of pyruvate. Acetyl CoA is also obtained from fats and proteins when they are used to generate energy, as we will see in Chapter 25. There are a total of 8 reactions in the citric acid cycle, which we can separate into two parts. In part 1, a two-carbon acetyl group bonds with four-carbon oxaloacetate to yield citrate. Then decarboxylation reactions remove two carbon atoms as CO_2, which produces a four-carbon compound. Then the reactions in part 2 convert the four-carbon compound back to oxaloacetate, which bonds with another acetyl CoA and goes through the cycle again. (See Figure 24.2.) In one turn of the citric acid cycle, four oxidation reactions provide hydrogen ions and electrons, which are used to reduce FAD and NAD^+ coenzymes.

LEARNING GOAL

Describe the oxidation of acetyl CoA in the citric acid cycle.

Stages of Metabolism

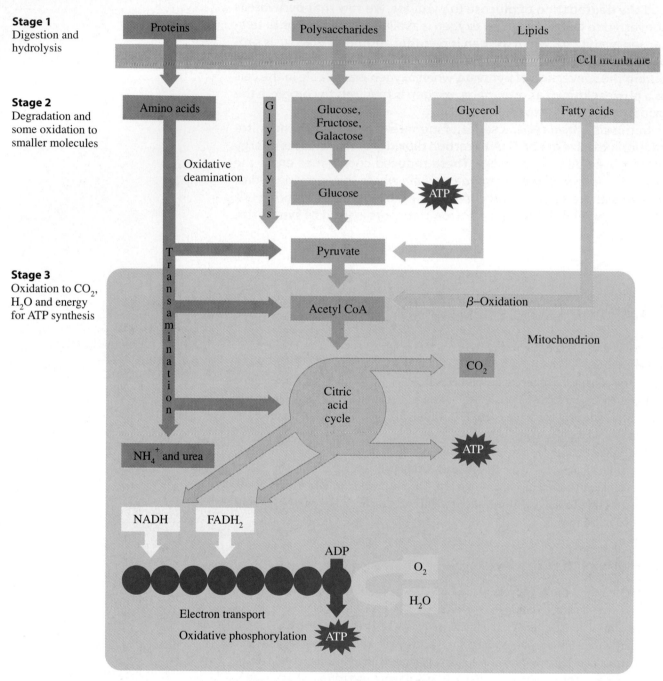

Figure 24.1 The citric acid cycle connects the catabolic pathways that begin with the digestion and degradation of foods in stages 1 and 2 with the oxidation of substrates in stage 3 that generates most of the energy for ATP synthesis.

Q *Why is the citric acid cycle called a central metabolic pathway?*

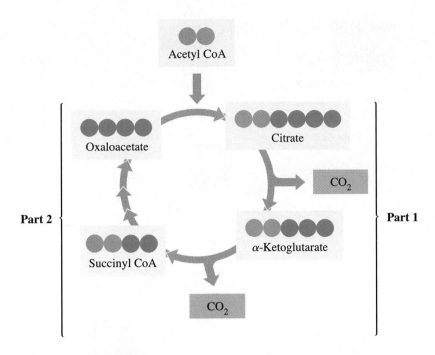

Figure 24.2 In part 1 of the citric acid cycle, two carbon atoms are removed as CO_2 from six-carbon citrate to give four-carbon succinyl CoA, which is converted in part 2 to four-carbon oxaloacetate.
Q *What is the difference in part 1 and part 2 of the citric acid cycle?*

Part 1   Decarboxylation Removes Two Carbon Atoms

Reaction 1   Formation of Citrate

In the first reaction, the two-carbon acetyl group in acetyl CoA bonds with the four-carbon oxaloacetate to yield citrate and CoA. The formation of citrate is an important regulation point in the cycle. (See Figure 24.3.)

$$CH_3-\overset{\overset{\displaystyle O}{\|}}{C}-S-CoA \;+\; \underset{\underset{\displaystyle COO^-}{\underset{\displaystyle |}{\underset{\displaystyle CH_2}{|}}}}{\overset{\overset{\displaystyle COO^-}{|}}{C}}{=}O \;+\; H_2O \;\xrightarrow[synthase]{Citrate}\; \underset{\underset{\displaystyle COO^-}{\underset{\displaystyle |}{\underset{\displaystyle CH_2}{|}}}}{HO-\overset{\overset{\displaystyle COO^-}{\overset{|}{CH_2}}}{C}-COO^-} \;+\; HS-CoA \;+\; H^+$$

Acetyl CoA   Oxaloacetate   Citrate

Reaction 2   Isomerization to Isocitrate

In order to continue oxidation, citrate undergoes isomerization to yield isocitrate. This is necessary because the secondary hydroxyl group (Chapter 14) in isocitrate can be oxidized in the next reaction, while the tertiary hydroxyl group in citrate cannot be oxidized. The isomerization actually consists of two steps. Citrate loses water (dehydration) to yield aconitate, which is rehydrated to form isocitrate. This reaction converts the tertiary hydroxyl ($-OH$) group to a secondary hydroxyl group, because a tertiary hydroxyl ($-OH$) group cannot be oxidized further.

Citrate $\xrightarrow[\text{Aconitase}]{H_2O}$ Aconitate $\xrightarrow[\text{Aconitase}]{H_2O}$ Isocitrate

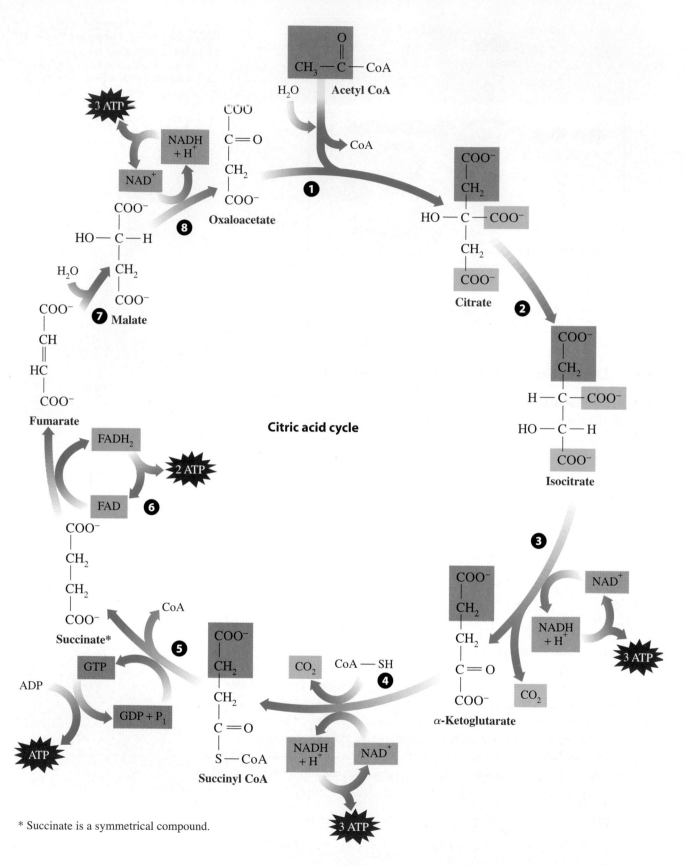

Figure 24.3 In the citric acid cycle, oxidation reactions produce two CO_2, reduced coenzymes NADH and $FADH_2$, and regenerate oxaloacetate.

Q *How many reactions in the citric acid cycle produce a reduced coenzyme?*

Reaction 3   First Oxidative Decarboxylation (CO_2)

This is the first time in the citric acid cycle that both an oxidation and a decarboxylation occur together. The oxidation converts the hydroxyl group to a ketone and the decarboxylation removes a carbon as a CO_2 molecule. That carbon is from a carboxylate group (COO^-) in the original oxaloacetate and not from the acetyl group. The loss of CO_2 shortens the carbon chain to yield a five-carbon α-ketoglutarate. The energy from the oxidation is used to transfer hydrogen ions and electrons to NAD^+. We can summarize the reactions as follows:

1. The hydroxyl group ($—OH$) is oxidized to a ketone ($C{=}O$).
2. NAD^+ is reduced to yield NADH.
3. A carboxylate group (COO^-) is removed as CO_2.

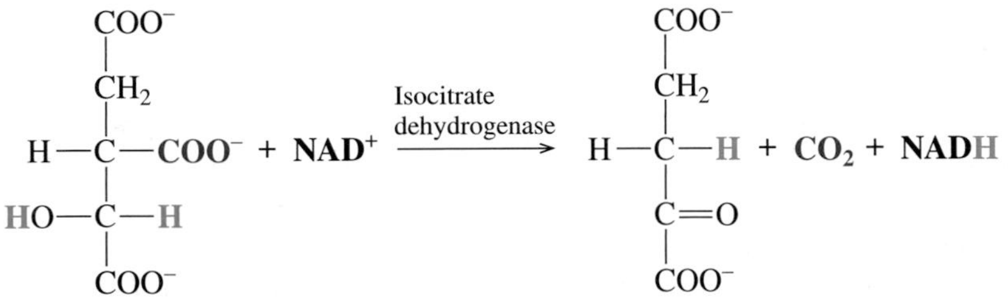

Physical Therapist

"We do all kinds of activities that are typical of the things kids would do," says Helen Tong, physical therapist. "For example, we play with toys to help improve motor control. When a child can't do something, we use adaptive equipment to help him or her do that activity in a different way, which still allows participation. In school, we learn how the body works, and why it does not work. Then we can figure out what to do to help a child learn new skills. For example, this child with Rett syndrome has motor difficulties. Although she has difficulty talking, she does amazing work at a computer using the switch. There has been a real growth in assisted technology for children, and it has changed our work tremendously."

$$
\begin{array}{c}
COO^- \\
| \\
CH_2 \\
| \\
H—C—COO^- \\
| \\
HO—C—H \\
| \\
COO^-
\end{array}
\; + \; NAD^+
\quad
\xrightarrow[\text{dehydrogenase}]{\text{Isocitrate}}
\quad
\begin{array}{c}
COO^- \\
| \\
CH_2 \\
| \\
H—C—H \\
| \\
C{=}O \\
| \\
COO^-
\end{array}
\; + \; CO_2 \; + \; NADH
$$

Isocitrate　　　　　　　　　　　　　　　　　α-Ketoglutarate

Reaction 4   Second Oxidative Decarboxylation (CO_2)

In this reaction, a second CO_2 is removed as α-ketoglutarate undergoes oxidative decarboxylation. The resulting four-carbon group combines with coenzyme A to form succinyl CoA and hydrogen ions and electrons are transferred to NAD^+. The two reactions that occur are as follows:

1. A second carbon is removed as CO_2.
2. NAD^+ is reduced to yield NADH.

$$
\begin{array}{c}
COO^- \\
| \\
CH_2 \\
| \\
CH_2 \\
| \\
C{=}O \\
| \\
COO^-
\end{array}
\; + \; NAD^+ \; + \; CoA{—}SH
\quad
\xrightarrow[\text{dehydrogenase}]{\alpha\text{-Ketoglutarate}}
\quad
\begin{array}{c}
COO^- \\
| \\
CH_2 \\
| \\
CH_2 \\
| \\
C{=}O \\
| \\
S{—}CoA
\end{array}
\; + \; CO_2 \; + \; NADH
$$

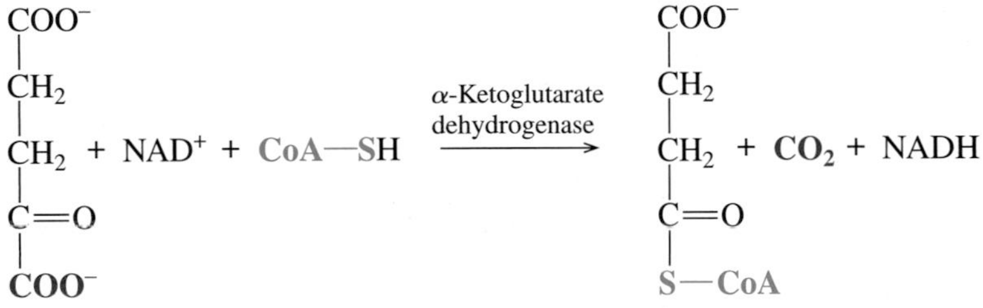

α-Ketoglutarate　　　　　　　　　　　　　Succinyl CoA

Part 2

Reaction 5   Hydrolysis of Succinyl CoA

The energy released by the hydrolysis of the thioester bond in succinyl CoA is used to add a phosphate group (P_i) directly to GDP (guanosine diphosphate). The products are succinate and GTP, which is a high-energy compound similar to ATP.

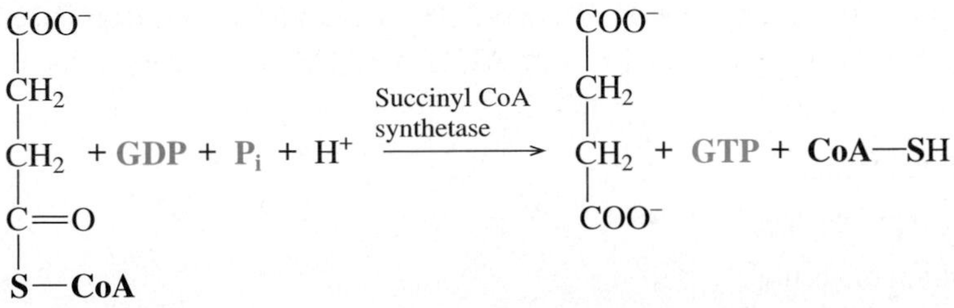

$$
\begin{array}{c}
COO^- \\
| \\
CH_2 \\
| \\
CH_2 \\
| \\
C{=}O \\
| \\
S{—}CoA
\end{array}
\; + \; GDP \; + \; P_i \; + \; H^+
\quad
\xrightarrow[\text{synthetase}]{\text{Succinyl CoA}}
\quad
\begin{array}{c}
COO^- \\
| \\
CH_2 \\
| \\
CH_2 \\
| \\
COO^-
\end{array}
\; + \; GTP \; + \; CoA{—}SH
$$

Succinyl CoA　　　　　　　　　　　　　　Succinate

The hydrolysis of GTP is used to add a phosphate group to ADP, which regenerates GDP for the citric acid cycle. This is the only time in the citric acid cycle that a direct substrate phosphorylation is used to produce ATP.

$$GTP + ADP \longrightarrow GDP + ATP$$

Reaction 6   Dehydrogenation of Succinate

In this oxidation reaction, hydrogen is removed from two carbon atoms in succinate, which produces fumarate, a compound with a trans double bond. This is the only place in the citric acid cycle where FAD is reduced to $FADH_2$.

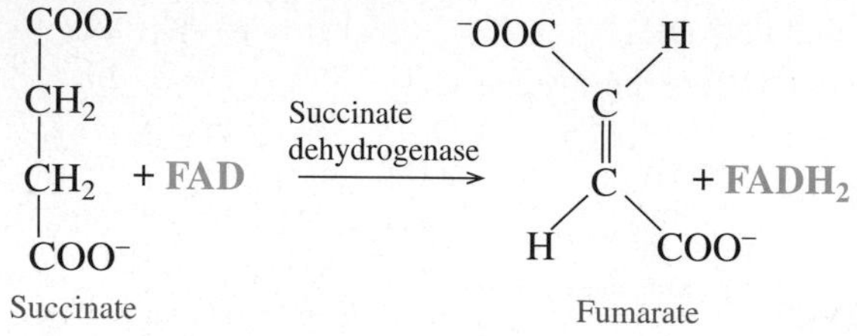

Succinate dehydrogenase

Succinate   Fumarate

Reaction 7   Hydration

In a hydration reaction, water adds to the double bond of fumarate to yield malate.

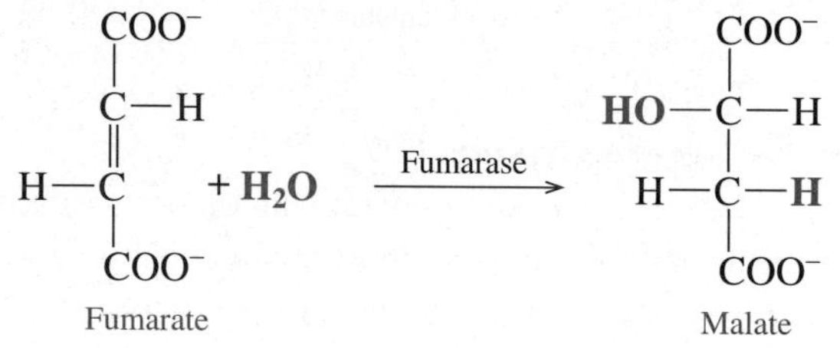

Fumarase

Fumarate   Malate

Reaction 8   Dehydrogenation Forms Oxaloacetate

In the last step of the citric acid cycle, the hydroxyl (—OH) group in malate is oxidized to yield oxaloacetate, which has a ketone group. The coenzyme NAD^+ is reduced to $NADH + H^+$.

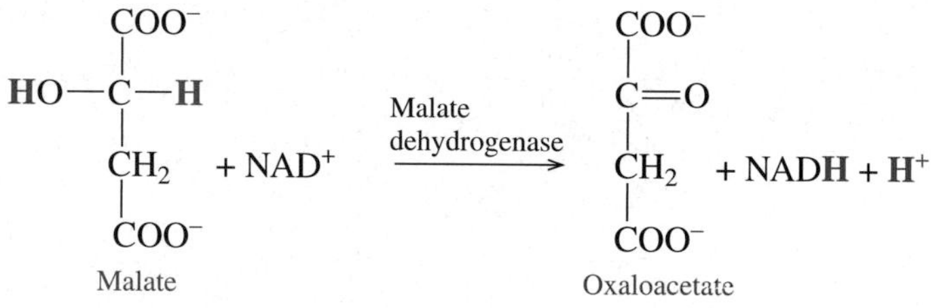

Malate dehydrogenase

Malate   Oxaloacetate

Summary of Products from the Citric Acid Cycle

We have seen that the citric acid cycle begins when a two-carbon acetyl group from acetyl CoA combines with oxaloacetate. In part 1 of the cycle, two carbon atoms are removed to yield two CO_2 and a four-carbon compound that undergoes more reactions in part 2 to regenerate oxaloacetate. In four oxidation reactions, energy is released that reduces three NAD^+ and one FAD. In one reaction, GTP produced by a direct phosphorylation is used to form ATP from ADP. A summary of the products from one turn of the citric acid cycle is as follows:

2 CO_2 molecules

3 NADH molecules

1 $FADH_2$ molecule

1 GTP molecule used to form ATP

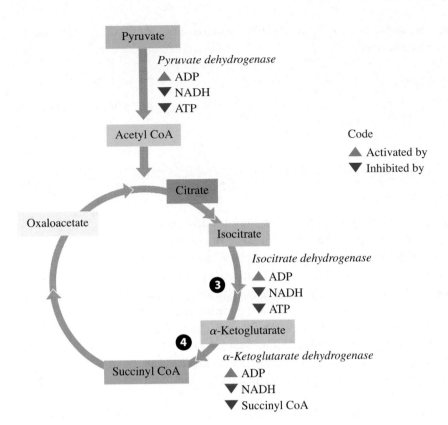

Figure 24.4 High levels of ADP activate enzymes for the production of acetyl CoA and the citric acid cycle, whereas high levels of ATP, NADII, and succinyl CoA inhibit enzymes in the citric acid cycle.
Q *How do high levels of ATP affect the rate of the citric acid cycle?*

We can write the overall chemical equation for one turn of the citric acid cycle as follows:

$$\text{Acetyl CoA} + 3\,\text{NAD}^+ + \text{FAD} + \text{GDP} + \text{P}_i + 2\text{H}_2\text{O} \longrightarrow$$

$$2\text{CO}_2 + 3\,\text{NADH} + 2\text{H}^+ + \text{FADH}_2 + \text{HS—CoA} + \text{GTP}$$

Regulation of Citric Acid Cycle

The primary function of the citric acid cycle is to produce high-energy compounds for ATP synthesis. When the cell needs energy, low levels of ATP stimulate the conversion of pyruvate to acetyl CoA, the fuel for the citric acid cycle. When ATP and NADH levels are high, there is a decrease in the production of acetyl CoA from pyruvate.

In the citric acid cycle, the enzymes that catalyze reactions 3 and 4 respond to allosteric activation and inhibition (Chapter 21). In reaction 3, isocitrate dehydrogenase is activated by high levels of ADP and inhibited by high levels of ATP and NADH. In reaction 4, α-ketoglutarate dehydrogenase is activated by high levels of ADP and inhibited by high levels of NADH and succinyl CoA. (See Figure 24.4.)

SAMPLE PROBLEM 24.1

Citric Acid Cycle

When one acetyl CoA completes the citric acid cycle, how many of each of the following are produced?

a. NADH   **b.** ketone group   **c.** CO_2

Solution

a. One turn of the citric acid cycle produces three molecules of NADH.

b. Two ketone groups form when the secondary alcohol groups in isocitrate and malate are oxidized by NAD^+.

c. Two molecules of CO_2 are produced by the decarboxylation of isocitrate and α-ketoglutarate.

Study Check

What substance is a substrate in the first reaction of the citric acid cycle and a product in the last reaction?

QUESTIONS AND PROBLEMS

The Citric Acid Cycle

24.1 What other names are used for the citric acid cycle?

24.2 What compounds are needed to start the citric acid cycle?

24.3 What are the products from one turn of the citric acid cycle?

24.4 What compound is regenerated in each turn of the citric acid cycle?

24.5 Which reactions of the citric acid cycle involve oxidative decarboxylation?

24.6 Which reactions of the citric acid cycle involve a dehydration reaction?

24.7 Which reactions of the citric acid cycle reduce NAD^+?

24.8 Which reactions of the citric acid cycle reduce FAD?

24.9 Where does a direct substrate phosphorylation occur?

24.10 What is the total NADH and total $FADH_2$ produced in one turn of the citric acid cycle?

24.11 Refer to the diagram of the citric acid cycle to answer each of the following:
a. What are the six-carbon compounds?
b. How is the number of carbon atoms decreased?
c. What are the five-carbon compounds?
d. Which reactions are oxidation reactions?
e. In which reactions are secondary alcohols oxidized?

24.12 Refer to the diagram of the citric acid cycle to answer each of the following:
a. What is the yield of CO_2 molecules?
b. What are the four-carbon compounds?
c. What is the yield of GTP molecules?
d. What are the decarboxylation reactions?
e. Where does a hydration occur?

24.13 Indicate the name of the enzyme for each of the following reactions in the citric acid cycle:
a. joins acetyl CoA to oxaloacetate
b. forms a carbon–carbon double bond
c. adds water to fumarate

24.14 Indicate the name of the enzyme for each of the following reactions in the citric acid cycle:
a. isomerizes citrate
b. oxidizes and decarboxylates α-ketoglutarate
c. adds P_i to GDP

24.15 State the acceptor for hydrogen or phosphate in each of the following reactions:
a. isocitrate $\longrightarrow$ α-ketoglutarate
b. succinyl CoA $\longrightarrow$ succinate   c. succinate $\longrightarrow$ fumarate

24.16 State the acceptor for hydrogen or phosphate in each of the following reactions:
a. malate $\longrightarrow$ oxaloacetate
b. α-ketoglutarate $\longrightarrow$ succinyl CoA
c. pyruvate $\longrightarrow$ acetyl CoA

24.17 What enzymes in the citric acid cycle are allosteric enzymes?
24.18 How does NADH affect the rate of the citric acid cycle?
24.19 How do high levels of ADP affect the rate of the citric acid cycle?
24.20 Why does the rate of the oxidation of pyruvate affect the rate of the citric acid cycle?

24.2 Electron Carriers

LEARNING GOAL
Describe the electron carriers involved in electron transport.

At this point, the metabolic cycles of glycolysis, oxidation of two pyruvate, and the citric acid cycle for two acetyl CoA would produce four ATP along with ten NADH and two FADH$_2$ from the degradation of glucose.

Glycolysis: 2 NADH and 2 ATP
Oxidation of 2 pyruvate: 2 NADH
Citric acid cycle (2 acetyl CoA):   6 NADH, 2 FADH$_2$, and 2 ATP

Now we will see how the oxidation of these reduced coenzymes provides the energy for the synthesis of considerably more ATP. In the **electron transport chain** or *respiratory chain*, hydrogen ions and electrons from NADH and FADH$_2$ are passed from one electron acceptor to the next until they combine with oxygen to form H$_2$O. The electron acceptors in this transport system are known as **electron carriers.** The energy released during electron transport is used to synthesize ATP from ADP and P$_i$, a process called *oxidative phosphorylation*. As long as oxygen is available for the mitochondria in the cell, electron transport and oxidative phosphorylation function to produce most of the ATP energy manufactured in the cell.

Oxidation and Reduction of Electron Carriers

The electron carriers in the electron transport include flavins, iron–sulfur proteins, coenzyme Q, and cytochromes. Each type of electron carrier contains a group or ion that is reduced and then oxidized as the electrons are accepted and then passed on. We can illustrate this process with two electron carriers A and B. As hydrogen ions and electrons are transferred from the reduced AH$_2$ to the oxidized carrier B, reduced BH$_2$ is formed along with the oxidized form of carrier A.

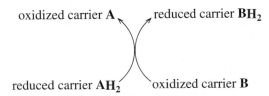

There are four types of electron carriers that make up the electron transport system.

1. **FMN (flavin mononucleotide)** is a coenzyme derived from riboflavin (vitamin B$_2$). FMN contains a flavin ring system that is also found in FAD. In riboflavin, the ring system is attached to ribitol, the sugar alcohol of ribose. (See Figure 24.5.) The reduced product is FMNH$_2$.

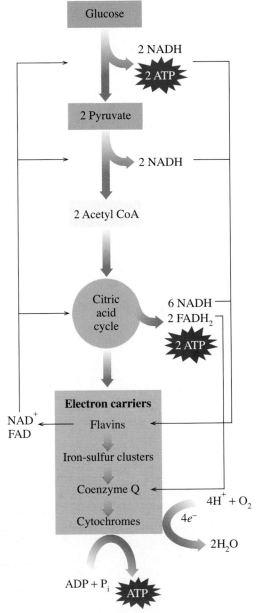

Figure 24.5 The electron carrier FMN consists of a flavin ring system containing the reactive center, ribitol, and a phosphate group.

Q *What part of the FMN molecule is reduced when hydrogen ions and electrons are accepted?*

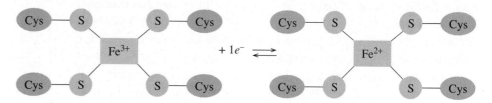

FMN (flavin in mononucletotide)   **FMNH₂**

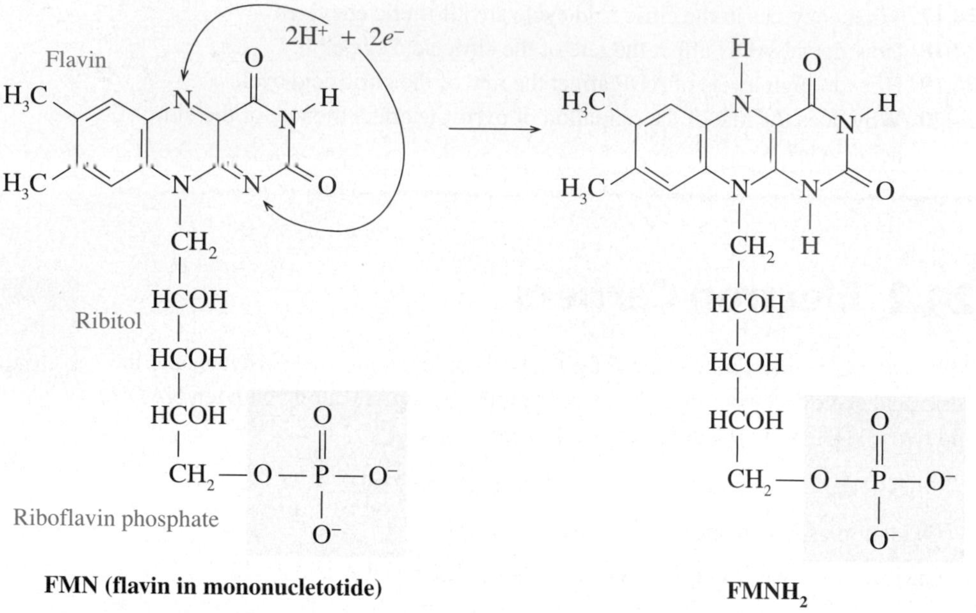

Figure 24.6 In a typical iron–sulfur cluster, an iron ion bonds to sulfur atoms in the thiol (—SH) groups of four cysteine groups in proteins.

Q *In an iron–sulfur cluster, what are the ionic charges of the oxidized and reduced iron ions?*

Typical Fe-S cluster

2. **Fe–S** clusters is the name given to a group of iron–sulfur proteins that contain iron–sulfur clusters embedded in the proteins of the electron transport chain. The clusters contain iron ions, inorganic sulfides, and several cysteine groups. The iron in the clusters is reduced to Fe^{2+} and oxidized to Fe^{3+} as electrons are accepted and lost. (See Figure 24.6.)

3. **Coenzyme Q (Q or CoQ)** is derived from quinone, which is a six-carbon cyclic compound with two double bonds and two keto groups attached to a long carbon chain. (See Figure 24.7.) Coenzyme Q is reduced when the keto groups of quinone accept hydrogen ions and electrons.

4. **Cytochromes (cyt)** are proteins that contain an iron ion in a heme group. The different cytochromes are indicated by the letters following the abbreviation for cytochrome (cyt): cyt *b*, cyt c_1, cyt *c*, cyt *a*, and cyt a_3. In each cytochrome, the Fe^{3+} accepts a single electron to form Fe^{2+}, which is oxidized back to Fe^{3+} when the electron is passed to the next cytochrome. (See Figure 24.8.)

$$Fe^{3+} + 1e^- \rightleftharpoons Fe^{2+}$$

SAMPLE PROBLEM 24.2

Electron Carriers

Give the abbreviation for each of the following carriers:

a. the oxidized form of flavin mononucleotide

b. the reduced form of coenzyme Q

Solution

a. FMN   **b.** QH_2

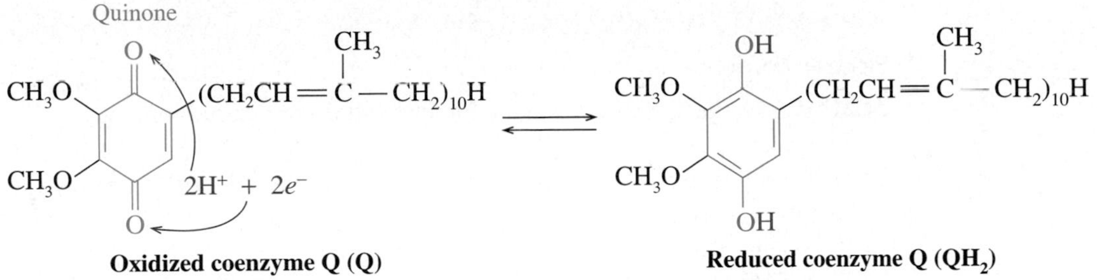

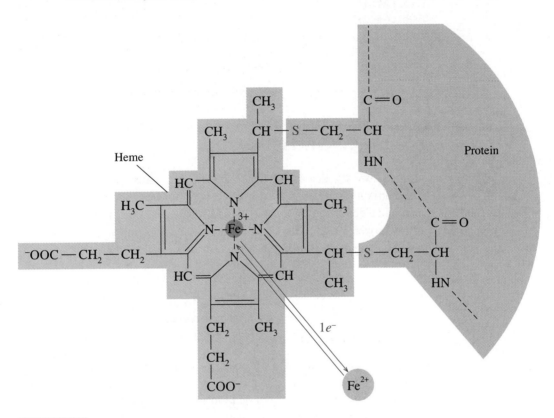

Figure 24.7 The electron carrier coenzyme Q accepts electrons from FADH$_2$ and FMNH$_2$ and passes them to the cytochromes.
Q *How does reduced coenzyme Q compare to the oxidized form?*

Figure 24.8 The iron-containing proteins known as cytochromes are identified as *b*, *c*, *c*$_1$, *a*, and *a*$_3$.
Q *What are the reduced and oxidized forms of the cytochromes?*

Study Check

What is the oxidized form of iron in cyt *a*$_3$?

SAMPLE PROBLEM 24.3

Oxidation and Reduction

Identify the following steps in the electron transport chain as oxidation or reduction:

a. FMN + 2H$^+$ + 2e^- $\longrightarrow$ FMNH$_2$
b. Cyt *c* (Fe^{2+}) $\longrightarrow$ cyt *c* (Fe^{3+}) + e^-

Solution

a. The gain of hydrogen is reduction.   **b.** The loss of electrons is oxidation.

Study Check

Identify each of the following as oxidation or reduction:

a. QH$_2$ $\longrightarrow$ Q   **b.** Cyt *b* (Fe^{3+}) $\longrightarrow$ cyt *b* (Fe^{2+})

QUESTIONS AND PROBLEMS

Electron Carriers

24.21 Is cyt b (Fe^{3+}) the abbreviation for the oxidized or reduced form of cytochrome b?

24.22 Is $FMNH_2$ the abbreviation for the oxidized or reduced form of flavin mononucleotide?

24.23 Identify the following as oxidation or reduction:
a. $FMNH_2 \longrightarrow FMN + 2H^+ + 2e^-$
b. $Q + 2H^+ + 2e^- \longrightarrow QH_2$

24.24 Identify the following as oxidation or reduction:
a. Cyt c (Fe^{3+}) $+ e^- \longrightarrow$ cyt c (Fe^{2+})
b. Fe^{2+} S cluster $\longrightarrow Fe^{3+}$S cluster $+ e^-$

LEARNING GOAL

Describe the role of the electron carriers in electron transport.

24.3 Electron Transport

In Chapter 23, we saw that a mitochondrion consists of inner and outer membranes with the matrix located between. Along the highly folded inner membrane are the enzymes and electron carriers required for electron transport. When these membranes are broken up, four distinct protein complexes are obtained. Within each complex are some of the electron carriers needed for electron transport.

Complex I	NADH dehydrogenase
Complex II	Succinate dehydrogenase
Complex III	Coenzyme Q-cytochrome c reductase
Complex IV	Cytochrome c oxidase

Two electron carriers, coenzyme Q and cytochrome c, are not firmly attached to the membrane. They function as mobile carriers shuttling electrons between the protein complexes that are tightly bound to the membrane. The mobile carrier coenzyme Q moves electrons from complexes I and II to complex III. The other mobile carrier, cytochrome c, transfers electrons from complex III to complex IV. (See Figure 24.9.)

Complex 1   NADH Dehydrogenase

At complex I, all of the NADH generated in the cell transfers hydrogen ions and electrons to the electron carrier FMN. The reduced $FMNH_2$ forms, while NADH is reoxidized to NAD^+, which returns to oxidative pathways such as the citric acid cycle to oxidize more substrates.

$$NADH + H^+ + FMN \longrightarrow NAD^+ + FMNH_2$$

Within complex I, the electrons are transferred to iron–sulfur (Fe–S) clusters and then to coenzyme (Q).

$$FMNH_2 + Q \longrightarrow QH_2 + FMN$$

The overall reaction sequence in complex I can be written as follows:

$$NADH + H^+ + Q \longrightarrow QH_2 + NAD^+$$

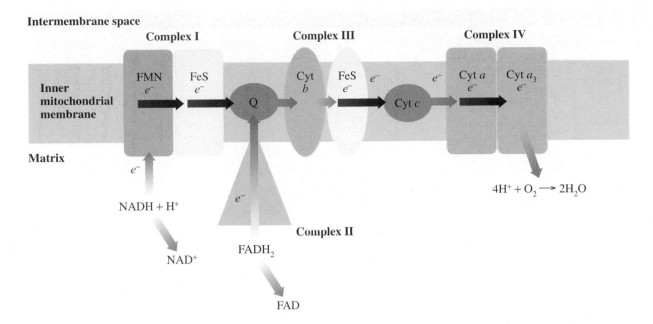

Complex II   Succinate Dehydrogenase

Complex II is specifically used when $FADH_2$ is generated by the conversion of succinate to fumarate in the citric acid cycle. The electrons from $FADH_2$ are transferred to coenzyme Q to yield QH_2. Because complex II is at a lower energy level than complex I, the electrons from $FADH_2$ enter electron transport at a lower energy level than those from NADH.

$$FAD\textbf{H}_2 \ + \ Q \ \longrightarrow \ FAD \ + \ Q\textbf{H}_2$$

Complex III   Coenzyme Q–Cytochrome *c* Reductase

The mobile carrier QH_2 transfers the electrons it has collected from NADH and $FADH_2$ to an iron–sulfur (Fe–S) cluster, and then to cytochrome *b*, the first cytochrome in complex III.

$$QH_2 \ + \ 2 \ cyt \ b \ (Fe^{3+}) \ \longrightarrow \ Q \ + \ 2 \ cyt \ b \ (Fe^{2+}) \ + \ 2H^+$$

From cyt *b*, the electron is transferred to an Fe–S cluster and then to cytochrome c_1 and then to cytochrome *c*. Each time an Fe^{3+} ion accepts an electron, it is reduced to Fe^{2+}, and then oxidized back to Fe^{3+} as the electron is passed on along the chain. Cytochrome *c* is another mobile carrier, it moves the electron from complex III to complex IV.

Complex IV   Cytochrome *c* Oxidase

At complex IV, electrons are transferred from cytochrome *c* to cytochrome *a*, and then to cytochrome a_3, the last cytochrome. In the final step of electron transport, electrons and hydrogen ions combine with oxygen (O_2) to form water.

$$4H^+ \ + \ 4e^- \ + \ O_2 \ \longrightarrow \ 2H_2O$$

Figure 24.9 Most of the electron carriers in the electron transport chain are found in protein complexes bound to the inner membrane of the mitochondria. Two are mobile carriers that carry electrons between the protein complexes.
Q *What is the function of the electron carriers coenzyme Q and cytochrome c?*

Toxins: Inhibitors of Electron Transport

There are several substances that inhibit the electron carriers in the electron transport chain. Rotenone, a product from a plant root used in South America to poison fish, blocks electron transport between the NADH dehydrogenase (complex I) and coenzyme Q. The barbiturates amytal and demerol also inhibit the NADH dehydrogenase complex. Another inhibitor is the antibiotic antimycin A, which blocks the flow of electrons between cytochrome b and cytochrome c_1 (complex III). Another group of compounds including cyanide (CN^-) and carbon monoxide inhibit cytochrome c oxidase (complex IV). The toxic nature of these compounds makes it clear that an organism relies heavily on the process of electron transport.

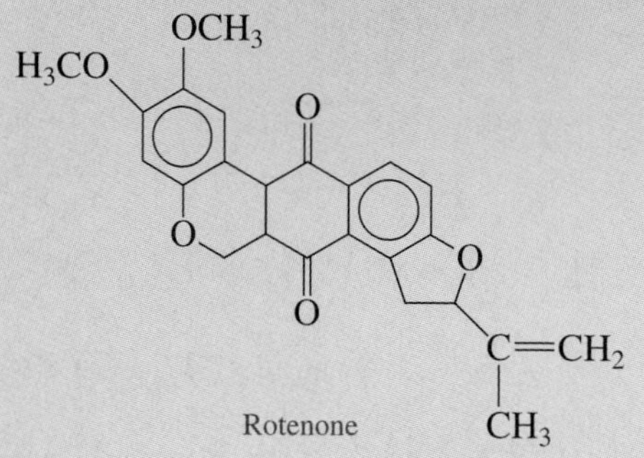

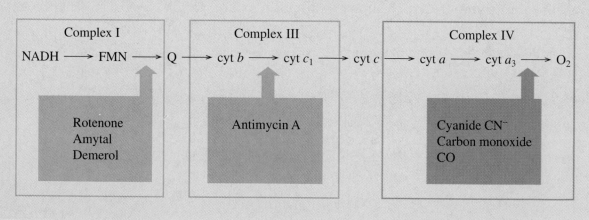

Rotenone

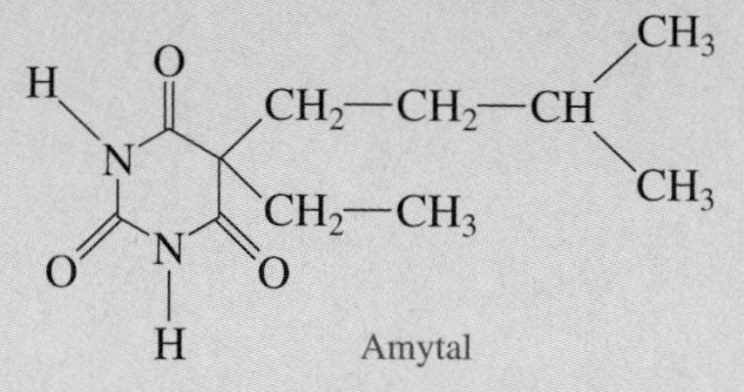

Amytal

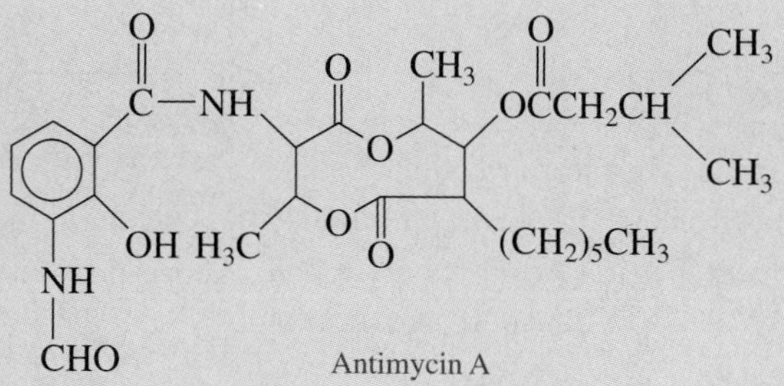

Antimycin A

When an inhibitor blocks a step in the electron transport chain, the carriers preceding that step are unable to transfer electrons and remain in their reduced forms. All the carriers after the blocked step remain oxidized without a source of electrons. Thus, any of these inhibitors shut down the flow of electrons through the electron transport chain.

Complex I	Complex III	Complex IV
NADH → FMN → Q	→ cyt b → cyt c_1 →	cyt c → cyt a → cyt a_3 → O_2
Rotenone Amytal Demerol	Antimycin A	Cyanide CN^- Carbon monoxide CO

SAMPLE PROBLEM 24.4

Electron Transport

Identify the following electron carriers that are mobile carriers.

a. Cyt c **b.** FMN   **c.** Fe–S clusters   **d.** Q

Solution

a. and **d.** Cyt c and Q are mobile carriers.

Study Check

What is the final substance that accepts electrons in the electron transport chain?

QUESTIONS AND PROBLEMS

Electron Transport

24.25 What reduced coenzymes provide the electrons for electron transport?

24.26 What happens to the energy level as electrons are passed along the electron transport chain?

24.27 Arrange the following in the order they appear in electron transport: cytochrome c, cytochrome b, $FADH_2$, and coenzyme Q.

24.28 Arrange the following in the order they appear in electron transport: O_2, NAD^+, cytochrome a_3, and FMN.

24.29 How are electrons carried from complex I to complex III?

24.30 How are electrons carried from complex III to complex IV?

24.31 How is NADH oxidized in electron transport?

24.32 How is $FADH_2$ oxidized in electron transport?

24.33 Complete the following reactions in electron transport:
a. $NADH + H^+ + \underline{\hspace{2cm}} \longrightarrow \underline{\hspace{1.5cm}} + FMNH_2$
b. $QH_2 + 2\ cyt\ b\ (Fe^{3+}) \longrightarrow \underline{\hspace{1.5cm}} + \underline{\hspace{1.5cm}} + 2H^+$

24.34 Complete the following reactions in electron transport:
a. $Q + \underline{\hspace{2cm}} \longrightarrow \underline{\hspace{1.5cm}} + FAD$
b. $2\ cyt\ a\ (Fe^{3+}) + 2\ cyt\ a_3\ (Fe^{2+}) \longrightarrow \underline{\hspace{1.5cm}} + \underline{\hspace{1.5cm}}$

24.4 Oxidative Phosphorylation and ATP

LEARNING GOAL

Describe the process of oxidative phosphorylation in ATP synthesis.

We have seen that energy is generated when electrons from the oxidation of substrates flow through the electron transport chain. Now we will look at how that energy is used coupled with the production of ATP for the cell, which is a process called **oxidative phosphorylation.**

Chemiosmotic Model

In 1978, Peter Mitchell received the Nobel Prize for his theory called the **chemiosmotic model,** which links the energy from electron transport to a proton gradient that drives the synthesis of ATP. In this model, three of the complexes (I, III, and IV) extend through the inner membrane with one end of each complex in the matrix and the other end in the intermembrane space. In the chemiosmotic model, each of these complexes act as a **proton pump** by pushing protons (H^+) out of the matrix and into the intermembrane space. This increase in protons in the intermembrane space lowers the pH and creates a proton gradient. Because protons are positively charged, both the lower pH and the electrical charge of the proton gradient make it an electrochemical gradient. (See Figure 24.10.)

To equalize the pH between the intermembrane space and the matrix, there is a tendency by the protons to return the matrix. However, protons cannot diffuse

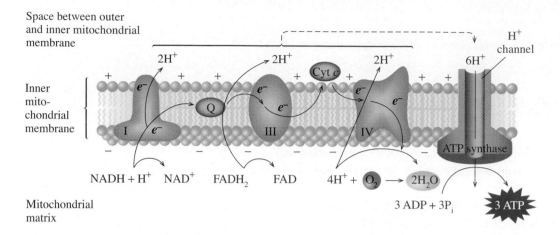

through the inner membrane. The only way protons can return to the matrix is to pass through a protein complex called **ATP synthase.** As the protons flow through ATP synthase, energy generated from the proton gradient is used to drive the ATP synthesis. Thus the process of oxidative phosphorylation couples the energy from electron transport to the synthesis of ATP from ADP and P_i.

Figure 24.10 In the electron transport chain, protein complexes oxidize and reduce coenzymes to provide electrons and protons that move into the intermembrane space where they create a proton gradient that drives ATP synthesis.

Q *What is the major source of NADH for the electron transport chain?*

$$ADP + P_i + energy \xrightarrow{\text{ATP synthase}} ATP$$

Details of ATP Synthase

ATP synthase consists of two enzyme complexes. (See Figure 24.11.) The F_0 complex contains the channel for the return of protons to the matrix. The F_1 section consists of a center subunit (γ) and three surrounding protein subunits, which have three active sites with different shapes or conformations known as loose (L), tight (T), and open (O). As the protons flow through the F_0 channel, the energy released turns the center subunit (γ). We might think of the flow of protons as a stream or river that turns a water wheel. As the center unit supplies energy to the three active sites, their shapes change. ATP synthesis begins when the substrates ADP and P_i enter a loose (L) active site. As the loose (L) site shape converts to a tight (T) shape, ATP is formed. However, the ATP is tightly bound to the active site (T). When energy turns the γ unit, the tight site (T) converts to an open (O) site, which releases the ATP. The open site will convert to an L site and accept new substrates ADP and P_i. According to Paul Boyer, who earned the 1997 Nobel Prize in chemistry for his work on ATP synthase, the formation of ATP is spontaneous, whereas its release from the synthase requires the energy supplied by the proton gradient. The following steps summarize the changes in the active site in the synthesis of ATP. (See Figure 24.12.)

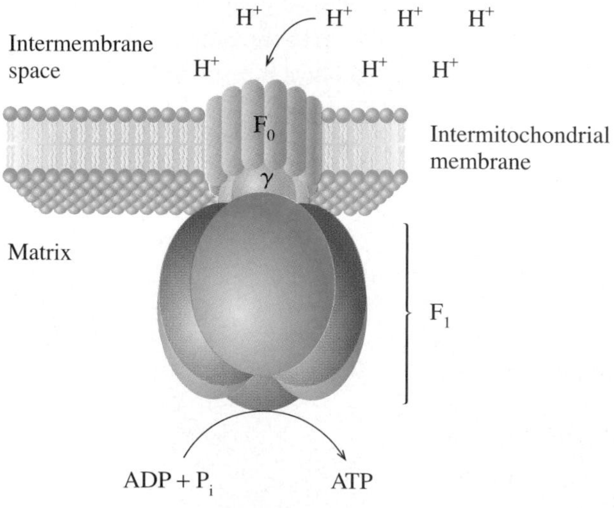

Figure 24.11 ATP synthase consists of two protein complexes. An F_0 section contains the channel for proton flow, and an F_1 section uses the energy from the proton gradient to drive the synthesis of ATP.

Q *What are the functions of the F_0 and F_1 sections of ATP synthase?*

Step 1   The synthesis of ATP begins when ADP and P_i bind to a loose (L) site.

Step 2   Energy from the γ center now converts all the sites: an L site becomes a tight (T) site, an O site changes to an L site, and a T site changes to an O site.

Step 3   Now the ADP and P_i—which are now in a tight (T) site—spontaneously form ATP, which is held tightly in this active site.

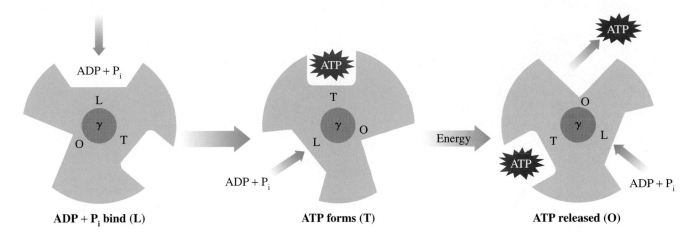

ADP + P$_i$ bind (L)   **ATP forms (T)**   **ATP released (O)**

Step 4   Another input of energy from proton flow changes all the active sites again. The ATP site changes to an open (O) site, which has little affinity for ATP, and ATP is released. This site is open once again and ready to accept more ADP and P$_i$.

In summary, the energy from protons flowing through F$_0$ turns the center γ unit in F$_1$, which causes a change in the shapes of the active sites from loose (L), where ADP, and P$_i$ bind, to tight (T) where ATP forms, and to open (O), which releases ATP. This process of oxidative phosphorylation continues as long as energy from the electron transport system is generated, which pumps the protons into the inner membrane space and produces the proton gradient to fuel ATP synthase.

Figure 24.12 In the F$_1$ ATP synthase, ATP is formed when an L active site containing ADP and P$_i$ converts to a T site. When energy from the proton flow in F$_0$ changes the site to an open (O) site, ATP is released.
Q *What shape of an active site in F$_1$ ATP synthase accepts the substrates, and which shape releases the ATP?*

Electron Transport and ATP Synthesis

We have seen that oxidative phosphorylation couples the energy from electron transport with the synthesis of ATP. Because NADH enters the electron transport chain at complex I, energy is released from the oxidation of NADH to synthesize three ATP. However, FADH$_2$, which enters the chain at a lower energy level at complex II, provides energy to drive the synthesis of only two ATP. (See Figure 24.13.) The overall equation for the oxidation of NADH and FADH$_2$ can be written as follows:

$$\boxed{\textbf{NADH + H}^+} + \tfrac{1}{2}O_2 + 3\,ADP + 3P_i \longrightarrow NAD^+ + H_2O + \boxed{\textbf{3 ATP}}$$

$$\boxed{\textbf{FADH}_2} + \tfrac{1}{2}O_2 + 2\,ADP + 2P_i \longrightarrow FAD + H_2O + \boxed{\textbf{2 ATP}}$$

Regulation of the Electron Transport Chain and Oxidative Phosphorylation

The electron transport chain is regulated by the availability of ADP, P$_i$, oxygen (O$_2$), and NADH. Low levels of any of these compounds will decrease the activity of the electron transport chain and formation of ATP. When a cell is active and ATP is consumed rapidly, the elevated levels of ADP will activate the synthesis of ATP. Therefore, the activity of the electron transport chain is strongly dependent on the levels of ADP for ATP synthesis.

High energy

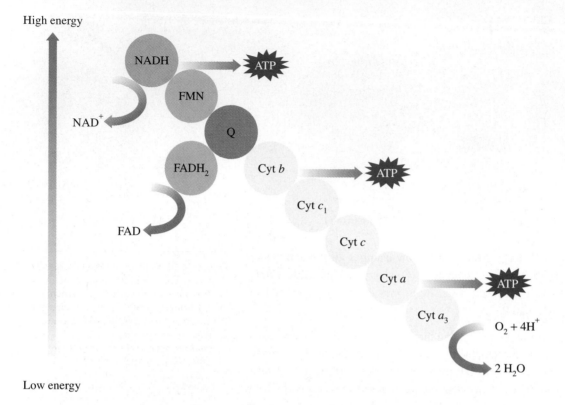

Low energy

Figure 24.13 As the energy levels decrease in the flow of electrons along the major electron carriers, three of the electron transfers release sufficient energy to drive ATP synthesis.
Q *Why does electron transfer from FADH₂ provide less energy than from NADH and H⁺?*

SAMPLE PROBLEM 24.5

ATP Synthesis

Why does the oxidation of NADH provide energy for the formation of three ATP whereas FADH₂ produces two ATP?

Solution

Electrons from the oxidation of NADH enter the electron chain at a higher energy level than FADH₂, providing energy to pump three pairs of protons to the inter-membrane. However, FADH₂ transfers electrons to Q that go through two sites that pump protons and therefore provide energy for two ATP.

Study Check

What complexes act as proton pumps?

QUESTIONS AND PROBLEMS

Oxidative Phosphorylation and ATP

24.35 What is meant by oxidative phosphorylation?

24.36 How is the proton gradient established?

24.37 According to the chemiosmotic theory, how does the proton gradient provide energy to synthesize ATP?

24.38 How does the phosphorylation of ADP occur?

24.39 How are glycolysis and the citric acid cycle linked to the production of ATP by the electron transport chain?

24.40 Why does FADH₂ have a yield of two ATP via the electron chain, but NADH yields three ATP?

24.41 What are the parts of ATP synthase?

ATP Synthase and Heating the Body

Some types of compounds called *uncouplers* separate the electron transport system from the ATP F_0F_1 synthase. They do this by disrupting the proton gradient needed for the synthesis of ATP. The electrons are transported to O_2 in electron transport, but ATP is not formed by ATP synthase.

Some uncouplers transport the protons through the inner membrane, which is normally impermeable to protons; others block the channel in the F_0 portion of ATP synthase. Compounds such as dicumarol and 2,4-dinitrophenol (DNP) are hydrophobic and bind with protons to carry them across the inner membrane. An antibiotic, oligomycin, binds to the F_0 complex, and blocks the channel, which does not allow any protons to return to the matrix. By removing protons or blocking the F_0 channel, there is no proton flow through the F_0 channel to generate energy for ATP synthesis.

Dicumarol   2,4-Dinitrophenol (DNP)

When there is no mechanism for ATP synthesis, the energy of electron transport is released as heat. Certain animals that are adapted to cold climates have developed their own uncoupling system, which allows them to use electron transport energy for

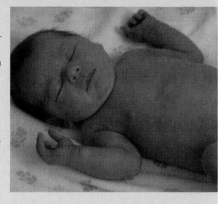

heat production. These animals have large amounts of a tissue called brown fat, which contains a high concentration of mitochondria. This tissue is brown because of the color of iron in the cytochromes of the mitochondria. The proton pumps still operate in brown tissue but a protein embedded in the cell wall allows the protons to bypass ATP synthase. The energy that would be used to synthesize ATP is released as heat. In newborn babies brown fat is used to generate heat because newborns have not stored much fat. The brown fat deposits are located near major blood vessels, which carry the warmed blood to the body. Infants have a small mass but a large surface area and need to produce more heat than the adult. Most adults have little or no brown fat, although someone who works outdoors in a cold climate will develop some brown fat deposits.

Plants also use uncouplers. Some plants use uncoupling agents to volatize fragrant compounds that attract insects to pollinate the plants. Skunk cabbage uses this system. In early spring, heat is used to warm early shoots of plants under the snow, which helps them melt the snow around the plants.

24.42 What is the role of each part of ATP synthase in ATP synthesis?

24.43 What type of active site in ATP synthase binds ADP and P_i?

24.44 How is the ATP released from ATP synthase?

24.5 ATP Energy from Glucose

Under aerobic conditions, the oxidation of glucose through glycolysis, oxidation to pyruvate, the citric acid cycle, the electron transport chain produces ATP from many NADH molecules and some $FADH_2$. Let's see how much ATP is associated with each of these metabolic cycles.

> **LEARNING GOAL**
>
> **Account for the ATP produced by the complete oxidation of glucose.**

ATP from Glycolysis

In glycolysis, the oxidation of glucose stores energy in two NADH molecules as well as two ATP molecules from direct substrate phosphorylation. However, glycolysis occurs in the cytoplasm, and the NADH produced cannot pass through the mitochondrial membrane to the electron transport chain.

Therefore the hydrogen ions and electrons from NADH in the cytoplasm are transferred to compounds that can enter the mitochondria. In this shuttle system, dihydroxyacetone phosphate, a compound in glycolysis, is reduced to glycerol-3-phosphate and NAD^+ is regenerated. After glycerol-3-phosphate crosses the mitochondrial membrane, the hydrogen ions and electrons are transferred to FAD. $FADH_2$ is produced along with glycerol-3-phosphate, which returns to the cytoplasm. The overall reaction for the glycerol-3-phosphate shuttle is:

$$NADH + H^+ + FAD \longrightarrow NAD^+ + FADH_2$$

Cytoplasm   Mitochondria

Therefore the transfer of electrons from NADH in the cytoplasm to $FADH_2$ produces only two ATP, rather than three. In glycolysis, four ATP from two NADH and two ATP from direct phosphorylation add up to six ATP from one glucose molecule.

Glucose $\longrightarrow$ 2 pyruvate $+$ 2 ATP $+$ 2 NADH ($\longrightarrow$ 2 $FADH_2$)

Glucose $\longrightarrow$ 2 pyruvate $+$ 6 ATP

ATP from the Oxidation of Two Pyruvate

Under aerobic conditions, pyruvate enters the mitochondria, where it is oxidized to give acetyl CoA, CO_2, and NADH. Because each glucose molecule yields two pyruvate, two NADH enter electron transport. The oxidation of two pyruvate molecules leads to the production of six ATP molecules.

2 Pyruvate $\longrightarrow$ 2 acetyl CoA $+$ 6 ATP

ATP from the Citric Acid Cycle

One turn of the citric acid cycle produces two CO_2, three NADH, one $FADH_2$, and one ATP by direct substrate phosphorylation. When the NADH and $FADH_2$ enter electron transport, three NADH produce a total of nine ATP molecules, and one $FADH_2$ produces two more ATP.

$$
\begin{aligned}
3 \text{ NADH} \times 3 \text{ ATP} &= 9 \text{ ATP} \\
1 \text{ FADH}_2 \times 2 \text{ ATP} &= 2 \text{ ATP} \\
\underline{1 \text{ GTP} \times 1 \text{ ATP}} &= \underline{1 \text{ ATP}} \\
\text{Total (one turn)} &= 12 \text{ ATP}
\end{aligned}
$$

Thus, one turn of the citric acid cycle generates energy for the synthesis of a total of 12 ATP molecules. Because two acetyl CoA molecules are produced from each glucose, two turns of the citric acid cycle produces 24 ATP.

Acetyl CoA $\longrightarrow$ $2CO_2$ $+$ 12 ATP (one turn of citric acid cycle)

2 Acetyl CoA $\longrightarrow$ $4CO_2$ $+$ 24 ATP (two turns of citric acid cycle)

ATP from the Complete Oxidation of Glucose

We can now determine the total ATP for the complete oxidation of glucose by combining the equations from glycolysis, oxidation of pyruvate, and citric acid cycle. (See Figure 24.14.) The details of the reaction pathways are given in Table 24.1.

ATP and Complete Oxidation of Glucose

Metabolic Pathway	Substrate(s) Oxidized	Products	ATP
Glycolysis	1 Glucose	2 Pyruvate, 2 ATP, 2 NADH	6 ATP
Oxidation	2 Pyruvate	2 Acetyl CoA, 2 NADH, $2CO_2$	6 ATP
Citric acid cycle	2 Acetyl CoA	6 NADH, 2 $FADH_2$, 2 ATP, $4CO_2$	24 ATP
Complete oxidation	Glucose $+$ $6O_2$ $\longrightarrow$ $6CO_2$ $+$ $6H_2O$ $+$ 36 ATP		36 ATP

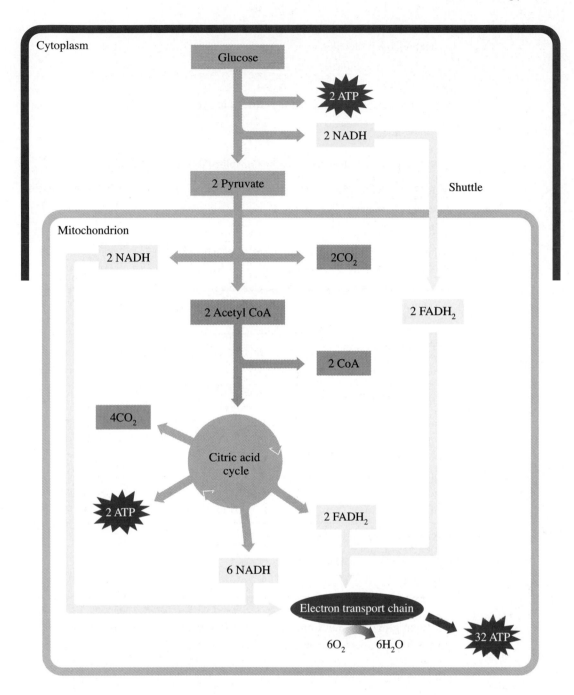

Cytoplasm

Glucose

2 ATP

2 NADH

2 Pyruvate

Shuttle

Mitochondrion

2 NADH

2CO$_2$

2 Acetyl CoA

2 FADH$_2$

2 CoA

4CO$_2$

Citric acid
cycle

2 ATP

2 FADH$_2$

6 NADH

Electron transport chain

6O$_2$ 6H$_2$O

32 ATP

SAMPLE PROBLEM 24.6

ATP Production

Indicate the amount of ATP produced by each of the following oxidation reactions:

a. pyruvate to acetyl CoA   **b.** glucose to acetyl CoA

Solution

a. The oxidation of pyruvate to acetyl CoA produces one NADH, which yields
three ATP.

b. Six ATP are produced from the oxidation of glucose to two pyruvate molecules.

Figure 24.14 The complete oxidation of
glucose to CO$_2$ and H$_2$O yields a total of
36 ATP.

Q *What metabolic pathway produces most
of the ATP from the oxidation of glucose?*

Another six ATP result from the oxidation of two pyruvate molecules to two acetyl CoA molecules.

Study Check

What are the sources of ATP in the citric acid cycle?

Table 24.1   ATP Produced by the Complete Oxidation of Glucose

Reaction Pathway	ATP for One Glucose
ATP from Glycolysis	
Activation of glucose	−2 ATP
Oxidation of glyceraldehyde-3-phosphate (2 NADH)	6 ATP
Transport of 2 NADH across membrane	−2 ATP
Direct ADP phosphorylation (two triose phosphate)	4 ATP
Summary: $C_6H_{12}O_6$ $\longrightarrow$ 2 pyruvate + $2H_2O$	6 ATP
Glucose	
ATP from Acetyl CoA	
2 Pyruvate $\longrightarrow$ 2 acetyl CoA (2 NADH)	6 ATP
ATP from Citric Acid Cycle	
Oxidation of 2 isocitrate (2 NADH)	6 ATP
Oxidation of 2 α-ketoglutarate (2 NADH)	6 ATP
2 Direct substrate phosphorylations (2 GTP)	2 ATP
Oxidation of 2 succinate (2 $FADH_2$)	4 ATP
Oxidation of 2 malate (2 NADH)	6 ATP
Summary: 2 Acetyl CoA $\longrightarrow$ $4CO_2$ + $2H_2O$	24 ATP
Overall ATP Production for One Glucose	

$C_6H_{12}O_6$ + $6O_2$ + 36 ADP + 36 P_i $\longrightarrow$ $6CO_2$ + $6H_2O$ + 36 ATP
Glucose

When glucose is not immediately used by the cells for energy, it is stored as glycogen in the liver and muscles. When the levels of glucose in the brain or blood become low, the glycogen reserves are hydrolyzed and glucose is released into the blood. If glycogen stores are depleted, some glucose can be synthesized from non-carbohydrate sources. It is the balance of all these reactions that maintains the necessary blood glucose level available to our cells and provides the necessary amount of ATP for our energy needs. (See Figure 24.15.)

QUESTIONS AND PROBLEMS

ATP Energy from Glucose

24.45 Why does the NADH produced in glycolysis yield only two ATP?

24.46 Under anaerobic conditions, what is the maximum number of ATP molecules that can be produced from one glucose molecule?

24.47 What is the energy yield in ATP molecules associated with each of the following?
   a. NADH $\longrightarrow$ NAD^+
   b. glucose $\longrightarrow$ 2 pyruvate

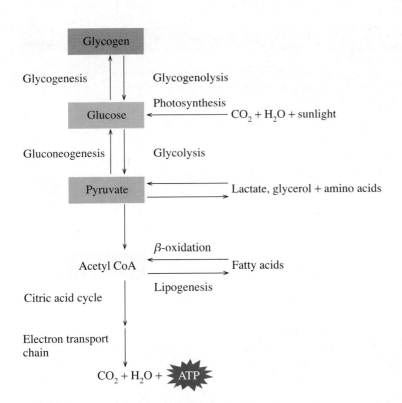

Figure 24.15 The ATP level is maintained by a balance of metabolic pathways that increase or decrease the glucose according to the energy requirements in the cell.
Q *What metabolic pathways are stimulated by low ATP levels?*

HEALTH NOTE

Efficiency of ATP Production

In a laboratory, a calorimeter is used to measure the heat energy from the combustion of glucose. In a calorimeter, one mole of glucose produces a total of 687 kcal (2870 kJ).

$$C_6H_{12}O_6 + 6O_2 \longrightarrow 6CO_2 + 6H_2O + 687 \text{ kcal (2870 kJ)/mole}$$

Let's compare the amount of energy produced from one mole of glucose in a calorimeter with the ATP energy produced in the mitochondria. We can use the energy of the hydrolysis of ATP, which is 7.3 kcal (31 kJ)/mole ATP. Because one mole of glucose generates energy for 36 moles of ATP, the total energy from the oxidation of one mole of glucose in the cells would be 263 kcal (1100 kJ)/per mole.

$$\frac{36 \text{ mole ATP}}{1 \text{ mole glucose}} \times \frac{7.3 \text{ kcal (31 kJ)}}{1 \text{ mole ATP}} = 263 \text{ kcal (1100 kJ) per 1 mole glucose}$$

Compared to the energy produced by burning glucose in a calorimeter, our cells are about 38% (263 kcal/687 kcal) efficient in converting the total available chemical energy in glucose to ATP.

$$\frac{263 \text{ kcal (cells)}}{687 \text{ kcal (calorimeter)}} \times 100 = 38\%$$

The rest of the energy from glucose produced during the oxidation of glucose in our cells is lost as heat.

Calorimeter	Cells
Energy produced by 1 mole of glucose (687 kcal)	Stored as ATP (263 kcal)
	Lost as heat (424 kcal)

---

c. 2 pyruvate $\longrightarrow$ 2 acetyl CoA + 2CO$_2$

d. acetyl CoA $\longrightarrow$ 2CO$_2$

24.48 What is the energy yield in ATP molecules associated with each of the following?

a. FADH$_2$ $\longrightarrow$ FAD

b. glucose + 6O$_2$ $\longrightarrow$ 6CO$_2$ + 6H$_2$O

c. glucose $\longrightarrow$ 2 lactate

d. pyruvate $\longrightarrow$ lactate

Chapter Review

24.1 The Citric Acid Cycle

In a sequence of reactions called the citric acid cycle, an acetyl group is combined with oxaloacetate to yield citrate. Citrate undergoes oxidation and decarboxylation to yield two CO_2, GTP, three NADH, and $FADH_2$ with the regeneration of oxaloacetate. The direct phosphorylation of ADP by GTP yields ATP.

24.2 Electron Carriers

Electron carriers that transfer hydrogen ions and electrons include FMN, iron–sulfur proteins, coenzyme Q, and several cytochromes. Both iron–sulfur proteins and cytochromes contain iron ions that are reduced to Fe^{2+} and reoxidized to Fe^{3+} as electrons are accepted and then passed to the next electron carrier.

24.3 Electron Transport

The reduced coenzymes NADH and $FADH_2$ from various metabolic pathways are oxidized to NAD^+ and FAD when their protons and electrons are transferred to the electron transport chain. The final acceptor, O_2, combines with protons and electrons to yield H_2O. At three transfer points in electron transport, the energy decrease provides the necessary energy for ATP synthesis.

24.4 Oxidative Phosphorylation and ATP

The protein complexes in electron transport act as proton pumps to move protons into the inner membrane space, which produces a proton gradient. As the protons return to the matrix by way of ATP synthase, energy is generated. This energy is used to drive the synthesis of ATP in a process known as oxidative phosphorylation. The available ADP and ATP levels in the cells control the activity of the electron transport chain.

24.5 ATP Energy from Glucose

With the exception of the NADH produced from glycolysis, the oxidation of NADH yields three ATP molecules, and $FADH_2$ yields two ATP. The energy from the NADH produced in the cytoplasm is used to form $FADH_2$. Under aerobic conditions, the complete oxidation of glucose yields a total of 36 ATP from the oxidation of the reduced coenzymes NADH and $FADH_2$ by electron transport, oxidative phosphorylation, and from some direct substrate phosphorylation.

Summary of Key Reactions

Citric Acid Cycle

$$\text{Acetyl CoA} + 3\,NAD^+ + FAD + GDP + P_i + 2H_2O \longrightarrow$$
$$2CO_2 + 3\,NADH + 3H^+ + FADH_2 + CoA + GTP$$

Electron Transport Chain

$$NADH + H^+ + 3\,ADP + 3\,P_i + \tfrac{1}{2}O_2 \longrightarrow NAD^+ + 3\,ATP + H_2O$$
$$FADH_2 + 2\,ADP + 2\,P_i + \tfrac{1}{2}O_2 \longrightarrow FAD + 2\,ATP + H_2O$$

Phosphorylation of ADP

$$ADP + P_i \longrightarrow ATP + H_2O$$

Complete Oxidation of Glucose

$$C_6H_{12}O_6 + 6O_2 + 36\,ADP + 36\,P_i \longrightarrow 6\,CO_2 + 6\,H_2O + 36\,ATP$$

Key Terms

ATP synthase (F_0F_1) An enzyme complex that links the energy released by protons returning to the matrix with the synthesis of ATP from ADP and P_i. The F_0 section contains the channel for proton flow, and the F_1 section uses the energy from the proton flow to drive the synthesis of ATP.

chemiosmotic model The conservation of energy from transfer of electrons in the electron transport chain by pumping protons into the intermembrane space to produce a proton gradient that provides the energy to synthesize ATP.

citric acid cycle A series of oxidation reactions in the mitochondria that convert acetyl CoA to CO_2 and yield NADH and $FADH_2$. It is also called the tricarboxylic acid cycle and the Krebs cycle.

coenzyme Q (CoQ, Q) A mobile carrier that transfers electrons from NADH and $FADH_2$ to cytochrome b in complex III.

cytochromes (cyt) Iron-containing proteins that transfer electrons from QH_2 to oxygen.

electron carriers A group of proteins that accept and pass on electrons as they are reduced and oxidized. Most of the carriers are tightly attached to the inner mito-

chondrial membrane, but two are mobile carriers, which move electrons between the complexes containing the other carriers.

electron transport chain A series of reactions in the mitochondria that transfer electrons from NADH and $FADH_2$ to electron carriers, which are arranged from higher to lower energy levels, and finally to O_2, which produces H_2O. Energy changes during three of these transfers provide energy for ATP synthesis.

Fe–S (iron–sulfur) clusters Proteins containing iron and sulfur in which the iron ions accept electrons from $FMNH_2$ and cytochrome b.

FMN (flavin mononucleotide) An electron carrier derived from riboflavin (vitamin B_2) that transfers hydrogen ions and electrons from NADH entering in the electron transport chain.

oxidative phosphorylation The synthesis of ATP from ADP and P_i using energy generated by the oxidation reactions in the electron transport chain.

proton pumps The enzyme complexes I, III, and IV that move protons from the matrix into the intermembrane space, creating a proton gradient.

Additional Problems

24.49 What is the main function of the citric acid cycle in energy production?

24.50 Most metabolic pathways are not considered cycles. Why is the citric acid cycle considered to be a metabolic cycle?

24.51 If there are no reactions in the citric acid cycle that used oxygen, O_2, why does the cycle operate only in aerobic conditions?

24.52 What products of the citric acid cycle are needed for the electron transport chain?

24.53 Identify the compounds in the citric acid cycle that have the following:
  a. six carbon atoms
  b. five carbon atoms
  c. a keto group

24.54 Identify the compounds in the citric acid cycle that have the following:
  a. four carbon atoms
  b. a hydroxyl group
  c. a double bond

24.55 In which reaction of the citric acid cycle does each of the following occur?
  a. a five-carbon keto acid is decarboxylated
  b. a double bond is hydrated
  c. NAD^+ is reduced
  d. a secondary hydroxyl group is oxidized

24.56 In which reaction of the citric acid cycle does each of the following occur?
  a. FAD is reduced
  b. a six-carbon keto acid is decarboxylated
  c. a carbon–carbon double bond is formed
  d. GDP undergoes direct phosphorylation

24.57 Indicate the coenzyme for each of the following reactions:

a. isocitrate $\longrightarrow$ α-ketoglutarate

b. α-ketoglutarate $\longrightarrow$ succinyl CoA

24.58 Indicate the coenzyme for each of the following reactions:

a. succinate $\longrightarrow$ fumarate

b. malate $\longrightarrow$ oxaloacetate

24.59 How do each of the following regulate the citric acid cycle?

a. high levels of NADH   **b.** high levels of ATP

24.60 How do each of the following regulate the citric acid cycle?

a. high levels of ADP   **b.** low levels of NADH

24.61 Identify each as part of the structure of one of the following components in the electron transport chain as (1) FMN, (2) Fe–S cluster (3) CoQ, or (4) cytochrome

a. a heme group   **b.** contains a ribitol group

24.62 Identify each as part of the structure of one of the following components in the electron transport chain as (1) FMN, (2) Fe–S cluster (3) CoQ, or (4) cytochrome

a. contains the three-ring system of flavins

b. a six-atom ring attached to a long-carbon chain

24.63 Identify each of the following electron carriers as part of a complex or as a mobile carrier? If part of a complex, indicate which one.

a. CoQ   **b.** Fe–S clusters   **c.** cyt a_3

24.64 Identify each of the following electron carriers as part of a complex or as a mobile carrier. If part of a complex, indicate which one.

a. cyt b **b.** cyt c **c.** FMN

24.65 Identify the complex where each of the following are oxidized or reduced and complete the equation:

a. $FADH_2 + Q \longrightarrow$

b. cyt a (Fe^{2+}) + cyt a_3 (Fe^{3+}) $\longrightarrow$

24.66 Identify the complex where each of the following are oxidized or reduced and complete the equation:

a. cyt c (Fe^{2+}) + cyt a (Fe^{3+}) $\longrightarrow$

b. NADH + FMN $\longrightarrow$

24.67 Complete the following by adding the substances that are missing:

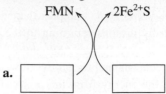

a.

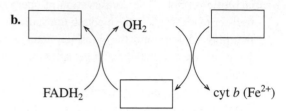

b.

24.68

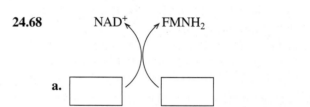

a.

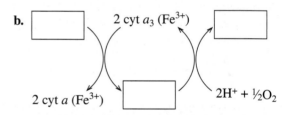

b.

24.69 At which parts of the electron transport system are protons pumped into the intermembrane space?

24.70 What is the effect of proton accumulation in the intermembrane space?

24.71 In the chemiosmotic model, how is energy provided to synthesize ATP?

24.72 In what part of the electron transport chain does the synthesis of ATP take place?

24.73 Why do protons tend to leave the intermembrane space and return to the matrix?

24.74 Why do the enzymes that pump protons extend across the membrane from the matrix to the intermembrane space?

24.75 How many ATP molecules are produced by energy generated when electrons flow from $FADH_2$ to oxygen (O_2)?

24.76 How many ATP molecules are produced by energy generated when electrons flow from NADH to oxygen (O_2)?

24.77 What part of the electron transport chain is inhibited by each of the following?
a. amytal and rotenone
b. antimycin A
c. cyanide and carbon monoxide

24.78 **a.** When an inhibitor blocks the electron transport chain, how are the coenzymes that precede the blocked site affected?
b. When an inhibitor blocks the electron transport chain, how are the coenzymes that follow the blocked site affected?

24.79 How many ATP are produced when glucose is oxidized to pyruvate compared to when glucose is oxidized to CO_2 and H_2O?

24.80 Why do the two NADH produced in glycolysis provide a net of two ATP and not three?

24.81 Using the value of 7.3 kcal/mole for ATP, how many kcal are conserved as ATP from one mole of glucose in each of the following?
a. glycolysis
b. oxidation of pyruvate to acetyl CoA

c. citric acid cycle
d. complete oxidation of glucose to CO_2 and H_2O

24.82 What percent of ATP energy is conserved from one mole of glucose in problem 24.81a–c?

24.83 What does it mean to say that the cell is 38% efficient in storing the energy from the complete combustion of glucose?

24.84 Considering the efficiency of ATP synthesis, how many kcal of energy would be conserved from the complete oxidation of four moles of glucose?

24.85 Where is ATP synthase for oxidative phosphorylation located in the cell?

24.86 How is a proton gradient developed in a cell?

24.87 How is the energy from the proton gradient utilized by ATP synthase?

24.88 How do the active sites on F_1 ATP synthase change during ATP production?

24.89 Why would a bear that is hibernating have more brown fat than one that is active?

24.90 A student is considering using a 2,4-dinitrophenol, which is an uncoupler, to lose weight. Explain how the uncoupler will affect the body temperature of the student. How would you advise the student if the dosage of DNP needed to lose weight is close to toxic levels?

25 Metabolic Pathways for Lipids and Amino Acids

"Occupational therapists teach children and adults the skills they need for the job of living," says occupational therapist Leslie Wakasa. "When working with the pediatric population, we are crucial in educating children with disabilities, their families, caregivers, and school staff in ways to help them be as independent as they can be in all aspects of their daily lives. It's rewarding when you can show children how to feed themselves, which is a huge self-esteem issue for them. The opportunity to help people become more independent is very rewarding."

A combination of technology and occupational therapy helps children who are nonverbal to communicate and interact with their environment. By leaning on a red switch, Alex is learning to use a computer.

LOOKING AHEAD

the **Chemistry** place

www.chemplace.com/college

Visit the URL above or use the CD-ROM in the book for extra quizzing, interactive tutorials, and career resources.

In previous chapters, we focused on carbohydrates because glucose is the primary fuel for the synthesis of ATP. However, lipids and proteins also play an important role in metabolism and energy production. In this chapter we will look at how the digestion of lipids produces fatty acids and glycerol and digestion of proteins gives amino acids. Almost all our energy is stored in the form of triacylglycerols in fat cells of **adipose tissue.** Many people go on diets after they discover that adipose tissue can store unlimited quantities of fat. This fact has become quite apparent in the large number of people in the U.S. that are considered obese. When our caloric intake exceeds the nutritional and metabolic needs of the body, excess carbohydrates and fatty acids are converted to triacylglycerols and added to our fat cells.

The digestion and degradation of dietary proteins as well as body proteins provides amino acids, which are needed to synthesize nitrogen-containing compounds in our cells such as new proteins and nucleic acids. Although amino acids are not considered a primary source of fuel, energy can be extracted from amino acids if glycogen and fat reserves have been depleted. However, when a person who is fasting or starving utilizes amino acids as the only source of energy, the breakdown of the body's own proteins eventually destroys essential body tissues, particularly the heart muscle.

25.1 Digestion of Triacylglycerols

LEARNING GOAL

Describe the sites and products obtained from the digestion of triacylglycerols.

Our adipose tissue is made of fat cells called *adipocytes,* which store triacylglycerols. (See Figure 25.1.) Let's compare the amount of energy stored in the fat cells to the energy from glucose, glycogen, and protein. A typical 70 kg (150 lb) person has about 135,000 kcal of energy stored as fat, 24,000 kcal as protein, 720 kcal as glycogen reserves, and 80 kcal as blood glucose. Therefore, the energy available from stored fats is about 85% of the total energy available in the body. Thus body fat is our major source of stored energy.

Digestion of Dietary Fats

The digestion of dietary fats begins in the small intestine, when the hydrophobic fat globules mix with bile salts released from the gallbladder. In a process called *emulsification,* the bile salts break the fat globules into smaller droplets called micelles. Then *pancreatic lipases* released from the pancreas hydrolyze the triacylglycerols

Figure 25.1 The fat cells (adipocytes) that make up adipose tissue are capable of storing unlimited quantities of triacylglycerols. **Q** *What are some sources of fats in our diet?*

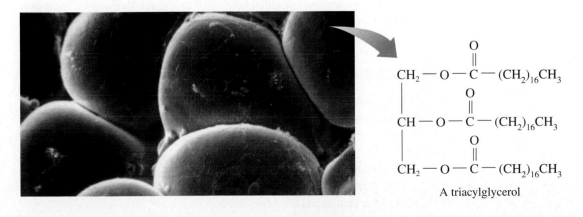

A triacylglycerol

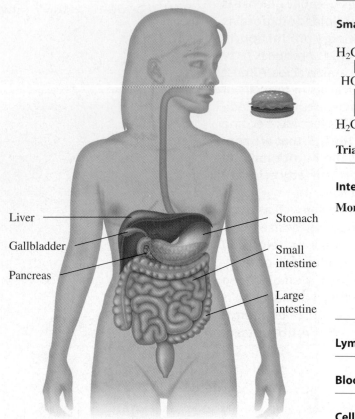

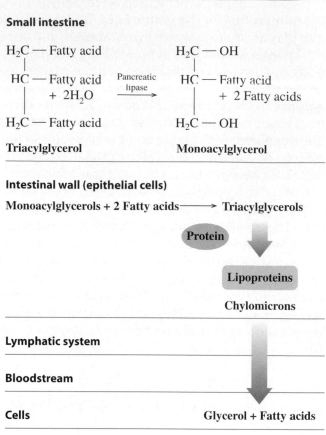

Small intestine

$$\begin{array}{ccc}
\text{H}_2\text{C} - \text{Fatty acid} & & \text{H}_2\text{C} - \text{OH} \\
| & & | \\
\text{HC} - \text{Fatty acid} & \xrightarrow[\text{lipase}]{\text{Pancreatic}} & \text{HC} - \text{Fatty acid} \\
| \quad + 2\text{H}_2\text{O} & & | \qquad + 2 \text{ Fatty acids} \\
\text{H}_2\text{C} - \text{Fatty acid} & & \text{H}_2\text{C} - \text{OH} \\
\textbf{Triacylglycerol} & & \textbf{Monoacylglycerol}
\end{array}$$

Intestinal wall (epithelial cells)

Monoacylglycerols + 2 Fatty acids ⟶ Triacylglycerols

Protein

Lipoproteins

Chylomicrons

Lymphatic system

Bloodstream

Cells   **Glycerol + Fatty acids**

Figure 25.2 The triacylglycerols that reform in the intestinal wall from the digestion products monoacylglycerols and fatty acids bind to proteins for transport through the lymphatic system and bloodstream to the cells.

Q *What kinds of enzymes are secreted from the pancreas into the small intestine to hydrolyze triacylglycerols?*

in the micelles to yield monoacylglycerols and free fatty acids. These digestion products are absorbed into the intestinal lining where they recombine to form triacylglycerols, which are coated with proteins to form lipoproteins called **chylomicrons.** The chylomicrons transport the triacylglycerols through the lymph system and into the bloodstream to be carried to cells of the heart, muscle, and adipose tissues. (See Figure 25.2.) We can write the overall equation for the digestion of triacylglycerols as follows:

$$\text{Triacylglycerols} + 2\text{H}_2\text{O} \xrightarrow{\text{Pancreatic lipase}} \text{Monoacylglycerols} + 2 \text{ Fatty acids}$$

In the cells, enzymes hydrolyze the triacylglycerols to yield glycerol and free fatty acids, which can be used for energy production. The preferred fuel of the heart is fatty acids that are oxidized to acetyl CoA units for ATP synthesis. However, the brain and red blood cells cannot utilize fatty acids. Fatty acids cannot diffuse across the blood-brain barrier, and red blood cells have no mitochondria, which is where fatty acids are oxidized. Therefore, glucose and glycogen are the only source of energy for the brain and red blood cells.

Mobilization of Fat Stores

When blood glucose is depleted and glycogen stores are low, the process of **fat mobilization** breaks down triacylglycerols in the adipose tissue to fatty acids and glycerol. The mobilization occurs when the hormones glucagon or epinephrine are secreted into the bloodstream and bind to receptors on the membrane of adipose cells. This activates enzymes within the fat cells that begin the hydrolysis of a triacylglycerol. A fatty acid is hydrolyzed from carbon 1 or carbon 3 followed by the hydrolysis of the second and third fatty acids.

EXPLORE YOUR WORLD

Digestion of Fats

Place some water and several drops of vegetable oil in a container with a top. Cap and shake. Observe. Add a few drops of liquid soap to the oil and water, cap the container again, and shake. Observe.

Questions

1. Why does the oil separate from the water?
2. How does the appearance of the oil change after soap is added?
3. How is soap like the bile salts in the digestion of fats?
4. Where does fat digestion occur in the body?
5. People without gallbladders take lipase medication. Why is this necessary?

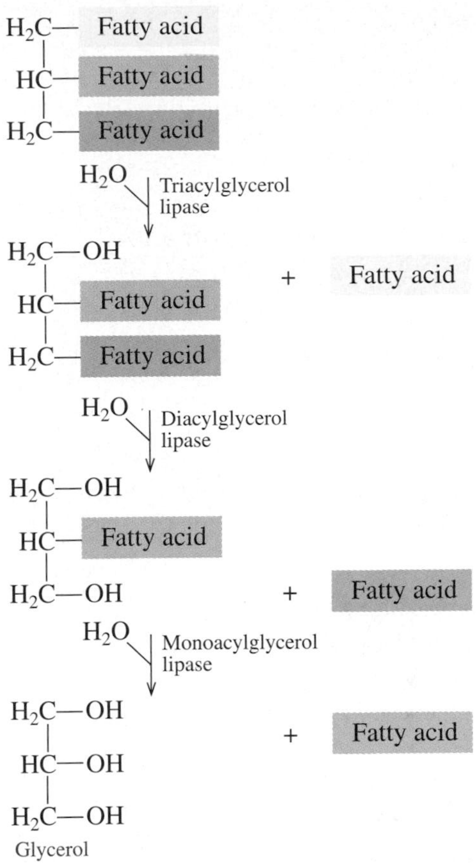

We can write the overall reaction for the mobilization of fats in fat cells as follows:

$$\text{Triacylglycerols} + 3H_2O \xrightarrow{\text{Lipases}} \text{Glycerol} + 3\ \text{Fatty acids}$$

The products of fat mobilization, glycerol and fatty acids, diffuse into the bloodstream and bind with plasma proteins (albumin) to be transported to the tissues. Most of the glycerol goes into the liver where it is converted to glucose.

Metabolism of Glycerol

Using two steps, enzymes in the liver convert glycerol to dihydroxyacetone phosphate. In the first step, glycerol is phosphorylated using ATP to yield glycerol-3-phosphate. In the second step, the hydroxyl group is oxidized to yield dihydroxyacetone phosphate, which is an intermediate in several metabolic pathways including glycolysis and gluconeogenesis. (See Chapter 23.)

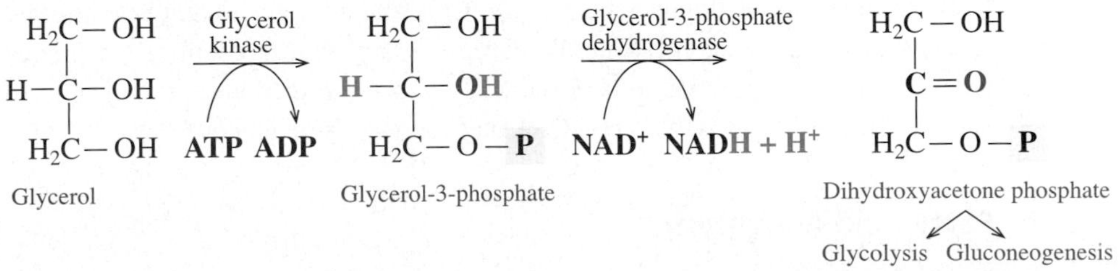

The overall reaction for the metabolism of glycerol is written as follows:

$$\text{Glycerol} + \text{ATP} + \text{NAD}^+ \longrightarrow \text{Dihydroxyacetone phosphate} + \text{ADP} + \text{NADH} + \text{H}^+$$

SAMPLE PROBLEM 25.1

Fats and Digestion

What are the sites, enzymes, and products for the digestion of triacylglycerols?

Solution

The digestion of triacylglycerols takes place in the small intestine where pancreatic lipase catalyzes their hydrolysis to monoacylglycerols and fatty acids.

Study Check

What happens to the products from the digestion of triacylglycerols in the membrane of the small intestine?

QUESTIONS AND PROBLEMS

Digestion of Triacylglycerols

25.1 What is the role of bile salts in lipid digestion?

25.2 How are insoluble triacylglycerols transported to the tissues?

25.3 When are fats released from fat stores?

25.4 What happens to the glycerol produced from the hydrolysis of triacylglycerols in adipose tissues?

25.5 How is glycerol converted to an intermediate of glycolysis?

25.6 How can glycerol be used to synthesze glucose?

LEARNING GOAL

Describe the metabolic pathway of β oxidation.

25.2 Oxidation of Fatty Acids

A large amount of energy is obtained when fatty acids undergo oxidation in the mitochondria to yield acetyl CoA. In stage 2 of fat metabolism, fatty acids undergo **beta oxidation (β oxidation),** which removes two-carbon segments, one at a time, from a fatty acid.

$$CH_3-(CH_2)_{14}-\underset{\beta}{CH_2}-\underset{\alpha}{CH_2}-\overset{\overset{\displaystyle O}{\|}}{C}-OH$$

β-Oxidation occurs here

Stearic acid

Each cycle in β oxidation produces acetyl CoA and a fatty acid that is shorter by two carbons. The cycle repeats until the original fatty acid is completely degraded to two-carbon acetyl CoA units. Each acetyl CoA can then enter the citric acid cycle in the same way as the acetyl CoA units derived from glucose.

Fatty Acid Activation

Before a fatty acid can enter the mitochondria, it undergoes activation in the cytosol. The activation process combines a fatty acid with coenzyme A to yield fatty acyl CoA. The energy released by the hydrolysis of two phosphate groups from ATP is used to drive the reaction. The products are AMP and pyrophosphate (PP_i), which hydrolyzes to yield two inorganic phosphates ($2 P_i$).

$$R-CH_2-\overset{\overset{\text{O}}{\|}}{C}-O^- \;+\; \text{ATP} \;+\; \text{HS}-\text{CoA} \xrightarrow[\text{synthetase}]{\text{Acyl CoA}} R-CH_2-\overset{\overset{\text{O}}{\|}}{C}-S-\text{CoA} \;+\; \text{AMP} \;+\; 2\,P_i$$

Fatty acid   Fatty acyl CoA

Transport of Fatty Acyl CoA

The long hydrocarbon chain prevents the fatty acyl CoA molecule from crossing the inner mitochondrial membrane. However, it is transported into the matrix when it binds with a charged carrier called *carnitine*. (See Figure 25.3.)

$$R-CH_2-\overset{\overset{\text{O}}{\|}}{C}-S-\text{CoA} + H-\underset{\underset{CH_2}{|}}{\overset{\overset{\overset{+}{N(CH_3)_3}}{|}}{\underset{|}{\overset{|}{C}}}}-OH \;\rightleftharpoons\; H-\underset{\underset{CH_2}{|}}{\overset{\overset{\overset{+}{N(CH_3)_3}}{|}}{\underset{|}{\overset{|}{C}}}}-O-\overset{\overset{\text{O}}{\|}}{C}-CH_2-R + \text{HS}-\text{CoA}$$

Fatty acyl CoA   Carnitine   Fatty acyl carnitine

After the fatty acyl group is released in the matrix, it recombines with coenzyme A, and the carnitine returns to the inner membrane. While it may seem like a complicated way to move fatty acyl CoA into the matrix, this transport system provides a way to regulate degradation (oxidation) and synthesis of fatty acids. When fatty acids are being synthesized in the cytosol, the transport of fatty acyl CoA into the matrix is blocked, which prevents fatty acid degradation.

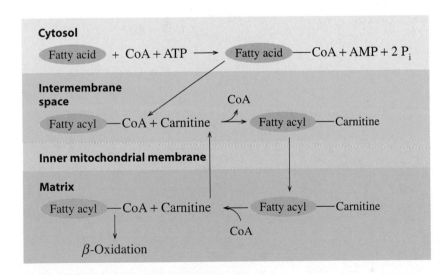

Figure 25.3 Fatty acids are activated and transported by carnitine through the inner mitochondrial membrane into the matrix.
Q *Why is carnitine used to transport a fatty acid into the matrix?*

Reactions of β-Oxidation Cycle

In the matrix, fatty acyl CoA molecules undergo β oxidation, which is a cycle of four reactions that convert the —CH_2— of the β carbon to a β-keto group. Once the β-keto group is formed, a two-carbon acetyl group can be split from the chain, which shortens the fatty acyl group.

β-Oxidation Pathway

Reaction 1   Oxidation (Dehydrogenation)

In the first reaction of β oxidation, the FAD coenzyme removes hydrogen atoms from the α and β carbons of the activated fatty acid to form a trans carbon–carbon double bond, and $FADH_2$.

Reaction 2   Hydration

A water molecule now adds across the trans double bond, which places a hydroxyl group (—OH) on the β carbon.

Reaction 3   Oxidation (Dehydrogenation)

The hydroxy group on the β carbon is oxidized to yield a ketone. The hydrogen atoms removed in the dehydrogenation reduce coenzyme NAD^+ to $NADH + H^+$. At this point, the β carbon has been oxidized to a keto group.

Reaction 4   Cleavage of Acetyl CoA

In the final step of β oxidation, the C_α—C_β bond splits to yield free acetyl CoA and a fatty acyl CoA molecule that is shorter by two carbon atoms. This shorter fatty acyl CoA is ready to go through the β-oxidation cycle again.

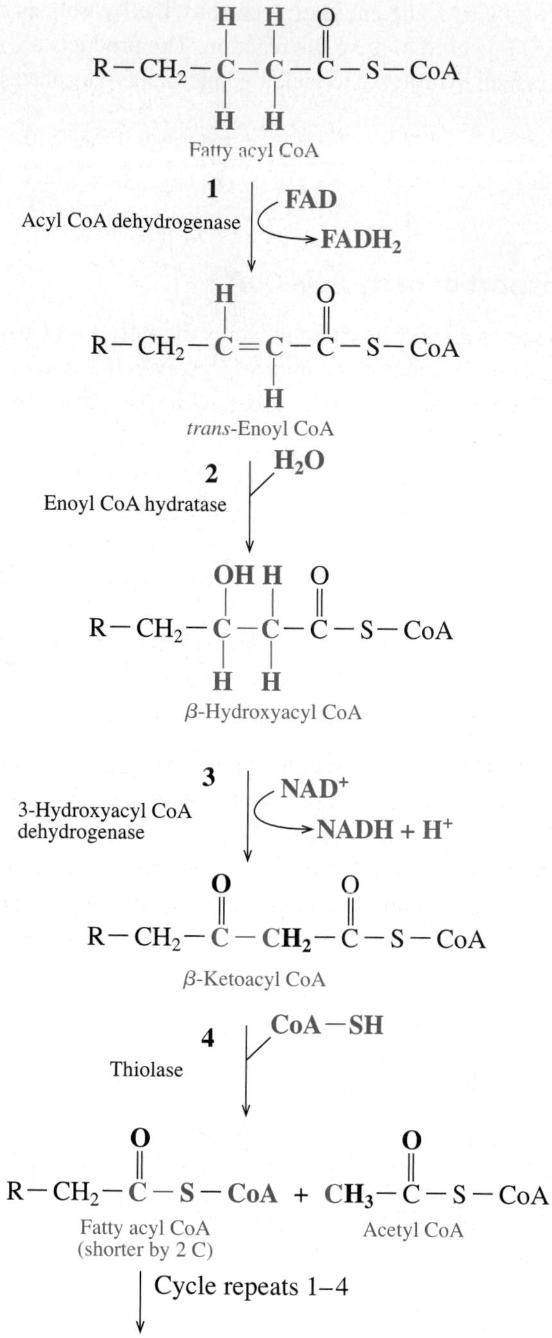

$$R-CH_2-\overset{\overset{H}{|}}{\underset{\underset{H}{|}}{C}}-\overset{\overset{H}{|}}{\underset{\underset{H}{|}}{C}}-\overset{\overset{O}{\|}}{C}-S-CoA$$

Fatty acyl CoA

1 Acyl CoA dehydrogenase   FAD → $FADH_2$

$$R-CH_2-\overset{\overset{H}{|}}{C}=\overset{\underset{H}{|}}{C}-\overset{\overset{O}{\|}}{C}-S-CoA$$

trans-Enoyl CoA

2 Enoyl CoA hydratase H_2O

$$R-CH_2-\overset{\overset{OH}{|}}{\underset{\underset{H}{|}}{C}}-\overset{\overset{H}{|}}{\underset{\underset{H}{|}}{C}}-\overset{\overset{O}{\|}}{C}-S-CoA$$

β-Hydroxyacyl CoA

3 3-Hydroxyacyl CoA dehydrogenase NAD^+ → $NADH + H^+$

$$R-CH_2-\overset{\overset{O}{\|}}{C}-CH_2-\overset{\overset{O}{\|}}{C}-S-CoA$$

β-Ketoacyl CoA

4 Thiolase   CoA—SH

$$R-CH_2-\overset{\overset{O}{\|}}{C}-S-CoA \;+\; CH_3-\overset{\overset{O}{\|}}{C}-S-CoA$$

Fatty acyl CoA (shorter by 2 C)   Acetyl CoA

Cycle repeats 1–4

The reaction for one cycle of β oxidation is written as follows:

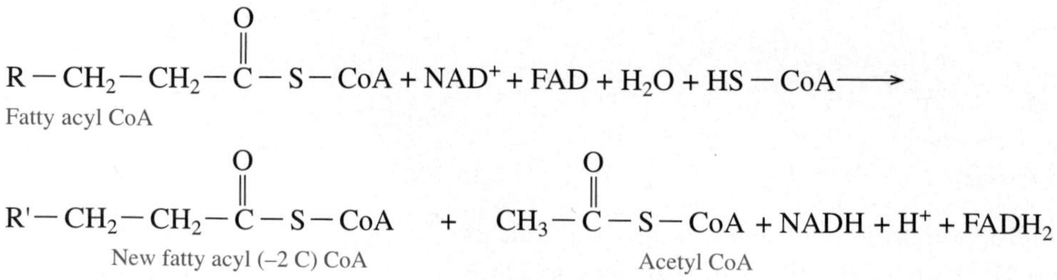

$$R-CH_2-CH_2-\overset{\overset{O}{\|}}{C}-S-CoA + NAD^+ + FAD + H_2O + HS-CoA \longrightarrow$$

Fatty acyl CoA

$$R'-CH_2-CH_2-\overset{\overset{O}{\|}}{C}-S-CoA \quad + \quad CH_3-\overset{\overset{O}{\|}}{C}-S-CoA + NADH + H^+ + FADH_2$$

New fatty acyl (–2 C) CoA   Acetyl CoA

Fatty Acyl Length Determines Cycle Repeats

The number of carbon atoms in a fatty acid determines the number of times the cycle repeats and the number of acetyl CoA units it produces. For example, the complete β oxidation of myristic acid (C_{14}) produces seven acetyl CoA groups, which is equal to one-half the number of carbon atoms in the chain. Because the final turn of the cycle produces two acetyl CoA groups, the total number of times the cycle repeats is one less than the total number of acetyl groups it produces. Therefore, the C_{14} fatty acid goes through the cycle six times. (See Figure 25.4.)

Fatty acid		Number of Acetyl CoA	β-Oxidation cycles
Myristic acid	C_{14}	7 acetyl CoA	6
Palmitic acid	C_{16}	8 acetyl CoA	7
Stearic acid	C_{18}	9 acetyl CoA	8

We can write an overall equation for the complete oxidation of myristyl-CoA as follows:

$$\text{Myristyl-CoA} + 6\,\text{CoA} + 6\,\text{FAD} + 6\,\text{NAD}^+ + 6\,H_2O \longrightarrow$$
$$7\,\text{Acetyl CoA} + 6\,\text{FADH}_2 + 6\,\text{NADH} + 6H^+$$

Oxidation of Unsaturated Fatty Acids

The β-oxidation sequence we have described applies to the most typical type of fatty acids, which are saturated with an even number of carbon atoms. However the fats in our diets, particularly the oils, contain unsaturated fatty acids, which have one or more cis double bonds. The hydration reaction adds water to trans double bonds, not cis. When the double bond in an unsaturated fatty acid is ready for hydration, an isomerase forms a trans double bond between the α and β carbons, which is the arrangement needed for the hydration reaction.

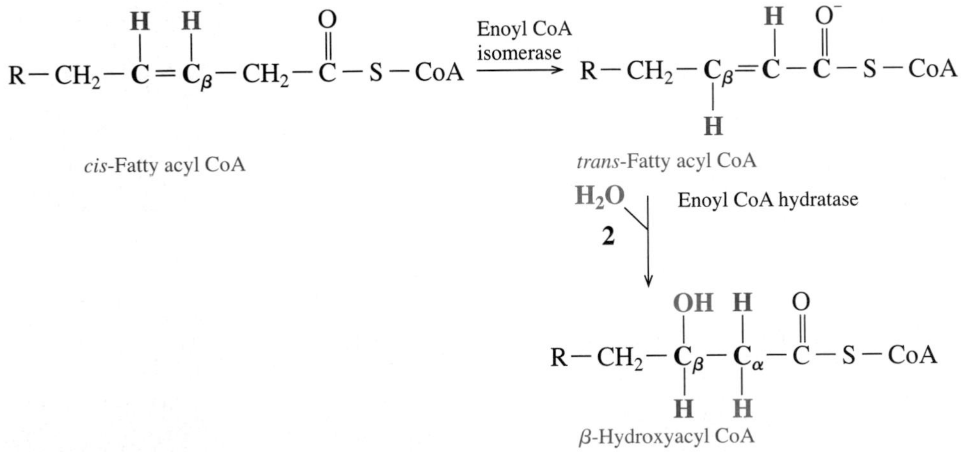

Because the isomerization provides the trans double bond for the hydration in reaction 2, it bypasses the first reaction. Therefore, the energy released by the β oxidation of an unsaturated fatty acid is slightly less since no $FADH_2$ is produced in that cycle.

SAMPLE PROBLEM 25.2

β Oxidation

Match each of the following with reactions in the β-oxidation cycle:

(1) first oxidation (2) hydration (3) second oxidation (4) cleavage

Figure 25.4 Myristic acid (C$_{14}$) undergoes six oxidation cycles that repeat reactions 1–4 to yield 7 acetyl CoA molecules.

Q *How many NADH and FADH$_2$ are produced in one turn of the fatty acid cycle of β oxidation?*

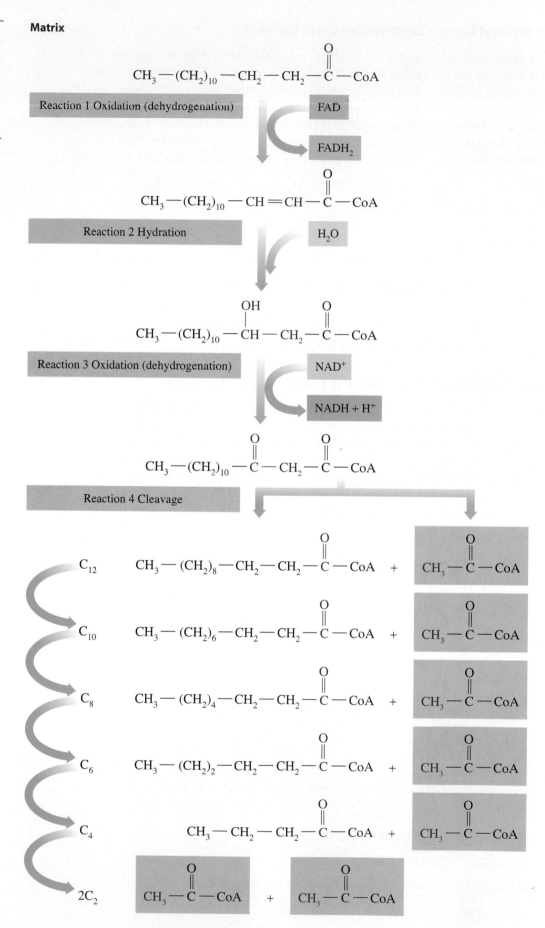

a. water is added to a trans double bond
b. an acetyl CoA is removed
c. FAD is reduced to FADH$_2$
d. bypassed during the oxidation of unsaturated fatty acids

Solution

a. (2) hydration   **b.** (4) cleavage
c. (1) first oxidation   **d.** (1) first oxidation

Study Check

Which coenzyme is needed in reaction 3 when a β-hydroxyl group is converted to a β-keto group?

QUESTIONS AND PROBLEMS

Oxidation of Fatty Acids

25.7 Where in the cell is a fatty acid activated?

25.8 What is the function of carnitine in the degradation of fatty acids?

25.9 What coenzymes are required for β oxidation?

25.10 When does an isomerization occur during the β oxidation of a fatty acid?

25.11 In each of the following acyl CoA molecules, identify the β carbon.

$$\textbf{a. } CH_3-CH_2-CH_2-CH_2-CH_2-CH_2-CH_2-\overset{\overset{\displaystyle O}{\|}}{C}-S-CoA$$

$$\textbf{b. } CH_3-(CH_2)_{14}-CH_2-CH_2-\overset{\overset{\displaystyle O}{\|}}{C}-S-CoA$$

$$\textbf{c. } CH_3-CH_2-CH=CH-CH_2-CH_2-CH_2-CH_2-CH_2-\overset{\overset{\displaystyle O}{\|}}{C}-S-CoA$$

25.12 Write the product when each of the following undergoes the indicated reaction in β oxidation:

a.

$$CH_3-(CH_2)_{12}-CH=CH-\overset{\overset{\displaystyle O}{\|}}{C}-S-CoA + H_2O \xrightarrow{\text{Enoyl CoA hydratase}}$$

b.

$$CH_3-(CH_2)_6-CH_2-CH_2-\overset{\overset{\displaystyle O}{\|}}{C}-S-CoA \xrightarrow{\text{Acyl CoA dehydrogenase}}$$

c.

$$CH_3-(CH_2)_4-\overset{\overset{\displaystyle O}{\|}}{C}-CH_2-\overset{\overset{\displaystyle O}{\|}}{C}-S-CoA + HS-CoA \xrightarrow{\text{Thiolase}}$$

25.13 Capric acid, $CH_3(CH_2)_8COOH$, is a C_{10} fatty acid.
  a. Write the formula of the activated form of capric acid.
  b. Indicate the α and β carbon atoms in the fatty acid.
  c. Write the overall equation for the first cycle of β oxidation for capric acid.
  d. Write the overall equation for the complete β oxidation of capric acid.

25.14 Arachidic acid, CH_3—$(CH_2)_{18}$—COOH, is a C_{20} fatty acid.
  a. Write the formula of the activated form of arachidic acid.
  b. Indicate the α and β carbon atoms in the fatty acid.
  c. Write the overall equation for the first cycle of β oxidation for arachidic acid.
  d. Write the overall equation for the complete β oxidation of arachidic acid.

LEARNING GOAL

Calculate the total ATP produced by the complete oxidation of a fatty acid.

25.3 ATP and Fatty Acid Oxidation

We can now determine the total energy yield from the oxidation of a particular fatty acid. The total ATP is the sum of the ATP produced from the $FADH_2$, NADH, and acetyl CoA units. In each β-oxidation cycle, one NADH, one $FADH_2$, and one acetyl CoA are produced. From Chapter 24, we know that hydrogen ions and electrons transferred from NADH to coenzyme Q in the electron transport chain generate sufficient energy to synthesize three ATP, whereas $FADH_2$ leads to the synthesis of two ATP. However, the greatest amount of energy produced from a fatty acid is generated by the production of the acetyl CoA units that enter the citric acid cycle. We saw in Chapter 24 that one acetyl CoA leads to the synthesis of a total of 12 ATP.

Let's continue with our example of myristic acid and calculate the total ATP it produces. So far we know that the C_{14} acid produces seven acetyl CoA units and goes through six turns of the cycle. We also need to remember that activation of the myristic acid requires the equivalent of two ATP. We can set up the calculation as follows.

ATP Production for Myristic Acid

Activation	−2 ATP
7 acetyl CoA	
7 acetyl CoA × 12 ATP/acetyl CoA	84 ATP
6 β-oxidation cycles	
6 $FADH_2$ × 2 ATP/$FADH_2$	12 ATP
6 NADH × 3 ATP/NADH	18 ATP
Total	112 ATP

The Energy Yield from Fats

Myristic acid, $C_{14}H_{28}O_2$, has a molar mass of 228 g/mole. We can calculate the ATP produced per gram of the fatty acid as follows:

$$\frac{112 \text{ moles ATP}}{1 \text{ mole myristic acid}} \times \frac{1 \text{ mole myristic acid}}{228 \text{ g myristic acid}} = 0.49 \text{ mole ATP per 1 g myristic acid (fat)}$$

In Chapter 24 we saw that the complete oxidation of glucose generated a total of 36 ATP. Glucose has the formula $C_6H_{12}O_6$ and a molar mass 180 g. We can calculate the ATP produced per gram of glucose as follows:

$$\frac{36 \text{ moles ATP}}{1 \text{ mole glucose}} \times \frac{1 \text{ mole glucose}}{180 \text{ g glucose}} = 0.20 \text{ mole ATP per 1 g glucose}$$

Stored Fat and Obesity

The storage of fat is an important survival feature in the lives of many animals. In hibernating animals, large amounts of stored fat provide the energy for the entire hibernation period, which could be several months. In camels, such as dromedary camels, large amounts of food are stored in the camel's hump, which is actually a huge fat deposit. When food resources are low, the camel can survive months without food or water by utilizing the fat reserves in the hump. Migratory birds preparing to fly long distances also store large amounts of fat. Whales are kept warm by a layer of body fat called "blubber" under their skin, which can be as thick as 2 feet. Blubber also provides energy when whales must survive long periods of starvation. Penguins also have blubber, which protects them from the cold and provides energy when they are sitting on a nest of eggs.

Humans also have the capability to store large amounts of fat, although they do not hibernate or usually have to survive for long periods of time without food. When humans survived on sparse diets that were mostly vegetarian, the fat content was about 20%. Today, a typical diet includes more dairy products and foods with high fat levels, which increases the daily fat intake to as much as 60% of the diet. The U.S. Public Health Service now estimates that in the United States, more than one-third of adults are obese. Obesity is defined as a body weight that is more than 20 percent over an ideal weight. Obesity is a major factor in health problems such as diabetes, heart disease, high blood pressure, stroke, and gallstones as well as some cancers and forms of arthritis.

At one time we thought that obesity was simply a problem of eating too much. However research now indicates that certain pathways in lipid and carbohydrate metabolism may cause excessive weight gain in some people. In 1995, scientists discovered that a hormone called *leptin* is produced in fat cells. When fat cells are full, high levels of leptin signal the brain to limit the intake of food. When fat stores are low, leptin production decreases, which signals the brain to increase food intake. Some obese persons have high levels of leptin, which means that leptin did not cause them to decrease how much they ate.

Research on obesity has become a major research field. Scientists are studying differences in the rate of leptin production, degrees of resistance to leptin, and possible combinations of these factors. After a person has dieted and lost weight, the leptin level drops. This decrease in leptin may cause an increase in hunger, slow metabolism, and increased food intake, which starts the weight-gain cycle all over again. Currently, studies are being made to assess the safety of leptin therapy following a weight loss.

From these calculations, we see that one gram of fat produces more than twice the ATP energy as 1 gram of glucose. This also means that we obtain more than double the number of nutritional calories from 1 g of fat (9 kcal/g fat) than we do from 1 g of carbohydrate (4 kcal/g). This is one reason why a low-fat diet is recommended when we are trying to lose weight.

SAMPLE PROBLEM 25.3

ATP Production from β Oxidation

How much ATP will be produced from the β oxidation of palmitic acid, a C_{16} saturated fatty acid?

Solution

A 16-carbon fatty acid will produce 8 acetyl CoA units and go through 7 β oxidation cycles. Each acetyl CoA can produce 12 ATP by way of the citric acid cycle. In the electron transport chain, each $FADH_2$ produces 2 ATP, and each NADH produces 3 ATP.

Fat Storage and Blubber

Obtain two medium-sized plastic baggies and a can of Crisco or other type of shortening used for cooking You will also need a bucket or large container with water and ice cubes. Place several tablespoons of the shortening in one of the baggies. Place the second baggie inside and tape the outside edges of the bag. With your hand inside the inner baggie, move the shortening around to cover your hand. With one hand inside the double baggie, submerge both your hands in the container of ice water. Measure the time it takes for one hand to feel uncomfortably cold. Experiment with different amounts of shortening.

Questions

1. How effective is the bag with "blubber" in protecting your hand from the cold?
2. How does "blubber" help an animal survive starvation?
3. How would twice the amount of shortening affect your results?
4. Why would animals in warm climates, such as camels and migratory birds, need to store fat?
5. If you placed 300 g of shortening in the baggie, how many moles of ATP could it provide if used for energy production (assume it produces the same ATP as myristic acid)?

ATP Production for Palmitic Acid ($C_{16}H_{32}O_2$)

Activation of palmitic acid to palmitoyl-CoA	−2 ATP
8 acetyl CoA × 12 ATP (citric acid cycle)	96 ATP
7 $FADH_2$ × 2 ATP (electron transport chain)	14 ATP
7 NADH × 3 ATP (electron transport chain)	21 ATP
Total	129 ATP

Study Check

Compare the percent of the total ATP from reduced coenzymes and from acetyl CoA in the β oxidation of palmitic acid.

QUESTIONS AND PROBLEMS

ATP and Fatty Acid Oxidation

25.15 Why is the energy of fatty acid activation from ATP to AMP considered the same as the hydrolysis of 2 ATP $\longrightarrow$ 2 ADP?

25.16 What is the number of ATP obtained from one acetyl CoA in the citric acid cycle?

25.17 Consider the complete oxidation of capric acid CH_3—$(CH_2)_8$—COOH, a C_{10} fatty acid.
   a. How many acetyl CoA units are produced?
   b. How many cycles of β oxidation are needed?
   c. How many ATPs are generated from the oxidation of capric acid?

25.18 Consider the complete oxidation of arachidic acid, CH_3—$(CH_2)_{18}$—COOH, a C_{20} fatty acid.
   a. How many acetyl CoA units are produced?
   b. How many cycles of β oxidation are needed?
   c. How many ATPs are generated from the oxidation of arachidic acid?

Describe the pathway of ketogenesis.

25.4 Ketogenesis and Ketone Bodies

When carbohydrates are not available to meet energy needs, the body breaks down body fat. However, the oxidation of large amounts of fatty acids can cause acetyl CoA molecules to accumulate in the liver. Then acetyl CoA molecules combine to form keto compounds called **ketone bodies** in a pathway known as **ketogenesis.** (See Figure 25.5.)

1. In ketogenesis, two molecules of acetyl CoA combine to form acetoacetyl CoA, which reverses the last reaction in β oxidation.
2. The hydrolysis of acetoacetyl CoA forms acetoacetate, a ketone body, which reacts further to produce two other ketone bodies.
3. Acetoacetate can be reduced to yield β-hydroxybutyrate, which is considered a ketone body even though it does not contain a keto group.
4. Acetone forms when acetoacetate undergoes decarboxylation and loses CO_2.

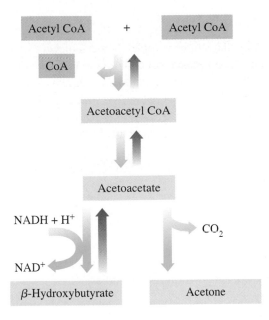

Figure 25.5 In ketogenesis, acetyl CoA molecules combine to produce ketone bodies: acetoacetate, β-hydroxybutyrate, and acetone.
Q *What condition in the body leads to the formation of ketone bodies?*

Ketogenesis

$$CH_3-\overset{O}{\overset{\|}{C}}-S-CoA + CH_3-\overset{O}{\overset{\|}{C}}-S-CoA$$

HS—CoA **1** Thiolase

$$CH_3-\overset{O}{\overset{\|}{C}}-CH_2-\overset{O}{\overset{\|}{C}}-S-CoA +$$
Acetoacetyl CoA

2

$$CH_3-\overset{O}{\overset{\|}{C}}-CH_2-\overset{O}{\overset{\|}{C}}-O^- + HS-CoA$$
Acetoacetate

β-Hydroxybutyrate dehydrogenase   **3**   **4**

NADH + H⁺

NAD⁺

CO₂

$$CH_3-\overset{OH}{\underset{|}{CH}}-CH_2-\overset{O}{\overset{\|}{C}}-O^- \qquad CH_3-\overset{O}{\overset{\|}{C}}-CH_3$$
β-Hydroxybutyrate   Acetone

Ketone bodies are produced mostly in the liver and transported to cells in the heart, brain, and skeletal muscle, where small amounts of energy can be obtained by converting acetoacetate or β-hydroxybutyrate back to acetyl CoA.

β-Hydroxybutyrate ⟶ acetoacetate ⟶ acetoacetyl-CoA ⟶ 2 acetyl CoA

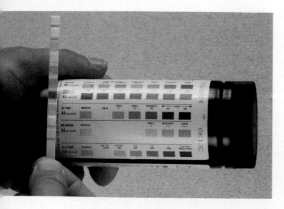

Ketosis

When ketone bodies accumulate, they may not be completely metabolized by the body. This may lead to a condition called **ketosis,** which is found in severe diabetes, diets high in fat and low in carbohydrates, and starvation. Because two of the ketone bodies are acids, they can lower the blood pH below 7.4, which is **acidosis,** a condition that often accompanies ketosis. A drop in blood pH can interfere with the ability of the blood to carry oxygen and cause breathing difficulties.

SAMPLE PROBLEM 25.4

Ketogenesis

When does ketogenesis take place in the liver?

Solution

When excess acetyl CoA cannot be processed by the citric acid cycle, acetyl CoA units enter the ketogenesis pathway where they form ketone bodies.

Study Check

What are the names of the three compounds that are called ketone bodies?

HEALTH NOTE

Ketone Bodies and Diabetes

The blood glucose is elevated within 30 minutes following a meal containing carbohydrates. The elevated level of glucose stimulates the secretion of the hormone insulin from the pancreas, which increases the flow of glucose into muscle and adipose tissue for the synthesis of glycogen. As blood glucose levels drop, the secretion of insulin decreases. When blood glucose is low, another hormone, glucagon, is secreted by the pancreas, which stimulates the breakdown of glycogen in the liver to yield glucose.

In *diabetes mellitus,* glucose cannot be utilized or stored as glycogen because insulin is not secreted or does not function properly. In Type I, *insulin-dependent diabetes,* which often occurs in childhood, the pancreas produces inadequate levels of insulin. This type of diabetes can result from damage to the pancreas by viral infections or from genetic mutations. In Type II, *insulin-resistant diabetes,* which usually occurs in adults, insulin is produced, but insulin receptors are not responsive. Thus a person with Type II diabetes does not respond to insulin therapy. *Gestational diabetes* can occur during pregnancy, but blood glucose levels usually

return to normal after the baby is born. Mothers with diabetes tend to gain weight and have large babies.

In all types of diabetes, insufficient amounts of glucose are available in the muscle, liver, and adipose tissue. As a result liver cells synthesize glucose from noncarbohydrate sources (gluconeogenesis) and break down fat, which elevates the acetyl CoA level. Excess acetyl CoA undergoes ketogenesis and ketone bodies accumulate in the blood. As the level of acetone increases, its odor can be detected on the breath of a person with uncontrolled diabetes who is in ketosis.

In uncontrolled diabetes, the concentration of blood glucose exceeds the ability of the kidney to reabsorb glucose, and glucose appears in the urine. High levels of glucose increase the osmotic pressure in the blood, which leads to an increase in urine output. Symptoms of diabetes include frequent urination and excessive thirst. Treatment for diabetes includes a change to a diet to limit carbohydrate intake, and may require medication such as a daily injection of insulin.

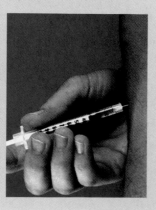

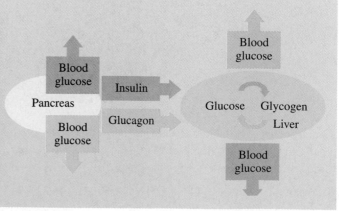

QUESTIONS AND PROBLEMS

Ketogenesis and Ketone Bodies

25.19 What is ketogenesis?

25.20 If a person is fasting, why would they have high levels of acetyl CoA?

25.21 What type of reaction converts acetoacetate to β-hydroxybutyrate?

25.22 How is acetone formed from acetoacetate?

25.23 What is ketosis?

25.24 Why do diabetics produce high levels of ketone bodies?

25.5 Fatty Acid Synthesis

LEARNING GOAL

Describe the biosynthesis of fatty acids from acetyl CoA.

When the body has met all its energy needs and the glycogen stores are full, acetyl CoA from the breakdown of carbohydrates and fatty acids is used to form new fatty acids. In the pathway called **lipogenesis,** two-carbon acetyl units are linked together to give a 16-carbon fatty acid, palmitic acid. Although the reactions appear much like the reverse of the reactions we discussed in fatty acid oxidation, the synthesis of fatty acids proceeds in a separate pathway with different enzymes. Fatty acid oxidation occurs in the mitochondria and uses FAD and NAD^+, whereas fatty acid synthesis occurs in the cytosol and uses the reduced coenzyme NADPH. NADPH is similar to NADH, except it has a phosphate group.

Acyl Carrier Protein (ACP)

In β oxidation, acetyl and acyl molecules are activated using coenzyme A (CoA). In fatty acid synthesis, acyl compounds also are activated, but by an acyl carrier protein (ACP—SH). In the ACP—SH molecule, the thiol and pantothenic acid found in CoA are attached to a protein.

ACP—SH

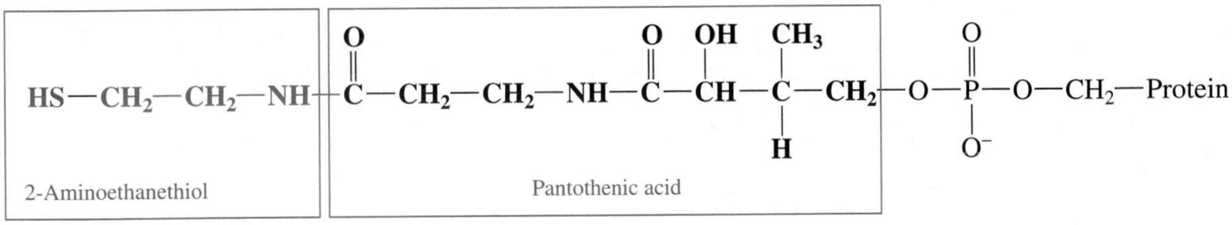

For fatty acid synthesis, the activated forms of malonyl-ACP and acetyl-ACP are produced by transferring the acetyl or acyl group from CoA—SH to ACP—SH.

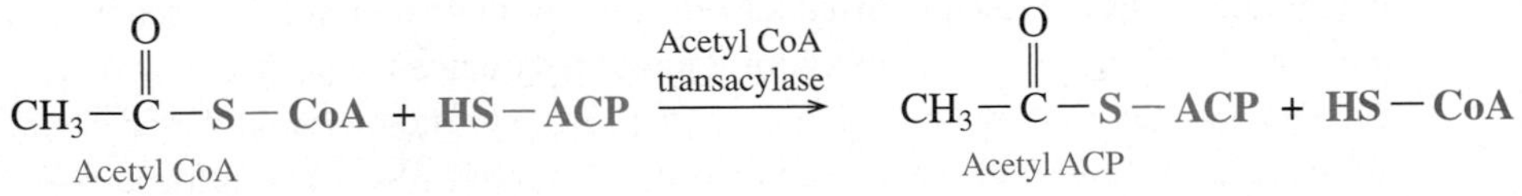

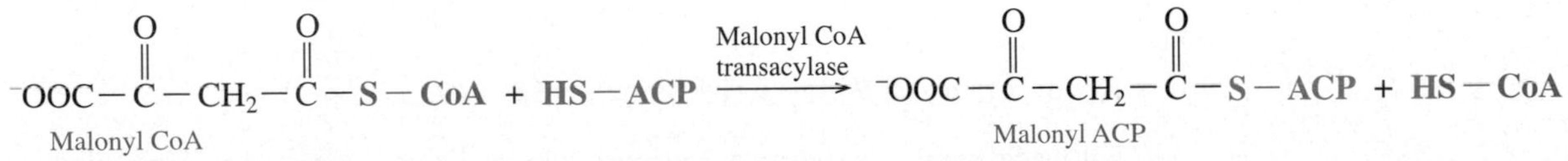

Synthesis of Malonyl CoA

Fatty acid synthesis begins when acetyl CoA combines with bicarbonate to form a three-carbon compound malonyl CoA. The hydrolysis of ATP provides the energy for the reaction.

$$CH_3\text{—}\overset{\displaystyle O}{\overset{\|}{C}}\text{—CoA} + HCO_3^- + ATP \xrightarrow[\text{carboxylase}]{\text{Acetyl CoA}} {}^-O\text{—}\overset{\displaystyle O}{\overset{\|}{C}}\text{—}CH_2\text{—}\overset{\displaystyle O}{\overset{\|}{C}}\text{—CoA} + ADP + P_i + H^+$$

$$\text{Acetyl CoA} \qquad\qquad\qquad\qquad\qquad\qquad\qquad\qquad\qquad \text{Malonyl ACP}$$

Synthesis of Palmitate

The next four reactions occur in a cycle that adds two-carbon acetyl groups to a carbon chain. (See Figure 25.6.)

Reaction 1   Condensation
Acetyl ACP and malonyl ACP condense to yield acetoacetyl-ACP and CO_2.

Reaction 2   Reduction
The keto group on the β carbon is reduced to a hydroxyl group using hydrogen from the reduced coenzyme NADPH.

Reaction 3   Dehydration
The alcohol is dehydrated to form a trans double bond in *trans*-enoyl-ACP.

Reaction 4   Reduction
NADPH reduces the double bond to a single bond, which forms butyryl-ACP, a saturated four-carbon compound.

Cycle Repeats

The cycle repeats as the longer four-carbon butyryl-ACP reacts with another malonyl-ACP to produce hexanoyl-ACP. After seven cycles, the product, C_{16} palmitoyl-ACP, is hydrolyzed to yield palmitate and free ACP.

The overall reaction starts with acetyl CoA and requires ATP and NADPH. We can write the overall equation for the synthesis of palmitate as follows:

$$8\,\text{Acetyl CoA} + 14\,\text{NADPH} + 14H^+ + 7\,\text{ATP} \longrightarrow$$

$$\text{palmitate} + 8\,\text{CoA} + 14\,\text{NADP}^+ + 7H_2O + 7\,\text{ADP} + 7\,P_i$$

Longer and Shorter Fatty Acids

Although we have looked at the synthesis of the fatty acid palmitate, shorter and longer fatty acids are also produced in cells. Shorter fatty acids are released before there are 16 carbon atoms in the chain. Longer fatty acids are produced with special enzymes that add two-carbon units to the carboxyl end of the fatty acid chain. An unsaturated cis bond can also be incorporated into a 10-carbon fatty acid followed by the same elongation reactions we have seen.

Regulation of Fatty Acid Synthesis

Fatty acid synthesis takes place primarily in the adipose tissue, where triacylglycerols are formed and stored. The hormone insulin stimulates the formation of fatty

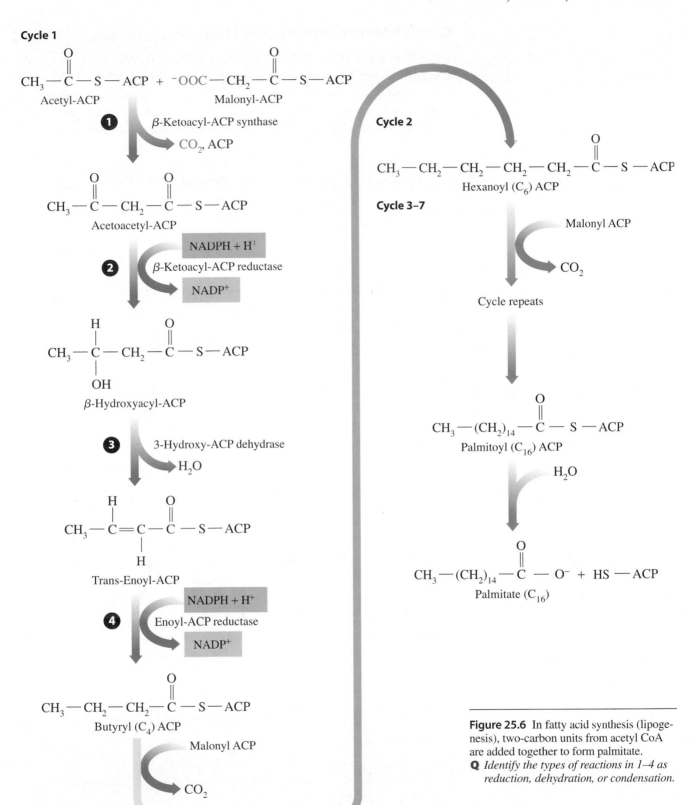

Figure 25.6 In fatty acid synthesis (lipogenesis), two-carbon units from acetyl CoA are added together to form palmitate.
Q *Identify the types of reactions in 1–4 as reduction, dehydration, or condensation.*

acids. When blood glucose is high, insulin moves glucose into the cells. In the cell, insulin stimulates glycolysis and the oxidation of pyruvate, thereby producing acetyl CoA for fatty acid synthesis. During lipogenesis, the production of malonyl CoA blocks the transport of fatty acyl groups into the matrix of the mitochondria, which prevents their oxidation.

Comparison of β Oxidation and Fatty Acid Synthesis

We have seen that many of the steps in the synthesis of palmitate are similar to those that occur in the β oxidation of palmitate. Synthesis combines two-carbon units, whereas β oxidation removes two-carbon units. Synthesis involves reduction and dehydration, whereas β oxidation involves oxidation and hydration. We can distinguish between the two pathways by comparing some of their features in Table 25.1.

Table 25.1   A Comparison of β Oxidation and Fatty Acid Synthesis

	β Oxidation	Fatty Acid Synthesis (lipogenesis)
Site	Mitochondrial matrix	Cytosol
Activated by	Glucagon Low blood glucose	Insulin High blood glucose
Activator	Coenzyme A (CoA)	Acyl carrier protein (ACP)
Initial subtrate	Fatty acid	Acetyl CoA $\longrightarrow$ Malonyl CoA
Coenzymes	FAD, NAD$^+$	NADPH, NADP$^+$
Types of Reaction	Oxidation Hydration Cleavage	Reduction Dehydration Condensation
Function	Cleaves two-carbon acyl group	Adds two-carbon acyl group
Final product	Acetyl CoA units	Palmitate (C_{16}) and other fatty acids

SAMPLE PROBLEM 25.5

Fatty Acid Synthesis

Malonyl ACP is required for the elongation of fatty acid chains. How is malonyl ACP formed from the starting material acetyl CoA?

Solution

Acetyl CoA combines with bicarbonate to form malonyl CoA, which reacts with ACP to form malonyl ACP.

Study Check

If malonyl ACP is a three-carbon acyl group, why are only two carbon atoms added each time malonyl ACP is combined with a fatty acid chain?

QUESTIONS AND PROBLEMS

Fatty Acid Synthesis

25.25 Where does fatty acid synthesis occur in the cell?

25.26 What compound is involved in the activation of acyl compounds in fatty acid synthesis?

25.27 What is the starting material for fatty acid synthesis?

25.28 What is the function of malonyl ACP in fatty acid synthesis?

25.29 Identify the reaction catalyzed by each of the following enzymes:
       (1) acetyl CoA carboxylase   (2) acetyl CoA transacylase
       (3) malonyl CoA transacylase
     a. converts malonyl CoA to malonyl ACP
     b. combines acetyl CoA with bicarbonate to give malonyl CoA

    c. converts acetyl CoA to acetyl ACP

25.30 Identify the reaction catalyzed by each of the following enzymes:
    (1) β-ketoacyl-ACP synthase   (2) β-ketoacyl-ACP reductase
    (3) 3-hydroxy-ACP dehydrase   (4) enoyl-ACP reductase
    a. catalyzes the dehydration of an alcohol
    b. converts a carbon–carbon double bond to a carbon–carbon single bond
    c. combines a two-carbon acyl group with a three-carbon acyl group accompanied by the loss of CO_2
    d. reduces a keto group to a hydroxyl group

25.31 Determine the number of each of the following involved in the synthesis of one molecule of capric acid, a fatty acid with 10 carbon atoms, $C_{10}H_{20}O_2$.
    a. HCO_3^- **b.** ATP   **c.** acetyl CoA
    d. malonyl ACP   **e.** NADPH   **f.** CO_2 removed

25.32 Determine the number of each of the following needed in the synthesis of one molecule of myristic acid, a fatty acid with 14 carbon atoms $C_{14}H_{28}O_2$.
    a. HCO_3^- **b.** ATP   **c.** acetyl ACP
    d. malonyl ACP   **e.** NADPH   **f.** CO_2 removed

25.6 Digestion of Proteins

> **LEARNING GOAL**
> Describe the hydrolysis of dietary protein and absorption of amino acids.

The major role of proteins is to provide amino acids for the synthesis of new proteins for the body and nitrogen atoms for the synthesis of compounds such as nucleotides. We have seen that carbohydrates and lipids are the major sources of energy, but when they are not available, amino acids are degraded to substrates that enter energy-producing pathways.

In stage 1, the digestion of proteins begins in the stomach, where hydrochloric acid (HCl) at pH 2 denatures proteins and activates enzymes such as pepsin that begin to hydrolyze peptide bonds. Polypeptides from the stomach move into the small intestine where trypsin and chymotrypsin complete the hydrolysis of the peptides to amino acids. The amino acids are absorbed through the intestinal walls into the blood stream for transport to the cells. (See Figure 25.7.)

SAMPLE PROBLEM 25.6

Digestion of Proteins

What are the sites and end products for the digestion of proteins?

Solution

Proteins begin digestion in the stomach and complete digestion in the small intestine to yield amino acids.

Study Check

What is the function of HCl in the stomach?

Protein Turnover

Our bodies are constantly replacing old proteins with new ones. The process of synthesizing proteins and breaking them down is called **protein turnover.** Many

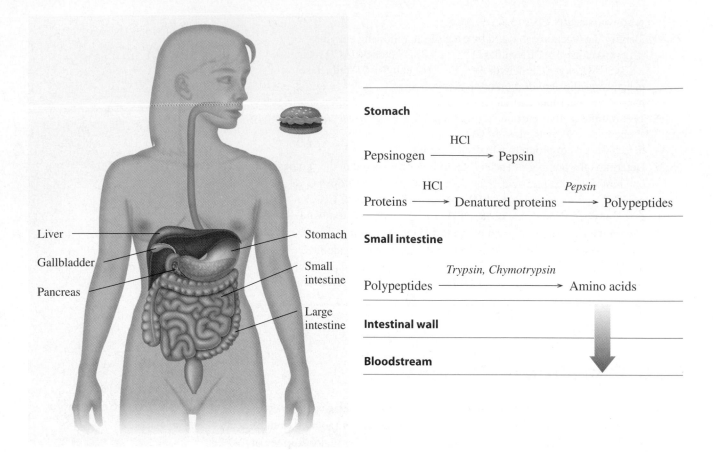

Stomach

$$\text{Pepsinogen} \xrightarrow{\text{HCl}} \text{Pepsin}$$

$$\text{Proteins} \xrightarrow{\text{HCl}} \text{Denatured proteins} \xrightarrow{\textit{Pepsin}} \text{Polypeptides}$$

Small intestine

$$\text{Polypeptides} \xrightarrow{\textit{Trypsin, Chymotrypsin}} \text{Amino acids}$$

Intestinal wall

Bloodstream

Figure 25.7 Proteins are hydrolyzed to polypeptides in the stomach and to amino acids in the small intestine.

Q *What enzyme, secreted into the small intestine, hydrolyzes peptides?*

types of proteins including enzymes, hormones, and hemoglobin are synthesized in the cells and then degraded. For example, the hormone insulin has a half-life of 10 minutes, whereas the half-lives of lactate dehydrogenase is about 2 days and hemoglobin is 120 days. Damaged and ineffective proteins are also degraded and replaced. While most amino acids are used to build proteins, other compounds also require nitrogen for their synthesis as seen in Table 25.2. (See Figure 25.8.)

Usually we maintain a nitrogen balance in the cells so that the amount of protein we break down is equal to the amount that is reused. A diet that is high in protein has a positive nitrogen balance because it supplies more nitrogen than we need. Because the body cannot store nitrogen, the excess is excreted as urea. A diet that does not provide sufficient nitrogen has a negative nitrogen balance, which is a condition that occurs during starvation and fasting.

Table 25.2 Nitrogen-Containing Compounds

Type of Compound	Example
Nonessential amino acids	Alanine, aspartate, cysteine, glycine
Proteins	Muscle protein, enzymes
Neurotransmitters	Acetylcholine, dopamine, serotonin
Amino alcohols	Choline, ethanolamine
Heme	Hemoglobin
Hormones	Thyroxine, epinephrine, insulin
Nucleotides (nucleic acids)	Purines, pyrimidines

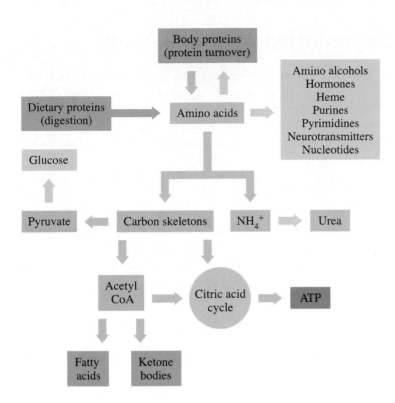

Figure 25.8 Proteins are used in the synthesis of nitrogen-containing compounds or degraded to carbon skeletons that enter other metabolic pathways and urea.
Q *What are some compounds that require nitrogen for their synthesis?*

Energy from Amino Acids

Normally, only a small amount (about 10%) of our energy needs is supplied by amino acids. However, more energy is extracted from amino acids in conditions such as fasting or starvation, when carbohydrate and fat stores are exhausted. If amino acids remain the only source of energy for a long period of time, the breakdown of body proteins eventually leads to a destruction of essential body tissues.

SAMPLE PROBLEM 25.7

In positive nitrogen balance, why are excess amino acids excreted?

Solution

Because the body cannot store nitrogen, amino acids that are not needed for the synthesis of proteins are excreted.

Study Check

Under what condition does the body have a negative nitrogen balance?

QUESTIONS AND PROBLEMS

Digestion of Proteins

25.33 Where do dietary proteins undergo digestion in the body?

25.34 What is meant by protein turnover?

25.35 What are some nitrogen-containing compounds that need amino acids for their synthesis?

25.36 What is the fate of the amino acids obtained from a high protein diet?

25.7 Degradation of Amino Acids

When dietary protein exceeds the nitrogen needed for protein synthesis, the excess amino acids are degraded in a similar way. The α-amino group is removed to yield a keto acid, which can be converted to an intermediate of other metabolic pathways. The carbon atoms from amino acids are used in the citric acid cycle as well as the synthesis of fatty acids, ketone bodies, and glucose. Most of the amino groups are converted to urea.

Transamination

The degradation of amino acids occurs primarily in the liver. In a **transamination** reaction, an α-amino group is transferred from an amino acid to an α-keto acid, usually α-ketoglutarate. A new amino acid and a new α-keto acid are produced. The enzymes for the transfer of amino groups are known as transaminases or aminotransferases.

$$R_1-\overset{\overset{+}{N}H_3}{\underset{|}{C}H}-COO^- \ + \ R_2-\overset{O}{\overset{||}{C}}-COO^- \ \xrightarrow{\text{Aminotransferase}} \ R_1-\overset{O}{\overset{||}{C}}-COO^- \ + \ R_2-\overset{\overset{+}{N}H_3}{\underset{|}{C}H}-COO^-$$

Amino acid   α-Keto acid   New α-Keto acid   New amino acid

We can write an equation to show the transfer of the amino group from alanine to α-ketoglutarate to yield glutamate, the new amino acid, and the α-keto acid pyruvate. The α-keto acid often used in transamination reactions is α-ketoglutarate, which is converted to glutamate.

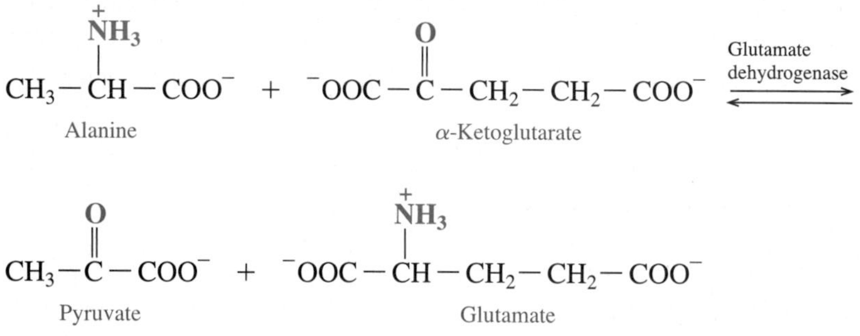

$$CH_3-\overset{\overset{+}{N}H_3}{\underset{|}{C}H}-COO^- \ + \ ^-OOC-\overset{O}{\overset{||}{C}}-CH_2-CH_2-COO^- \ \underset{\text{dehydrogenase}}{\overset{\text{Glutamate}}{\rightleftharpoons}}$$

Alanine   α-Ketoglutarate

$$CH_3-\overset{O}{\overset{||}{C}}-COO^- \ + \ ^-OOC-\overset{\overset{+}{N}H_3}{\underset{|}{C}H}-CH_2-CH_2-COO^-$$

Pyruvate   Glutamate

SAMPLE PROBLEM 25.8

Transamination

Write the formula of the amino acid and α-keto acid produced from the transamination of oxaloacetate by alanine.

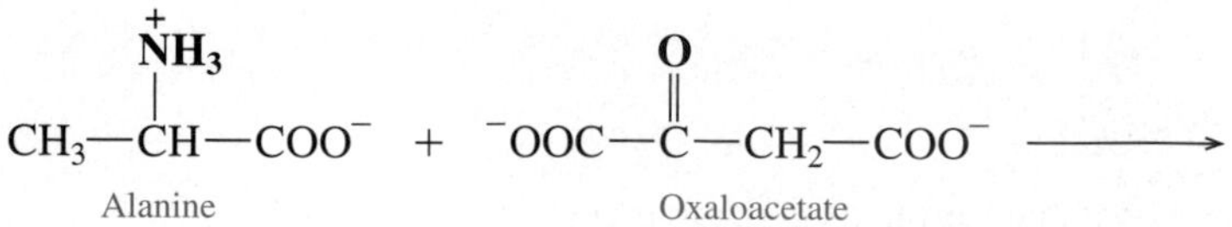

$$CH_3-\overset{\overset{+}{N}H_3}{\underset{|}{C}H}-COO^- \ + \ ^-OOC-\overset{O}{\overset{||}{C}}-CH_2-COO^- \ \longrightarrow$$

Alanine   Oxaloacetate

Solution

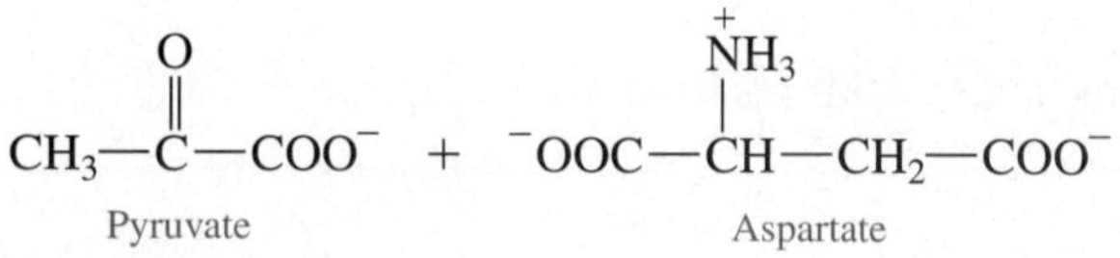

$$CH_3-\overset{O}{\overset{||}{C}}-COO^- \ + \ ^-OOC-\overset{\overset{+}{N}H_3}{\underset{|}{C}H}-CH_2-COO^-$$

Pyruvate   Aspartate

What is a possible name for the enzyme that catalyzes this reaction?

Oxidative Deamination

In a process called **oxidative deamination,** the amino group in glutamate is removed as an ammonium ion NH_4^+. This reaction is catalyzed by *glutamate dehydrogenase,* which uses either NAD^+ or $NADP^+$ as a coenzyme.

$$\overset{\overset{+}{N}H_3}{\underset{|}{^-OOC-CH}}-CH_2-CH_2-COO^- + H_2O + NAD^+\ (or\ NADP^+) \xrightarrow{\substack{Glutamate \\ dehydrogenase}}$$

Glutamate

$$\overset{\overset{O}{\|}}{^-OOC-C}-CH_2-CH_2-COO^- + NH_4^+ + NADH\ (or\ NADPH) + H^+$$

α-Ketoglutarate

Therefore the amino group from any amino acid can be used to form glutamate, which undergoes oxidative deamination converting the amino group to an ammonium ion. Then the ammonium ion is converted to urea, which we will discuss in the next section.

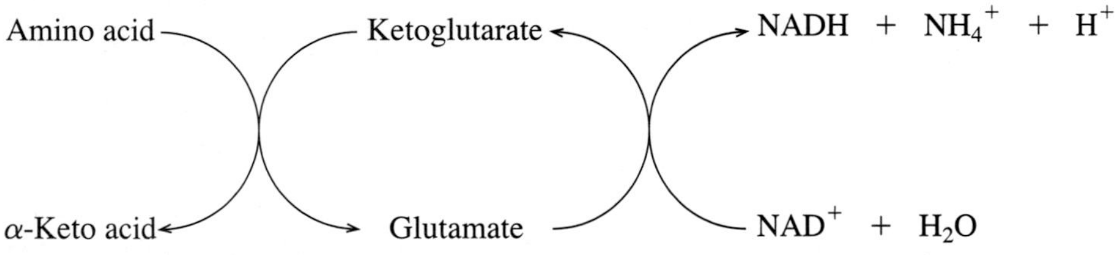

Amino acid — Ketoglutarate ← NADH + NH$_4^+$ + H$^+$

α-Keto acid ← Glutamate → NAD$^+$ + H$_2$O

SAMPLE PROBLEM 25.9

Transamination and Oxidative Deamination

Indicate whether each of the following represents a transamination or an oxidative deamination:
a. Glutamate is converted to α-ketoglutarate and NH_4^+.
b. Alanine and α-ketoglutarate react to form pyruvate and glutamate.
c. A reaction is catalyzed by glutamate dehydrogenase, which requires NAD^+.

Solution

a. oxidative deamination
b. transamination
c. oxidative deamination

How is α-ketoglutarate regenerated to participate in more transamination reactions?

QUESTIONS AND PROBLEMS

Degradation of Amino Acids

25.37 What are the reactants and products in transamination reactions?

25.38 What types of enzymes catalyze transamination reactions?

25.39 Write the structure of the α–keto acid produced from each of the following in transamination:

          $\overset{+}{N}H_3$ $\overset{+}{N}H_3$

a. 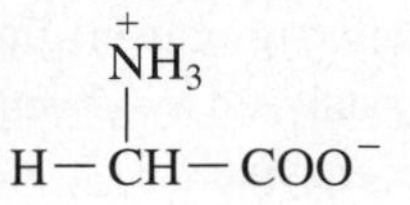 $H - CH - COO^-$ glycine   **b.** 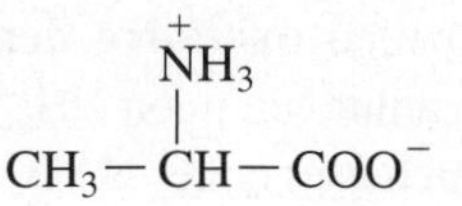 $CH_3 - CH - COO^-$ alanine

               $CH_3 \quad \overset{+}{N}H_3$

c. 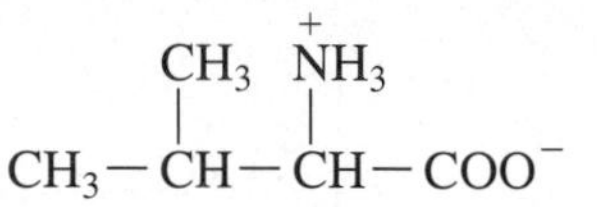 $CH_3 - CH - CH - COO^-$ valine

25.40 Write the structure of the α–keto acid produced from each of the following in transamination:

         $\overset{+}{N}H_3$ $CH_3 \quad \overset{+}{N}H_3$

a. 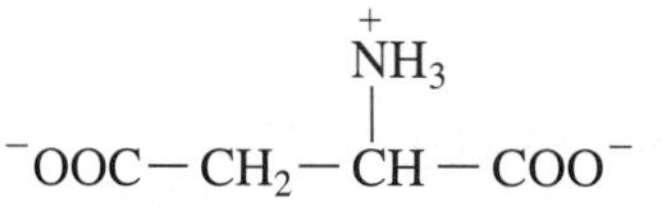 $^-OOC - CH_2 - CH - COO^-$ aspartate   **b.** 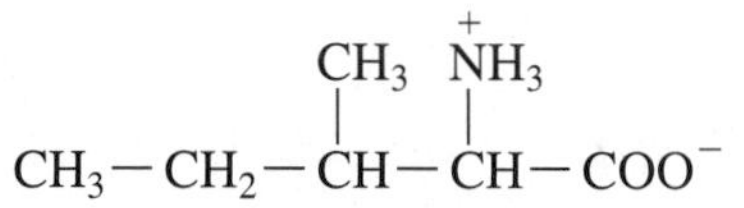 $CH_3 - CH_2 - CH - CH - COO^-$ isoleucine

          $CH_3 \quad \overset{+}{N}H_3$

c. 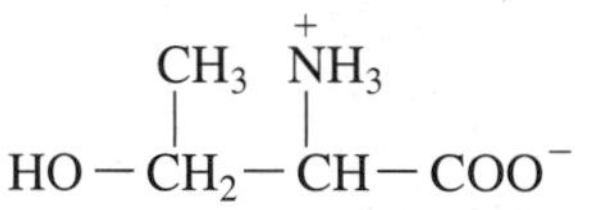 $HO - CH_2 - CH - COO^-$ serine

25.41 Write the reaction for the oxidative deamination of glutamate.

25.42 How do the amino groups from all 20 amino acids produce ammonium ions when glutamate undergoes oxidative deamination?

LEARNING GOAL

Describe the formation of urea from ammonium ion.

25.8 Urea Cycle

The ammonium ion, which is the end product of amino acid degradation, is toxic if it is allowed to accumulate. Therefore, a pathway called the **urea cycle** detoxifies ammonium ions by converting them to urea, which is transported to the kidneys to form urine.

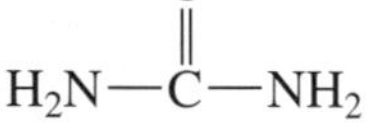

$$\underset{\text{Urea}}{H_2N - \overset{\overset{\textstyle O}{\|}}{C} - NH_2}$$

In one day, a typical adult may excrete about 25–30 g of urea in the urine. This amount increases when a diet is high in protein. If urea is not properly excreted, it builds up quickly to a toxic level. To detect renal disease, the blood urea nitrogen (BUN) level is measured. If the BUN is high, protein intake must be reduced, and hemodialysis may be needed to remove toxic nitrogen waste from the blood.

Urea Cycle

The urea cycle in the liver cells consists of reactions that take place in both the mitochondria and cytosol. (See Figure 25.9.) In preparation for the urea cycle, the ammonium ions react with carbon dioxide from the citric acid cycle and two ATP to yield carbamoyl phosphate.

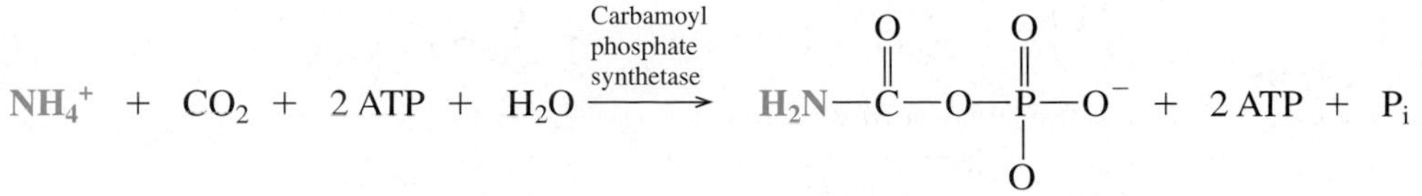

$$NH_4^+ + CO_2 + 2\,ATP + H_2O \xrightarrow{\text{Carbamoyl phosphate synthetase}} \underset{\text{Carbamoyl phosphate}}{H_2N-\overset{\overset{\displaystyle O}{\|}}{C}-O-\overset{\overset{\displaystyle O}{\|}}{\underset{\underset{\displaystyle O}{|}}{P}}-O^-} + 2\,ATP + P_i$$

Reaction 1   Transfer of Carbamoyl Group

In the mitochondria, the carbamoyl group is transferred from carbamoyl phosphate to ornithine to yield citrulline, which is transported across the mitochondrial membrane into the cytosol. The hydrolysis of the phosphate bond provides the energy to drive the reaction.

Reaction 2   Condensation with Aspartate

In the cytosol, citrulline condenses with the amino acid aspartate to form argininosuccinate. The hydrolysis of ATP to AMP and two inorganic phosphates provides the energy for the reaction. The nitrogen atom in aspartate becomes the other nitrogen atom in the urea that is produced in the final reaction.

Reaction 3   Cleavage of Fumarate

The argininosuccinate undergoes a cleavage to yield fumarate, a citric acid cycle intermediate, and arginine.

Reaction 4   Hydrolysis to Form Urea

The hydrolysis of arginine yields urea and ornithine, which returns to the mitochondria to repeat the cycle.

Summary of Urea Formation

The formation of urea starts with two nitrogen atoms, one from NH_4^+ and one from aspartate, and a carbon atom from CO_2. In the urea cycle, a total of four phosphate bonds are hydrolyzed to provide the energy for the reactions. We can write an overall reaction starting with ammonium ion as follows:

$$NH_4^+ + CO_2 + 3\,ATP + \text{aspartate} + 2H_2O \longrightarrow \text{Urea} + 2\,ADP + AMP + 4P_i + \text{fumarate}$$

SAMPLE PROBLEM 25.10

Urea Cycle

Indicate the reaction in the urea cycle where each of the following compounds is a reactant.

a. aspartate   **b.** ornithine   **c.** arginine

Solution

a. Aspartate condenses with citrulline in reaction 2.
b. Ornithine accepts the carbamoyl group in reaction 1.
c. Arginine is cleaved in reaction 4.

Study Check

Name the products of each reaction in problem 25.10.

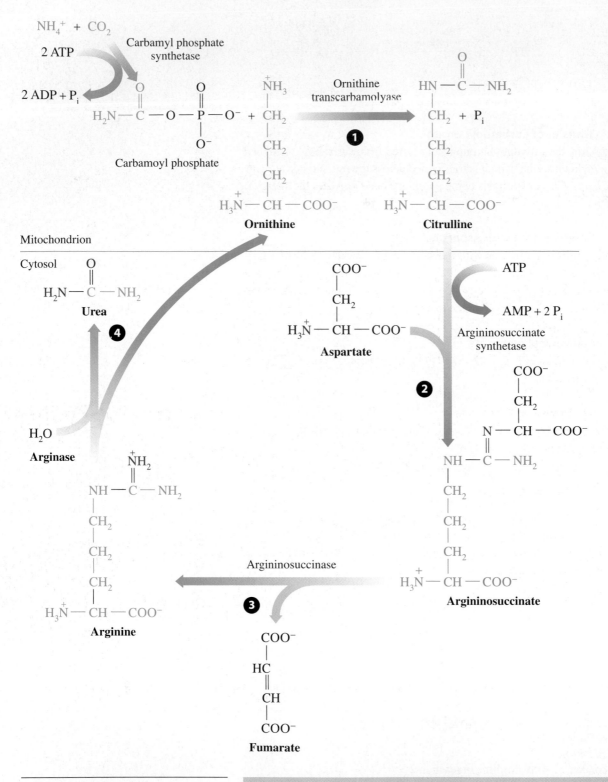

Figure 25.9 In the urea cycle, urea is formed from a carbon and nitrogen (blue) from carbamoyl phosphate (initially an ammonium ion from oxidative deamination) and a nitrogen from aspartate (pink). **Q** *Where in the cell is urea formed?*

QUESTIONS AND PROBLEMS

Urea Cycle

25.43 Why does the body convert NH_4^+ to urea?

25.44 Where is the energy source for the formation of urea?

25.45 What is the structure of urea?

25.46 What is the structure of carbamoyl phosphate?

25.47 What is the source of carbon in urea?

25.48 How much ATP energy is required to drive one turn of the urea cycle?

25.9 Fates of the Carbon Atoms from Amino Acids

LEARNING GOAL

Describe where carbon atoms from amino acids enter the citric acid cycle or other pathways.

The carbon skeletons from the transamination of amino acids are used as intermediates of the citric acid cycle or other metabolic pathways. We can classify the amino acids according to the number of carbon atoms in those intermediates. (See Figure 25.10.) The amino acids with three carbons or those that are converted into carbon skeletons with three carbons are converted to pyruvate. The four-carbon group consists of amino acids that are converted to oxaloacetate, and the five-carbon group provides α-ketoglutarate. Some amino acids are listed twice because they can enter different pathways to form citric acid cycle intermediates.

A **glucogenic amino acid** (orange) generates pyruvate or oxaloacetate, which can be converted to glucose by gluconeogenesis. A **ketogenic amino acid** (green) produces acetoacctyl CoA or acetyl CoA, which can enter the ketogenesis pathway to form ketone bodies or fatty acids.

Carbon Skeletons from the Three-Carbon Amino Acids

The carbon atoms from alanine, serine, and cysteine are converted to pyruvate in one step or several steps. Alanine is converted by a simple transamination.

Alanine + α-ketoglutarate $\longrightarrow$ pyruvate + glutamate

Glycine is converted to serine and then to pyruvate. When tryptophan is degraded, the three carbon atoms that are not part of the ring system form alanine, which goes to pyruvate. Although it has more than three carbon atoms, threonine can be degraded to glycine and acetaldehyde. The oxidation of acetaldehyde gives acetyl-CoA.

Carbon Skeletons from the Four-Carbon Amino Acids

Oxaloacetate is the α-keto acid that forms when the four-carbon aspartate undergoes transamination.

Aspartate + α-ketoglutarate $\longrightarrow$ oxaloacetate + glutamate

Asparagine hydrolyzes to give NH_4^+ and aspartate, which goes to oxaloacetate. The degradation pathways of other four-carbon amino acids, isoleucine, methionine, and valine, produce succinyl CoA, another intermediate of the citric acid cycle.

Carbon Skeletons from the Five-Carbon Amino Acids

The five-carbon amino acids glutamine, glutamate, proline, arginine, and histidine are converted to glutamate, which undergoes oxidation–deamination to yield α-ketoglutarate and ammonium ion.

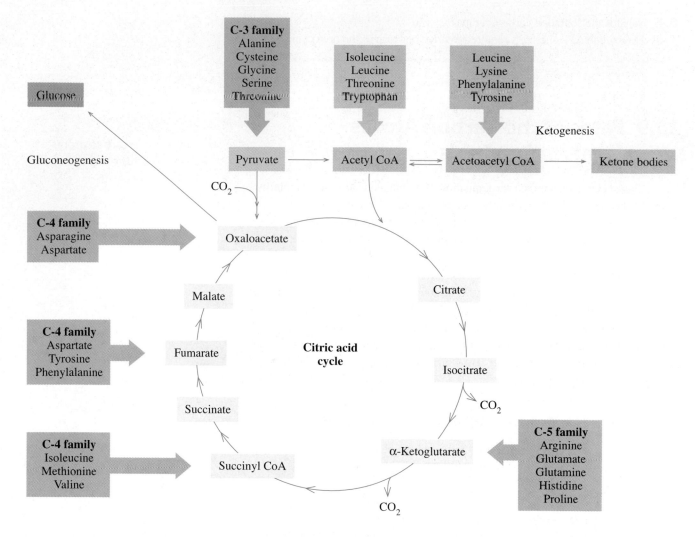

Figure 25.10 Carbon atoms from degraded amino acids are converted to the intermediates of the citric acid cycle or other pathways. Glucogenic amino acids (orange boxes) produce carbon skeletons that can form glucose, and ketogenic amino acid (green boxes) can produce ketone bodies.

Q *Why is aspartate glucogenic, but leucine is ketogenic?*

Degradation of Ketogenic Amino Acids

Two ketogenic amino acids, leucine and lysine, require a series of reactions that degrade them to acetoacetate, which is an intermediate in the production of ketone bodies. Lysine also produces acetyl CoA. Some of the carbon atoms of the aromatic amino acids phenylalanine and tyrosine are converted to acetoacetate as well as fumarate.

SAMPLE PROBLEM 25.11

What degradation products of amino acids are citric acid intermediates?

Solution

Carbon atoms from amino acids enter the citric acid cycle as acetyl CoA, α-ketoglutarate, succinyl CoA, fumarate, or oxaloacetate.

Study Check

Which amino acids provide carbon atoms that enter the citric acid as α-ketoglutarate?

QUESTIONS AND PROBLEMS

Fates of the Carbon Atoms from Amino Acids

25.49 What is a glucogenic amino acid?

25.50 What is a ketogenic amino acid?

25.51 What metabolic substrate(s) can be produced from the carbon atoms of each of the following amino acids?
  a. alanine   **b.** aspartate
  c. valine   **d.** glutamine

25.52 What metabolic substrate(s) can be produced from the carbon atoms of each of the following amino acids?
  a. leucine   **b.** asparagine
  c. cysteine   **d.** arginine

25.10 Synthesis of Amino Acids

LEARNING GOAL

Illustrate how some nonessential amino acids are synthesized from intermediates in the citric acid cycle and other metabolic pathways.

Plants and bacteria such as *E. coli* produce all of their amino acids using NH_4^+ and NO_3^-. However, humans can synthesize only 10 of the 20 amino acids found in their proteins. The **nonessential amino acids** are synthesized in the body, whereas the **essential amino acids** must be obtained from the diet. (See Table 25.3.) Two amino acids, arginine and histidine, are essential in diets for children due to their rapid growth requirements, but not adults.

Some Pathways for Amino Acid Synthesis

There are a variety of pathways involved in the synthesis of nonessential amino acids. When the body synthesizes nonessential amino acids, the α-keto acid skeletons are obtained from the citric acid cycle or glycolysis and converted to amino acids by transamination. (See Figure 25.11.)

Some of the amino acids are formed from a simple transamination. For example, the transfer of the amino group from glutamate to pyruvate, a three-carbon α-keto acid, produces alanine, an amino acid with three carbons.

Table 25.3   Essential and Nonessential Amino Acids in Humans

Essential		Nonessential	
Arginine*	Methionine	Alanine	Glutamine
Histidine*	Phenylalanine	Asparagine	Glycine
Isoleucine	Threonine	Aspartate	Proline
Leucine	Tryptophan	Cysteine	Serine
Lysine	Valine	Glutamate	Tyrosine

*Essential for children only

$$CH_3-\overset{O}{\overset{\|}{C}}-COO^- \ + \ {}^-OOC-\overset{\overset{+}{NH_3}}{\overset{|}{CH}}-CH_2-CH_2-COO^- \ \xrightarrow{\text{Glutamate pyruvate transaminase}}$$

Pyruvate   Glutamate

$$CH_3-\overset{\overset{+}{NH_3}}{\overset{|}{CH}}-COO^- \ + \ {}^-OOC-\overset{O}{\overset{\|}{C}}-CH_2-CH_2-COO^-$$

Alanine α-Ketoglutarate

In another transamination using glutamate, the four-carbon oxaloacetate from the citric acid cycle is converted to aspartate.

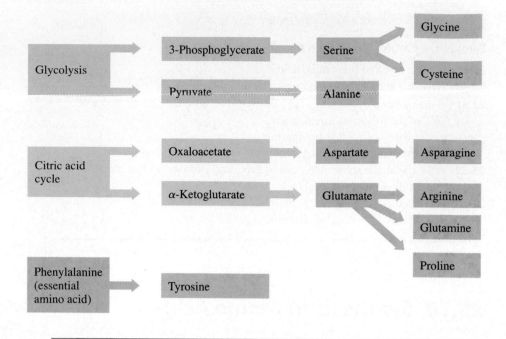

Figure 25.11 Nonessential amino acids are synthesized from intermediates of glycolysis and the citric acid cycle.

Q *How is alanine formed from pyruvate formed in glycolysis?*

$$^-OOC-CH_2-\overset{\overset{\displaystyle O}{\|}}{C}-COO^- \;+\; ^-OOC-\overset{\overset{\displaystyle \overset{+}{N}H_3}{|}}{CH}-CH_2-CH_2-COO^- \;\xrightarrow{\text{Glutamate oxaloacetate transaminase}}$$

Oxaloacetate $\qquad\qquad$ Glutamate

$$^-OOC-CH_2-\overset{\overset{\displaystyle \overset{+}{N}H_3}{|}}{CH}-COO^- \;+\; ^-OOC-\overset{\overset{\displaystyle O}{\|}}{C}-CH_2-CH_2-COO^-$$

Aspartate $\qquad\qquad$ α-Ketoglutarate

These two transaminases are abundant in the cells of the liver and heart, but are present only in low levels in the bloodstream. When an injury or disease occurs, they are released from the damaged cells into the bloodstream. The elevated levels of *serum glutamate-pyruvate transaminase* (SGPT) and *serum glutamate-oxaloacetate transaminase* (SGOT) provide a means to diagnose the extent of damage to the liver or heart.

Synthesis of Glutamine

The synthesis of the other nonessential amino acids requires several other reactions in addition to transamination. For example, glutamine is synthesized when a second amino group is added to glutamate using the energy from the hydrolysis of ATP.

$$^-OOC-\overset{\overset{\displaystyle \overset{+}{N}H_3}{|}}{CH}-CH_2-CH_2-COO^- \;+\; NH_3 \;\xrightarrow[\text{ATP}\quad\text{ADP + P}_i]{\text{Glutamine synthetase}}\; ^-OOC-\overset{\overset{\displaystyle \overset{+}{N}H_3}{|}}{CH}-CH_2-CH_2-\overset{\overset{\displaystyle O}{\|}}{C}-NH_2$$

Glutamate $\qquad\qquad\qquad\qquad\qquad\qquad\qquad\qquad\qquad$ Glutamine

Synthesis of Serine and Cysteine

In the synthesis of serine, three steps are required starting with 3-phosphoglycerate from glycolysis. In this pathway, the —OH group of glycerate is oxidized to give an α-keto acid, which undergoes transamination by glutamate accompanied by the loss of the phosphate group.

$$^-OOC-\overset{\overset{\displaystyle OH}{|}}{CH}-CH_2-O-\overset{\overset{\displaystyle O}{\|}}{\underset{\underset{\displaystyle O^-}{|}}{P}}-O^- \;+\; NAD^+ \;+\; Glutamate \;+\; H_2O \longrightarrow$$

3-Phosphoglycerate

$$^-OOC-\overset{\overset{\displaystyle \overset{+}{N}H_3}{|}}{CH}-CH_2-OH \;+\; NADH \;+\; H^+ \;+\; \alpha\text{-Ketoglutarate} \;+\; P_i$$

Serine

Once serine is formed, its —OH group is replaced by —SH from a reaction with homocysteine.

$$^-OOC-\overset{\overset{\displaystyle \overset{+}{N}H_3}{|}}{CH}-CH_2-OH \;+\; ^-OOC-\overset{\overset{\displaystyle \overset{+}{N}H_3}{|}}{CH}-CH_2-CH_2-SH \;\xrightarrow[H_2O]{Cysteine\ synthase}$$

Serine   Homocysteine

$$^-OOC-\overset{\overset{\displaystyle \overset{+}{N}H_3}{|}}{CH}-CH_2-SH \;+\; ^-OOC-\overset{\overset{\displaystyle O}{\|}}{C}-CH_2-CH_3$$

Cysteine α-Ketobutyrate

Synthesis of Tyrosine

Tyrosine, an aromatic amino acid with a hydroxyl group, is formed from phenylalanine, an essential amino acid.

$$\bigcirc\!\!\!\!\!\hexagon-CH_2-\overset{\overset{\displaystyle \overset{+}{N}H_3}{|}}{CH}-COO^- \;+\; O_2 \;\xrightarrow{Phenylalanine\ hydroxylase}\; HO-\bigcirc\!\!\!\!\!\hexagon-CH_2-\overset{\overset{\displaystyle \overset{+}{N}H_3}{|}}{CH}-COO^- \;+\; H_2O$$

Phenylalanine   Tyrosine

HEALTH NOTE

Homocysteine and Coronary Heart Disease

In the body, the amino acid methionine is degraded to homocysteine. We obtain most of our methionine, an essential amino acid, from the proteins in meat. In turn, homocysteine can be used to synthesize methionine in a process that requires folic acid and vitamin B_{12} (cobalamin). In a study at Harvard in the 1960s, children suffering from a genetic disorder, *homocystinuria*, were found to have high homocysteine levels. They were also found to have advanced atherosclerosis, which led to strokes and heart attacks early in life. This was one of the first indications of a link between elevated homocysteine levels and heart disease. In the past few years, additional clinical research has indicated that elevated blood levels of homocysteine are associated with increased risk of coronary heart disease, which can also lead to stroke and myocardial infarction (heart attack). It was also found that low levels of folic acid accompanied the elevated levels of homocysteine. This finding suggests that inadequate levels of folic acid limit the synthesis of methionine from homocysteine causing an accumulation of homocysteine.

When the body has adequate amounts of vitamin B_6, B_{12}, and folic acid, the synthesis of methionine maintains proper levels of homocysteine, which we need for healthy tissues. However, a deficiency of any one of these vitamins such as folic acid can lead to increased homocysteine levels, which may be damaging to the heart. Folic acid is recommended as supplement for pregnant women to avoid folic acid deficiencies during the growth of a fetus.

HEALTH NOTE

Phenylketonurea (PKU)

In the genetic disease phenylketonurea (PKU), a person cannot convert phenylalanine to tyrosine because the gene for an enzyme in the conversion is defective. As a result, large amounts of phenylalanine accumulate. In a different pathway, phenylalanine undergoes transamination to form phenylpyruvate, which is decarboxylated to phenylacetate. Large amounts of these compounds are excreted in the urine.

In infants, high levels of phenylpyruvate and phenylacetate cause severe mental retardation. However, the defect can be identified at birth, and all newborns are now tested for PKU. By detecting PKU early, retardation is avoided by using a diet with proteins that are low in phenylalanine and high in tyrosine. It is also important to avoid the use of sweeteners and soft drinks containing aspartame, which contains phenylalanine as one of two amino acids in its structure. In adulthood, persons with PKU can eat a nearly normal diet as long as they are checked for phenylpyruvate periodically.

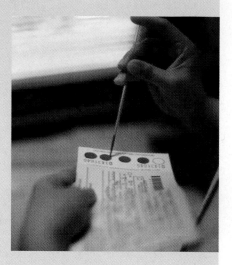

$$\overset{+}{N}H_3$$

Phenylalanine $\xrightarrow{\text{Transanimation}}$ Phenylpyruvate

Phenylalanine → Phenylalanine hydroxylase

Tyrosine

CO_2

Phenylacetate

SAMPLE PROBLEM 25.12

Synthesis of Amino Acids

What compound is most often the source of amino groups when transamination is used to synthesize nonessential amino acids?

Solution

Glutamate is the usual source of amino groups in the synthesis of nonessential amino acids.

Study Check

What are the sources of the substrates used to synthesize nonessential amino acids?

Overview of Metabolism

In these chapters, we have seen that catabolic pathways degrade large molecules to small molecules that are used for energy production via the citric acid cycle and the electron transport chain. We have also looked at the anabolic pathways that lead to the synthesis of larger molecules in the cell. In the overall view of metabolism, there are several branch points from which compounds may be degraded for energy or used to synthesize larger molecules. For example, glucose can be degraded to acetyl CoA for the citric acid cycle to produce energy or converted to glycogen for storage. When glycogen stores are depleted, fatty acids are degraded for energy. Amino acids normally used to synthesize nitrogen-containing compounds in the

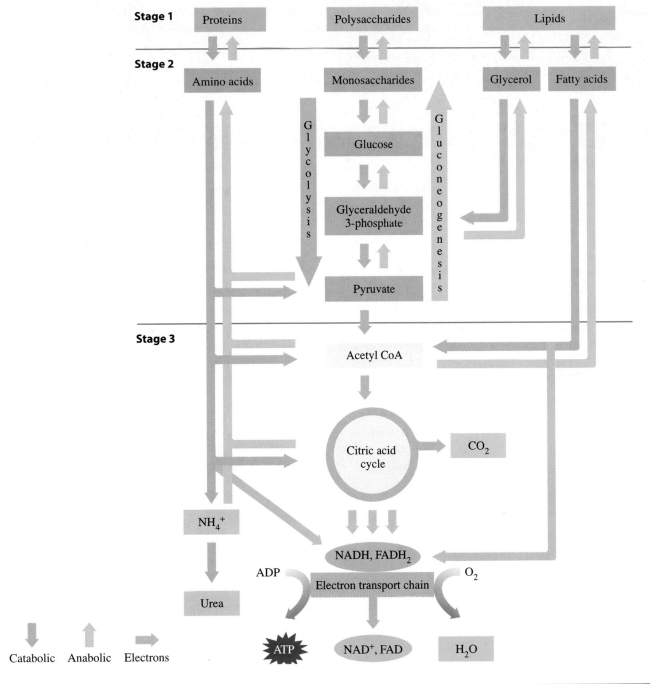

Figure 25.12 Catabolic and anabolic pathways in the cells provide the energy and necessary compounds for the cells.
Q *Under what conditions in the cell are amino acids degraded for energy?*

cells can also be used for energy after they are degraded to intermediates of the citric acid cycle. In the synthesis of nonessential amino acids, α-keto acids of the citric acid cycle enter a variety of reactions that convert them to amino acids through transamination by glutamate. (See Figure 25.12.)

QUESTIONS AND PROBLEMS

Synthesis of Amino Acids

25.53 What do we call the amino acids that humans can synthesize?

25.54 How do humans obtain the amino acids that cannot be synthesized in the body?

25.55 How is glutamate converted to glutamine?

25.56 What amino acid is needed for the synthesis of tyrosine?

25.57 What do the letters PKU mean?

25.58 How is PKU treated?

Chapter Review

25.1 Digestion of Triacylglycerols
Triacylglycerols are hydrolyzed in the small intestine to monoacylglycerol and fatty acids, which enter the intestinal wall and form new triacylglycerols. They bind with proteins to form chylomicrons, which transport them through the lymphatic system and bloodstream to the tissues.

25.2 Oxidation of Fatty Acids
When needed as an energy source, fatty acids are linked to coenzyme A and transported into the mitochondria where they undergo β oxidation. The fatty acyl chain is oxidized to yield a shorter fatty acid, acetyl CoA, and the reduced coenzymes NADH and $FADH_2$.

25.3 ATP and Fatty Acid Oxidation
Although the energy from a particular fatty acid depends on its length, each oxidation cycle yields 5 ATP with another 12 ATP from the acetyl CoA that enter the citric acid cycle.

25.4 Ketogenesis and Ketone Bodies
When high levels of acetyl CoA are present in the cell, they enter the ketogenesis pathway forming ketone bodies such as acetoacetate, which cause ketosis and acidosis.

25.5 Fatty Acid Synthesis
When there is an excess of acetyl CoA in the cell, the two-carbon acetyl CoA units link together to synthesize palmitate, which is converted to triacylglycerols and stored in the adipose tissue.

25.6 Digestion of Proteins
The digestion of proteins, which begins in the stomach and continues in the small intestine, involves the hydrolysis of peptide bonds by proteases to yield amino acids that are absorbed through the intestinal wall and transported to the cells.

25.7 Degradation of Amino Acids
When the amount of amino acids in the cells exceeds that needed for synthesis of nitrogen compounds, the process of transamination converts them to α-keto acids and glutamate. Oxidative deamination of glutamate produces ammonium ions and α-ketoglutarate.

25.8 Urea Cycle
Ammonium ions from oxidative deamination combine with bicarbonate and ATP to form carbamoyl phosphate, which is converted to urea.

25.9 Fates of the Carbon Atoms from Amino Acids
The carbon atoms from the degradation of glucogenic amino acids enter the citric acid cycle or glucose synthesis, whereas ketogenic amino acids provide acetyl CoA or acetoacetate for ketogenesis.

25.10 Synthesis of Amino Acids
Nonessential amino acids are synthesized when amino groups from glutamate are transferred to an α-keto acid obtained from glycolysis or the citric acid cycle.

Summary of Key Reactions

Digestion of Triacylglycerols

$$\text{Triacylglycerols} + 2H_2O \xrightarrow{\text{Pancreatic lipase}} \text{monoacylglycerols} + 2 \text{ fatty acids}$$

Mobilization of Fats

$$\text{Triacylglycerols} + 3H_2O \xrightarrow{\text{Lipases}} \text{glycerol} + 3 \text{ fatty acids}$$

Metabolism of Glycerol

Glycerol $+$ ATP $+$ NAD$^+$ $\longrightarrow$ dihydroxyacetone phosphate $+$ ADP $+$ NADH $+$ H$^+$

β Oxidation of Fatty Acid

Myristyl-CoA $+$ 6 CoA $+$ 6 FAD $+$ 6 NAD$^+$ $+$ 6 H$_2$O $\longrightarrow$

7 acetyl CoA $+$ 6 FADH$_2$ $+$ 6 NADH $+$ 6H$^+$

Fatty Acid Synthesis

8 Acetyl CoA $+$ 14 NADPH $+$ 14H$^+$ $+$ 7 ATP $\longrightarrow$

Palmitate (C$_{16}$) $+$ 8 CoA $+$ 14 NADP$^+$ $+$ 7 H$_2$O $+$ 7 ADP $+$ 7 P$_i$

Transamination

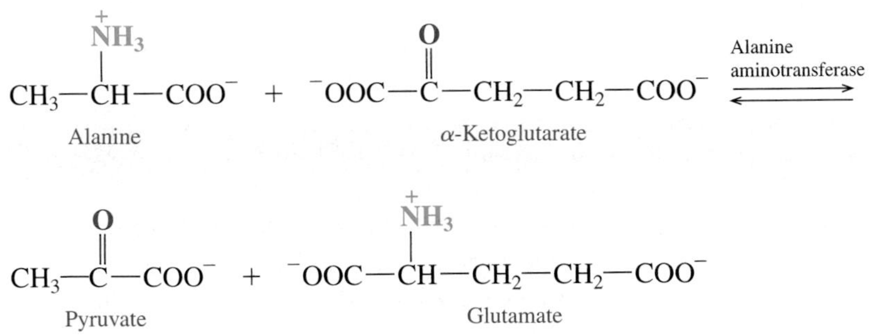

$$
\overset{\overset{+}{NH_3}}{\underset{|}{CH_3-CH-COO^-}} + \ ^-OOC-\overset{O}{\overset{\|}{C}}-CH_2-CH_2-COO^-
\underset{\longleftarrow}{\overset{\text{Alanine}}{\overset{\text{aminotransferase}}{\longrightarrow}}}
$$

Alanine α-Ketoglutarate

$$
CH_3-\overset{O}{\overset{\|}{C}}-COO^- + \ ^-OOC-\overset{\overset{+}{NH_3}}{\underset{|}{CH}}-CH_2-CH_2-COO^-
$$

Pyruvate   Glutamate

Oxidative Deamination

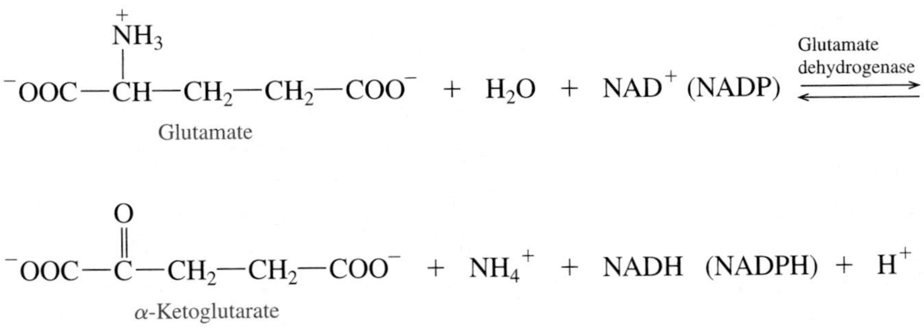

$$
^-OOC-\overset{\overset{+}{NH_3}}{\underset{|}{CH}}-CH_2-CH_2-COO^- + H_2O + NAD^+ (NADP)
\underset{\longleftarrow}{\overset{\text{Glutamate}}{\overset{\text{dehydrogenase}}{\longrightarrow}}}
$$

Glutamate

$$
^-OOC-\overset{O}{\overset{\|}{C}}-CH_2-CH_2-COO^- + NH_4^+ + NADH (NADPH) + H^+
$$

α-Ketoglutarate

Urea Cycle

NH$_4^+$ $+$ CO$_2$ $+$ 3 ATP $+$ aspartate $+$ 2H$_2$O $\longrightarrow$ Urea $+$ 2 ADP $+$ AMP $+$ 4 P$_i$ $+$ fumarate

Key Terms

acidosis   Low blood pH resulting from the formation of acidic ketone bodies.

adipose tissue   Tissues containing cells in which triacylglycerols are stored.

beta (β) oxidation   The degradation of fatty acids that removes two-carbon segments from an oxidized β carbon in the chain.

chylomicron   Lipoproteins formed by coating triacylglycerols with proteins for transport in the lymph and bloodstream.

essential amino acid   An amino acid that must be obtained from the diet because it cannot be synthesized in the body.

fat mobilization   The hydrolysis of triacylglycerols in the adipose tissue to yield fatty acids and glycerol for energy production.

glucogenic amino acid   An amino acid that provides carbon atoms for the synthesis of glucose.

ketogenesis   The pathway that converts acetyl CoA to four-carbon acetoacetate and other ketone bodies.

ketogenic amino acid   An amino acid that provides carbon atoms for the fatty acid synthesis or ketone bodies.

ketone bodies   The products of ketogenesis: acetoacetate, β-hydroxybutyrate, and acetone.

ketosis   A condition in which high levels of ketone bodies cannot be metabolized, leading to lower blood pH.

lipogenesis   The synthesis of fatty acid in which two-carbon acetyl units link together to yield palmitic acid.

nonessential amino acid   An amino acid that can be synthesized by reactions including transamination of α-keto acids in the body.

oxidative deamination   The loss of ammonium ion when glutamate is degraded to α-ketoglutarate.

protein turnover   The amount of protein that we break down from our diet and utilize for synthesis of proteins and nitrogen-containing compounds.

transamination   The transfer of an amino group from an amino acid to an α-keto acid.

urea cycle   Ammonium ions from the degradation of amino acids and CO_2 form carbamoyl phosphate, which is converted to urea.

Additional Problems

25.59   How are dietary triacylglycerols digested?

25.60   What is a chylomicron?

25.61   Why are the fats in the adipose tissues of the body considered as the major form of stored energy?

25.62   How are fatty acids obtained from stored fats?

25.63   Why doesn't the brain utilize fatty acids for energy?

25.64   Why don't red blood cells utilize fatty acids for energy?

25.65   A triacylglycerol is hydrolyzed in fat cells of adipose tissues and the fatty acid is transported to the liver.
   a. What happens to the glycerol?
   b. Where in the liver cells is the fatty acid activated for β oxidation?
   c. What is the energy cost for activation of the fatty acid?
   d. What is the purpose of activating fatty acids?

25.66   Consider the β oxidation of a saturated fatty acid.
   a. What is the activated form of the fatty acid?
   b. Why is the oxidation called β oxidation?
   c. What reactions in the fatty acid cycle require coenzymes?
   d. What is the yield in ATP for one cycle of β oxidation?

25.67   Lauric acid, CH_3—$(CH_2)_{10}$—COOH, is a saturated fatty acid.
   a. Write the formula of the activated form of lauric acid.

   b. Indicate the α and β carbon atoms in the fatty acyl molecule.
   c. Write the overall equation for the complete β oxidation for lauric acid.
   d. How many acetyl CoA units are produced?
   e. How many cycles of β oxidation are needed?
   f. Account for the total ATP yield from β oxidation of lauric acid (C_{12} fatty acid) by completing the following calculation:

activation	$\longrightarrow$	-2 ATP
____acetyl CoA	$\longrightarrow$	____ATP
____FADH$_2$	$\longrightarrow$	____ATP
____NADH	$\longrightarrow$	____ATP
Total		____ATP

25.68   Calculate the total ATP produced in the complex oxidation of caproic acid, $C_6H_{12}O_2$, with the total ATP produced from the oxidation of glucose, $C_6H_{12}O_6$.

25.69   Identify each of the following as involved in β oxidation or in fatty acid synthesis:
   a. NAD$^+$
   b. occurs in the mitochondrial matrix
   c. malonyl ACP
   d. cleavage of two-carbon acyl group
   e. acyl protein carrier
   f. acetyl CoA carboxylase

25.70   Identify each of the following as involved in β oxidation or in fatty acid synthesis:
   a. NADPH   **b.** takes place in the cytosol

c. FAD   d. oxidation of a hydroxyl group
e. coenzyme A   f. hydration of a double bond

25.71 The metabolism of triacylglycerols and carbohydrates is influenced by the hormones insulin and glucagon. Indicate the results of each of the following as stimulating (1) fatty acid oxidation or (2) the synthesis of fatty acids.
    a. low blood glucose   **b.** glucagon secreted

25.72 The metabolism of triacylglycerols and carbohydrates is influenced by the hormones insulin and glucagon. Indicate the results of each of the following as stimulating (1) fatty acid oxidation or (2) the synthesis of fatty acids.
    a. high blood glucose   **b.** insulin secreted

25.73 Why is ammonium ion produced in the liver converted immediately to urea?

25.74 What compound is regenerated in the urea cycle?

25.75 Indicate the corresponding reactant in the urea cycle for each of the following compounds:
    a. aspartate   **b.** ornithine

25.76 Indicate the products in the urea cycle for each of the following compounds:
    a. arginine   **b.** argininosuccinate

25.77 What metabolic substrate(s) can be produced from the carbon atoms of each of the following amino acids?
    a. serine   **b.** lysine
    c. methionine   **d.** glutamate

25.78 What metabolic substrate(s) can be produced from the carbon atoms of each of the following amino acids?
    a. leucine   **b.** isoleucine
    c. cysteine   **d.** phenylalanine

25.79 How much ATP can the degradation product of serine provide?

25.80 A camel hump contains 14 kg of triacylglycerols. Using the value of 0.49 moles of ATP per gram of fat, how many mole of ATP could be produced by the fat in the camel's hump?

Answers

Chapter 11

Answers to Study Checks

11.1 The structure needs more hydrogen atoms to complete four bonds to carbon and three bonds to nitrogen.

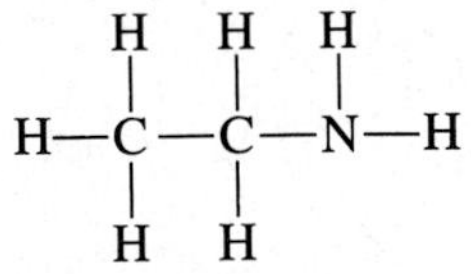

H—C—C—N—H (with H atoms)

11.2 In CBr$_4$, the shape that allows the bonds to be as far apart as possible is a tetrahedron.

11.3 The CH$_3$F molecule is polar because the stronger C—F dipole is not canceled by the C—H bonds.

11.4 Octane is not soluble in water; it is an organic compound.

11.5 CH$_3$CH$_2$—O—CH$_3$ contains the functional group C—O—C; it is an ether.

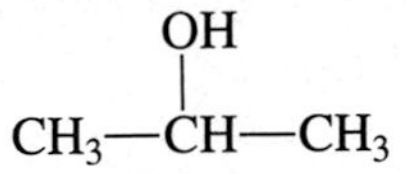

$$CH_3-\overset{\overset{\textstyle OH}{|}}{CH}-CH_3$$

contains the —OH functional group; it is an alcohol.

11.6 A primary amine has one carbon atom attached to nitrogen. A secondary amine has two carbon atoms attached to nitrogen.

11.7 Because the two compounds are constitutional isomers, they probably have different boiling points.

11.8 CH$_3$—CH$_2$—NH$_2$ and CH$_3$—$\overset{\overset{\textstyle H}{|}}{N}$—CH$_3$

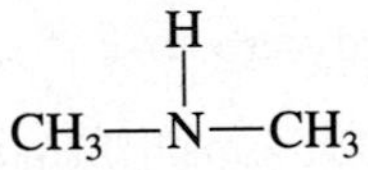

Answers to Selected Questions and Problems

11.1 a. 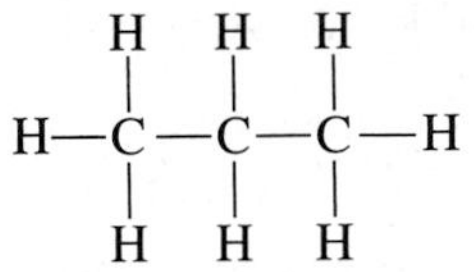 H—C—C—C—H (with H atoms, all single bonds)

b. H—C—C—O—H (with H atoms)

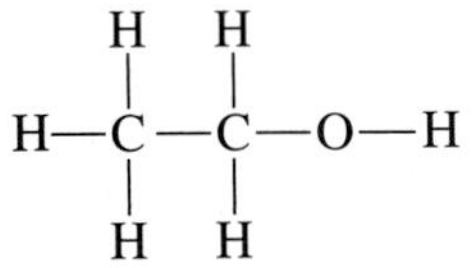

c. H—C=C—N—H (with H atoms)

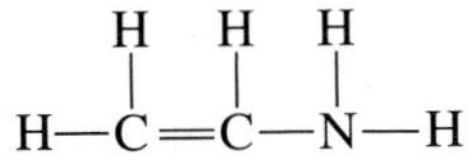

11.3 a. incorrect
b. incorrect
c. correct
d. correct

11.5 VSEPR theory predicts that the four bonds in CH$_4$ will be as far apart as possible, which means that the hydrogen atoms are at the corners of a tetrahedron.

11.7 a. nonpolar
b. nonpolar
c. polar
d. polar

11.9 a. inorganic
b. organic
c. organic
d. inorganic
e. inorganic
f. organic

11.11 a. inorganic
b. organic
c. organic
d. inorganic

11.13 a. alcohol
b. alkene
c. aldehyde
d. ester

11.15 a. ether
b. alcohol
c. ketone
d. carboxylic acid
e. amine

11.17 a. constitutional isomers
b. identical compounds
c. constitutional isomers
d. constitutional isomers
e. constitutional isomers
f. different compounds

11.19

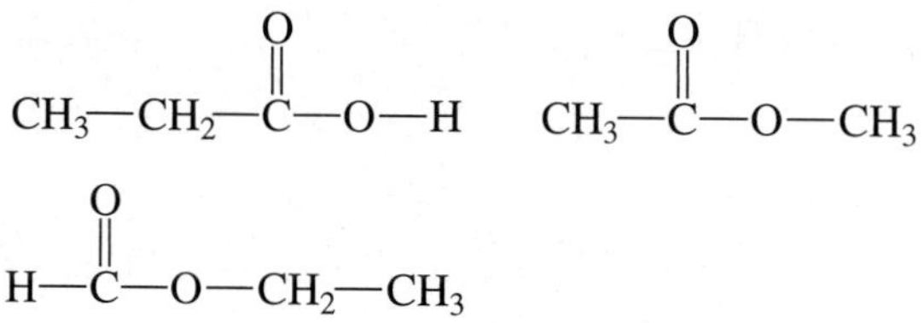

11.21 a. Organic compounds have covalent bonds; inorganic compounds have ionic as well as polar covalent and a few have nonpolar covalent bonds.
b. Most organic compounds are insoluble in water; many inorganic compounds are soluble in water.
c. Most organic compounds have low melting points; inorganic compounds have high melting points.
d. Most organic compounds are flammable; inorganic compounds are not flammable.

11.23 a. butane   **b.** butane
c. potassium chloride   **d.** potassium chloride
e. butane

11.25

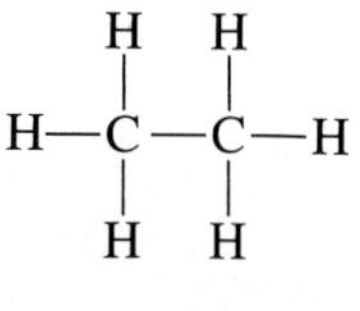

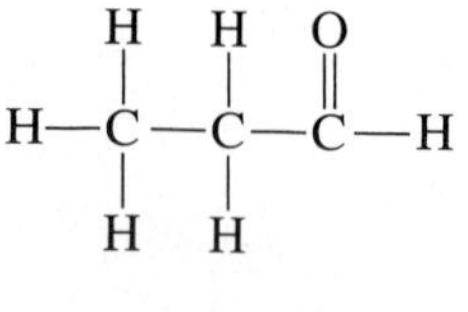

d.

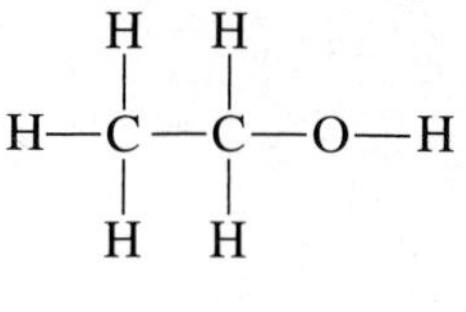

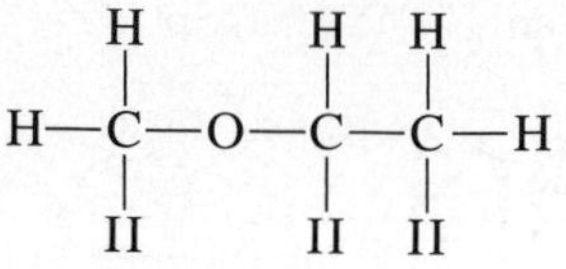

11.27 a. A hydroxyl group is $-OH$; a carbonyl group is $C=O$.
b. An alcohol contains the hydroxyl ($-OH$) functional group; an ether contains a $C-O-C$ functional group.
c. A carboxylic acid contains the COOH functional group; an ester contains a $-COOC-$ functional group.

11.29 a. constitutional isomers
b. different compounds
c. constitutional isomers
d. different compounds
e. constitutional isomers
f. identical compounds
g. different compounds

11.31 a. polar   **b.** polar
c. nonpolar   **d.** polar
e. polar   **f.** polar

11.33 a. alcohol   **b.** unsaturated hydrocarbon
c. aldehyde   **d.** saturated hydrocarbon
e. carboxylic acid   **f.** amine
g. tetrahedral   **h.** constitutional isomers
i. ester   **j.** hydrocarbon
k. ether   **l.** unsaturated hydrocarbon
m. functional group   **n.** ketone

Chapter 12

Answers to Study Checks

12.1 $CH_3-CH_2-CH_2-CH_2-CH_2-CH_2-CH_3$

12.2 This is a constitutional isomer. There is a five-carbon chain with a methyl group bonded to the middle (third) carbon.

12.3 3-ethyl-5-methylheptane

12.4

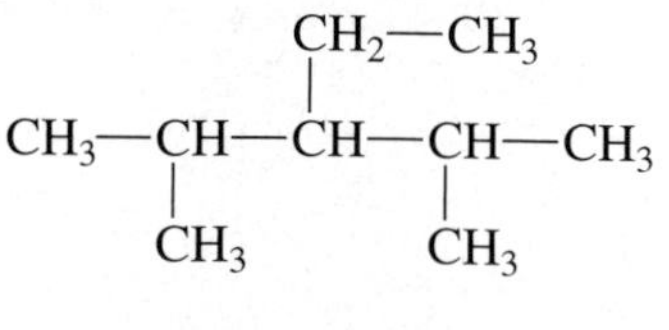

12.5 2,2-dimethylpentane; 3,3-dimethylpentane
2,3-dimethylpentane; 2,4-dimethylpentane

12.6 1-bromo-1-chloro-2,2,2-trifluoroethane

12.7 1-bromo-2-chloro-3-methylcyclobutane

12.8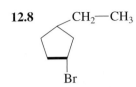

12.9

H₃C   CH₃

12.10 Cyclohexane

12.11

$$CH_3—CH—CH_2—CH_3 = C_5H_{12}$$
(with CH_3 on the second carbon)

$$C_5H_{12} + 8O_2 \longrightarrow 5CO_2 + 6H_2O$$

12.12

$$CH_3—CH—Br \qquad Br—CH_2—CH_2—Br$$
(with Br on the second carbon)

Answers to Selected Questions and Problems

12.1 a. C_7H_{16} **b.** 12   **c.** 4

12.3

a.   H—C—C—C—H (each carbon with H above and below)

b. $CH_3—CH_2—CH_2—CH_2—CH_2—CH_3$

c.

12.5 a. pentane
b. heptane
c. hexane

12.7 a. two different conformations
b. constitutional isomers
c. constitutional isomers

12.9 a. propyl
b. isopropyl
c. butyl
d. methyl

12.11 a. 2-methylpropane
b. 2-methylpentane
c. 4-ethyl-2-methylhexane
d. 2,3,4-trimethylheptane
e. 5-isopropyl-3-methyloctane

12.13 a. 2,3-dimethylpentane
b. 3-ethyl-5-methylheptane
c. 3-isopropyl-2,4-dimethylpentane

12.15

a. $CH_3—CH—CH_2—CH_3$ (with CH_3 above the second carbon)

b. $CH_3—CH_2—C—CH_2—CH_3$ (with CH_3 above and below the central carbon)

c. $CH_3—CH—CH—CH_2—CH—CH_3$ (with CH_3 above carbons 2, 3, and 5)

d. $CH_3—CH—CH—CH_2—CH—CH_2—CH_2—CH_3$ (with CH_3 above C2, $CH_2—CH_3$ above C3, CH_3 above C5)

e. $CH_3—CH—CH_2—CH—CH_2—CH_2—CH_3$ (with CH_3 above C2, and $CH—CH_3$ with CH_3 above C4)

f. $CH_3—CH_2—CH_2—CH—CH_2—CH_2—CH_2—CH_2—CH_3$ (with $CH_2—CH_2—CH_3$ above C4)

12.17 a. 2-methylheptane; 3-methylheptane;
4-methylheptane
b. 2,2-dimethylpentane; 3,3-dimethylpentane;
2,3-dimethylpentane; 2,4-dimethylpentane;
3-ethylpentane

12.19 a. bromoethane; ethyl bromide
b. 1-fluoropropane; propyl fluoride
c. 2-chloropropane; isopropyl chloride
d. trichloromethane; chloroform

12.21 a. 2-bromo-3-methylbutane
b. 3-bromo-2-chloropentane
c. 2-fluoro-2-methylbutane

12.23

a. 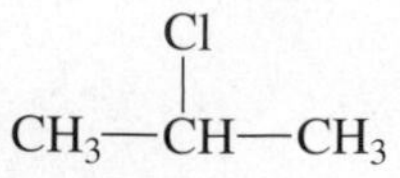 CH$_3$—CH—CH$_3$ (with Cl on the middle CH)

b. 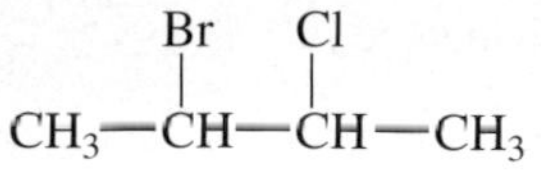 CH$_3$—CH—CH—CH$_3$ (with Br and Cl)

c. CH$_3$Br

d. CH$_3$—CH$_2$—CH$_2$—CH$_2$—Cl

e. 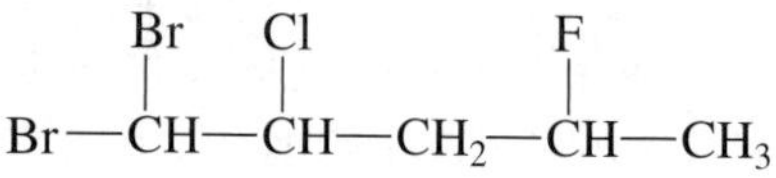 Br—CH—CH—CH$_2$—CH—CH$_3$ (with Br, Cl, F)

f. CBr$_4$

12.25 methyl chloride is CH$_3$Cl; ethyl chloride is CH$_3$CH$_2$Cl.

12.27

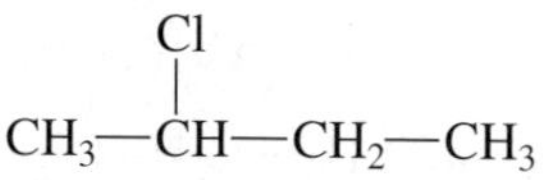

Cl—CH$_2$—CH$_2$—CH$_2$—CH$_3$ CH$_3$—CH—CH$_2$—CH$_3$ (Cl on 2nd C)

1-chlorobutane   2-chlorobutane

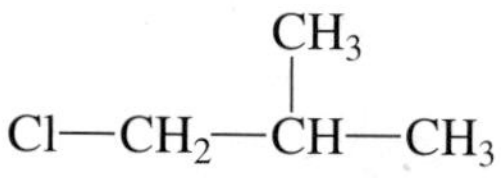

Cl—CH$_2$—CH—CH$_3$ (CH$_3$ branch)   CH$_3$—C—CH$_3$ (CH$_3$ up, Cl down)

1-chloro-2-methylpropane 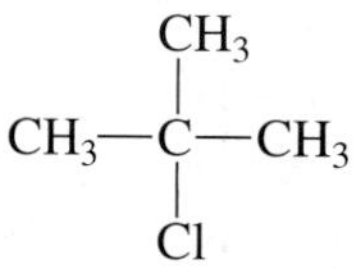   2-chloro-2-methylpropane

12.29 **a.** C$_5$H$_{10}$
b. 8
c. 6

12.31 **a.** cyclobutane
b. chlorocyclopentane
c. methylcyclohexane
d. 1-bromo-3-methylcyclobutane
e. 1-bromo-2-chlorocyclopentane
f. 1,3-dibromo-5-methylcyclohexane

12.33 **a.** 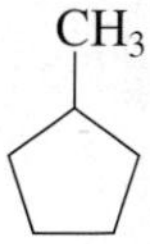 CH$_3$ (cyclopentane)   **b.** 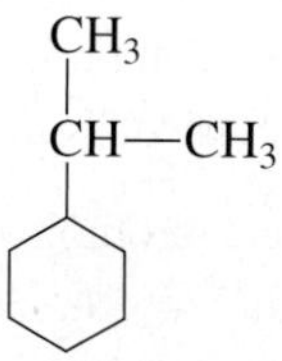 CH$_3$—CH—CH$_3$ (cyclohexane)

c. 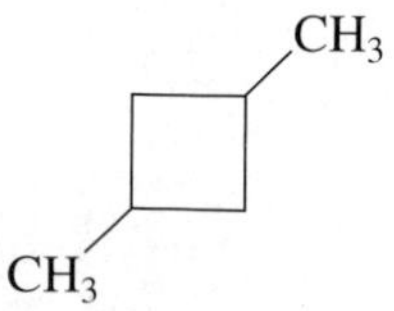 CH$_3$ / CH$_3$ (cyclobutane)   **d.** 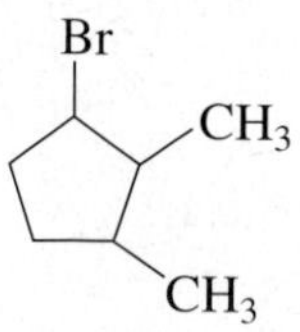 Br / CH$_3$ / CH$_3$ (cyclopentane)

12.35 Choice of three: 1,1-dimethylcyclopropane, 1,2-dimethylcyclopropane, ethylcyclopropane, methyl-cyclobutane

12.37 **a.** Does not have cis-trans isomers
b. *trans*-1,2-dimethylcyclopropane
c. *cis*-1,2-dichlorocyclobutane
d. *trans*-1,3-dimethylcyclopentane

12.39 **a.** 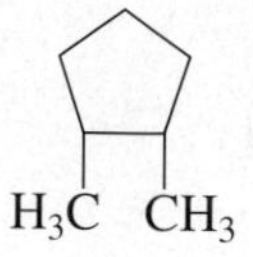 (cyclopentane) H$_3$C   CH$_3$

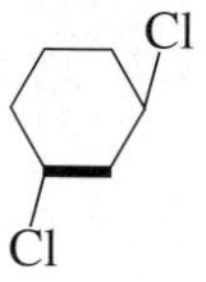

b. (cyclohexane) Cl / Cl

c. (cyclobutane) Br   CH$_3$

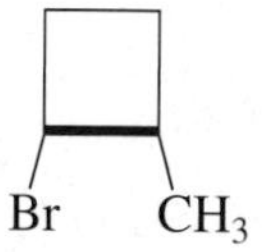

12.41 **a.** CH$_3$—CH$_2$—CH$_2$—CH$_2$—CH$_2$—CH$_2$—CH$_3$
b. liquid
c. insoluble in water
d. float
e. lower

12.43 **a.** heptane   **b.** cyclopropane   **c.** hexane

12.45 **a.** 2C$_2$H$_6$ + 7O$_2$ $\longrightarrow$ 4CO$_2$ + 6H$_2$O
b. 2C$_3$H$_6$ + 9O$_2$ $\longrightarrow$ 6CO$_2$ + 6H$_2$O
c. 2C$_8$H$_{18}$ + 25O$_2$ $\longrightarrow$ 16CO$_2$ + 18H$_2$O
d. C$_6$H$_{12}$ + 9O$_2$ $\longrightarrow$ 6CO$_2$ + 6H$_2$O

12.47 **a.** CH$_3$—CH$_2$—Cl

b. 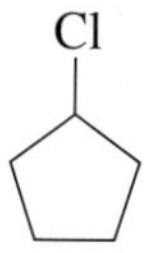 Cl (cyclopentane)

c. 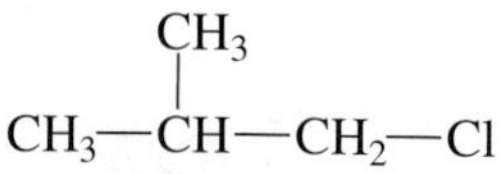 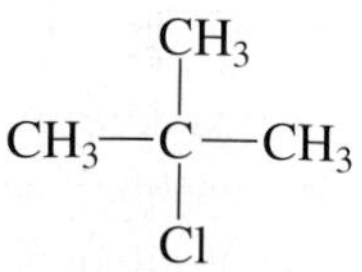 CH$_3$—CH—CH$_2$—Cl (with CH$_3$ branch)   CH$_3$—C—CH$_3$ (with CH$_3$ up, Cl down)

12.49 **a.** constitutional isomers
b. constitutional isomers
c. same molecule
d. constitutional isomers

12.51 a. methyl
 b. propyl
 c. isopropyl

12.53 a. 2,2-dimethylbutane
 b. chloroethane
 c. 2-bromo-4-ethylhexane
 d. 1,1-dibromocyclohexane

12.55

a. 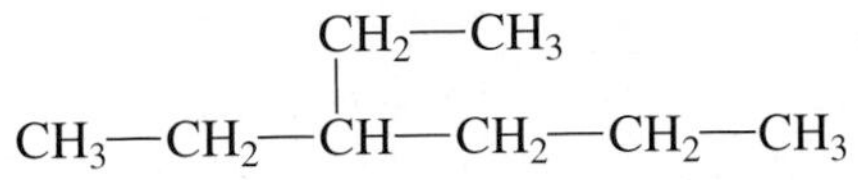 CH$_3$—CH$_2$—CH—CH$_2$—CH$_2$—CH$_3$ (with CH$_2$—CH$_3$ branch)

b. 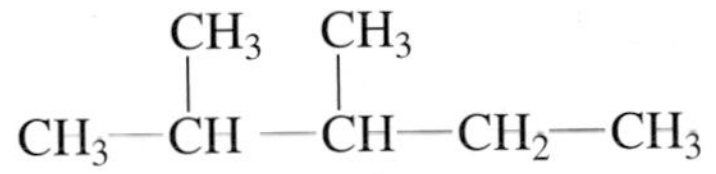 CH$_3$—CH—CH—CH$_2$—CH$_3$ (with CH$_3$, CH$_3$ branches)

c. 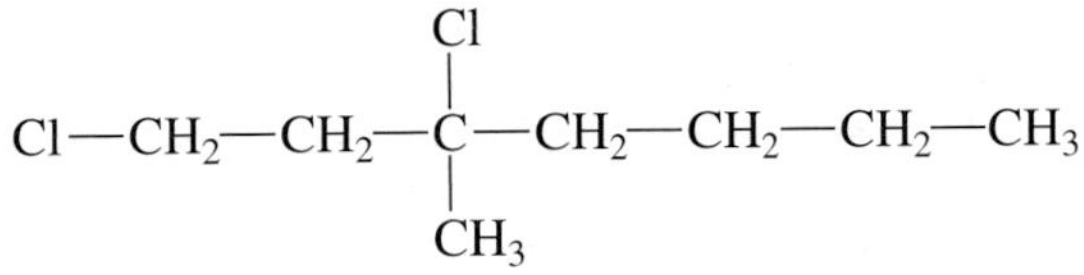 Cl—CH$_2$—CH$_2$—C—CH$_2$—CH$_2$—CH$_2$—CH$_3$ (with Cl above and CH$_3$ below)

d. 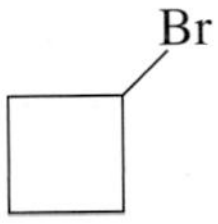 (cyclobutane with Br)

e. 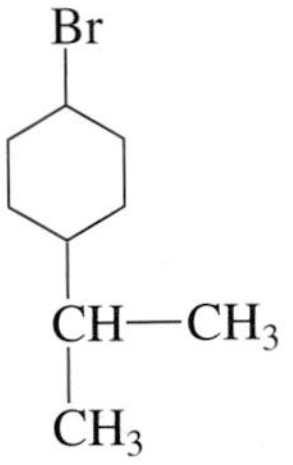 (cyclohexane with Br and CH—CH$_3 / CH_3$)

f. 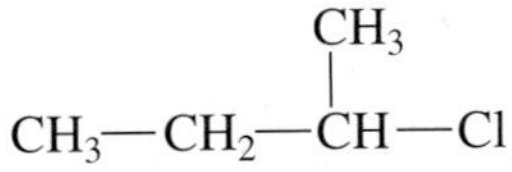 CH$_3$—CH$_2$—CH—Cl (with CH$_3$ branch)

12.57 a.    **b.**

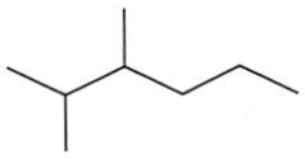

 c. 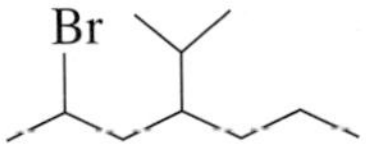 (Br and isopropyl)

12.59 a. You should have two of the following four possibilities.

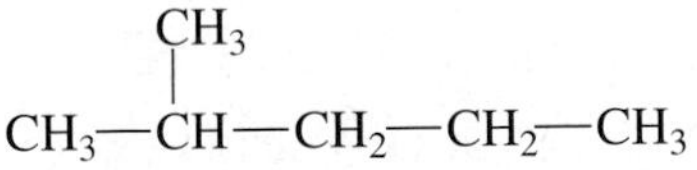

CH$_3$—CH—CH$_2$—CH$_2$—CH$_3$ 2-methylpentane (with CH$_3$ branch)

CH$_3$—CH$_2$—CH—CH$_2$—CH$_3$ 3-methylpentane (with CH$_3$ branch)

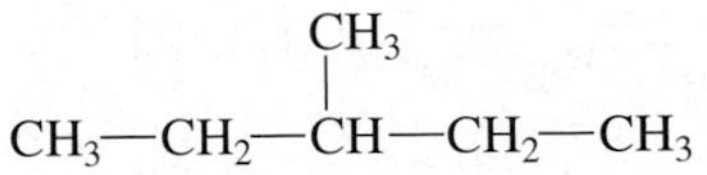

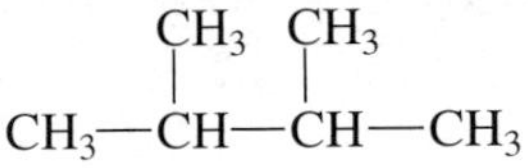

CH$_3$—CH—CH—CH$_3$ 2,3-dimethylbutane (with CH$_3$, CH$_3$ branches)

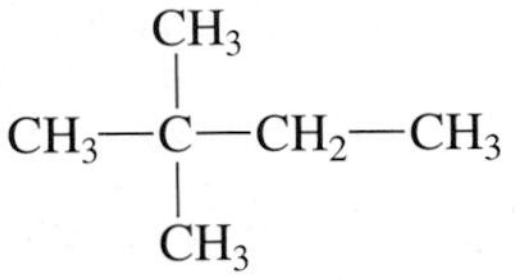

CH$_3$—C—CH$_2$—CH$_3$ 2,2-dimethylbutane (with CH$_3$ above and CH$_3$ below)

b. The molecular formula C$_7$H$_{16}$ would be the formula for the following constitutional isomers:

CH$_3$—CH$_2$—CH$_2$—CH$_2$—CH$_2$—CH$_2$—CH$_3$ heptane

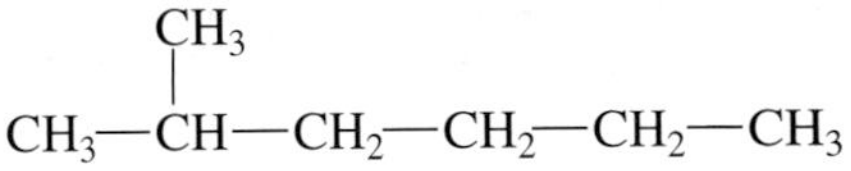

CH$_3$—CH—CH$_2$—CH$_2$—CH$_2$—CH$_3$ 2-methylhexane (with CH$_3$ branch)

CH$_3$—CH$_2$—CH—CH$_2$—CH$_2$—CH$_3$ 3-methylhexane (with CH$_3$ branch)

CH$_3$—C—CH$_2$—CH$_2$—CH$_3$ 2,2-dimethylpentane (with CH$_3$ above and CH$_3$ below)

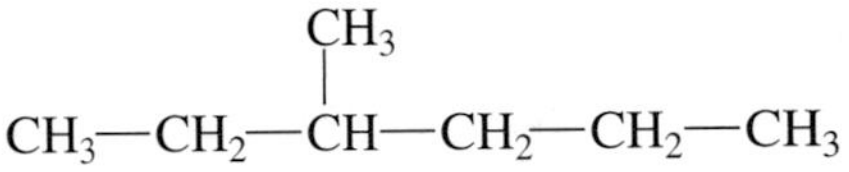

CH$_3$—CH$_2$—C—CH$_2$—CH$_3$ 3,3-dimethylpentane (with CH$_3$ above and CH$_3$ below)

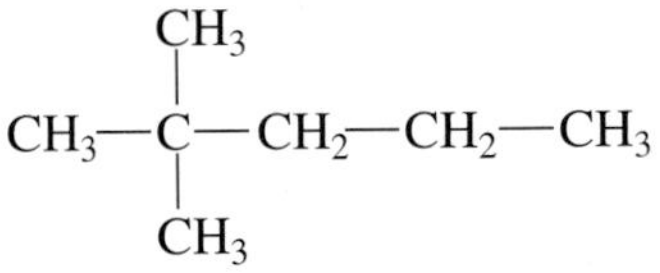

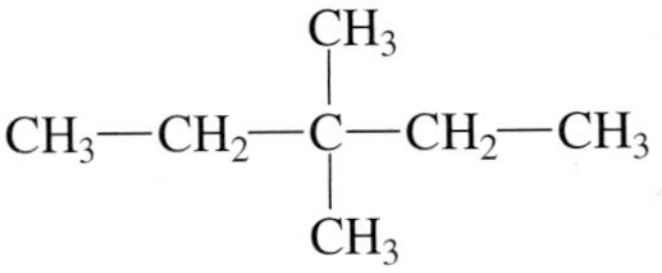

CH$_3$—CH—CH—CH$_2$—CH$_3$ 2,3-dimethylpentane (with CH$_3$, CH$_3$ branches)

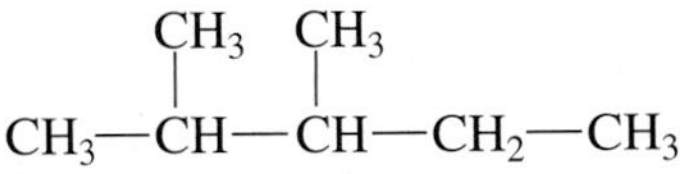

CH$_3$—CH—CH$_2$—CH—CH$_3$ 2,4-dimethylpentane (with CH$_3$, CH$_3$ branches)

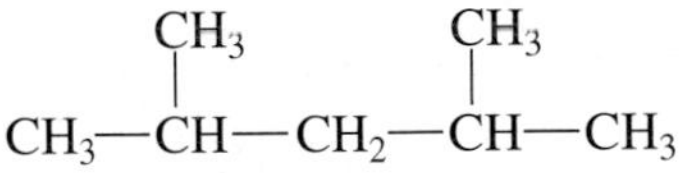

CH$_3$—C—CH—CH$_3$ 2,2,3-trimethylbutane (with CH$_3$, CH$_3$ above and CH$_3$ below)

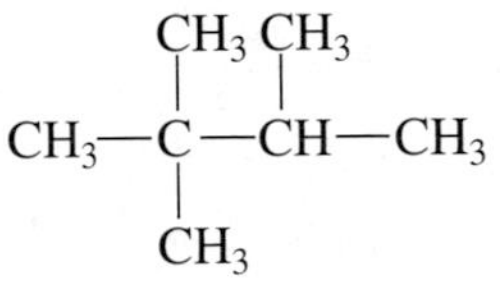

c. CH₃—CH₂—CH₂—CH₂—CH₂—CH₃   hexane

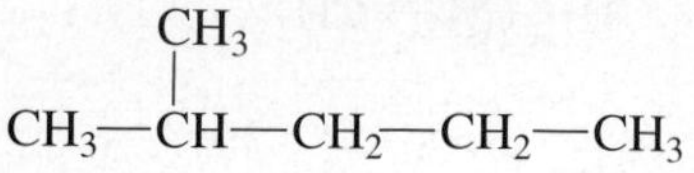

CH₃—CH—CH₂—CH₂—CH₃   2-methylpentane
　　│
　　CH₃

CH₃—CH₂—CH—CH₂—CH₃   3-methylpentane
　　　　　│
　　　　　CH₃

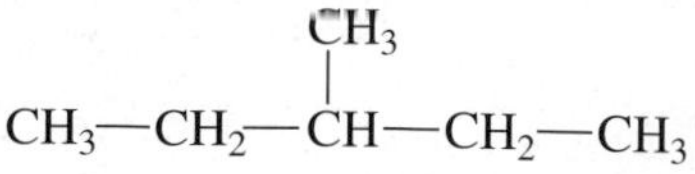

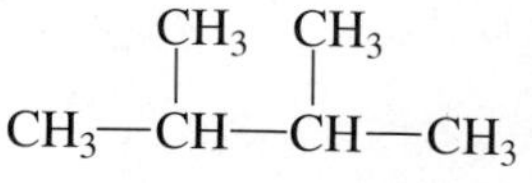

CH₃—CH—CH—CH₃   2,3-dimethylbutane
　　│　　│
　　CH₃　CH₃

d.

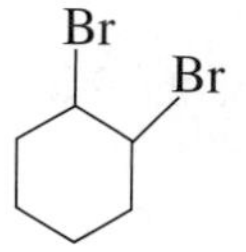

1,2-dibromo-
cyclohexane

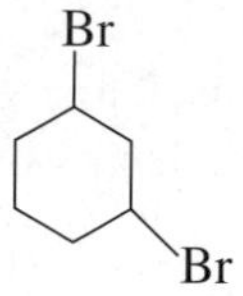

1,3-dibromo-
cyclohexane

1,4-dibromo-
cyclohexane

12.61 a. 2-methylbutane
b. 2,3-dimethylpentane
c. 1,3-dibromocyclohexane
d. hexane

12.63 CH₃—CH₂—CH₂—CH₂—CH₂—Cl

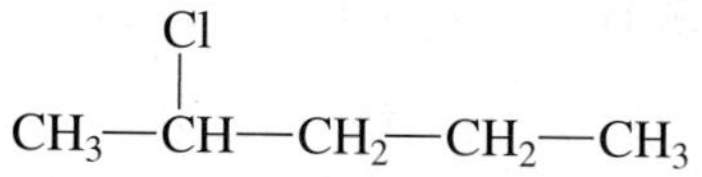

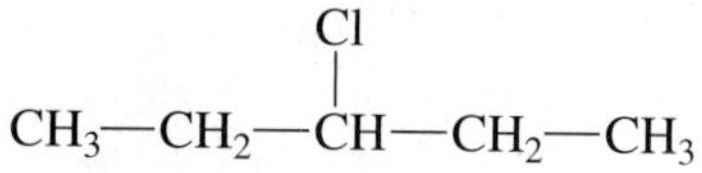

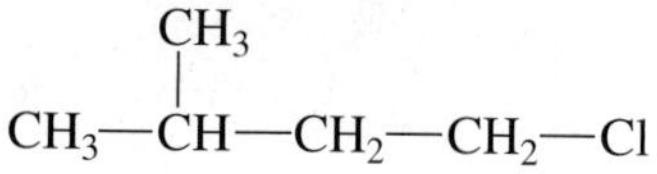

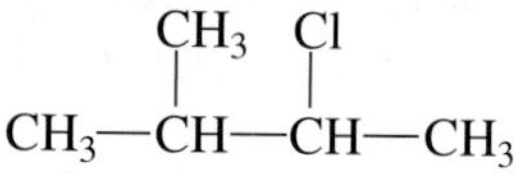

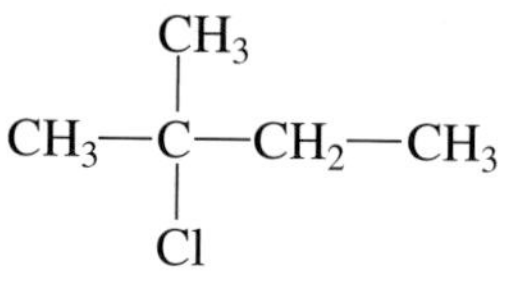

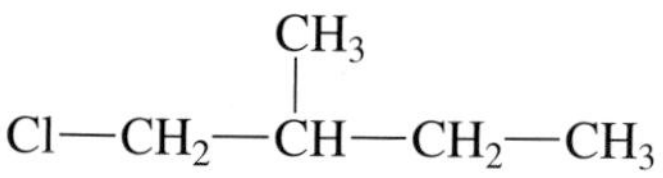

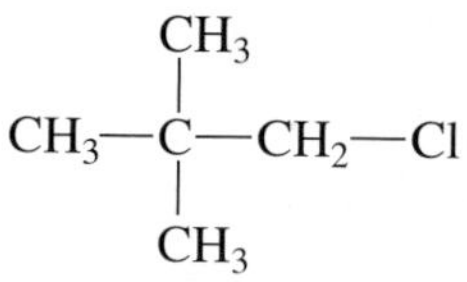

CH₃—CH—CH₂—CH₂—CH₃
　　│
　　Cl

CH₃—CH₂—CH—CH₂—CH₃
　　　　　│
　　　　　Cl

CH₃—CH—CH₂—CH₂—Cl
　　│
　　CH₃

CH₃—CH—CH—CH₃
　　│　　│
　　CH₃　Cl

CH₃—C—CH₂—CH₃
　　│
　　Cl
（CH₃ above C）

Cl—CH₂—CH—CH₂—CH₃
　　　　　│
　　　　　CH₃

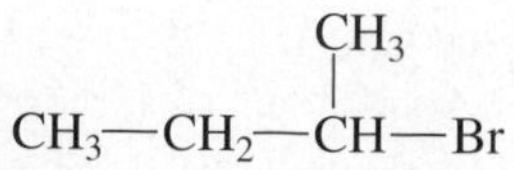

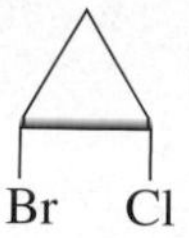

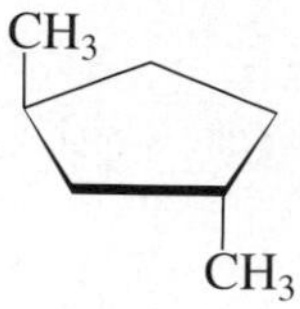

CH₃—C—CH₂—Cl
　　│
　　CH₃
（CH₃ above C）

12.65

a. CH₃—CH₂—CH—Br
　　　　　│
　　　　　CH₃

b.
Br   Cl

c.
CH₃
（cyclopentane with two CH₃）
CH₃

d.
（cyclobutane with two Br）
Br   Br

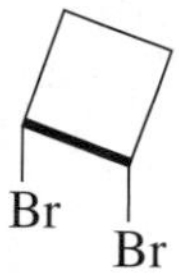

12.67 Condensed structural formula

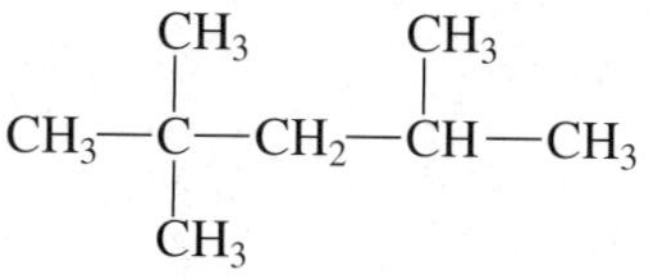

CH₃—C—CH₂—CH—CH₃
（CH₃ above and below left C, CH₃ above the CH）

molecular formula of C₈H₁₈

The combustion reaction: 2C₈H₁₈ + 25O₂ ⟶ 16CO₂ + 18H₂O

12.69 a. heptane   b. cyclopentane
c. hexane   d. cyclohexane

12.71 a. C₃H₈ + 5O₂ ⟶ 3CO₂ + 4H₂O
b. C₅H₁₂ + 8O₂ ⟶ 5CO₂ + 6H₂O
c. C₄H₈ + 6O₂ ⟶ 4CO₂ + 4H₂O
d. 2C₈H₁₈ + 25O₂ ⟶ 16CO₂ + 18H₂O

12.73 a. CH₃—CH₂—Cl

b. CH₃—CH₂—CH₂—Cl   and   CH₃—CH—CH₃
　　　　　　　　　　　　　　　　　│
　　　　　　　　　　　　　　　　　Cl

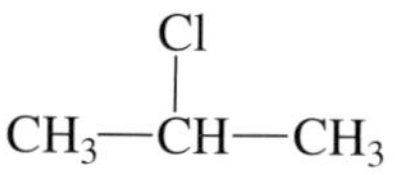

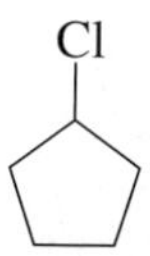

c.
（cyclopentane with Cl）

12.75 a. C₅H₁₂ + 8O₂ ⟶ 5CO₂ + 6H₂O
b. 72.0 g/mole
c. 2.8 × 10⁴ kcal
d. 3700 L

Chapter 13

Answers to Study Checks

13.1 This is cyclobutene, a cycloalkene.

13.2. In an alkene, the three bonded atoms are as far apart as possible at 120°. In an alkyne, the two bonded atoms are as far apart as possible at 180°.

13.3 **a.** $CH_3C≡CCH_2CH_3$ **b.**

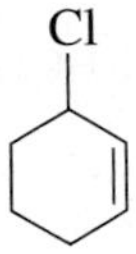

13.4

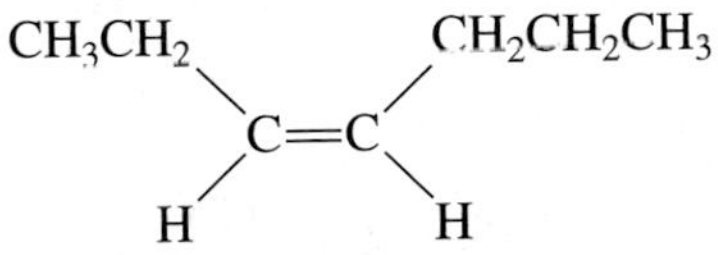

cis-3-Heptene

13.5

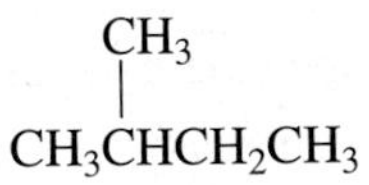

$CH_3CHCH_2CH_3$

13.6 1,2-dichlorobutane

13.7 CH_3 Br 1-bromo-1-methylcyclopentane

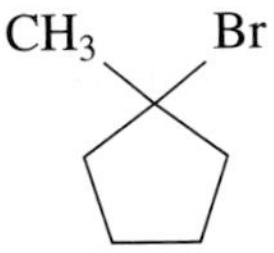

13.8 $CH_2=CHCH_2CH_2CH_3$ + HCl ⟶

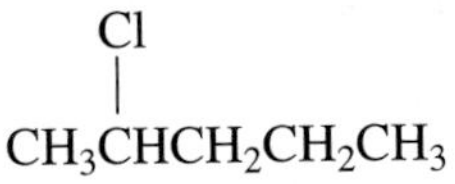

$CH_3CHCH_2CH_2CH_3$

13.9 CH_3

$CH_3CCH_2CH_3$

OH

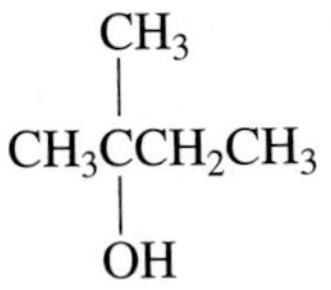

13.10 The monomer of PVC, polyvinyl chloride, is chloroethene:

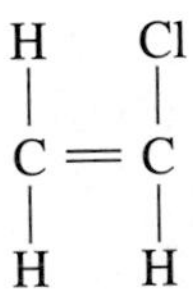

13.11 1,3-diethylbenzene; *m*-diethylbenzene

13.12 Chlorobenzene can be prepared from benzene and chlorine, using $FeCl_3$ as a catalyst.

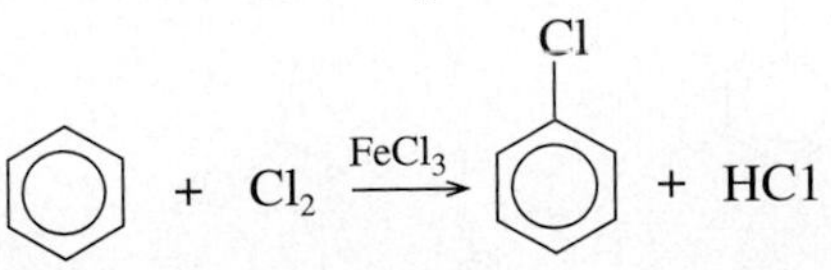

+ Cl_2 →(FeCl₃)→ + HCl

Answers to Selected Questions and Problems

13.1 **a.** An alkane has only sigma bonds.
b. An alkyne has a sigma bond and two pi bonds.
c. An alkene has the groups on the carbon atom arranged at 120°.

13.3 **a.** An alkane has the general formula C_nH_{2n+2}.
b. An alkene has a double bond.
c. An alkyne has a triple bond.
d. An alkene has a double bond.
e. A cycloalkene has a double bond in a ring.

13.5. $CH_2=CH—CH_3$ Propene

Cyclopropane

13.7 **a.** Propene contains a double bond, and propyne has a triple bond.
b. Cyclohexane is a saturated cycloalkane, and cyclohexene contains a double bond and is unsaturated.

13.9 **a.** ethene **b.** 2-methyl-1-propene
c. 4-bromo-2-pentyne **d.** cyclobutene
e. 4-ethylcyclopentene **f.** 4-ethyl-2-hexene

13.11 **a.** $CH_3CH=CH_2$ **b.** $CH_2=CHCH_2CH_2CH_3$

e. $CH_2=CCH_2CH_3$ **d.**
CH_3

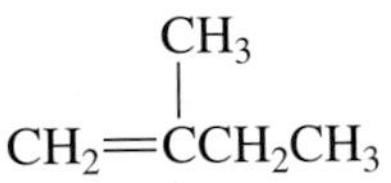

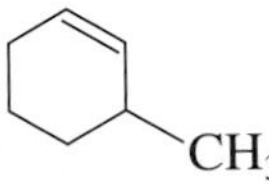

c. $CH_3—CH—C≡C—CH_2—CH_3$
Cl

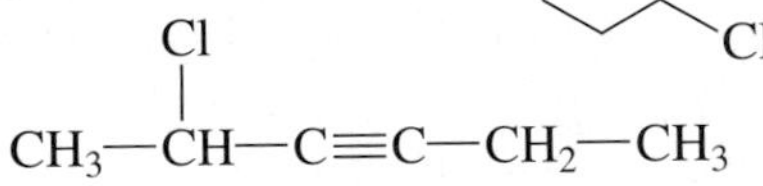

13.13 Constitutional isomers have the same molecular formula, but different arrangements of the atoms. Geometric isomers in alkenes have different geometric relationship of the groups attached to a double bond.

13.15 There are four constitutional isomers with the mo-

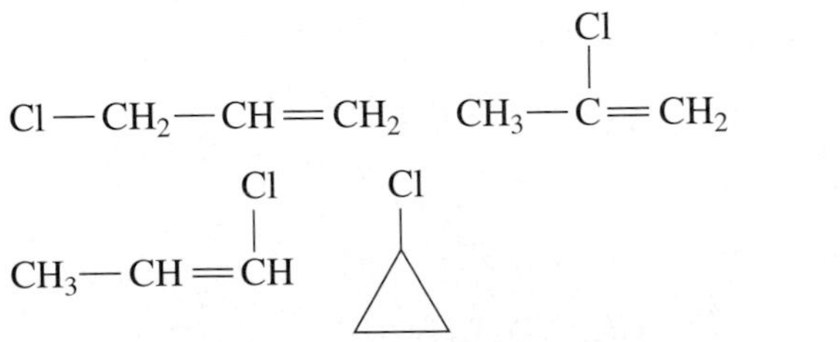

$Cl—CH_2—CH=CH_2$ $CH_3—C=CH_2$
Cl

$CH_3—CH=CH$
Cl Cl

13.17 **a.** There are no cis–trans isomers.
b. This alkene has cis–trans isomers.
c. There are no cis–trans isomers.

13.19 **a.** *cis*-2-butene **b.** *trans*-3-octene
c. *cis*-3-heptene

13.21 a.

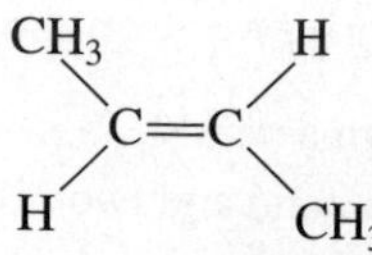

b.

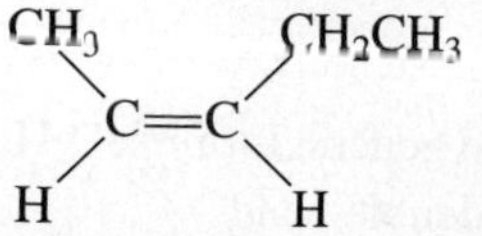

c.

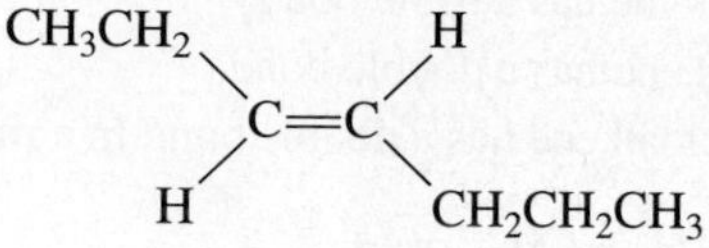

13.23 a. $CH_3-CH_2-CH_2-CH_2-CH_3$ Pentane

b.

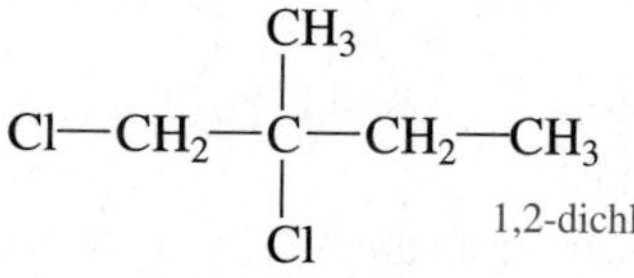

1,2-dichloro-2-methylbutane

c.

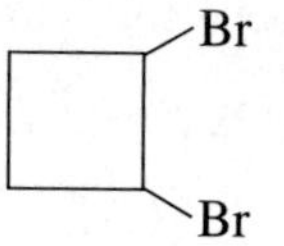

1,2-dibromocyclobutane

d.

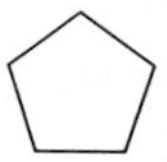

Cyclopentane

e.

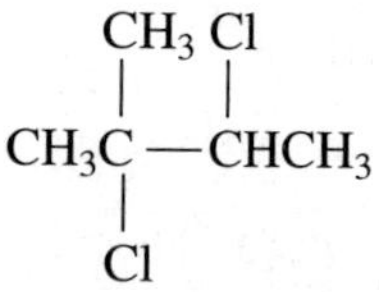

2,3-dichloro-2-methylbutane

f. $CH_3-CH_2-CH_2-CH_2-CH_3$ Pentane

13.25 a.

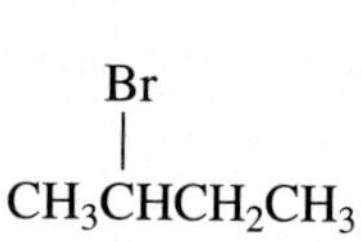

b.

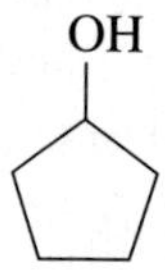

c.

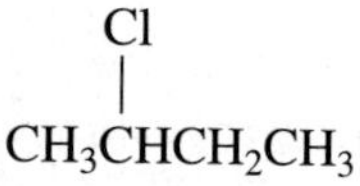

d.

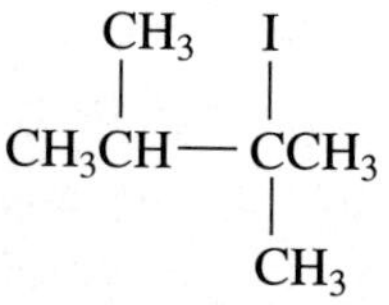

e.

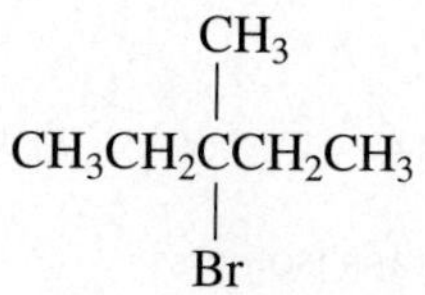

f.

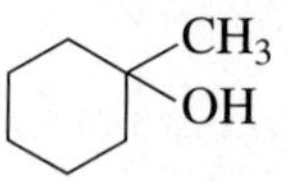

13.27

a.

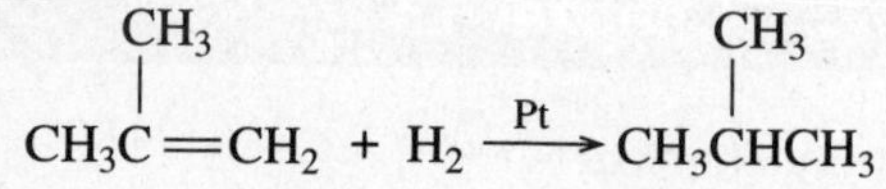

b.

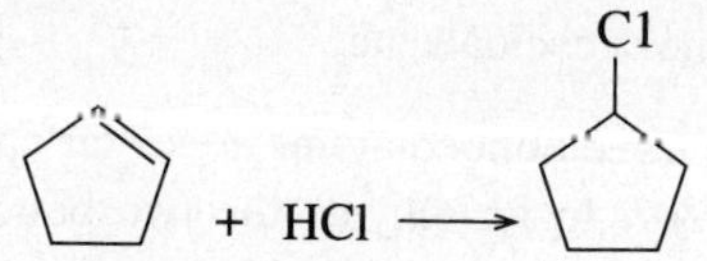

c. $CH_3CH = CHCH_2CH_3 + Br_2 \longrightarrow$

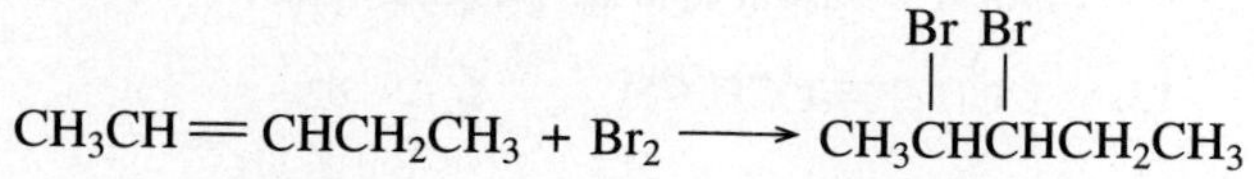

d. $CH_2 = CHCH_3 + HOH \xrightarrow{H^+}$

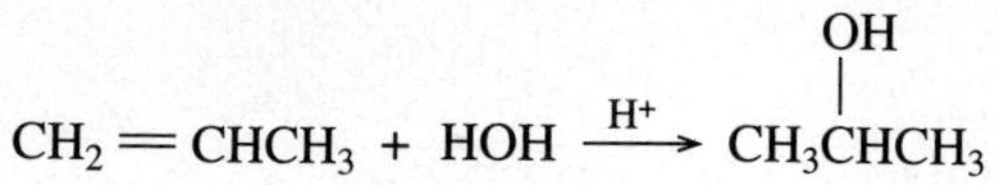

e. $CH_3-C\equiv C-CH_3 + 2Cl_2 \longrightarrow$

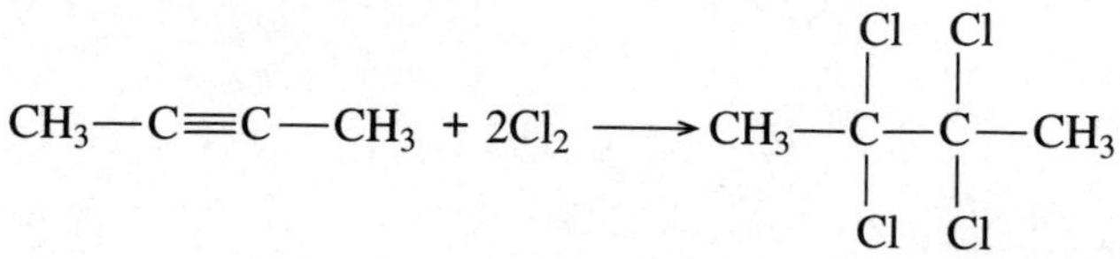

13.29   A polymer is a very large molecule composed of small units that are repeated many times.

13.31

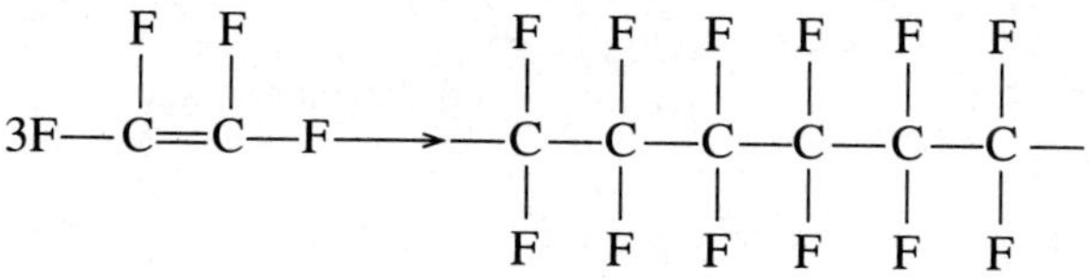

13.33

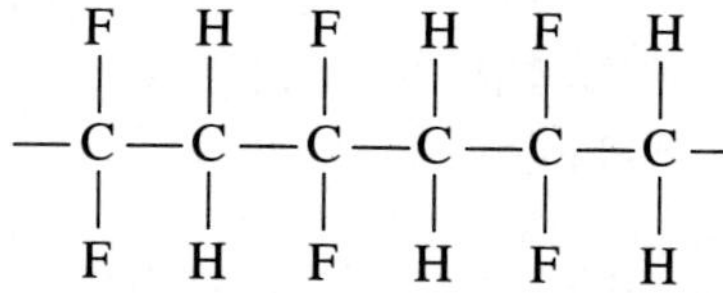

13.35   Cyclohexane, C_6H_{12}, is a cycloalkane in which six carbon atoms in a ring are linked by single bonds. In benzene, C_6H_6, alternating single and double bonds link the six carbon atoms.

13.37 a. 1-chloro-2-methylbenzene; o-chloromethylbenzene; o-chlorotoluene; 2-chlorotoluene
  b. ethylbenzene
  c. 1,3,5-trichlorobenzene
  d. m-xylene; m-methyltoluene;1,3-dimethylbenzene; 3-methyltoluene
  e. 1-bromo-3-chloro-5-methylbenzene; 3-bromo-5-chlorotoluene
  f. isopropyl benzene

13.39

a.

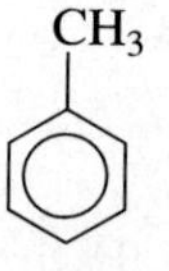

b.

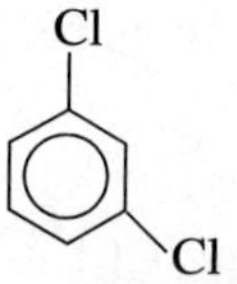

c. CH₂CH₃   **d.** CH₃

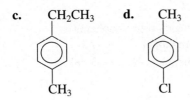

13.41 Benzene undergoes substitution reactions because a substitution reaction allows benzene to retain the stability of the aromatic system.

13.43 a. no reaction

b. Cl   **c.** NO₂

13.45 All the compounds have three carbon atoms. Propane is a saturated alkane, and cyclopropane is a saturated cyclic hydrocarbon. Both propene and propyne are unsaturated hydrocarbons, but propene has a double bond and propyne has a triple bond.

13.47 a. chlorocyclopentane
b. 2-chloro-4-methylpentane
c. 2-methyl-1-pentene
d. 2-pentyne
e. 1-chlorocyclopentene
f. *trans*-2-pentene
g. 1,3-dichlorocyclohexene

13.49 a. constitutional isomers **b.** cis-trans isomers
c. identical   **d.** constitutional isomers

13.51 CH₃   CH₃   CH₃   CH₂

13.53
a. CH₃   H
         C=C   *trans*-2-pentene
       H   CH₂CH₃

       CH₃   CH₂CH₃
         C=C   *cis*-2-pentene
       H   H

b. CH₃CH₂   H
          C=C   *trans*-3-hexene
        H   CH₂CH₃

       CH₃CH₂   CH₂CH₃
          C=C   *cis*-3-hexene
        H   H

c. CH₃   H
         C=C   *trans*-2-butene
       H   CH₃

       CH₃   CH₃
         C=C   *cis*-2-butene
       H   H

d. CH₃   H
         C=C   *trans*-2-hexene
       H   CH₂CH₂CH₃

       CH₃   CH₂CH₂CH₃
         C=C   *cis*-2-hexene
       H   H

13.55

a. ⬡   **b.** CH₃CH=CHCH₂CH₃

c. CH₂=CHCH₃   **d.** ⬠

13.57

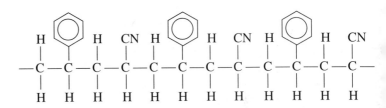

13.59 a. methylbenzene; toluene
b. 1-chloro-2-methylbenzene; *o*-chlorotoluene; 2-chlorotoluene
c. 1-ethyl-4-methylbenzene; *p*-ethylmethylbenzene; *p*-ethyltoluene
d. 1,3-diethylbenzene; *m*-diethylbenzene

13.61 a. chlorobenzene
b. *o*-bromotoluene, *m*-bromotoluene, *p*-bromotoluene
c. benzenesulfonic acid
d. no products

Chapter 14

Answers to Study Checks

14.1 3°; The carbon attached to the hydroxyl group is bonded to three carbon atoms.

14.2 3-bromo-4-methylcyclopentanol

14.3

$$CH_3CH_2\overset{\overset{\displaystyle OH}{|}}{CH}-\overset{\overset{\displaystyle CH_3}{|}}{CH}-\overset{\overset{\displaystyle Cl}{|}}{CH}CH_2CH_3$$

14.4 CH_3CH_2-SH

14.5 Ethanol

14.6 ethyl phenyl ether

14.7 1-butanol, 2-butanol, methoxy-1-propane, ethoxy ethane.

14.8 Both are unsaturated cyclic ethers, but furan has five atoms in the ring and pyran has six atoms.

14.9 Ethanol molecules can hydrogen bond with each other, but ether molecules cannot. Thus, a higher temperature is required to break the hydrogen bonds between ethanol molecules.

14.10 cyclopentene

14.11 2-methyl-1-propanol, 2-methyl-2-propanol

14.12 cyclohexyl alcohol

14.13 Propene adds water in the presence of an acid to form 2-propanol, which loses two hydrogen atoms to form propanone (acetone).

$$CH_2{=}CHCH_3 + HOH \xrightarrow{H^+} CH_3\overset{\overset{\displaystyle OH}{|}}{CH}CH_3 \xrightarrow{[O]}$$

Propene   2-Propanol

$$CH_3\overset{\overset{\displaystyle O}{\|}}{C}CH_3$$

Propanone

Answers to Selected Questions and Problems

14.1 a. 1°   **b.** 1°   **c.** 3°
    d. 2°   **e.** 3°

14.3 a. ethanol   **b.** 2-butanol
    c. 3-hexanol   **d.** 3-methyl-1-butanol
    e. 3,4-dimethylcyclohexanol
    f. 3,5,5-trimethyl-1-heptanol

14.5

a. $CH_3CH_2CH_2OH$ **b.** CH_3OH

c. $CH_3CH_2\overset{\overset{\displaystyle OH}{|}}{CH}CH_2CH_3$ **d.** $CH_3\overset{\overset{\displaystyle OH}{|}}{\underset{\underset{\displaystyle CH_3}{|}}{C}}CH_2CH_3$

e. (cyclohexanol with OH)   **f.** $HOCH_2CH_2CH_2CH_2OH$

14.7 a. phenol
    b. 2-bromophenol, *o*-bromophenol
    c. 3,5-dichlorophenol
    d. 3-bromophenol, *m*-bromophenol

14.9 a. (phenol with Br)   **b.** (phenol with Cl)

    c. (phenol with two Cl)   **d.** (biphenol)

    e. (phenol with CH_2CH_3)

14.11 a. methanethiol   **b.** 2-propanethiol
    c. 2,3-dimethyl-1-butanethiol
    d. cyclobutanethiol

14.13 a. ethanol   **b.** menthol   **c.** *ortho*-phenylphenol

14.15 a. methoxyethane, ethyl methyl ether
    b. methoxycyclohexane, cyclohexyl methyl ether
    c. ethoxycyclobutane, cyclobutyl ethyl ether
    d. 1-methoxypropane, methyl propyl ether

14.17 a. $CH_3CH_2-O-CH_2CH_2CH_3$

    b. CH_3CH_2-O-

    c. OCH_3 (on cyclopentane)

    d. $CH_3CH_2-O-CH_2\overset{\overset{\displaystyle CH_3}{|}}{CH}CH_2CH_3$

    e. $CH_3\overset{\overset{\displaystyle OCH_3}{|}}{CH}\underset{\underset{\displaystyle OCH_3}{|}}{CH}CH_2CH_3$

14.19 a. Constitutional isomers ($C_5H_{12}O$)
    b. Different compounds
    c. Constitutional isomers ($C_5H_{12}O$)

14.21 a. tetrahydrofuran
b. 3-methylfuran
c. 5-methyl-1,3-dioxane

14.23 a. methanol
b. 1-butanol
c. 1-butanol

14.25 a. yes, hydrogen bonding
b. yes; hydrogen bonding
c. no; long carbon chain diminishes effect of —OH group
d. no; alkanes are nonpolar
e. yes: —OH ionizes

14.27 a. CH₃CH₂CH=CH₂ **b.**

c. 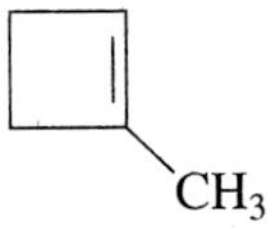   **d.** CH₃CH₂CH=CHCH₃

14.29 a. CH₃OCH₃
b. CH₃CH₂CH₂OCH₂CH₂CH₃

14.31 a. CH₃CH₂OH   **b.** CH₃OH + CH₃CH₂OH
c. 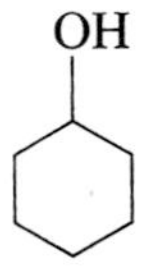 OH

14.33 a. CH₃CH₂CH₂CH₂CH (=O) **b.** CH₃CH₂CCH₃ (=O)

c. (cyclohexanone)   **d.** CH₃CCH₂CHCH₃ (with O and CH₃)

e. CH₃CHCH₂CH (with CH₃ and O)

14.35 a. CH₃OH   **b.** 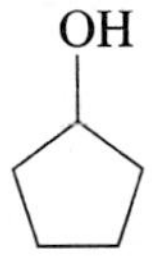 (cyclopentanol, OH)

c. CH₃CH₂CHCH₃ (OH)   **d.** 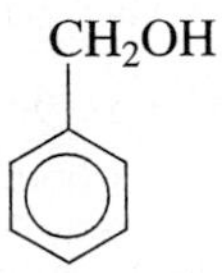 (benzene with CH₂OH)

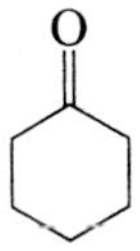

e. 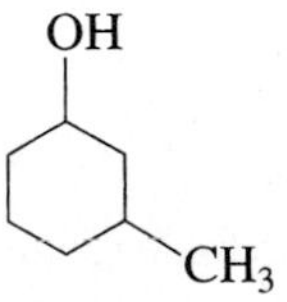 (cyclohexane with OH and CH₃)

14.37 a. 2°   **b.** 1°   **c.** 1°
d. 2°   **e.** 1°   **f.** 3°

14.39 a. alcohol   **b.** ether   **c.** thiol
d. alcohol   **e.** ether   **f.** cyclic ether
g. alcohol   **h.** phenol

14.41 a. 2-chloro-4-methylcyclohexanol
b. methyl phenyl ether
c. 2-propanethiol
d. 2,4-dimethyl-2-pentanol
e. methyl propyl ether   **f.** 3-methyl furan
g. 4-bromo-2-pentanol   **h.** *meta*-cresol

14.43

a. 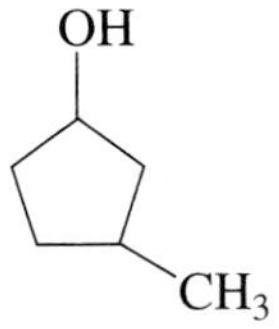 (cyclopentane with OH and CH₃)

b. (benzene with OH and Cl)

c. H₃C—CH—CH—CH₂—CH₃ (with CH₃ and OH)

d. 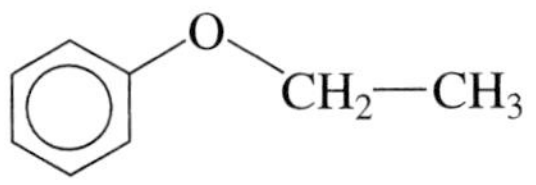 (benzene—O—CH₂—CH₃)

e. CH₃—CH₂—CH—CH₂—CH₃ (with SH)   **f.** 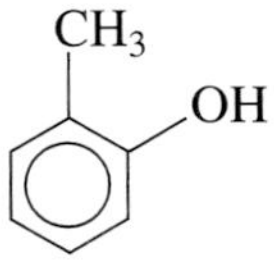 (benzene with CH₃ and OH)

g. 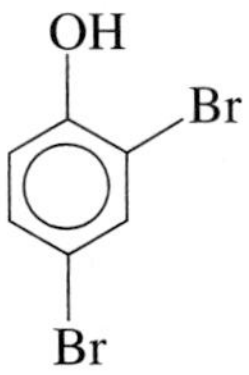 (benzene with OH, Br, and Br)

14.45 a. glycerol
b. 1,2-ethanediol; ethylene glycol
c. ethanol

14.47 CH₃—CH₂—CH₂—CH₂—OH

CH₃—CH—CH₂—OH (with CH₃)

CH₃—CH—CH₂—CH₃ (with OH)

CH₂—C—CH₃ (with OH and CH₃)

14.49 **a.** 1-propanol; hydrogen bonding
b. 1-propanol; hydrogen bonding
c. 1-butanol; higher molar mass

14.51 **a.** soluble; hydrogen bonding
b. soluble; hydrogen bonding
c. insoluble; long carbon chain diminishes effect of polar —OH on hydrogen bonding

14.53 **a.** $CH_3-CH=CH_2$

b. $CH_3-CH_2-\overset{\overset{\displaystyle O}{\|}}{C}-H$

c. $CH_3-CH=CH-CH_3$

d. $CH_3-CH_2-\overset{\overset{\displaystyle O}{\|}}{C}-CH_3$

e. $CH_3-CH_2-CH_2-O-CH_2-CH_2-CH_3$

f. 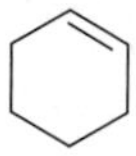   **g.**

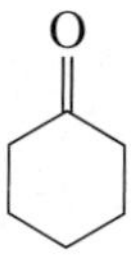

14.55

a. $CH_3-CH_2-CH_2-OH \xrightarrow{H^+, \text{ heat}} CH_3-CH=CH_2 + HCl \longrightarrow CH_3-\overset{\overset{\displaystyle Cl}{|}}{CH}-CH_3$

b. $CH_3-\overset{\overset{\displaystyle OH}{|}}{\underset{\underset{\displaystyle CH_3}{|}}{C}}-CH_3 \xrightarrow{H^+, \text{ heat}} CH_3-\overset{\overset{\displaystyle}{}}{\underset{\underset{\displaystyle CH_3}{|}}{C}}=CH_2 + H_2 \xrightarrow{Pt} CH_3-\overset{}{\underset{\underset{\displaystyle CH_3}{|}}{CH}}-CH_3$

c. $CH_3-CH_2-CH_2-OH \xrightarrow{H^+, \text{ heat}} CH_3-CH=CH_2 + H_2O \xrightarrow{H^+, \text{ heat}} CH_3-\overset{\overset{\displaystyle OH}{|}}{CH}-CH_3$

$\xrightarrow{[O]} CH_3-\overset{\overset{\displaystyle O}{\|}}{C}-CH_3$

14.57 **a.** cycloalkene, ketone, alcohol

14.59

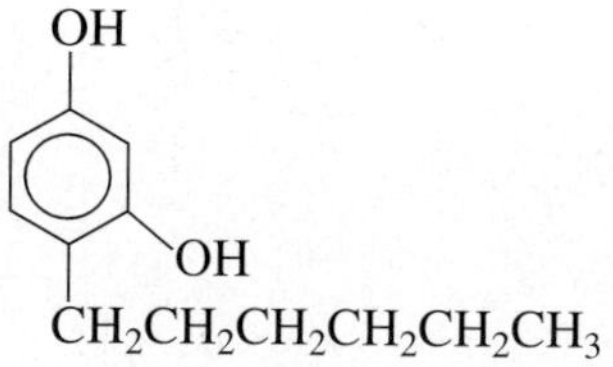

$CH_2CH_2CH_2CH_2CH_2CH_3$

14.61 **a.**

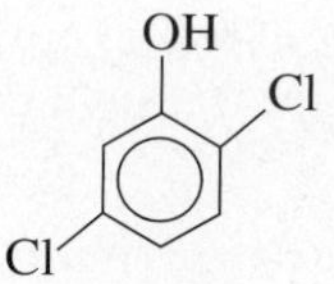

b. $CH_3-\overset{\overset{\displaystyle CH_3}{|}}{CH}-CH_2-CH_2-SH$

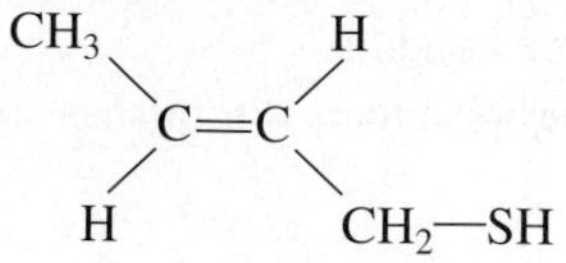

c.

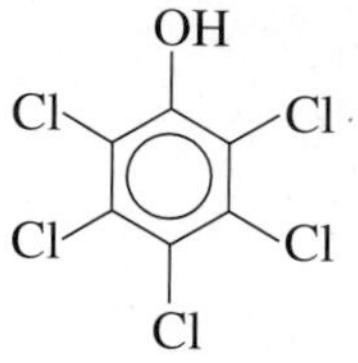

Chapter 15

Answers to Study Checks

15.1 $CH_3-CH_2-\overset{\overset{\displaystyle O}{\|}}{C}-CH_2-CH_3$

15.2 The oxygen atom is much more electronegative than the carbon atom in the carbonyl group.

15.3 ethyl propyl ketone

15.4 propanal (IUPAC), propionaldehyde (common)

15.5 $HO-CH_2-\overset{\overset{\displaystyle O}{\|}}{C}-CH_2-OH$

15.6 The oxygen atom in the carbonyl group of acetone hydrogen bonds with water molecules.

15.7 achiral. A bowling pin has a symmetrical shape.

15.8

$HO-CH_2-\overset{\overset{\displaystyle OH}{|}}{\underset{}{CH}}-\overset{\overset{\displaystyle OH}{|}}{\underset{}{CH}}-\overset{\overset{\displaystyle O}{\|}}{C}-H$

15.9

CHO	CHO
H——OH	HO——H
CH$_3$	CH$_3$
D-2-hydroxypropanal	L-2-hydroxypropanal

15.10 cyclohexanol

15.11 The oxidation of benzaldehyde reduces Ag^+ to metallic silver, which forms a silvery coating on the walls of the test tube.

15.12 1-propanol

15.13 hemiacetal

15.14 $CH_3-\overset{\overset{\displaystyle OCH_3}{|}}{\underset{\underset{\displaystyle OCH_3}{|}}{C}}-CH_3$

Answers to Selected Problems

15.1 a. ketone   **b.** aldehyde
     c. ketone   **d.** aldehyde

15.3 a. 1   **b.** 1   **c.** 2

15.5 a. propanal   **b.** 2-methyl-3-pentanone
     c. 3-hydroxybutanal   **d.** 2-pentanone
     e. 3-methylcyclohexanone   **f.** 4-chlorobenzaldehyde

15.7 a. acetaldehyde
     b. methyl propyl ketone
     c. formaldehyde

15.9   a. $CH_3-\overset{\overset{\displaystyle O}{\|}}{C}-H$

   b. $CH_3-\overset{\overset{\displaystyle O}{\|}}{C}-CH_2-\overset{\overset{\displaystyle OH}{|}}{CH}-CH_3$

   c. $CH_3-\overset{\overset{\displaystyle Br}{|}}{CH}-\overset{\overset{\displaystyle Br}{|}}{CH}-\overset{\overset{\displaystyle O}{\|}}{C}-H$

d. $CH_3-\overset{\overset{\displaystyle O}{\|}}{C}-CH_2-CH_2-CH_2-CH_3$

e. $CH_3-CH_2-\overset{\overset{\displaystyle CH_3}{|}}{CH}-CH_2-\overset{\overset{\displaystyle O}{\|}}{C}-H$

f. $CH_3-CH_2-\overset{\overset{\displaystyle O}{\|}}{C}-CH_2-\overset{\overset{\displaystyle O}{\|}}{C}-H$

15.11   CHO

   OCH$_3$

15.13 a. benzaldehyde
      b. acetone; propanone
      c. formaldehyde

15.15 a. $CH_3-CH_2-\overset{\overset{\displaystyle O}{\|}}{C}-H$ has a polar carbonyl group.
      b. pentanal has a higher molar mass and thus a higher boiling point
      c. 1-butanol because it can hydrogen bond with other 1-butanol molecules

15.17 a. $CH_3-\overset{\overset{\displaystyle O}{\|}}{C}-\overset{\overset{\displaystyle O}{\|}}{C}-CH_3$; more hydrogen bonding
      b. acetaldehyde; more hydrogen bonding
      c. acetone; lower number of carbon atoms

15.19 No. The long carbon chain diminishes the effect of the carbonyl group.

15.21 a. achiral

      b. chiral $CH_3-\overset{\overset{\displaystyle Br}{|}\text{ chiral carbon}}{CH}-CH_2CH_3$

      c. chiral $CH_3-\overset{\text{chiral carbon}\,\overset{\displaystyle Br}{|}}{CH}-\overset{\overset{\displaystyle O}{\|}}{C}-H$

      d. achiral

15.23
a. $CH_3-\overset{\overset{\displaystyle CH_3}{|}}{C}=CH-CH_2-CH_2-\overset{\overset{\displaystyle CH_3}{|}\text{chiral carbon}}{CH}-CH_2-CH_2-OH$

b. $H_2N-\overset{\overset{\displaystyle CH_3}{|}}{CH}-\overset{\overset{\displaystyle O}{\|}}{C}-OH$
chiral carbon

15.25

a.

$$\underset{\overset{|}{CH_3}}{\overset{\overset{H}{|}}{HO-\!\!\!-Br}}$$

b.

$$\underset{\overset{|}{OH}}{\overset{\overset{CH_3}{|}}{Cl-\!\!\!-Br}}$$

c.

$$\underset{\overset{|}{CH_2CH_3}}{\overset{\overset{CHO}{|}}{HO-\!\!\!-H}}$$

15.27 a. identical   **b.** enantiomers
     c. enantiomers   **d.** enantiomers

15.29 a. CH_3OH

b. cyclopentanol with OH

c. $CH_3CH_2\underset{\overset{|}{OH}}{CH}CH_3$

d. benzyl alcohol (CH_2OH on benzene ring)

e. 3-methylcyclohexanol (ring with OH and CH_3)

15.31 a. $CH_3CH_2CH_2CH_2\overset{\overset{O}{\|}}{C}H$

b. $CH_3CH_2\overset{\overset{O}{\|}}{C}CH_3$

c. cyclohexanone

d. $CH_3\overset{\overset{O}{\|}}{C}CH_2\overset{\overset{CH_3}{|}}{C}HCH_3$

e. $CH_3\overset{\overset{CH_3}{|}}{C}HCH_2\overset{\overset{O}{\|}}{C}H$

15.33 a. $CH_3CH_2CH_2CH_2OH$

b. $CH_3\overset{\overset{OH}{|}}{C}HCH_3$

c. $CH_3CH_2CH_2\overset{\overset{Br}{|}}{C}HCH_2CH_2OH$

d. $CH_3\overset{\overset{CH_3}{|}}{C}H\underset{\overset{|}{OH}}{C}HCH_2CH_3$

15.35 a. $CH_3-\underset{\overset{|}{OH}}{\overset{\overset{OH}{|}}{C}}-H$

b. $H-\underset{\overset{|}{OH}}{\overset{\overset{OH}{|}}{C}}-H$

15.37 a. hemiacetal   **b.** hemiacetal   **c.** acetal
     d. hemiacetal   **e.** acetal

15.39 a. $CH_3-\underset{\overset{|}{OH}}{\overset{\overset{OCH_3}{|}}{C}}H$

b. $CH_3-\underset{\overset{|}{OH}}{\overset{\overset{OCH_3}{|}}{C}}-CH_3$

c. cyclopentane with HO and OCH_3

d. $CH_3-CH_2-CH_2-\underset{\overset{|}{OH}}{\overset{\overset{OCH_3}{|}}{C}}H$

15.41 a. $CH_3-\underset{\overset{|}{OCH_3}}{\overset{\overset{OCH_3}{|}}{C}}H$

b. $CH_3-\underset{\overset{|}{OCH_3}}{\overset{\overset{OCH_3}{|}}{C}}-CH_3$

c. cyclopentane with CH_3O and OCH_3

d. $CH_3-CH_2-CH_2-\underset{\overset{|}{OCH_3}}{\overset{\overset{OCH_3}{|}}{C}}H$

15.43 The carbonyl group consists of a sigma bond and a pi bond, which is an overlapping of the *p* orbitals of the carbon and oxygen atom.

15.45 $CH_3-CH_2-CH_2-\overset{\overset{O}{\|}}{C}-H$ $CH_3-\overset{\overset{CH_3}{|}}{C}H-\overset{\overset{O}{\|}}{C}-H$

$CH_3-CH_2-\overset{\overset{O}{\|}}{C}-CH_3$

15.47 a. 2-bromo-4-chlorocyclopentanone
     b. 4-chloro-3-hydroxybenzaldehyde
     c. 3-chloropropanal
     d. 5-hydroxy-3-hexanone
     e. 2-chloro-3-pentanone
     f. 3-methylcyclohexanone

15.49

a. [structure: 3-methylcyclopentanone]

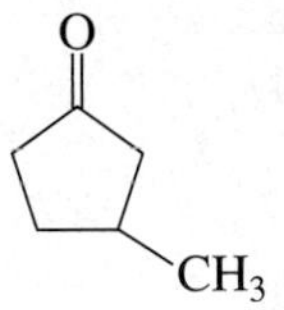

b. [structure: 4-chlorobenzaldehyde, CHO on benzene ring with Cl]

c. Cl—CH₂—CH₂—C(=O)—H

d. CH₃—CH₂—C(=O)—CH₃

e. CH₃—CH₂—CH₂—CH(CH₃)—CH₂—C(=O)—H

f. CH₃—CH₂—CH₂—CH₂—CH₂—C(=O)—CH₃

15.51 b, c, and d

15.53 **a.** CH₃—CH₂—OH   **b.** CH₃—CH₂—C(=O)—H

c. CH₃—CH₂—CH₂—OH

15.55

a. H—C(Cl)(Cl)—C(Cl)(Cl)—O—H   **b.** none   **c.** none

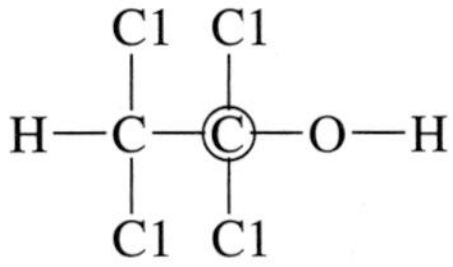

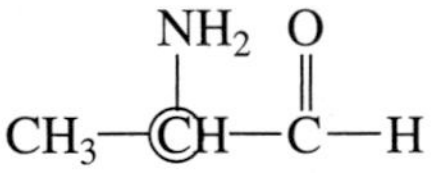

d. CH₃—CH(NH₂)—C(=O)—H

e. CH₃—CH₂—CH(Br)—CH₂—CH₂—CH₃

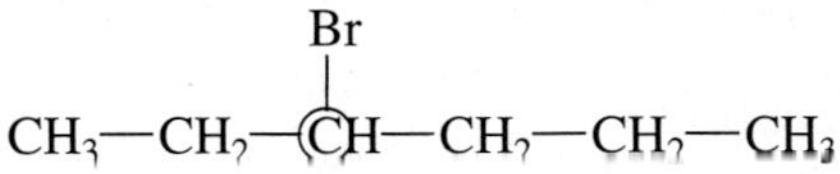

f. none

15.57 **a.** identical   **b.** enantiomers
c. enantiomers (turn 180°)   **d.** identical

15.59 **a.** CH₃—CH₂—C(=O)—H →(Further oxidation)→ CH₃—CH₂—C(=O)—OH

b. CH₃—C(=O)—CH₂—CH₂—CH₃

c. CH₃—CH₂—CH₂—C(=O)—OH

d. [structure: cyclohexanone]

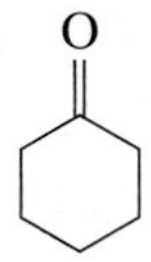

15.61 **a.** CH₃—CH(OH)—CH₃

b. [structure: benzene ring with CH₂—CH₂—OH]

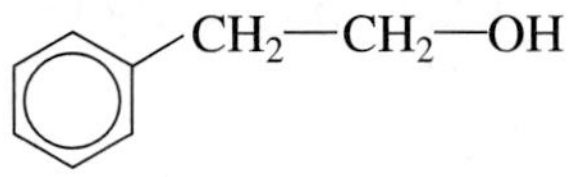

c. CH₃—CH(CH₃)—CH₂—CH(OH)—CH₃

15.63

a. 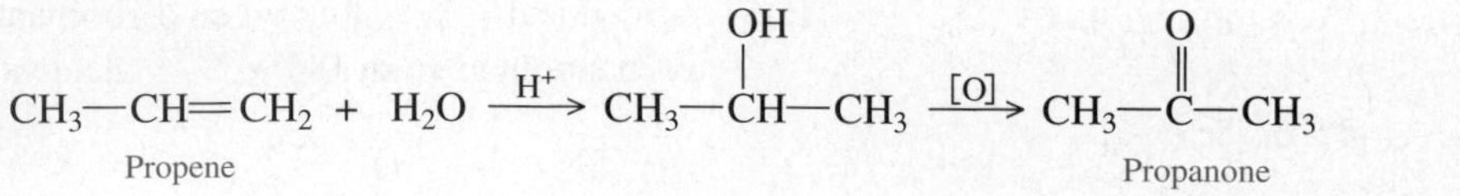 $CH_3-CH=CH_2 + H_2O \xrightarrow{H^+} CH_3-\overset{\overset{\displaystyle OH}{|}}{CH}-CH_3 \xrightarrow{[O]} CH_3-\overset{\overset{\displaystyle O}{||}}{C}-CH_3$

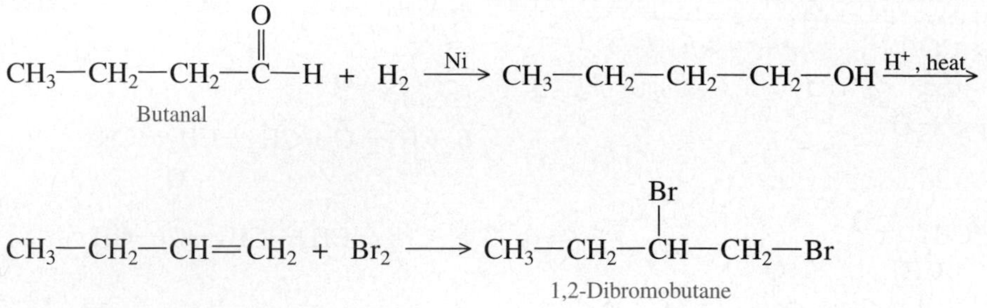

Propene   Propanone

b. $CH_3-CH_2-CH_2-\overset{\overset{\displaystyle O}{||}}{C}-H + H_2 \xrightarrow{Ni} CH_3-CH_2-CH_2-CH_2-OH \xrightarrow{H^+, heat}$

Butanal

$CH_3-CH_2-CH=CH_2 + Br_2 \longrightarrow CH_3-CH_2-\overset{\overset{\displaystyle Br}{|}}{CH}-CH_2-Br$

1,2-Dibromobutane

c. $CH_3-CH_2-CH_2-\overset{\overset{\displaystyle O}{||}}{C}-H + H_2 \xrightarrow{Ni} CH_3-CH_2-CH_2-CH_2-OH \xrightarrow{H^+, heat}$

Butanal

$CH_3-CH_2-CH=CH_2 + H_2O \longrightarrow CH_3-CH_2-\overset{\overset{\displaystyle OH}{|}}{CH}-CH_3 \xrightarrow{[O]}$

$CH_3-CH_2-\overset{\overset{\displaystyle O}{||}}{C}-CH_3$

Butanone

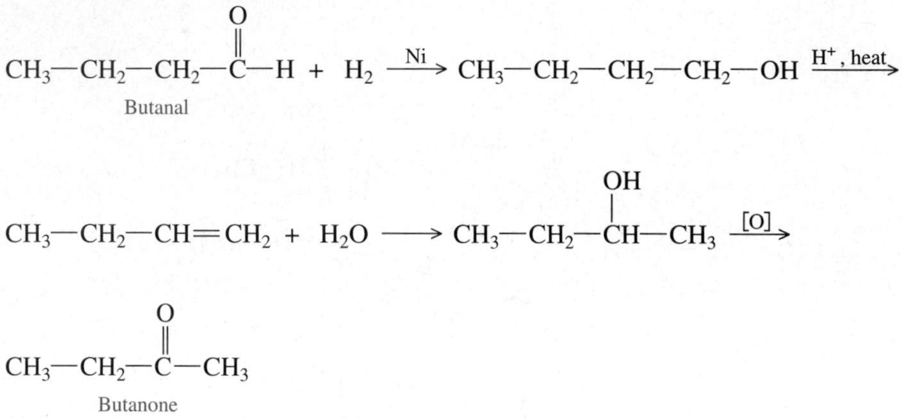

15.65 a. acetal; propanal and methanol
b. hemiacetal; butanone and ethanol
c. acetal; cyclohexanone and ethanol

Chapter 16

Answers to Study Checks

16.1 a monosaccharide

16.2
$$\begin{array}{c} CH_2OH \\ | \\ C=O \\ | \\ CH_2OH \end{array}$$

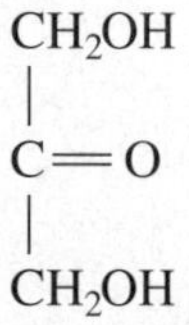

16.3

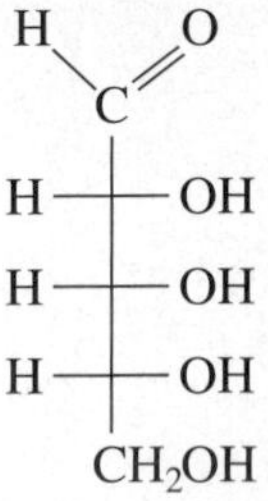

16.4 Ribulose is a ketopentose.

16.5

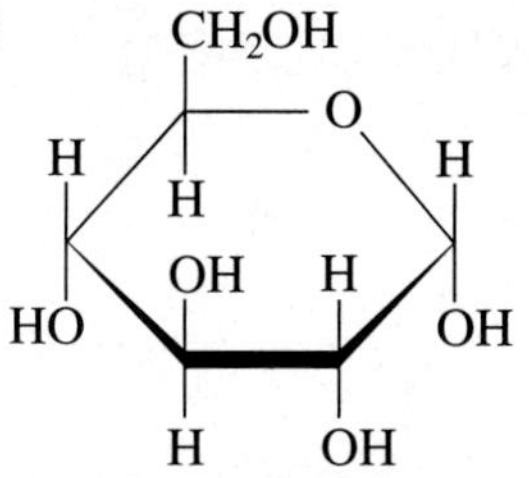

16.6 This indicates a high level of reducing sugar (probably glucose) in the urine. One common cause of this condition is diabetes mellitus.

16.7

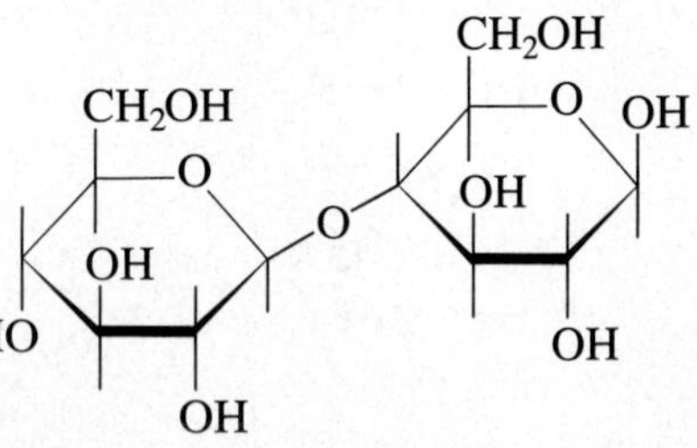

16.8 Cellulose contains glucose units connected by β-1,4-glycosidic bonds, whereas the glucose units in amylose are connected by α-1,4-glycosidic bonds.

Answers to Selected Questions and Problems

16.1 Photosynthesis requires CO_2, H_2O, and the energy from the sun. Respiration requires O_2 from the air and glucose from our foods.

16.3 A monosaccharide is a simple sugar composed of three to six carbon atoms. A disaccharide is composed of two monosaccharide units.

16.5 Hydroxyl groups are found in all monosaccharides along with a carbonyl on the first or second carbon.

16.7 A ketopentose contains hydroxyl and ketone functional groups and has five carbon atoms.

16.9 **a.** ketose   **b.** aldose   **c.** ketose
       d. aldose   **e.** aldose

16.11 A Fischer projection is a two-dimensional representation of the three-dimensional structure of a molecule.

16.13 **a.** D   **b.** D   **c.** L   **d.** D

16.15 **a.**

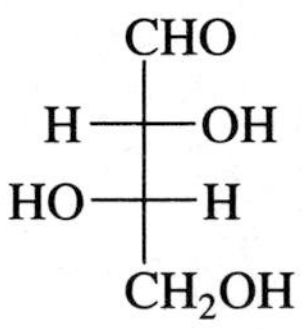

```
      CHO
  H ——— OH
 HO ——— H
     CH2OH
```

b.

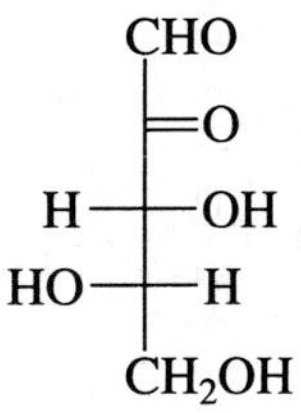

```
     CHO
      ==O
  H —— OH
 HO —— H
    CH2OH
```

c.

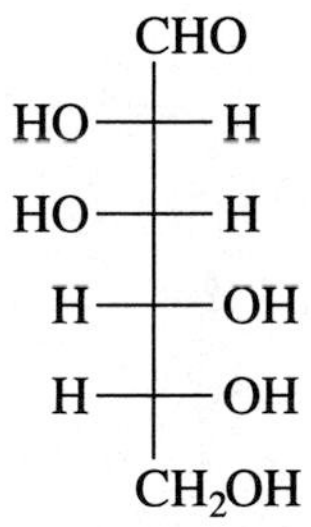

```
      CHO
 HO —— H
 HO —— H
  H —— OH
  H —— OH
    CH2OH
```

d.

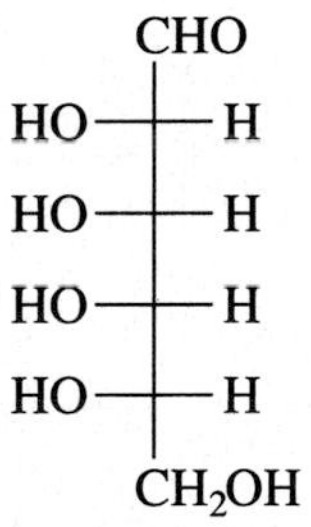

```
      CHO
 HO —— H
 HO —— H
 HO —— H
 HO —— H
    CH2OH
```

16.17

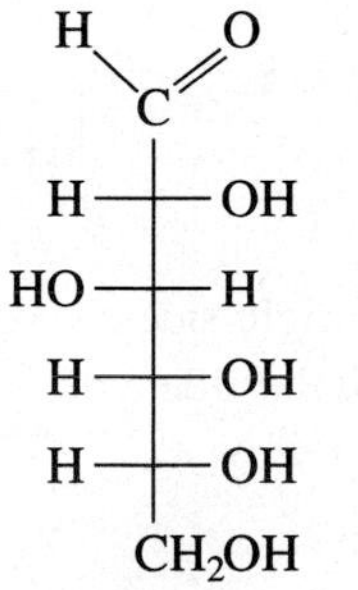

```
   H   O
    \ //
     C
  H —— OH
 HO —— H
  H —— OH
  H —— OH
    CH2OH
   D-Glucose
```

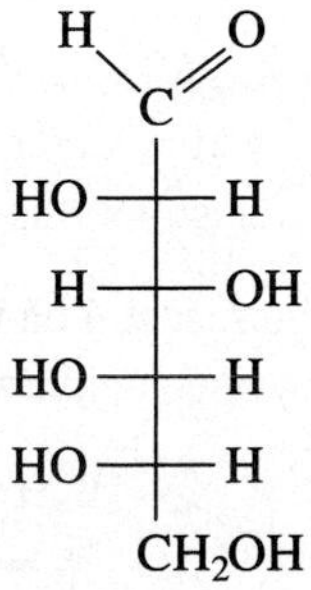

```
   H   O
    \ //
     C
 HO —— H
  H —— OH
 HO —— H
 HO —— H
    CH2OH
   L-Glucose
```

16.19 In D-galactose the hydroxyl on carbon four extends to the left. In glucose this hydroxyl goes to the right.

16.21 **a.** glucose   **b.** galactose   **c.** fructose

16.23 In the cyclic structure of glucose, there are five carbon atoms and an oxygen.

16.25

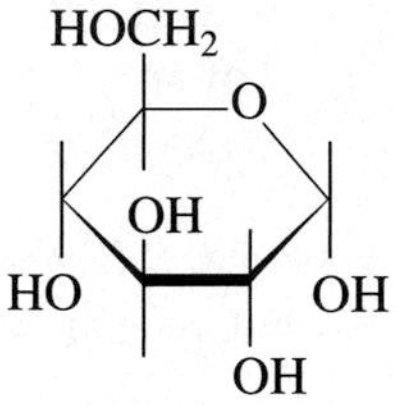

α-D-Glucose

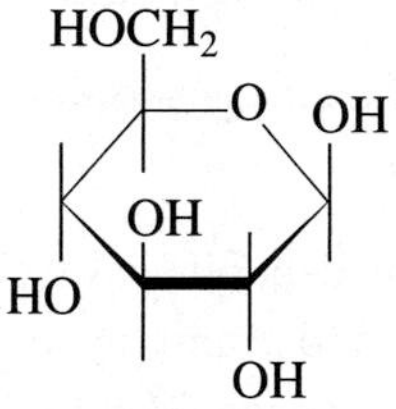

β-D-Glucose

16.27 **a.** α-anomer   **b.** α-anomer

16.29

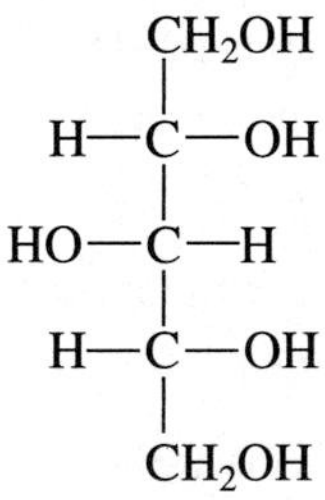

```
     CH2OH
  H —— C —— OH
 HO —— C —— H
  H —— C —— OH
     CH2OH
     Xylitol
```

16.31 Oxidation product:

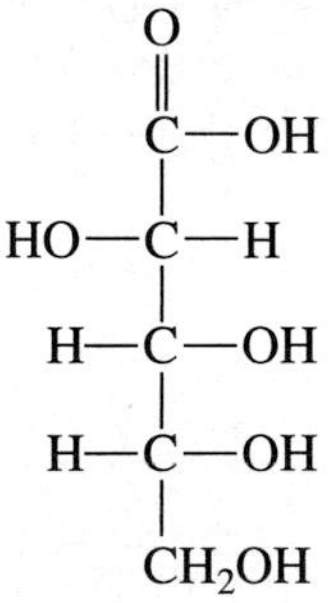

```
      O
      ||
      C —— OH
 HO —— C —— H
  H —— C —— OH
  H —— C —— OH
     CH2OH
```

Reduction product (sugar alcohol):

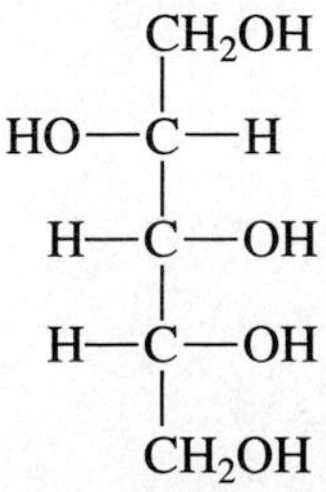

```
     CH2OH
 HO —— C —— H
  H —— C —— OH
  H —— C —— OH
     CH2OH
    D-Arabitol
```

16.33

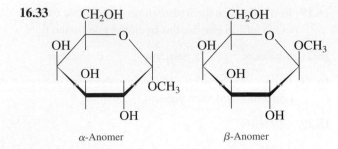

α-Anomer   β-Anomer

16.35 a. galactose and glucose; β-1,4 bond; β-lactose
b. glucose and glucose; α-1,4 bond; α-maltose

16.37 a. Will undergo mutarotation; can be oxidized
b. Will undergo mutarotation; can be oxidized

16.39 a. sucrose   **b.** lactose
c. maltose   **d.** lactose

16.41 a. Amylose is an unbranched polymer of glucose units joined by α-1,4 bonds; amylopectin is a branched polymer of glucose joined by α-1,4 and α-1,6 bonds.
b. Amylopectin, which is produced in plants, is a branched polymer of glucose, joined by α-1,4 and α-1,6 bonds. Glycogen, which is produced in animals, is a highly branched polymer of glucose, joined by α-1,4 and α-1,6 bonds.

16.43 a. cellulose   **b.** amylose, amylopectin
c. amylose   **d.** glycogen

16.45 They differ only at carbon 4 where the —OH in D-glucose is on the right side and in D-galactose it is on the left side.

16.47 D-galactose is the mirror image of L-galactose. In D-galactose, the —OH group on carbon 5 is on the right side whereas in L-galactose, the —OH group on carbon 5 is on the left side.

16.49
a.

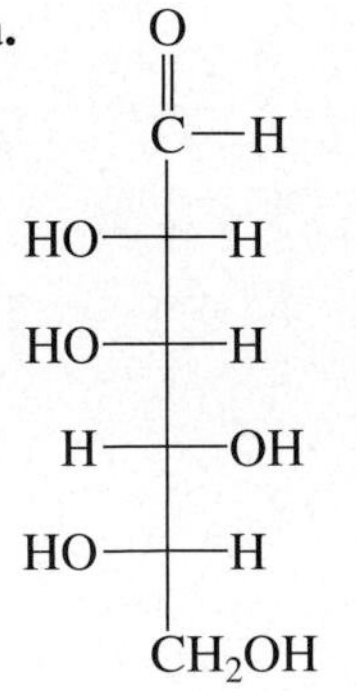

```
        O
        ‖
        C—H
  HO ——┼—— H
  HO ——┼—— H
   H ——┼—— OH
  HO ——┼—— H
       CH₂OH
```
L-Gulose

b.

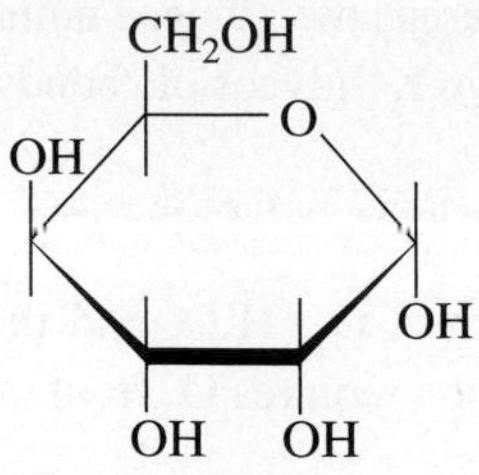

α-D-Gulose   β-D-Gulose

16.51

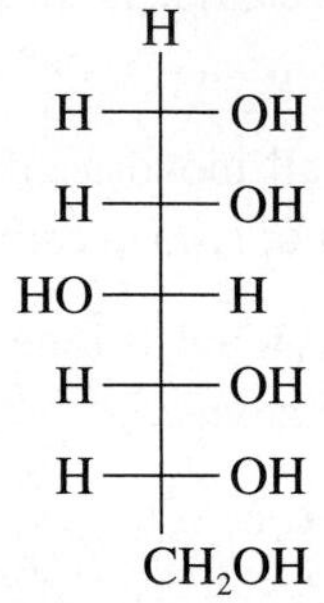

```
        H
   H ——┼—— OH
   H ——┼—— OH
  HO ——┼—— H
   H ——┼—— OH
   H ——┼—— OH
       CH₂OH
```

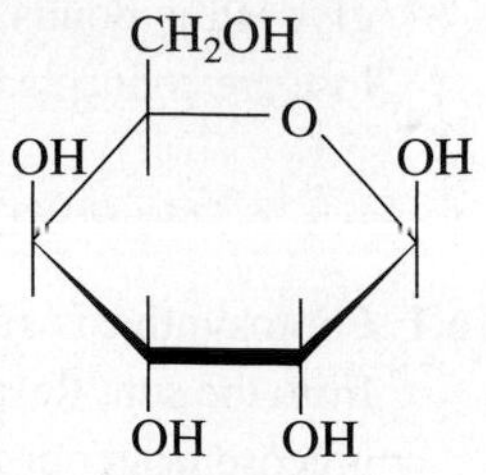

16.53 The α-galactose forms an open-chain structure, and when the chain closes, it can form both α- and β-galactose.

16.55

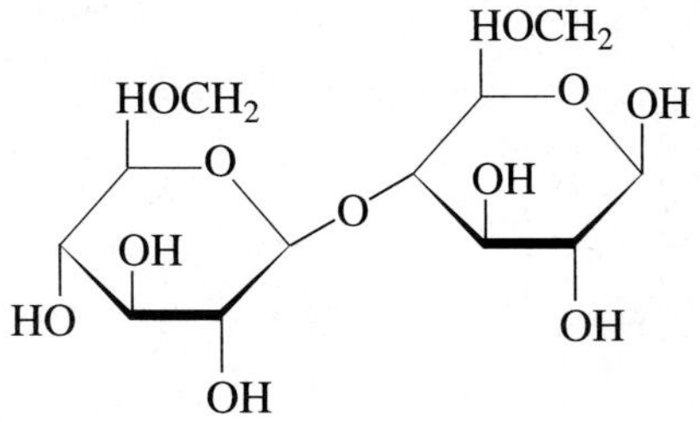

16.57 a.

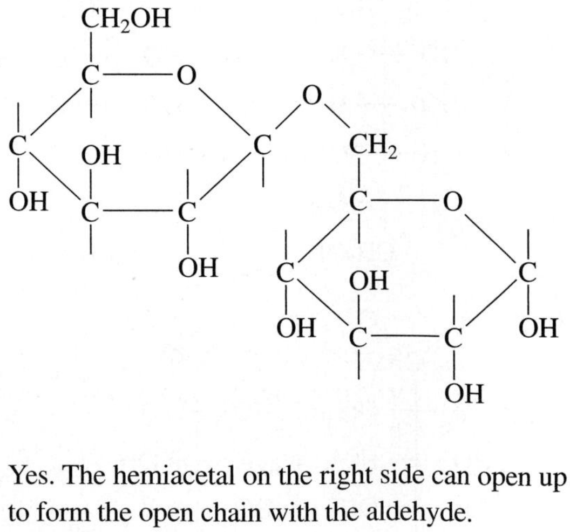

b. Yes. The hemiacetal on the right side can open up to form the open chain with the aldehyde.

Chapter 17

Answers to Study Checks

17.1

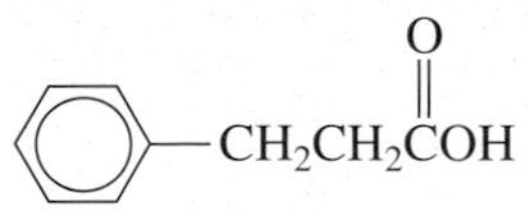

(benzene ring)—CH$_2$CH$_2$COH with C=O

17.2

CH$_3$CH$_2$CH$_2$COH (with C=O)

17.3 Two carboxylic acid molecule hydrogen bond forming a dimer, which has a mass twice that of the single acid molecule.

17.4 HCOH + H$_2$O $\rightleftharpoons$ HCO$^-$ + H$_3$O$^+$ (with C=O on both)

17.5 butanoic acid, butyric acid

17.6 propanoic (propionic) acid and 1-pentanol

17.7 CH$_3$C—OCH$_2$CH$_2$CH$_2$CH$_2$CH$_3$ (with C=O)

17.8 propanoic (propionic) acid and ethanol

17.9

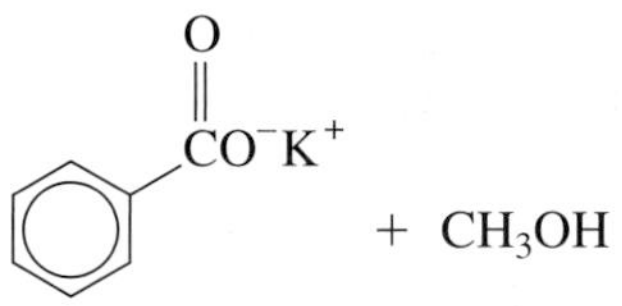

(benzene ring)—CO$^-$K$^+$ (with C=O)   + CH$_3$OH

Answers to Selected Problems

17.1 methanoic acid (formic acid)

17.3 Each compound contains three carbon atoms. They differ because propanal, an aldehyde, contains a carbonyl group bonded to a hydrogen. In propanoic acid, the carbonyl group connects to a hydroxyl group, forming a carboxyl group.

17.5
a. ethanoic acid (acetic acid)
b. butanoic acid (butyric acid)
c. 2-chloropropanoic acid (α-chloropropionic acid)
d. 3-methylhexanoic acid
e. 3,4-dihydroxybenzoic acid
f. 4-bromopentanoic acid

17.7 a. CH$_3$—CH$_2$—C—OH (with C=O)   b. (benzene ring)—C—OH (with C=O)

17.9

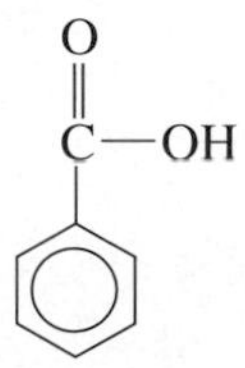

c. Cl—CH$_2$—C—OH (with C=O)

d. HO—CH$_2$—CH$_2$—C—OH (with C=O)

e. CH$_3$—CH$_2$—CH—C—OH (with C=O), with CH$_3$ on the CH

f. CH$_3$—CH$_2$—CH—CH$_2$—CH—CH$_2$—C—OH (with C=O), with Br on each CH

17.9 a. HCOH (with C=O)   b. CH$_3$—C—OH (with C=O)

c. CH$_3$—CH—CH$_2$—C—OH (with C=O), with CH$_3$ on the CH

d. (cyclopentane ring)—CH$_2$—C—OH (with C=O)

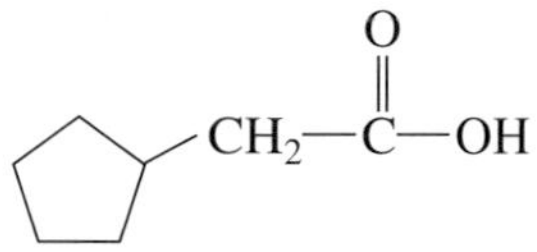

17.11
a. Butanoic acid has a higher molar mass and would have a higher boiling point.
b. Propanoic acid can form more hydrogen bonds and would have a higher boiling point.
c. Butanoic acid can form hydrogen bonds and would have a higher boiling point.

17.13
a. acetone, propanol, propanoic acid because of the greater number of hydrogen bonds.
b. butanoic acid, propanoic acid, acetic acid because of the decrease in molar mass
c. propane, ethanol, acetic acid because of the greater number of hydrogen bonds.

17.15

a. H—C—OH + H$_2$O $\rightleftharpoons$ H—C—O$^-$ + H$_3$O$^+$ (with C=O)

b. CH$_3$—CH$_2$—C—OH + H$_2$O $\rightleftharpoons$ CH$_3$—CH$_2$—C—O$^-$ + H$_3$O$^+$ (with C=O)

c. CH$_3$—C—OH + H$_2$O $\rightleftharpoons$ CH$_3$—C—O$^-$ + H$_3$O$^+$ (with C=O)

17.17 **a.** $H-\overset{\overset{\displaystyle O}{\|}}{C}-OH + NaOH \longrightarrow H-\overset{\overset{\displaystyle O}{\|}}{C}-O^-Na^+ + H_2O$

b. $CH_3-CH_2-\overset{\overset{\displaystyle O}{\|}}{C}-OH + NaOH \longrightarrow CH_3-CH_2-\overset{\overset{\displaystyle O}{\|}}{C}-O^-Na^+ + H_2O$

c. (benzene ring)$-\overset{\overset{\displaystyle O}{\|}}{C}-OH + NaOH \longrightarrow$ (benzene ring)$-\overset{\overset{\displaystyle O}{\|}}{C}-O^-Na^+ + H_2O$

17.19 a. sodium methanoate, sodium formate
 b. sodium propanoate, sodium propionate
 c. sodium benzoate

17.21 a. aldehyde   **b.** ester
 c. ketone   **d.** carboxylic acid

17.23 a. $CH_3-\overset{\overset{\displaystyle O}{\|}}{C}-O-CH_3$

b. $CH_3-CH_2-CH_2-\overset{\overset{\displaystyle O}{\|}}{C}-O-CH_3$

c. $\overset{\overset{\displaystyle O}{\|}}{C}-O-CH_3$ (attached to benzene ring)

17.25 a. $CH_3-CH_2-\overset{\overset{\displaystyle O}{\|}}{C}-O-CH_2-CH_2-CH_3$

b. $CH_3-CH_2-CH_2-CH_2-\overset{\overset{\displaystyle O}{\|}}{C}-O-\overset{\overset{\displaystyle CH_3}{|}}{CH}-CH_3$

17.27 a. formic acid (methanoic acid) and methyl alcohol (methanol)
 b. acetic acid (ethanoic acid) and methyl alcohol (methanol)
 c. butyric acid (butanoic acid) and methyl alcohol (methanol)
 d. β-methylbutyric acid (3-methylbutanoic acid) and ethyl alcohol (ethanol)

17.29 a. methyl formate (methyl methanoate)
 b. methyl acetate (methyl ethanoate)
 c. methyl butyrate (methyl butanoate)
 d. ethyl-β-methyl butyrate (ethyl-3-methyl butanoate)

17.31 a. $CH_3-\overset{\overset{\displaystyle O}{\|}}{C}-O-CH_3$

b. $H-\overset{\overset{\displaystyle O}{\|}}{C}-O-CH_2-CH_2-CH_2-CH_3$

c. $CH_3-CH_2-CH_2-CH_2-\overset{\overset{\displaystyle O}{\|}}{C}-O-CH_2-CH_3$

d. $CH_3-CH_2-\overset{\overset{\displaystyle O}{\|}}{C}-O-CH_2-\overset{\overset{\displaystyle Br}{|}}{CH}-CH_3$

17.33 a. pentyl ethanoate (pentyl acetate)
 b. octyl ethanoate (octyl acetate)
 c. pentyl butanoate (pentyl butyrate)
 d. isobutyl methanoate (isobutyl formate)

17.35 a. $CH_3-\overset{\overset{\displaystyle O}{\|}}{C}-OH$

b. $CH_3-CH_2-CH_2-CH_2-OH$

c. $CH_3-O-\overset{\overset{\displaystyle O}{\|}}{C}-CH_3$

17.37 The products of the acid hydrolysis of an ester are an alcohol and a carboxylic acid.

17.39

a. $CH_3CH_2\overset{\displaystyle O}{\overset{\|}{C}}O^-Na^+$ and CH_3OH

b. $CH_3\overset{\displaystyle O}{\overset{\|}{C}}OH$ and $CH_3CH_2CH_2OH$

c. $CH_3CH_2CH_2\overset{\displaystyle O}{\overset{\|}{C}}OH$ and CH_3CH_2OH

d. (benzene ring)$\overset{\displaystyle O}{\overset{\|}{C}}OH$ and CH_3CH_2OH

e. (benzene ring)$\overset{\displaystyle O}{\overset{\|}{C}}O^-Na^+$ and CH_3CH_2OH

17.41 **a.** 3-methylbutanoic acid; β-methylbuyric acid
b. ethylbenzoate
c. ethyl propanoate; ethylpropionate
d. 2-chlorobenzoic acid; *ortho*-chlorobenzoic acid
e. 4-hydroxypentanoic acid
f. 2-propyl ethanoate; isopropyl acetate

17.43

$CH_3-CH_2-CH_2-CH_2-\overset{\displaystyle O}{\overset{\|}{C}}-OH$ $CH_3-CH_2-\overset{\displaystyle CH_3}{\overset{\displaystyle |}{C}H}-\overset{\displaystyle O}{\overset{\|}{C}}-OH$

$CH_3-\overset{\displaystyle CH_3}{\overset{\displaystyle |}{C}H}-CH_2-\overset{\displaystyle O}{\overset{\|}{C}}-OH$ $CH_3-\overset{\displaystyle CH_3}{\underset{\displaystyle CH_3}{\overset{\displaystyle |}{\underset{\displaystyle |}{C}}}}-\overset{\displaystyle O}{\overset{\|}{C}}-OH$

17.45 **a.** $CH_3-O-\overset{\displaystyle O}{\overset{\|}{C}}-CH_3$

b. (benzene ring with COOH top and Cl bottom)

c. $Cl-CH_2-CH_2-\overset{\displaystyle O}{\overset{\|}{C}}-OH$

d. $CH_3-CH_2-O-\overset{\displaystyle O}{\overset{\|}{C}}-CH_2-CH_2-CH_3$

e. $CH_3-CH_2-\overset{\displaystyle CH_3}{\overset{\displaystyle |}{C}H}-CH_2-\overset{\displaystyle O}{\overset{\|}{C}}-OH$

f. $\overset{\displaystyle O}{\overset{\|}{C}}-O-CH_2-CH_3$ (benzene ring)

17.47 a. $CH_3-\overset{\displaystyle O}{\overset{\|}{C}}\;OH$

b. $CH_3-CH_2-\overset{\displaystyle O}{\overset{\|}{C}}-OH$

c. $CH_3-CH_2-CH_2-\overset{\displaystyle O}{\overset{\|}{C}}-OH$

17.49 The presence of two polar groups in the carboxyl group allows hydrogen bonding and the formation of a dimer that doubles the effective molar mass.

17.51 b, c, d, and e are all soluble in water

17.53

(benzene ring)$-\overset{\displaystyle O}{\overset{\|}{C}}-OCH_3 + KOH \longrightarrow$

(benzene ring)$-\overset{\displaystyle O}{\overset{\|}{C}}-O^-K^+ + CH_3OH$

A soluble salt, potassium benzoate, is formed. When acid is added, the salt is converted to insoluble benzoic acid.

17.55 a. hydroxyl and carboxylic acid

b. (benzene ring)$-O-\overset{\displaystyle O}{\overset{\|}{C}}-CH_3$ and $-\overset{\displaystyle}{\underset{\displaystyle O}{\overset{\|}{C}}}-OH$

c. (benzene ring)$-OH$ and $-\overset{\displaystyle}{\underset{\displaystyle O}{\overset{\|}{C}}}-OCH_3$

17.57

a. $CH_3-CH_2-\overset{\displaystyle O}{\overset{\|}{C}}-O^- + H_3O^+$

b. $CH_3-CH_2-\overset{\displaystyle O}{\overset{\|}{C}}-O^-K^+ + H_2O$

c. $CH_3-CH_2-\overset{\displaystyle O}{\overset{\|}{C}}-O-CH_3 + H_2O$

d.

$$\underset{\overset{\displaystyle O}{\overset{\displaystyle \|}{}}}{C}-O-CH_2-CH_3 \ + \ H_2O$$

(benzene ring)

17.59 a. 3-methylbutanoic acid and methanol
   b. 3-chlorobenzoic acid and ethanol
   c. hexanoic acid and methanol

17.61

a. $CH_3-CH_2-\overset{\overset{\displaystyle O}{\displaystyle \|}}{C}-OH$ and $HO-\overset{\overset{\displaystyle CH_3}{|}}{CH}-CH_3$

b. $CH_3-\overset{\overset{\displaystyle CH_3}{|}}{CH}-\overset{\overset{\displaystyle O}{\displaystyle \|}}{C}-O^-Na^+$ and $HO-CH_2-CH_2-CH_3$

17.63

a. $CH_2{=}CH_2 \ + \ H_2O \xrightarrow{H^+} CH_3CH_2OH \xrightarrow{[O]} CH_3\overset{\overset{\displaystyle O}{\displaystyle \|}}{C}OH$

b. $CH_3-CH_2-CH_2-CH_2-OH \xrightarrow{[O]} CH_3-CH_2-CH_2-\overset{\overset{\displaystyle O}{\displaystyle \|}}{C}-OH$

Chapter 18

Answers to Study Checks

18.1 a glycolipid

18.2 a. 16   **b.** unsaturated   **c.** liquid

18.3

$$CH_2-O-\overset{\overset{\displaystyle O}{\displaystyle \|}}{C}-(CH_2)_{12}-CH_3$$
$$CH-O-\overset{\overset{\displaystyle O}{\displaystyle \|}}{C}-(CH_2)_{12}-CH_3$$
$$CH_2-O-\overset{\overset{\displaystyle O}{\displaystyle \|}}{C}-(CH_2)_{12}-CH_3$$

18.4 tristearin

18.5 Triacylglycerols contain three fatty acids connected by ester bonds to glycerol. Glycerophospholipids con-

tain glycerol esterified to two fatty acids and a phosphate connected to an amino alcohol.

18.6 Cerebrosides contain only one monosaccharide, and gangliosides contain two or more monosaccharide units.

18.7 Cholesterol is not soluble in water; it is classified with the lipid family.

18.8 Testosterone and nandrolone both contain a steroid nucleus with one double bond and a ketone group in the first ring, and a methyl and alcohol group on the five-carbon ring. Nandrolone does not have the second methyl group at the first and second ring fusion that is seen in the structure of testosterone.

18.9 Protein channels allow electrolytes to flow in and out of the cell through the lipid bilayer.

Answers to Selected Questions and Problems

18.1 Lipids provide energy and protection and insulation for the organs in the body.

18.3 Because lipids are not soluble in water, a polar solvent, they are nonpolar molecules.

18.5 All fatty acids contain a long chain of carbon atoms with a carboxylic acid group. Saturated fats contain only carbon-to-carbon single bonds; unsaturated fats contain one or more double bonds. More saturated fats are found in animal fats, whereas vegetable oils contain more unsaturated fats.

18.7 a. palmitic acid

(zig-zag chain ending in)—COOH

   b. oleid acid

(zig-zag chain with one double bond ending in)—COOH

18.9 a. saturated   **b.** unsaturated
   c. unsaturated   **d.** saturated

18.11 In a cis fatty acid, the hydrogen atoms are on the same side of the double bond, which produces a bend in the carbon chain. In a trans fatty acid, the hydrogen atoms are on opposite sides of the double bond, which gives a carbon chain without any bend.

18.13 In an omega-3 fatty acid, there is a double bond on carbon 3 counting from the methyl group, whereas in an omega-6 fatty acid, there is a double bond beginning at carbon 6 counting from the methyl group.

18.15 Arachidonic acid contains four double bonds and no side groups. In PGE$_2$, a part of the chain forms cyclopentane and there are hydroxyl and ketone functional groups.

18.17 Prostaglandins affect blood pressure and stimulate contraction and relaxation of smooth muscle.

18.19

$$CH_3(CH_2)_{14}\overset{O}{\underset{\|}{C}}O(CH_2)_{29}CH_3$$

18.21

$$CH_2O\overset{O}{\underset{\|}{C}}(CH_2)_{16}CH_3$$
$$CHO\overset{O}{\underset{\|}{C}}(CH_2)_{16}CH_3$$
$$CH_2O\overset{O}{\underset{\|}{C}}(CH_2)_{16}CH_3$$

18.23

$$CH_2O\overset{O}{\underset{\|}{C}}(CH_2)_{14}CH_3$$
$$CHO\overset{O}{\underset{\|}{C}}(CH_2)_{14}CH_3$$
$$CH_2O\overset{O}{\underset{\|}{C}}(CH_2)_{14}CH_3$$

18.25 Safflower oil contains fatty acids with two or three double bonds; olive oil contains a large amount of oleic acid, which has only one (monounsaturated) double bond.

18.27 Although coconut oil comes from a vegetable, it has large amounts of saturated fatty acids and small amounts of unsaturated fatty acids.

18.29

$$CH_2O\overset{O}{\underset{\|}{C}}(CH_2)_7CH{=}CH(CH_2)_7CH_3$$
$$CHO\overset{O}{\underset{\|}{C}}(CH_2)_7CH{=}CH(CH_2)_7CH_3 + H_2 \xrightarrow{Pt}$$
$$CH_2O\overset{O}{\underset{\|}{C}}(CH_2)_7CH{=}CH(CH_2)_7CH_3$$

$$CH_2O\overset{O}{\underset{\|}{C}}(CH_2)_{16}CH_3$$
$$CHO\overset{O}{\underset{\|}{C}}(CH_2)_{16}CH_3$$
$$CH_2O\overset{O}{\underset{\|}{C}}(CH_2)_{16}CH_3$$

18.31 a. Some of the double bonds in the unsaturated fatty acids have been converted to single bonds.
b. It is mostly saturated fatty acids.

18.33 a.

$$CH_2O\overset{O}{\underset{\|}{C}}(CH_2)_{12}CH_3$$
$$CHO\overset{O}{\underset{\|}{C}}(CH_2)_{12}CH_3 + 3H_2O \xrightarrow{H^+}$$
$$CH_2O\overset{O}{\underset{\|}{C}}(CH_2)_{12}CH_3$$

$$CH_2OH$$
$$CHOH + 3CH_3(CH_2)_{12}\overset{O}{\underset{\|}{C}}OH$$
$$CH_2OH$$

b.

$$CH_2O\overset{O}{\underset{\|}{C}}(CH_2)_{12}CH_3$$
$$CHO\overset{O}{\underset{\|}{C}}(CH_2)_{12}CH_3 + 3\,NaOH \longrightarrow$$
$$CH_2O\overset{O}{\underset{\|}{C}}(CH_2)_{12}CH_3$$

$$CH_2OH$$
$$CHOH + 3CH_3(CH_2)_{12}\overset{O}{\underset{\|}{C}}O^-Na^+$$
$$CH_2OH$$

18.35 A triacylglycerol is composed of glycerol with three hydroxyl groups that form ester links with three long-chain fatty acids. In olestra, six to eight long-chain fatty acids form ester links with the hydroxyl groups on sucrose, a sugar. The olestra cannot be digested because our enzymes cannot break down the large olestra molecule.

18.37

$$CH_2O{-}\overset{O}{\underset{\|}{C}}{-}(CH_2)_{16}CH_3$$
$$CHO{-}\overset{O}{\underset{\|}{C}}{-}(CH_2)_{16}CH_3$$
$$CH_2O{-}\overset{O}{\underset{\|}{C}}{-}(CH_2)_{16}CH_3$$

18.39 A triacylglycerol consists of glycerol and three fatty acids. A glycerophospholipid consists of glycerol, two fatty acids, a phosphate group, and an amino alcohol.

18.41

$$CH_2O\overset{O}{\overset{||}{C}}(CH_2)_{14}CH_3$$
$$CHO\overset{O}{\overset{||}{C}}(CH_2)_{14}CH_3$$
$$CH_2OP OCH_2CH_2\overset{+}{N}H_3$$
$$\overset{|}{O^-}$$

This is a cephalin

18.43 This phospholipid is a cephalin. It contains glycerol, oleic acid, stearic acid, a phosphate, and ethanolamine.

18.45 A sphingolipid contains the amino alcohol sphingosine (instead of glycerol) and only one fatty acid. A glycerophospholipid consists of glycerol, two fatty acids, a phosphate group, and an amino alcohol.

18.47

$$CH_3(CH_2)_{12}CH=CHOH$$
$$CHNH\overset{O}{\overset{||}{C}}(CH_2)_{14}CH_3$$

HOCH₂ O O—CH₂
HO   OH   OH

18.49

18.51 Bile salts act to emulsify fat globules, allowing the fat to be more easily digested.

18.53 Lipoproteins are large, spherically shaped molecules that transport lipids in the bloodstream. They consist of an outside layer of phospholipids and proteins surrounding an inner core of hundreds of nonpolar lipids and cholesterol esters.

18.55 Chylomicrons have a lower density than VLDLs. They pick up triacylglycerols from the intestine, whereas VLDLs transport triacylglycerols synthesized in the liver.

18.57 "Bad" cholesterol is the cholesterol carried by LDLs that can form deposits in the arteries called plaque, which narrow the arteries.

18.59 Both estradiol and testosterone contain the steroid nucleus and a hydroxyl group. Testosterone has a ketone group, a double bond, and two methyl groups. Estradiol has a benzene ring, a hydroxyl group in place of the ketone, and a methyl group.

18.61 Testosterone is a male sex hormone.

18.63 Phospholipids with smaller amounts of glycolipids and cholesterol.

18.65 The lipid bilayer in a cell membrane surrounds the cell and separates the contents of the cell from the external fluids.

18.67 The peripheral proteins in the membrane emerge on the inner or outer surface only, whereas the integral proteins extend through the membrane to both surfaces.

18.69 The carbohydrates as glycoproteins and glycolipids on the surface of cells act as receptors for cell recognition and chemical messengers such as neurotransmitters.

18.71 Substances move through cell membrane by simple transport, facilitated transport, and active transport.

18.73 Beeswax and carnauba are waxes. Vegetable oil and capric triglyceride are triacylglycerols.

$$CH_2O\overset{O}{\overset{||}{C}}(CH_2)_8CH_3$$
$$CHO\overset{O}{\overset{||}{C}}(CH_2)_8CH_3$$
$$CH_2O\overset{O}{\overset{||}{C}}(CH_2)_8CH_3$$
Capric triacylglycerol

18.75 **a.** A typical fatty acid has a cis double bond.
b. A trans fatty acid has a trans double bond.
c.

$$\underset{CH_3(CH_2)_6CH_2}{H}C=C\underset{H}{CH_2(CH_2)_6\overset{O}{\overset{||}{C}}OH}$$

18.77

$$CH_2O\overset{O}{\overset{||}{C}}(CH_2)_{16}CH_3$$
$$CHO\overset{O}{\overset{||}{C}}(CH_2)_{16}CH_3$$
$$CH_2O\overset{O}{\overset{||}{C}}(CH_2)_{16}CH_3$$
Glyceryl stearate

$$CH_2OC(CH_2)_{14}CH_3$$
(structure with carbonyl O above first carbon)

$$CHOC(CH_2)_{14}CH_3$$

$$CH_2OPOCH_2CH_2\overset{+}{N}(CH_3)_3$$
$$O^-$$

Lecithin

(Note: This lecithin structure is shown with choline, but ethanolamine or serine are also possible amino alcohols in lecithin.)

18.79 Stearic acid is a fatty acid. Sodium stearate is a soap. Glyceryl tripalmitate, safflower oil, whale blubber, and adipose tissue are triacylglycerols. Beeswax is a wax. Lecithin is a glycerophospholipid. Sphingomyelin is a sphingolipid. Cholesterol, progesterone, and cortisone are steroids.

18.81 **a.** 5   **b.** 1, 2, 3, 4   **c.** 2
       d. 1, 2   **e.** 1, 2, 3, 4   **f.** 2, 3, 4, 6

18.83 **a.** 4   **b.** 3   **c.** 1
       d. 4   **e.** 4   **f.** 3
       g. 2   **h.** 1

Chapter 19

Answers to Study Checks

19.1 tertiary (3°)

19.2 $CH_3CH_2NHCH_2CH_2CH_3$

19.3 Hydrogen bonding makes amines with six or fewer carbon atoms soluble in water.

19.4
$$CH_3 - \overset{\overset{CH_3}{|+}}{\underset{\underset{CH_3}{|}}{N}} - H\ Cl^-$$

19.5 pyrmidine

19.6 piperidine

19.7
$$\overset{O}{\overset{\|}{HCOH}}\quad \text{and}\quad \overset{CH_3}{\overset{|}{HNCH_3}}$$

19.8
Cl—(benzene ring)—$\overset{O}{\overset{\|}{C}}$—$\overset{\overset{CH_3}{|}}{N}$—CH$_3$

19.9 $CH_3CH_2CH_2\overset{O}{\overset{\|}{C}}OH$ and $CH_3NH_3{}^+Br^-$

Answers to Selected Problems

19.1 In a primary amine, there is one alkyl group (and two hydrogens) attached to a nitrogen atom.

19.3 **a.** primary   **b.** secondary
       c. primary   **d.** tertiary   **e.** tertiary

19.5 **a.** ethylamine; ethanamine
       b. methylpropylamine; N-methyl-1-propanamine
       c. diethylmethylamine; N-methyl-N-ethylethanamine
       d. Isopropylamine; 2-propanamine

19.7 **a.** 2-butanamine
       b. 2-chloroaniline
       c. 3-aminopropanal
       d. N-ethylaniline

19.9 **a.** $CH_3CH_2NH_2$ **b.** (benzene ring with NHCH$_3$)

c. $CH_3CH_2CH_2CH_2 - \overset{\overset{H}{|}}{N} - CH_2CH_2CH_3$

d. $CH_3 - \overset{\overset{NH_2}{|}}{CH} - CH_2 - CH_2 - CH_3$

19.11 **a.** $CH_3 - CH_2 - OH$
        b. $CH_3 - CH_2 - CH_2 - NH_2$
        c. $CH_3 - CH_2 - CH_2 - NH_2$

19.13 As a primary amine, propylamine can form two hydrogen bonds, which gives it the highest boiling point. Ethylmethylamine, a secondary amine, can form one hydrogen bond, and butane cannot form hydrogen bonds. Thus butane has the lowest boiling point of the three compounds.

19.15 **a.** yes   **b.** yes   **c.** no   **d.** yes

19.17

a. $CH_3NH_2 + H_2O \rightleftharpoons CH_3-NH_3^+ + OH^-$

b. $CH_3-NH-CH_3 + H_2O \rightleftharpoons CH_3-\overset{+}{N}H_2-CH_3 + OH^-$

c.

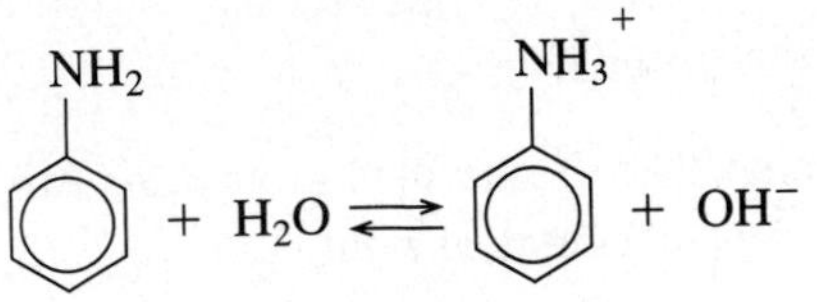

$\underset{\text{(benzene ring with } NH_2)}{} + H_2O \rightleftharpoons \underset{\text{(benzene ring with } \overset{+}{N}H_3)}{} + OH^-$

19.19   **a.** $CH_3-NH_3^+ Cl^-$

b. $CH_3-\overset{+}{N}H_2-CH_3 Cl^-$

c. $NH_3^+ Cl^-$ (benzene ring)

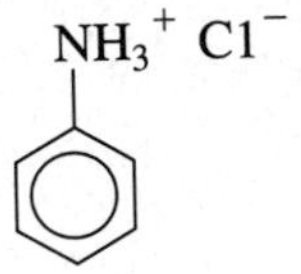

19.21

a. $H_2N-\underset{\text{(benzene ring)}}{}-\overset{O}{\overset{\|}{C}}-O-CH_2-CH_2-\overset{\overset{CH_2CH_3}{|}}{\underset{\underset{CH_2CH_3}{|}}{N^+}}-H\ Cl^-$

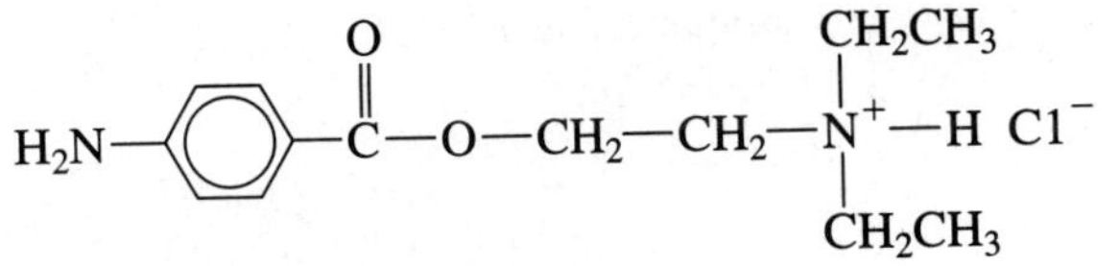

b. Amine salts are soluble in body fluids.

19.23   **a.** amine   **b.** amine
c. heterocyclic amine   **d.** heterocyclic amine

19.25   **c.** pyrimidine   **d.** pyrrole

19.27   pyrrole

19.29   **a.** $CH_3\overset{O}{\overset{\|}{C}}-NH_2$

b. $CH_3\overset{O}{\overset{\|}{C}}-NHCH_2CH_3$

c. $\underset{\text{(benzene ring)}}{}-\overset{O}{\overset{\|}{C}}-\overset{\overset{H}{|}}{N}-CH_2CH_2CH_3$

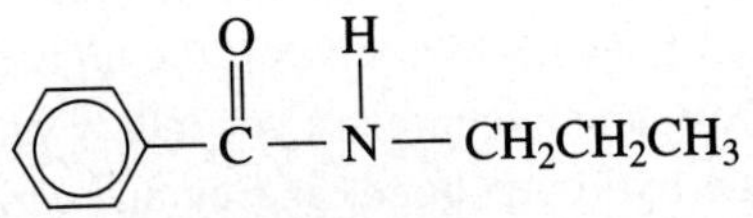

19.31   **a.** *N*-methylethanamide (*N*-methylacetamide)
b. butanamide (butyramide)
c. methanamide (formamide)
d. *N*-methylbenzamide

19.33

a. $CH_3CH_2\overset{O}{\overset{\|}{C}}-NH_2$

b. $CH_3CH_2CH_2\overset{\overset{CH_3}{|}}{CH}-\overset{O}{\overset{\|}{C}}-NH_2$

c. $H\overset{O}{\overset{\|}{C}}NH_2$

d. $\underset{\text{(benzene ring)}}{}-\overset{O}{\overset{\|}{C}}-\overset{\overset{H}{|}}{N}-CH_2CH_3$

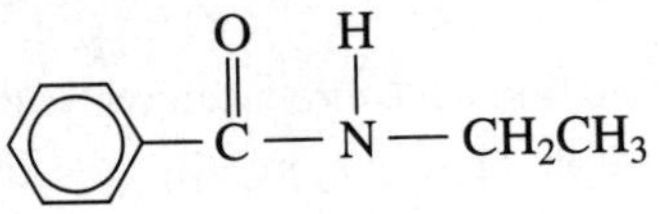

e. $CH_3CH_2CH_2\overset{O}{\overset{\|}{C}}-\overset{\overset{H}{|}}{N}-CH_2CH_3$

19.35   **a.** acetamide   **b.** propionamide
c. *N*-methylpropanamide

19.37

a. $CH_3\overset{O}{\overset{\|}{C}}OH + NH_4^+Cl^-$

b. $CH_3CH_2\overset{O}{\overset{\|}{C}}OH + NH_4^+Cl^-$

c. $CH_3CH_2CH_2\overset{O}{\overset{\|}{C}}OH + CH_3NH_3^+Cl^-$

d. $\underset{\text{(benzene ring)}}{}-\overset{O}{\overset{\|}{C}}OH + NH_4^+Cl^-$

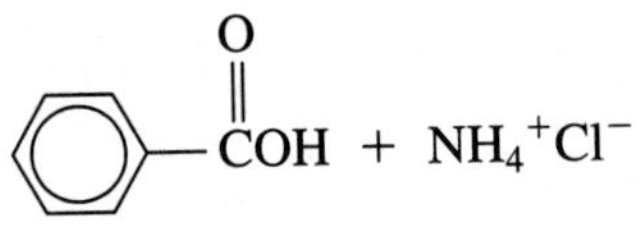

e. $CH_3CH_2CH_2CH_2\overset{O}{\overset{\|}{C}}OH + CH_3CH_2-NH_3^+Cl^-$

19.39 $CH_3-CH_2-CH_2-NH_2$
Propylamine 1°

$CH_3-CH_2-NH-CH_3$
Ethymethylamine 2°

$CH_3-\overset{\overset{CH_3}{|}}{N}-CH_3$
Trimethylamine 3°

$CH_3-\overset{\overset{CH_3}{|}}{CH}-NH_2$
Isopropylamine 1°

19.41 a.
$$\underset{\overset{|}{NH_2}}{CH_3-CH_2-CH-CH_2-CH_3}$$

b.
$$\underset{\text{cyclohexane with } NH_2}{\bigcirc}$$ (cyclohexane ring with NH₂)

c.
$$CH_3-\overset{\overset{\displaystyle CH_3}{|}}{N}H_2^+ \ Cl^-$$

d.
$$CH_3-CH_2-\overset{\overset{\displaystyle CH_2-CH_3}{|}}{N}-CH_2-CH_3$$

e.
$$CH_3-\overset{\overset{\displaystyle OH}{|}}{CH}-\overset{\overset{\displaystyle NH_2}{|}}{CH}-CH_2-CH_2-CH_3$$

f.
$$CH_3-\overset{\overset{\displaystyle CH_3}{|}}{\underset{\underset{\displaystyle CH_3}{|}}{N^+}}-CH_3 \ Br^-$$

g.
$$CH_3\diagdown \overset{\displaystyle CH_3}{\underset{N}{\diagup}}$$ (N-dimethyl aniline, with benzene ring)

19.43 a. ethylamine   **b.** trimethylamine
c. butylamine
d. $NH_2-CH_2CH_2CH_2CH_2CH_2-NH_2$

19.45 a. quinine   **b.** nicotine
c. caffeine   **d.** morphine, codeine

19.47 a. $CH_3-CH_2-NH_3^+OH^-$

b. $CH_3-CH_2-NH_3^+Cl^-$

c. $CH_3-CH_2-\overset{+}{N}H_2-CH_3 \ OH^-$

d. $CH_3-CH_2-\overset{+}{N}H_2-CH_3 \ Cl^-$

e. $CH_3-CH_2-CH_2-NH_2 + NaCl + H_2O$

f. $CH_3-CH_2-\overset{\overset{\displaystyle CH_3}{|}}{N}H + NaCl + H_2O$

19.49 carboxylate salt, aromatic, amine, haloaromatic

19.51 a. aromatic, amine, amide, carboxylic acid, cycloalkene
b. aromatic, ether, alcohol, amine
c. aromatic, carboxylic acid
d. phenol, amine, carboxylic acid
e. aromatic, ether, alcohol, amine, ketone
f. aromatic, amine

Chapter 20

Answers to Study Checks

20.1 hormones

20.2 a. nonpolar   **b.** polar

20.3 In the Fischer projection of D-serine, the —NH₂ group is on the right side.

20.4
$$\underset{\overset{+}{N}H_3-CH-\overset{\overset{\displaystyle O}{\|}}{C}-OH}{\overset{\overset{\displaystyle OH}{|}}{\underset{\displaystyle CH_2}{|}}}$$

20.5
$$\underset{H_3\overset{+}{N}-CH-\overset{O}{\underset{\|}{C}}-N-\overset{}{\underset{}{CH}}-\overset{O}{\underset{\|}{C}}-O^-}{\overset{\overset{\displaystyle O \diagdown \diagup OH}{C}}{\underset{\displaystyle CH_2}{|}} \quad \overset{\displaystyle \bigcirc}{\underset{\displaystyle CH_2}{|}}}$$

20.6 threonylleucylphenylalanine

20.7 a triple helix

20.8 Both are nonpolar and would be found on the inside of the tertiary structure.

20.9 quaternary

20.10 The heavy metal Ag⁺ denatures the protein in bacteria that cause gonorrhea.

Answers to Selected Questions and Problems

20.1 a. transport   **b.** structural
c. structural   **d.** enzyme

20.3 All amino acids contain a carboxylic acid group and an amino group on the α carbon.

20.5

a.

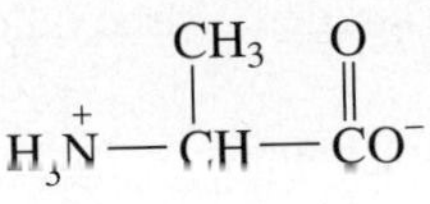

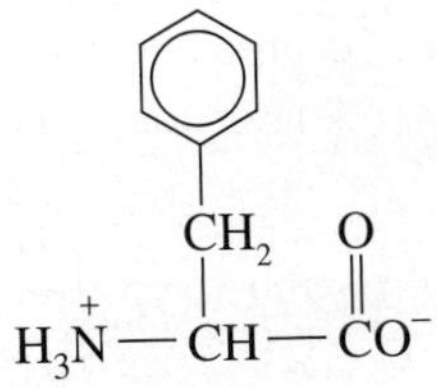

b.

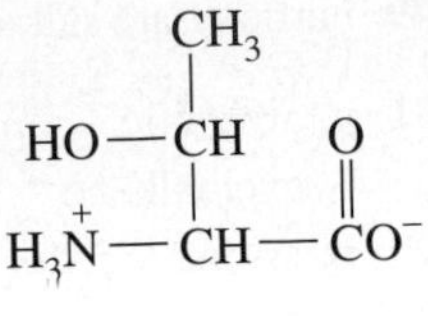

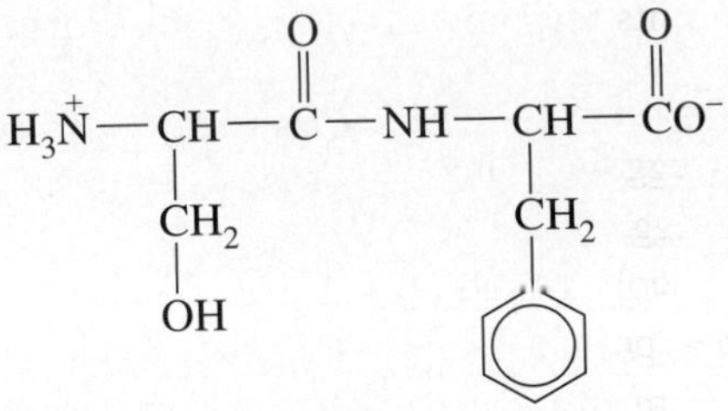

c.

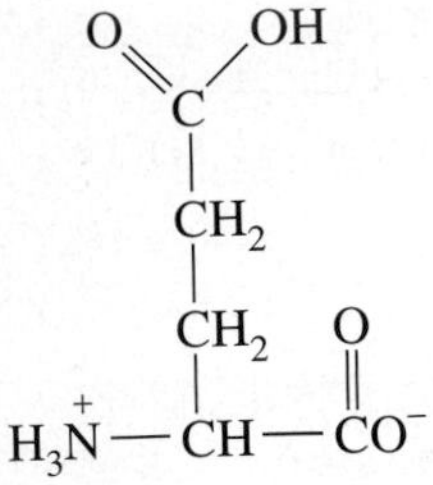

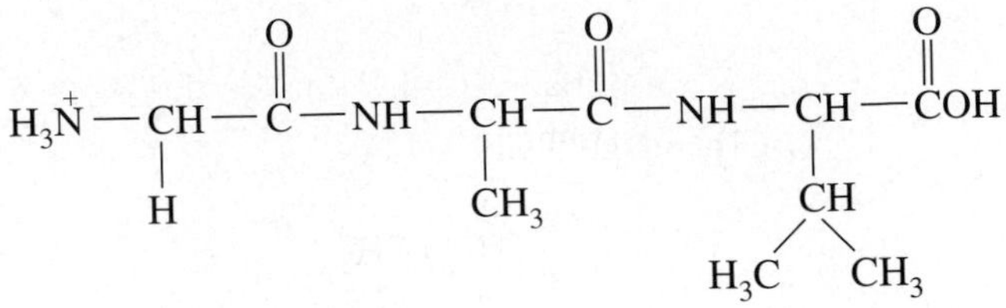

d.

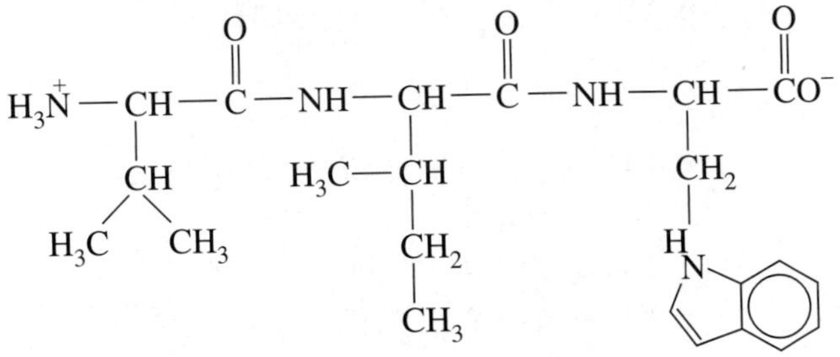

20.7 **a.** hydrophobic nonpolar
c. acidic
b. hydrophilic polar
d. hydrophobic nonpolar

20.9 **a.** alanine
c. lysine
b. valine
d. cysteine

20.11

20.13

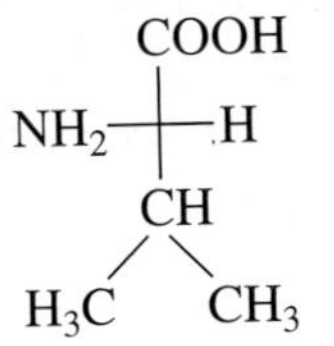

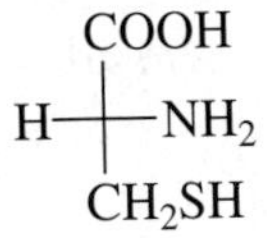

20.15

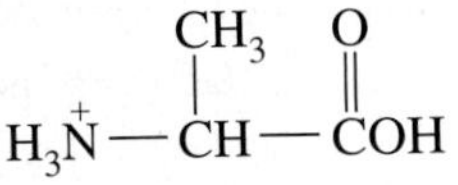

20.17 **a.** above pI
b. below pI
c. at pI

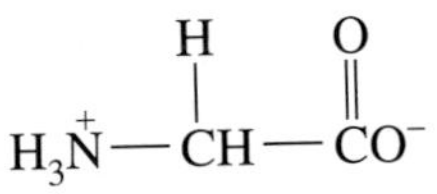

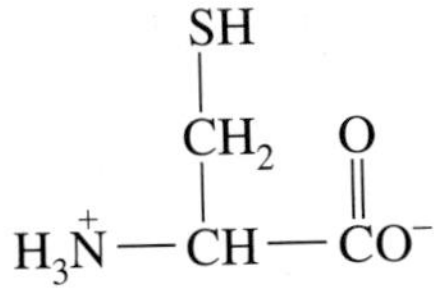

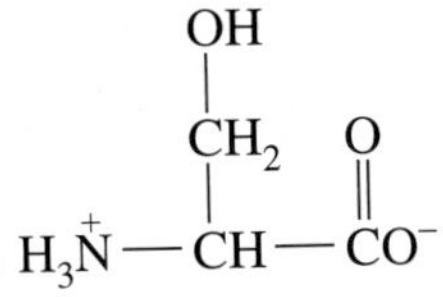

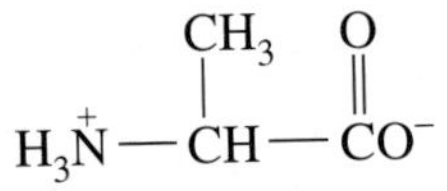

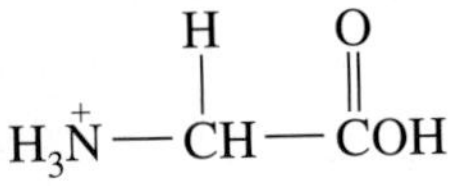

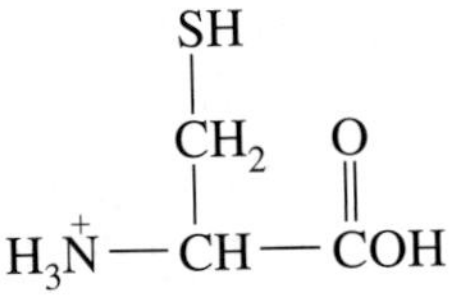

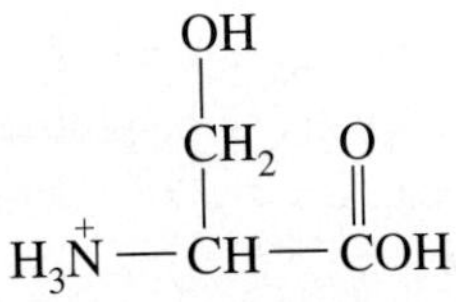

20.19

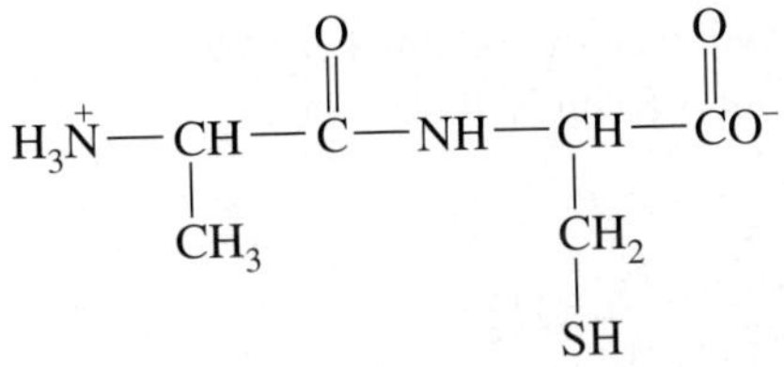

20.21 Amide bonds form to connect the amino acids that make up the protein.

20.23 Val-Ser-Ser, Ser-Val-Ser, or Ser-Ser-Val

20.25 The primary structure remains unchanged and intact as hydrogen bonds form between carbonyl oxygen atoms and amino hydrogen atoms.

20.27 In the α-helix, hydrogen bonds form between the carbonyl oxygen atom and the amino hydrogen atom of the fourth amino acid in the sequence. In the β-pleated sheet, hydrogen bonds occur between parallel peptides or across sections of a long polypeptide chain.

20.29 **a.** a disulfide bond
b. salt bridge
c. hydrogen bond
d. hydrophobic interaction

20.31 **a.** cysteine
b. Leucine and valine will be found on the inside of the protein because they are hydrophobic.
c. The cysteine and aspartic acid would be on the outside of the protein because they are polar.
d. The order of the amino acids (the primary structure) provides the R groups, whose interactions determine the tertiary structure of the protein.

20.33 The products would be the amino acids glycine, alanine, and serine.

20.35 His-Met; Met-Gly; Gly-Val

20.37 Hydrolysis splits the amide linkages in the primary structure.

20.39 **a.** Placing an egg in boiling water coagulates the protein of the egg.
 b. Using an alcohol swab coagulates the protein of any bacteria present.
 c. The heat from an autoclave will coagulate the protein of any bacteria on the surgical instruments.
 d. Heat will coagulate the surrounding protein to close the wound.

20.41 **a.** yes   **b.** yes   **c.** no
 d. yes   **e.** no   **f.** yes

20.43 **a.** The secondary structure of a protein depends on hydrogen bonds to form a helix or a pleated sheet; the tertiary structure is determined by the interaction of R groups and determines the three-dimensional structure of the protein.
 b. Nonessential amino acids can be synthesized by the body; essential amino acids must be supplied by the diet.
 c. Polar amino acids have hydrophilic side groups, whereas nonpolar amino acids have hydrophobic side groups.
 d. Dipeptides contain two amino acids, whereas tripeptides contain three.
 e. An ionic bond is an interaction between a basic and acidic side group; a disulfide bond links the sulfides of two cysteines.
 f. Fibrous proteins consist of three to seven α helixes coiled like a rope. Globular proteins form a compact spherical shape.
 g. The α-helix is the secondary shape like a staircase or corkscrew. The β-pleated sheet is a secondary structure that is formed by many protein chains side by side like a pleated sheet.
 h. The tertiary structure of a protein is its three-dimensional structure. The quaternary structure involves the grouping of two or more peptide units for the protein to be active.

20.45 **a.** α-keratins are fibrous proteins that provide structure to hair, wool, skin, and nails.
 b. α-keratins have a high content of cysteine.

20.47

a.

b. This segment contains polar R groups, which would be found on the surface of a globular protein where they can hydrogen bond with water.

20.49 **a.** β-pleated sheet   **b.** α-helix

20.51 Serine is a polar amino acid, whereas valine is nonpolar. Valine would be in the center of the tertiary structure. However, serine would pull that part of the chain to the outside surface of the protein where it forms hydrophilic bonds with water.

20.53

a.

b.

c.

Chapter 21

Answers to Study Checks

21.1 Enzymes act as catalysts in biological systems because they lower the activation energy of chemical reactions in the cells.

21.2 hydrolase

21.3 In the lock-and-key model, the shape of a substrate fits the shape of the active site exactly. In the induced-fit model, the substrate induces the active site to adjust its shape to fit the substrate.

21.4 At a pH lower than the optimum pH, denaturation will decrease the activity of urease.

21.5 HCN is a noncompetitive inhibitor.

21.6 Pepsin hydrolyzes proteins in the foods we ingest. It is synthesized as a zymogen, pepsinogen, to prevent its digestion of the proteins that make up the organs in the body

21.7 vitamin B_6

21.8 Water-soluble vitamins are easily destroyed by heat.

Answers to Selected Questions and Problems

21.1 The chemical reactions can occur without enzymes, but the rates are too slow. Catalyzed reactions, which are many times faster, provide the amounts of products needed by the cell at a particular time.

21.3 **a.** oxidation–reduction
b. transfer of a group from one substance to another
c. hydrolysis (splitting) of molecules with the addition of water

21.5 **a.** hydrolase   **b.** oxidoreductase
c. isomerase   **d.** transferase

21.7 **a.** lyase   **b.** transferase

21.9 **a.** succinate oxidase
b. fumarate hydrase
c. alcohol dehydrogenase

21.11 **a.** enzyme   **b.** enzyme–substrate complex
c. substrate

21.13 **a.** $E + S \rightleftharpoons ES \longrightarrow E + P$
b. The active site is a region or pocket within the tertiary structure of an enzyme that accepts the substrate, aligns the substrate for reaction, and catalyzes the reaction.

21.15 Isoenzymes are slightly different forms of an enzyme that catalyze the same reaction in different organs and tissues of the body.

21.17 A doctor might run tests for the enzymes CK, LDH, and AST to determine if the patient had a heart attack.

21.19 **a.** The rate would decrease.
b. The rate would decrease.
c. The rate would decrease.
d. The rate would increase if [S] < [E].

21.21 pepsin, pH 2; urease, pH 5; trypsin, pH 8

21.23 **a.** competitive   **b.** noncompetitive
c. competitive   **d.** noncompetitive
e. competitive

21.25 **a.** methanol, CH_3OH; ethanol, CH_3CH_2OH
b. Ethanol has a similar structure to methanol and could compete for the active site.
c. Ethanol is a competitive inhibitor of methanol oxidation.

21.27 Digestive enzymes are proteases and would digest the proteins of the organ where they are produced if they were active immediately upon synthesis.

21.29 In feedback inhibition, the product binds to the first enzyme in a series and changes the shape of the active site. If the active site can no longer bind the substrate effectively, the reaction will stop.

21.31 When a regulator molecule binds to an allosteric site, the shape of the enzyme is altered, which makes the active site more reactive or less reactive and thereby increases or decreases the rate of the reaction.

21.33 **a.** 3; negative regulator
b. 4; allosteric enzyme
c. 1; zymogen

21.35 **a.** an enzyme that requires a cofactor
b. an enzyme that requires a cofactor
c. a simple enzyme

21.37 **a.** THF   **b.** NAD^+

21.39 **a.** pantothenic acid (vitamin B_5)   **b.** folic acid
c. niacin (vitamin B_3)

21.41 **a.** vitamin D or cholecaliferol
b. ascorbic acid or vitamin C
c. niacin or vitamin B_3

21.43 Vitamin B_6 is a water-soluble vitamin, which means that each day any excess of vitamin B_6 is eliminated from the body.

21.45 The side chain $-CH_2OH$ on the ring is oxidized to $-CHO$, and the other $-CH_2OH$ forms a phosphate ester.

21.47 The many different reactions that take place in cells require different enzymes because enzymes react with only a certain type of substrate.

21.49 When exposed to conditions of strong acids or bases, or high temperatures, the protein of enzymes is denatured rapidly causing a loss of tertiary structure and catalytic activity.

21.51 **a.** The reactant is lactose and the products are glucose and galactose.
b.

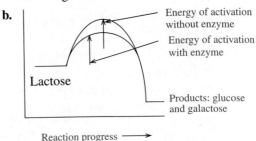

c. By lowering the energy of activation, the enzyme furnishes a lower energy pathway by which the reaction can take place.

21.53 Indicate whether each of the following would be a substrate (S) or an enzyme (E):
a. S   **b.** E
c. E   **d.** E
e. S   **f.** E

21.55 a. urea   **b.** lactose
c. aspartate   **d.** tyrosine

21.57 a. transferase   **b.** oxidoreductase
c. hydrolase   **d.** lyase

21.59 a. oxidoreductase   **b.** hydrolase   **c.** lyase

21.61 Sucrose fits the shape of the active site in sucrase, but lactose does not.

21.63 A heart attack may be the cause.

21.65 a. saturated   **b.** unsaturated

21.67 In a reversible inhibition, the inhibitor can dissociate from the enzyme, whereas in irreversible inhibition, the inhibitor forms a strong covalent bond with the enzyme and does not dissociate. Irreversible inhibitors act as poisons to enzymes.

21.69 a. oxidoreductase
b. At high concentration, ethanol, which acts as a competitive inhibitor of ethylene glycol, would saturate the alcohol dehydrogenase enzyme to allow ethylene glycol to be removed from the body without producing oxalic acid.

21.71 a. Antibiotics such as amoxicillin are irreversible inhibitors.
b. Antibiotics inhibit enzymes needed to form cell walls in bacteria, not humans.

21.73 a. When pepsinogen enters the stomach, the low pH cleaves a peptide from its protein chain to form pepsin.
b. An active protease would digest the proteins of the pancreas rather than the proteins in the foods entering the stomach.

21.75 An allosteric enzyme contains sites for regulators that alter the enzyme and speed up or slow down the rate of the catalyzed reaction.

21.77 This would be a negative regulator because the end product of the reaction pathway binds to the enzyme to decrease or stop the first reaction in the reaction pathway.

21.79 a. requires a cofactor
b. simple enzyme
c. requires a cofactor (coenzyme)

21.81 a. pantothenic acid (B_5); coenzyme A
b. niacin (B_3); NAD^+
c. biotin; biocytin

21.83 A vitamin combines with an enzyme only when the enzyme and coenzyme are needed to catalyze a reaction. When the enzyme is not need, the vitamin dissociates for use by other enzymes in the cell.

21.85 a. niacin; pellagra
b. vitamin A; night blindness
c. vitamin D; weak bone structure

Chapter 22

Answers to Study Checks

22.1 a. Guanine is found in both RNA and DNA.
b. Uracil is found only in RNA.

22.2 deoxycytidine 5'—monophosphate (dCMP)

22.3

22.4 5'—C—C—A—A—T—T—G—G—3'

22.5 In DNA, the lagging strand is synthesized in the 3'–5' direction in short sections that are joined by DNA ligase.

22.6 Each type of tRNA matches a specific codon to a specific amino acid.

22.7 3'—CCCAAATTT—5'

22.8 A regulatory gene produces mRNA for the production of a protein repressor, which binds to the operator and blocks protein synthesis.

22.9 UGA is a stop codon that signals the termination of translation.

22.10 at UAG

22.11 If the substitution of an amino acid in the polypeptide affects an interaction essential to functional structure on the binding of a substrate, the resulting protein could be less effective or nonfunctional.

22.12 Gene cloning is the process by which recombinant DNA technology inserts the DNA of a gene into the plasmid of *E. coli* bacteria that multiply rapidly to make many copies of the gene.

22.13 A protease inhibitor modifies the three-dimensional structure of protease, which prevents the enzymes from synthesizing the proteins needed to produce more viruses.

Answers to Selected Questions and Problems

22.1 a. pyrimidine   **b.** pyrimidine

22.3 a. DNA   **b.** both DNA and RNA

22.5 deoxyadenosine 5'—monophosphate (dAMP), deoxythymidine 5'—monophosphate (dTMP), deoxycytidine 5'—monophosphate (dCMP), and deoxyguanosine 5'—monophosphate (dGMP)

22.7 a. nucleoside   **b.** nucleoside
c. nucleoside   **d.** nucleotide

22.9

22.11 The nucleotides in nucleic acids are held together by phosphodiester bonds between the 3'—OH of a sugar (ribose or deoxyribose) and a phosphate group on the 5'—carbon of another sugar.

22.13

22.15 The two DNA strands are held together by hydrogen bonds between the bases in each strand.

22.17 a. 3'—TTTTTT—5'
b. 3'—CCCCCC—5'
c. 3'—TCAGGTCCA—5'
d. 3'—GACATATGCAAT—5'

22.19 The enzyme helicase unwinds the DNA helix to prepare the parent DNA strand for the synthesis of daughter DNA strands.

22.21 The DNA strands separate and the DNA polymerase pairs each of the bases with its complementary base and produces two exact copies of the original DNA.

22.23 Ribosomal RNA, messenger RNA, and transfer RNA

22.25 A ribosome, which is about 65% rRNA and 35% protein, consists of a small subunit and a large subunit.

22.27 In transcription, the sequence of nucleotides on a DNA template (one strand) is used to produce the base sequences of a messenger RNA.

22.29 5'—GGCUUCCAAGUG—3'

22.31 In eukaryotic cells, genes contain sections called exons that code for protein and sections called introns that do not code for protein.

22.33 An operon is a section of DNA that regulates the synthesis of one or more proteins.

22.35 When the lactose level is low in *E. coli*, a repressor produced by the mRNA from the regulatory gene binds to the operator blocking the synthesis of mRNA from the genes and preventing the synthesis of protein.

22.37 A three-base sequence in mRNA that codes for a specific amino acid in a protein.

22.39 a. leucine   **b.** serine
         c. glycine   **d.** arginine

22.41 When AUG is the first codon, it signals the start of protein synthesis. Thereafter, AUG codes for methionine.

22.43 A codon is a base triplet in the mRNA. An anticodon is the complementary triplet on a tRNA for a specific amino acid.

22.45 Initiation, translocation, and termination.

22.47 a. —Lys—Lys—Lys—
         b. —Phe—Pro—Phe—Pro—
         c. —Tyr—Gly—Arg—Cys—

22.49 The new amino acid is joined by a peptide bond to the peptide chain. The ribosome moves to the next codon, which attaches to a tRNA carrying the next amino acid.

22.51 a. 5'—CGA—AAA—GUU—UUU—3'
         b. GCU, UUU, CAA, AAA
         c. Using codons in mRNA: Arg—Lys—Val—Phe

22.53 A base in DNA is replaced by a different base.

22.55 If the resulting codon still codes for the same amino acid, there is no effect. If the new codon codes for a different amino acid, there is a change in the order of amino acids in the polypeptide.

22.57 The normal triplet TTT forms a codon AAA, which codes for lysine. The mutation TTC forms a codon AAG, which also codes for lysine. There is no effect on the amino acid sequence.

22.59 a. —Thr—Scr—Arg—Val—
         b. —Thr—Thr—Arg—Val—
         c. —Thr—Ser—Gly—Val—
         d. —Thr—STOP   Protein synthesis would terminate early. If this occurs early in the formation of the polypeptide, the resulting protein will probably be nonfunctional.
         e. The new protein will contain the sequence —Asp—Ile—Thr—Gly—.
         f. The new protein will contain the sequence —His—His—Gly—.

22.61 a. GCC and GCA both code for alanine.
         b. A vital ionic cross-link in the tertiary structure of hemoglobin cannot be formed when the polar glutamine is replaced by valine, which is nonpolar. The resulting hemoglobin is malformed and less capable of carrying oxygen.

22.63 *E. coli* bacterial cells contain several small circular plasmids of DNA that can be isolated easily. After the recombinant DNA is formed, *E. coli* multiply rapidly, producing many copies of the recombinant DNA in a relatively short time.

22.65 *E. coli* are soaked in a detergent solution that dissolves the cell membrane and releases the cell contents including the plasmids, which are collected.

22.67 When a gene has been obtained using restriction enzymes, it is mixed with the plasmids that have been opened by the same enzymes. When mixed together in a fresh *E. coli* culture, the sticky ends of the DNA fragments bond with the sticky ends of the plasmid DNA to form a recombinant DNA.

22.69 In DNA fingerprinting, restriction enzymes cut a sample DNA into fragments, which are sorted by size by gel electrophoresis. After tagging the DNA fragments with a radioactive isotope, a piece of x-ray film placed over the gel is exposed by the radioactivity to give a pattern of dark and light bands known as a DNA fingerprint.

22.71 DNA or RNA, but not both.

22.73 a. A viral RNA is used to synthesize a viral DNA to produce the proteins for the protein coat, which allows the virus to replicate and leave the cell.
         b. retrovirus

22.75 Nucleoside analogs such as AZT and ddI are similar to the nucleosides required to make viral DNA in reverse transcription. However, they interfere with the ability of the DNA to form and thereby disrupt the life cycle of the HIV-1 virus.

22.77 a. pyrimidine   **b.** purine   **c.** pyrimidine
         d. pyrimidine   **e.** purine

22.79 a. thymine and deoxyribose
         b. adenine and ribose
         c. cytosine and ribose
         d. guanine and deoxyribose

22.81 They are both pyrimidines, but thymine has a methyl group.

22.83

22.85 They are both polymers of nucleotides connected through phosphodiester bonds between alternating sugar and phosphate groups with bases extending out from each sugar.

22.87 28% T, 22% G, and 22% C

22.89 a. two   **b.** three

22.91 a. 3'—CTGAATCCG—5'
     b. 5'—ACGTTTGATCGT—3'
     c. 3'—TAGCTAGCTAGC—5'

22.93 DNA polymerase synthesizes the leading strand continuously in the 5' to 3' direction. The lagging strand is synthesized in small segments called Okazaki fragments because it must grow in the 3' to 5' direction.

22.95 One strand of the parent DNA is found in each of the two copies of the daughter DNA molecule.

22.97 a. tRNA.   **b.** rRNA   **c.** mRNA

22.99 a. ACU, ACC, ACA, and ACG
     b. UCU, UCC, UCA, UCG, AGU, and AGC
     c. UGU and UGC

22.101 What is the amino acid for each of the following codons?
     a. lysine   **b.** isoleucine   **c.** arginine

22.103 start—Tyr—Gly—Gly—Phe—Leu—stop

22.105 a. UCG   **b.** AUA   **c.** GGU

22.107 Three nucleotides are needed for each amino acid plus a start and stop triplet, which makes a minimum total of 33 nucleotides.

22.109 A DNA virus attaches to a cell and injects viral DNA that uses the host cell to produce copies of DNA to make viral RNA. A retrovirus injects viral RNA from which complementary DNA is produced by reverse transcription.

Chapter 23

Answers to Study Checks

23.1 The cytoplasm is all the cellular material including cytosol and organelles between the plasma membrane and the nucleus. The cytosol, the aqueous part of the cytoplasm, is a solution of electrolytes and enzymes.

23.2 Adenine, ribose, and three phosphate groups.

23.3 $FADH_2$

23.4 The digestion of amylose begins in the mouth when salivary amylase hydrolyzes some of the glycosidic bonds. In the small intestine, pancreatic amylase hydrolyzes more glycosidic bonds, and finally maltose is hydrolyzed by maltase to yield glucose.

23.5 In the initial reactions of glycolysis, energy in the form of two ATP is invested to activate glucose and convert it to fructose-1,6-bisphosphate.

23.6

23.7 Only liver and kidney cells contain the phosphatase enzyme that converts glucose-6-phosphate to free glucose.

23.8 The reaction catalyzed by hexokinase in glycolysis is irreversible

23.9 When a cell needs a supply of ribose-5-phophsate for nucleotide synthesis.

Answers to Selected Questions and Problems

23.1 The digestion of polysaccharides takes place in stage 1.

23.3 In metabolism, a catabolic reaction breaks apart large molecules, releasing energy.

23.5 a. (3) smooth endoplasmic reticulum
     b. (1) lysosome
     c. (2) Golgi complex

23.7 The phosphoric anhydride bonds (P—O—P) in ATP release energy that is sufficient for energy-requiring processes in the cell.

23.9 a. PEP + H_2O $\longrightarrow$ pyruvate + P_i + 14.8 kcal /mole
     b. ADP + P_i + 7.3 kcal/mole $\longrightarrow$ ATP

c. Coupled: PEP + ADP $\longrightarrow$ ATP + pyruvate + 7.5 kcal /mole

23.11 a. coenzyme A   **b.** NAD^+ **c.** FAD

23.13 a. NADH   **b.** FAD

23.15 FAD

23.17 Hydrolysis is the main reaction involved in the digestion of carbohydrates.

23.19 a. lactose
  b. glucose and fructose
  c. glucose

23.21 glucose

23.23 ATP is required in phosphorylation reactions.

23.25 glyceraldehyde-3-phosphate and dihydroxyacetone phosphate

23.27 ATP is produced in glycolysis by transferring a phosphate from 1,3-bisphosphoglycerate and from phosphoenolpyruvate directly to ADP.

23.29 a. hexokinase   **b.** phosphokinase

23.31 a. 1 ATP required
  b. 1 NADH is produced for each triose.
  c. 2 ATP and 2 NADH

23.33 a. In reaction 1, a hexokinase uses ATP to phosphorylate glucose.
  b. In reactions 7 and 10, phosphate groups are transferred from 1,3-bisphosphoglycerate and phosphoenolpyruvate directly to ADP to produce ATP.
  c. In reaction 4, the six-carbon molecule fructose-1,6-bisphosphate is split into two three-carbon molecules, glyceraldehyde-3-phosphate and dihydroxyacetone phosphate.

23.35 Galactose reacts with ATP to yield galactose-1-phosphate, which is converted to glucose-6-phosphate, an intermediate in glycolysis. Fructose reacts with ATP to yield fructose-1-phosphate, which is cleaved to give dihydroxyacetone phosphate and glyceraldehyde. Dihydroxyacetone phosphate isomerizes to glyceraldehyde-3-phosphate, and glyceraldehyde is phosphorylated to glyceraldehyde-3-phosphate, which is an intermediate in glycolysis.

23.37 a. activate **b.** inhibit

23.39 Aerobic (oxygen) conditions are needed.

23.41 The oxidation of pyruvate converts NAD^+ to NADH and produces acetyl CoA and CO_2.

Pyruvate + NAD^+ + CoA $\longrightarrow$
Acetyl CoA + CO_2 + NADH + H^+

23.43 When pyruvate is reduced to lactate, NAD^+ is produced, which can be used for the oxidation of more glyceraldehyde-3-phosphate in glycolysis. This recycles NADH.

23.45 Carbon dioxide, CO_2

23.47 Glycogenesis is the synthesis of glycogen from glucose molecules.

23.49 Muscle cells break down glycogen to glucose-6-phosphate, which enters glycolysis.

23.51 Glycogen phosphorylase cleaves the glycosidic bonds at the ends of glycogen chains to remove glucose monomers as glucose-1-phosphate.

23.53 When there are no glycogen stores remaining in the liver, gluconeogenesis synthesizes glucose from noncarbohydrate compounds such as pyruvate and lactate.

23.55 Phosphoglucoisomerase, aldolase, triosephosphate isomerase, glyceraldehyde-3-phosphate dehydrogenase, phosphoglycerokinase, phosphoglyceromutase, and enolase.

23.57 a. activates   **b.** activates   **c.** inhibits

23.59 Metabolism includes all the reactions in cells provide energy and material for cell growth.

23.61 Stage 1

23.63 Eukaryotic cell

23.65 Adenosine triphosphate

23.67 ATP + H_2O $\longrightarrow$ ADP + P_i + 7.3 kcal (31kJ)/mole

23.69 Flavin adenine dinucleotide

23.71 Nicotinamide adenine dinucleotide

23.73 a. $FADH_2$ **b.** NADH + H^+

23.75 Lactose undergoes digestion in the mucosal cells of the small intestine to yield galactose and glucose.

23.77 Galactose and fructose are converted in the liver to glucose phosphate compounds that can enter the glycolysis pathway.

23.79 Glucose is the reactant and pyruvate is the product of glycolysis.

23.81 Reaction 1 and 3 involve phosphorylation of hexoses with ATP, and reactions 7 and 10 involve direct substrate phosphorylation that generates ATP.

23.83 Reaction 4, which converts fructose-1,6-bisphosphate into two triose phosphates, is catalyzed by aldolase.

23.85 Glucose-6-phosphate

23.87 Pyruvate is converted to lactate when oxygen is not present in the cell (anaerobic) to regenerate NAD^+ for glycolysis.

23.89 Phosphofructokinase is an allosteric enzyme that is activated by high levels of AMP and ADP because the cell needs to produce more ATP. When ATP levels are high due to a decrease in energy needs, ATP inhibits phosphofructokinase, which reduces its catalysis of fructose-6-phosphate.

23.91 The rate of glycogenolysis increases when blood glucose levels are low and glucagon has been secreted, which accelerate the breakdown of glycogen.

23.93 Glucose

23.95 The cells in the liver, but not skeletal muscle, contain a phosphatase enzyme needed to convert glucose-6-phosphate to free glucose that can diffuse through cell membranes into the blood stream. Glucose-6-phosphate, which is the end product of glycogenolysis in muscle cells, cannot diffuse easily across cell membranes.

23.97 Insulin increases the rate of glycogenolysis and glycolysis and decreases the rate of glycogenesis. Glucagon decreases the rate glycogenolysis and glycolysis and increases the rate of glycogenesis.

23.99 The Cori cycles is a cyclic process that involves the transfer of lactate from muscle to the liver where glucose is synthesized, which can be used again by the muscle.

23.101 **a.** increase   **b.** decrease
  c. increase   **d.** decrease

23.103 **a.** decrease   **b.** decrease
  c. increase   **d.** decrease

Chapter 24

Answers to Study Checks

24.1 oxaloacetate

24.2 Fe^{3+}

24.3 **a.** oxidation   **b.** reduction

24.4 Oxygen (O_2) is the last substance that accepts electrons.

24.5 The protein complexes I, III, and IV pump protons from the matrix to the intermembrane space.

24.6 Three NADH provide nine ATP, one $FADH_2$ provides two ATP, and one direct phosphorylation provides one ATP.

Answers to Selected Questions and Problems

24.1 Krebs cycle and tricarboxlyic acid cycle

24.3 $2CO_2$, 3 NADH + $3H^+$, $FADH_2$, GTP (ATP), and HS—CoA.

24.5 Two reactions, 3 and 4, involve oxidative decarboxylation.

24.7 NAD^+ is reduced in reactions 3, 4, and 8 of the citric acid cycle.

24.9 In reaction 5, GDP undergoes a direct substrate phosphorylation.

24.11 **a.** citrate and isocitrate
  b. In decarboxylation, a carbon atom is lost as CO_2.
  c. α-ketoglutarate
  d. isocitrate $\longrightarrow$ α-ketoglutrate; α-ketoglutarate $\longrightarrow$ succinyl CoA; succinate $\longrightarrow$ fumarate; malate $\longrightarrow$ oxaloacetate
  e. steps 3, 8

24.13 **a.** citrate synthase
  b. succinate dehydrogenase and aconitase
  c. fumarase

24.15 **a.** NAD^+ **b.** GDP   **c.** FAD

24.17 Isocitrate dehydrogenase and α-ketoglutarate dehydrogenase are allosteric enzymes

24.19 High levels of ADP increase the rate of the citric acid cycle.

24.21 oxidized

24.23 **a.** oxidation   **b.** reduction

24.25 NADH and $FADH_2$

24.27 FAD, coenzyme Q, cytochrome b, cytochrome c

24.29 The mobile carrier Q transfers electrons from complex I to III.

24.31 NADH transfers electrons to FMN in complex I to give NAD⁺.

24.33 **a.** $NADH + H^+ + FMN \longrightarrow NAD^1 + FMNH_2$
b. $QH_2 + 2 \text{ cyt } b \text{ Fe}^{3+} \longrightarrow Q + 2 \text{ cyt } b (Fe^{2+}) + 2H^+$

24.35 In oxidation phosphorylation, the energy from the oxidation reactions in the electron transport chain is used to drive ATP synthesis.

24.37 As protons return to the lower energy environment in the matrix, they pass through ATP synthase where they release energy to drive the synthesis of ATP.

24.39 The oxidation of the reduced coenzymes NADH and $FADH_2$ by the electron transport chain generates energy to drive the synthesis of ATP.

24.41 ATP synthase consists of two protein complexes known as F_0 and F_1.

24.43 The loose (L) site in ATP synthase binds ADP and P_i.

24.45 Glycolysis takes place in the cytoplasm, not in the mitochondria. Because NADH cannot cross the mitochondrial membrane, one ATP is hydrolyzed to transport the electrons from NADH to FAD. The resulting $FADH_2$ produces only 2 ATP for each NADH produced in glycolysis.

24.47 **a.** 3 ATP   **b.** 2 ATP   **c.** 6 ATP   **d.** 12 ATP

24.49 The oxidation reactions of the citric acid cycle produce a source of reduced coenzymes for the electron transport chain and ATP synthesis.

24.51 The oxidized coenzymes NAD⁺ and FAD needed for the citric acid cycle are regenerated by the electron transport chain.

24.53 **a.** citrate, isocitrate
b. α-ketoglutarate
c. α-ketoglutarate, succinyl CoA, oxaloacetate

24.55 **a.** In reaction 4, α-ketoglutarate, a five-carbon keto acid, is decarboxylated.
b. In reaction 1 and reaction 7, double bonds in aconitate and fumarate are hydrated.
c. NAD⁺ is reduced in reactions 3, 4, and 8.
d. In reactions 3 and 8, a secondary hydroxyl group in isocitrate and malate is oxidized.

24.57 **a.** NAD⁺   **b.** NAD⁺ and CoA

24.59 **a.** High levels of NADH inhibit isocitrate dehydrogenase and α-ketoglutarate dehydrogenase to slow the rate of the citric acid cycle.
b. High levels of ATP inhibit the citric acid cycle.

24.61 **a.** (4) cytochrome   **b.** (1) FMN

24.63 **a.** CoQ is a mobile carrier
b. Fe–S clusters are found in complex I, III, and IV
c. cyt a_3 is part of complex IV

24.65 **a.** complex II; $FADH_2 + Q \longrightarrow FAD + QH_2$
b. complex IV; cyt a (Fe²⁺) + cyt a_3 (Fe³⁺) $\longrightarrow$ cyt a (Fe³⁺) + cyt a_3 (Fe²⁺)

24.67

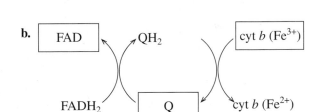

24.69 complex I, III, and IV

24.71 Energy released as protons flow through ATP synthase back to the matrix is utilized for the synthesis ATP.

24.73 Protons flow into the matrix where H⁺ concentration is lower.

24.75 Two ATP molecules are produced from $FADH_2$

24.77 **a.** NADH dehydrogenase (complex I)
b. electron flow from cyt b to cyt c_1 (complex III)
c. cytochrome c oxidase (complex IV)

24.79 The oxidation of glucose to pyruvate produces 6 ATP whereas the oxidation of glucose to CO_2 and H_2O produces 36 ATP.

24.81 **a.** 6 ATP × 7.3 kcal/mole = 44 kcal
b. 6 ATP × 7.3 kcal/mole = 44 kcal (2 pyruvate to 2 acetyl CoA)
c. 24 ATP × 7.3 kcal/ mole = 175 kcal (2 acetyl CoA citric acid cycle)
d. 36 × 7.3 kcal/mol = 263 (complete oxidation of glucose to CO_2 and H_2O)

24.83 In a calorimeter, the complete combustion of glucose gives 687 kcal. Using the value for glucose in problem 24.81, $283/687 \times 100 = 38.3\%$

24.85 The ATP synthase extends through the inner mitochondrial membrane with the F_0 part in contact with the proton gradient in the intermembrane space, while the F_1 complex is in the matrix.

24.87 As protons from the proton gradient move through the ATP synthase to return to the matrix, energy is released and used to drive ATP synthesis at F_1 ATP synthase.

24.89 A hibernating bear has stored fat as brown fat, which can be used during the winter for heat rather than ATP energy.

Chapter 25

Answers to Study Checks

25.1 In the membrane of the small intestine, monoacylglycerols and fatty acids recombine to form new triacylglycerols that bind with proteins to form chylomicrons for transport to the lymphatic system and the bloodstream.

25.2 NAD^+

25.3

$$\text{8 acetyl CoA} - \text{2 ATP} \qquad \frac{94\ \text{ATP}}{129\ \text{ATP}} \times 100 = 73\%$$

$$\text{35 ATP from NADH and FADH}_2 \qquad \frac{35\ \text{ATP}}{129\ \text{ATP}} \times 100 = 27\%$$

25.4 The ketone bodies are acetoacetate, β-hydroxybutyrate, and acetone.

25.5 In each cycle of fatty acid synthesis, a two-carbon atom acyl group from malonyl ACP adds to a fatty acid chain and one carbon atom forms CO_2.

25.6 HCl denatures proteins and activates enzymes such as pepsin.

25.7 In conditions such as fasting or starvation, a diet insufficient in protein leads to a negative nitrogen balance.

25.8 Alanine oxaloacetate transaminase or transaminotransferase.

25.9 The process of oxidative deamination regenerates α-ketoglutarate from glutamate.

25.10 **a.** argininosuccinate
b. citrulline
c. urea and ornithine

25.11 Arginine, glutamate, glutamine, histidine, and proline provide carbon atoms for α-ketoglutarate.

25.12 Substrates for the synthesis of nonessential amino acids are obtained from glycolysis and the citric acid cycle.

Answers to Selected Questions and Problems

25.1 The bile salts emulsify fat to give small fat globules for lipase hydrolysis.

25.3 When blood glucose and glycogen stores are depleted

25.5 Glycerol is converted to glycerol-3-phosphate, which is converted to dihydroxyacetone phosphate, an intermediate of glycolysis.

25.7 In the cytosol at the outer mitochondrial membrane

25.9 FAD and NAD^+

25.11

a. $CH_3-CH_2-CH_2-CH_2-CH_2-\overset{\beta}{C}H_2-CH_2-\overset{O}{\overset{\|}{C}}-S-CoA$

b. $CH_3-(CH_2)_{14}-\overset{\beta}{C}H_2-CH_2-\overset{O}{\overset{\|}{C}}-S-CoA$

c. $CH_3-CH_2-CH=CH-CH_2-CH_2-CH_2-\overset{\beta}{C}H_2-CH_2-\overset{O}{\overset{\|}{C}}-S-CoA$

25.13

a. and b. $CH_3-(CH_2)_6-\underset{\beta}{CH_2}-\underset{\alpha}{CH_2}-\overset{\displaystyle O}{\overset{\|}{C}}-S-CoA$

c. $CH_3-(CH_2)_8-\overset{\displaystyle O}{\overset{\|}{C}}-S-CoA + NAD^+ + FAD + H_2O + SH-CoA \longrightarrow$

$CH_3-(CH_2)_6-\overset{\displaystyle O}{\overset{\|}{C}}-S-CoA + CH_2-\overset{\displaystyle O}{\overset{\|}{C}}-S-CoA + NADH + H^+ + FADH_2$

d. $CH_3-(CH_2)_8-COOH + 4CoA + 4FAD + 4NAD^+ + 4H_2O \longrightarrow$
$5Acetyl\ CoA + 4FADH_2 + 4NADH + 4H^+$

25.15 The hydrolysis of ATP to AMP hydrolyzes ATP to ADP, and ADP to AMP, which provides the same amount of energy as the hydrolysis of 2 ATP to 2 ADP.

25.17 **a.** 5 acetyl CoA units
b. 4 cycles of β oxidation
c. 60 ATP from 5 acetyl CoA (citric acid cycle) + 12 ATP from 4 NADH + 8 ATP from 4 FADH$_2$ −2 ATP (activation) = 80 −2 = 78 ATP

25.19 Ketogenesis is the synthesis of ketone bodies from excess acetyl CoA from fatty acid oxidation, which occurs when glucose is not available for energy, particularly in starvation, fasting, and diabetes.

25.21 Acetoacetate undergoes reduction using NADH + H$^+$ to yield β-hydroxybutyrate.

25.23 High levels of ketone bodies lead to ketosis, a condition characterized by acidosis (a drop in blood pH values), excessive urination, and strong thirst.

25.25 in the cytosol of cells in liver and adipose tissue

25.27 acetyl CoA, HCO$_3^-$, and ATP

25.29 **a.** (3) malonyl CoA transacylase
b. (1) acetyl CoA carboxylase
c. (2) acetyl CoA transacylase

25.31 **a.** 4 HCO$_3^-$ **b.** 4 ATP
c. 5 acetyl CoA   **d.** 4 malonyl ACP
e. 8 NADPH   **f.** 4 CO$_2$ removed

25.33 The digestion of proteins begins in the stomach and is completed in the small intestine.

25.35 Hormones, heme, purines and pyrimidines for nucleotides, proteins, nonessential amino acids, amino alcohols, and neurotransmitters require nitrogen obtained from amino acids.

25.37 The reactants are an amino acid and an α-keto acid, and the products are a new amino acid and a new α-keto acid.

25.39

a. $H-\overset{\displaystyle O}{\overset{\|}{C}}-COO^-$ **b.** $CH_3-\overset{\displaystyle O}{\overset{\|}{C}}-COO^-$

c. $CH_3-\underset{\underset{\displaystyle CH_3}{|}}{CH}-\overset{\displaystyle O}{\overset{\|}{C}}-COO^-$

25.41

$\overset{\overset{\displaystyle \overset{+}{N}H_3}{|}}{{}^-OOC-CH-CH_2-CH_2-COO^-} + H_2O + NAD^+ (NADP^+) \xrightarrow{\text{Glutamate dehydrogenase}}$

Glutamate

$^-OOC-\overset{\displaystyle O}{\overset{\|}{C}}-CH_2-CH_2-COO^- + NH_4^+ + NADH\ (NADPH) + H^+$

α-Ketoglutarate

25.43 NH_4^+ is toxic if allowed to accumulate in the liver.

25.45

$$H_2N-\overset{\overset{\displaystyle O}{\|}}{C}-NH_2$$

25.47 CO_2 from the citric acid cycle

25.49 Glucogenic amino acids produce compounds used to synthesis glucose.

25.51 **a.** pyruvate   **b.** oxaloacetate, fumarate
      c. succinyl CoA   **d.** α-ketoglutarate

25.53 nonessential amino acid

25.55 Glutamine synthetase catalyzes the addition of an amino group to glutamate using energy from the hydrolysis of ATP.

25.57 phenylketournia

25.59 Triacylglycerols are hydrolyzed to monoacylglycerols and fatty acids in the small intestine, which reform triacylglycerols in the intestinal lining for transport as lipoproteins to the tissues.

25.61 Fats can be stored in unlimited amounts in adipose tissue compared to the limited storage of carbohydrates as glycogen.

25.63 The fatty acids cannot diffuse across the blood brain barrier.

25.65 **a.** Glycerol is converted to glycerol-3-phosphate and to dihydroxyacetone phosphate, which enters glycolysis or gluconeogensis.
     b. Activation of fatty acids occurs on the outer mitochondrial membrane.
     c. The energy cost is equal to 2 ATP.
     d. Only fatty acyl CoA can move into the intermembrane space for transport by carnitine into the matrix.

25.67 Lauric acid, $CH_3-(CH_2)_{10}-COOH$, is a fatty acid.
     a. and **b.**

$$CH_3-(CH_2)_8-\underset{\beta}{CH_2}-\underset{\alpha}{CH_2}-\overset{\overset{\displaystyle O}{\|}}{C}-CoA$$

     c. Lauryl-CoA + 5 CoA + 5 FAD + 5 NAD^+ + 5 $H_2O \longrightarrow$
        6 Acetyl CoA + 5 $FADH_2$ + 5NADH + $5H^+$
     d. Six acetyl CoA units are produced
     e. Five cycles of β oxidation are needed
     f. activation $\longrightarrow$ -2 ATP

6 acetyl CoA x 12	$\longrightarrow$	72 ATP
5 $FADH_2$	$\times$ 2 $\longrightarrow$	10 ATP
5 NADH	$\times$ 3 $\longrightarrow$	15 ATP
	Total	95 ATP

25.69 **a.** β oxidation   **b.** β oxidation
     c. Fatty acid synthesis   **d.** β oxidation
     e. Fatty acid synthesis   **f.** Fatty acid synthesis

25.71 **a.** (1) fatty acid oxidation
     b. (2) the synthesis of fatty acids.

25.73 Ammonium ion is toxic if allowed to accumulate in the liver.

25.75 **a.** citrulline   **b.** carbamoyl phosphate

25.77 **a.** pyruvate   **b.** acetoacetyl-CoA, acetyl CoA
     c. succinyl-CoA   **d.** α-ketoglutarate

25.79 Serine is degraded to pyruvate, which is oxidized to acetyl CoA. The oxidation produces NADH + H^+, which provides 3 ATP. In one turn of the citric acid cycle, the acetyl CoA provides 12 ATP. Thus, serine can provide a total of 15 ATP.

Credits

Unless otherwise acknowledged, all photographs are the property of Benjamin Cummings.

Chapter 11

p. 376   Paul Hanna/Reuters NewMedia Inc./Corbis.

Chapter 12

p. 396   © Stephen Kline/Bruce Coleman, Inc.
p. 397   Total Ozone Mapping Spectrometer/NASA.
p. 405   Natalie Fobes/Corbis
p. 407   © 1997 PhotoDisc/K. Knudson/Photolink.

Chapter 13

p. 417   © 1997 PhotoDisc/Don Tremain.
p. 420   © 1997 PhotoDisc/Tim Hall.
p. 422   Ian O'Leary/Stone.
p. 425   bottom: Alastair Shay/Papilio/Corbis.
p. 435   top: Courtesy of NASA.
p. 441   © Dianora Niccolini/Medical Images, Inc.

Chapter 14

p. 463   © Tom Pantages.
p. 468   © Al Assid/The Stock Market.
p. 473   Owen Franken/Corbis.
p. 475   Shelley Gazin/Corbis.

Chapter 16

p. 525   John Wilson White/Addison Wesley Longman.
p. 533   © Custom Medical Stock Photo.
p. 543   left: Richard Hamilton Smith/Corbis.
         right: David Toase/PhotoDisc.

Chapter 17

p. 550   Robert and Linda Mitchell/Robert and Linda Mitchell Photography.
p. 558   left: Warren Morgan/Corbis

Chapter 18

p. 582   Lawson Wood/Corbis.
p. 584   Ralph A. Clevenger/Corbis.
p. 586   George D. Lepp/Corbis.
p. 591   Lori Adamski Peek/Corbis.
p. 599   © C. Raines/Visuals Unlimited.
p. 601   Courtesy of the National Heart, Lung, and Blood Institute/National Institutes of Health.
p. 604   Custom Medical Stock Photo.
p. 606   AFP Photo/Paul Jones/Corbis.

Chapter 19

p. 624   bottom: Dr. Morley Read/Science Photo Library/Photo Researchers, Inc.
p. 628   top left: © Ed Drews/Photo Researchers, Inc.
p. 629   Michael S. Yamashita/Corbis.

Chapter 20

p. 643   left: D. Robert & Lorri Franz/Corbis.
         top right: D. Robert & Lorri Franz/Corbis.
         bottom right: Richard Hamilton Smith/Corbis.
p. 658   Jilly Wendell/Stone.
p. 662   left: Richard Hamilton Smith/Corbis.
         middle: Eric and David Hosking/Corbis.
         right: Bruce Wilson/Stone.
p. 665   © Lewin/Royal Free Hospital/Photo Researchers, Inc.

Chapter 22

p. 716 © 1997 PhotoDisc/M. Freeman/Photolink.
p. 736 Morton Beebe, S.F./Corbis.

Chapter 23

p. 757 Lori Adamski Peek/Stone.
p. 772 top: Joel W. Rogers/Corbis.
p. 777 Neal Preston/Corbis.

Chapter 24

p. 805 Philip James Corwin/Corbis.

Chapter 25

p. 815 Hossler/Custom Medical Stock Photo.
p. 825 left: Gunter Marx Photography/Corbis.
right: Chase Swift/Corbis.
p. 828 bottom left: Custom Medical Stock Photo.
bottom right: SIU/Custom Medical Stock Photo.
p. 846 Custom Medical Stock Photo.

Glossary/Index